Spatio-Temporal Patterns in Nonequilibrium Complex Systems

Spatio-Temporal Patterns in Nonequilibrium Complex Systems

NATO Advanced Research Workshop

Editors

P. E. Cladis
AT&T Bell Laboratories
Murray Hill, NJ 07974

P. Palffy-Muhoray
Kent State University
Kent, OH 44242

Proceedings Volume XXI

Santa Fe Institute
Studies in the Sciences of Complexity

Addison-Wesley Publishing Company
The Advanced Book Program

Reading, Massachusetts   Menlo Park, California   New York
Don Mills, Ontario   Wokingham, England   Amsterdam   Bonn
Sydney   Singapore   Tokyo   Madrid   San Juan
Paris   Seoul   Milan   Mexico City   Taipei

Publisher: *David Goehring*
Production Manager: *Michael Cirone*
Production Supervisor: *Lynne Reed*

Director of Publications, Santa Fe Institute: *Ronda K. Butler-Villa*
Publications Assistant, Santa Fe Institute: *Della L. Ulibarri*

This volume was typeset using TEXtures on a Macintosh II computer. Camera-ready output from a Hewlett Packard Laser Jet 4M Printer.

ISBN 0-201-40984-4 (Hardcover)
ISBN 0-201-40987-9 (Paperback)

1 2 3 4 5 6 7 8 9 10-MA-97969594
First printing, November 1994

About the Santa Fe Institute

The *Santa Fe Institute* (SFI) is a multidisciplinary graduate research and teaching institution formed to nurture research on complex systems and their simpler elements. A private, independent institution, SFI was founded in 1984. Its primary concern is to focus the tools of traditional scientific disciplines and emerging new computer resources on the problems and opportunities that are involved in the multidisciplinary study of complex systems—those fundamental processes that shape almost every aspect of human life. Understanding complex systems is critical to realizing the full potential of science, and may be expected to yield enormous intellectual and practical benefits.

All titles from the *Santa Fe Institute Studies in the Sciences of Complexity* series will carry this imprint which is based on a Mimbres pottery design (circa A.D. 950–1150), drawn by Betsy Jones. The design was selected because the radiating feathers are evocative of the outreach of the Santa Fe Institute Program to many disciplines and institutions.

Santa Fe Institute
Studies in the Sciences of Complexity

Contributors to This Volume

Y. Adachi, Kyushu Institute of Technology

K. I. Agladze, Max-Plank-Institut für molekulare Physiologie

G. Ahlers, University of California, Santa Barbara

K. R. Amundson, AT&T Bell Laboratories

I. Aranson, Bar-Ilan University

C. B. Arnold, University of Pennsylvania

O. Avidan, Tel-Aviv University

J. Bechhoefer, Simon Fraser University

M. Ben Amar, Ecole Normale Supérieure

E. Ben-Jacob, Tel-Aviv University

D. Berry, The University of Iowa

A. Bösch, Institut für Festkörperforschung

P. Borckmans, Université Libre de Bruxelles

H. R. Brand, Universität Bayreuth

E. Brener, Institute for Solid State Physics

J.-M. Buchlin, Von Karman Institute

A. Buka, Universität Bayreuth

D. S. Cannell, University of California, Santa Barbara

O. Cardoso, Ecole Normale Supérieure

E. F. Carr, University of Maine

H. Chaté, Centre d Etudes de Saclay

P.E. Cladis, AT&T Bell Laboratories

V. Croquette, LPS, ENS, URA D 1306 CNRS

A. Czirók, Institute for Technical Physics, Budapest

S. H. Davis, Northwestern University

J. R. de Bruyn, Memorial University of Newfoundland

M. Decré, Von Karman Institute

M. Dennin, University of California, Santa Barbara

G. Dewell, Université Libre de Bruxelles

C. R. Doering, Clarkson University

J. M. Flesselles, INLN, UMR 129 CNRS

M. Frey, Bucknell University

H. Furukawa, Yamaguchi University

A. M. Gabrielov, Cornell University

E. Gailly, Von Karman Institute

B. J. Gluckman, University of Pennsylvania

H. F. Goldstein, University of California, Berkeley

J. P. Gollub, University of Pennsylvania

M. Grinfeld, Rutgers University

T. Hashimoto, Kyoto University

A. Hernández-Machado, Universitat de Barcelona

H. Herzel, Humboldt University

N. A. Hill, The University, Leeds

H. Honjo, Kyushu University

W. Horsthemke, Southern Methodist University

G. Huber, University of California, Lawrence Berkeley Laboratory
T. Ihle, Institut für Festkörperforschung
B. Janiaud, LPS, ENS, URA D 1306 CNRS
M. H. Jensen, Niels Bohr Institute
O. Jensen, The Technical University of Denmark
S. Jucquois, LPS, ENS, URA D 1306 CNRS
S. Kai, Kyushu Institute of Technology
K. Kawasaki, Kyushu University
J. O. Kessler, University of Arizona
J. Y. Kim, Kent State University
E. Knobloch, University of California, Berkeley
T. Koga, Kyushu University
L. Kramer, Physikalisches Institut der Universität Bayreuth
K. Krischer, Fritz Haber Institute, Max Planck Society
T. Kyu, University of Akron
L. Lam, San Jose State University
R. G. Larson, AT&T Bell Laboratories
S. J. Linz, Universität Augsburg
N. V. Madhusudana, Raman Research Institute
J. V. Maher, University of Pittsburgh
D. Marteau, Ecole Normale Supérieure
M. Matsushita, Chuo University
T. Matsuyama, Niigata University
A. D. May, University of Toronto and Ontario Laser Lightwave Research
J. P. McClymer, University of Maine
M. M. Millonas, Los Alamos National Laboratory and Santa Fe Institute
T. Mizuguchi, Kyoto University
S. Morris, University of California, Santa Barbara
E. Mosekilde, The Technical University of Denmark
H. Müller-Krumbhaar, Institut für Festkörperforschung
S. C. Müller, Max-Plank-Institut für molekulare Physiologie
M. Mustafa, University of Akron
S. Nasuno, Kyushu University
A. C. Newell, University of Arizona
W. I. Newman, University of California, Los Angeles
P. Nozières, Institut Laue-Langevin
S. Ohta, Kyushu University
P. Palffy-Muhoray, Kent State University
L. Pan, Memorial University of Newfoundland
V. O. Pannbacker, The Technical University of Denmark
S. L. Phoenix, Cornell University
R. D. Pochy, San Jose State University
R. Pratibha, Raman Research Institute
M. Rabaud, Ecole Normale Supérieure
M. I. Rabinovich, University of California, San Diego
D. Raitt, Northwestern University
I. Rehberg, Universität Bayreuth

H. Richter, Universität Bayreuth
H. Riecke, Northwestern University
D. M. Ronis, McGill University
Y. Saito, Keio University
M. Sano, Tohoku University
S.-I. Sasa, Kyoto University
R. Schmitz, IFF, Forschungszentrum Julich
H. Shehadeh, University of Maine
K. Shiraishi, Keio University
O. Shochet, Tel-Aviv University
E. Simiu, National Institute of Standards and Technology
K. Sneppen, Niels Bohr Institute
O. Steinbock, Max-Plank-Institut für molekulare Physiologie
P. Tabeling, Ecole Normale Supérieure
M. Takenaka, Kyoto University
A. Tenenbaum, Tel-Aviv University
I. Titze, The University of Iowa
L. S. Tsimring, University of California, San Diego
D. L. Turcotte, Cornell University
W. van Saarloos, University of Leiden
M. C. Veinott, San Jose State University
T. Vicsek, Eötvös University
J.-I. Wakita, Chuo University
D. Walgraef, Free University of Brussels
A. Warda, Max-Plank-Institut für molekulare Physiologie
J. A. Warren, National Institute of Standards and Technology
A. Weber, Physikalisches Institut der Universität Bayreuth
J.-Y. Yuan, McGill University
H. Zhao, University of Pittsburgh
W. Zimmermann, IFF, Forschungszentrum Julich

Contents

Electrohydrodynamic Convection in Liquid Crystals

Phase Separation in Complex Systems   379

Chemical and Materials Instabilities   423

Noise, Turbulence, Chaos, and Intermittency　　　519

Pattern Formation in Biological Systems　　　599

P. E. Cladis† and P. Palffy-Muhoray‡
†AT&T Bell Laboratories Murray Hill, NJ 07974
‡The Liquid Crystal Institute, Kent State University, Kent, OH 44242

Preface

The enthusiasm of the participants and the quality of their contributions to the NATO Advanced Research Workshop on *Spatio-Temporal Patterns in Nonequilibrium Complex Systems* held on April 13-17, 1993 in Santa Fe, NM, are two measures of the importance and excitement of the new field of nonequilibrium physics of complex systems. Although many delegates could not be adequately funded from our slim budget, 120 researchers from 14 different countries participated. On the basis of the above metrics, then, this workshop was an outstanding success: links were formed across disciplinary boundaries and there was a pervading sense that something new was emerging. We thank all participants for their contributions, particularly those whose contributions are presented here.

Given the intense study and advances in nonequilibrium phenomena in "simple systems" over the past several years, Pattern Formation in "simple" systems appears practically mature in comparison to the more diffuse activity in complex systems. The first workshop to address nonequilibrium phenomena in Liquid Crystals (Bayreuth, Germany, 1989) was a small workshop organized by Lorenz Kramer and Walter Zimmermann. The second workshop (Kitakyushu, Japan, 1991), broadening the scope to "Complex Systems," was organized by Shoichi Kai. The Proceedings of Kai's conference have been published.[4] We view this workshop as the third in this series where, again, the goal was to bring together experimentalists working with pattern-forming systems in complex materials, such as liquid crystals, polymers, and biological systems, and theorists knowledgeable and interested in a physical understanding of the many, sometimes startlingly beautiful, patterns exhibited by these complex materials.

We believe that pattern formation in complex systems is important because it stimulates the growth of new physics and therefore new technologies. The technological relevance touched on in these chapters ranges from laser design, the development of high-strength alloys, polymer processing, flat-panel liquid crystal displays, and living systems. As descriptions in terms of order parameters revolutionized the understanding of phase transitions, the hope is that descriptions via amplitude equations and phase dynamics will give rise to "simple" descriptions of pattern formation in a wide variety of complex systems. To realize this hope, clearly, much remains to be done.

The complex systems we wish to understand are often complicated, and the equations, initial and boundary conditions, are often not known. Nevertheless, many different systems show similar patterns that can be modeled by solutions to amplitude equations. Our understanding is that here details of the system enter only through the coefficients in the relevant amplitude equations. These coefficients can, at least in principle, be experimentally determined, and the behavior of the system then compared to theory.

While the usefulness of amplitude equations to describe macroscopic spatio-temporal features of dynamical systems was established long ago by Newell,[6] to our knowledge, there are relatively few dissipative systems where the connection between amplitude equations and the basic hydrodynamic equations governing each system has been made. This was first done for Rayleigh-Benard convection in simple fluids[5,8] and for electroconvection in nematic liquid crystals.[2]

In both these systems, length scales in the pattern are set by the container height. However, in a large variety of patterns this is not the case. So, the question still looms: what is the physics behind spatio-temporal patterns in extended (no boundaries) systems? In traveling nematic/cholesteric liquid crystal interfaces (Section II), a good candidate is the nonlinear elasticity that is a consequence of the broken continuous symmetries of these states of matter.

Because of their seminal nature and their "macroscopic" point of view, then, Section I deals with an overview and some detailed predictions of amplitude equations. Subsequent chapters, with system complexity and the difficulty of materials characterization monotonically increasing from chapter to chapter, give experimental studies and theoretical interpretations of pattern-forming phenomena in a variety of systems. We mention here only a few of the many points to emerge in the Chapters contributed to this Workshop Proceedings.

In Section II, dealing with *Pattern Formation at Traveling Phase Boundaries and Other Interfaces*, the first point to emerge is that technologically relevant and/or simple materials such as helium, tend to be well characterized. Consequently, these systems are attractive for well-controlled experiments because they enable quantitative comparison with theory. Because of their importance in flat panel displays, one example of a well-characterized complex system is provided by nematic/cholesteric liquid crystals and the "simplicity" of the patterns they exhibit. This leads to the second point, with again liquid crystals leading the way, that when length and time scales are amenable to well-controlled experiments, the likelihood of discovering qualitatively new physics increases.

Section III opens with another example of the power of a macroscopic physical point of view. Without reference to specific material parameters, Brand argues that the "second order fluids" description frequently used in current theories of viscoelastic materials violates the second law of thermodynamics!

Section IV highlights the well-characterized aspect of liquid crystals in electrohydrodynamic convection as well as the novel thin film geometry described by Morris et al.

Phase separation and phase separated systems are important in the production of new materials and new display technologies. Section V describes recent fundmental results in this area.

In Section VI, Walgraef points out that Turing patterns have only recently been observed[3,7] after almost 40 years of theoretical modeling of pattern formation in reaction diffusion equations. The experimentalists have finally caught up!

Recently, many investigators concerned with patterns in nonequilibrium systems have turned their attention to the problem of spatio-temporal chaos. In Section VII, we have collected several contributions on the role of noise, turbulence, transitions to chaos, and intermittency. As pointed out by Tabeling et al., it appears that despite considerable effort, neither theoretical nor numerical results are in agreement with their experimental observations of turbulent flows in thin electrolyte layers.

Section VIII deals with pattern formation in biological systems. These are clearly the most complicated, the least well-characterized, and the most important of all known nonequilibrium systems. It is encouraging to see that despite daunting difficulties, experimentalists and theorists are developing new tools to expand our understanding of these systems. One ambitious goal of studying pattern formation in biological systems is to provide information useful to medicine.

An even more ambitious effort was outlined in the lively and entertaining banquet address by Brian Arthur.[1] His talk on *Pattern Formation in the Economy* combined humor with a description of modeling economic indicators at the Santa Fe Institute. We thank him for taking time out of his busy schedule to give this banquet address to people interested in "simple" models for complex systems.

The technological relevance of pattern formation in complex systems was addressed in a panel discussion. This session was chaired by Alan Newell. Points to emerge here were: (1) nonlinear, nonequilibrium physics addresses real-life issues often in a "win- win" scenario. Examples were given. (2) Perhaps the organization of our academic institutions along the traditional lines of engineering, chemistry, physics, etc. (e.g., our current university departments) is not well suited to stimulate cross-disciplinary interactions needed to understand complex phenomena; that is, it closes rather than opens doors. (3) The curiosity, enthusiasm, and talent of investigators in the field know no bounds nor boundaries.

We thank the sponsoring organizations for their support: the NATO Special Program on Chaos, Order and Patterns; the U.S. Office of Naval Research; the Center for Nonlinear Studies (Los Alamos National Laboratories); the National Science Foundation (U.S.A.); the NEC Research Institute (New Jersey, U.S.A.); Sandia National Laboratories; the Santa Fe Institute; AT&T Bell Laboratories; and the Liquid Crystal Institute (KSU NSF STC ALCOM). We also thank the staff of the Santa Fe Hilton and the Santa Fe Institute for organizational support and hospitality. A key contribution was the outstanding organizational effort of Ms. Brenda Buck of the Liquid Crystal Institute, KSU.

Finally, we thank again the workshop participants for their scientific contributions and open-minded, spirited, and generous sharing of information, insights, and questions. A unifying theme of this workshop was the similarities exhibited by patterns in many different complex systems out of equilibrium, and the impressive efforts of the participants to rise to the task of extracting from the complexity of their systems, and sharing with us, what they thought were "significant" and what they thought were "irrelevant" to help advance *our discovery and understanding...of the truly significant simplicity of the basic laws of nature embedded in the amazing complexity of natural phenomena.*[9]

REFERENCES

1. Arthur, W. B. "Positive Feedbacks in the Economy." *Sci. Am.* **Feb.** (1990): 92.
2. Bodenschatz, E., W. Zimmermann, and L. Kramer. "On Electrically Driven Pattern-Forming Instabilities in Planar Nematics." *J. Physique* **49** (1988): 1875.
3. Castets, V., E. Dulos, J. Boissonade, and P. de Kepper. "Experimental Evidence of a Sustained Standing Turing-Type Nonequilibrium Chemical Pattern." *Phys. Rev. Lett.* **64** (1990): 2953.
4. Kai, S., (ed). *Pattern Formation in Complex Dissipative Systems.* Singapore, World Scientific, 1991.
5. Newell, A. C., and J. A. Whitehead. "Finite Bandwidth Finite Amplitude Convection." *J. Fluid Mech.* **38** (1969): 279.
6. Newell, A. C. "Envelope Equations." In *Lectures in Applied Mathematics* **15** (1974): 157.
7. Ouyang, Q., and H. L. Swinney. "Transition from a Uniform State to Hexagonal and Striped Turing Patterns." *Nature* **352** (1991): 610.
8. Segel, L. A. "Distant Sidewalls Cause Slow Amplitude Modulation of Cellular Convection." *J. Fluid Mech.* **38** (1969): 203.

9. van Hove, L. "Concluding Remarks." In *Order and Fluctuations in Equilibrium and Nonequilibrium Statistical Mechanics*, edited by G. Nicolis, G. Dewel, and J. W. Turner, 367. New York: John Wiley and Sons, 1981.

Chapter 1
Amplitude Equations

Alan C. Newell
Arizona Center for Mathematical Sciences, University of Arizona, Tucson, AZ 85721

Patterns in Nonlinear Optics: A Paradigm

In this brief overview, we introduce the reader to the ideas and methods involved in the macroscopic description of patterns using the unidirectional propagation of light in an amplifying medium as a paradigm. This model has several advantages: it is simple and easy to understand; it is nontrivial; the computations are not unnecessarily burdensome: the territory is relatively unexplored and the results are interesting from the technological as well as scientific points of view. Some of the results, particularly the introduction of the canonical complex Swift-Hohenberg equation with dispersion (CSHD) as an asymptotically valid approximation, are new.

Spatio-Temporal Patterns, Ed. P. E. Cladis and P. Palffy-Muhoray,
SFI Studies in the Sciences of Complexity, Addison-Wesley, 1995

3

---

1. INTRODUCTION

Patterns of an almost periodic nature appear all over the place: the sand ripples on flat beaches and desert dunes, the buckles on a thin elastic shell under compression, the convection cells in a nematic liquid crystal stressed by an alternating electric field. Indeed, most spatially extended, continuous, dissipative systems, when subjected to an external stress, undergo symmetry-breaking phase transitions or bifurcation. At first, the patterns that occur are periodic, in space and/or time, breaking translational symmetry (spatial and/or temporal). As the stress is increased, however, the patterns become more complicated, often through defect formation, and eventually exhibit spatio-temporal turbulence. The goal of theory is to understand patterns from a macroscopic point of view that both unifies and simplifies classes of problems that are unrelated at the microscopic level.

A key step in achieving a macroscopic description is the identification of a suitable order parameter (or order parameters) for the system and writing down equations that capture its space-time evolution. These equations have universal properties. While it is true that the coefficients depend on the particular microscopic system under consideration, the shape of the equation is determined by the nature of the symmetries, such as translation and rotation, which the original microscopic system enjoys.

The starting point is the simplest state, usually the spatially uniform, time-independent solution, from whose loss of stability one may determine the dominant shape (or shapes; often symmetries lead to degeneracies) of the first pattern to emerge. The equations for the relevant order parameters near onset are of the Newell-Whitehead-Segel (NWS) (both with real and complex coefficients) and complex Ginzburg-Landau types.[13] The order parameters near onset are the envelopes (space- and time-dependent amplitudes) of the dominant shapes. These are complex numbers containing information on both amplitude and phase. Far from onset (actually "far" means that the stress parameter is supercritical by an amount greater than the square of the inverse of the aspect ratio $\epsilon = \lambda/l$, λ, the pattern wavelength; l, the horizontal dimension, e.g., container size, over which one expects the pattern wavevector to vary by an order one amount), the amplitude of the order parameter becomes slaved to the phase and then one has a single order parameter, that satisfies the Cross-Newell phase diffusion equation.[4] The latter is ill-posed for certain ranges of wavenumber and must be regularized. This has just recently been done.[19]

One interesting and special feature of these order parameter equations is that they contain singular or weak solutions at points and along curves that describe defects (disclinations, dislocations) and phase grain boundaries, respectively. Thus the macroscopic description is a field-particle theory that contains both smooth and singular components which is obtained by averaging over the almost everywhere, periodic structure of a pure microscopic field.

We now give concrete manifestation to these ideas in the context of an almost planar, unidirectional light beam that travels in an amplifying medium of two-level atoms.

2. THE EQUATIONS

In nondimensional form, the relevant equations are[16]

$$\frac{\partial A}{\partial z} + \frac{\partial A}{\partial t} - ia\nabla^2 A = -\sigma A + \sigma P, \tag{1}$$

$$\frac{\partial P}{\partial t} + (1 + i\Omega)P = (r - n)A, \tag{2}$$

$$\frac{\partial n}{\partial t} + bn = \frac{1}{2}(AP^* + A^*P). \tag{3}$$

TABLE 1

Symbol	Description, Comments	Dimensional Factor
A	Electric field envelope of the carrier wave $\exp(i\omega/cz - i\omega t)$. Eq. (1) is Maxwell's Equation in which $\partial^2 A/\partial z^2$, $\partial^2 A/\partial t^2$ are ignored and $\partial^2 P/\partial t^2$ is approximated by $-\omega^2 P$.	$(i\hbar/2p)\gamma_{12}$
a	Inverse Fresnel number. The term $ia\nabla^2 A$, $\nabla^2 = \partial^2/\partial x^2 + \partial^2/\partial y^2$ describes the transverse diffraction of the beam.	$a = c^2/2\omega\gamma_{12}w_0^2$
σ	Electric field attenuation. In a laser, this is mainly due to imperfect mirror reflection.	$\sigma = \kappa/\gamma_{12}$
Ω	Mismatch between energy level difference $\Delta E = h\omega_{12}$ and the frequency ω of the electric field or laser cavity.	$\Omega = (\omega_{12} - \omega)/\gamma_{12}$

TABLE 2

Symbol	Description, Comments	Dimensional Factor
P	Dipole Polarization induced by electric field.	$(\hbar\gamma_{12}\kappa\epsilon_0)/\omega p$
r	External pumping. The amount by which the population of atoms in the higher energy state exceeds that in the lower energy state.	$(\omega p^2 |N_0|)/(2\epsilon_0 \hbar\kappa\gamma_{12})$
n	A measure of the population inversion N.	$N - N_0 = (2\epsilon_0\kappa\hbar\gamma_{12}/\omega p^2)n$
b	Decay of population inversion of spontaneous emission.	γ_{11}/γ_{12}
t	Time.	$1/\gamma_{12}$
$\vec{x}(x,y)$	Distance in transverse direction.	w_0
z	Distance in propagation direction.	c/γ_{12}

The parameters are: h, Planck's constant; ε_0, vacuum dielectric constant in MKS units; p, dipole polarization strength; γ_{12}, homogeneous broadening, or the loss in polarization due to spontaneous emission; γ_{11}, population decay due to spontaneous emission of population inversion; c, speed of light; κ, attenuation of electric field; and w_0, beam width.

In addition, if the medium is also inhomogeneously broadened (because the atoms are in motion or that the frequency mismatch Ω is, due to Doppler shift, a random variable with distribution $g(\Omega) = (\beta/\pi)\exp(-(\Omega - \Omega_0)^2/\beta^2)$ or $\beta/\pi(\beta^2 + (\Omega - \Omega_0)^2)$, for examples), then one replaces the P in Eq. (1) by $\langle P \rangle = \int g(\Omega)P(\Omega, x, y, t)d\Omega$.

Observe that in the absence of diffraction (i.e., $a = 0$), Eqs. (1)–(3) are the complex Lorenz equations and, for zero-frequency mismatch $\Omega = 0$, the fields A and P may be taken real so that they are exactly the Lorenz equations. We shall see, however, that the limit $a \to 0$ is singular and, indeed, that the most interesting transverse structures occur in the large Fresnel number, low a, limit. A modification to Eqs. (1)–(3), the so-called Raman laser, with nontrivial consequences,[8] is found by adding $i\delta_1 An$ and $i\delta_2|A^2|P$ (A.C. Stark shift) to the right-hand sides of Eq. (1) and Eq. (2), respectively, and by adding $r\delta_3$ to Ω. The reader may wish to repeat the calculations of this review for the Raman laser as an exercise.

3. LINEAR STABILITY ANALYSIS OF THE NONLASING SOLUTION

We test the stability of the nonlasing solution,

$$A = P = n = 0, \tag{4}$$

by setting $(A, P) = (\hat{A}, \hat{P}) \exp(i\vec{k} \cdot \vec{x} + \lambda t)$, $\lambda = \mu - i\nu$ and $n = \hat{n} \exp(\mu t)$ from which we find,

$$\hat{L} \begin{pmatrix} \hat{A} \\ \hat{P} \\ \hat{n} \end{pmatrix} = \begin{pmatrix} \lambda + iak^2 + \sigma & -\sigma & 0 \\ -r & \lambda + 1 + i\Omega & 0 \\ 0 & 0 & \mu + b \end{pmatrix} \begin{pmatrix} \hat{A} \\ \hat{P} \\ \hat{n} \end{pmatrix} = 0, \tag{5}$$

giving two relations $\mu = \mu(k^2, r)$ and $\nu = \nu(k^2, r)$ for the growth rate and frequency of the perturbation. On the neutral curve, $\mu(k^2, r) = 0$, we obtain

$$\nu = \frac{\sigma\Omega + ak^2}{\sigma + 1}, \tag{6}$$

$$r_c(k^2) = 1 + \frac{(\Omega - ak^2)^2}{(\sigma + 1)^2}. \tag{7}$$

If inhomogeneous broadening is included, then we simply replace the 1 by $1 + \beta$ in the denominators of Eqs. (6) and (7), where β^{-1} is the Lorentzian line width, and multiply the RHS of Eq. (7) by $1 + \beta$.

We observe now a very special and intuitively reasonable result. If $\Omega < 0$, then $r_c(k^2)$ is minimized at $k^2 = 0$ and $\nu = \sigma\Omega/(1 + \sigma)$ so that the total frequency is (in dimensional units) $\omega + \nu = (\gamma_{12}\omega + \kappa\omega_{12})/(\gamma_{12} + \kappa)$, which for $\sigma = \kappa/\gamma_{12} \ll \gamma_{12}$ is nearly ω. Namely, the electric field forces the medium to operate at its frequency or, in the case of a laser, at the cavity frequency of the longitudinal TEM_{00} mode closest to ω_{12}. On the other hand, for positive mismatch $\Omega > 0$, the critical pumping value $r_0 = \min r_c(k^2)$ can be lowered by choosing a nontrivial transverse pattern with $k^2 = \Omega/a$ so that $r_0 = 1$. Then, the total frequency $\omega + \nu$, in dimensional units, is exactly ω_{12}, so that the electric field tunes itself to the medium's natural frequency in order to optimize its output. In a laser cavity, the same tuning occurs. For a given longitudinal mode TEM_{00}, the frequencies of the higher transverse TEM_{rs} modes are situated above the frequency ω_c of the TEM_{00} mode. If no account must be taken for the additional losses incurred by the higher order transverse modes, then the system will choose to go unstable in the shape of that particular mode whose frequency is closest to $\omega_{12} > \omega_c$. In our case, we have not imposed (yet) any horizontal boundary conditions and we are allowing for an infinitely large transverse domain ($\Gamma = \epsilon^{-1}$, the aspect ratio, is infinite), so that k is continuous. In a finite system, we must be concerned about losses when the electromagnetic wave meets the real horizontal boundary. Note that the group velocity $\vec{v}_g = ((2ak_0/\sigma + 1)\cos\varphi, (2ak_0/\sigma + 1)\sin\varphi)$, where $\vec{k} = (k_0\cos\varphi, k_0\sin\varphi)$ is nonzero when $k_0 = \sqrt{\Omega/a} \neq 0$ so that power is carried to the boundaries. The situation resembles, therefore, convection in binary fluid mixtures of finite length when the Soret effect inhibits ordinary convection and gives rise to oscillatory and travelling wave convection.[5] We will avoid these considerations here by assuming that we have an annular domain so that periodic boundary conditions obtain.

The right and left eigenvectors of $\hat{L}$ at $\mu = 0$ and at $k^2 = 0 (\Omega < 0)$ or $k^2 = \frac{\Omega}{a}(\Omega > 0)$ are respectively:

$$\Omega < 0: \quad (1, x_0 = 1 - \frac{i\Omega}{\sigma + 1}, 0)^T, \quad (1, y_0 = \frac{\sigma x_0}{r_0}, 0) \tag{8}$$

$$\Omega > 0: \quad (1, 1, 0)^T, \quad (1, \sigma, 0). \tag{9}$$

We note that, for r close to but greater than r_0, any initial value of $(A, P, n)^T$ evolves quickly onto the unstable manifolds $(1, x_0, 0)^T$ when $\Omega < 0$ or $(1, 1, 0)^T$ when $\Omega > 0$.

4. NEAR ONSET: WEAKLY NONLINEAR ANALYSIS

The last observation suggests that, for

$$r = r_0 + R, R \ll 1, \tag{10}$$

one can approximate (A, P, n) to leading order by

$$\begin{pmatrix} A \\ P \\ n \end{pmatrix} = \sqrt{R}\sum \begin{pmatrix} 1 \\ z_0 \\ 0 \end{pmatrix} Be^{i\vec{k}_0\cdot\vec{x} - i\nu t} + (*), \tag{11}$$

where $z_0 = x_0, \vec{k}_0 = 0, \nu = \sigma\Omega/(\sigma + 1)$ when $\Omega < 0$, and there is only one term in the sum and $z_0 = 1, k_0 = \sqrt{\Omega/a}, \nu = \Omega$ when $\Omega > 0$, and the sum is over all $\vec{k}_0$ such that $|\vec{k}_0| = \sqrt{\Omega/a}$, each with a different envelope B. The choice of magnitude $\sqrt{R}$ reflects the fact that equilibrium can only be achieved through a balance of linear growth RA and the saturating nonlinear feedback $-nA$ where n is generated by $Rl(AP^*)$ and is therefore of order amplitude A squared. The balance give an amplitude A of order $\sqrt{R}$. I will write down the evolution equations for both cases $\Omega > 0$ and $\Omega < 0$ but will now derive in detail what happens in the transition case,

$$\Omega = \sqrt{R}\Omega_1. \tag{12}$$

5. WHEN THE FREQUENCY MISMATCH IS SMALL AND OF EITHER SIGN

In this case, the curve $r_c(k^2)$ versus k is quartic near $r = r_0, k = 0$ so that for $r > r_0 = 1$, there is an order $R^{1/4}$ bandwidth of modes excited in both transverse directions. Thus we may take, to leading order,

$$\begin{pmatrix} A \\ P \\ n \end{pmatrix} = \begin{pmatrix} 1 \\ 1 \\ 0 \end{pmatrix} \sqrt{R} B(X = R^{1/4}x, \ Y = R^{1/4}y, t), \tag{13}$$

where B, the envelope order parameter, is a slowly varying function of X, Y, and t. For bookkeeping purposes, it is convenient to write,

$$\frac{\partial B}{\partial t} = \sqrt{R} \frac{\partial B}{\partial T_1} + R \frac{\partial B}{\partial T_2}, \ T_1 = \sqrt{R} t_1, T_2 = Rt. \tag{14}$$

Now, set

$$\begin{pmatrix} A \\ P \\ n \end{pmatrix} = \sqrt{R} \begin{pmatrix} A \\ P \\ n \end{pmatrix}_0 + R \begin{pmatrix} A \\ P \\ n \end{pmatrix}_1 + R^{3/2} \begin{pmatrix} A \\ P \\ n \end{pmatrix}_2 + \dots \tag{15}$$

and determine $\partial B/\partial T_1$, $\partial B/\partial T_2$, namely $\partial B/\partial t$ from the solvability of the first and second iterates.

At order $R^{1/2}$, Eqs. (1), (2), and (3) are satisfied exactly by Eq. (13). At order R, we have

$$L \begin{pmatrix} A \\ P \\ n \end{pmatrix}_1 = \begin{pmatrix} \partial_t - ia\nabla^2 + \sigma & -\sigma & 0 \\ -1 & \partial_t + 1 & 0 \\ 0 & 0 & \partial_t + b \end{pmatrix} \begin{pmatrix} A \\ P \\ n \end{pmatrix}_1$$
$$= \begin{pmatrix} -\partial_{T_1} B & + & ia\nabla^2 B \\ \partial_{T_1} B & - & i\Omega_1 B \\ & 0 & \end{pmatrix} + \begin{pmatrix} 0 \\ 0 \\ BB^* \end{pmatrix}, \tag{16}$$

where the first term on the RHS of Eq. (16) comes from ∂_t and ∇^2 and $\Omega = \sqrt{R}\Omega_1$ acting on $(A, P, n)_0^T$ and the second from the nonlinear terms $(0, -n_0 A_0, 1/2(A_0 P_0^* + A_0^* P_0))^T$. The ∇^2 on the RHS of Eq. (16) is $\partial^2/\partial X^2 + \partial^2/\partial Y^2$, the slow Laplacian, whereas on the left in L, it is $\partial^2/\partial x^2 + \partial^2/\partial y^2$. As far as (A_1, P_1, n_1) is concerned, the RHS is constant and thus must be orthogonal to the left eigenvector because

$$(1, \sigma, 0) \begin{pmatrix} \sigma & -\sigma & 0 \\ -1 & 1 & 0 \\ 0 & 0 & \partial_t + b \end{pmatrix} = 0.$$

This condition gives

$$(\sigma + 1)\frac{\partial B}{\partial T_1} - ia\nabla_X^2 B + i\Omega_1 B = 0. \tag{17}$$

Using Eq. (17), we may now solve Eq. (16). We find

$$\begin{pmatrix} A \\ P \\ n \end{pmatrix}_1 = \frac{BB^*}{b}\begin{pmatrix} 0 \\ 0 \\ 1 \end{pmatrix} - \begin{pmatrix} 0 \\ 1 \\ 0 \end{pmatrix}\left(\frac{\partial B}{\partial T_1} + i\Omega_1 B\right). \tag{18}$$

At order $R^{3/2}$,

$$L\begin{pmatrix} A \\ P \\ n \end{pmatrix} = -\partial_{T_2}B\begin{pmatrix} 1 \\ 1 \\ 0 \end{pmatrix} + \begin{pmatrix} 0 \\ 1 \\ 0 \end{pmatrix}B + \left(\frac{\partial}{\partial T_1} + i\Omega_1\right)^2 B\begin{pmatrix} 0 \\ 1 \\ 0 \end{pmatrix}$$
$$- \frac{B^2B^*}{b}\begin{pmatrix} 0 \\ 1 \\ 0 \end{pmatrix}, \tag{19}$$

where the first term comes from L acting on

$$\begin{pmatrix} A \\ P \\ n \end{pmatrix}_0,$$

the third from the fact that $r = 1 + R$ and the R part acts on

$$\begin{pmatrix} A \\ P \\ n \end{pmatrix}_0,$$

the second from the quantity $\partial/\partial T_1 + i\Omega_1$ acting on the second term in Eq. (18), and the fourth from the nonlinear product $n_1 A_0$. The solvability condition is

$$(\sigma + 1)\frac{\partial B}{\partial T_2} = \sigma\left(\frac{\partial}{\partial T_1} + i\Omega_1\right)^2 B + \sigma B - \frac{\sigma}{b}B^2 B^*,$$

which, when combined with Eq. (17), and upon reintroducing $R = r - 1$, $x = R^{1/4}X$, $y = R^{1/4}Y$, $\Omega = R^{-1/2}\Omega_1$ and calling $\psi = \sqrt{R}B$ gives

$$(\sigma - 1)\psi_t - ia\nabla^2\psi + i\sigma\Omega\psi + \frac{\sigma}{(\sigma + 1)^2}(a\nabla^2 + \Omega)^2\psi = \sigma(r - 1)\psi - \frac{1}{b}\psi^2\psi^*. \tag{20}$$

For obvious reasons, we call Eq. (20) the CSHD (complex Swift-Hohenberg with dispersion) equation. For $a = \Omega = 0$ and ψ real, it is exactly the Swift-Hohenberg equation. It is interesting that nonlinear optics provides the first context for which the SH equation is a legitimate asymptotic approximation, just as it does the Lorenz equations. In the fluid context for which both the Lorentz and the Swift-Hohenberg equations were first suggested, neither approximation is quantitatively valid in any asymptotic sense!

For $\Omega < 0$ and finite,

$$\begin{pmatrix} A \\ P \\ n \end{pmatrix} = \begin{pmatrix} 1 \\ x_0 \\ 0 \end{pmatrix} \psi \exp(-i\frac{\sigma\Omega}{\sigma+1}t) + \cdots \tag{21}$$

and ψ satisfies $(r_0 = 1 + \Omega^2/(\sigma+1)^2 = x_0 x_0^*)$

$$(1 + \frac{\sigma x_0^2}{r_0})\frac{\partial\psi}{\partial t} - ia\nabla^2\psi = \frac{\sigma x_0^2}{r_0}\psi\left(r - r_0 - \frac{1}{b}\psi\psi^*\right). \tag{22}$$

In the limit, $\Omega \to 0$, Eqs. (20) and (22) agree.

For $\Omega > 0$ and finite, we take the one-dimensional case of two oppositely travelling waves, where

$$\begin{pmatrix} A \\ P \\ n \end{pmatrix} = \begin{pmatrix} 1 \\ 1 \\ 0 \end{pmatrix} \left(\psi_1 e^{ik_o x} + \psi_2 e^{-ik_o x}\right) e^{-i\Omega t} + \cdots \tag{23}$$

and find

$$(\sigma+1)\frac{\partial\psi_1}{\partial t} + 2ak_0\frac{\partial\psi_1}{\partial x} - ia\nabla^2\psi_1 + \frac{\sigma a^2}{(\sigma+1)^2}(2ik_0\frac{\partial}{\partial x} + \frac{\partial^2}{\partial y^2})^2\psi_1$$

$$= \sigma(r-1)\psi_1 - \frac{1}{b}(|\psi_1|^2 + 2|\psi_2|^2)\psi_1, \tag{24}$$

$$(\sigma+1)\frac{\partial\psi_2}{\partial t} - 2ak_0\frac{\partial\psi_2}{\partial x} - ia\nabla^2\psi_2 + \frac{\sigma a^2}{(\sigma+1)^2}(-2ik_0\frac{\partial}{\partial x} + \frac{\partial^2}{\partial y^2})^2\psi_2$$

$$= \sigma(r-1)\psi_2 - \frac{1}{b}(2|\psi_1|^2 + |\psi_2|^2)\psi_2. \tag{25}$$

Again, in the limit $\Omega \to 0$ through positive values, Eqs. (20) and (24), Eq. (25) agree.

Equation (20), the CSHD, is a new and canonical equation for patterns and is worthy of study in its own right. When the Raman laser terms, mentioned earlier, are included, the changes are that $\Omega \to \Omega + \delta|\psi|^2$ and the coefficient of $\psi^2\psi^*$ is complex and can take either sign. Thus, new and nontrivial behavior of the Benjamin-Feir turbulence type with dislocation nucleation, can occur.

The limit b small (of order R, say,) is also of much interest. In that case, the field $n(0, 0, 1)$ must be introduced as an additional order parameter and one obtains coupled equations between n and ψ. Lega and Meunier are currently working on this.

We will now study Eq. (20), as a legitimate model for Eqs. (1), (2), and (3), in the case where $r - 1$ is not small.

6. FAR FROM ONSET

It is not hard to show that in a competition between planforms, travelling waves, standing waves, or multiple combinations associated with different wavevectors $\vec{k}$, all lying on the critical circle, $|\vec{k}|^2 = \Omega_0/a$, the single travelling wave solution is the only stable one and therefore is the locally preferred planform. We now study these travelling waves, and in particular modulations thereof, as the stress parameter r is increased. To do this, we will use as a model a modification of Eq. (20) which covers both the case of two level atoms and the Raman laser medium.

First, observe that Eq. (20) admits simple periodic solutions $\psi = f(\theta; A) = A \exp i\theta$, $\theta = \vec{k} \cdot \vec{x} - \omega t$ which correspond to travelling plane waves. We next seek to find a description for a class of solutions that, almost everywhere, are modulations of plane wave solutions in which the wavevector $\vec{k}$ is no longer constant but varies slowly over some distance much greater than the pattern wavelength λ. The small parameter in this theory is $\epsilon = \lambda/l$, the inverse aspect ratio, rather than $R = r - r_0$, the distance from onset. The dividing line between the theories is the ratio $\sqrt{R}/\epsilon$, which we take to be order one or larger. In order to make the problem a little more interesting, we will examine the equation,

$$W_t - ia\nabla^2 W + (\nabla^2 + k_0^2)^2 W = \alpha W - \beta W^2 W^* \qquad (26)$$

α, β complex and $a, \alpha_r, \beta_r > 0$. The exact, periodic travelling waves are particularly simple as they contain only one harmonic. We now seek solutions of Eq. (26) in the form

$$W = f(\theta = \frac{\Theta(X = \epsilon x, Y = \epsilon y, T = \epsilon t)}{\epsilon}, A) + \epsilon W_1 + \cdots, \qquad (27)$$

where $f(\theta)$ is 2π-periodic and has the form of an exact travelling wave. The difference this time, however, is that $\vec{k} = \nabla_x \theta = \nabla_X \Theta$ is not constant but slowly varying. Consequently $f(\theta)$ is not an exact solution of Eq. (26) and corrections $W_1, W_2, \cdots$ must be sought. Because of translational invariance in both space and time, if $f(\theta)$ is a solution, so is $f(\theta + \theta_0)$ so that $\partial f/\partial \theta_0 = \partial f/\partial \theta$ satisfies the homogeneous equations for the infinitesimal variation of W about $f(\theta)$. Hence, the nonhomogeneous equations for $W_1, W_2, \cdots$, have solvability conditions. These solvability conditions give rise to the phase diffusion equation and its higher order corrections.

The calculation is particularly simple in the present context because all solutions $f, W_1, W_2, \cdots$ have the same form $A \exp i\theta$ and thus we may simply substitute $W = A \exp i\theta$. The later iterates can be found by expanding A and θ, but it is more correct and more useful[4] to consider the equations for A and θ without re-expanding. We obtain, to order ϵ,

$$\beta_r A^2 = \alpha_r - (k^2 - k_0^2)^2 - \epsilon a \frac{A_T + 2\vec{k} \cdot \nabla A + A \nabla \cdot \vec{k}}{A} + O(\epsilon^2), \qquad (29)$$

$$A^2(\Theta_T + \omega) - 2\epsilon \nabla \cdot \vec{k} A^2 (k^2 - k_0^2) + 0(\epsilon^2) = 0, \qquad (29)$$

where

$$\omega(k^2, A^2) = ak^2 - \alpha_i + \beta_i A^2. \tag{30}$$

Now, if $a = \alpha_i = \rho_i = 0, \omega = 0$, then the relevant time scale is $T_2 = \epsilon T = \epsilon^2 t$, the horizontal diffusion time scale, namely the time necessary for diffusion influences to move across the container. If $\beta_i, a \neq 0$, we need to account for group velocity advection which takes place on the ϵ^{-1} time scale. In the former case, we may replace A^2 in Eq. (29) by

$$A_0^2 = \frac{1}{\beta_r} \left(\alpha_r - (k^2 - k_0^2)^2 \right), \tag{31}$$

which is accurate to order ϵ^2 and then obtain the Cross-Newell CN phase diffusion equation,

$$\tau(k)\Theta_{T_2} + \nabla \cdot \vec{k}B(k) = 0, \tag{32}$$

with

$$\tau(k) = A_0^2(k), \ B(k) = A_0^2 \frac{dA_0^2}{dk^2}, \tag{33}$$

relevant for the complex Swift-Hohenberg equation with real coefficients. When dispersion and wavelike behaviors are present, however, one must include the $0(\epsilon)$ correction to Eq. (31) because the ω in Eq. (29) contains A^2 and will lead to an order ϵ correction comparable in magnitude to the diffusion term $\nabla \cdot \vec{k}A^2(k^2 - k_0^2)$. We write

$$A^2 = A_0^2 - \frac{\epsilon a}{2\beta_r} \frac{(A_0^2)_T + \nabla \cdot \vec{k}A_0^2}{A_0^2} + \theta(\epsilon^2) \tag{34}$$

and

$$\omega = \omega_0 - \frac{\epsilon a\beta_i}{2\beta_r} \frac{(A_0^2)_T + \nabla \cdot \vec{k}A_0^2}{A_0^2} + 0(\epsilon^2) \tag{35}$$

where ω_0 is Eq. (30) with A^2 replaced by A_0^2. Now,

$$\frac{\partial A_0^2}{\partial T} = \frac{dA_0^2}{dk^2} 2\vec{k} \cdot \vec{k}_T = -2\frac{dA_0^2}{dk^2} \vec{k} \cdot \nabla_X \omega_0 + 0(\epsilon), \tag{36}$$

The phase diffusion equation is then (to order ϵ),

$$A_0^2(\Theta_T + \omega_0) + \epsilon \nabla \cdot \vec{k}(B - \frac{a\beta_i}{2\beta_r}A_0^2) + \frac{\epsilon a\beta_i}{2\beta_r}(A_0^2)'\vec{k} \cdot \nabla k^2 = 0, \tag{37}$$

where,

$$(A_0^2)' = \frac{dA_0^2}{dk^2}, B = A_0^2(A_0^2)', \quad \omega_0 = ak^2 - \alpha_i + \beta_i A_0^2, \tag{38}$$

Observe that, because of the last term, it is not in flux divergence form which has major nontrivial consequences (it cannot be written in gradient form; the shock conditions when the order ϵ terms are hyberbolic in X, Y, may depend on the form of regularization) which we discuss elsewhere.[19] In fact, even when no dispersion and traveling wave effects are present, one is still not guaranteed that the spatial derivatives in Eq. (32) have flux divergence form with $B(k)$ analytic in k for all k within the neutral stability boundaries $k_l < k < k_r$, $A(k_l) = A(k_r) = 0$. This is a nontrivial point that I do not have room to discuss here. The interested reader should consult Passot and Newell.[19]

The long wave instability boundaries (the Eckhaus and zig-zag instabilities) are found by linearizing about $\Theta = kX$, that is by setting $\Theta = kX - \omega_0 T + \Theta'$ and dropping terms quadratic in Θ'. We find

$$A_0^2 \Theta'_T + \epsilon \left\{ \frac{d}{dk} \left(B - \frac{a\beta_i}{2\beta_r} A_0^2 \right) + \frac{2a\beta_i}{\beta_r} (A_0^2)' k^2 \frac{\partial \omega_0}{\partial k^2} \right\} \Theta'_{XX}$$
$$+ \epsilon \left\{ B - \frac{a\beta_i}{2\beta_r} A_0^2 \right\} \Theta'_{YY} = 0, \tag{39}$$

from which we see that both boundaries

$$k = k_Z, B - \frac{a\beta_i}{2\beta_r} A_0^2 = 0 \qquad (\text{zig} - \text{zag}), \tag{40}$$

and

$$k = k_E, \frac{d}{dk} k \left(B - \frac{a\beta_i}{2\beta_r} A_0^2 \right) + \frac{2a\beta_i}{\beta_r} (A_0^2)' k^2 \frac{\partial \omega_0}{\partial k^2} = 0 \qquad (\text{Eckhaus}), \tag{41}$$

are shifted from their dispersionless ($a = 0$) positions. Observe that when $k = k_0$, the travelling wave is unstable to perturbations depending on the coordinate along the crests of the waves when $\alpha\beta_i < 0$. This is nothing other than the Benjamin-Feir instability.[1] Indeed, it is possible, I have not checked, that there is no Busse balloon (the stable band between $k_Z < k < k_{Er}$, k_{Er}, the right Eckhaus boundary) at all in certain parameter ranges. On the other hand, for a defocusing medium, $\beta_i > 0$, the zig-zag instability is weakened.

7. TRAVELLING SHOCKS, DEFECTS, PHASE GRAIN BOUNDARIES, AND OTHER SINGULARITIES

The first type of weak solution which emerges from Eq. (37) is due entirely to the fact that ω_0 is nonlinear in k and that to leading order

$$\vec{k}_T + (\vec{V}_g \cdot \nabla)\vec{k} = 0, \qquad \vec{V}_g = \frac{d\omega_0}{dk} \hat{k}. \tag{42}$$

Equation (42) is a two-dimensional form of Burgers' equation. In one dimension, $\vec{k} = \vec{k}(X, T)$ only and,

$$k_T + \frac{\partial \omega_0}{\partial k} k_X = 0, \tag{43}$$

is Burgers' equation. Because of the regularizing influence of the order ϵ terms (and this of course depends critically on whether the wavenumbers involved lie in the Busse balloon), the weak solutions that are realized are shocks that travel with speeds $S = (\omega_0(k_+) - \omega_0(k_-))/(k_+ - k_-)$, with k_+ and k_- being constant wavenumber states before and after the shock. Solutions of this type have been examined by Nozaki et al.[18] in the context of the complex Ginzburg-Landau equation, and by Kopell and Howard[7] in the context of excitable media.

Another class of singular solutions arise because the local wavenumber k is ouside of the Busse balloon so that either the zig-zag instability Eq. (40) or the Eckhaus instability Eq. (41) is triggered. In the former case, the diffusion equation is regularized by including order ϵ^2 terms containing more spatial derivations of the phase. The upshot, however, is that, in the limit of small ϵ, the solution is again of simple wave type, in the sense that there is a phase grain boundary (separating zigs from zags) along which there is a jump in wavevector. If the discontinuity is only one of direction, then the shock is stationary and either absorbs or is a source for travelling waves.

For wavenumbers which lie outside the right Eckhaus boundary, however (and I have indicated that in certain parameter ranges this may mean all wavenumbers), the instability is not always saturated in a nearby state. (It can be; see Weber et al.[22]) In most cases, the pattern attempts to adjust its wavelength by forming point defects, pairs of dislocations, which move so as to remove roll pairs. When many defects are formed or when defects form everywhere (it can be that the far field wavenumber selected by the dislocation lies outside the stable band), the pattern develops a spatio-temporal turbulent character.

Finally, we mention the fundamental point singularities, disclinations. They are the building blocks of all other kinds of point singularities (for example, a dislocation is two bound pairs of convex and concave disclinations). They are not present in isolation in model Eq. (26) because of the $W \to W \exp i\theta$ symmetry it enjoys. However, should we break the symmetry by adding terms such as γW^*, W_0 constant, which corresponds in the light context to a pump electric field with a combination of the first and second harmonics of the natural medium frequency (or a frequency close to it). The bound pairs of disclinations can separate and isolated disclinations can occur. They first occur at the ends of phase grain boundaries so that the total pattern looks like a patchwork quilt of stationary rolls or travelling waves (depending on whether ω_0 is zero or nonzero), where domains of constant wavevector are enclosed by convex simplices, the edges of which are phase grain boundaries and the corners of which are disclinations. One final and key point. Isolated disclinations can only be captured by an order parameter which is real valued.

The reader will find more detailed discussions of these ideas in a couple of papers soon to be published.

ACKNOWLEDGMENTS

Research support from the Arizona Center for Mathematical Sciences, sponsored by AFOSR Contract 90-0021 with the University Research Initiative Program at the University of Arizona, and NSF INT 8914313, are gratefully acknowledged. I also enjoyed many useful discussions with Roland Ribotta and Pierre Coullet. Much of the work reported here was done in collaboration with Thierry Passot, Joceline Lega, Jerry Moloney, Nick Ercolani, and Rob Indik.

REFERENCES

1. Benjamin, T. B., and J. E. Feir. "The Disintegration of Wave Trains on Deep Water, Part 1: Theory." *J. Fluid Mech.* **27** (1967): 417–430.
2. Bernoff, A. "Slowly-Varying, Fully Nonlinear Wavetrains in the Ginzburg-Landau Equation." *Physica D* **30** (1988): 363–391.
3. Brand, H. R., P. S. Lomdahl, and A. C. Newell. "Benjamin-Feir Turbulence in Convective Binary Mixtures." *Physica D* **23** (1986): 345–361.
4. Cross, M. C., and A. C. Newell. "Convection Patterns in Large Aspect Ratio Systems." *Physica D* **10** (1984): 299–328.
5. Cross, M. C. "Traveling and Standing Waves in Binary-Fluid Convection in Finite Geometries." *Phys. Rev. Lett.* **57** (1986): 2935–2938.
6. Cross, M. C., and P. C. Hohenberg. "Pattern Formation Outside of Equilibrium." *Rev. Mod. Phys.* (1993).
7. Howard, L. N., and N. Kopell. "Slowly Varying Waves and Shocks Structures in Reaction-Diffusion Equations." *Stud. Appl. Math.* **56** (1977): 95–145.
8. Jakobsen, P. K., J. Lega, Q. Feng, M. Staley, J. V. Moloney, and A. C. Newell. *Nonlinear Transverse Modes of Large Aspect Ratio Homogeneously Broadened Lasers: I. Analysis and Numerics*, in press.
9. Kolodner, P., A. Passner, C. M. Surko, and R. W. Walden. "Onset of Oscillatory Convection in a Binary Fluid Mixture." *Phys. Rev Lett.* **56** (1986): 2621–2624.
10. Kolodner, P. "Stable and Unstable Pulses of Travelling Wave Convection." *Phys. Rev. A* **43** (1991): 2827–2832.

11. Lega J., P. K. Jakobsen, J. V. Moloney, and A. C. Newell. *Nonlinear Transverse Modes of Large Aspect Ratio Homogeneously Broadened Lasers: II. Pattern Analysis Near and Beyond Threshold,* in press.
12. Newell, A. C., and J. A. Whitehead. "Review of the Finite Bandwidth Concept." In *Proc. IUTAM 1969 Symposium on Instability of Continuous Systems,* edited by H. Leipholz, 284–289. Berlin: Springer-Verlag, 1971.
13. Newell, A. C., and J. A. Whitehead. "Finite Bandwidth, Finite Amplitude Convection." *J. Fluid Mech.* **38** (1969): 279–303.
14. Newell, A. C. "Envelope Equations." In *Lectures in Applied Mathematics, Nonlinear Wave Motion,* Vol. 15, 157–163. Providence, RI: AMS, 1974.
15. Newell, A. C., T. Passot, and M. Souli. "The Phase Diffusion and Mean Drift Equations for Convection at Finite Rayleigh Numbers in Large Containers." *J. Fluid Mech.* **220** (1990): 187–252.
16. Newell, A. C., and J. V. Moloney. *Nonlinear Optics.* Reading, MA: Addison-Wesley, 1992.
17. Newell, A. C., T. Passot, and J. Lega. "Order Parameter Equations for Patterns." *Ann. Rev. Fluid Mech.* **25** (1993): 399–653.
18. Nozaki, K., and N. Bekki. "Pattern Selection and Spatio-Temporal Transition to Chaos in the Ginzburg-Landau Equation." *Phys. Rev. Lett.* **51** (1983): 2171–2173.
19. Passot, T., and A. C. Newell. "Towards a Universal Theory of Patterns." *Physica D* (1994): to appear.
20. Pomeau, Y., and P. Manneville. "Stability and Fluctuations of a Spatially Periodic Convective Flow." *J. Phys. Lett.* **40** (1979): 609–612.
21. Segel, L. A. "Distant Sidewalls Cause Slow Amplitude Modulation of Cellular Convection." *J. Fluid Mech.* **38** (1969): 203–224.
22. Weber, A., L. Kramer, I. S. Aranson, and J. E. Aranson. "Stability Limits of Traveling Waves and the Transition to Spatio-Temporal Chaos in the Complex Ginzburg-Landau Equation." Preprint, 1992
23. Whitham, G. B. *Linear and Nonlinear Waves.* Chichester, UK: Wiley-Interscience, 1974.

W. van Saarloos
Instituut Lorentz, University of Leiden, P. O. Box 9506, 2300 RA Leiden, The Netherlands

The Complex Ginzburg-Landau Equation for Beginners

SCOPE OF THIS CHAPTER

Several systems discussed at this workshop on *Spatio-Temporal Patterns in Non-equilibrium Complex Systems* have been related to or analyzed in the context of the so-called Complex Ginzburg-Landau equation (CGL). What is the difference between the physics underlying the usual amplitude description for stationary patterns and the one underlying the CGL? Why are there many more stable *coherent structures* [pulses, sources (holes), sinks] possible in systems described by the CGL than in systems exhibiting a stationary bifurcation, and what is their relation, if any, to the chaotic behavior that is characteristic of the CGL in some parameter regimes? The organizers of this workshop have asked me to try to provide some answers to these questions for the *non-expert*, someone with an interest in pattern formation but who has not had an introduction to the CGL before or who has not followed the recent developments in this field. Since there are several very recent review papers on this subject[3,8,9,13] where a more thorough and detailed discussion

Spatio-Temporal Patterns, Ed. P. E. Cladis and P. Palffy-Muhoray,
SFI Studies in the Sciences of Complexity, Addison-Wesley, 1995   **19**

can be found,[1] I will confine myself here to a brief low-level introduction, in which I try to paint some of the main ideas with broad strokes. I stress that I do not pretend to give a balanced review—this chapter is extremely sketchy and colored by my own interests, and I urge the reader interested in learning more about this line of research to consult the papers cited above and the references therein. A more detailed introduction aimed at graduate students is given by van Hecke et al.[12]

THE CGL AS AN AMPLITUDE EQUATION

Amplitude equations describe the slow modulation in space and time near the threshold for an instability. Imagine we consider a system that initially is in a basic state which is homogeneous (e.g., the purely conductive state of a Rayleigh-Bénard cell), and that exhibits a finite-wave-length instability when one of the parameters is varied. We can always define an appropriately rescaled control parameter ε, so that this instability occurs when ε becomes positive. The existence of a finite wave length instability means that when we consider the evolution of a Fourier mode $\exp[-i\Omega t + ikx]$, the growth rate $\text{Im}\,\Omega$ of each mode k behaves as sketched in Figure 1. For $\varepsilon < 0$, all modes are decaying ($\text{Im}\,\Omega < 0$), so the basic state is stable. If k_c denotes the wave number of the mode whose growth rate is zero at $\varepsilon = 0$, then we see that there is a narrow band of width around k_c where the growth rate is slightly positive (of order ε) for $\varepsilon > 0$. Let us assume that the transition to patterns corresponds to a supercritical bifurcation, meaning that the nonlinearities are saturating so that patterns slightly above threshold have a small amplitude. We then also expect that the patterns slightly above threshold,[2] i.e., for $0 < \varepsilon \ll 1$, will have a wave length close to $2\pi/k_c$. However, due to the fact that the width of the band of unstable modes is nonzero and of order $\varepsilon^{1/2}$, there can be slow modulations of the patterns on length scales of order $1/\varepsilon^{1/2}$.[3] Close to threshold, a typical physical field (the temperature field in a Rayleigh-Bénard cell, say) will therefore look like that sketched in Figure 2.

---

[1] Note that van Saarloos and Hohenberg's paper[13] was not written as an introductory review on the complex CGL *approach*, but as a research paper on the CGL equation; its main emphasis is on systematizing many (new) results for coherent structures in the one-dimensional equation.

[2] Of course, if we start with special initial conditions, the wave number can be different during initial transients.

[3] Just like in radio transmission a low-frequency signal is transmitted by modulating a high-frequency signal within a narrow band.

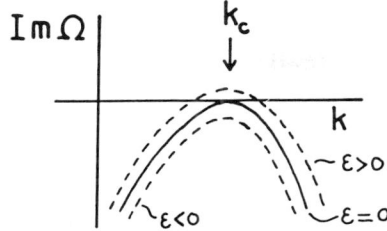

FIGURE 1 Behavior of the growth rate as a function of wave number k.

FIGURE 2 Slow modulation of the critical modes for $\varepsilon \ll 1$. The envelope is indicated with the dashed line.

If $\mathrm{Re}\,\Omega = 0$, the amplitude of modes near k_c is growing in time for $\varepsilon > 0$, but each mode is stationary in space. This situation arises near a bifurcation to stationary patterns. If $\mathrm{Re}\,\Omega \neq 0$, each mode corresponds to a traveling wave, so we will generally have $\mathrm{Re}\,\Omega \neq 0$ at a transition to traveling wave patterns. Let us first consider the case of a stationary bifurcation ($\mathrm{Re}\,\Omega = 0$). The separation of the dynamics of the patterns close to threshold in terms of a fast component (with the length scale set by the critical wave number) and an envelope that varies slowly in space and time, can then be formulated by writing the relevant field(s) close to threshold as follows:

$$\text{physical fields} \propto A(x,t)e^{ik_c x} + A^*(x,t)e^{-ik_c x} + \text{higher harmonics} , \qquad (1)$$

where $A(x,t)$ is the complex amplitude or envelope, and where "higher harmonics" stands for terms proportional to $\exp(2ik_c x)$. For simplicity, we will consider here only one dimension. To lowest order in ε, A then obeys an equation of the form

$$\tau_0 \frac{\partial A}{\partial t} = \xi_0^2 \frac{\partial^2 A}{\partial x^2} + \varepsilon A - g|A|^2 A , \quad (g > 0) . \qquad (2)$$

For a given problem, the constants τ_0, ξ_0, and g can in principle be calculated from the starting equations. As they only set the scales of time, of length, and of the size of the amplitude, we will rescale these here and write the equation in the rescaled form

$$\frac{\partial A}{\partial t} = \frac{\partial^2 A}{\partial x^2} + \varepsilon A - |A|^2 A . \qquad (3)$$

A few remarks concerning this result are in order:

i.   With an additional rescaling $x \to \varepsilon^{-1/2}x$, $t \to \varepsilon^{-1}t$, $A \to \varepsilon^{1/2}A$, ε can be scaled out of Eq. (3). This confirms that patterns evolve on long time scales ε^{-1} and large length scales $\varepsilon^{-1/2}$, and that the amplitude of the patterns grows as $\varepsilon^{1/2}$ as we had anticipated. Nevertheless, we prefer to keep ε explicit in Eq. (3), so as to avoid a control parameter dependent rescaling. Moreover, this will make it easier to consider what happens when ε changes through zero.

ii. Although the equation and the coefficients τ_0, etc. can be calculated from the full equations describing the physical problem under study, this equation arises naturally near any stationary supercritical bifurcation if the system is translation invariant and reflection symmetric ($x \to -x$). The latter symmetry dictates that the second-order term $\partial^2/\partial x^2$ arises as the lowest order spatial derivative, while the form of the cubic term is prescribed by the requirement that the equation is invariant upon multiplying A by an arbitrary phase factor $\exp(i\phi)$: according to Eq. (1) this corresponds to translating the pattern by a distance ϕ/k_c, so translation invariance implies that the equation for A has to be invariant under $A \to Ae^{i\phi}$.[4]

iii. Note that for $\varepsilon > 0$ Eq. (3) has stationary solutions of the form $A = a_0 e^{iqx}$, with $q^2 = \varepsilon - a_0^2$. According to Eq. (1), these so-called phase winding solutions describe steady-state periodic patterns with total wave number slightly bigger ($q > 0$) or slightly smaller ($q < 0$) than k_c. We will come back to the stability of these solutions below.

iv. It is easy to check that Eq. (3) can be written in the form

$$\frac{\partial A}{\partial t} = -\frac{\delta \mathcal{F}}{\delta A^*} \,, \qquad \text{with } \mathcal{F} = \int dx \left[\left|\frac{\partial A}{\partial x}\right|^2 - \varepsilon|A|^2 + 1/2|A|^4 \right] \,, \qquad (4)$$

from which it follows that $d\mathcal{F}/dt < 0$. Thus, $\mathcal{F}$ plays the role of a "free energy" or Lyapunov functional, and many aspects of the dynamics of patterns can simply be understood in terms of the tendency of patterns to evolve towards the "lowest free-energy" state. In this sense, the dynamics of Eq. (3) is very thermodynamic-like.

v.   Equation (3) has the form of the Ginzburg-Landau equation for superconductivity in the absence of a magnetic field. To distinguish it from the amplitude equation for traveling waves given below, we will refer to it as the real Ginzburg-Landau equation, since the coefficients in this equation are real.

In the case in which a traveling wave mode of the form $\exp(-i\Omega_c t + ik_c x)$ becomes unstable for $\varepsilon = 0$, the separation of the pattern into the critical mode

---

[4] The invariance under a change of the phase implies that the phase is a slow variable, whose dynamics is governed by a diffusion-type equation.

and a slowly varying amplitude can then be done in essentially the same way as before; we now write[5]

$$\text{physical fields} \propto A(x,t)e^{-i\Omega_c t + i k_c x} + A^*(x,t)e^{i\Omega_c t - i k_c x} + \text{higher harmonics} . \quad (5)$$

In this case, the equation obeyed by A reads, after properly rescaling space, time, and amplitude,[6]

$$\frac{\partial A}{\partial t} = (1 + ic_1)\frac{\partial^2 A}{\partial x^2} + \varepsilon A - (1 - ic_3)|A|^2 A . \quad (6)$$

This equation will be referred to as the Complex Ginzburg-Landau (CGL) equation. How can we understand that the amplitude equation for traveling waves has complex coefficients in front of the second derivative and cubic term? To see this, note that for Eq. (3) we found a band of stationary phase winding solutions $A = a_0 e^{iqx}$; these correspond to a band of stationary patterns of the physical system that are periodic in space. For traveling waves, there is also a band of traveling wave solutions $A = a_0 e^{-i\omega t + iqx}$ with $\text{Im}\,\omega = 0$ to exist. Just like q measures the difference between the wave number of the pattern and the critical wave number, so does ω measure the difference between the frequency of the pattern and the frequency of the critical mode, Ω_c—after all, there is in general no reason why the frequency should not depend on the wave number or on the amplitude of the pattern. Indeed, if we substitute the Ansatz $A = a_0 e^{-i\omega t + iqx}$ with $\text{Im}\,\omega = 0$ into Eq. (6), we obtain

$$\omega = c_1 q^2 - c_3 a_0^2 , \qquad q^2 = \varepsilon - a_0^2 . \quad (7)$$

The expression for ω illustrates that c_1 is the coefficient that measures the strength of the linear dispersion, i.e., the dependence of the frequency of the waves on the wave number, while c_3 is a measure of the nonlinear dispersion. We will see below that these terms can dramatically affect the stability of the phase winding solutions.

Let us also make a number of brief comments on the CGL:

i.   For traveling waves, the group velocity $v_{gr} \equiv \partial\omega/\partial k$ is generally nonzero. As it stands, Eq. (6) is written in the frame moving with the group velocity. In the laboratory frame, there is an additional term $v_{gr}\partial A/\partial x$ on the left-hand side of the equation. This distinction is particularly important for determining whether the instability is convective or absolute.

ii. We have written only one CGL equation for a single amplitude. When the underlying system is symmetric under reflection[7] ($x \to -x$), both left- and

---

[5] Note that $\text{Im}\,\Omega = 0$ for $k = k_c$.

[6] There are many different conventions for the imaginary parts of the coefficients in Eq. (6); here I follow the convention of van Saarloos and Hohenberg.[13]

[7] This happens, e.g., for traveling waves in binary mixtures, but not in the rotating Rayleigh-Bénard cells discussed by Ecke and co-workers at this workshop.

right-moving traveling waves can exist. For such systems, one actually obtains two coupled CGL equations, one for the amplitude of the left-moving waves and one for the amplitude of the right-moving waves. Depending on the nonlinear interaction terms, one can either have a situation in which standing waves are favored, or one in which one wave completely suppresses the other. In the latter case, one can effectively use a single CGL equation like Eq. (6).

iii. For $c_1, c_3 \neq 0$, the CGL can *not* be derived from a Lyapunov functional or "free energy." As a result, it displays a much richer type of dynamical behavior than the real Ginzburg-Landau equation (3). In fact, in the limit $c_1, c_3 \rightarrow \infty$ the equation reduces to the nonlinear Schrödinger equation, which is not only Hamiltonian but also integrable (it has the well-known soliton solutions). The fact that the CGL reduces to a relaxational equation in one limit and to a Hamiltonian equation in another limit makes the equation very interesting from a theoretical point of view. In addition, these two limits can be exploited as starting points for perturbation theories.

STABILITY OF PHASE WINDING SOLUTIONS

We saw above that both the real and the complex Ginzburg-Landau equation admit phase winding solutions with wave-vector $-\varepsilon^{1/2} < q < \varepsilon^{1/2}$. For the real equation these are stationary solutions that correspond to stationary periodic patterns in the band $[k_c - \varepsilon^{1/2}, k_c + \varepsilon^{1/2}]$; in the complex case, these correspond to spatially periodic traveling wave solutions. What is the stability of these solutions? The linear stability analysis of these solutions is quite straightforward, and can be found in the various reviews cited. Here we just want to give the reader some intuitive understanding of the fact that, in the region $c_1 c_3 > 1$, all periodic solutions are linearly unstable (so-called Benjamin-Feir unstable).

For the real case ($c_1, c_3 = 0$), the qualitative answer is well known; consider the left part of Figure 3(a). For a given $\varepsilon > 0$, the values of q for which steady-state solutions exist fall within the solid line. However, only the solutions within the dashed lines are stable—solutions corresponding to values of q in the hatched region, close to the edge of the band, are linearly unstable (sometimes called Eckhaus or Benjamin-Feir unstable). Intuitively, one may understand this as follows. A wave number q close to the left edge of the band corresponds to a smaller total wave number of the pattern and, hence, a larger wave length. When the wave length is too small, as illustrated for a roll pattern in a Rayleigh-Bénard cell on the left of Figure 3(b), the pattern is unstable because a roll is so wide that it will split into three. Likewise, a q near the right edge amounts to a pattern that is unstable because three narrow rolls like those in the right part of Figure 3(b) merge into one. Only patterns with wave length close to the critical one (those in the center of Figure 3(b) with $q \approx 0$) are stable. Now, in the Rayleigh-Bénard example of

Figure 3(b), the phase difference of $Ae^{ik_c x}$ between two points divided by 2π is equal to the number of pairs of rolls between these two points. Thus, when three rolls merge into one or when one roll splits into three, the number of phase windings of A over a certain distance changes by one. But since the phase of A is well defined and continuous whenever $|A|$ is nonzero, the only way the number of phase windings can change discontinuously in a localized region is if at some point in time and space $|A| = 0$. At that point the phase is undefined, and so can "slip" by 2π. These points are called phase slip centers. In Figure 4 we illustrate the rapid variation of the phase and the decrease in modulus $|A|$ that leads to such behavior.

We note that the analogous process is well known for the superconductivity of one-dimensional wires (thin enough that the magnetic field generated may be neglected). In Ginzburg-Landau theory, the supercurrent is proportional to the gradient of the phase, so phase winding solutions are current-carrying solutions. The wave number corresponding to a value on the dashed line of Figure 3(a) and beyond which the pattern is unstable, corresponds to the critical current: beyond the critical current, these current-carrying solutions are unstable and lead to the

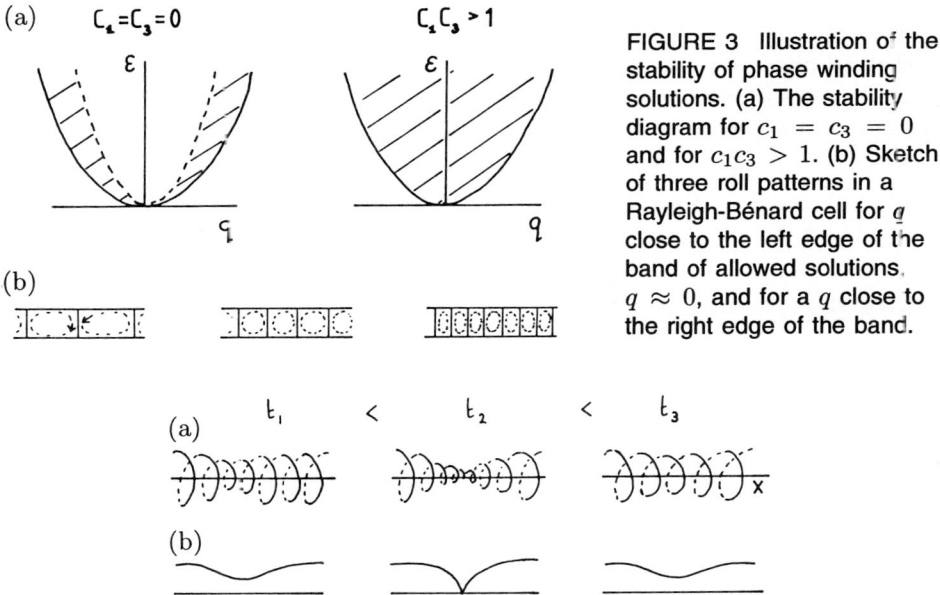

FIGURE 3 Illustration of the stability of phase winding solutions. (a) The stability diagram for $c_1 = c_3 = 0$ and for $c_1 c_3 > 1$. (b) Sketch of three roll patterns in a Rayleigh-Bénard cell for q close to the left edge of the band of allowed solutions, $q \approx 0$, and for a q close to the right edge of the band.

FIGURE 4 Illustration of the dynamical process by which phase winding solutions with too large $|q|$ go unstable. In (a) the complex envelope A is plotted as a function of x for three different times. Note that the plane perpendicular to the x-axis is a complex plane. In (b) the dynamics of $|A|$ is sketched; at time t_2 the phase slip center occurs.

generation of phase slips. In superconducting wires, there is also a small dissipation for currents slightly smaller than the critical current due to phase slips that are generated by thermal fluctuations—see Langer and Ambegaokar[6] for details.

When the parameters c_1 and c_3 in the CGL are nonzero but small, the stability of phase winding solutions is qualitatively the same as for the case $c_1 = c_3 = 0$ discussed above, with the band of stable solutions narrowing when c_1 and c_3 are of the same sign and increasing. However, when the product $c_1 c_3$ equals one, the width of the band of stable solutions is zero, so that for $c_1 c_3 > 1$ none of the phase winding solutions is stable, as illustrated on the right-hand side of Figure 3(b). To get some feeling for why this happens,[8] note that we can combine the two Eqs. (6) to

$$\omega = \varepsilon c_3 + (c_1 + c_3)q^2 \ . \tag{8}$$

This equation shows that, if we imagine a distortion of a phase winding solution in the case c_1, $c_3 \neq 0$ so that the local wave number[9] q (or, because of Eq. (7), the local modulus) varies with x, we see that regions with a different local wave number will rotate with a different frequency, and the phase gradients tend to increase, as if the spring-like pattern of Figure 4(a) is wound up more tightly. We can imagine that when c_1 and c_3 are large enough the difference in local frequencies tends to drive every perturbation unstable [and possibly towards the generation of a phase slip event illustrated in Figure (4)]. This is indeed what happens for $c_1 c_3 > 1$. The fact that all phase winding solutions are unstable in this regime of the CGL leads to various types of chaotic behavior, and is one of the examples of the richness of behavior encountered in this equation—see, e.g., the contribution by Chaté to this workshop.

COHERENT STRUCTURES

Another aspect of the CGL is that there is a larger variety of "coherent structures." These are solutions that are either themselves localized or that consist of domains of regular patterns connected by localized defects or interfaces. In two dimensions,

[8] The following argument only partially "explains" why this happens. In particular, it would suggest that only the combination $c_1 + c_3$ is important, whereas in reality the product $c_1 c_3$ determines the transition to the regime where all phase winding solutions are unstable: both linear and nonlinear dispersion need to be present. Another caveat: the argument below is not meant to suggest that phase slips *always* occur if all phase winding solutions are unstable. In the "phase chaos" regime discussed by Chaté (see also Shraiman et al.[11]), one is indeed in the Benjamin-Feir unstable regime, but no phase slips are dynamically generated.

[9] This is allowed since the Benjamin-Feir instability we are considering is essentially a long-wavelength instability.

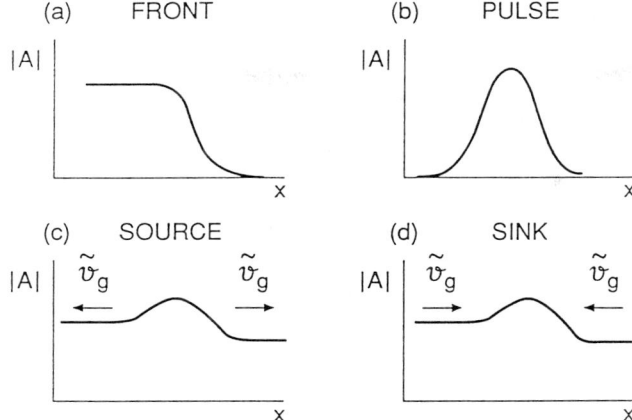

FIGURE 5 Four possible coherent structures in the one-dimensional CGL

a well-known example is a spiral,[10] but we will confine ourselves here to summarizing the most important features of the one-dimensional structures illustrated in Figure 5.

SOURCES AND SINKS

Let us first consider solutions that connect one phase winding solution on the left to another phase winding solution on the right. These are domain wall or shock-type solutions. Since a traveling wave has a nonzero group velocity v_{gr}, there are several possibilities depending on whether the group velocity of each phase winding solution points away from or towards the localized structure connecting the two asymptotic states. It is useful to use the group velocity $\tilde{v}_{gr}$ relative to the velocity v of this localized structure: $\tilde{v}_{gr} \equiv v_{gr} - v$. Thus $\tilde{v}_{gr}$ is the velocity with which a small perturbation to a phase winding solution travels in the frame moving with the domain wall between the two states we consider. If the relative group velocity points away from it on both sides, the localized structure has the properties of a *source*, while if $\tilde{v}_{gr}$ points inwards on both sides, we will call the domain wall a *sink*.[11] In principle, solutions with the relative group velocity pointing in the same direction on both sides of the wall are also possible, but one can show[13] that such solutions do not exist in the cubic CGL (6). They should, however, exist in higher-order extensions, like the quintic extension of the CGL mentioned below. In the real Ginzburg-Landau equation, most of the dynamics of domain walls can be understood in terms of the tendency of walls to move in the direction of the

[10] See the contribution by Huber at this workshop.
[11] See, for example, Coullet, Frisch, and Plaza[4] and references therein, and van Saarloos and Hohenberg.[13]

lowest "free energy" state [see the discussion after Eq. (4)]. The behavior of sinks and sources in the CGL is more complicated.

Sinks are, in a sense, relatively dull objects, since the very fact that the relative group velocities point inwards means that the waves it connects must come from some other regions in space. One then tends to focus on the dynamics in these regions. Nevertheless, the velocity of sinks is important during transients by determining which region shrinks and which one expands. For example, in two dimensions the arms of competing spirals form a sink-type solution when they meet. In this case, it is known that the sink moves in the direction of the spiral with the smallest frequency. In this sense, the sink does play a role in determining which spiral survives. In the one-dimensional CGL, there is typically a two-parameter family of sinks for not too large velocities.[13] This means that if we select two arbitrary asymptotic phase winding solutions on the left and right with group velocities pointing inwards, there is always a sink solution that connects these two states. I would intuitively expect such solutions to be stable, but their stability has, to my knowledge, not been studied in detail.

Sources send out waves, and so may determine the large time asymptotic dynamics. On the basis of simple counting arguments,[13] one generically expects the existence of a $v = 0$ source, as well as a *discrete set* of $v \neq 0$ solutions.[12] Surprisingly, however, Bekki and Nozaki[2] found a *family* of exact source solutions of Eq. (6); Hohenberg and I have taken this as a hint that there might be some hidden symmetry or some accidental nongenericity in the cubic CGL. Support for this point of view comes from recent work by Aranson et al.,[1] who discovered that if a small perturbation is added to the CGL (6), the stability of these solutions depend sensitively on the sign and strength of this perturbation.[13] For certain ranges of the parameters c_1 and c_3, these source solutions play an important role in the chaotic dynamics of the CGL—see Chaté's contribution or Aranson et al.[1] for details.

PULSES

The work on pulse-type solutions of the type sketched in Figure 5 was motivated largely by observations of localized convective regions in experiments on binary mixtures by Niemela, Ahlers, and Cannell, and by Kolodner.[14] Since the instability to traveling waves in this system actually corresponds to a subcritical bifurcation, this motivated a number of groups to study the following quintic extension of the CGL

$$\frac{\partial A}{\partial t} = (1 + ic_1)\frac{\partial^2 A}{\partial x^2} + \varepsilon A + (1 - ic_3)|A|^2 A - (1 - ic_5)|A|^4 A . \tag{9}$$

[12] In other words, one expects that in addition to the $v = 0$ solutions there are only solutions with particular values of the velocity and asymptotic wave numbers.

[13] The stability of the Bekki-Nozaki solutions was recently studied by Sasa and Iwamoto, and by Manneville and Chaté. See Aranson et al.[1] for references.

[14] See references in Riecke's paper.[10]

Both perturbation expansions about the relaxational limit ($c_i \to 0$) and about the Hamiltonian limit[15] ($c_i \to \infty$) have shown that there exist stable pulse solutions with zero velocity in large subcritical ($\varepsilon < 0$) ranges of the $\varepsilon, c_1, c_3, c_5$ parameter space. Pulse solutions can go unstable by splitting into two fronts that move apart. Since they move apart, the long time properties of each front is given by that of a single front like that shown in Figure 5(a), so one can get information on the range of existence of pulses by analyzing the dynamics of a single front. Pulse solutions of Eq. (9) are stationary in the frame moving with the group velocity of the traveling waves; in principle, if one considers an amplitude expansion near a weakly first-order subcritical bifurcation, in the same order as the quintic term $|A|^4 A$, other nonlinear terms arise that give pulses a drift velocity slightly different from the group velocity. Experimentally, however, pulses are found to have a drift velocity much smaller than the group velocity.[5] Although this can be accounted for on an ad hoc basis by taking some of the parameters in the extension of Eq. (9) large, a more fundamental analysis[10] attributes the small drift velocity of pulses to the coupling with the concentration field.

FRONTS

The dynamics of fronts in the cubic CGL (6) for $\varepsilon > 0$ is relatively well understood in terms of marginal stability selection criteria (see van Saarloos and Hohenberg[13] and references therein). Because of their importance for the stability of pulse solutions, the dynamics of fronts both for $\varepsilon < 0$ and for $\varepsilon > 0$ has recently been studied in great detail.[13] It turns out that an *exact* nonlinear front solution can be found whose dynamics plays an important role in the selection of patterns. Together with a set of rules and conjectures, a fairly complete picture of the stability of pulses and of the dynamics of fronts emerges. Some of the surprising findings are:

i.   In some ranges of the parameters, pulses can remain stable in the limit $\varepsilon \uparrow 0$. In these regions of parameter space, fronts only advance into the state $A = 0$ for $\varepsilon > 0$, and dynamically the distinction between the supercritical and subcritical case seems to have blurred.

ii.   For fronts to propagate into the state $A = 0$ the nonlinear dispersion has to be relatively small.

iii. It is possible to have fronts that propagate with the linear marginal stability speed but that are *not* uniformly translating.

iv.   There are subcritical regions of parameter space where chaotic "slugs" spread.

v.   There are regions of parameter space in the limit $c_i \to \infty$ where there are dynamically important front solutions that cannot be obtained perturbatively.

---

[15] In the relaxational limit no stable pulse solutions exists; one then perturbs about the point where a front like that in Figure 5(a) has zero velocity. In the Hamiltonian limit, a two-parameter family of moving pulse solutions can be used as a starting point for a perturbation expansion.

CONCLUDING REMARK

In this brief and extremely elementary contribution, I have tried to introduce the CGL in relatively nontechnical terms. Although the coherent structures discussed do illustrate that the CGL exhibits much richer and complicated behavior than the real equation, this overview does not do justice to its full richness, nor does it do justice to the many other interesting results obtained by a number of groups. A full discussion of coherent structures should include the work by Coullet and co-workers[4] on transitions in domain wall motion and on defects. Moreover, most of the current work is focussed on patterns in two dimensions and on the chaotic behavior in one and two dimensions; though shocks and Bekki-Nozaki holes do appear to play a role in some of the chaotic regimes, the extent to which they drive the chaotic dynamics is far from understood.

REFERENCES

1. Aranson, I., L. Kramer, S. Popp, O. Stiller, and A. Weber. "Localized Hole Solutions and Spatio-Temporal Chaos in the 1-D Complex Ginzburg-Landau Equation." *Phys. Rev. Lett.* **70** (1993): 3880.
2. Bekki, N., and K. Nozaki. "Formalations of Spatial Patterns and Holes in the Generalized Ginzburg-Landau Equation." *Phys. Lett.* **A110** (1985): 133.
3. Cross, M. C., and P. C. Hohenberg. "Pattern Formation Outside of Equilibrium." *Rev. Mod. Phys.* **65** (1993): 851.
4. Coullet, P., T. Frisch, and F. Plaza. "Sources and Sinks of Wave Patterns." *Physica* **62D** (1993): 75.
5. Kolodner, P. "Drifting Pulses of Traveling-Wave Convection." *Phys. Rev. Lett.* **66** (1991): 1165.
6. Langer, J. S., and V. Ambegaokar. "Intrinsic Resistive Transition in Narrow Superconducting Channels." *Phys. Rev.* **164** (1967): 498.
7. Newell, A. C. "Envelope Equations." In *Nonlinear Wave Motion.* Lectures in Applied Mathematics, Vol. 15, 157. Providence, RI: American Mathematical Society, 1974.
8. Newell, A. C., T. Passot, and J. Lega. "Order Parameter Equations for Patterns." *Ann. Rev. Fluid Mech.* **25** (1993): 399.
9. Passot, T., and A. C. Newell. "Towards a Universal Theory for Natural Patterns." Preprint. See also the chapter in these proceedings.
10. Riecke, H. "Self-Trapping of Traveling Wave Pulses in Binary Fluid Mixture Convection." *Phys. Rev. Lett.* **68** (1992): 301.

11. Shraiman, B. I., A. Pumir, W. van Saarloos, P. C. Hohenberg, H. Chaté, and M. Holen. "Spatiotemporal Chaos in the One-Dimensional Complex Ginzburg-Landau Equation." *Physica* **D57** (1992): 241.
12. van Hecke, M. L., P. C. Hohenberg, and W. van Saarloos. "Amplitude Equations for Pattern Forming Systems." In *Fundamental Problems in Statistical Mechanisms VIII*, edited by H. van Beijjeren. Amsterdam: North Holland, 1994.
13. van Saarloos, W., and P. C. Hohenberg. "Fronts, Pulses, Sources and Sinks in Generalized Complex Ginzburg-Landau Equations." *Physica* **D56** (1992): 303.

Hugues Chaté
Service de Physique de l'Etat Condensé, Centre d'Etudes de Saclay, 91191 Gif-sur-Yvette, France

Disordered Regimes of the One-Dimensional Complex Ginzburg-Landau Equation

I review recent work on the "phase diagram" of the one-dimensional complex Ginzburg-Landau equation for system sizes at which chaos is extensive. Particular attention is paid to a detailed description of the spatiotemporally disordered regimes encountered. The nature of the transition lines separating these phases is discussed, and preliminary results are presented which aim at evaluating the phase diagram in the infinite-size, infinite-time, thermodynamic limit.

1. A PROTOTYPE FOR STUDYING SPATIOTEMPORAL CHAOS

The complex Ginzburg-Landau equation (CGL),

$$\partial_t A = A + (1 + ib_1)\Delta A - (b_3 - i)|A|^2 A$$

Spatio-Temporal Patterns, Ed. P. E. Cladis and P. Palffy-Muhoray,
SFI Studies in the Sciences of Complexity, Addison-Wesley, 1995   **33**

where A is a complex field, $x \in [0, L]$ and $b_3 > 0$, is of considerable importance to everyone interested in spatially extended nonequilibrium systems.[8] It accounts for the slow modulations, in space and time, of the oscillatory state in a physical system which has undergone a Hopf bifurcation.[7,13,14] As such, the CGL is closely related to numerous experimental situations.[8] The resulting universality and genericity are accompanied by specific features which make this model interesting for its own sake, simply as a prototype of spatially extended dynamical systems.

There exist two important limits to CGL: the dissipative, relaxational "real Ginzburg-Landau" equation ($b_1 = 0, b_3 \to \infty$) and the dispersive, integrable nonlinear Schrödinger equation (NLS) ($b_3 = 0, b_1 \to \infty$). Studying CGL in a comprehensive manner, one encounters dynamical regimes where the relative importance of dissipation and dispersion can be tuned at will in a nonintegrable, nonvariational system. This has been done recently for the one-dimensional case, uncovering the "phase diagram" presented in Figure 1 and discussed below.[6,16,18]

The relative simplicity of Figure 1 stems mostly from the existence of a "thermodynamic limit" for the spatiotemporally chaotic regimes observed. As a matter of fact, away from the intricacy of the bifurcation diagrams at small sizes ($L \lesssim 50$), there exists a large-size limit beyond which chaos becomes extensive and

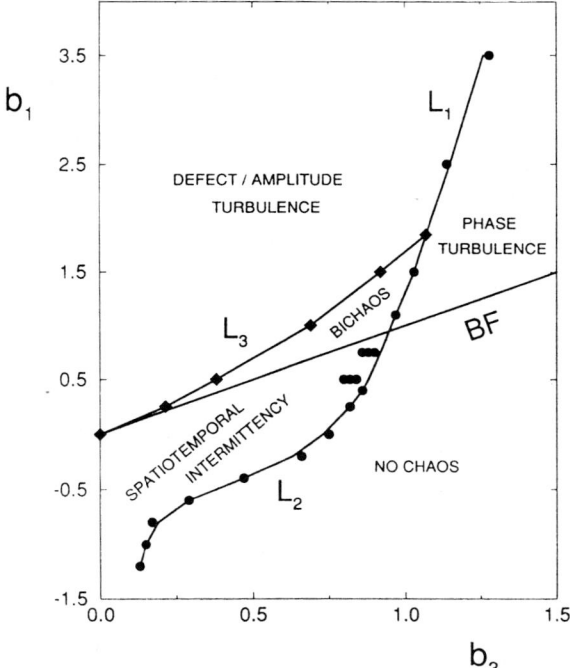

FIGURE 1 Phase diagram in the (b_1, b_3) parameter plane. Lines L_1 and L_3 were determined by Shraiman et al.[18] and Pumir et al.,[16] line L_2 by Chaté.[6] See Chaté's paper[6] for numerical details.

can be characterized by intensive quantities independent of system size, boundary conditions, and, to a large extent, initial conditions. Perhaps the most convincing illustration, for CGL, of this essential property of spatiotemporal chaos can be found in a recent paper by Egolf and Greenside[9] where they show that the Lyapunov dimension is proportional to the system size L.

Here, I review the various disordered "phases" observed in the one-dimensional case in the large-size limit and the nature of the transitions leading to them. Emphasis is put on the physical-space structures and objects that compose the spatiotemporal chaos, as they could well play a crucial role in building a statistical analysis of the disordered phases. This should also hopefully provide a clearer picture of the "elementary mechanics" at play in these regimes.

All the results reported and presented here were obtained using a pseudo-spectral code with periodic boundary conditions and second-order accuracy in space and time. Spatial resolution was typically 1024 modes for a domain of size $L = 1000$. Time steps varied like $1/\sqrt{b_3}$, with typical values as indicated in the figure captions.

2. PHASES

Early work on CGL has dealt with the problem of the linear stability of its family of plane-wave solutions $A = a_k \exp i(kx + \omega_k t)$ with $a_k^2 = (1 - k^2)/b_3$ and $\omega_k = 1/b_3 - (b_1 + 1/b_3)k^2$. All these solutions are unstable for $b_1 > b_3$, a condition which defines the so-called "Benjamin-Feir line" (BF). For $b_1 < b_3$, plane-wave solutions with $k^2 < k^2_{\text{Eckhaus}} = (b_3 - b_1)/(3b_3 - b_1 + 2/b_3)$ are linearly stable.[8] The BF line thus provides a first separation of the (b_1, b_3) plane, indicating where the constant-modulus, phase-winding solutions might play a role.

I now give a brief account of what is known about the disordered phases. For all these spatiotemporal chaos regimes, chaos is extensive: the Lyapunov dimension (and the other quantities derived from the Lyapunov spectrum such as the entropy or the number of positive exponents) scales linearly with the system size L.[3]

2.1 DEFECT TURBULENCE

Above the BF line and to the left of L_1, a strongly disordered phase is observed, best defined by a finite density of space-time zeros of $|A|$. This "defect turbulence," also named "amplitude turbulence," is characterized by a quasi-exponential decay of space and time correlation functions, and slower-than-exponential tails of the pdf of the phase gradient ϕ_x.[16,18]

One elementary process is the key ingredient of this strong spatiotemporal chaos: pulses of $|A|$ grow under the effect of the dispersion term; this "self-focusing" is stopped by the action of dissipation, breaking the pulse. Such events can be seen in the spatiotemporal plots presented in Figure 2; regions of almost zero amplitude are present, another indication that pulses are the relevant objects to consider when approaching the NLS limit. As one gets closer to the NLS limit (increasing b_1), the pulses get sharper and higher, whereas closer to the BF line, dissipation dominates and $|A|$ rarely overpasses the "saturation" value $1/\sqrt{b_3}$.

(a)

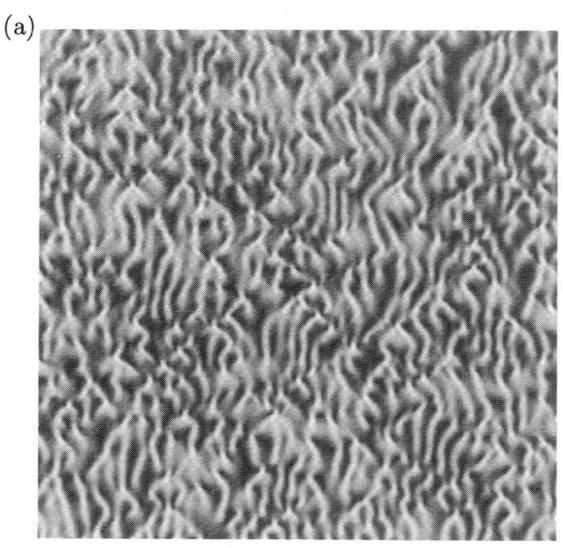

(b)

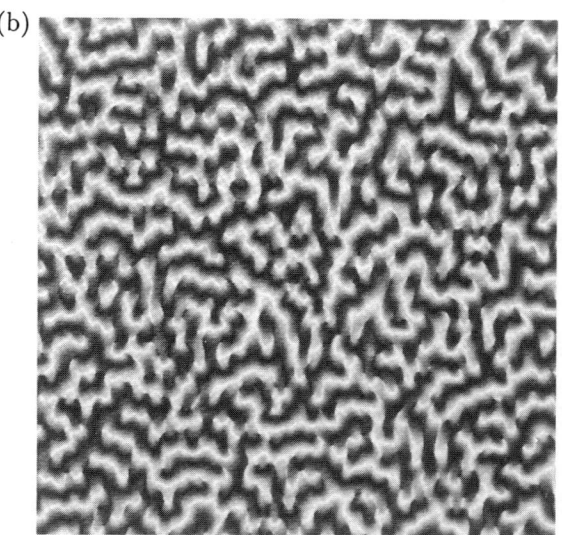

FIGURE 2   Spatiotemporal representation of a defect turbulence regime in a system of size $L = 250$ for $b_1 = 3.5$ and $b_3 = 0.5$ (periodic boundary conditions, timestep $\delta t = 0.02$). Time is running upward for $\Delta T = 82.5$ (transient discarded). Part (a) shows the evolution of $|A|$ (grey scale between $|A| = 0$/black and $|A| = 2$/white). Part (b) shows the corresponding evolution of $\phi = \arg(A)$ (grey scale between 0/white and π/black).

This disordered phase is reached from (almost) all initial conditions for parameter values to the left of L_3. In the triangular region delimited by BF, L_1, and L_3, it coexists with "phase turbulence," which is described below.

2.2 PHASE TURBULENCE

Phase turbulence is a weakly disordered regime observed in the region of parameter space above the BF line to the right of L_1, as well as in the triangular region delimited by BF, L_1, and L_3. It is best defined by the absence of space-time defects or, equivalently, by the bounded character of the pdf of ϕ_x and $|A|$ (see below for a discussion of this point). The absence of phase singularities implies that the "winding number" $\nu = \int_0^L \phi_x dx$ is a conserved quantity equal to a multiple of 2π which classifies the different attractors reached at given values of b_1 and b_3.

Chaos is very weak, as shown by the slow decay of space and time correlation functions, indicative of diffusive or sub-diffusive modes.[16,18] This is corroborated by the flat shoulder of the spatial power spectrum of $|A|$ at low wavenumbers, reminiscent of the Kuramoto-Sivashinsky equation (KS). The dynamics is, in fact, very similar to that of KS, which is not surprising since this equation was originally derived to describe the phase dynamics of CGL near the BF line.[7,13,14] The close relationship with KS suggests, in turn, that the Kardar-Parisi-Zhang interface equation (KPZ),[12] which have been argued to account for the large-scale, long-time, properties of KS,[15,19,20,21] might provide the correct description of the long-wavelength limit of phase turbulence. In such a setting, one expects exponential decay of spatial correlations and stretched-exponential decay of temporal correlations.[12]

Spatiotemporal diagrams (see Figure 3) reveal that the objects involved are "shocks" of $|A|$ (similar graphs for ϕ_x show localized modulations of the wavenumber, and look essentially the same). Close to the BF line, the dynamics is indeed very much like in KS (Figure 3(a)), while away from it trains of propagative shocks become more and more frequent (Figure 3(b)). The overall effect of ν is to introduce a general drift on the localized shocks of $|A|$ which increases with abs(ν) (Figure 3(c)). The phase turbulence regimes thus look very different depending on ν and the distance, in the parameter plane, from (b_1, b_3) to the BF line. Increasing this distance, the characteristic scales decrease (the difference of the time scales between Figures 3(a) and 3(b) is only partly due to the variation of the basic frequency like $\sqrt{b_3}$). The other important effect, the appearance of trains of propagative shocks, is an indication that KS is *not* a valid description away from the BF line and that other terms, incorporating odd derivatives of ϕ, are present in the "effective phase equation" describing the corresponding phase turbulence regimes. Whether the connection to KPZ still holds in such circumstances and what is then the valid phase equation replacing KS, both are open problems under investigation.

(a)

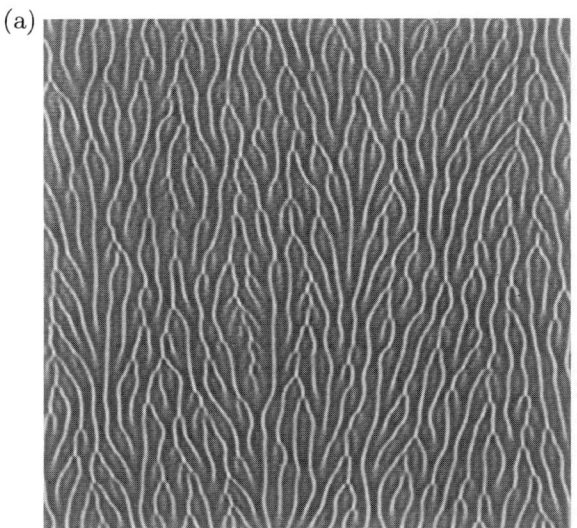

(b)

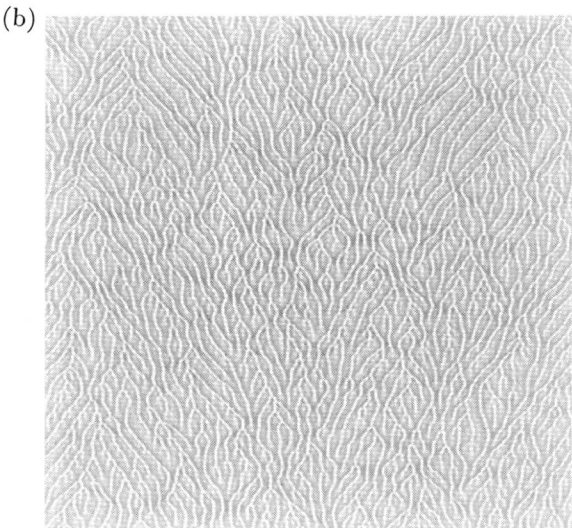

FIGURE 3 Spatiotemporal evolution of $|A|$ in phase turbulence regimes in a system of size $L = 1000$ for $b_1 = 3.5$ (periodic boundary conditions, timestep $\delta t = 0.16$, transient discarded, time running upward, grey scale with white corresponding to maximum value of $|A|$). Part (a): close to the BF line ($\nu = 0$, $b_3 = 2$, $\Delta T = 3300$, $0.68 \leq |A| \leq 0.75$). Part (b): close to L_1 ($\nu = 0$, $b_3 = 1.4$, $\Delta T = 1650$, $0.69 \leq |A| \leq 1.01$). (continued)

It must be mentioned, at this point, that it is still a controversial and unresolved issue to know whether phase turbulence "exists" in the thermodynamic limit, i.e., whether it is a transient phenomenon or a finite-size artifact. In Sections 3.1 and 3.3 below, I discuss this point in relationship with the nature of the transition lines

(c)

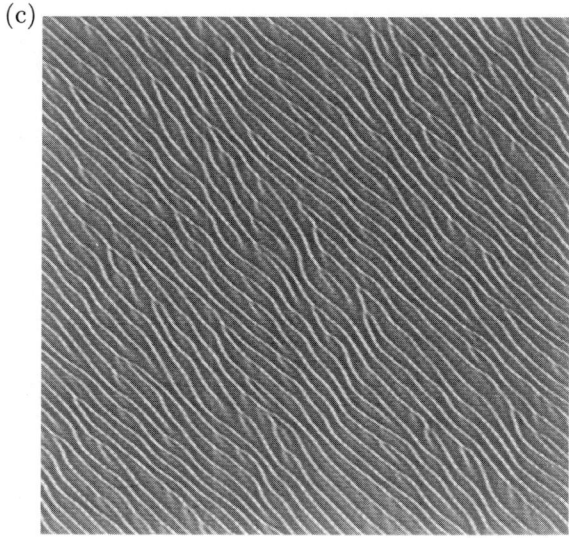

FIGURE 3 (continued) Part
(c): same as (a) but with
$\nu = -10\pi$.

delimiting the phase turbulence region in Figure 1. In any case, even if there always exists a very small (essentially undetectable) density of defects, the spacetime regions between those rare events are large enough so as to justify the study of phase turbulence for its own sake.

2.3 SPATIOTEMPORAL INTERMITTENCY

Below the BF line, plane-wave solutions of wavenumber $k^2 \leq k^2_{\text{Eckhaus}}$ are linearly stable. This does not preclude the existence, to the left of L_2, of other, mostly chaotic, solutions which are easily reached for initial conditions outside the basin of attraction of the stable plane waves.[6] These disordered regimes are *spatiotemporal intermittency* regimes: they consist of space-time regions of stable plane waves separated by localized objects evolving and interacting in a complex manner (see Figure 4). The plane waves constitute the passive, "absorbing" state while the localized objects carry the spatiotemporal disorder. The simplest of these objects are members of the family of exact solutions found by Nozaki and Bekki.[1,2] These propagating "holes" of $|A|$ are not defects, strictly speaking, even though $|A|$ might remain very small in their core.[6] Defects (zeros of A) do occur, but at scattered points in spacetime. This makes the density of defects difficult to estimate and rather worthless, since, for example, the average number (in space) of hole-like objects is probably a better characterisation of the dynamics. Likewise, space and time correlation functions are not the easiest quantities to measure when evaluating the coherence scales in these regimes, due to the intermittent character of the disorder. As usual with spatiotemporal intermittency,[3,5] one can take advantage of the intrinsic binary structure of the spatiotemporal dynamics and evaluate, say, the distributions of sizes and lifetimes of the patches of plane wave solutions. These distributions are roughly exponential, yielding characteristic coherence scales.[6]

(a)

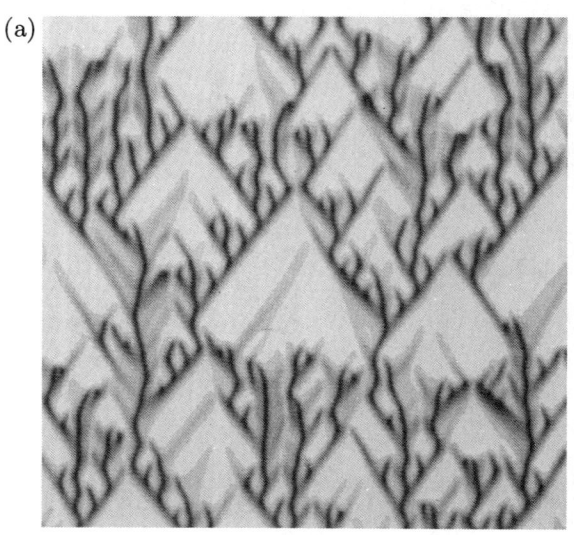

(b)

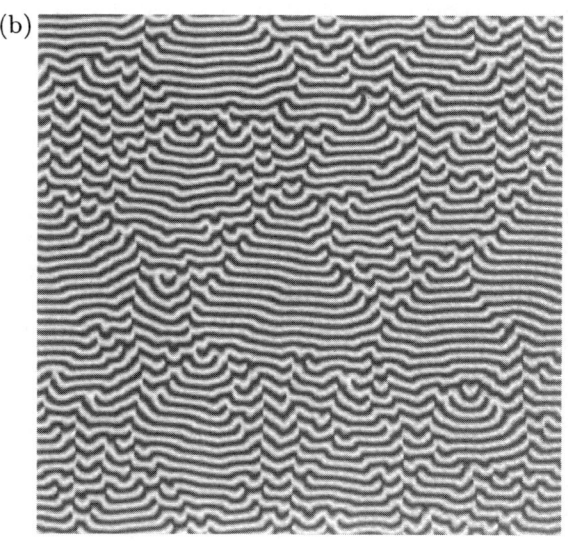

FIGURE 4   Spatiotemporal intermittency regime in a system of size $L = 250$ (periodic boundary conditions, timestep $\delta t = 0.04$) for $b_1 = 0$ and $b_3 = 0.5$. Time is running upward for $\Delta T = 165$ (transient discarded). Same representation as in Figure 2, but $0 \leq |A| \leq 1.413$ in part (a).

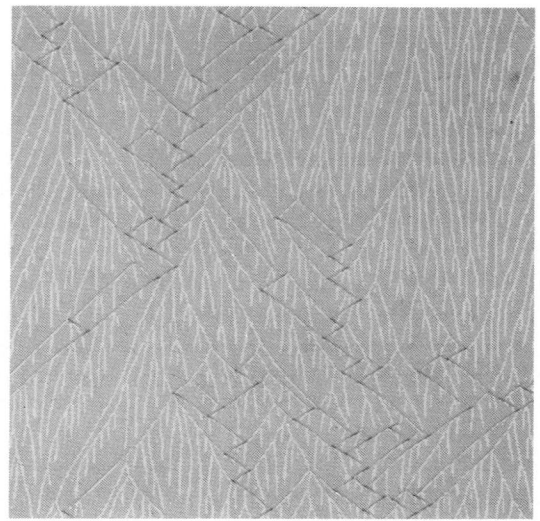

FIGURE 5 Defect turbulence in the bichaos region ($L = 1000$, $b_1 = 1.5$, $b_3 = 1$, $\Delta T = 1312$, periodic boundary conditions, timestep $\delta t = 0.04$, and transient discarded). Spatiotemporal representation of $|A|$ in grey scale from $|A| = 1.12$ (white) to $|A| = 0$ (black); time is running upward. Note that the hole-like propagating and branching objects typical of the defect phase do not appear spontaneously in the phase turbulent medium.

The spatiotemporal intermittency regimes do exhibit space-time defects. To that extent, they may be considered as defect turbulence regimes. The key difference with the region of defect turbulence described above is that here the defects do not appear spontaneously, they are produced by the localized objects carrying the disorder.

2.4 THE "BICHAOS" REGION

The domain of parameter space delimited by the BF, L_1, and L_3 lines deserves further comment. Depending on the initial conditions, one can reach one of two spatiotemporally chaotic regimes: phase turbulence or defect turbulence. A closer look at the defect turbulence reveals that it consists of localized propagating and branching hole-like objects which separate more quiescent space-time regions (see Figure 5). These objects do not appear spontaneously. The quiescent regions are nothing else than patches of phase turbulence, with their characteristic shocks of $|A|$. Defect turbulence is thus a spatiotemporal intermittency regime, with the absorbing state being the phase turbulence regime.

3. TRANSITIONS

I now examine the various transition lines between the phases and discuss to what extent they can be considered as phase transitions.

3.1 THE L_1 LINE

Line L_1 separates defect turbulence from phase turbulence for values of $b_1 \gtrsim 1.85$. As such, its best definition is given by the parameter values for which the density of defects n_D goes to zero. It was shown[16,18] that, for $b_1 = 3.5$, $n_D \sim (b_3 - b_3^*)^2$ with $b_3^* \simeq 1.29$. No hysteresis has been detected. This transition thus appears continuous, even though one cannot exclude a crossover scenario where n_D would remain small beyond b_3^*.

The situation is not as clear from the point of view of the correlation length ξ deducted from the exponential decay of the two-point correlation function: approaching L_1, ξ increases but does not diverge at the transition point.[10,16,18] Although difficult to measure in the phase turbulence regimes, indirect arguments imply that ξ remains finite. Only a change of behavior is observed, apparently accompanied by a discontinuity of $d\xi/db_3$. A similar situation holds for the variation of the Lyapunov dimension and the other quantities derived from the Lyapunov spectrum when crossing L_1, as shown by Egolf and Greenside.[9] In other words, no critical behavior seems to be present at the transition, even though it is continuous.

This remark is valid for the experimental conditions within which the above results were obtained. In particular, the system sizes used ensure the extensivity of chaos, but no assessment of finite-size effects has been made in order to evaluate the infinite-size, infinite-time, "thermodynamic" limit. The main question in this respect is that of the position of L_1 as $L \to \infty$. If L_1 remains away from the BF line, the existence of phase turbulence in the thermodynamic limit is assured and the transition at L_1 does not show any critical behavior. Another hypothesis consists in L_1 moving toward BF as L increases, with the correlation length at threshold increasing and finally diverging for an infinite-size system. In this case, the phase turbulence region disappears in the thermodynamic limit and the BF line is the asymptotic boundary of defect turbulence.

Preliminary results (see Figure 6) indicate a clear displacement of L_1 toward BF as L increases, making the second hypothesis more likely. A similar trend was recently found by Egolf and Greenside.[10] Extensive numerical work is under completion to obtain a quantitative extrapolation of the position of L_1 in the infinite-size limit.

3.2 THE L_2 LINE

The line L_2, best defined as the limit of existence of spatiotemporal disordered states, is difficult to determine in a precise manner.[6] This is mostly due to the appearance, in the transition region, of "frozen" states, i.e., spatially disordered arrangements of localized objects with trivial time dependence (see Figure 7). The coherence scales of the spatiotemporal intermittency regimes increase when approaching L_2, but their divergence is not observed; the system "falls" on a frozen

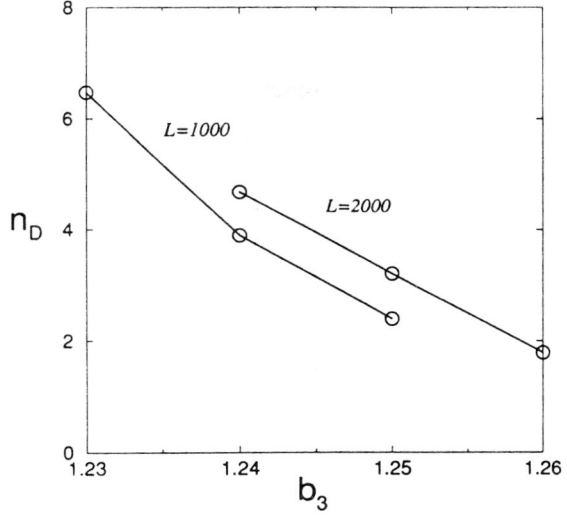

FIGURE 6 Variation of the defect density n_D ($\times 10^5$ on the graph) with b_3 near L_1. These results were obtained from runs of duration $\Delta T \sim$ 20000 after transients on systems of size $L = 1000$ and 2000 with periodic boundary conditions and timestep $\delta t = 0.04$. Although these results are still not very precise (at such low densities, much longer runs would be necessary), the general trend is significant: higher densities are reached for larger sizes.

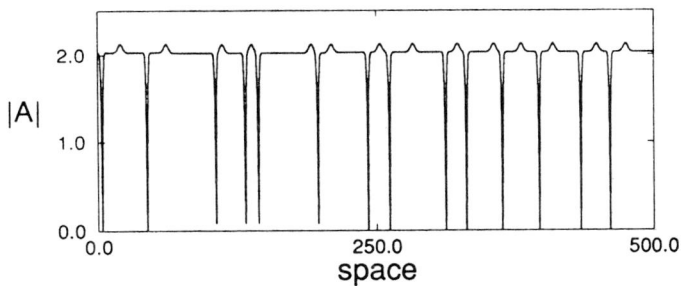

FIGURE 7 Frozen state observed near L_2 in a system of size $L = 500$ with $b_1 = -0.9$ and $b_3 = 0.18$ (periodic boundary conditions). This spatially disordered state is made of zero-velocity Nozaki-Bekki holes and shocks. $|A|$ is stationary, while ϕ winds regularly along time.

state. As often in spatiotemporal intermittency situations,[4] the deterministic features of the system "mask" the directed-percolation-like phase transition. The line L_2 is thus determined only crudely, awaiting further progress. In particular, two directions seem worth investigating: a "local" approach based on the study of the interactions between the localized objects at the origin of the spatiotemporal disorder, and a "global" approach based on Lyapunov analysis of the disordered regimes.

3.3 THE L_3 LINE

The L_3 line delimits the parameter space region where phase turbulence ceases to coexist with defect turbulence. Here I give a preliminary account of ongoing work on the nature of the breakdown of phase turbulence occurring when crossing L_3 (say, by decreasing b_3), for a winding number $\nu = 0$.

The transition is hysteretic: defect turbulence persists when crossing L_3 back, up to line L_2. The breakdown of phase turbulence[11] occurs *via* the nucleation of a first defect, followed by the quasi-deterministic invasion of the phase turbulent state by the defect phase (see Figure 8). The speed of invasion is always finite and rather large; it is given by the average propagation speed v_D of the localized objects composing the defect phase in this region of parameter space. The precursor of the first defect is apparently a local event: one of the shocks of $|A|$ composing the phase turbulence accelerates, and a depression of $|A|$ develops at its tail. When the velocity reaches v_D, the first hole-like object typical of the defect phase is effectively created, and quickly generates the first space-time defect. It is difficult to determine what triggers the initial acceleration of one of the shocks of $|A|$. As mentioned above, more and more trains of propagating shocks are observed in the phase turbulence regimes as one goes away, in parameter space, from the BF line. Simultaneously, the average velocity v_S of these propagative shocks increases. A possible interpretation of the nucleation is that, for parameter values beyond L_3, the fluctuations around v_S bring the speed of one shock past a critical value, after which the shock is attracted to the hole-like solution characteristic of the defect phase. Such a critical value would be related to coreponding critical values of the local minimum of $|A|$ and local extremum of ϕ_x, since as the velocity of the shock increases, the local depression of $|A|$ at its tail deepens, and the local phase gradient increases. Ongoing work aims at making those mostly qualitative statements more quantitative.

As described above, the breakdown of phase turbulence strongly resembles a first-order phase transition. So far, L_3 has been determined numerically by merely checking whether the breakdown occurs within a fixed (very long) integration time.[12] However crude, this methodology produced a sharp transition line, all the more so as no size effects could be detected.

A quantitative estimate of the probability of nucleation is needed to improve this situation. Preliminary results on the study of the pdfs of $|A|$ and ϕ_x in the phase turbulence regimes provide the beginning of an answer (see Figure 9). Away from L_3 (Figures 9(a) and 9(c)), these distributions have either Gaussian or strictly bounded tails. Close to L_3 (Figures 9(b) and 9(d)), these tails are exponential, signalling a clear change of behavior on which the definition of L_3 can be firmly based. Nevertheless, the question of the Gaussian/bounded character of the tails away from L_3 remains a crucial one since it determines whether the probability of nucleation can be strictly zero or not. A plausible scenario would consist of

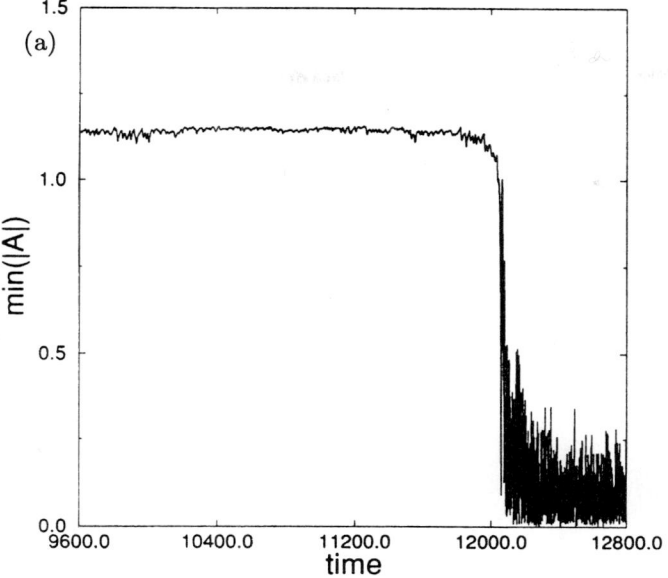

(a)

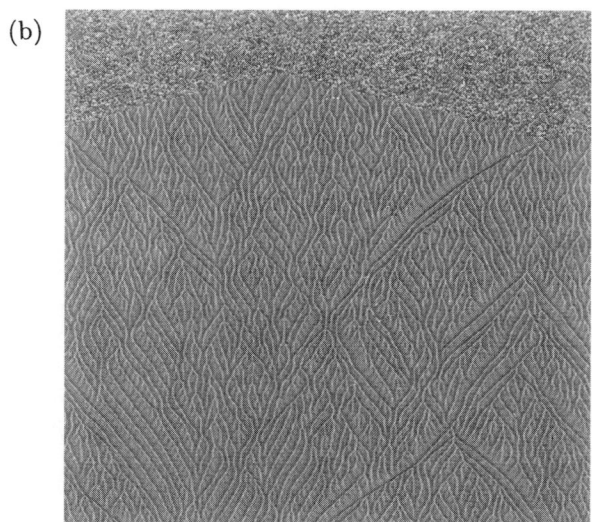

(b)

FIGURE 8 Breakdown of phase turbulence past the L_3 line in a system of size $L = 1000$ for $b_1 = 1$, $b_3 = 0.675$ and $\nu = 0$ (periodic boundary conditions, timestep $\delta t = 0.16$). Initial condition was a phase turbulent state for $b_3 = 0.7$. The breakdown occurred near $t = 12000$ as shown by the time series of $|A|$ (part (a)). Part (b) is the spatiotemporal evolution of $|A|$ for $t \in [9600, 12800]$ (grey scale from $|A| = 1.29$/white to $|A| = 1.1$/black, time running upward); at this resolution, the details of the defect phase are not visible (continued)

(c)

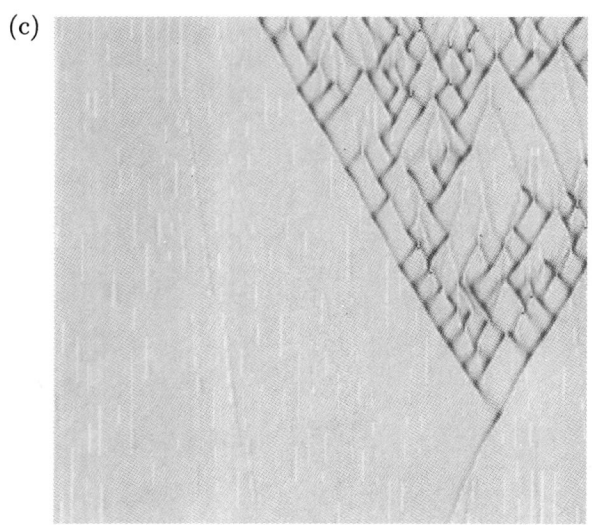

(d)

FIGURE 8 (continued) Part (c): zoom of (b) with a grey scale and a resolution adapted to the defect phase ($x \in [500, 1000]$, $t \in [12550, 12800]$, $0 \leq |A| \leq 1.29$); the nucleation of the first defect is obvious. Part (d): as in (b) but for the locus of the minimum (in space) of $|A|$.

exponential tails truncated by a finite bound moving to infinity (resp. zero) for ϕ_x (resp. $|A|$) when approaching L_3. Whereas this scenario would assure the existence of the phase turbulence in the infinite-size, infinite-time limit, the other scenario of a crossover from Gaussian to exponential tails would not. Current numerical efforts aim at providing data to decide this.

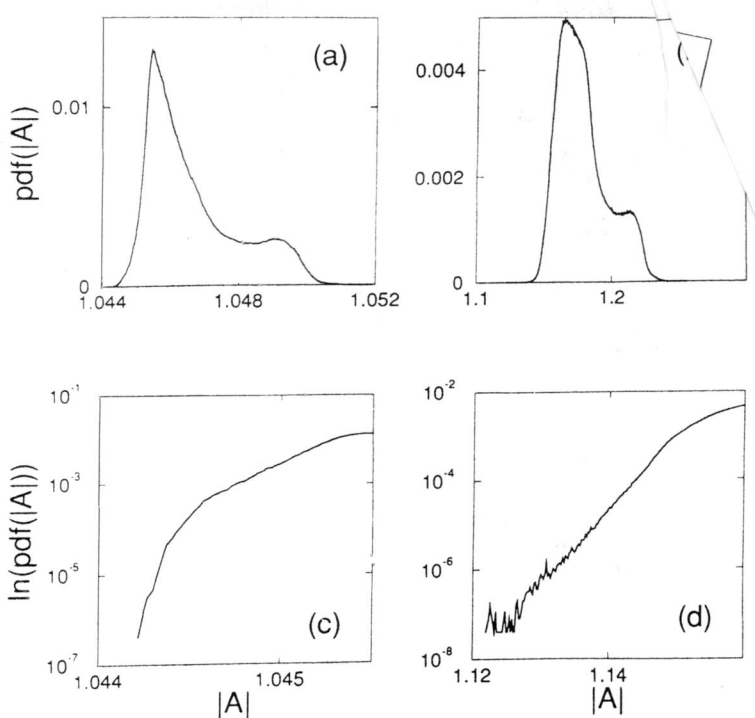

FIGURE 9 Probability distribution functions of $|A|$ for phase turbulence regimes in the bichaos region. These distributions were obtained from the simulation of a system of size $L = 4000$ during $\Delta T = 11500$ sampled every 4 timesteps $\delta t = 0.16$. (periodic boundary conditions, transient discarded). Parts (a) and (c): $\nu = 0$, $b_1 = 1$, $b_3 = 0.9$. Parts (b) and (d): $\nu = 0$, $b_1 = 1$, $b_3 = 0.7$.

In the same spirit, another important question deals with the estimation of size effects in this region of parameter space. So far, results for sizes $L = 1000$ and $L = 4000$ have not shown any significant difference, strengthening the case for a first-order transition.

Finally, the overall effect of a non-zero winding number ν is to give rise to earlier breakdown of the phase turbulence regime, moving L_3 to the right, as expected from the general additional drift of the shocks in this case.

4. OPEN PROBLEMS

The "phase diagram" presented in Figure 1 was obtained for systems sizes L large (for the extensivity of chaos is ensured) but finite. From the above discussion (for L_1 and L_3, it appears that an estimation of the finite-size effects is necessary, even crucial, since the existence of phase turbulence depends on the position and nature of these transition lines in the infinite-size, infinite-time, "thermodynamic" limit. For line L_3, a detailed study of the pdfs of interest in the phase turbulence regimes should be able to provide an objective criterion which can help to determine precisely the breakdown of phase turbulence. In this respect, preliminary results (see Figure 9) are encouraging. For the line L_2 delimiting the domain of existence of spatiotemporal intermittency, the key point lies in finding a quantifier which can be used to determine its position efficiently. Quantities related to the Lyapunov spectrum might constitute such a quantifier.

Apart from these "global," statistical quantities, approaches based on the study of the local objects involved in the spatiotemporally disordered dynamics ("shocks" for phase turbulence, "pulses" and "defects" for defect turbulence, "sources" and "sinks" for spatiotemporal intermittency) might also help to clarify some of the problems left open.

At any rate, and whatever the situation is in the thermodynamic limit, the "finite-size" phase diagram of Figure 1 is of interest to all "experimentalists" (i.e., anybody working with finite-size systems observable during a finite time). In this respect, it shows how varied and complex statiotemporally chaotic regimes can emerge from such a simple equation as CGL. I hope the above review will also provide guidelines for dealing with similar cases, in particular when the question of what to measure arises.

ACKNOWLEDGMENTS

I thank D. A. Egolf and H. S. Greenside for stimulating exchanges and the communication of their unpublished results, and P. Manneville for many discussions and comments.

REFERENCES

1. Bekki, N., and K. Nozaki. "Exact Solutions of the Generalized Landau Equation." *J. Phys. Soc. Jap.* **53** (1984): 1581–1582.

2. Bekki, N., and K. Nozaki. "Formation of Spatial Patterns and Ho Generalized Ginzburg-Landau Equation." *Phys. Lett. A* **110** (1985

3. Chaté, H., and P. Manneville. "Spatiotemporal Intermittency in Co Map Lattices." *Physica D* **32** (1988): 409–422.

4. Chaté, H., and P. Manneville. "Role of Defects in the Transition to T lence via Spatiotemporal Intermittency." *Physica D* **37** (1989): 33–41.

5. Chaté, H. "Subcritical Bifurcations and Spatiotemporal Intermittency." *Spontaneous Formation of Space-Time Structures and Criticality*, edited T. Riste and D. Sherrington, 273–311. Dordrecht: Kluwer, 1991.

6. Chaté, H. "Spatiotemporal Intermittency Regimes of the One-Dimensional Complex Ginzburg-Landau Equation." *Nonlinearity* **7**(1994): 185–204.

7. Coullet, P., L. Gil, and J. Lega. "Defect Mediated Turbulence." *Phys. Rev. Lett.* **62** (1989): 1619–1622.

8. Cross, M. C., and P. C. Hohenberg. "Pattern Formation Outside of Equilibrium," and references therein. *Rev. Mod. Phys.* **65** (1993): 851–1112.

9. Egolf, D. A. and H. S. Greenside. "Complexity Versus Disorder for Spatiotemporal Chaos." Preprint available from the LANL database "chao-dyn@xyz.lanl.gov" under number # 9307010. See also the abstract of a poster by the same authors in this volume.

10. Greenside, H.S., private communication.

11. Janiaud, B., A. Pumir, D. Bensimon, V. Croquette, H. Richter, and L. Kramer. "The Eckhaus Instability for Travelling Waves." *Physica D* **55** (1992): 259–269.

12. Kardar, M., G. Parisi and Y. C. Zhang. "Dynamic Scaling of Growing Interfaces". *Phys. Rev. Lett.* **56** (1986) 889–892.

13. Kuramoto, Y. *Chemical Oscillations, Waves and Turbulence*. Tokyo: Springer, 1984.

14. Lega, J. "Défauts Topologiques Associés à la Brisure de l'Invariance de Translation dans le Temps." Thèse de Doctorat, Université de Nice, 1989.

15. Procaccia, I., M. H. Jensen, V. S. L'vov, K. Sneppen, and R. Zeitak. "Surface Roughening and the Long-Wavelength Properties of the Kuramoto-Sivashinsky Equation." *Phys. Rev. A.* **46** (1992): 3220–3224.

16. Pumir, A., B.I. Shraiman, W. van Saarloos, P. C. Hohenberg, H. Chaté, and M. Holen. "Phase vs. Defect Turbulence in the One-Dimensional Complex Ginzburg-Landau Equation." In *Ordered and Turbulent Patterns in Taylor-Couette Flows*, edited by C. D. Andereck. New York: Plenum Press, 1992.

17. Sakaguchi, H. "Breakdown of the Phase Dynamics." *Prog. Theor. Phys.* **84** (1990): 792–800.

, B.I., A. Pumir, W. van Saarloos, P. C. Hohenberg, H. Chaté,
18. Shraiben. "Spatiotemporal Chaos in the One-Dimensional Complex
a.rg-Landau Equation." *Physica D* **57** (1992): 241–248.

en, K., J. Krug, M. H. Jensen, C. Jayaprakash and T. Bohr. "Dynamic
ng and Crossover Analysis for the Kuramoto-Sivashinsky Equation."
ys. *Rev. A* **46** (1992): 7351–7354.

akhot, V. "Large-Scale Properties of Unstable Systems Governed by the
Kuramoto-Sivashinsky Equation." *Phys. Rev. A.* **24** (1981): 642–644.

Zaleski, S. "A Stochastic Model for the Large Scale Dynamics of Some Fluctuating Interfaces." *Physica D* **34** (1989): 427–438.

Greg Huber
Department of Physics, and Department of Mathematics, University of California, Lawrence
Berkeley Laboratory, Berkeley, California 94720 USA

Vortex Solids and Vortex Liquids in a Complex Ginzburg-Landau System

1. INTRODUCTION

Order and disorder in oscillatory media are closely connected with the presence of topological defects. In a large class of two-dimensional chemical, optical, and hydrodynamic systems, the dynamics is dominated by point defects (or vortices) and the extended spiral wave patterns they can generate. The goal in this chapter is to present some of the phenomena seen in large-scale, parallel computer experiments. For this, a discrete form of the complex Ginzburg-Landau (CGL) equation based on a coupled-map lattice[1] is used to explore the defect dynamics. This algorithm defines a new Ginzburg-Landau system, but one whose properties can be brought into arbitrarily close correspondence with those of the complex Ginzburg-Landau equation. Details of this particular construction,[2,3] and of our simulation parameters[4] have appeared elsewhere.

Spatio-Temporal Patterns, Ed. P. E. Cladis and P. Palffy-Muhoray,
SFI Studies in the Sciences of Complexity, Addison-Wesley, 1995   **51**

2. THE CGL EQUATION

The prototype for modeling defect dynamics in oscillatory media is the complex Ginzburg-Landau equation,

$$\dot{A}(\mathbf{x}, t) = A(\mathbf{x}, t) - (1 + i\alpha)|A(\mathbf{x}, t)|^2 A(\mathbf{x}, t) + (1 + i\beta)\nabla^2 A(\mathbf{x}, t), \qquad (1)$$

which describes the behavior of a slowly varying complex amplitude field $A(\mathbf{x}, t)$ in the neighborhood of a Hopf bifurcation.[5,6] The parameters α and β are real numbers, and one research goal is to map out the asymptotic solutions in an α-β phase diagram (see Figure 1).[7,4] The global, homogeneous solution $A(t) = \exp(-i\alpha t)$ is linearly stable when the condition $\alpha\beta + 1 > 0$ is satisfied (the Benjamin-Feir

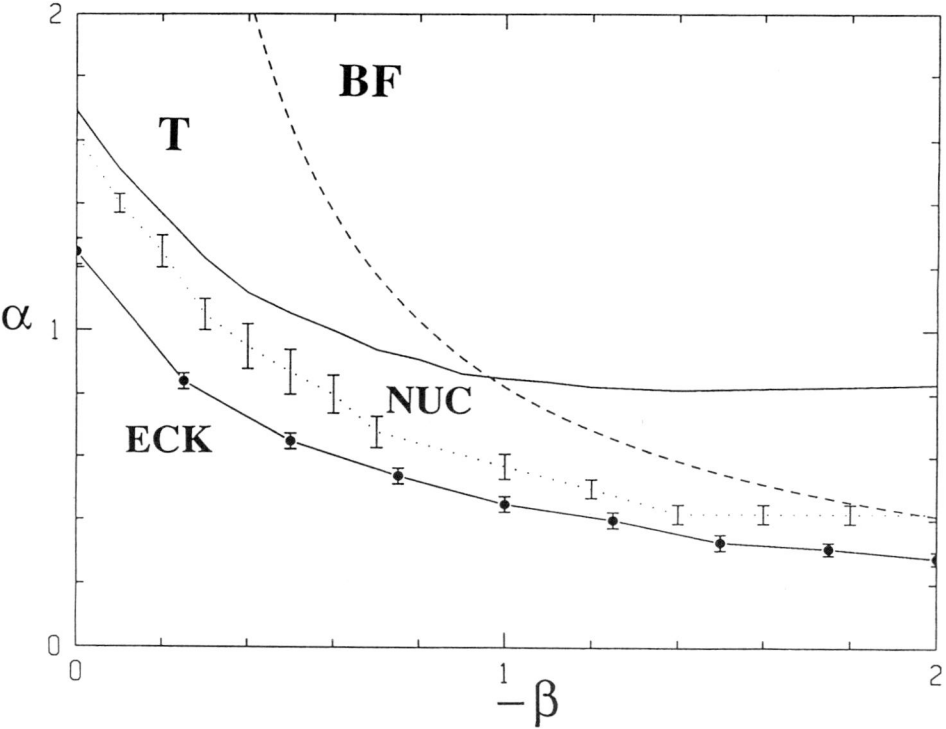

FIGURE 1   Phase diagram for $0 \leq \alpha \leq 2$ and $-2 \leq \beta \leq 0$. Curves are labeled (BF) Benjamin-Feir, (T) turbulence transition, (NUC) nucleation limit, (ECK) Eckhaus instability.

condition). This result can be reproduced in the coupled-map formalism, but due to our choice of discrete time steps, the analogous stability condition is $\alpha\beta + \gamma > 0$, where $\gamma = 0.824$. This stability limit is labeled "BF" in Figure 1.

3. VORTEX PHASES

Qualitatively different from the homogeneous state, are solutions that allow the field A to assume phase-less points. In the two-dimensional context, these defects are referred to as *vortices*, since the phase undergoes a complete rotation on a closed loop encircling one. If the loop direction and induced phase rotation have the same sense, then the defect is properly called a vortex; if opposite senses, then an *antivortex*. Viewed as topological excitations, the vortices and antivortices carry positive and negative *topological charge*, respectively. (Where the distinction between vortices and antivortices is irrelevant, we refer to them collectively as vortices.) At the phase singularity, $|A| = 0$. Amplitude portraits clearly show the vortices as amplitude minima; also, one-dimensional structures (*domain walls*) corresponding to amplitude maxima are found separating the vortices.

If the initial conditions are random (this and periodic boundary conditions apply to all the simulations described here), the subsequent evolution in case $\alpha = \beta = 0$ does resemble the relaxation of a thermodynamic system following a deep quench.[8] In general, however, the system does *not* settle, even at very long times, into the homogeneous state; rather it finds a many-vortex state. This can be either a "frozen" (or "solid") state of stationary vortices, or a turbulent ("liquid") state[5] characterized by a high density of vortices, and vortex-antivortex creation and annihilation. In the α-β diagram of Figure 1, the curve $\alpha_0(\beta)$ marking the transition from frozen to turbulent states (labeled "T") was found point by point, by fixing β and carefully charting the time history of the total vortex number at various α values. In a recent paper,[4] this transition was found to have many properties of a first-order phase transition. Furthermore, plots of vortex density versus time provide a tool for directly tracing the different dynamical regimes. As an example representing the variety of dynamical changes, consider the log-log density trace in Figure 2. Four distinct dynamical behaviors can be identified: initial decay, transient turbulence, decay to frozen state ("Zamboni clean-up"), and frozen state. This is the generic dynamics in the parameter region lying between the curves T and NUC (see Figure 1). The curve NUC marks the appearance of transient turbulence and subsequent nucleation phenomena. By comparison, the boundary of the Eckhaus instability (ECK), associated with the observed spiral waves, states lies somewhat lower in the α-β diagram. The gap between these curves is not completely understood.[7,4])

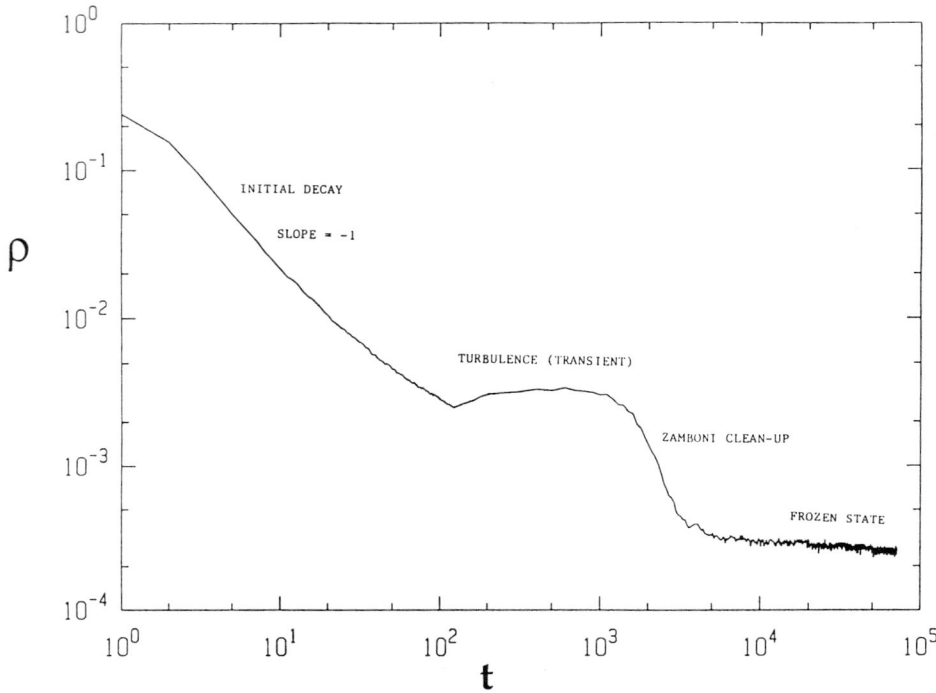

FIGURE 2 Decay of vortex density in time, from random initial conditions, for $\alpha = 0.75, \beta = -1$, lattice size 512×512. Four dynamical regimes are indicated.

4. NUCLEATION AND GROWTH

We consider first the vicinity of the "knee" in Figure 2. It is helpful to see the system as it decays out of the turbulent transient, as in Figure 3(a)–(c). (These are actually taken from three different runs, at three successive times, using the same parameter values.) The phenomenon observed in these amplitude portraits is the nucleation and growth of spiral waves. (Although the spiral waves can only be seen in the phase portrait, amplitude waves of corresponding frequency and wavenumber *can* be seen in Figure 3.) These *vortex droplets* nucleate out of the turbulent sea, and grow until the entire space is covered by the frozen phase. Here the turbulent state is unstable with respect to the finite-sized droplets; in other words, the transient state can be viewed as a *metastable* state. It was found[4] that the average lifetime T of the metastable state is dependent on the distance to the turbulence transition line $\Delta\alpha$ $(= \alpha_0(\beta) - \alpha)$ according to the simple law,

$$T \sim e^{\text{const}/(\Delta\alpha)^2}. \tag{2}$$

The divergence of the lifetime is one means of locating the transition curve $\alpha_0(\beta)$.
The relatively rapid decay of the vortex density toward the frozen density is due
to the steady growth of the vortex droplets at the expense of the turbulent fraction,

(a) (b)

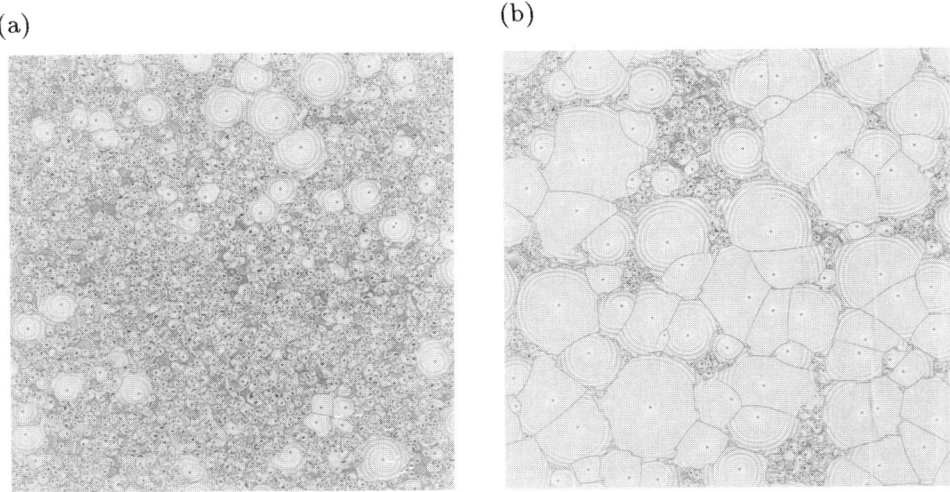

(c)

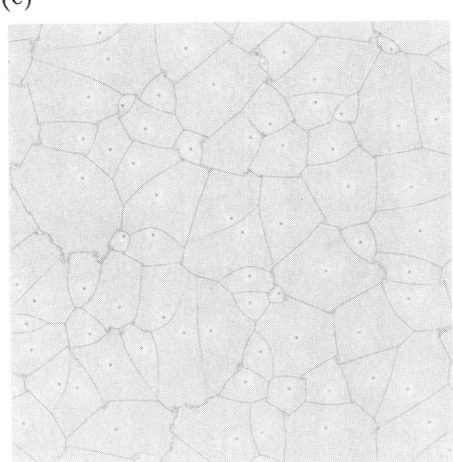

FIGURE 3 Nucleation sequence.
Dark points are vortices (amplitude
minima), dark lines are domain
walls (amplitude maxima). $\alpha =
0.79, \beta = -1$, 1024×1024 lattice.
(a) Early, (b) intermediate and (c)
late stages.

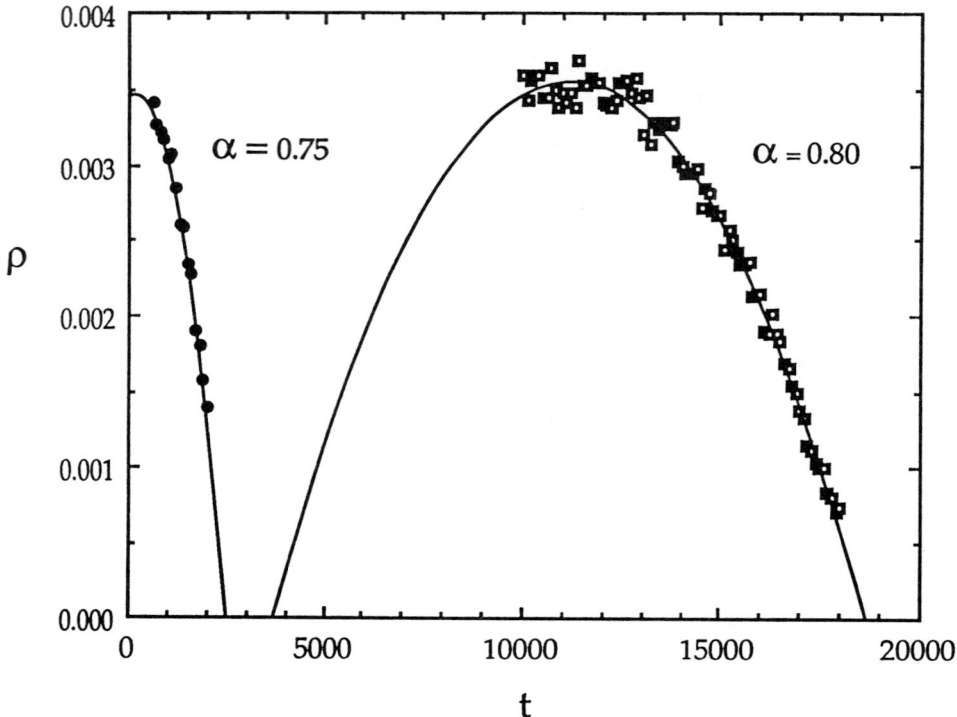

FIGURE 4 Vortex density vs. time for the "Zamboni" clean-up, i.e., fast decay to the frozen state. Two parameter values, $\beta = -1$ for both. Curves are best quadratic fits.

labeled the "Zamboni clean-up"[1] in Figure 2. There is a general decay law lurking in that figure, but it is hidden by the log-log axes. Plotting the same regime for two different α values makes clear the parabolic nature of the decay in density (see Figure 4). A model consisting of circular regions with radii growing linearly in time, together with a few other reasonable assumptions, leads to a density decreasing like $-t^2$, and accounts for the parabola law.

[1] *Zamboni* (after its inventor, F. J. Zamboni) is a popular brand of machine that levels and resurfaces the ice during the intermissions of ice hockey games. Since the machines render the disordered surface smooth, and does so in a spiral pattern, it is somehow appropriate in the context of nucleating spiral-wave droplets. (See Montville.[9])

5. VORTEX SOLID: DOMAIN WALLS, POLYCRYSTALS, EDGE VORTICES

The amorphous frozen state seen in Figure 3(c) is a fascinating structure, about which much remains to be learned. The subtle degrees of order and disorder evident here are a topic of on-going work. Note that some information on the vortex state is absent from the amplitude picture in Figure 3, namely the relative phase and direction of the spiral waves. The vortex domains are typically evolving toward four or five-sided polygons, often distinctly different from the Voronoi polygons calculated from the vortex centers. For starters, the domain walls are not straight line segments, but show various curvatures. Numerically, their shape is well approximated by a hyperbola. One simple explanation of this follows from the nucleation model. Let Δt_0 be the interval of time between the nucleation of two neighboring vortex droplets. Then, considering the two frozen vortices as the sources

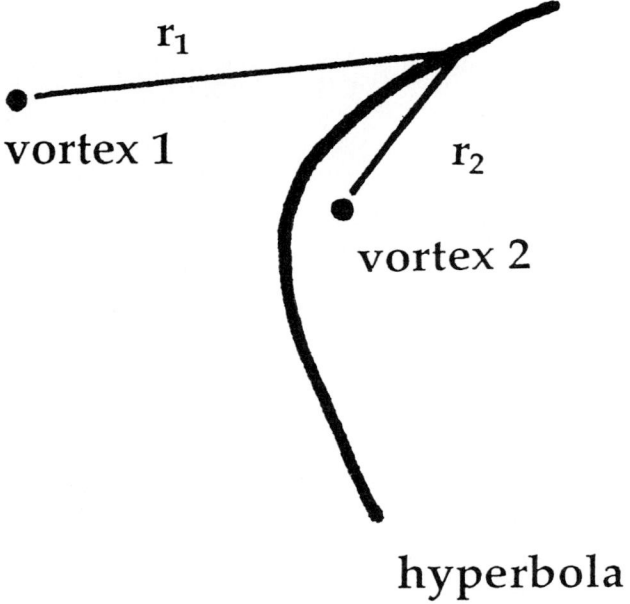

FIGURE 5 Domain walls are nearly hyperbolic, since the path difference $r_1 - r_2$ between two frozen vortices is a constant, being proportional to the time interval between their times of nucleation.

(a)

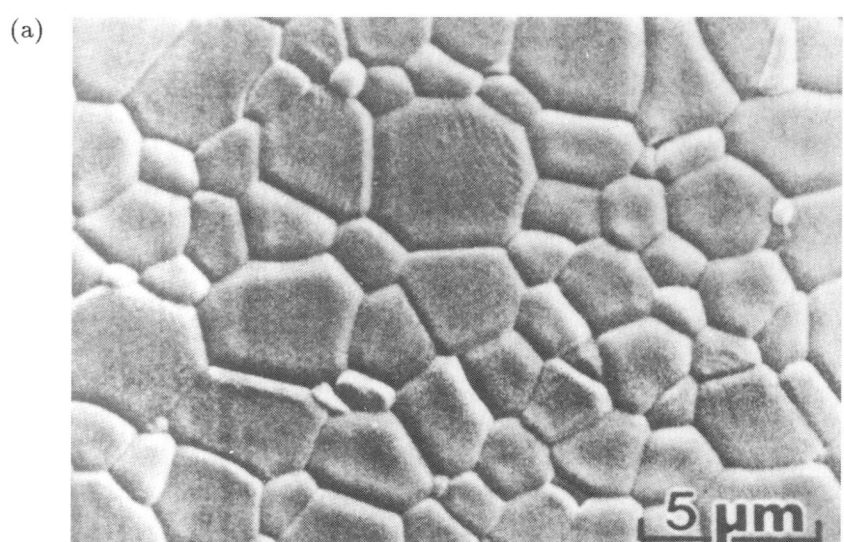

(b)

FIGURE 6   Polycrystalline structures. (a) Grains of polycrystalline alumina (MgO doped and hot pressed). (Courtesy of A. M. Glaeser and J. Rödel.) (b) Spherulites (5 cells) of polyethylene crystallized from the melt, as observed between crossed polarizers. Horizontal scale of the photograph is 130 μm. (Courtesy of P. H. Geil.)

of two annihilating, nearly-plane waves (the spiral waves), the interface between them can be described by the radial path difference from the two vortex cores $r_1 - r_2$ (see Figure 5). We have the usual relation,

$$r_1 - r_2 = \frac{2\pi\omega}{q}\Delta t_0. \tag{3}$$

Since Δt_0 is fixed, and the vortices have the same values of wavenumber q and frequency ω, the right-hand side of Eq. (3) is constant—this is just the equation for a hyperbola. Only when all the vortices nucleate at the same time ($\Delta t_0 = 0$), do we come back to the Voronoi picture of straight, bisecting domain walls. The explanation falls short if the cells undergo any "equilibration"—which does occur—and especially if the frozen cells extend over only a few wavelengths. Here, a more general picture of domain walls is needed.

Although the positions of the vortex centers are disordered, the field A within a vortex cell can be thought of as a perfect spiral wave, like a spatiotemporal crystal. In fact, the array of polygonal vortex cells strongly resembles the microstructure of polycrystalline materials. We illustrate the similarity in Figure 6 with two diverse examples. In Figure 6(a) is shown a scanning-tunneling micrograph of the grain structure of hot-pressed alumina; in Figure 6(b) a photograph of spherulites of low-density polyethylene grown from a thin-film melt, and viewed in polarized light (hence the Maltese-cross pattern). These cellular arrays of metallic grains, or of polymer spherulites, are not isolated examples—polycrystals are a common structural motif in nature. The analog of the disorder in relative phase of the spirals is present as the random orientation of the crystal axes within a grain. And, likewise, crystal grain boundaries are the analog of our spiral-wave domain walls.

Analogies to the formation and growth of polycrystalline patterns are likewise intriguing. The spherulite example shows striking parallels. Nucleation and growth of spherulites is the characteristic mode of polymer crystallization from the melt, and the parabola law is found there as well.[10] Moreover, it is interesting to note that the helically varying orientation of the crystal creates a tightly wound spiral pattern in polarized light. This makes photographs of nucleating spherulites eerily reminiscent of nucleating spiral waves.[11,10]

The existence of *edge* vortices, which do not generate extended spiral waves, is another suprise revealed by the simulations. In a sense, they control the dynamics of an otherwise frozen set of vortex cells. They move along the domain walls, braiding the walls roughly one wavelength around them, only coming to rest at the intersection of two or three domain walls. In doing so, they act somewhat like zippers, "unzipping" or changing the phase to a new value along the domain wall. As they diffuse around the network of domain walls, their numbers decrease due to vortex-antivortex annihilations at the interstices. A number of long-lived edge-vortex arrangements have been observed in the simulations. A common configuration is a bound vortex-antivortex pair. Triplets, quadruplets, and higher-order

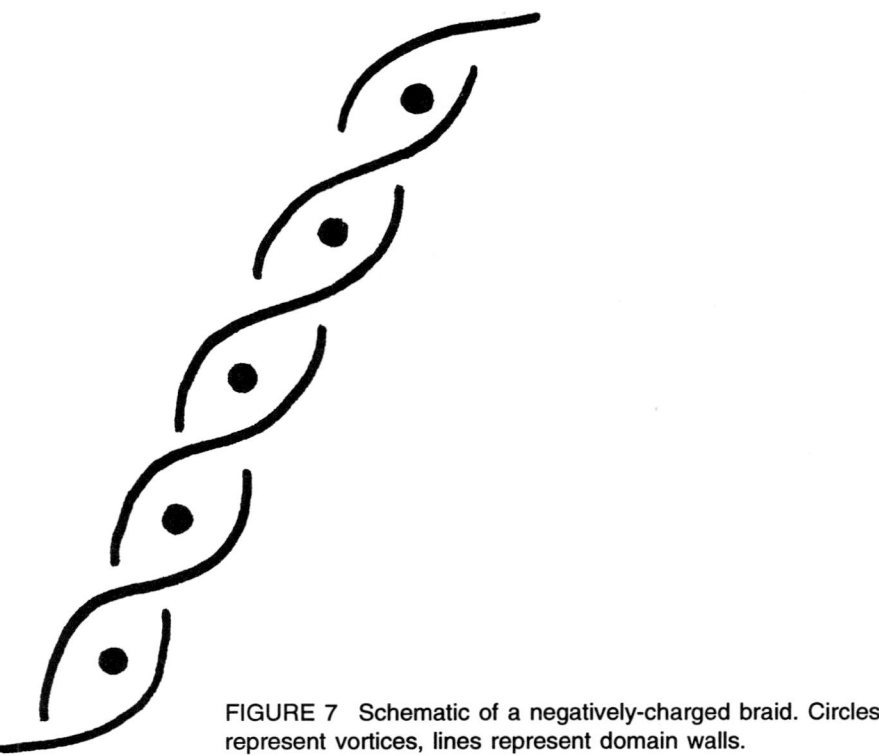

FIGURE 7 Schematic of a negatively-charged braid. Circles represent vortices, lines represent domain walls.

combinations are also seen. One beautiful configuration of same-sign edge vortices is the *vortex braid*. In Figure 7 is a schematic of a vortex braid, really an antivortex braid, redrawn from an amplitude portrait. +-charged vortex braid looks like the mirror image of this drawing.

6. QUESTIONS

To conclude, we present some unanswered questions. The presumed glasslike order of the vortex positions motivates one to look for a *bona fide* glass transition. Is there a vortex solid with a density just slightly less than the high liquid density? And is there an associated second-order transition to the vortex-liquid phase? Since the metastability (NUC) and turbulent transition (T) lines seem to meet somewhere near the $\beta = 0$ axis, future work in this region of the phase diagram is called for.

Are the vortex solids described here really "frozen"; in other words, are they truly asymptotic, or are they unstable, finally relaxing to a single vortex-antivortex pair (or even the homogeneous state) in the limit of long times and large system sizes? On the one hand, there seems no way to single out a natural length scale corresponding to a final density. On the other hand, our simulations show no definite trend toward a zero-density state, but the possibility of dynamics on geological time scales cannot be eliminated. The addition of an explicit noise term to Eq. (1) would be a step towards settling the question of further equilibration. There may exist both algebraic- and exponential-time behaviors, with the present simulations probing only the former.

Finally, the analogous vortex-line phases in three dimensions are almost completely unexplored. We end with a teaser in the form of Color Plate 1. It shows two sets of amplitude isosurfaces, one corresponding to vortex filaments, the other to domain walls. At the top of the cube, a two-dimensional slice shaded by amplitude shows cross-sections of some of the filaments and walls threading through the volume.

ACKNOWLEDGMENTS

Discussions with P. Alstrøm, T. Bohr, L. Kramer, and S. Strogatz are gratefully acknowledged. This work is supported by the Applied Mathematical Sciences Subprogram of the Office of Energy Research, U. S. Department of Energy under Contract DE-AC03-76SF00098.

REFERENCES

1. Aranson, I. S., L. B. Aranson, L. Kramer, and A. Weber. *Phys. Rev. A* **46** (1992): R2992–R2995.
2. Bohr, T., A. W. Pedersen, and M. H. Jensen. *Phys. Rev. A* **42** (1990): 3626–3629.
3. Bohr, T., A. W. Pedersen, M. H. Jensen, and D. A. Rand. In *New Trends in Nonlinear Dynamics and Pattern Forming Phenomena: The Geometry of Nonequilibrium*, edited by P. Coullet and P. Huerre, 185–192. New York: Plenum, 1989.
4. Geil, P. H. *Polymer Single Crystals.* New York: Interscience, 1963.
5. Gil, L., J. Lega, and J. L. Meunier. *Phys. Rev. A* **41** (1990): 1138–1141.
6. Huber, G., and P. Alstrøm. *Physica A* **195** (1993): 448–456.
7. Huber, G., P. Alstrøm, and T. Bohr. *Phys. Rev. Lett.* **69** (1992): 2380–2383.

8. Keller, A. In *Polymers, Liquid Crystals, and Low-Dimensional Solids*, edited by N. March and M. Tosi, 33–69. New York: Plenum, 1984.

9. Kuramoto, Y. *Chemical Oscillations, Waves, and Turbulence*. Berlin: Springer-Verlag, 1984.

10. Montville, L. *Sports Ill.* **66** 1987): 38–45.

11. Newell, A. C., and J. A. Whitehead. *J. Fluid Mech.* **38** (1969): 279–303.

12. Oono, Y., and S. Puri. *Phys. Rev. Lett.* **58** (1987): 836–839.

Chapter 2
Pattern Formation at Phase Boundaries and Other Interfaces

P. Nozières
Institut Laue-Langevin, B.P. 156, 38042 Grenoble Cedex 9, France

The Grinfeld Instability of Stressed Crystals

A crystal subject to a nonhydrostatic stress is unstable towards mass transport at the surface. Such a transport may be due to melting and growth at a solid liquid interface, or to surface diffusion. As a result, a planar surface spontaneously corrugates. The origin of the effect, the nature of the bifurcation, and the nonlinear behavior are briefly discussed. Recent experimental observations in various fields are surveyed.

1. INTRODUCTION

Consider a pure crystal subject to a nonhydrostatic stress, and an interface at which it is in contact with either its melt or with vacuum. At interface, the stress tensor σ_{ij} must obey mechanical equilibrium

$$\sigma_{nn} = -p_L, \quad \sigma_{nt} = 0 \tag{1}$$

where n and t stand for normal and tangential. In contrast the tangential stress σ_{tt} is free. When $\sigma_{tt} \neq \sigma_{nn}$ a planar interface is *unstable* against corrugations due to melting and growth (or to atomic diffusion at a free surface). The effect is monitored by elasticity, but the primary mechanism is a transport of matter, not an elastic displacement. Note that growth is supposed to respect the underlying strain, a condition met only for crystals.

This effect was first described by Asaro and Tiller in 1972[1] as a possible mechanism for stress corrosion cracking. Its generality was not fully appreciated at that time and it was rediscovered in 1985 by Grinfeld.[2,3] It is presently known as the Grinfeld instability and it has been widely studied recently. We shall survey its physical nature and its applications.

2. QUALITATIVE ARGUMENTS

Stressing a solid increases its energy: it is less competitive as compared to the liquid and thus *it melts*. Consider a planar (xy) interface and assume that

$$\sigma_{xx} = \sigma_{zz} + \sigma_0, \quad u_{yy} = 0 .$$

According to Gibbs, melting equilibrium implies

$$\frac{f_S - \sigma_{zz}}{\rho_S} = \mu_S^{\text{eff}} = \mu_L$$

where f_s is the free energy per unit volume of the solid and μ_L the liquid chemical potential. The extra uniaxial stress σ_0 shifts the equilibrium pressure by δP_L. Melting equilibrium is maintained if

$$\begin{aligned}
\delta\mu_S^{\text{eff}} &= \frac{(1-\sigma^2)\sigma_0^2}{2\rho_S E} + \frac{\delta P_L}{\rho_S} \\
&= \delta_{\mu_L} = \frac{\delta P_L}{\rho_L}
\end{aligned} \tag{2}$$

(σ and E are respectively Poisson's coefficient and Young's modulus). In a gravity field g the interface goes down by an amount

$$h = \frac{(1-\sigma^2)\sigma_0^2}{2Eg[\rho_S - \rho_L]} .$$

The origin of the instability is then straightforward. Assume that the interface corrugates: the extra stress σ_0 *decreases in the hills* (which feel the substrate less), *while it increases in the troughs*. As a result the hills will melt less and the troughs will melt more—the corrugation will *grow*! Clearly the instability is countered by gravity and by capillarity—as usual it is a matter of wavelength.

One can devise an equivalent energetic argument. Assume that melting-growth creates a corrugation at *constant* stress (strain) in the solid. Ignoring gravity and capillarity, that does not change the free enthalpy, since the two phases were originally in equilibrium. But it clearly breaks mechanical equilibrium at interface (the normal has turned; see Figure 1). If we want to restore that equilibrium we must exert a surface force

$$dF_i = \sigma_{ij}dS_j + P_L dS_i \Rightarrow dF_x = \sigma_0 dS_x \,.$$

If we don't, the solid relaxes elastically, which always *lowers* the energy. Corrugation implies a *negative* elastic energy, leading to the instability.

QUANTITATIVE CALCULATIONS

The planar interface is displaced by an amount $h(x)$ (through growth). We consider only a one-dimensional geometry. The stress field acquires a correction $\sigma^{(1)}$ due to the corrugation that must obey the boundary conditions in Eq. (1). From that we infer an elastic energy of the solid phase

$$E_{el}[h(x)] = -\frac{1}{2}\int d^3r u_{ij}^{(1)}\sigma_{ij}^{(1)} \,. \tag{3}$$

(The minus sign in Eq. (3) comes from a compensation with the cross term involving σ_0: the total change in energy is the elastic energy stored in the bulk plus the work of surface forces.)

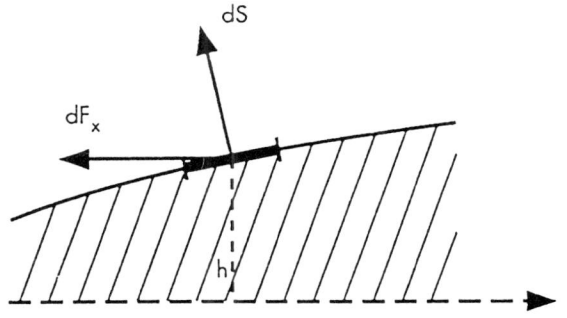

FIGURE 1 Force equilibrium at a displaced interface.

A calculation to order h^2 is enough to discuss linear stability. For a periodic distortion $h(x) = \alpha \cos kx$ it is easily found that

$$E_{el} = -\frac{1-\sigma^2}{2E}\sigma_0 k\alpha^2 . \tag{4}$$

The origin of the factor k is clear: the instability is monitored by the slope of the interface, i.e., by kh. But the elastic perturbation penetrates over a wavelength $1/k$. The elastic contribution Eq. (4) is balanced by contributions due to gravity and capillarity

$$E_{gr} = (\rho_S - \rho_L)g\frac{\alpha^2}{4}, \quad E_{cap} = \gamma k^2\frac{\alpha^2}{4} .$$

Gravity is stabilizing at long wavelength, capillarity at short wavelength. Altogether we have

$$E_{tot} = \frac{\alpha^2}{4}\lambda(k), \quad \lambda(k) = (\rho_S - \rho_L)g + \gamma k^2 - 2\left(\frac{1-\sigma^2}{E}\right)\sigma_0^2 k .$$

The bifurcation diagram is shown on Figure 2. The instability appears at the capillary wavevector k^* past a critical stress $\sigma_0 = \sigma^*$,

$$k^* = \sqrt{\frac{(\rho_S - \rho_L)g}{\gamma}}, \quad \sigma^* = \sqrt{\frac{\gamma E k^*}{1-\sigma^2}} .$$

Well beyond threshold gravity becomes negligible: all modes are unstable up to a wavevector

$$k_m = \frac{2\sigma_0^2}{E\gamma} .$$

Such a behavior is in fact familiar for any surface instability monitored by a bulk energy (ferrofluids in a magnetic field, charged interfaces in an electric field). The bifurcation is always the same, differences appearing when one goes beyond linear stability.

In practice one can "control gravity" using a vertical temperature gradient: what matters is the local distance to the phase equilibrium curve $p_L(T)$. If J is the heat flow through the interface, κ_L and κ_S the thermal conductivities of the two phases, it is easily shown that

$$g_{\text{eff}} = g - \frac{2J}{\rho_L(\kappa_S + \kappa_L)}\frac{dp_L}{dT} .$$

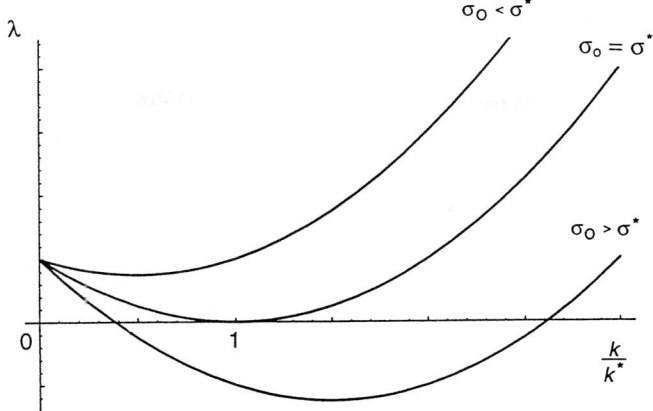

FIGURE 2   Bifurcation diagram of the Grinfeld instability: $\lambda(k)$ for various values of σ_0.

The next question is the nature of the bifurcation: subcritical or supercritical? For that we must push the amplitude expansion to order h^4. As usual in such problems the quartic terms in the energy are modified by the occurrence of a second harmonic in the pattern $h(x)$. We thus write h as

$$h(x) = \alpha \cos kx + \beta \cos 2kx .$$

The energy contains terms of order $\alpha^2, \beta^2, \alpha^4, \alpha^2\beta$. The latter generate β upon minimization. A detailed calculation is straightforward,[4] but extremely cumbersome (matching equations must be written at the displaced interface). The final result is

$$E_{el} = \frac{1-\sigma^2}{2E}\sigma_0^2 k \left[-\alpha^2 - 2\beta^2 + \frac{3}{4}k^2\alpha^4 + 2k\alpha^2\beta \right] ,$$

$$E_{gr} = \frac{(\rho_S - \rho_L)g}{4}[\alpha^2 + \beta^2] ,$$

$$E_{cap} = \frac{\gamma k^2}{4}\left[\alpha^2 + 4\beta^4 - \frac{3}{16}k^2\alpha^4 \right] .$$

Minimization with respect to β yields the amount of second harmonic, the final amplitude expansion being

$$E_{tot} = \frac{1}{4}[\lambda(k)\alpha^2 + v(k)\alpha^4 + \ldots] .$$

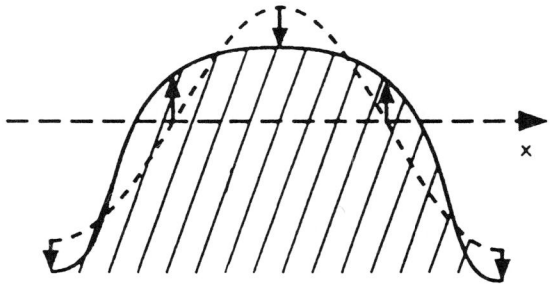

FIGURE 3 A negative second harmonic generates grooves on the solid side.

At bifurcation $k = k^*, \sigma = \sigma^*$ we find

$$\beta = -2k\alpha^2, v = -\frac{43}{16}(\rho_S - \rho_L)gk^2 \,.$$

The bifurcation is *subcritical* and β is negative, which means sharp grooves on the solid side of the interface (see Figure 3).

What happens beyond fourth order can only be inferred from a numerical calculation. One thing is sure: the interface cannot overcome the position it would have for $\sigma_0 = 0$. Far above threshold it should flatten in the front and develop deep grooves in the back. This is substantiated in preliminary calculations of Kassner and Misbah.[5]

The generalization to a three-dimensional geometry is more complicated (three-dimensional elasticity cannot rely on Airy stress functions). $\sigma_{xx}, \sigma_{xy}, \sigma_{yy}$ are now independent control parameters, which should allow a rich phase diagram (roll structures, hexagonal lattices, etc....)

---

3. EXPERIMENTAL OBSERVATION

Perhaps the first observation of the effect was the spontaneous corrugation of a solid He layer deposited on a metallic substrate when the temperature is quenched (in Moscow). Differential thermal expansion creates a tangential stress—hence, the instability. A similar effect may occur in the growth of alloy grains: radial concentration profiles create radial stresses and the grain takes strange shapes.[6] Such evidences were only qualitative and not conclusive.

The first clear demonstration was obtained on the solid-superfluid interface of ⁴He, for which the kinetics is practically instantaneous. In the experiment of Thiel et.al.[7] the stress is created by a piezobimorph, while Torii and Balibar[8] use pistons driven by piezoceramics. The latter geometry is better controlled and we focus on its results. As expected the interface goes *down* by an amount δh when a tangential stress is applied. The result shown on Figure 4 is in fair agreement with the above theory. When the applied strain exceeds a threshold h_c corrugations appear (which depend on crystal orientation). The threshold σ^* is approximately correct. The transition displays *hysteresis,* which means a subcritical bifurcation. Sharp grooves develop downward (on the solid side): all of that is in agreement with the above amplitude expansion. (As a byproduct, the distortion quickly disappears—typically 10 seconds at $1,2°K$—which implies fast plastic relaxation).

The demonstration is convincing, but a more quantitative analysis is needed. Another evidence comes in a very different context, at the surface of polymerized polydiacetylene.[9] This material is originally crystallized in its monomer form, and surface polymerization is induced by electron bombardment. Varying the energy of those electrons one can control the *thickness d* of the polymerized layer. That layer has a lower equilibrium lattice spacing: since it is epitaxied on the substrate it is subject to a large tangential stress—hence the possibility of a Grinfeld instability if surface diffusion allows for mass transfer along the surface (here there is no liquid phase to act as a reservoir).

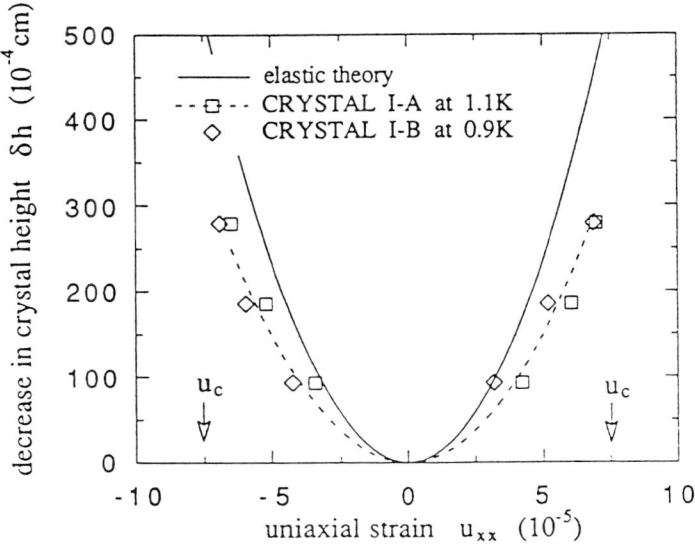

FIGURE 4   The lowering of interface as a function of applied stress (from Torii and Balibar[8]). In that experiment the basal plane is parallel to the c-axis.

The geometry is here that of a *finite* layer and the preceding calculation should be refined.[10] Clearly results are unchanged at short wavelength, $kd \gg 1$. Results are again unchanged whatever d if the elastic coefficients of the layer and substrate are the same. The first-order response to surface forces is not affected by the preexisting layer stress so as long as we stay within *linear* elasticity. If elasticity is the same in the two media, the response is that of a semiinfinite medium. More generally a very thin layer will play a negligible role if it sits on a deformable substrate: it is a thin foil with no resistance to stress.

The situation is very different in the opposite limit of an undeformable substrate, in which case the layer elastic deformation is severely reduced when $kd \ll 1$. The situation is similar to waves in shallow water: the elastic energy picks up an extra factor kd. The calculation is simple when $d \to 0$. The elastic boundary conditions are

$$u_{xx}^{(1)} = 0, \quad u_z^{(1)} = 0 \quad \text{at } z = 0; \quad \sigma_{zz}^{(1)} = 0, \quad \sigma_{zx}^{(1)} = -\sigma_0 h' \quad \text{at } z = d.$$

We notice that

$$\frac{\partial u_{xx}}{\partial z} = \frac{\partial}{\partial x}\left[\frac{\partial u_x}{\partial z}\right] = \frac{2\partial u_{zx}}{\partial x} - \frac{\partial^2 u_z}{\partial x^2}.$$

The last term is of order d, thus negligible. It follows that at the surface

$$\sigma_{tt} = \sigma_0 \left[1 - \frac{2k^2 dh}{1 - \sigma}\right].$$

Disregarding gravity the net energy is

$$E_{tot} = \frac{\alpha^2}{4}\left[\gamma k^2 - \frac{2\sigma_0^2(1+\sigma)}{E}k^2 d\right].$$

All long wavelength modes turn unstable together when the thickness d exceeds

$$d^* = \frac{\gamma E}{2\sigma_0^2(1+\sigma)}. \tag{5}$$

In the case of polydiacetylene the substrate *is* deformable: we face a standard Grinfeld instability. Indeed corrugations do occur when d exceeds $\approx 2000\text{Å}$, with a wavelength of order 2μ. They quickly evolve towards deep grooves, and eventually *lines of fracture* at the base of these grooves. The qualitative behavior is correct but we have no theory for singularities and fracture.

4. RELEVANCE TO EPITAXIAL GROWTH

Very frequently an epitaxial layer has a lattice mismatch

$$u_0 = \frac{a_2 - a_1}{a_1}$$

which creates a tangential stress in the layer. Various growth scenarios are then possible:

- *Frank-van der Merwe*: dislocations appear past a critical thickness d^*. They act to release the strain u_o.
- *Vollmer-Weber*: The epitaxial layer breaks into nonwetting droplets on the substrate.
- *Stranski-Krastanow*: The layer first grows flat; islands appear past a minimum thickness d^*. Subsequent growth leads to "kinetic amplification" of the islands, which turn into larges blobs at the surface.

A specific example of that latter regime is epitaxial growth of Ge on a (100)Si substrate. The mismatch is $u_o = .04$, while d^* is roughly 3 atomic layers. Can one explain such a threshold behavior?

The Grinfeld instability has been put forward as a candidate—and it indeed it would work fine if the substrate were rigid: the critical thickness would be given by Eq. (5). Alas that is not realistic as elasticities of Si and Ge are very similar. An alternate possibility is to invoke a *nonlocal Hooke's law*: the sharp discontinuity of u_{xx} at interface would then build up $\sigma_{xx} = \sigma_0$ progressively, threshold being reached only past a critical d. Again it is not reasonable to invoke nonlocality extending to three lattice spacings!

Another possibility is to rely on the van der Waals interaction with the substrate to play the role of gravity. A thin layer δd a distance d from the substrate feels an *extra* attraction (per unit area)

$$\frac{A_{12} - A_{22}}{2\pi d^4} \delta d = \rho g_{\text{eff}} \delta d \,.$$

A_{12}, A_{22} are the Hamaker constants, the interface is stabilized if $A_{12} > A_{22}$. It is convenient to measure that interaction in terms of a characteristic length ℓ_w defined as

$$\rho_L g_{\text{eff}} = \gamma \frac{\ell_w^2}{d^4}, \quad \ell_w^2 = \frac{A_{12} - A_{22}}{2\pi \gamma} \,.$$

The threshold for a Grinfeld instability then corresponds to

$$k^* = \frac{\ell_w}{d^2}, \quad \sigma^* = \left[\frac{E}{1 - \sigma^2} \frac{\gamma \ell_w}{d^2} \right]^{1/2} \,.$$

It corresponds to a thickness

$$d^* = \sqrt{\frac{b\ell_w}{u_o}} \text{ with } b = (1 - \sigma^2)\frac{\gamma}{E}.$$

b is comparable to the atomic spacing. Orders of magnitude are not stupid—except that the signs are wrong for Ge(Si): $A_{12} < A_{22}$! The van der Waals interaction is destabilizing! The situation thus remains completely open.

5. CONCLUSION

This brief survey shows that the Grinfeld instability is a very common phenomenon, far more than could have been surmised at the time of its inception by Asaro and Tiller.[1] It deserves a more systematic study. On the fundamental side its nonlinear behavior looks singular and unusual; moreover it allows a very flexible study of the transition between one-dimensional and two-dimensional structures. On the practical side, it applies to very different systems, both on a macroscopic and on an atomic scale. The main message is that nonhydrostatic stresses can lead to all sorts of instabilities and pattern formations.

ACKNOWLEDGMENTS

The author wishes to thank V. I. Marchenko for calling his attention to the work of Grinfeld in January 1987. His own contribution to that problem stems from that discussion.

REFERENCES

1. Asaro, I., and W. A. Tiller. *Metall. Trans.* **3** (1972): 1972.
2. Grinfeld, M. Ya. *Sov. Phys. Dokl.* **31** (1986): 831. A recent review is published in *J. Nonlinear Sci.* **3** (1993): 35.
3. Srolowitz, D. J. *Acta Metall.* **37** (1989): 621.
4. Noziéres, P. *J. Phys. I. France* **3** (1993): 681.
5. Kassner, K., and C. Misbah. To be published.
6. Caroli, B., C. Caroli, B. Roulet, and P. W. Voorhees. *Acta Metall.* **37** (1989): 257.
7. Thiel, M., A. Willibald, P. Evers, A. Levchenko, P. Leiderer, and S. Balibar. *Europhys. Lett.* **26** (1992): 707.
8. Torii, R., and S. Balibar. *J. Low Temp. Phys.* **89** (1992): 391.
9. Berrehar, J., C. Caroli, C. Lapersonne-Meyer, and M. Schott. *Phys. Rev. B* **46** (1992): 13487.
10. Spencer, B. J., P. W. Voorhees, and S. H. Davis. *Phys. Rev. Lett.* **67** (1991): 3696.

Michael Grinfeld
Department of Mathematics, Rutgers University, New Brunswick, NJ 08903, USA

Patterns of Stress-Driven Islanding in Solid Epitaxial and He-4 Films

It was demonstrated in a series of publications that in the absence of surface tension a flat boundary of nonhydrostatically stressed elastic solids is always unstable with respect to "mass rearrangement." The instability is pure thermodynamic in nature and does not depend on the specific mechanisms of the rearrangement that can be different, for instance, (a) melting-freezing or vaporization-sublimation processes at liquid-solid or vapor-solid phase boundaries, (b) surface diffusion of particles along free or interface boundaries, (c) adsorption-desorption of atoms in epitaxial crystal growth, etc.... We discuss the role of this instability in the problems of epitaxy and low temperature physics.

We believe that the instability allows to explain some recent experiments on the He-4 crystals under stress and also that it delivers new opportunities for realistic explanation of recently discovered phenomenon of the *dislocation-free Stranski-Krastanov pattern of epitaxial growth*. This phenomenon cannot be interpreted in the framework of traditional viewpoints since, according to the classical theory, the Stranski-Krastanov pattern develops as the result of proliferation of misfit dislocations appearing at the interface "crystalline film-substrate." Our approach leads to a new formula

Spatio-Temporal Patterns, Ed. P. E. Cladis and P. Palffy-Muhoray,
SFI Studies in the Sciences of Complexity, Addison-Wesley, 1995

of the critical thickness for the exchange of stability from a planar (layer-by-layer) growth to a growth mode showing long parallel trenches with periodic spacing. We discuss the circumstances of the existence of the second critical thickness. When the film exceeds the second critical thickness the mode of fastest growth of perturbations of an unstable isotropic planar film corresponds to a two-dimensional superlattice of rectangular islands rather than a one-dimensional lattice of trenches.

1. INTRODUCTION

For many decades, various morphological instabilities of interfaces remain to be the field of intensive studies. Traditionally, it is a topic of primary interest in crystal growth, metallurgy, fracture, geology, petrology, etc. In past years, it has become even more attractive because of the needs of relatively new disciplines like epitaxy, electronic packaging, tribology, biology, etc. (see modern reviews by Godréche[10] and references therein).

Recently, active efforts of many researchers provided a rapid progress and much deeper understanding of the specific "rearrangement" instability of the interfaces and free boundaries in solids. This stress-driven instability occurs at the interfaces of nonhydrostatically stressed solids, the material particles of which are capable to rearrange their equilibrium positions in the lattice. At present, the rearrangement instabilities of free boundaries and of the phase boundaries "stressed solid-melt" have attracted a particular attention because of their the most promising and obvious applications. The studies and lectures of Noziéres[23] have made the topic absolutely transparent; in particular, Noziéres gave amazingly general and elementary argument clearly showing that a flat boundary of a nonhydrostatically stressed solid of any symmetry (both linear and nonlinear) is always unstable with respect to "mass rearrangement." This conceptually rigorous proof does not require even a single formula!

Because of its universal nature, the stress-driven "rearrangement" instability and the instability "stressed solid-melt," in particular, delivers new insights and provides new opportunities in different branches of materials and other sciences, a part of which was discussed by Grinfeld.[11,14,15] The list of the applications is rapidly growing at present (see, for instance, discussions of the instability "stressed solid-melt" in the problems of pattern formation in metallurgy,[5,22] theory of gels,[25] low-temperature physics,[2,4,24,29,30] and geology[20]; also, see the publications on the stress driven "rearrangement" instability of isolated crystals relating to the problems of epitaxy[5,7,17,27,28] and fracture[1,8,9]). In particular, it has already allowed to predict the stress-driven corrugations of thin solid films of He-4, experimentally evidenced by Theil et al.[29] and Torii and Balibar,[30] and to propose a new approach to the phenomenon of dislocation-free Stranski-Krastanov pattern of growth of epitaxial films of GaAs on Si substrates.[6,19,21] Key roles of misfit stresses and of the rearrangement in this phenomena were discussed in the past by several authors exploiting different approaches. One of the most interesting parameters in the problems of epitaxy is the magnitude of the critical thickness of the so-called pseudomorphic growth.

2. INTUITIVE JUSTIFICATION OF THE MECHANISM OF THE STRESS-DRIVEN REARRANGEMENT INSTABILITY

Let us consider the processes of crystallization or sublimation at at free surface of the uniaxially prestressed deformable crystalline solid layer (see Figure 1). We take into account three following constituents of accumulated energy: (a) "gravity," (b) elasticity, and (c) surface energy. The stresses can be generated artificially (like in the recent experiments. with He-4 described by Thiel et al.[29] and Torii and Balibar[30]) or to be of natural origin (like "mismatch" intrinsic stresses in epitaxy). Let us consider the two-dimensional picture and assume that an unstressed elementary lattice cell of the crystalline substance has square shape. Under uniaxial lateral stress, the cells of completely filled ad-layers take on a rectangular shape. When new particle **A** sticks to the uniform ad-layer, its bottom stretches more intensively than the top, the square distorts, becoming a trapezia. Consider now the preferences of the particle **B** fooling around the particle **A**. The "normal" gravity pushes the particle **B** to occupy as low position as possible (the positions 1, 2, 3, and 4 are all equivalent from this point of view and they are preferential as compared with the position atop the particle **A**). The surface energy dictates the particle B to have as many solid neighbors as possible—this makes the position (2, 3) the most advantageous among $(1, \ldots, 4, \mathbf{B})$. The both discussed agents favor the positions that make the interface as smooth as possible. At the same time the influence of elasticity is absolutely opposite. Intuitively it looks obvious that the particle **B** accumulates less elastic energy if it sticks atop the particle A as compared with the positions $1, \ldots, 4$ since the particle **A** has already compensated partially the mismatch of the uniformly stressed ad-layer and the free particles. Thus, elasticity favors corrugating of a smooth surface while gravity and surface energy oppose this process.

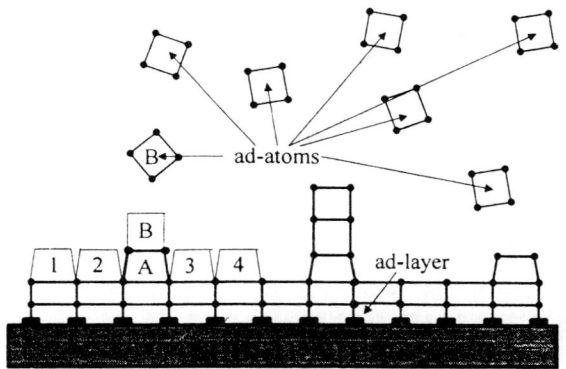

FIGURE 1   The mechanism of the stress-driven rearrangement instability

---

3. THE ENERGY BALANCE OF ELASTIC SYSTEMS WITH REARRANGEMENT

Let us consider a thin solid film of a thickness H, that is uniformly stressed and attached to a solid substrate. The uniformly-stressed film accumulates the energy E^{reg}. We use notation E_g, E_b, and E_s for the contributions of the potential energy, due to the presence of the external force field (which is assumed one-dimensional with the acceleration g), of the bulk (elastic) energy and of the surface energy, respectively. The surface energy is assumed proportional, with a surface tension coefficient σ, to the elemental area of the free boundary in the unstressed (reference) configuration. A corrugated film of the same film accumulates another amount of energy E^{irreg}. Here we limit ourselves with the case of small surface corrugations (see the paper by Noziéres[24] dealing with the nonlinear effects associated with the corrugations of finite sizes). In what follows, we denote by $\epsilon\Delta(x^a)$ the corrugations height at the point with the lateral coordinates $x^a (\epsilon \ll 1, \Delta \sim H)$. The in-plane indexes $a, b, c, \ldots$ run values $1, 2$ while the spatial ones $i, j, k, \ldots$ run values $1, 2, 3$. For the films and substrates consisting of isotropic elastic substances (with the shear modules μ_c, μ_s and the Poisson ratios ν_c, ν_s, respectively) a difference of the energies $E^{\text{irreg}} - E^{\text{reg}}$ (which determines the stability of the flat film with respect to spontaneous corrugations of its free surface) can be computed explicitly and written in the following Fourier form:

$$E^{\text{irreg}} - E^{\text{reg}} = \epsilon^2 \int_{-\infty}^{+\infty} dk_1 \int_{-\infty}^{+\infty} dk_2 \mathbf{K}(\mathbf{k}, H)\chi(\mathbf{k})\chi^*(-\mathbf{k}) \,. \tag{1}$$

Here $\chi(k_1, k_2)$ is the Fourier component of the surface corrugation $\Delta(x^1, x^2)$ with the in-plane wavevector $\mathbf{k} = (k_1, k_2)$. The kernel $\mathbf{K}(\mathbf{k}, H)$ includes the same three constituents associated with

i. the external potential energy K_g,

ii. the bulk (elastic) energy K_b, and

iii. the surface energy K_s,

to within insignificant positive multiplier it can be written as follows:

$$\mathbf{K}(\mathbf{k}, H) = K_g + K_b + K_s = g\rho_v(\gamma - 1) - |\mathbf{k}|J_b(\mathbf{e}, h) + \sigma|\mathbf{k}|^2, \tag{2}$$

where $\gamma = \rho_s/\rho_v$ is the density ratio of the solid film and the melt or vapor, and $h = |\mathbf{k}|H$ and $\mathbf{e} = \mathbf{k}|\mathbf{k}|^{-1}$ are the dimensionless length and the unidirectional vector of $\mathbf{k}$. In the case of epitaxial film, we plug $g\rho_s$ for $g\rho_v(\gamma - 1)$ to account for van der Waals forces[24]; it should be underlined that these forces depend strongly on the distance from the substrate.

In contrast to the external and surface energies, the bulk energy term $J_b(\mathbf{e}, h)$ depends on the mechanical properties of the film and of the substrate, and of their thicknesses. In what follows, we consider two following situations: (a) an isotropic compressible film of thickness H attached to a rigid substrate; and (b) an isotropic incompressible film of a thickness H coherently attached to infinitely thick incompressible isotropic substrate. In particular, in the case of isotropic prestressed film with Poisson ratio n coherently attached to a rigid substrate we arrive at the formula[16]:

$$J_b(\mathbf{e}, h) = \frac{T_n^2}{\mu} \frac{(1 - \nu)[h + (3 - 4n)\sinh(h)\cosh(h)]}{4(1 - n)^2 + h^2 + (3 - 4\nu)\sinh^2 h} + \frac{T_t^2}{\mu} \frac{\sinh(h)}{\cosh(h)}. \tag{3}$$

The function $J_b(\mathbf{e}, h)$ depends on the directional vector $\mathbf{e}$ implicitly via the normal T_n and tangential T_t components of the in-plane stress $\mathbf{T_k}$ acting at the cross-section orthogonal to the wave-vector $\mathbf{k}$ (see Figure 2). Introducing the principal in-plane prestresses (the "misfit" stresses) Tab can present the above quantities as

$$T_n = T_1 e_1^2 + T_2 e_2^2 = T_1 \cos^2\theta + T_2 \sin^2\theta, \quad T_t = T_1 e_1 q_1 + T_2 e_2 q_2 = (T_1 - T_2)\sin\theta\cos\theta,$$

where θ is the angle between $\mathbf{e}$ and the principal direction of the in-plane stress T_1; e_a, q_a are the components of the unit in-plane vectors $\mathbf{e}$ and $\mathbf{q} \perp \mathbf{e}$, respectively, with respect to the in-plane axes coinciding with the principal directions of the misfit stresses.

(a)

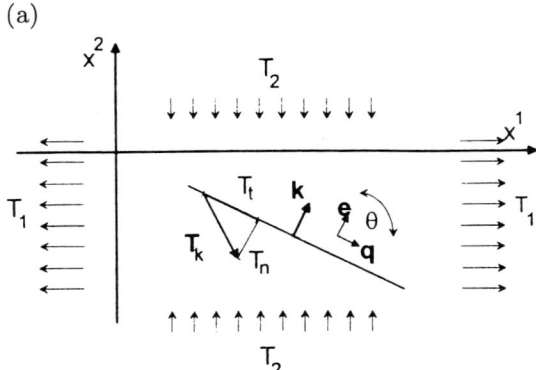

(b)

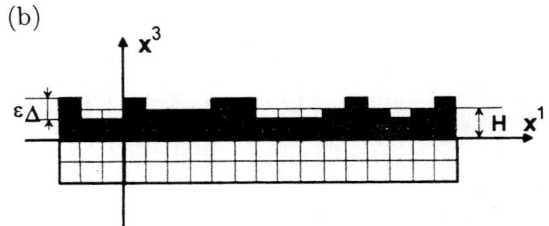

FIGURE 2 The geometry of
the corrugated film. (a) top
view, (b) side view.

---

4. MORPHOLOGICAL PATTERNS OF PRESTRESSED FILMS WITH THE REARRANGEMENT. THE CRITICAL THICKNESSES AND THE PATTERNS OF CORRUGATION.

We call *stable, neutral, and destabilizing* the corrugations with the wave-vector $\mathbf{k}$ satisfying the following relations, respectively:

$$\mathbf{K}(\mathbf{k}, H) > 0, \quad \mathbf{K}(\mathbf{k}_{ne}, H) = 0, \quad \mathbf{K}(\mathbf{k}, H) < 0. \tag{4}$$

We call *locally stable* the configuration having stable corrugations only. We call *unstable* the configuration having, at least, one unstable corrugation. We call *critical* the thickness of the uniform film without unstable corrugations for which, at least, one mode of the surface corrugations appears to be neutral.

Combining Eqs. (3) and (4), we arrive at the following equation of the neutral corrugations of the isotropic prestressed film attached to the rigid substrate[17]:

$$|\mathbf{k}| - 1\Gamma^* + |\mathbf{k}|\Sigma^* - (\cos^2\theta + \alpha\sin^2\theta)^2 \frac{(1-\nu)[h + (3 - 4\nu)\sinh(h)\cosh(h)]}{4(1-n)2 + h2 + (3 - 4n)\sinh 2h}$$

$$- (1 - \alpha)^2 \sin^2\theta \cos^2\theta \frac{\sinh(h)}{\cosh(h)} = 0 \tag{5}$$

where $\alpha = T_2/T_1$ is the ratio of the two in-plane principal "misfit" stresses whereas the parameters Γ^* and Σ^* are defined as follows:

$$\Gamma^* = \frac{g\rho_v(\gamma - 1)\mu}{T_1^2} = \frac{g\rho_s(\gamma - 1)\mu}{\gamma T_1^2}, \quad \Sigma^* = \frac{\sigma\mu}{T_1^2}. \tag{6}$$

Let us begin with the case of negligible mass forces. Then, Eqs. (3) and (4) lead, in particular, to the following value of the critical film thickness:

$$H_{\mathrm{crit}} = \frac{\mu\sigma}{T_{\mathrm{max}}^2} \tag{7}$$

where T_{max} is the greater (in the absolute value) of the two principal in-plane misfit stresses T_1, T_2. The investigation of Eqs. (4) and (5) shows that at $H < H_{\mathrm{crit}}$, the film is stable with respect to arbitrary surface corrugations. At H exceeding H_{crit} it becomes unstable. At this initial stage (i.e., at the thicknesses slightly exceeding H_{crit}) the most unstable (and the fastest) corrugations appear to be of the shape of "trenches" parallel to the direction of the minimal in-plane principal stress T_{min}. While the film becomes thicker, the most unstable mode changes. The scenario of the evolution depends drastically on the misfit stresses and, in particular, on the dimensionless parameter $s \equiv (T_1 + T_2)/(T_1 - T_2)$. Say, in the case of uniaxial misfit stresses, ($|s| = 1$) for each thickness; the most unstable corrugations take on the shape of the array of trenches with the direction orthogonal to the stress, the distance between the trenches diminishes with the increase of the thickness. In the case of the purely shearing misfit stresses, (when $T_1 = -T_2$ and $s = 0$) the most unstable mode corresponds to the array of the squares the size of which diminishes with increase of the thickness.

At the final stage of growth, when the film is infinitely thick, the corrugations with the fastest rate of growth can form two different ultimate surface morphologies depending on the stresses and material parameters: (a) the trenches-like morphology; or (b) the islands-like morphology. The trenches-like morphology corresponds to the array of trenches of the permanent direction (but with specific changes of the shape of the trenches and of the distance between them); this morphology dominates at the final stage providing the inequality

$$|s| > \frac{1 - \nu}{\nu} \tag{8}$$

is valid.

If the inequality Eq. (8) violates the "islands"-like pattern develops and the final surface morphology appears to become the array of rectangular "islands." The ratio Ra of the legs of elementary cell of the island is the following:

$$Ra = tg\,\theta = \sqrt{\frac{1 - \nu - s\nu}{1 - \nu + s\nu}}. \tag{9}$$

In particular, if the film is incompressible ($n = 1/2$), the ration Ra is determined by the very simple and illuminating formula:

$$Ra = tg\,\theta = \sqrt{\frac{1-s}{1+s}} = \sqrt{-\frac{T_2}{T_1}}. \tag{10}$$

We recall that at the initial stage of the instability the morphology of the corrugations with the fastest rate of growth is always the trenches-like (with a single exception of the misfit stress of the pure shear: $T_1 = -T_2$). Thus, in the case (b) at the certain (second!) critical thickness there appears the dramatic change in the pattern of growth from the trenches-like to the islands-like morphology.

---

5. THE STABILITY CRITERIA OF THE TRENCHES-LIKE PATTERN OF CORRUGATION. THE $\Phi\alpha$-NUMBERS, CRITICAL WAVE-LENGTHS AND THICKNESSES.

The trenches-like pattern of growth leads to the simplest key formulas and criteria, and it can be investigated in the framework of two-dimensional theory of elasticity. All these criteria are formulated in terms of different but similar dimensionless parameters which we call the $\Phi\alpha$-numbers. All these numbers reflect the competition of a destabilizing influence of the shear stresses on one hand, and of stabilizing influence the (isotropic) surface tension and gravity on the other. To distinguish the "rearrangement" stress driven instability from various other bulk instabilities, we call it shortly the $\Phi\alpha$-bulk instability.

For a prestressed elastic half-plane in the absence of mass forces the stability criterion can be written as follows:

$$\Phi\alpha^\sigma < \frac{1}{1-\nu}, \tag{11}$$

where n is the Poisson ratio of the isotropic half-space; the $\Phi\alpha^\sigma$-number is defined as

$$\Phi\alpha^\sigma = \frac{\tau^2}{\sigma\mu_c k}. \tag{12}$$

Here k is the tangential wavenumber of the surfacial corrugations (when dealing with epitaxial film in two-dimensional approach we use notation $\tau = T_1$ for the uniaxial misfit stress appearing in the flat film). According to Eqs. (11) and (12), the elastic half-space is stable with respect to the disturbances with sufficiently large tangential wavenumbers and (in the absence of gravity) it is unstable with respect to the disturbances with sufficiently small wavenumbers. The critical value of the wavenumber is given by the formula

$$k_{ne} = \frac{\tau^2}{\sigma \mu^c}(1 - \nu).$$ (13)

In the case of a solid film of finite (undisturbed) thickness H, it is convenient to introduce another dimensionless $\Phi\alpha$-number

$$\Phi\alpha^H = \frac{\tau^2 H}{\sigma \mu^c}.$$ (14)

For the fixed magnitude of $\Phi\alpha$ the flat, traction-free boundary can be destabilized by those and only those components of the corrugation which obey the inequality

$$\Phi\alpha^H > J_\nu(h); \quad J_\nu(h) = h\frac{4(1 - \nu)^2 + h^2 + (3 - 4\nu)\sinh^2 h}{(1 - \nu)[h + (3 - 4\nu)\sinh h \cosh h]}.$$ (15)

The neutral dimensionless wavenumber h_{ne} obeys the equation $J_\nu(h_{ne}) = \Phi\alpha^H$. For $h \gg 1$ (i.e., for relatively short horizontal wavelengths), the function $J_\nu(h)$ approaches the value $h/(1 - \nu)$, and Eq. (15) leads to the criterion equivalent to Eq. (13).

In the opposite asymptotic case of very small values of dimensionless wavenumber $h \ll 1$ (or relatively lengthy "corrugations"); in Eq. (15), we show that $J\nu$ is equal to 1. Thus, according to Eq. (15), the a flat upper boundary of the layer is stable with respect to the "corrugations" of very large relative wavelengths if the number $\Phi\alpha^H$ is less than $\Phi\alpha^H_{\text{crit}}$ given by the formula

$$\Phi\alpha^H_{\text{crit}} = 1.$$ (16)

In view of the properties of function $J\nu(h)$, the stability at h tending to zero implies the stability with respect to the corrugations of arbitrary wavelengths. Equation (16) leads to the critical film thickness given by Eq. (7). At $H < H_{\text{crit}}$, the film is stable with respect to arbitrary surface corrugations even without mass forces.

In the absence of mass forces, the neutral wavevector k_{ne} appears to be equal to zero. Eqation (5) allows one to determine the critical thickness and neutral wavevector for the arbitrary function $g(H)$. In two-dimensional case, Eq. (5) reduces to the following:

$$L(k) = R(kH)$$ (17)

where

$$L(k) = k^{-1}\Gamma^* + k\Sigma^*, \quad R(kH) = \frac{(1 - \nu)[kH + (3 - 4\nu)\sinh(kH)\cosh(kH)]}{4(1 - \nu)^2 + (kH)^2 + (3 - 4\nu)\sinh^2(kH)}.$$ (18)

For positive $\Gamma^*(H)$, the function $L(k)$ has a single minimum $2\sqrt{\Gamma^*\Sigma^*}$ achieved at $k = \sqrt{\Gamma^*\Sigma^{*-1}}$ (see the sketch in Figure 3). The RHS of Eq. (17) depends on both quantities k and H. The function $R(kH)$ is shown in Figure 3 for several values of H (shrinking the graph for $H = 1$ parallel to the k-axis with the coefficient H, we obtain the corresponding graph for arbitrary value of H). The critical thickness H_{crit} corresponds to the magnitude of H at which $L(k)$ and $R(kH_{\text{crit}})$ touch at a single point. Abscissa of the touch-point gives the corresponding neutral wavenumber k_{ne}. Thus, we arrive at the following system for the H_{crit} and k_{ne}:

$$L(k_{ne}) = R(k_{ne}H_{\text{crit}}), L'(k_{ne}) = H_{\text{crit}}R'(k_{ne}H_{\text{crit}}). \tag{19}$$

For small g independent of H the constant Γ^* is small also. In this case, the study of the system Eq. (19) leads to the following asymptotic formulas:

$$H_{\text{crit}} = \Sigma^* + \sqrt[2]{\frac{2}{3}\frac{3+4\nu}{1-\nu}\Gamma^*\Sigma^{*3}}, \quad k_{ne} = \sqrt[4]{\frac{6(1-\nu)}{3+4\nu}\frac{\Gamma^*}{\Sigma^{*3}}}. \tag{20}$$

In the case of incompressible film, $n = 1/2$ the Eq. (20) reduce to:

$$H_{\text{crit}} = \Sigma^* + \sqrt[2]{\frac{20}{3}\Gamma^*\Sigma^{*3}}, \quad k_{ne} = \sqrt[4]{\frac{3}{5}\frac{\Gamma^*}{S^{*3}}}. \tag{21}$$

Thus, the correction of the critical thickness for a small mass force is of the order of $g^{1/2}$, whereas k_{ne} is of the order of $g^{1/4}$ (for infinitely thick film $k_{ne} \sim g^{1/2}$).[24]

According to the experiment of Torii and Balibar,[30] $n = 1/3$, $\Delta\rho \sim 0.2$ g/cm³, $\sigma \sim 0.2$ erg/cm², $\mu \sim 1.1 \times 10^6$ erg/cm³, $T \sim 105$ cgs (10^{-1} bar); for these parameters, the formulas in Eq. (20) give $k_{ne} = 64.6$ cm^{-1}($\lambda_{ne} = 2\pi/k_{ne} = 0.01$ cm), $\Sigma^* = 0.23 \times 10^{-2}$ cm, $H_{\text{crit}} = \Sigma^* + 0.23 \times 10^{-3}$ cm.

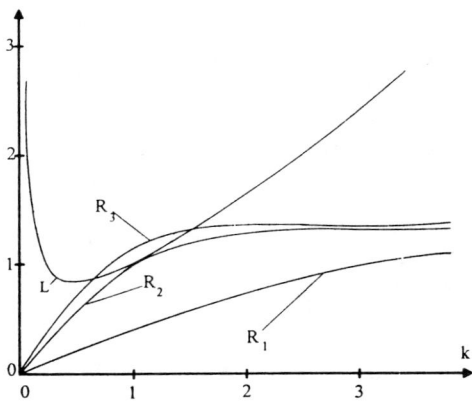

FIGURE 3 The graphs of the LHS (L) and of the RHS (R) of Eq. (17) for rigid substrate (R_1, R_2, and R_3 are the graphs for the subcritical, critical and transcritical thicknesses, respectively).

Let us dwell on the case of the isotropic incompressible film of a thickness H coherently attached to an infinitely thick incompressible isotropic substrate. In the absence of mass forces, we arrive at the following equation for the dimensionless neutral corrugations h_{ne}[15]:

$$L^\chi(h_{ne}) = \Phi\alpha^H;$$

$$L^\chi(h) = h\frac{(\chi^2 - 1)h^2 - (\chi \sinh h + \cosh h)^2}{(\chi^2 - 1)h - (\sinh h + \chi \cosh h)(\cosh h + \chi \sinh h)}, \qquad (22)$$

where $\chi = \mu_c/\mu_s$ is the ratio of shear modules of the film and the substrate. At $\chi = 1$ function $L^\chi(h)$ reduces to h, and we arrive at the formula $h_{ne} = \Phi\alpha^H$, which is equivalent to Eq. (13) of the neutral dimensional wavenumber. Function $L^\chi(h)$ behaves like χh at $h \ll 1$ and it tends to zero at h approaching zero in contradistinction to the function $J_\nu(h)$ which tends to a finite value at h approaching zero. In view of that asymptotic behavior, in the absence of mass forces, the crystalline layer coherently attached to a *deformable* substrate can always be destabilized by means of the "corrugations" containing sufficiently lengthy components. Equation (22) leads to the following values of the neutral wavenumber h_{ne}^χ (provided $h_{ne}^\chi \ll 1$):

$$h_{ne}^\chi = \chi\Phi\alpha^H = \chi h_{ne}^1 \quad \text{or} \quad \frac{2\pi}{h_{ne}^\chi} = \frac{\mu_s}{\mu_c}\frac{2\pi}{h_{ne}^1}. \qquad (23)$$

Equations (23) show that the neutral wavelength of "corrugation" increases in direct proportion to the rigidity of the substrate.

Thus, without mass forces, there is no critical thickness of the solid films growing atop deformable substrates: at arbitrary nonzero thickness the laterally unbounded films appear to be unstable with respect to sufficiently large corrugations. This circumstance causes some conceptual difficulties in the explanation of the Stranski-Krastanov pattern of growth of epitaxial films. The mass forces, however, suppress the unstable corrugations in that very domain of the spectrum providing the chance of making the critical thickness non-vanishing. Thinking of the experimental results[6,19,21,26] we assume that the first corrugations correspond to the neutral wavenumber k_{ne} and the critical thickness H_{crit} such that $\exp(k_{ne}H_{\text{crit}}) \sim 1$. Then, using Eq. (5), one can easily establish the following formulae:

$$\rho_s g(H_{\text{crit}}) = \frac{T^4\chi^2}{16\mu_c^2\sigma}, \quad k_{ne} = \sqrt{\frac{\rho_s g}{\sigma}} = \frac{T^2\chi}{4\mu_c\sigma}. \qquad (24)$$

First, Eq. (24) has to be treated as the equation with respect to the unknown thickness H_{crit}. Solutions of this equation depend on the function $\rho_s g(H)$ giving a magnitude of the mass forces at the distance H from the substrate. A typical relation for van der Waals force is $\rho_s g(H) = WH^{-4}$, where W is a constant. In this case, we arrive at the following formula of the critical thickness which generalization the formula of critical thickness of overlayers atop attractive deformable substrates established by Nozières in his lectures on Stranski-Krastanov pattern of epitaxial growth for the case of the film and substrate with equal elastic modules:

$$H_{\text{crit}} = 2W^{1/4}\mu_s^{1/2}\sigma^{1/4}|T|^{-1}; \ (k_{ne}H_{\text{crit}} = 2^{-1}W^{1/4}\mu_s^{-1/2}\sigma^{-3/4}|T|). \quad (25)$$

The second of the formulae in Eq. (25) allows one to check the applicability of the formula of the critical thickness (see the above-mentioned assumption on the relative wavelength).

The study of the $\Phi\alpha$-bulk instability of isolated films has shown the existence of two different unstable modes: (a) symmetrical and (b) asymmetrical. In the absence of gravity, the corresponding equations for the dimensionless wavenumber of neutral corrugations are of the following form[14]:

$$(a) \quad \Phi\alpha_s^H = \frac{1}{1-\nu}(h\tanh h - h^2\tanh^2 h + h^2)$$

$$(26)$$

$$(b) \quad \Phi\alpha_{as}^H = \frac{1}{1-\nu}(h\coth h - h^2\coth^2 h + h^2)$$

where the $\Phi\alpha_s^H = \Phi\alpha_{as}^H = \frac{2H\tau^2}{\sigma\mu_c}$, $h = kH$; H is a half-thickness of the films (given $\Phi\alpha_s^H$, $\Phi\alpha_{as}^H$ the roots of Eqs. (26) allow one to compute magnitude of the neutral corrugations h_{ne}).

---

6. CONCLUSIONS

In the absence of surface tension, a flat boundary of nonhydrostatically stressed elastic solids is always unstable with respect to "mass rearrangement." The instability is pure thermodynamic in nature and does not depend on the specific mechanisms of mass transport. The instability allows to explain some recent experiments on the He-4 crystals under stress.[29,30] Also it delivers new opportunities in the theory of dislocation-free Stranski-Krastanov pattern of epitaxial growth.

There may exit two critical thickness for the prestressed elastic films attached to solid substrate. The first one corresponds to the exchange of stability from a planar (layer-by-layer) growth to a growth mode showing long parallel trenches with periodic spacing. For some prestresses there exists also the second critical thickness. When the film exceeds the second critical thickness the mode of fastest growth of perturbations of an unstable isotropic planar film corresponds to a two-dimensional superlattice of rectangular islands rather than a one dimensional lattice of trenches.

Attractive mass forces (those with positive acceleration g) always increase the static critical thickness of stressed solid films attached to the substrates. Both the critical thickness and its correction for the gravity are detectable in the experiments like those of Thiel et al.[29] and Torii and Balibar.[30] Combining the mechanism of stress-driven rearrangement instability, stabilizing effects of surface tension, and van der Waals forces, one obtains rather flexible treatment of the phenomenon of dislocation free Stranski-Krastanov growth of heteroepitaxial solid films attached to attractive substrates.

REFERENCES

1. Asaro, R. J., and Tiller, W. A. "Interface Morphology Development During Stress Corrosion Cracking." *Metall. Trans.* (1972): 1789.
2. Balibar, S., D. O. Edwards, and W. F. Saam. "The Effect of Heat Flow and Non-Hydrostatic Strain on the Surface of Helium Crystals." *J. Low Temp. Phys.* **82** (1991): 119.
3. Berrehar, J., C. Caroli, C. Lapersonne-Meyer, and M. Schott. "Surface Patterns on Single-Crystal Films Under Uniaxial Stress: Experimental Evidence for the Grinfeld Instability." *Phys. Rev. B* **46** (1992): 13487.
4. Bowley, R. M. "Instabilities of the Liquid Solid Interface." *J. Low Temp. Phys.* **89** (1992): 401.
5. Caroli, B., C. Caroli, B. Roulet, and P. W. Voorhees. "Effect of Elastic Stresses on the Morphological Stability of a Solid Sphere Growing from a Supersaturated Melt." *Acta Metall.* **37** (1989): 257.
6. Eaglesham, D. J., and M. Cerullo. "Dislocation-Free Stranski-Krastanov Growth of Ge on Si(100)." *Phys. Rev. Lett.* **64** (1990): 1943.
7. Freund, L. B., and F. Jonsdottir. "Instability of Biaxially Stressed Thin Film on a Substrate due to Material Diffusion over Its Free Surface." *J. Mech. Phys. Solids* (1993): in press.
8. Gao, H. "Stress Concentration at Slightly Undulating Surfaces." *J. Mech. Phys. Solids* **39** (1991): 443.
9. Gao, H. "Surface Roughening and Branching Instabilities in Dynamic Fracture." *J. Mech. Phys. Solids* **41** (1991): 457.
10. Godréche, C., ed. *Solids Far From Equilibrium.* Cambridge: Cambridge University Press, 1991.
11. Grinfeld, M. A. "Instability of the Separation Boundary Between a Nonhydrostatically Stressed Elastic Body and a Melt." *Sov. Phys. Doklady* **31** (1986): 831.
12. Grinfeld, M. A. "On the Instability of Equilibrium of the Nonhydrostatically Stressed Body and Its Melt." *Fluid Dynamics* **22** (1987): 169.
13. Grinfeld, M. A. "The Stability of Interphase Boundaries in Solid Elastic Media." *Appl. Math. Mech.* **51** (1987): 489.

14. Grinfeld, M. A. *Thermodynamic Methods in the Theory of Heterogeneous Systems.* Sussex: Longman, 1991.

15. Grinfeld, M. A. "The Stress Driven Instabilities in Crystals: Mathematical Models and Physical Manifestations." *J. Nonlinear Sci.* **3** (1993): 35.

16. Grinfeld, M. A. "On Morphology of the Stress Driven Corrugations of the Phase Boundary Between the Solid and its Melt." *J. Phys. (Cond. Matt.)* **4** 1992: 647.

17. Grinfeld, M. A. "The Stress Driven 'Rearrangement' Instability in Crystalline Films." *J. Intellig. Mater. Syst. Struct.* **4** (1993): 76.

18. Grinfeld, M. A. "The Influence of Mass Forces on the Critical Thickness of Pre-Stressed He-4 and Solid Epitaxial Films." *Europhys. Lett.* (1993): in press.

19. Guha, S., A. Madhukar, and K. C. Rajkumar, "Onset of Incoherency and Defect Introduction in the Initial Stages of Molecular Beam Epitaxial Growth of Highly Strained $In_xGa1_{-x}As$ on GaAs(100)." *Appl. Phys. Lett.* **57** (1990): 2110.

20. Heidug, W. "A Thermodynamic Analysis of the Conditions of Equilibrium at Nonhydrostatically Stressed and Curved Solid-Fluid Interfaces." *J. Geophys. Res.* **21** (1991): 909.

21. LeGoues, F. K., M. Copel, R. M. Tromp, "Microstructure and Strain Relief on Ge Films Grown Layer by Layer on Si(001)." *Phys. Rev. B* **42** (1990): 11690.

22. Leo, P. H., and R. F. Sekerka. "The Effect of Surface Stress on Crystal-Melt and Crystal-Crystal Equilibrium." *Acta Metall.* **37** (1989): 3119.

23. Noziéres, P. "Growth and Shape of Crystals." In: *Solids Far From Equilibrium.* edited by C. Godréche, 1–153. Cambridge: Cambridge University Press, 1991.

24. Noziéres, P. "Amplitude Expansion for the Grinfeld Instability due to Uniaxial Stress at a Solid Surface." *J. Phys.* **3** (1993): 681.

25. Onuki, A. "Theory of Pattern Formation in Gels: Surface Folding in Highly Compressible Elastic Bodies." *Phys. Rev. A* **39** (1989): 5932 .

26. Snyder, C. W., B. G. Orr, D. Kessler, and L. M. Sander. "Effect of Strain on Surface Morphology in Highly Strained InGaAs films." *Phys. Rev. Lett.* (1991): 66, 3032.

27. Spencer, B. J., P. W. Voorhees, and S. H. Davis. "Morphological Instability in Epitaxially Strained Dislocation-Free Solid Films." *Phys. Rev. Lett.* **67** (1991): 3696.

28. Srolovitz, D. J. "On the Instability of Surfaces of Stressed Solids." *Acta Metall.* **37** (1989): 621.

29. Thiel, M., A. Willibald, P. Evers, A. Levchenko, P. Leiderer, and S. Balibar. "Stress-Induced Melting and Surface Instability of He-4 Crystals." *Europhys. Lett.* **20** (1992): 707.

30. Torii, R. H., and S. Balibar. "Helium Crystals Under Stress: the Grinfeld Instability." *J. Low Temp. Phys.* **89** (1992): 391.

James A. Warren
National Institute of Standards and Technology, Gaithersburg, MD 20899

Selected Spacings During Directional Solidification of a Binary Alloy

1. INTRODUCTION

The casting of alloys is an industrially important problem which has generated a great deal of theoretical interest—see Kurz and Fisher[5] and Langer[6,7] for general reviews. The patterns observed during the solidification of an alloy are quite intricate, and although much theoretical progress has been made, a satisfactory quantitative description is still lacking. In this review we shall attempt to explain a specific phenomena that occurs during the directional solidification of a binary alloy, but it should be pointed out that the principles upon which the analysis is based are quite general. In particular we wish to emphasize the importance of *transients* in pattern formation.

Recently, Warren and Langer (WL)[10,11] provided a simple model to explain observed spacings in experments by Trivedi and Somboonsuk (TS).[9] Since then, there has been work by Caroli, Caroli, and Ramírez-Piscina (CCR)[1] and by Karma[3,4] expanding on the earlier work, and placing the approximations made by WL on a rigorous footing. Given these developments, it is worthwhile to present a unified discussion and synthesis of the WL, Karma1, Karma2, and CCR papers, in order to emphasize the accuracy of the results, and pave the way for future investigations.

Spatio-Temporal Patterns, Ed. P. E. Cladis and P. Palffy-Muhoray,
SFI Studies in the Sciences of Complexity, Addison-Wesley, 1995   **91**

2. THE EXPERIMENT OF TRIVEDI AND SOMBOONSUK

The TS experiment involved the directional solidification of a 5.5% molar solution of acetone in succinonitrile. The temperature-concentration phase diagram for this system in the limit of small concentrations is given in Figure 1. The two lines are called the solidus and the liquidus, and they represent the equilibrium concentrations of solute at a given temperature in the solid and liquid respectively. We assume, mostly for simplicity sake, that the slopes of the solidus and liquidus are constant, and define two parameters

$$m = \left| \frac{dT}{dc_l} \right|, \tag{1}$$

the liquidus slope, and

$$K = \frac{c_s}{c_l}, \tag{2}$$

the partition coefficient. Note that with this form for the phase diagram solute will build up in front of the liquid-solid boundary as solidification occurs.

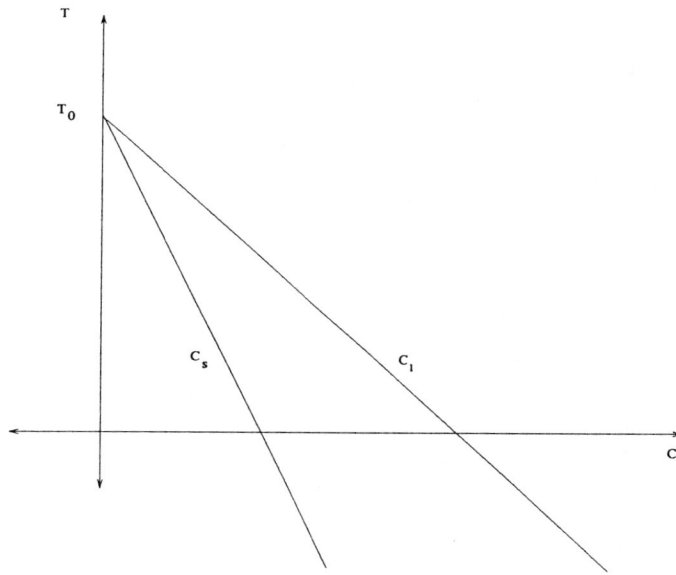

FIGURE 1   Phase Diagram for a typical binary alloy, for dilute concentrations of solute. The quantities c_l and c_s are the equilibrium concentrations in the liquid and solid, respectively.

A directional solidification experimental setup and observations are schematically represented in Figures 2(a)–2(d), and Figure 3. Initially the alloy is at rest between two glass plates in a constant, applied, temperature gradient G in the z direction (TS used a gradient of $G = 67$ K/cm). The glass plates are assumed close enough that the governing equations are correct in their two-dimensional forms. We choose an origin for the z-axis by defining

$$T = T_0 + Gz, \tag{3}$$

where T_0 is the melting point for pure succinonitrile. It is worthwhile to note that this choice of origin always places the interface at points where $z < 0$. We assume that Eq. (3) determines the temperature everywhere in the system, which is reasonable if thermal diffusion occurs much more quickly than solute diffusion, that the latent heats are small, and that the thermal conductivities in the liquid and solid are nearly equal. All of these assumptions are quite good for the succinonitrile-acetone system. In Figure 2(a) the initial equilibrium position of the interface, and the associated concentration profile are shown. We see that initially the concentration is just a step function across the interface of the form

$$c_0(z) = \begin{cases} c_\infty, & z > z_0, \\ Kc_\infty, & z < z_0, \end{cases} \tag{4}$$

where z_0 is the position of the interface. Given the initial state described by Eq. (4), a motor is then turned on which pulls the sample through the temperature gradient at the pulling speed v_p. This is illustrated in Figure 2(b). We should note that the solid-liquid interface is initially at rest, and will only move at the final pulling speed after a transient period of acceleration. It is our argument that the understanding of this transient period is *crucial* to understanding the patterns that are observed experimentally. In Figure 2(c) we suggest what a transient concentration profile might look like, while in Figure 2(d) we represent the $t = \infty$ steady-state exponential concentration profile when the interface has achieved the pulling speed v_p, of the form

$$c_0(z, t = \infty) = c_\infty + \left(\frac{c_\infty}{K} - c_\infty\right) \exp\left(-\frac{2}{\ell_0}\left(\frac{z - z_\infty}{K}\right)\right), \tag{5}$$

where $\ell_0 = 2D/v_p$ is called the diffusion length, c_∞ is the concentration far in front of the interface, z_∞/K is the equilibrium position of the interface, and D the diffusivity of the solute in the liquid. It is important to realize, however, that Eq. (5) is only valid, and therefore interesting, if the interface remains flat throughout the history of the solidification process, which is not the case. We will shortly examine the consequences of this change in shape.

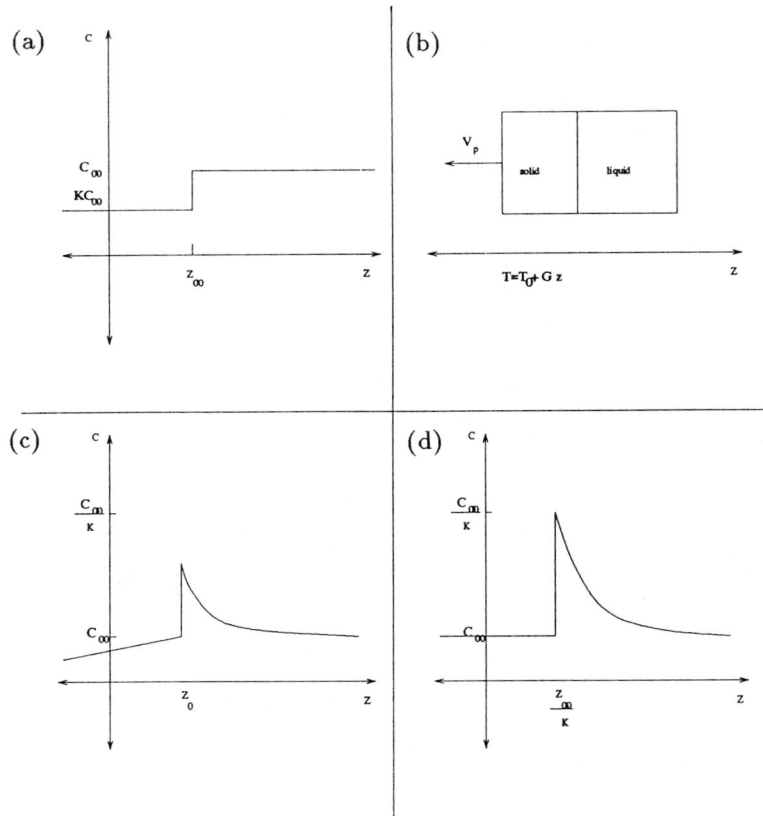

FIGURE 2 The history of a directional solidification experiment. In (a) we show the initial concentration profile, before the pulling motor is turned on. In (b) we have a schematic of the sample being pulled. In (c) and (d) we see transient and steady-state concentration profiles.

Equation (5) is a steady–state solution to the diffusion equation

$$\frac{\partial c}{\partial t} = D\nabla^2 c + v_p \frac{\partial c}{\partial z} \tag{6}$$

in the frame moving at the pulling speed v_p (the gradient frame). We have assumed that all transport is diffusive in the liquid (no convection), and that there is no diffusion in the solid. There are several associated boundary conditions which fix the solution to the diffusion equation. We assume local equilibrium at the interface, which when combined with Eq. (3) requires that

$$c(z_0, t) = -\frac{G}{m}z_0 - d_0 \mathcal{K}, \tag{7}$$

where z_0 is now the time-dependent position of the interface in the gradient frame. The second term on the right-hand side of Eq. (7) is called the Gibbs-Thomson curvature correction and is nonzero whenever the interface is nonplanar. $\mathcal{K}$ is the curvature of the interface, and

$$d_0 = \frac{\gamma T_0}{Lm} \tag{8}$$

is called the capillary length, where γ is the surface energy, and L is the latent heat per unit volume. The other boundary condition is conservation of solute at the interface, which requires that

$$D\frac{\partial c}{\partial z}\bigg|_{z_0} = v_0(1 - K)c(z_0, t). \tag{9}$$

In Eq. (9) we have used $v_0 = v_p + \dot{z}_0$, which is the velocity of the interface in the frame fixed in the solidifying material (the "material frame").

As we have previously noted, the equilibrium state represented by Eq. (5) and Figure 2(d) is rarely observed. In Figure 3 we show schematically what the interface

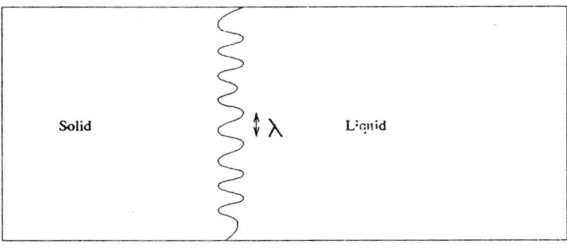

FIGURE 3   A schematic illustration of an interface which has become morphologically unstable. The quantity λ represents the mean distance between "bumps."

truly looks like some time after the pulling begins. A morphological instability has deformed the interface, such that there is a specific wavelength describing the nearly uniform space λ between "bumps" on the interface. We define this point in the evolution of the interface as that time when the deformations are approximately the same width as height. Beyond this time we may consider the "bumps" as nascent cells or dendrites. It is often the case that this change in shape will occur long before the interface achieves its final velocity, and the steady-state solute profile is established. We will now examine a simple yet illustrative attempt to explain this spacing, and then outline a better method of solution.

3. A PREVIOUS ATTEMPT TO EXPLAIN EXPERIMENT

When Trivedi and Somboonsuk reported their observations, they also included an attempt at a theoretical explanation.[9] The theory did not perform very well, unfortunately, but they recognized that their analysis did not take into account the acceleration of the interface or the concomitant buildup of solute. In this section we will briefly describe the method they employed, in order to help motivate our own analysis.

The basic assumption of TS was that the interface could be assumed to be moving at the pulling speed v_p, and that the concentration profile was that given by Eq. (5). One is then able to use the now classic analysis of Mullins and Sekerka[8] to explain the observed patterns.

In order to determine the selected wavelength at the interface, a linear stability analysis upon the interface concentration profile was performed. One defines the perturbed position of the interface $\bar{z}$

$$\bar{z}(t) = \frac{z_\infty}{K} + \delta z e^{i\mathbf{k}\cdot\mathbf{x} + \omega_k t}, \tag{10}$$

and the resultant concentration field

$$c(z,t) = c_0 + \delta c e^{i\mathbf{k}\cdot\mathbf{x} - q(z - \bar{z}) + \omega_k t}. \tag{11}$$

Substituting Eqs. (10) and (11) into Eqs. (5) and (6)–(9) yields an expression for ω_k, the amplification rate of a perturbation of wavelength $2\pi/k$. The wavenumber k for which ω_k is largest is then computed, and it is postulated that such a mode, being the most unstable, is the wavelength selected by the the system. In Figure 4 we show the results of this theoretical calculation, along with the experimental values of Trivedi and Somboonsuk. We immediately see how poorly this calculation performs, erring by more than a factor of eight at high velocities. The necessity for an improved calculation is clear, and so we must search for a more accurate method.

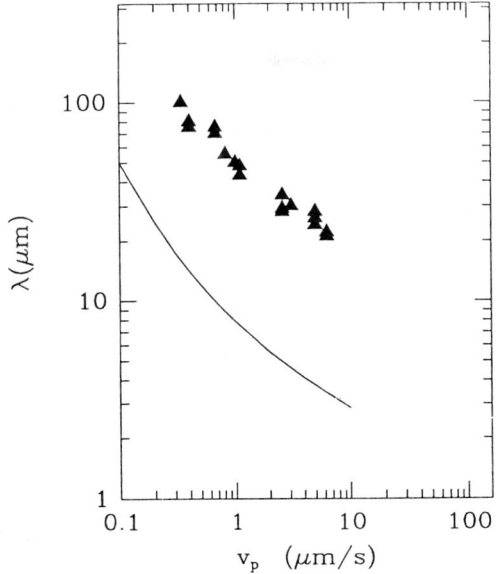

FIGURE 4   A plot of the theoretical predictions (dark line) versus experimental results (black triangle). Reprinted from *Acta Metallica*, **33**, Trivedi and Somboonsuk,[9] "Pattern Formation During the Directional Solidification of Binary Systems," 1061. Copyright © 1985, with kind permission from Elsevier Science Ltd, The Boulevard, Langford Lane, Kidlington OX5 1GB, UK.

4. THE MOVING FLAT INTERFACE AND MORPHOLOGICAL INSTABILITIES

In order to accurately predict the wavelength selected by our system, we must model the dynamic development of the liquid-solid interface. Our first task towards this end is to model the acceleration of the interface from rest, after the gradient has begun to move relative to the material. There is no known analytic solution to the diffusion equation, and its boundary conditions, for the accelerating flat interface problem (a version of the classic Stefan problem); hence, we must resort to either exact numerical solutions or approximate model equations. We choose the later, and follow the recent work of Warren and Langer.

The WL model requires two elements: the time-dependent position of the interface, and the thickness of the boundary layer of solute, which builds up as the interface moves. With these ideas in mind, we propose an exponential profile for the concentration similar to Eq. (5), but now with a time-dependent decay length $\ell(t)$, as well as a time-dependent position $z_0(t)$. Our ansatz is

$$c(z,t) = c_\infty + (c(z_0,t) - c_\infty)e^{-2(z-z_0(t))/\ell(t)}, \tag{12}$$

where we must recognize that this is *not* a solution to the diffusion equation. Instead, we find the zeroth moment of Eq. (6) by integrating over the liquid volume, and substitute Eq. (12) into that expression. This yields

$$v_0(c_\infty - Kc(z_0, t)) = \frac{\partial}{\partial t}\left[\frac{\ell(c(z_0, t) - c_\infty)}{2}\right],$$

(13)

which is just a statement of mass conservation in the sample. A second equation

$$v_0(1 - K)c(z_0, t) = \frac{2D}{\ell}(c(z_0, t) - c_\infty)$$

(14)

is found by inserting Eq. (12) into Eq. (9). We recall that since $c(z_0, t) = -Gz_0/m$, and $v_0 = \dot{z}_0 + v_p$, Eqs. (13–14) are simply two first-order nonlinear differential equations that completely describe the evolution of both the position of the interface z_0 and the thickness of the boundary layer ℓ. We will not concern ourselves with the inaccurate long time behavior of this system, because, as we will show shortly, the interface becomes unstable long before the equilibrium flat interface is realized.

Caroli, Caroli, and Ramírez-Piscina have investigated the accuracy of the above approximations, by direct numerical integration of the diffusion equation. They found, in the limit of relatively short and intermediate times, surprisingly good agreement with the numerics, in fact far better than one could expect by any systematic expansion in higher moments.[1] At higher velocities and longer times, this boundary layer model is not as good.

In order to calculate when the flat interface becomes unstable, we continue in the spirit of Warren and Langer, and perform a quasi-static Mullins-Sekerka stability analysis on the moving interface. We assume that

$$z_0(t) \rightarrow z_0(t) + \zeta_{\mathbf{k}}(t)e^{i\mathbf{k}\cdot\mathbf{x}},$$

(15)

where

$$\frac{d\zeta_k}{dt} = \omega(k, t)\zeta_{\mathbf{k}}(t),$$

(16)

and

$$c(z, t) \rightarrow c(z, t) + c_{\mathbf{k}}(t)e^{i\mathbf{k}\cdot\mathbf{x} - qz},$$

(17)

where again

$$\frac{dc_k}{dt} = \omega(k, t)c_{\mathbf{k}}(t),$$

(18)

and calculate $\omega(k, t)$, by linearizing the diffusion equation and its boundary conditions. We note that ω is now a function of time, because it is now dependent on the position of the interface $z_0(t)$. Figure 5 is a plot of $\omega(k, t)$, showing values of ω vs. k with later times higher on the plot, and it displays several noteworthy features. First, we see that initially all modes are stable, so one cannot find the most unstable

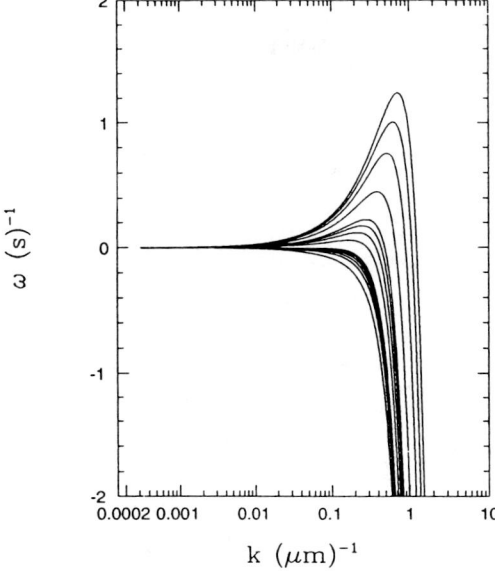

FIGURE 5 A plot of $\omega(k, t)$ $v_p = 10$ μm/s, with time decreasing as one moves from the top to the bottom of the graph. The times are $t = 0, 2, 4, 6, 8, 10, 30, 50, 90, 200, 400, 600,$ and 800 seconds.

mode, as was done in the previous section. Indeed, an unstable mode does not exist initially and then changes with time. Second, there is only a narrow band of modes that eventually become unstable. This will be important when we introduce noise into the problem, as it will render our calculation fairly insensitive to the precise nature of the noise.

CCR also numerically examined the quasi-static Mullins-Sekerka instability of the moving flat interface, and found that the interface always becomes unstable long before the WL model, described by Eqs. (13–14), becomes unphysical. Hence, we may use the WL model with a great deal of confidence to describe the growth of linear perturbations to the flat interface.

5. THERMAL NOISE AS AN INITIAL CONDITION

Equation (16) describes the evolution of a specific sinusoidal perturbation to the interface. We would like to introduce noise as the source of these perturbations, and then evolve those modes forward in time to find out which one is the most strongly selected. We will cut the calculation off when the "bumps" are approximately as large as their spacing, and then compare our results with the experimental results of Trivedi and Somboonsuk. As we have already noted, the form of our noise source is relatively unimportant, as $\omega(k, t)$ is sharply peaked about a narrow band of wavenumbers. However, we shall here reproduce much of the WL calculation, based on ground-breaking work by Cherapenova,[2] to introduce thermal noise into our system, and also include some of the work done by Karma [3] which greatly simplifies a critical portion of the calculation.

We start by introducing a stochastic version of Eq. (16)

$$\frac{d\zeta_{\mathbf{k}}}{dt} = \omega(k,t)\zeta_{\mathbf{k}} + \eta_{\mathbf{k}}(t), \tag{19}$$

where $\eta_{\mathbf{k}}(t)$ is a fluctuating force. We may formally solve Eq. (19) to find

$$\zeta_{\mathbf{k}}(t) = \int_{-\infty}^{t} dt' \exp\left[\int_{t'}^{t} \omega(k,t'')dt''\right]\eta_{\mathbf{k}}(t'). \tag{20}$$

We next assume that the $\eta_{\mathbf{k}}$ are statistically distributed such that

$$\langle\eta_{\mathbf{k}}(t)\eta_{\mathbf{k}'}(t')\rangle = \Gamma_0(k)(2\pi)^2\delta^2(\mathbf{k}+\mathbf{k}')\delta(t-t'), \tag{21}$$

where Γ_0 is assumed independent of time. We may then compute the correlation function

$$\langle\zeta_{\mathbf{k}}(t)\zeta_{\mathbf{k}'}(t)\rangle = \Gamma_0(k)(2\pi)^2\delta^2(\mathbf{k}+\mathbf{k}')\int_{-\infty}^{t} dt_1 \exp\left[2\int_{t_1}^{t}\omega(k,t'')dt''\right]. \tag{22}$$

The object of this calculation is to compute the average height of deformations on the surface as a function of time. The height-height correlation function is simply the Fourier transform of Eq. (22), and can be written as

$$\langle\zeta(\mathbf{r},t)\zeta(\mathbf{r}',t)\rangle = \int\frac{d^2k}{(2\pi)^2}\Gamma(k,t)e^{i\mathbf{k}\cdot(\mathbf{r}-\mathbf{r}')}, \tag{23}$$

where

$$\Gamma(k,t) = \Gamma_0(k)\int_{-\infty}^{t} dt_1 \exp\left[2\int_{t_1}^{t}\omega(k,t')dt'\right]. \tag{24}$$

In the limit that $\mathbf{r} \to \mathbf{r}'$, Eq. (23) gives us the mean height squared of a surface instability, which we see will grow or shrink exponentially depending on the value of $\omega(k,t)$. It is this behavior which makes our calculation relatively insensitive to the particular form of Γ_0. Here we will use thermal fluctuations to drive our system, but any other "reasonable" distribution should work as well.

In order to find the form for $\Gamma_0(k)$ we turn to the recent work of Karma. First, we note that when $t < 0$ the interface is not moving, the system is in stable equilibrium, and therefore $\omega(k, t < 0) \equiv \omega^{(eq)}(k) < 0$ for all k. Hence, when $t < 0$ we may write the time-independent expression

$$\langle\zeta_{\mathbf{k}}\zeta_{\mathbf{k}'}\rangle_{\mathrm{equil}} \equiv (2\pi)^2\delta^2(\mathbf{k}+\mathbf{k}')\Gamma_0(k)\frac{1}{2\left|\omega^{(eq)}(\mathbf{k})\right|}. \tag{25}$$

We may calculate $\langle \zeta_{\mathbf{k}} \zeta_{\mathbf{k}'} \rangle_{\text{equil}}$ in a straightforward manner by first reminding ourselves that the probability distribution for a state α in thermodynamic equilibrium is

$$p(\alpha) = \frac{1}{Z} \exp \left[-\frac{W(\alpha)}{k_B T_E} \right], \tag{26}$$

where $W(\alpha)$ is the work needed to bring about a reversible change in the variable α, and Z is the partition function. Here we are interested in the work done in changing the shape of the interface by some height function $\zeta(\mathbf{r})$, which is

$$W = \int_0^{\delta V} (P_L - P_S) dV + \gamma \delta A, \tag{27}$$

where $P_L - P_S$ is the pressure jump across the interface, and $\gamma \delta A$ is the change in energy do to the deformation of the surface. For small changes in surface area

$$\delta A = \frac{1}{2} \int d^2 r |\nabla_\perp \zeta(\mathbf{r})|^2, \tag{28}$$

while the jump in pressure is simply

$$P_L - P_S = \left[\left(\frac{\partial P_L}{\partial T} \right)_{\mu_L} - \left(\frac{\partial P_S}{\partial T} \right)_{\mu_S} \right] Gz = \frac{L}{T_E} Gz. \tag{29}$$

Inserting Eq. (29) into Eq. (27), as Karma found,

$$W(\zeta) = \int d^2 r \left(\frac{LG}{T_E} \int_0^{\zeta(\mathbf{r})} z \, dz + \frac{\gamma}{2} |\nabla_\perp \zeta(\mathbf{r})|^2 \right), \tag{30}$$

and therefore

$$p = \frac{1}{Z} \exp \left[-\frac{1}{k_B T_E} \int d^2 r \left(\frac{LG}{T_E} \zeta(\mathbf{r})^2 + \frac{\gamma}{2} |\nabla_\perp \zeta(\mathbf{r})|^2 \right) \right]. \tag{31}$$

We now compute $\langle \zeta_{\mathbf{k}} \zeta_{\mathbf{k}'} \rangle_{\text{equil}}$ using Eq. (31) and find

$$\langle \zeta_{\mathbf{k}} \zeta_{\mathbf{k}'} \rangle_{\text{equil}} = (2\pi)^2 \delta^2 (\mathbf{k} + \mathbf{k}') \frac{m}{(1-K) n_\infty (G + m d_0 k^2)}, \tag{32}$$

where we have remove T_E from the expression by using the Clausius-Clapeyron relation

$$\frac{k_B T_E^2}{L} = \frac{m}{(1-K) n_\infty}, \tag{33}$$

and n_∞ is the total number of molecules per unit volume in the fluid. We should note that Eq. (32) differs from the approximate calculation of WL by a factor of two, but, because of the exponential character of $\Gamma(k,t)$, the correction has little effect on the final result.

We now have enough information to compute $\Gamma_0(k)$ and, therefore, we may compute the rms height $h(t)$ of the deformations

$$h^2(t) = \langle \zeta^2(\mathbf{r},t) \rangle = \int \frac{d^2k}{(2\pi)^2} \Gamma(k,t). \tag{34}$$

$\Gamma(k,t)$ is plotted in Figure 6, and we clearly see how sharply peaked it is in k. We may associate the average wavelength of our interface with the value of $k = k_{\max}$ at the peak, and terminate the calculation at a time $t = t_0$ where the height is the same as the wavelength

$$\frac{2\pi}{k_{\max}(t_0)} \equiv \lambda \approx h(t_0). \tag{35}$$

Note that we only require an approximate equality, as $h(t)$ is growing exponentially, while $k_{\max}$ is changing quite slowly. A factor of 2 or more in Eq. (35) would make less than a percent change in the final wavelength selected.

In Figure 7 we have plotted the selected values of λ as a function of the pulling speed v_p versus the experimental values from Trivedi and Somboonsuk. The integral in Eq. (34) was computed numerically using a Gaussian approximation about the peak in $\Gamma(k,t)$. It should be noted that, for our choice of a noise source, Eq. (34)

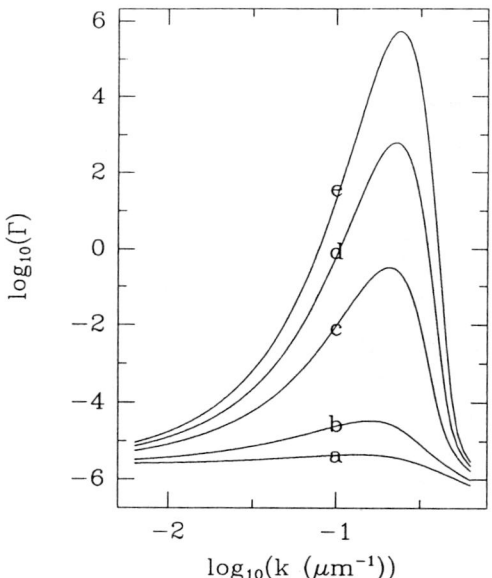

FIGURE 6   Plots of $\Gamma(k,t)$, for $v_p = 10 \ \mu$m/s, where times labeled "a–e". Time "a" corresponds to 10 s, "b" to 30 s, "c" to 70 s, "d" to 90 s, and "e" to 105 s. Point "e" identifies the crossover from linear instability of the flat interface and nonlinear dendritic behavior, and the cutoff in our calculation.

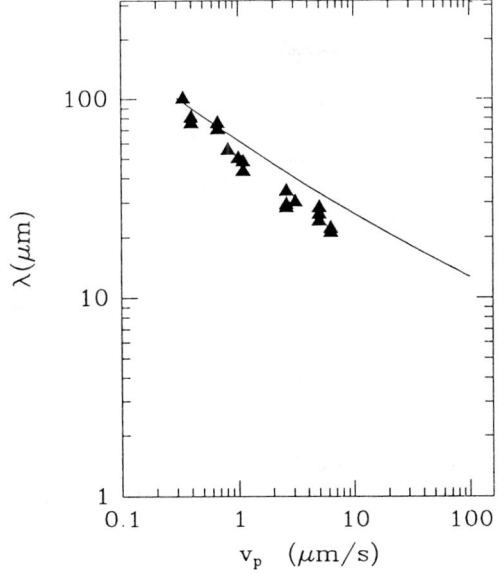

FIGURE 7 Theoretical values of λ (dark line) versus the experimental results of TS.

logarithmically diverges unless one introduces a physical cutoff, but even cutoffs as small as 3Å introduce a negligible correction. We see pleasingly good agreement between theory and experiment in this case, with the best correspondence at lower velocities, when the model for flat interface motion is more accurate,[1] while for higher velocities there is perhaps a 15% disagreement.

6. CONCLUSION

While the calculation presented above is not free from approximations, it presents a credible explanation of the TS experimental results. The discrepancies we have observed are possibly due to several sources of inaccuracy. First, our boundary layer model is only accurate for shorter times and at low speeds, this would explain the disagreement at high velocities. An improved model, or direct numerical integration of the one-dimensional problem, could test this hypothesis. Second, if the real noise distribution was more sharply peaked in k than our assumed thermal noise, it could explain some of the disagreement, but this seems less likely than errors due to the boundary layer approximations. Last, since there is some ambiguity in TS's definition of this wavelength, as it is a difficult quantity to measure, the experiment may need to be done with our cutoff criterion in mind. Overall, however, the results are encouraging.

In general, pattern-forming problems must take into account the importance of transients. In many experiments a rapid change (quenching) in one variable is introduced, and the system then undergoes a transition, due to an instability. In our case, to attempt to ignore the changing state of the system when calculating the pattern leads to a entirely incorrect answer to the problem, which does not even qualitatively appear to be accurate. If one hopes to understand such phenomena better, one needs to recognize the necessity of describing the history of such processes, and not just focus on the initial and final states. It is the dynamics of most systems that determines their structure.

ACKNOWLEDGMENTS

I would like to thank Jim Langer, who helped me do much of the analysis herein. I would also like to thank conference organizers P. E. Cladis and P. Palffy-Muhoray for inviting me to the NATO Advanced Research Workshop to present this work.

REFERENCES

1. Caroli, B., C. Caroli, and L. Ramírez-Piscina. "Initial Front Transients in Directional Solidification of Dilute Alloys." Preprint, 1993.
2. Cherapanova, T. A. "Fluctuation Mechanism of the Growth Instability of Crystal Faces." *Dokl. Akad. Nauk SSSR* **226** (1976): 1066. *Sov. Phys.-Dokl.* **21** (1976): 109.
3. Karma, A. "Fluctuations in Solidification." Preprint, 1993.
4. Karma, A. "Langevin Formalism for Solidification." Preprint, 1993.
5. Kurz, W., and D. J. Fisher. *Fundamentals of Solidification.* Switzerland: Trans Tech Publications, 1984.
6. Langer, J. S. "Chance and Matter." In *Lectures on the Theory of Pattern Formation* (Les Houches Summer School), edited by J. Souletie, J. Vannimenus, and R. Stora, 629–711. New York: North-Holland, 1987.
7. Langer, J. S. "Instabilities and Pattern Formation in Crystal Growth." *Rev. Mod. Phys.* **52** (1980): 1.
8. Mullins, W. W., and R. F. Sekerka. "Stability of a Planar Interface During Solidification of a Dilute Binary Alloy." *J. Appl. Phys.* **35** (1964): 444.
9. Trivedi, R., and K. Somboonsuk. "Pattern Formation During the Directional Solidification of Binary Systems." *Acta Metall.* **33** (1985): 1061.
10. Warren, J. A., and J. S. Langer. "Prediction of Dendritic Spacing in a Directional Solidification Experiment." *Phys. Rev. E* **47** (1993): 2702.

11. Warren, J. "A Dynamic Model of Wavelength Selection During Directional Solidification of a Binary Alloy." Ph.D. Thesis, University of California at Santa Barbara, 1992.

John Bechhoefer
Deptartment of Physics, Simon Fraser University, Burnaby, British Columbia, V5A 1S6, Canada

Nonequilibrium Phenomena in Liquid Crystals

The briefest glance through the literature on nonequilibrium phenomena shows that complex fluids, particularly liquid crystals, are often favored for experimental investigations. This might seem surprising in that complex fluids, as befits their name, are difficult materials: experiments require subtle tricks to prepare reproducible samples; theoretical descriptions lead to notoriously messy equations. Given the prejudice of physicists towards simple, well-controlled systems, and given the success that studies of simple fluids have enjoyed, why use complex fluids to study nonequilibrium phenomena? In this chapter, I shall offer two answers, one practical and obvious, the other more fundamental and subtle.

The obvious, practical reason is that the dynamics of complex fluids display a variety of interesting "effects" that have been—and will continue to be—exploited for gain. Indeed, the use of liquid crystals in flat-screen displays is perhaps the best known and most widely exploited of such special effects.

The very first observations of liquid crystals[21,26] noted that although clearly fluid, they were uniaxially birefringent, a property that had been associated only with solids. Even more interestingly, the optical axis could be aligned along an imposed electric or magnetic field; soon after, Mauguin[22] and Grandjean[17] discovered that suitably prepared solid surfaces would also align the optical axes of nematics.

In the 1930s, Freedericksz and Tsvetkov[15] and Zöcher[35] put these two effects together in an experiment illustrated in Figure 1. In the Freedericksz experiment, the surface and external field tend to align the molecules in the nematic phase along different directions. For small electric field, the orientation imposed by the surfaces wins out; but above some threshold E^*, the molecules align along the imposed electric field. The transition is really a supercritical bifurcation from one uniform state to a second, stationary state. This is perhaps the simplest of bifurcations, and one can immediately expect to see all of the universal behavior associated with such transitions. (For example, the maximum deflection of molecules grows as the square root of the distance above threshold, as suggested in Figure 3(a).)

In 1971, Schadt and Helfrich[27] modified the Freedericksz experiment slightly by rotating one plate 90° with respect to the other—thereby twisting the molecules in the sample—and by adding crossed polarizers. In this configuration, as shown in Figure 2, the transmitted light-intensity curve follows that of the molecular distortion. The configuration was the basis for the first commercially successful liquid crystal display and is still extensively used for small displays where a limited amount of information is to be shown.

In 1982, it was found that if the twist angle is increased past 270°, the bifurcation becomes subcritical.[28,32] (See Figure 3.) Although the resultant hysteresis causes difficulties for display switching, the limiting case of a 270° twist angle is useful. As Figure 3(c) shows, the transmission curves switches more abruptly for such "tricritical" bifurcations than for supercritical bifurcations. (An elementary analysis shows that the intensity now rises as the distance from threshold to the 1/4 power.[1]) This "supertwist" display is the dominant one used for the large flat-screen displays found in notebook computers.

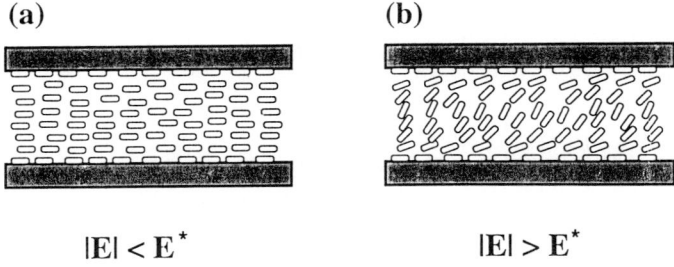

(a)   **(b)**

$|E| < E^*$ $|E| > E^*$

FIGURE 1   The Freedericksz transition. Glass plates are treated to align molecules in the nematic phase horizontally. (a) For small fields, the molecules lie flat throughout the sample. (b) Above a critical field strength E^*, the molecules tilt to align themselves along the electric field. The maximum distortion is at the midplane of the sample; the boundaries still force the molecules to lie flat.

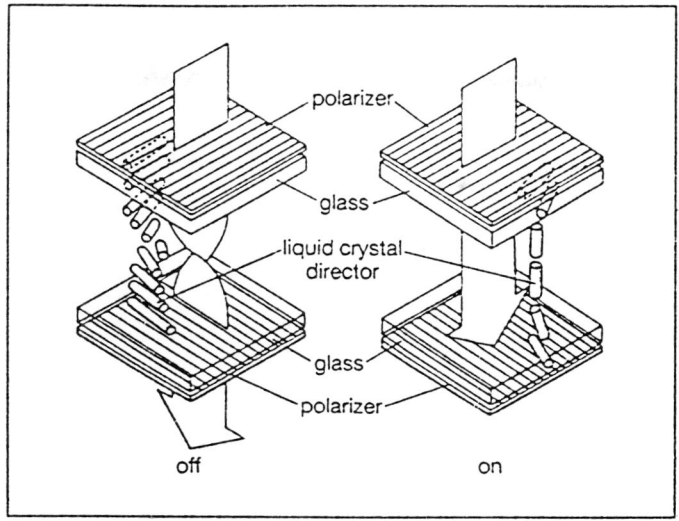

FIGURE 2 The twisted nematic display. The configuration is similar to that of the Freedericksz transition, but the bottom plate is rotated 90°, imposing a twist through the sample. Crossed polarizers are added to top and bottom. (Left.) With no field applied, the plane of polarization follows the nematic molecules adiabatically and light is transmitted through the display. The changing length of molecules represents rotation out of the plane of the illustration. (Right.) With a field applied, the polarization is no longer rotated and the analyzer blocks all light transmission.

I have outlined the history of liquid-crystal displays in some detail because—at least in hindsight—simple ideas from nonequilibrium science are relevant. A good display requires a sharp transition from the "off" state to the "on" state. Thus, it makes sense to use a supercritical transition, as opposed to a design in which the intensity is an analytic function of the control parameter. Changing the bifurcation from supercritical to sub- or tricritical further speeds the switching.

Simple ideas from nonequilibrium science can thus be combined with the special properties of complex fluids (birefringence, electric-field alignment of the optical axis) to create useful devices. The large markets for such devices—well over 3 billion per year for liquid-crystal displays[24]—certainly justifies continued research into understanding and further cataloguing of analogous special effects. Other special effects I could have cited include drag reduction in turbulent flows by adding small amounts of polymer,[13] which has been used to make fire hoses shoot farther and submarines move faster; the giant swelling transition in gels,[25] which promises robotic "fingers" that can grasp delicate parts without damage; and electrorheological fluids,[18] which are being tested in active automobile suspensions. In this

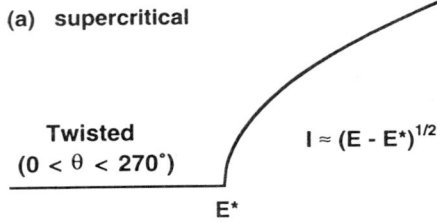

(a) **supercritical**

Twisted
(0 < θ < 270°)

$I \approx (E - E^*)^{1/2}$

E*

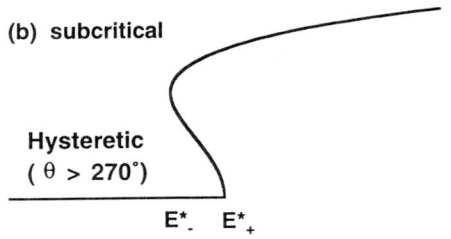

(b) **subcritical**

Hysteretic
(θ > 270°)

E*_−   E*_+

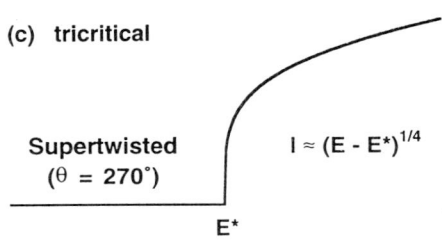

(c) **tricritical**

Supertwisted
(θ = 270°)

$I \approx (E - E^*)^{1/4}$

E*

FIGURE 3 Transmitted light intensity curves for liquid-crystal displays. (a) Ordinary twisted nematic cell. The bifurcation is supercritical; $I(E)$ is continuous and rises as $(E - E^*)^{1/2}$. (b) Twist-angle exceeds $270°$. The bifurcation is subcritical, and there is hysteresis in the switching. (c) "Supertwist display," with a twist angle of $270°$. The bifurcation is tricritical, and the intensity increases above threshold as $(E - E^*)^{1/4}$.

conference, K. Amundson,[2] R. Larson,[20] and H. R. Brand[7] have discussed other interesting polymer effects. I could go on, but I hope the point is clear.

In addition to the "bestiary" of special effects, there is a second, more fundamental reason to study nonequilibrium phenomena in complex fluids. Nonequilibrium science can loosely be characterized as the systematic exploration of systems as some "stress" is increased. And, simply put, complex fluids are easier to drive out of equilibrium than simple ones.

To understand this remark, consider what I shall call—with no disapproval implied—the "conventional" view of the progression of nonequilibrium phenomena. This view, largely shaped by work in fluid dynamics, is sketched in Figure 4: unstressed or lightly stressed systems are in a simple "lamellar" state. As the stress is increased, the system undergoes a sequence of bifurcations that results in a time-dependent, chaotic state with limited temporal but full spatial coherence. As one

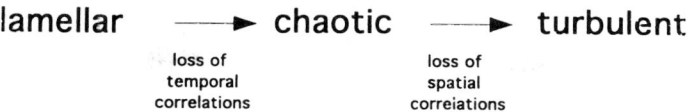

FIGURE 4 "Conventional" view of nonlinear phenomena as the driving "stress" is increased.

further increases the stress, a second series of transitions—less well understood—progressively destroys the spatial coherence of the system and results in a fully turbulent flow. Well-studied examples that illustrate this progression include Rayleigh-Bénard convection and Taylor vortex flow,[11] where "stress" is measured by the Rayleigh and the Reynolds numbers, respectively.

At first glance, the behavior in complex fluids would seem to parallel that of simple fluids. For example, when the Freedericksz experiment is performed on a nematic that tries to align perpendicularly to the applied field, convective motion is observed. (See, for example, W. Zimmerman's contribution to these proceedings.[34]) I want to suggest, though, that there is an important difference between the behavior of complex and simple fluids when driven out of equilibrium: In simple fluids, for reasonable driving stresses, the fluid is always in local—but not global—thermodynamic equilibrium. For simple fluids, this observation has a number of consequences. If, during an experiment on simple fluids (e.g., Rayleigh-Bénard convection using water), you were to sample the fluid used, you would find its material properties to be the same as in equilibrium. Moreover, at the end of the day, when you switched off the experiment, the fluid would settle down to its equilibrium state. Water that has been churned about at Reynolds numbers of 10^5 cannot be distinguished from water that has spent all day sitting at rest in a glass. Such observations—trivial as they may be—stand in contrast to the case of complex fluids where, I shall argue, modest driving forces can push a system out of equilibrium on length and time scales comparable to the microscopic scales that characterize the structure of the fluid.

Rather than discuss fluid dynamics, I want to focus on a phenomenon that is equally rich and about which I have personal experience: solidification. As is well-known, freezing fronts are often unstable to shape undulations. (See Color Plate 2.) This instability was first analyzed in detail by Mullins and Sekerka[23] and is relevant whenever front growth is controlled by diffusive processes (typically, these are either the diffusion of latent heat or chemical impurities away from the interface). If one freezes more rapidly, however, one finds another regime, the kinetics-limited regime, where front behavior is controlled by local ordering processes at the interface itself. As we shall see, the velocity separating the diffusion- from the kinetics-limited regimes, v_0, sets the scale for nonequilibrium phenomena. Fronts moving with $v \ll v_0$ are nearly in equilibrium, while fronts moving with $v \gtrsim v_0$ are strongly out of equilibrium. I shall call the former regime one of slow solidification and the latter regime one of rapid solidification.

To understand why v_0 sets the scale for nonequilibrium "stress" in solidification, we need to recall two facts: On the one hand, fronts have a finite thickness ℓ. This means that an interface moving at velocity v will take a time $t_p = \ell/v$ to pass a given observation point. On the other hand, a front may be viewed as an "ordering wave" that propagates through the fluid. As the front passes through an observation point, fluid molecules that were formerly in a disordered state now have to order. The ordering takes time—call it t_0. If the ordering time $t_0 \ll t_p$, then we have slow solidification, since the front has ample time to order. If $t_0 \gtrsim t_p$, then the front will have already passed through the observation point before the ordering is complete, and one may expect new phenomena to be observed. Equating the two time scales gives the velocity $v_0 \sim \ell/t_0$ described above.

The characteristic solidification speed of a front, v_0, is the ratio of a microscopic length, ℓ, to a microscopic ordering time, t_0. For simple fluids, this scale velocity turns out to be roughly the sound speed, and one can imagine that concocting a controlled experiment on fronts moving a kilometer a second is not easy! It turns out, though, that in a complex fluid, v_0 can be dramatically reduced, so that controlled experiments become feasible. This is then the second reason that complex fluids are useful in the exploration of nonequilibrium phenomena.

To understand where this reduction of v_0 comes from, let us first consider a **simple fluid**—nice examples include the noble elements, such as krypton and xenon—where the molecules (or atoms) are small and spherical and where interactions are short-ranged and isotropic. For such fluids, the interface width is roughly equal to an atomic diameter, so $\ell \approx 10^{-8}$cm. Since all atoms are identical and spherical, the ordering time is set by the time it takes to remove energy (heat) from the interface. This is given by the heat diffusion time, so that $t_0 \sim \ell^2/D_h \approx 10^{-16}cm/10^{-3}$cm^2/sec $\approx 10^{-13}$ sec. This gives $v_0 \sim D_h/\ell \sim 10^{-3}/10^{-8} \sim 10^{+5}$cm/sec. ($10^3$ m/sec), which is roughly the velocity of sound in a simple fluid.

Next, consider a **simple alloy**, made of a mixture of two simple fluids. The fundamental length scale is still about an angstrom ($\ell \approx 10^{-8}$cm), but now the solid phase is formed with an additional constraint: not only must energy be removed form the interface, but also the A and B molecules must be arranged in a precise pattern in the solid phase. In addition, the relative concentration of B and A molecules will differ in the two phases. Thus, freezing an alloy requires rearranging atoms, so that the time scale is set by mass diffusion and not by heat diffusion. Since the mass diffusivity $D \approx 10^{-5}$cm^2/sec is a hundred times smaller than the heat diffusivity, we expect $t_0 \approx 10^{-11}$ sec. and $v_0 \approx 10^{-5}/10^{-8} \approx 10^3$cm/sec (10 m/sec). Indeed, rapid solidification experiments on metallic alloys do show interesting phenomena when fronts move faster than about 10 m/sec.[8]

Notice that the microscopic time scale t_0 determining v_0 is set by the slower of the two relaxational processes (heat and mass diffusion). This is a general feature of

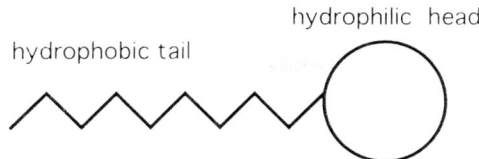

hydrophilic head

hydrophobic tail

FIGURE 5 Sketch of an amphiphilic molecules. When mixed in high concentration with water, molecules such as these order in lyotropic liquid crystalline phases.

complex fluids: the slowest relaxational process sets the microscopic ordering time scale. Notice, too, that although the length and time scales both increase as we go from a simple pure fluid to a simple alloy, the ratio v_0 decreases. This, too, is general.

Next, we consider **thermotropic liquid crystals**, which are pure materials made up of rigid, anisotropic molecules. In most cases, the molecules are rod-shaped, but disk-shaped molecules also form liquid-crystal phases.[9] The small dimension measures 5 Å across typically and the large dimension about 30 Å. Motions on the scale of the *large* dimension—the slowest process—set the length scale and the diffusion time scale. Although we once again have a pure fluid, the transition from an isotropic to a nematic state requires orientation alignment, so that one must consider rotational diffusivities in addition to heat diffusion. Using $\ell \approx 10^{-7}$cm and $D \approx 10^{-7}$cm^2/sec, we obtain $t_0 \approx 10^{-7}$sec and $v_0 \approx 1$cm/sec.

Lyotropic liquid crystal phases[10] are formed when large amounts of amphiphilic molecules are forced into an aqueous or oily solvent. (Amphiphilic molecules have two parts, one hydrophilic, the other hydrophobic. Examples include soaps and phospholipids, the constituents of biomembranes. See Figure 5.) A large variety of phases—lamellar, cubic, hexagonal, and others—can be observed for different temperatures and amphiphile concentrations. Here, the repeat distances are larger ($\ell \approx 50$ Å). Diffusivities vary greatly, ranging from 10^{-7} to 10^{-10}cm^2/sec. The small values occur because phase transitions can require topology changes in the amphiphile sheets that are the building blocks of the different configurations. Using $D \approx 10^{-8}$cm^2/sec, we expect $t_0 \approx 10^{-6}$ sec and $v_0 \approx 1$ mm/sec.

My final example is the ordering of **diblock copolymers**, which consist of a chain of A monomers joined covalently to a chain of B monomers.[4] (See Figure 6.) At high temperatures, the A and B chains are miscible and form a disordered solution. Below a critical temperature, the A and B chains phase separate. In contrast to a polymer blend, the phase separation must remain local, since the A and B chains remain joined together. Depending on the relative lengths of A and B chains, the microscopic ordering will vary. The phases that are formed have structures similar to those found in lyotropics. In contrast to liquid crystals, polymers—diblock or normal—are highly flexible molecules. There are new relaxational processes corresponding to the intricate meshing and disentangling of long polymer strands. To estimate this time scale theoretically, we use de Gennes's "reptation model," in which the polymer molecule is assumed to be confined to a tube enclosing the molecule.[12] This gives $t_0 \sim \tau N^3$, where $\tau \approx 10^{-11}$cm is the time scale of the monomer (assumed to be a simple molecule of size $a \sim 10$ Å). Alternatively, t_0 may be estimated experimentally from rheological measurements. (I thank Karl Amundson for pointing this out.) The appropriate length scale is the radius of gyration of the molecule, which in a random-walk model is simply $\ell \sim aN^{1/2}$. The diffusion constant then is $D \sim \ell^2/t_0 \sim (a^2/\tau)N^{-2}$ and the scale velocity for front growth is $\ell/t_0 \sim (a/\tau)N^{-5/2}$. Clearly, for large enough N, the velocity scale can be as small as one wishes. To get reasonable values, one might want to look at short molecules. For $N = 150$, we estimate $\ell \approx 120$ Å, $t_0 \approx 3 \times 10^{-5}$ sec, $D \approx 4 \times 10^{-8}$cm^2/sec, and $v \approx 300$ μ/sec.

The scales for the five examples discussed above are collected in Table 1, where it is immediately clear that increasing the complexity of the fluid dramatically reduces the velocity scale for rapid solidification. (For lyotropics, we selected a middle value, $D \approx 10^{-8}$cm^2/sec, and for the diblocks, we chose $N = 150$.) Notice that liquid crystals—both thermotropic and lyotropic—have convenient values of v_0. Simple fluids and alloys have v_0's so high that fronts cannot be followed during an experiment. Polymers, by contrast, have scales that are painfully slow, except perhaps for very short-chained molecules.

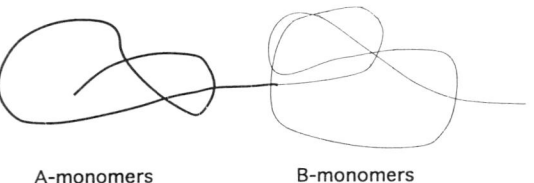

A-monomers   B-monomers

FIGURE 6   Sketch of a diblock copolymer. These are the polymer equivalent of amphiphilic molecules and form phases of similar structure to lyotropics.

TABLE 1 Microscopic scales of simple and complex fluids

	length scale ℓ (cm)	diffusion constant D (cm^2/sec)	time scale $t_0 =$ ℓ^2/D (sec)	velocity scale scale $v_0 =$ ℓ/t_0 (cm/sec)
simple fluid	10^{-8}	10^{-3}	10^{-13}	10^5
binary alloy	10^{-8}	10^{-5}	10^{-11}	10^3
thermotropic liq. cryst.	10^{-7}	10^{-7}	10^{-7}	1
lyotropic liq. cryst.	10^{-7}	10^{-8}	10^{-6}	10^{-1}
diblock copolymer	10^{-6}	10^{-8}	10^{-5}	10^{-2}

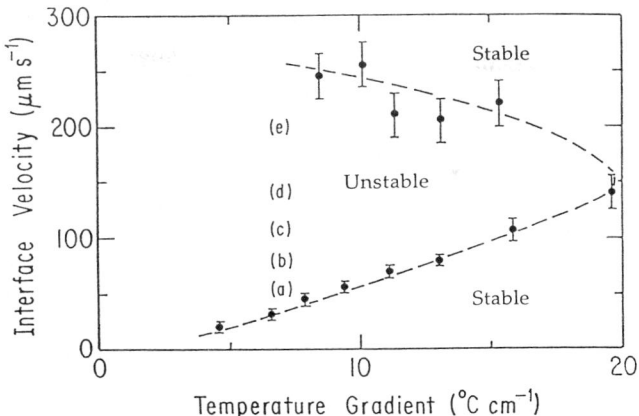

FIGURE 7 Linear stability of a flat interface under different combinations of front velocity and the temperature gradient normal to the interface (after Flesselles et al.[14]).

In my own work, I have studied the solidification of thermotropic liquid crystals with Patrick Oswald, Adam Simon, and Albert Libchaber.[5] Our directional solidification apparatus allowed a maximum speed of about 300 μm/sec. This is still somewhat slower than the scale speed of $v_0 \approx$ 1cm/sec, but already interesting phenomena were observed. In particular, we observed that in addition to a velocity threshold above which a flat interface destabilized, there was a second threshold above which the flat interface reappeared. In fact, the original study of a flat interface had predicted that for large freezing velocities and for large thermal gradients, the front would restabilize. The front restabilization velocity is indirectly linked to v_0 and occurs at about 300 μm/sec for the nematic-isotropic interface of a thermotropic liquid crystal lightly doped with ordinary impurities (*i.e.,* impurities that are themselves simple molecules). A typical stability curve is shown in Figure 7. These observations were significant in that the restabilization velocity of simple alloys is on the order of meters/sec. We were thus able to explore the entire bifurcation diagram, while previous experiments had probed just a small piece of it. We tested the linear stability analysis in the restabilization regime and also found a number of interesting secondary instabilities in the interior of the bifurcation diagram (parity breaking, traveling waves, breathing modes, etc.).[14,29]

One answer, then, to the question "why use liquid crystals and other complex fluids to study nonequilibrium phenomena" is that they can facilitate the study of instabilities that were already known in the context of simpler fluids. A second answer is that they allow access to the locally nonequilibrium regime. What can one expect to see here? In contrast to the usual nonlinear regime, much less is known, and I can only suggest what is to be learned. If we consider the case of solidification, we see that if we were to freeze a liquid instantaneously, the disorder of the fluid would be quenched in and produce a glassy state. One possibility, then, is that in the kinetics-limited regime, the ordered state will be progressively disrupted as the velocity is increased. The defect density in the ordered phase would then be a smoothly increasing function of the freezing velocity.[30]

Another—and to my mind, more interesting—possibility is that the route from the ordered state of near-equilibrium freezing to the glassy state of extremely rapid solidification will be marked by a series of transitions analogous to the phase transitions of equilibrium physics or the bifurcations of weakly nonlinear dynamics. With my colleagues Laurette Tuckerman and Hartmut Löwen, I have studied a simple theoretical model of solidification that displays such behavior.[6,31] As illustrated in Figure 8, we have proposed that a rapidly moving front can split into two separately moving fronts, one dividing the disordered phase (phase 0) from a new metastable phase (phase 1), the second dividing this metastable phase from the ordered, thermodynamically stable phase (phase 2). A necessary condition for the front to split is that the velocity of the leading edge v_{10} exceed that of the trailing edge v_{21}. If this condition is met and if reasonable initial conditions favor splitting, then an ever-widening region of phase 1 will be created. Our description of this transition turns out to be mathematically equivalent to surface melting and wetting transitions, so that one may view the appearance of phase 1 as being equivalent to the condensation of a liquid at a solid-vapor interface. Because the mathematics are the same, one expects to observe a pretransitional thickening of the 20 interface (logarithmic or power-law divergence, depending on the nature of the interactions). In addition, one can show that the transition can be continuous, hysteretic, or finite amplitude. An important difference from, say, surface melting, is that the transition need not occur in the vicinity of a special point on the equilibrium phase diagram (for example, the triple point), but can occur even if phase 1 is metastable at all temperatures. We require only that its free energy not greatly exceed that of the stable phase 2 and that it should somehow "resemble" the ordered phase. (For example, one phase might have an FCC lattice, the other a BCC or simple cubic.)

Referring to the list of complex fluids in Table 1, one might expect that lyotropic liquid crystals would be good candidates to search for such behavior. Not only is the scale velocity v_0 modest, but also lyotropics display a large variety of phases with weak first-order transitions separating them. Such experiments are currently being started in Lyon under Patrick Oswald and at Simon Fraser University, with Nancy Tamblyn and Anand Yethiraj. So far, these transitions have yet to be observed, but the experiments are still preliminary.

In the meantime, poor man's versions of the splitting transition have been observed in thermotropic liquid crystals. The transition is not between two thermodynamically distinct phases but between two configurations of the nematic phase. In Figure 9, I show a side view of the meniscus of the nematic-isotropic (NI) interface discussed above. The glass plates are treated to align surface molecules

Splitting Transition

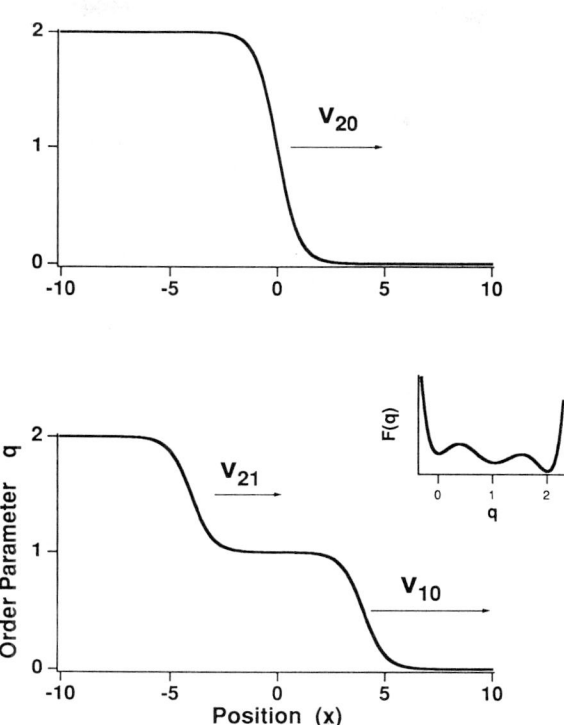

FIGURE 8 Schematic illustration of a splitting transition. In (a) and (b), we plot spatial profiles of a nonconserved order parameter that distinguishes the two phases. For example, in a solid-liquid transition, q could be the amplitude of one of the Fourier amplitudes of a reciprocal lattice vector. It is nonzero in the solid but zero in an isotropic fluid. For low velocities, the front between phases 2 and 0 propagates normally. For high velocities, the 20 front splits into a 21 and 10 fronts. The 10 front moves faster than the 21 front, leaving a widening region of the new metastable phase 1. The dependence of the free energy f on the order parameter q is sketched at right.

perpendicular to the plates (homeotropic orientation). There is another. globally incompatible condition at the NI interface itself. The resulting frustration forces a singularity in the nematic phase. (See Figure 9(a).) Topologically, the defect can either be next to the interface or be deep in the nematic phase. (See Figure 9(b).)

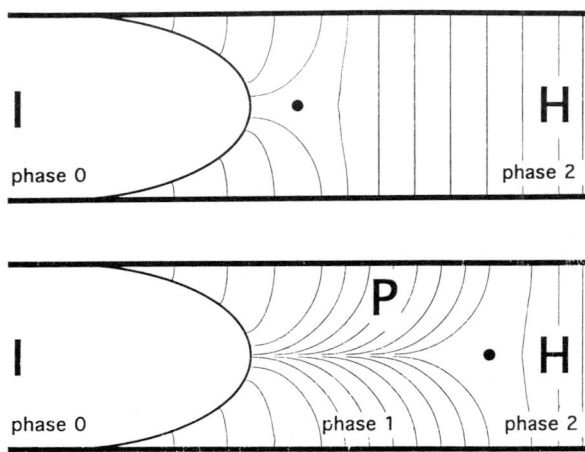

FIGURE 9   Side view of the nematic-isotropic meniscus spanning the gap between two plates. (a) Conflicting boundary conditions at the sidewalls and at the NI interface imply frustration in the nematic, leading to a defect (disclination line, denoted by the large black dot) in the nematic phase (here denoted by "H" for homeotropic orientation). (b) The defect may detach from the interface, creating a region of planar-oriented nematic (denoted "P").

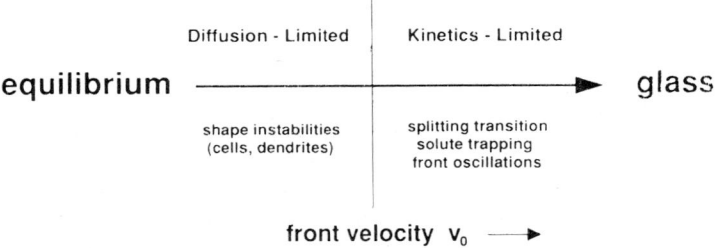

FIGURE 10   A "different" view of front behavior.

In the latter situation, the twisted region has a higher elastic energy than the homeotropic region. The defect line will then move back towards the NI interface at a velocity v_{defect} set by the nematic's viscosity and elastic constants. However, if the isotropic phase is moving faster than the defect line, the defect cannot catch up and we have the splitting transition described above. In this case, the isotropic is phase 0, the homeotropic phase 2, and the new (planar) orientation of the nematic is the metastable phase 1. If the freezing velocity v is low, we expect to see a homeotropic-isotropic interface (20 interface). For $v > v_{defect}$, we would expect to see the defect line peel back, creating a widening region of phase 1.

In fact, something slightly different happens. (See Color Plate 3.) The defect line detaches only when v substantially exceeds v_{defect} and then only when the interface passes through a dust particle. The interface detaches locally, and a planar region spreads out, creating a triangular shape that is a record of the space-time history of the new domain. Note that in Color Plate 3 there are simultaneously 20 and 10 interfaces present. This means that the splitting transition here is hysteretic. Finally, while physicists tend to be intrigued by the triangular shape of the domain, metallurgists are distinctly unimpressed: in the rapid casting of metal alloys, they see these shapes all the time.

Summing up, in Figure 10 is shown what the complete spectrum of behavior of a front might be as the driving force is systematically increased. In the near-equilibrium regime, the front is unstable to undulations whose size decreases with velocity. Above, v_0, one can expect to see front splitting and, eventually, disordering of the low-temperature phase. For lack of time, my discussion of rapid solidification has been incomplete, and I regret not talking about oscillatory instabilities[19] and solute trapping.[3,33] Moreover, my focus on solidification was purely for personal convenience; someone else could have easily rephrased this talk in terms of the Taylor-Couette experiment, where interesting features—including metastable phase formation—have been observed for complex fluids undergoing shear.

I began my discussion by saying that there were two reasons for using liquid crystals and other complex fluids to study nonequilibrium phenomena. The first was that there are a number of special effects that have great practical application, and I reviewed the history of liquid-crystal displays by way of illustration. The second point was the alteration of microscopic length, time, and velocity scales to values that are convenient experimentally. In the end, these two reasons happily do not separate as neatly as that. The metastable states that can result from strongly nonequilibrium processes are themselves new materials, and they may have useful properties. Indeed, metallurgists during the past 30 years have created thousands of new alloys through rapid solidification, and some of these are widely manufactured. A very old example is martensitic steel, which is significantly harder than the equilibrium austenite steel that is formed at slower cooling rates. Thus, although the more fundamentally minded scientist may wish to focus on strongly nonequilibrium phenomena, the result may be a better understanding of how to make new materials.

ACKNOWLEDGMENTS

Parts of this work were supported by an AT&T Bell Laboratories Graduate fellowship, an NSERC operating grant, and an Alfred P. Sloan fellowship. I thank Albert Libchaber, Patrick Oswald, Adam Simon, Laurette Tuckerman, and Hartmut Löwen for their contributions to the work on which this chapter was drawn. I thank Mike Cross for useful comments about the oral version of this chapter.

REFERENCES

1. Ahlers, G. "Experiments on Bifurcations and One-Dimensional Patterns in Nonlinear Systems Far from Equilibrium." In *Lectures in the Sciences of Complexity,* edited by D. L. Stein. Santa Fe Institute Studies in the Sciences of Complexity, Lect. Vol. I, 175–224. Redwood City, CA: Addison-Wesley, 1989.

2. Amundson, K. "Alignment of Block Copolymer Microstructure in an Orienting External Field." This volume.

3. Aziz, M. J. "Model for Solute Redistribution during Rapid Solidification." *J. Appl. Phys.* **53** (1982): 1158–1168.

4. Bates, F. S., and G. H. Fredrickson. "Block Copolymer Thermodynamics: Theory and Experiment." *Ann. Rev. Phys. Chem.* **41** (1990): 525–557.

5. Bechhoefer, J., A. J. Simon, and A. Libchaber. "Destabilization of a Flat Nematic-Isotropic Interface." *Phys Rev. A* **40** (1989): 2042–2056.

6. Bechhoefer, J., H. Löwen, and L. S. Tuckerman. "Dynamical Mechanism for the Formation of Metastable Phases." *Phys. Rev. Lett.* **67** (1991): 1266–1269.

7. Brand, H. R. "Pattern Formation in Polymers and Magnetic Liquids." This volume.

8. Cahn, R. W. "Recent Developments in Rapidly Melt-Quenched Crystalline Alloys." *Ann. Rev. Mat. Sci.* **12** (1982): 51–63.

9. Chandrasekhar, S. *Liquid Crystals.* 2nd ed. Cambridge: Cambridge University Press, 1992. [This is a good, recent reference on fundamental properties of thermotropic liquid crystals.]

10. Charvolin, Jean. "Lyotropic Liquid Crystals: Structures and Phase Transitions." In *Phase Transitions in Soft Condensed Matter,* edited by T. Riste and D. Sherrington. NATO ASI Series B: Physics, Vol. 211, 95–112. New York: Plenum, 1989.

11. Cross, M. C. and P. C. Hohenberg. "Pattern Formation Outside of Equilibrium." *Rev. Modern Phys.* (1993): in press.

12. De Gennes, P. G. *Scaling Concepts in Polymer Physics.* 2nd corrected printing. Ithaca: Cornell University Press, 1985, 223–230. [Note that use of the

reptation time implies that each polymer chain is well-entangled among its neighbors. For small N, this may not be the case. Also, the microscopic time scale will slow down greatly if the order-disorder temperature approaches the glass-transition temperature. The numbers given in the text, thus, are very crude estimates. The real point is the continous dependence on N.]

13. De Gennes, P. G. "Towards a Scaling Theory of Drag Reduction." *Physica* **140A** (1986): 9–25.
14. Flesselles, J. M., A. J. Simon, and A. J. Libchaber. "Dynamics of One-Dimensional Interfaces: An Experimentalist's View." *Advances in Physics* **40** (1991): 1–51.
15. Freedericksz, V. and V. Tsvetkov. *Trans. Faraday Soc.* **29** (1933): 919.
16. Fredrickson, G. H., and K. Binder. "Kinetics of Metastable States in Block Copolymer Melts." *J. Chem. Phys.* **91** (1989): 7265–7275.
17. Grandjean, F. *Bull. Soc. Franç. Mineral* **39** (1916): 71.
18. Halsey, T. C. "Electrorheological Fluids." *Science* **258** (1992): 761–766.
19. Karma, A. and A. Sarkissian. "Interface Dynamics and Banding in Rapid Solidification." *Phys. Rev. E* **47** (1993): 513.
20. Larson, R. "Patterns in Sheared Liquid Crystalline Polymers." This volume.
21. Lehmann, O. "Über fließende Krystalle." *Z. Phys. Chem.* **4** (1889): 462.
22. Mauguin, C. "Sur les Cristaux Liquides de Lehmann." *Bull. Soc. Franç. Mineral* **34** (1911): 71–117.
23. Mullins, W. W., and R. F. Sekerka. "Stability of a Planar Interface during Solidification of a Dilute Binary Alloy." *J. Appl. Phys.* **35** (1964): 444–451.
24. "Flat Panel Displays '93." Nikkei Business Publications,1993.
25. Osada, Y. and S. B. Ross-Murphy. "Intelligent Gels." *Sci.Am.* **268** (1993): 82–87.
26. Reinitzer, F. *Montsh. Chem.* **9** (1888): 421.
27. Schadt, M., and W. Helfrich. "Voltage-Dependent Optical Activity of a Twisted Nematic Liquid Crystal." *Appl. Phys. Lett.* **18** (1971): 127–128.
28. Scheffer, T., and J. Nehring. "Twisted Nematic and Supertwisted Nematic Mode LCDs." In *Liquid Crystals: Applications and Uses,* edited by Birendra Bahadur, Vol. I, 231–274. Singapore: World Scientific, 1990. [The three-volume set edited by Bahadur is also a good reference on other applications of thermotropic liquid crystals.]
29. Simon, A. J., J. Bechhoefer, and A. Libchaber. "Solitary Modes and the Eckhaus Instability in Directional Solidification." *Phys. Rev. Lett.* **61** (1988): 2574–2577.
30. Tiller, W. A. *The Science of Crystallization: Macroscopic Phenomena and Defect Formation.* Cambridge: Cambridge University Press, 1991.
31. Tuckerman, L. S., and J. Bechhoefer. "Dynamical Mechanism for the Formation of Metastable Phases: The Case of Two Nonconserved Order Parameters." *Phys. Rev. A* **46** (1992): 3178–3192.
32. Waters, C. M., and E. P. Raynes. British Patent 2, 123, 163B (1982).

33. Wheeler, A. A., W. J. Boettinger, and G. B. McFadden. "A Phase-Field Model of Solute Trapping during Solidification." *Phys. Rev. E* **47** (1993): 1893–1909.
34. Zimmerman, W. "Electroconvection in Nematics: Complex Patterns and Mechanisms." This volume.
35. Zöcher, H. *Trans. Faraday Soc.* **29** (1933): 945.

P. E. Cladis† **and H. R. Brand‡**
†AT&T Bell Laboratories, Murray Hill, NJ 07974 U.S.A.
‡Theoretical Physics III, University of Bayreuth, D95440 Bayreuth, F.R.G.

Nonequilibrium Phase Winding and Its Breakdown at a Chiral Interface

We describe a nonequilibrium phase winding of the liquid crystal director at a traveling flat cholesteric-isotropic interface. The physical origin of the driving force is a symmetry allowed coupling between an equilibrium helix structure and gradients in impurity concentration setup at a phase-transition boundary forced to move by displacing the sample through a temperature gradient. These observations enable us to determine a first estimate for the coupling constant. A second transition is observed at higher displacement speeds to an orientationally disordered state.

INTRODUCTION

Pattern formation at traveling phase boundaries is one of the major areas of nonequilibrium physics attracting considerable attention.[8] While most of the early work concentrated on the moving solid-liquid interface,[10] or pattern formation in simple systems,[7] now the focus has changed to phase transitions characterized by broken continuous symmetries such as the traveling nematic-isotropic interface.[12] Our interest is in the effect of chirality, the broken symmetry of life (no reflection symmetry), on moving phase boundaries involving broken continuous symmetries. We find that chirality has a profound effect on all patterns exhibited at these interfaces. Indeed, our observations are the first to reveal that chirality is sufficient to confer a frequency on a nonequilibrium system even at the first bifurcation.[4,5]

In Figure 1, we show the control parameter space; the interface speed v, and the temperature gradient $\mathbf{G}\|\mathbf{v}$, for the traveling cholesteric-isotropic interface, the chiral analogue of the nematic-isotropic interface. Patterns observed in the region marked A have both a wavelength, $2\pi/q$ and a frequency, $\omega/2\pi$.[4,5] In contrast, in the nonchiral case, only the region marked "A" is observed and the patterns have no characteristic frequency.[8] Here we describe the larger region where a pattern is created behind a moving flat interface ($q = 0$) by the director rotating in the interface (phase winding),[2] the *Lehmann Effect*.

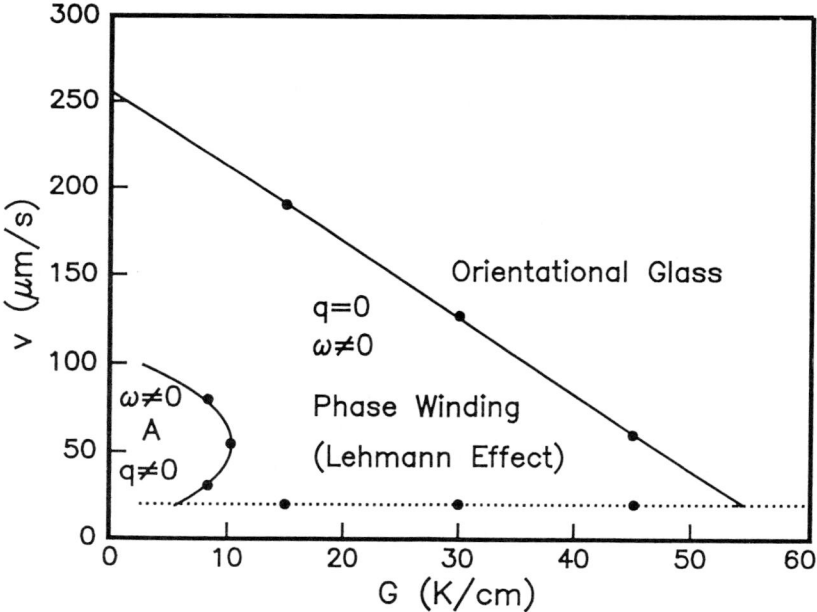

FIGURE 1   Patterns in the $G-v$ parameter space. Equilibrium is a $G = v = 0$.

COMPLEX MATERIALS: NEMATIC AND CHOLESTERIC LIQUID CRYSTALS"

Liquid crystals refers to states of matter characterized by long-range orientational order that occur between the isotropic liquid and solid state. Of the approximately 25 known liquid crystal states exhibited by organic compounds, the nematic and cholesteric states are most familiar because of their technological relevance to Flat Panel Displays.[9] Well-characterized materials are commercially available and a great deal is known about their physics under equilibrium conditions.[9] On the other hand, while it is known that living systems are also liquid crystalline in nature, relatively little is known about this and other applications of the physics of liquid crystals under nonequilibrium conditions.

Nematic liquid crystals break the continuous rotational symmetry of the isotropic liquid at a transition temperature, T_{NI}. We use materials in the nematic state formed by "small" molecules about $25\text{\AA}$ long and $5\text{\AA}$ in diameter. While the orientational ordering is long range, the microscopic difference between the nematic and isotropic liquid is subtle. As the direction chosen for orientation, called a director, $\mathbf{n}$, is infinitely degenerate, it may be controlled by weak external forces such as surface treatment and small electric fields. The director, a macroscopic variable, for which $\mathbf{n}$ and $-\mathbf{n}$ are equivalent, gives rise to an optic axis: when $\mathbf{n}$ is parallel to the direction of one of the crossed polarizers, the field of view is darkest and when $\mathbf{n}$ is at $45°$ to them, it is brightest. The power of nematics in visual displays resides in the fact that a continuous grey scale can be achieved both by different orientations of $\mathbf{n}$ relative to the polarizers as well as in the third direction perpendicular to the field of view.

Indeed, as their orientational order is long range, liquid crystals make beautiful pictures when observed with a polarizing light microscope and relatively weak magnification (i.e., classical optics). Recent advances in image analysis have enabled us to learn qualitatively new physics from pictures of liquid crystals under nonequilibrium conditions as a function of time.[4,5]

Nematic liquid crystals are a special case of a larger class called cholesteric (N^*) liquid crystals that have a characteristic length called the pitch, $p_o = 2\pi/q_o$. In cholesterics, $\mathbf{n}$ has a helix structure (see Figure 2). Nematics are the special case of a cholesteric with $q_o = 0$. Mixing an amount c of chiral material with a nematic results in a cholesteric with $q_o \propto c$: the larger the c, the tighter the pitch with an infinite pitch (the nematic) at $c = 0$. The transition temperature to the isotropic liquid also changes linearly with c. In our case, the larger the c, the greater the decrease in the transition temperature, T_{ChI}, relative to T_{NI} at $c = 0$. Thus, T_{ChI} is linearly related to q_o. We have quantitatively determined these relations for our system under equilibrium conditions and work with a $c_\infty = 9\%$ mixture giving an equilibrium pitch of $p_o = 38\mu m$. But, as they are liquids, weak external forces can change the pitch, indeed, even unwind the helix,[9] leading to a cholesteric-nematic transition. However, in the absence of such forces, the cholesteric ground state is one for which $\mathbf{q}_o = \mathbf{constant}$.

Cholesterics relax from a higher energy state to their ground state by a nonlinear elastic process called *orientational diffusion*[4,5] with diffusion constant, $D_o = 2 \times 10^{-7}$ cm^2/s. This means that there is an intrinsic frequency, $\omega_{e\ell}/2\pi \equiv D_o q_o^2$ also associated with the cholesteric state under nonequilibrium conditions.[4,5] For the mixture used here, $\omega_{e\ell}/2\pi = 2$ s^{-1}.

THE LEHMANN EFFECT

The cholesteric helix structure is characterized by a *handedness* perpendicular to $\mathbf{n}$.[6] In Figure 2, we have drawn a right-handed helix. In a mirror, a right-handed helix becomes a left-handed one, i.e., $q_o \rightarrow -q_o$. Specifically, for a helix with $\mathbf{q}_o$ oriented along $\hat{\mathbf{z}}$ with $\phi = q_o z + \phi_o$, $\mathbf{n}$ has the form $\mathbf{n} = (\cos\phi, \ \sin\phi, 0)$. Then ϕ is the director phase relative to a fixed direction in the x-y plane and ϕ_o is an arbitrary, constant phase shift. At fixed z and under stationary conditions, ϕ is a constant. The operation $z \rightarrow -z$, changes a right-handed helix into a left-handed one.

This symmetry allows terms in the dynamical equation[1] coupling q_o to gradients of a scalar, i.e., terms such as $\dot{\phi} \sim q_o \nabla T$ and $\dot{\phi} \sim q_o \nabla c$. $\dot{\phi} \neq 0$ means that the director rotates in time at fixed z in a process we call *phase winding.*

In these experiments, we will see that $\ell_D \sim 2$ μm, the dynamic length scale for impurity gradients, is much smaller than $\ell_T \sim 100$ μm, the length scale for thermal gradients, so, is the dominant effect to account for the observed phase winding.

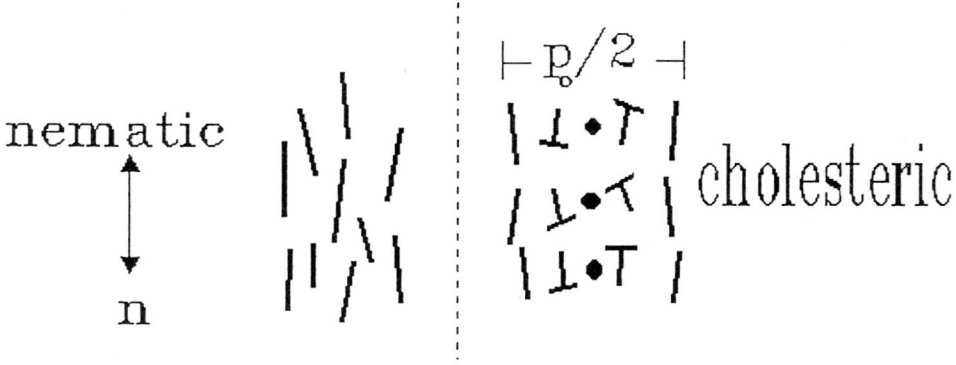

FIGURE 2   Equilibrium nematic and cholesteric liquid crystals.

Incorporating q_o into a *Lehmann coefficient* for impurity gradients, ν_c, we have $\gamma_1\dot\phi = \nu_C/\ell_D$, with γ_1 the director rotational viscosity. de Gennes[9] suggested that $\nu_C \sim K_2 q_o \sim 3 \times 10^{-4}$ dynes/cm². K_2 is the twist elastic constant related to the orientational diffusion constant by $D_o = K_2/\gamma_1$.

SET-UP FOR WELL-DEFINED TRAVELING INTERFACES

To obtain a traveling interface under well-defined conditions, we use the technique invented by Jackson and Hunt[10] known as "directional solidification." As details of our particular experimental set-up have been previously discussed,[3,5] we only outline here its features relevant to this chapter. The samples we use are all constructed from a mixture of 9% chiral impurity in a well-known nematic liquid crystal and held between glass plates prepared so that **n** is well oriented in the plane of each plate.

In this technique, a sample is placed in a well-defined, constant temperature gradient, **G**, chosen so that the interface between the two states is visible in a polarizing microscope/videocamera/image analysis arrangement. The first advantage of this technique is that it maps temperature onto a spatial dimension: knowing G, we can relate length scales in the field of view to a temperature difference and temperature differences in, e.g., the sample, to a length scale. For example, the width of the two-phase region for this mixture was found to be $\Delta T \sim 0.1$K giving a characteristic thermal length at $G = 7.5$ K/cm for this system of $\ell_T \equiv \Delta T/G = 133$ μm.

The sample is then mechanically displaced towards colder temperatures at constant velocity (known as the *pulling speed*) $-\mathbf{v}\|\mathbf{G}$. This displacement lowers the temperature at a fixed rate by a constant amount along the sample parallel to the temperature gradient, so that the cholesteric-isotropic interface at T_{ChI} is forced to move with a velocity $+\mathbf{v}$ towards the hot contact to maintain itself at T_{ChI}. Another advantage[3] of this technique is that the pulling speed can be controlled by a stepping motor/frequency generator arrangement to 1 part in 10^6.

As the interface advances, it expels impurities, here the chiral impurity we put into a nematic liquid crystal at an equilibrium concentration $c_\infty = 9\%$, into the isotropic liquid. The impurities diffuse away from the interface to c_∞ far from the interface in the isotropic liquid following linear diffusion. It turns out[4,5] that this is a relatively fast process in this system with a characteristic time, $\tau_D \equiv D_I/v^2 \sim 0.2$s for $D_I = 4 \times 10^{-7}$ cm²/s and $v \sim 20\mu$ m/s $\sim v_c$. Nevertheless, an impurity concentration gradient is set up in advance of the moving interface that exponentially decays from a maximum at the interface to c_∞ far in the isotropic liquid with a characteristic dynamic diffusion length, $\ell_D \equiv D_I/v \sim 2$ μm again for $v = 20$ μm/s, the smallest speed for these observations.

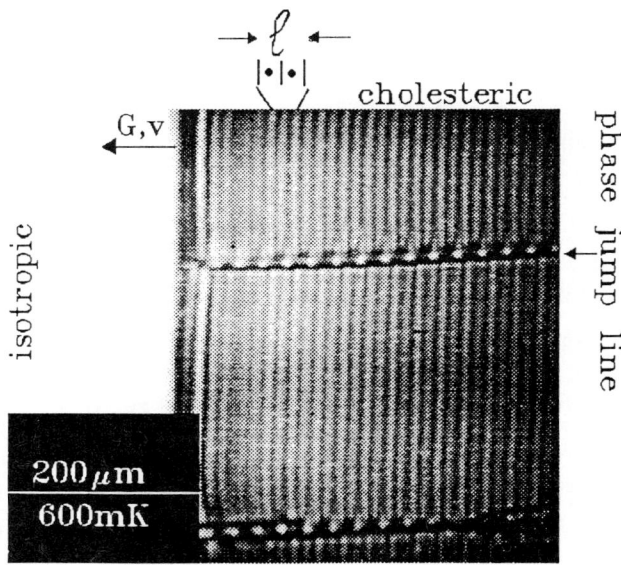

FIGURE 3   Pattern at a flat interface from forced phase winding (Lehmann Effect).

Above a critical speed, v_c, the destabilizing force of the concentration gradient ($\ell_D \sim 2\mu$m) overcomes the stabilizing influence of the temperature gradient ($\ell_t \sim 133\mu$) and elastic forces so that patterns, characterized by both a well-defined wavenumber, $q < q_o$, and frequency, $\omega/2\pi$, form at the interface (region A in Figure 1).[4,5] When $q \sim q_o$ in region A, or, if G is too large, the interface is flat ($q = 0$) but the director now winds at the flat interface (i.e., $q = 0$ but $\omega \neq 0$) leaving behind in the cholesteric state a pattern *perpendicular* to the interface of wavelength, ℓ (see Figure 3).

We attribute the forced winding of the director at this traveling interface to the previously discussed Lehmann Effect.

FORCED PHASE WINDING AND THE ORIENTATIONAL GLASS TRANSITION

We measured the pattern wavelength (ℓ in Figure 3) in the force phase-winding regime for three different sample thicknesses: 37 μm ($\bullet,\blacksquare$), 63 μm ($\times$), and 80 μm ($\bigcirc$), and three different temperature gradients: 15 K/cm ($\bullet, \times$), 30 K/cm ($\blacksquare$), and 45 K/cm ($\bigcirc$) (see Figure 4). In terms of Figure 1, the experimental procedure was to set G, then measure ℓ at fixed pulling speed, v, after the interface has reached a steady state temperature.

With increasing v, the extent of the spatial correlations in the director phase parallel to the interface, as measured by the distance between the π and sometimes 2π phase jump lines seen perpendicular to the interface (e.g., Figure 3), decreased so that measurements of even an average ℓ could no longer be made and there was a transition to an orientationally disordered quenched state (see Figure 5) that we call an orientational glass. This is seen on Figure 4 by the data points (highlighted by arrows) deviating markedly from the solid curve.

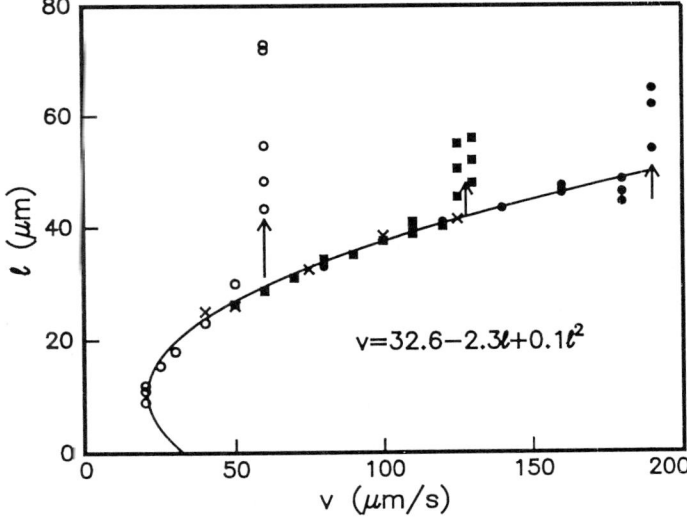

FIGURE 4 Pattern wavelength, ℓ, behind the flat interface (see Figure 3) vs. pulling speed.

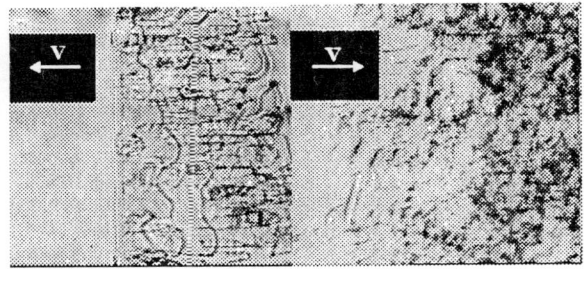

a   b

FIGURE 5 The orientational glass.

In Figure 1, we show that the pulling speed at the onset to the orientational glass state is linear in G. We have not yet formulated a simple reason why this should be. We believe that by measuring how spatial correlations in the director phase decrease as pulling speed is increased should give more hints to help understand this simple observation.

Apart from the data points in the vicinity of the orientational glass transition, all other data points appear to fall on a universal curve (i.e., independent of sample thickness and temperature gradient). A parabolic fit to these remaining data points gave the following fit (the solid curve in Figure 4):

$$v = 32.6 - 2.3\ell + 0.1\ell^2. \tag{1}$$

The good fit of the data to Eq. (1), suggests that the bifurcation to forced phase winding is transcritical.[11]

To arrive at a model that can be compared to the experimental results, we start with the terms in the hydrodynamic equations for the director:

$$\nu_C \frac{v}{D_I} = \gamma_1 \frac{d\phi}{dt} - \beta K_2 q_o \frac{2\pi}{\ell}, \tag{2}$$

and assume that the director phase rotates in the interface at a fixed rate so we can average Eq. 2 over phase changes of 2π. For $d\phi/dt$, we get two terms that are replaced under this assumption by $\partial\phi/\partial t \to 2\pi/\tau$ and $\partial\phi/\partial z \to 2\pi/\ell$. The term on the left-hand side of Eq. (2) is the Lehmann driving term for which the dominant contribution is expected to be from the concentration gradient, ℓ_D. The second term on the right-hand side is a simple order of magnitude estimate for the elastic torques that must be considered for cholesterics. This is a complicated task to calculate exactly for our geometry, so we pay the price for a simple order of magnitude estimate with an additional fit parameter, β. We then expand the velocity, v, as a quadratic polynomial in ℓ to get the final expression:

$$v = \beta D_o q_o - \frac{1}{\tau}\ell - \frac{\nu_C}{2\pi\gamma_1 D_I \tau}\ell^2. \tag{3}$$

Equation (3) contains three fit parameters: β, τ, ν_C. All other parameters are known. Comparing the expressions for the coefficients of ℓ in Eq. (3) with the fit to the experimental data (Eq. (1)): $\beta = 2.7$, $1/\tau = 2.3$ s^{-1} and $\nu_C = -3 \times 10^{-4}$ dynes/cm.

As expected, β is of order unity. The value for $1/\tau$ is close to the elastic frequency $\omega_{e\ell}/2\pi = 2$ s^{-1} and the value for the Lehmann coefficient coincides remarkably well with the order of magnitude estimate.[9]

CONCLUSIONS AND PERSPECTIVES

Forced phase winding and its breakdown at a well-controlled traveling chiral interface are examples of qualitatively new physics exhibited by a well-known liquid crystal system. A frequency is associated with all patterns in this system that has both a chirality determined intrinsic length and a frequency for response to perturbations of its equilibrium structure. For the forced phase-winding regime, we have proposed a simple model that is in reasonably good agreement with known material parameters and gives a first estimate of the Lehmann coefficient for the coupling between the intrinsic length determined by chirality to the concentration gradient in advance of the traveling interfaces. To improve our understanding of the orientational glass state, measurements are in progress to determine how spatial correlations in the phase winding regime decrease as a function of pulling speed.

ACKNOWLEDGEMENTS

Work partially supported by NATO CRG 890777 and a 1993 Guggenheim Fellowship (PEC).

REFERENCES

1. Brand, H. R., and H. Pleiner. "New Theoretical Results for the Lehmann Effect in Cholesteric Liquid Crystals." *Phys. Rev. A* **37** (1988): R2736–R2738.
2. Brand, H. R., and P. E. Cladis "Nonequilibrium Phase Winding and its Breakdown at a Chiral Interface." *Phys. Rev. Lett.* **72** (1994): 104–107.
3. Cladis, P. E., J. T. Gleeson, and P. L. Finn. "Experimental Results on the Analogy Between Saffman-Taylor Fingers and the Cellular Pattern in Directional Solidification of an Alloy." *Phys Rev. A* **44** (1991): R6173–R6176.
4. Cladis, P. E., J. T. Gleeson, P. L. Finn, and H. R. Brand. "Breathing Mode in a Pattern Forming System with Two Connecting Lengths." *Phys. Rev. Lett.* **67** (1991): 3239–3242.
5. Cladis, P. E. "Pattern Formation at the Cholesteric-Isotropic Interface." In *Pattern Formation in Complex Dissipative Systems,* edited by S. Kai, 3–13. Singapore: World Scientific, 1992.
6. Cladis, P. E. "Essay 17: The Liquid State of Materials or, How Does the Liquid Crystal Display (LCD) in my Wristwatch Work?" In *Fundamentals of Physics,* Extended Fourth Edition, edited by D. Halliday, R. Resnick, and J. Walker, 1230, E17-1–E17-5. New York: John Wiley, 1992.

7. Cross, M. C., and P. Hohenburg. "Pattern Formation Outside of Equilibrium." *Rev. Mod. Phys* **65** (1993): 851–1112.

8. Flesselles, J. M., A. J. Simon, and A. J. Libchaber. "Dynamics of One-Dimensional Interfaces: An Experimentalists' View." *Adv. Phys.* **40** (1991): 1–51.

9. de Gennes, P. G. *Physics of Liquid Crystals,* Third Edition. Oxford, UK: Clarendon, 1982.

10. Jackson, K. A., and J. D. Hunt. "Transparent Compounds that Freeze like Metals." *Acta Metall.* **13** (1965): 1212–1215.

11. Manneville, P. *Dissipative Structures and Weak Turbulence.* New York: Academic Press, 1990.

12. Oswald, P., J. Bechhoefer, and F. Melo. "Pattern Formation During the Growth of Liquid Crystal Phases." *MRS Bulletin* **XVI** (1991): 38–45.

Martine Ben Amar† and **Efim Brener†‡**

†Laboratoire de Physique Statistique de L'Ecole Normale Supérieure, associé au C.N.R.S. et aux Universités Paris VI et Paris VII, 24 rue Lhomond, 75231 Paris Cedex 05, France
‡Permanent address: 142432 Institute for Solid State Physics, Chernoglovka, Russia

Theory of Dendritic Growth in Three Dimensions

We study the selection of the shape and growth velocity of three dimensional dendritic crystals in cubically anisotropic materials. In the framework of asymptotics beyond all orders, we derive the inner boundary-layer equation for the non-axisymmetric shape correction to the Ivantsov paraboloid shape. The solvability condition for this equation provides selection for both the velocity and the shape. The comparison with available numerical and experimental results is reasonably good.

The problem of velocity selection for two-dimensional needle-crystal growth has been solved both numerically[11,12,21] and analytically.[1,3,5,7,8,9] Contradicting naive physical intuition, it turns out that the steady solutions exist only if the crystalline anisotropy of surface energy is taken into account. The predictions of this steady-state analysis agree with the numerical time-dependent simulations[27] which take into account sidebranching behavior of the tail of the dendrite. Details of this work may be found in papers by Brener and Melnikov,[10] Kessler et al.,[13] Langer,[17] and Pomeau and Ben Amar.[26] Most experiments, however, are three-dimensional and it is important to extend the previous theory to the three-dimensional case. A simple extrapolation of the two-dimensional case, in which the surface energy is averaged in the azimuthal direction (axisymmetric approach[1,6]), yields interesting qualitative predictions. But a physical anisotropy, say, with an underlying cubic symmetry, will give rise to a non-axisymmetric crystal shape, in agreement with experimental observations.[20] Nevertheless, this makes the problem more difficult to solve, and even some doubts about the solvability mechanism were expressed as an explanation for the dynamic behavior of this system.[19] An alternative and opposite point of view including time-dependent perturbations has been developed by Pines and Taylor[28] and Pines et al.[25] It is why it seems to us of great importance to bring some new material in the understanding of this theoretical problem which concerns a prototype of growth process. Moreover, we want to stress that this is the first theoretical treatment of a fully three-dimensional pattern. The treatment itself leads to an interesting puzzle. Roughly speaking, in the two-dimensional case or in the axisymmetric case, the velocity selection is given by the solvability condition associated with the smoothness of the dendrite tip. In the three-dimensional non-axisymmetric case, we must therefore satisfy a solvability condition for each of the azimuthal harmonics. We know only one attempt to solve a non-axisymmetric problem. This is the numerical calculation by Kessler and Levine.[13] They made several approximations and finally performed only a two-mode calculation, but the crucial point of their analysis is that they found enough degrees of freedom to satisfy the solvability conditions.

The main aim of this paper is to develop an analytic theory for the three-dimensional dendritic growth with an underlying cubic anisotropy. Using the general structure of the solvability theory beyond all orders, we derive the inner boundary-layer equation in the complex plane for the non-axisymmetric shape. This ordinary differential equation contains derivatives with respect to the fast variable in the boundary layer. Because we deal with a function of two variables the coefficients of the equation should depend also on the smooth variable. The remarkable property of the equation is that all dependences on the smooth variable are combined into a single common factor. The solvability condition requires this factor to be a pure constant. This will provide the selection of the velocity and the anisotropic shape corrections, which are in fact the additional degrees of freedom found in papers by Kessler and Levine[13] and Langer and Müller-Krumbhaar.[16] We would also like to stress an additional problem. In the two-dimensional case, the selected crystal shape is close to the Ivantsov[17] parabola since the anisotropy and surface tension effects are assumed to be small. In the three-dimensional case, the anisotropic shape correction becomes larger than the underlying Ivantsov solution as we move away from the dendrite tip. We will discuss this point at the end of this paper.

Let us study the problem of a free dendrite growing in its undercooled melt. The control parameter is the dimensionless undercooling $\Delta = (T_M - T_\infty)c_p/L$, where T_M is the melting temperature, L the latent heat, and c_p the specific heat. The temperature field satisfies the diffusion equation with the interface, moving with normal velocity v_n, acting as a source of magnitude $v_n L/c_p$. Together with the Gibbs-Thomson condition at the interface, this leads to a rather complicated integral-differential evolution equation.[10,14,17,26] It is possible to verify[23] that the

Ivantsov paraboloid of revolution $\zeta = -r^2/(2\rho)$ (ρ is the tip radius of curvature), which moves with an arbitrary constant velocity v, satisfies the steady-state equation when the Gibbs-Thomson shift is set equal to zero. The Péclet number $p = v\rho/(2D)$ (D is the thermal diffusivity) is related to the undercooling by the three-dimensional Ivantsov formula $\Delta = p\exp(p)E_1(p)$, where $E_1(p)$ is the exponential integral function. Adding the Gibbs-Thomson effect, one can expect to have velocity selection.[10,14,17,26] The usual analytical approach to this problem is to linearize the integral term in the evolution equation around the Ivantsov paraboloid, assuming the Gibbs-Thomson effect to be small. This gives[2,13]

$$-\Delta_\mu[\zeta] = \frac{C}{2\pi} \int_0^\infty r_1 dr_1 \int_{-\pi}^{\pi} \frac{d\phi_1(r_1^2 - r^2)[u(r_1, \phi_1) - u(r, \phi)]}{[r_1^2 + r^2 - 2rr_1\cos(\phi - \phi_1) + (r^2 - r_1^2)^2]^{3/2}}. \tag{1}$$

Here u is the correction to the Ivantsov shape, all lengths are reduced by 2ρ; $C = 8p\rho/d_0$ is an eigenvalue parameter; $d_0 = \tilde{\gamma}T_M c_p/L^2$ is a capillary length proportional to the isotropic part of the surface energy $\tilde{\gamma}$; $\Delta_\mu[\zeta]$ is the well-known[17] Gibbs-Thomson shift from the equilibrium melting point:

$$\Delta_\mu[\zeta] = \frac{1}{R_1}\left(\gamma + \frac{\partial^2\gamma}{\partial\Theta_1^2}\right) + \frac{1}{R_2}\left(\gamma + \frac{\partial^2\gamma}{\partial\Theta_2^2}\right), \tag{2}$$

where R_1, R_2 are the local principal radii of curvature of the surface, Θ_1, Θ_2 are the angles between the normal **n** and the local principal directions on the surface, and $\gamma(\mathbf{n})$ is the dimensionless anisotropic surface energy.[2,13]

$$\gamma(\theta, \tilde{\phi}) = 1 + 4\epsilon[\cos^4\theta + \sin^4\theta(\cos^4\tilde{\phi} + \sin^4\tilde{\phi})]$$
$$= 1 + 4\epsilon(\cos^4\theta + .75\sin^4\theta + .25\sin^4\theta\cos 4\tilde{\phi}). \tag{3}$$

Here θ and $\tilde{\phi}$ are the Euler angles of the normal vector to the interface. The parameter ϵ measures the strength of the cubic anisotropy, giving a maximum of surface energy in the (100) crystal direction. The axisymmetric approximation consists in dropping the last term in Eq. (3). A convenient alternative expression for the Gibbs-Thomson shift is given by Eq. (10) from Kessler[13]:

$$\begin{aligned}
-\Delta_\mu = {} & \gamma\nabla \cdot \left(\frac{\nabla\zeta}{\sqrt{1 + (\nabla\zeta)^2}}\right) + \frac{\partial^2\gamma}{\partial\theta^2}\frac{\nabla\theta \cdot \nabla\zeta}{\sqrt{1 + (\nabla\zeta)^2}|\nabla\zeta|} + \frac{\partial^2\gamma}{\partial\theta\partial\tilde{\phi}}\frac{\nabla\tilde{\phi} \cdot \nabla\zeta}{\sqrt{1 + (\nabla\zeta)^2}|\nabla\zeta|} \\
& - \frac{\partial^2\gamma}{\partial\tilde{\phi}^2}\frac{(\nabla\tilde{\phi} \times \nabla\zeta) \cdot \hat{\zeta}}{|\nabla\zeta|^2}\sqrt{1 + (\nabla\zeta)^2} - \frac{\partial^2\gamma}{\partial\tilde{\phi}\partial\theta}\frac{(\nabla\theta \times \nabla\zeta) \cdot \hat{\zeta}}{|\nabla\zeta|^2}\sqrt{1 + (\nabla\zeta)^2} \\
& + \frac{\partial\gamma}{\partial\theta}\frac{\nabla \cdot (\nabla\zeta/|\nabla\zeta|)}{\sqrt{1 + (\nabla\zeta)^2}} - \frac{\partial\gamma}{\partial\tilde{\phi}}\frac{\hat{\zeta} \cdot \nabla\zeta \times \nabla(1/|\nabla\zeta|^2)}{\sqrt{1 + (\nabla\zeta)^2}},
\end{aligned}$$
$$\tag{4}$$

where ∇ is the two-dimensional gradient operator and $\tan\theta = |\nabla\zeta|$.

Let us look for a solution of Eq. (1) of the form $\zeta = \zeta_0 + u(r, \phi)$, where

$$\zeta_0 = -r^2 + \sum A_m r^m \cos(m\phi). \tag{5}$$

This is possible, at least formally, because none of the terms $r^m \cos(m\phi)$ contribute to the integral in Eq. (1).[13,17] Kessler and Levine[13] proposed that these degrees of freedom are exactly those needed to satisfy the solvability conditions for the smoothness of the dendrite tip. They truncated the linear version of Eq. (1) to the first non-axisymmetric mode $\cos(4\phi)$ and solved this equation numerically. They found the parameter C to be close to the value of the axisymmetric version and also found the coefficient A_4 to be small but not zero. Of course, Eq. (5) cannot describe the true needle-crystal solution. It is clear that the linear approximation must break down eventually as we move away from the tip because the shift $r^m \cos(m\phi)$ grows at a faster rate that the underlying solution.

From the point of view of the general structure of the analytic theory[3,15,17] which has been developed only for the two-dimensional and three-dimensional axisymmetric cases, we must look, first of all, at the regular theory of perturbation for Eq. (1) with respect to the small parameter $1/C$. It is possible to prove that, for each mode, the regular theory of perturbation exists. Because the starting point of the perturbation theory implies to drop $u(r, \phi)$ in the Gibbs-Thomson shift $\Delta_\mu[\zeta]$, then one has to deal with linear inhomogeneous integral equations. A necessary condition for the existence of a solution is that the inhomogeneous term must be orthogonal to the null eigenvector of the adjoint operator. The crucial point here is that the solution of the adjoint integral operator simply does not exist. According to the general theorem of existence of the solution for the singular integral operator,[22] we reach the conclusion that the solution of the regular theory of perturbation exists for each mode without any additional conditions and, moreover, the solutions are smooth at the origin.

We turn now to the problem of the selection of the parameters C and A_m. As we mentioned before, the solvability condition appears only beyond all orders; to handle it properly, we have to derive the local equation near the singularities in the complex plane. The location of the singularities of this Gibbs-Thomson shift is given by the condition

$$1 + (\nabla \zeta_0)^2 = 0. \tag{6}$$

This equation describes the singularities line in the complex plane: $r = r_s(\phi)$. The function ζ_0 is given by Eq. (5) and, for example, for the pure Ivantsov shape we have $r_s = i/2$. We will derive the inner equation for $u(r, \phi)$ by imposing a smoothness condition in ζ_0. We do not assume that this function is close to the pure Ivantsov solution. In the close vicinity of the line of the singularities we can keep only the local singular contribution in the integral term and the most singular piece in the Gibbs-Thomson shift which contains the factor $[1 + (\nabla\zeta)^2]^{-3/2}$. This greatly simplifies Eq. (1). Let us start with the Gibbs-Thomson shift. The most singular contribution comes from the first and second terms in Eq. (4). Using the definition $\tan\theta = |\nabla\zeta|$, we find

$$-\Delta_\mu \approx \left(\gamma + \frac{\partial^2\gamma}{\partial\theta^2}\right)\frac{\nabla\zeta\cdot\nabla(|\nabla\zeta|)}{[1+(\nabla\zeta)^2]^{3/2}|\nabla\zeta|} \qquad (7)$$

and

$$\gamma + \frac{\partial^2\gamma}{\partial\theta^2} \approx 1 - \frac{7\alpha(1+(1/7)\cos 4\tilde\phi)}{[1+(\nabla\zeta)^2]^2}, \qquad (8)$$

where $\alpha = 15\epsilon$ and ϵ is the anisotropy parameter in Eq. (3) (assumed to be small). The function ζ contains two pieces. The first one, ζ_0, is smooth and the second one, u, is assumed to be small, but varies rapidly with respect to the fast variable: $t = 1 + i|\nabla\zeta_0|$ (we can choose, for example, ϕ as a smooth variable because r and ϕ are connected to each other by Eq. (6)). More precisely, and this will be checked self-consistently later, we assume that $u \sim \alpha$, the derivative $u_t \sim t \sim \sqrt{\alpha}$ and $u_{tt} \sim 1$. In principle, the ensuing calculations are quite straightforward; all we need to do is to keep the main term with respect to the small parameter α. Using the fact that $\nabla u \approx u_t \nabla t$ and $\nabla u_t \approx u_{tt} \nabla t$, we find, after some algebra,

$$1 + (\nabla\zeta)^2 \approx 2t + 2\nabla\zeta_0 \cdot \nabla u = 2(t - qu_t) \qquad (9)$$

and

$$\nabla\zeta \cdot \nabla(|\nabla\zeta|)|\nabla\zeta|^{-1} \approx q(1 - qu_{tt}), \qquad (10)$$

where

$$q = \nabla\zeta_0 \cdot \nabla(|\nabla\zeta_0|)|\nabla\zeta_0|^{-1}. \qquad (11)$$

Next, we perform a stretching transformation for both the variable $t = \tilde\alpha^{1/2}\tau$ and the function $u = -\tilde\alpha f/q$, and combine everything into the final formula for the Gibbs-Thomson shift:

$$-\Delta_\mu = \frac{q}{2\sqrt{2}\tilde\alpha^{3/4}}\left(1 - \frac{1}{(\tau + f_\tau)^2}\right)\frac{1 + f_{\tau\tau}}{(\tau + f_\tau)^{3/2}}, \qquad (12)$$

where $\tilde\alpha = (7\alpha/4)(1 + (1/7)\cos 4\tilde\phi)$ .Using the definition of $\tilde\phi$ as a spherical angle made by the normal vector to the interface, we find $\tilde\phi = \phi + \beta$, where

$$\sin\beta = -\frac{\partial\zeta/\partial\phi}{r|\nabla\zeta|} \approx \frac{i}{r}\frac{\partial\zeta_0}{\partial\phi}. \qquad (13)$$

We note that the structure of the Gibbs-Thomson shift (Eq. (12)) is very close to the one found in the two-dimensional and three-dimensional axisymmetric cases[3,17] except for the prefactor which depends now on the smooth variable.

Let us discuss the behavior of the integral term in Eq. (1) in the singular region. Because the function u is singular, we need keep only the local contribution. Therefore, we can write the integral term, which we call J, in the form: $J = -CBu/2$.

In the non-axisymmetric case a careful analysis , which takes into account both the pole term and the set of logarithmic singularities, gives

$$B = \frac{1}{1 + 4r^2} + \frac{2ir}{(1 + 4r^2)\sqrt{1 + \frac{1+4r^2}{r^2}(\frac{dr_s}{d\phi})^2}}, \tag{14}$$

which transforms into the usual[3,17] expression if the location of the singularities r_s is ϕ independent. Finally the local inner equation (which, in fact, is simply $-\Delta_\mu = J$) can be written as

$$\left(1 - \frac{1}{(\tau + f_\tau)^2}\right) \frac{1 + f_{\tau\tau}}{(\tau + f_\tau)^{3/2}} - f\lambda = 0, \tag{15}$$

where λ is given by:

$$\lambda = \sqrt{2}C\tilde{\alpha}^{7/4}Bq^{-2}. \tag{16}$$

Here, λ is to be evaluated on the line of singularities $r = r_s(\phi)$, and this depends on the smooth variable ϕ only. The crucial point is that Eq. (15) has precisely the same structure as in the two-dimensional case[3,17] and the dependence on the smooth variable is entirely contained in the single factor λ. Therefore, we can formulate the solvability condition which requires the parameter λ to be a pure constant and to belong to the discrete spectrum.[3,17] Equation (16) provides now the selection of the parameters C and A_m. First, we note that C obeys the usual two-dimensional scaling relation: $C = a\alpha^{-7/4}$. Then, Eq. (16) does not contain any small parameters, which means that α and A_m should be pure numbers. The axisymmetric approach to this complicated nonlinear equation simply consists of neglecting the dependence on ϕ in Eq. (8). In this case we can set A_m equal to zero and find the usual discrete spectrum of the parameter a; only the solution which corresponds to the smallest value of a is a stable one.[7,11] Let us consider the axisymmetric approach as an approximation of zero order with respect to the "small" parameter $\delta = 1/7$ (see Eq. (8)). Because of this small parameter, we can expect that for the stable solution, a should be only slightly different from its value in the pure axisymmetric approach. It is clear that the correction to a appears only at the second order with respect to δ (a change of sign of δ means simply the change of the origin of ϕ). This might explain that the numerics[13] give a selected value of C which is close to the value found in the axisymmetric approach. In the linear approximation we can satisfy the solvability condition by the correction $A_4 r^4 \cos(4\phi)$ to the Ivantsov shape. A simple but tedious calculation, which involves linearization of Eq. (16) with respect to δ and A_4, gives: $A_4 = 7\delta/11 = 1/11$. To leading order in α, the tip shape is given by:

$$\zeta_0 = -r^2 + \frac{1}{11}r^4 \cos(4\phi). \tag{17}$$

So, in the limit of small α, we predict that the shape correction is independent of the anisotropy parameter α, once written in units of ρ contrary to the tip radius

ρ itself. We note that this universal shape correction for cubic material is close to the only published experimental determination by Maurer et al.[20] Concerning the numerics, this number is about a factor of 2 smaller than the number given by Kessler and Levine.[13] This difference can be explained by the fact that several different approximations have been made in both calculations. Unfortunately, in Kessler and Levine's paper,[13] the calculation has been performed only for a single value of the anisotropy parameter, so we cannot make a comparison at this point. We hope that our result will suggest new experiments and numerical simulations on the three-dimensional dendrite shape. In principle, we can go further in the expansion (5) and see, for example, that A_{4m} should be small at least like δ^m. But what is more important is the fact that the number of degrees of freedom available are precisely those we need to satisfy the solvability condition.

We discussed the growth in the direction (100), which corresponds to the maximum of the surface energy (3) for positive ϵ. If, on the other hand, the parameter ϵ is negative, then the direction (100) corresponds to the minimum and the direction (111) corresponds to the maximum of the surface energy. In this case the theory predicts the dendrite to grow in the direction (111) for the used model of the anisotropy of surface energy. Only we need to replace Eq. (14) by

$$\tilde{\alpha} = \frac{7|\alpha|}{6}\left[1 + \frac{4i}{7}(\cos 3\tilde{\phi} - \sin 3\tilde{\phi})\right]. \qquad (18)$$

The main anisotropic correction to the shape in the tip region should be written as

$$A_3 r^3(\cos 3\tilde{\phi} - \sin 3\tilde{\phi}) \qquad (19)$$

with $A_3 \approx -8/21$.

As we mentioned before, the linear approximation breaks down eventually as we move away from the tip because the shift vector $r^m \cos m\phi$ grows at a faster rate than the underlying Ivantsov solution. This means that our approximation of linearizing the integral term, is valid in the tip region only. This is the crucial difference between the three-dimensional non-axisymmetric case and the two-dimensional case where the selected needle-crystal shape is close to the Ivantsov parabola everywhere if the the anisotropy is small. What does it mean? We think that the complete treatment of the three-dimensional dendritic shape requires two different steps. Our paper is concerned with the first one: the selection mechanism by a fully anisotropic surface tension and the determination of the tip shape. Since the shape correction cannot be extended to all distances, a further analysis is required to complete the description of the needle-crystal. A correct treatment requires a matching between the tip and the tail via an intermediate range of r values where the nonlinear effects cannot be neglected. From a purely theoretical point of view, this analysis will differ completely from the selection problem which is concerned with short distances near the tip and where an analytical extension in the complex plane is useful. Once the shape correction is established, one can always include time-dependent side-branching non-axisymmetric modes as in the paper by Langer.[18] Perhaps these modes are important in the matching treatment but we think, as does Langer,[18] that they are irrelevant to the selection process itself.

APPENDIX

The main contribution from the integral term (Eq. (1)) near the singularity has a local form. The pure local term gives the well-known[5,7] piece

$$J_l = -\frac{C_u}{2(1 + 4r^2)}. \tag{20}$$

The local behavior of the nonlocal part of the integral J_r is dominated by the vicinity of the point (r, ϕ). So we expand $u(r_1, \phi_1)$ when ϕ_1 is close to ϕ_1, which gives for J_r

$$J_r = \frac{C}{2\pi} \int_0^\infty r_1 dr_1 \frac{r_1^2 - r^2}{(2rr_1)^{3/2}} \left[u(r_1, \phi) I_0(\beta) + \sum_{n=1}^\infty \frac{\partial^n u(r_1, \phi)}{\partial \phi^n} I_n(\beta) \right] \tag{21}$$

with, for every n,

$$I_n = \frac{1}{n!} \int_{-\pi}^\pi \frac{(\phi_1 - \phi)^n}{[\beta - \cos(\phi_1 - \phi)]^{3/2}} \tag{22}$$

and

$$\beta - 1 = \frac{(r - r_1)^2 (1 + (r + r_1)^2)}{2rr_1}. \tag{23}$$

All the integrals J_n with odd n vanish; all the others are singular when β is close to 1 and ϕ_1 close to ϕ. For n equal zero, we obtain

$$I_0(\beta) = \frac{2\sqrt{2}}{\beta - 1} - \frac{1}{2\sqrt{2}} \ln(\beta - 1) + \ldots + B_l(\beta - 1)^l \ln(\beta - 1). \tag{24}$$

This series is mentioned in Langer's lecture.[7] For the axisymmetric shape only, the pole term gives the local contribution. Now we have to examine the terms that contain the derivatives with respect to ϕ. The point is that the derivatives with respect to ϕ can be transformed to the derivatives with respect to fast variable t and, in combination with less singular behavior of I_n, then gives rise to the additional local contribution. By expanding I_{2k} near $\beta = 1$ up to the $(k-1)$ order, we catch the logarithmic divergency which gives the main contribution for each k

$$I_{2k} = A_k(\beta - 1)^{k-1} \ln(\beta - 1) \tag{25}$$

and

$$A_k = (-1)^k 2^{3/2} \frac{(2k - 1)!!}{(k - 1)!(2k)!}. \tag{26}$$

We can always transform the weaker logarithmic singularities into a pole one by several $(2k)$ integration by part. Let us introduce the formal operator

$$\frac{\partial}{\partial \phi} \int dr = \frac{\partial t/\partial \phi}{\partial t/\partial r}, \tag{27}$$

where t is the fast variable. Combining all pole terms together, we get, after some tedious algebra,

$$J_r = -\frac{Cu}{2}\frac{2ir}{1+4r^2}\sum(-1)^k\frac{(2k-1)!!}{k!2^k}\frac{(1+4r^2)^k}{r^{2k}}\left(\frac{dr_s}{d\phi}\right)^{2k},\qquad(28)$$

where $r_s(\phi)$ is the line of singularities. The final summation gives Eq. (14) in the text. We note that summation gives the square-root dependence not accidentally, but with a deep origin: the local Mullins-Sekerka spectrum.

ACKNOWLEDGMENTS

We thank M. Adda Bedia, R. Combescot, V. Hakim, S. Iordanskii, D. Kessler, H. Levine, P. Pelcé, Y. Pomeau, and W.-J. Rappel for discussions. One of us (E.B.) acknowledges financial support from a grant of the D.R.E.T. (France).

REFERENCES

1. Barbieri, A., D. Hong, and J. S. Langer. *Phys. Rev. A* **36** (1987): 5353.
2. Barbieri, A., and J. S. Langer. *Phys. Rev. A* **39** (1989): 5314.
3. Ben Amar, M., and Y. Pomeau. *Europhys. Lett.* **2** (1986): 307.
4. Ben Amar, M., and B. Moussallam. *Physica D* **25** (1987): 155.
5. Ben Amar, M. *Physica D* **31** (1988): 409.
6. Ben Amar, M. *Phys. Rev. A* **41** (1990): 2080.
7. Bensimon, D., P. Pelcé, and B. Shariman. *J. Phys.* Paris **48** (1987): 2081.
8. Brener, E. A., S. E. Esipov, and V. I. Melnikov. *JETP Lett* **45** (1987): 759.
9. Brener, E. A., S. V. Iordanskii, and V. I. Melnikov. *Soviet Phys. JETP* **67** (1988): 2574.
10. Brener, E. A., and V. I. Melnikov. *Adv. Phys.* **40** (1991): 53.
11. Kessler, D., and H. Levine. *Phys. Rev. B* **33** (1986): 7678.
12. Kessler, D., and H. Levine. *Phys. Rev. Lett.* **57** (1986): 3069.
13. Kessler, D., and H. Levine. *Acta metall.* **36** (1988): 2693.
14. Kessler, D., J. Koplik, and H. Levine. *Adv. Phys.* **37** (1988): 255.
15. Kruskal, D., and H. Segur. *Stud. Appl. Math.* **85** (1991): 129.
16. Langer, J. S., and H. Müller-Krumbhaar. *Acta metall.* **26** (1978): 1689.
17. Langer, J. S. In *Chance and Matter*. Les Houches Lectures XLVI. New York: Elsevier, 1987.
18. Langer, J. S. *Phys. Rev. A* **36** (1987): 3350.
19. Langer, J. S. Special issue: Complexity and Materials Research. *Physics Today* **45** (1992): 24.
20. Maurer, J., B. Perrin, and P. Tabeling. *Europhys. Lett.* **14** (1991): 575.
21. Meiron, D. *Phys. Rev. A* **33** (1986): 2704.
22. Muskhelishvili, N. I. *Singular Integral Equations.* Holland: Groningen, 1953.
23. Pelcé, P., and Y. Pomeau. *Stud. Appl. Math.* **74** (1986): 245.
24. Pines, V. K., and P. L. Taylor. *Phys. Rev. A* **41** (1990): 1006.
25. Pines, V. K., P. L. Taylor, and M. A. Zlatkovski. *Phys. Rev. A* **41** (1990): 1021.
26. Pomeau, Y., and M. Ben Amar. In *Solid Far from Equilibrium,* edited by C. Godréche. Cambridge: Cambridge University Press, 1992.

27. Saito, Y., G. Goldbeck-Wood, and H. Müller-Krumbhaar. *Phys. Rev. Lett.* **58** (1987): 1541.

Shonosuke Ohta and Haruo Honjo
Department of Physics, College of General Education, Kyushu University, Ropponmatsu, Fukuoka 810, Japan

An Experimental Study of Periodic Tip-Splitting Process in NH4Cl Dendrite

1. INTRODUCTION

Dendritic crystal growth is an anisotropic pattern formed in a nonequilibrium diffusion field deeply affected by its own shape, and the growth process at the surface plays an important role for the anisotropy of dendritic pattern.[5,10,16] The growth surface is unstable by means of the Mullins-Sekerka instability arising from a nonlinearity of the diffusion equation. Therefore, the diversity of crystal shapes is strongly dependent on the boundary conditions at the surface, and also physically nontrivial patterns are formed by the effects of externally applied conditions. In this chapter, an attractive crystal pattern of a periodic tip-splitting dendrite formed by two anisotropies in the $\langle 100 \rangle$ and $\langle 110 \rangle$ directions, which is a unique property of the dendritic crystal growth in NH4Cl, will be presented.

In dendritic crystal growth for which external randomness is large enough to overcome the anisotropy and the surface tension, tip surfaces grow irregularly with a random tip-splitting process. Such a growth process leads the global pattern to a diffusion-limited aggregation[17] (DLA). Actually, a DLA pattern of dendrites has been observed experimentally in cells controlled by large randomness,[6,7,13] and a computer simulation supports the formation of a homogeneous DLA relying on the above process.[14,15] Since a DLA pattern was initially found in a system of electrochemical deposition,[11] the DLA formation is now recognized as common in the diffusion field.[17]

Recently, much attention has been paid to another common growth form called dense-branching morphology[17] (DBM), which has been observed in higher nonequilibrium conditions than that of DLA growth. The DBM pattern forms a randomly and densely tip-splitting structure, and the interfacial envelope is stable. Which tip-growth mechanism makes the DBM envelope stabilize against the intrinsic instability in the surfaces and the screening effects in the diffusion field, is an essential question of DBM pattern formation.

Let us return to the subject of anisotropic dendritic crystal growth. Recent advances in the theory of dendritic pattern formation apply a solvability condition of a nonlinear equation to find a steady state of needle form.[10] The stable form needs anisotropy, and experimental results are discussed in comparison with theory.[12]

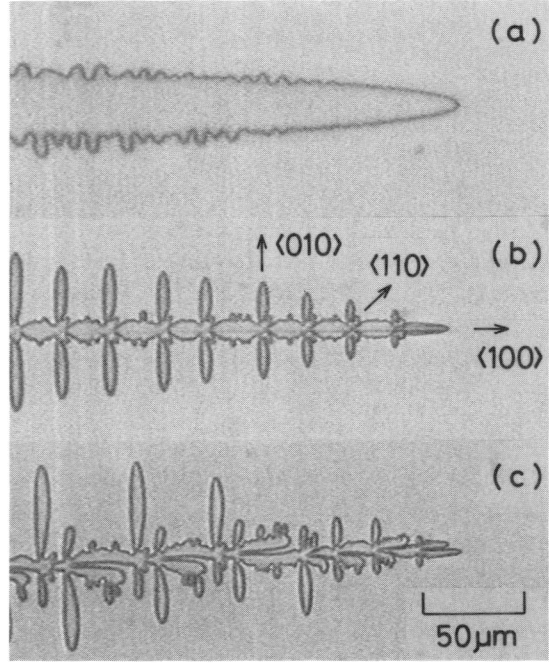

FIGURE 1 Normal dendrite (a), regular tip-splitting dendrite (b), and irregular tip-splitting dendrite (c) of NH_4Cl growing in a quasi-two-dimensional cell. Scale and crystal direction are equivalent for every picture.

The formation of sidebranches can be characterized by noise at the tip.[3,10] In the system of a supersaturated aqueous solution of NH$_4$Cl, a normal dendrite grows in the $\langle 100 \rangle$ direction due to the anisotropy of surface tension. According to the observation in a quasi-two-dimensional cell without any space,[5] the tip of the main stem has a parabolic form and the sidebranches are randomly produced in the $\langle 010 \rangle$ direction as shown in Figure 1(a). This is a common dendritic form in a system that has rough surfaces via the kinetic roughening phenomena and has an anisotropic surface tension.

An anisotropic DBM pattern of NH$_4$Cl crystal has been found, in our recent experiment,[8] in a cell of 300 μm thickness, in which the pattern has many self-originating tips because a stem bifurcates repeatedly into two stems—hereafter called the multiple tip-splitting (MTS) process—and these tips grow coherently toward the $\langle 100 \rangle$ direction. Although some tips break while growing, the front envelope grows continuously with a constant velocity and the envelope area increases gradually. The growth velocity of the envelope decreases with the increase of supersaturation, from 65 μm/sec to 52 μm/sec, just like a negative impedance of an Esaki diode in the relation between current and voltage. Such characteristics shed light on the interfacial stability on DBM pattern formation. Therefore, precise experimental research at nearby conditions becomes important in the study of such a negative tip-growth property.

The DBM pattern can be observed in the supersaturation of the transition region between $\langle 100 \rangle$ and $\langle 110 \rangle$ growths of NH$_4$Cl dendrites. From the previous experiment without any spacer, a regular tip-splitting (RTS) dendrite,[9] Figure 1(b), has been observed in the lower supersaturation of the transition region. The tip-splitting process of the RTS dendrite occurs periodically in space and time. The $\langle 110 \rangle$ arms grow closely to the $\langle 100 \rangle$ tip, and then the $\langle 110 \rangle$ arms stop growing abruptly. The well-developed $\langle 010 \rangle$ sidebranches, synchronized with the tip-splitting phenomena, are remarkable characteristics of the RTS dendrites. The growth velocity of this RTS dendrite is 43 μm/sec, which is smaller than the DBM velocities mentioned above. It seems that dimension, in this case on the order of a micronmeter. strongly affects the growth velocity of the dendrite, whereas, in intermediate supersaturation, an irregular tip-splitting (ITS) dendrite, Figure 1(c), has been observed. The $\langle 110 \rangle$ arms of ITS dendrite shift the growth direction from $\langle 110 \rangle$ to $\langle 100 \rangle$ after the tip-splitting processes has begun. And then, only one tip survives the competition of a few $\langle 100 \rangle$ tips, and a new tip splitting occur, from this tip. The ITS process enables the global growth direction to shift slightly from the $\langle 100 \rangle$ direction which is different from the rigid growth direction of the RTS dendrite. In the higher supersaturation of the transition region, a few $\langle 100 \rangle$ tips, produced by an ITS process, can survive together, i.e., an MTS process, and thus an DBM pattern appears.

Chan et al.[2] have experimentally found that the transition region of NH$_4$Cl dendrites is between 7.5% and 10.5% on the supersaturation of an aqueous solution at 25°C, and they have observed that the growth velocity of this region is a constant value of about 55 μm/sec. Moreover, they reported only an ITS morphology, unlike our previous observation. Such circumstances point out that the new experimental inspection of dendritic growth in the transition region is very significant. Brener[1] has theoretically studied the occurrence of the transition on the growth direction of dendrites caused by kinetics. If the new and old growths belonging to different growth directions coexist, and if a concentrational interaction exists between them, the new growth could be affected by the old growth through the diffusion field. Then, the new growth must be stopped because of the decreasing local supersaturation, and thus the transition must not abruptly occur when macroscopic supersaturation is increased. Therefore, the tip-splitting process is an extremely interesting problem coupled with the transition and geometry of the growth direction.

2. EXPERIMENTS

In the present experiment two significant points are taken into account: One is arranging the experiment to avoid the boundary effects of the sample cell. A cover glass of 200 μm thickness is put over the left side of a sample cell of 400 μm thickness. A dendrite that moves to the right along the surface of the the cover glass, after the tip passes by the edge of cover glass, grows freely in the open space without affecting the wall. A variation of velocity and a schematic drawing for a growing $\langle 100 \rangle$ normal dendrite are shown in Figure 2, which shows the case of maximum velocity, $v_{\max} = 81 \pm 1$ μm/sec, observed in the present experiments. The remarkable difference between the growth velocities of 60 $\mu m/sec$ on the boundary wall and of 81 μm/sec in the open space interesting, and the boundary effects for the growth velocity appears clearly within the boundary length of 40 μm from the wall. In this case the diffusion length $l_d = 2D/v$ is calculated l_d=64 μm and is compatible with the boundary length, where $D = 2.6 \times 10^{-5}$ cm^2/sec is the diffusion constant of NH$_4$Cl in water. The above results give an important suggestion for the diffusion length in the dendritric experiments.

The second point is the method for the control of supersaturation. All seven samples are prepared by the saturated aqueous solution of NH$_4$Cl at 40°C. From such a sample we can observe all morphologies of the transition region by varing temperature from 28°C to 30°C. In order to get more precise supersaturation and to avoid the complexity resulting from a lot of sample cells, the supersaturations $\sigma = 7 \sim 9.5\%$ are obtained from the temperature control within an error of ± 10mK, which control the water flow from a heat bath within an error of ± 1mK; the heater is driven by a thermistor thermometer and an electric circuit is used. As a result of our experiment, the mutual error of σ for the seven samples is estimated $\Delta \sigma = \pm 0.2\%$. Every picture of the dendrite observed by a microscope is recorded by a VTR every 1/30 sec, and then the VTR pictures are digitized through an image processor and are analyzed with an accuracy of 1 pixel = 1.932 μm.

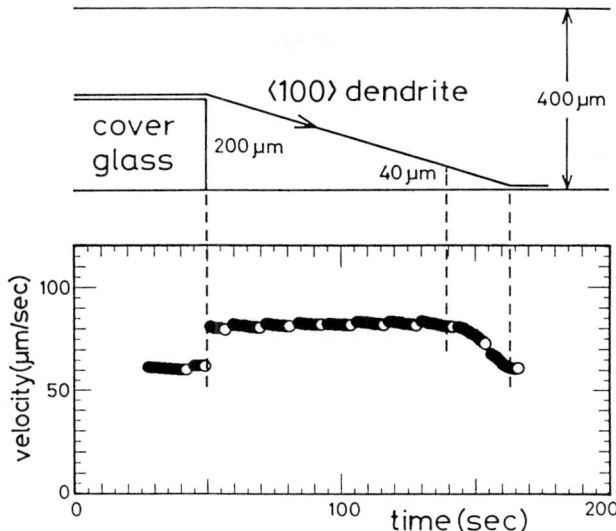

FIGURE 2 Schematic drawing and velocity variation of a $\langle 100 \rangle$ normal dendrite growing in a cell with a cover glass.

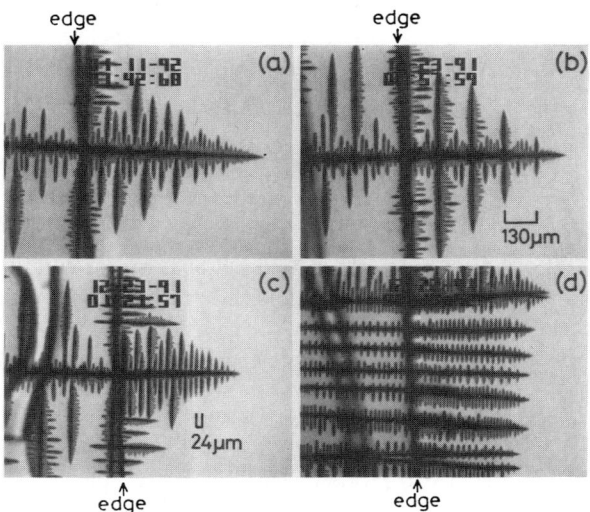

FIGURE 3 Typical examples of a normal dendrite (a), regular tip-splitting dendrites (b) and (c), and dense-branching morphology (d), where the edge line of the cover glass is shown. Supersaturations are (a) 7.33%, (b) 7.59%, (c) 8.45%, and (d) 9.22%. Scale is equivalent for every picture.

Typical examples of observed dendrites are shown in Figure 3, where Figure 3(a) is a normal dendrite, Figure 3(b) and 3(c) are RTS dendrites with $\lambda = 130$ and 24 μm respectively, and Figure 3(d) is a DBM dendrite. Here, λ is the mean length for the tip-splitting intervals of the RTS dendrite. The edge line of cover glass located on left side is shown. Tip splitting of RTS dendrites forms symmetrically well-developed sidebranches as seen in Figure 3(b) and 3(c). This implies a peculiar sidebranching mechanism in the RTS process different from the ordinary one in the normal dendrite. Paying attention to the sidebranching morphology, in the RTS dendrites of Figure 3(b) and 3(c) one can see the shapes of the normal dendrite when tips grow on the cover glass, and also the randomly sidebranching shapes in the intermediate range between well-developed sidebranches in the RTS dendrite of Figure 3(b). On the contrary, in the normal dendrite of Figure 3(a), a well-developed sidebranch can be formed as the tip crosses the edge of cover glass. When an RTS dendrite comes close to the glass boundary, λ is increasing, and then the morphology changes into a normal dendrite. These results indicate that the cell boundary has a great influence on the dendritic morphology. In the DBM dendrite of Figure 3(d), different tip-splitting shapes coexist, and this is where the RTS growths with various tip-splitting intervals are observed. The small shift of growth direction of main stems can be performed by ITS processes. The ITS processes are sometimes observed among the RTS processes in the supersaturation above $\sigma \sim 8.4\%$. An MTS process which is essential in the formation of a DBM pattern can be observed in the top stem of Figure 3(d). The spatial and temporal features of RTS dendrites observed in the open space, 14 μm$\leq \lambda \leq$450 μm, are analyzed in following discussion.

3. RESULTS AND DISCUSSION

The growth features of an RTS dendrite at $\sigma = 7.59\%$ are drawn in Figure 4, where Figure 4(a) is an overlapped picture of digitized tips snapped every 1/30 sec; the tip curvature clearly becomes small when tip splitting occurs as shown in Figure 4(a) and 4(b). The deviation of the location of the growing tip $\Delta L = L(t){-}vt$ versus time shows a clear sawtooth wave as seen in Figure 4(c), where $L(t)$ is the real tip location and v is the average velocity calculated from the data. This feature indicates that the growth velocity behaves just like a square pulse, as shown in Figure 4(d). For all data of RTS dendrites, both the curvature and the growth velocity yield above-periodic binary structure with two growth states. Such peculiar growth behavior points out that the processes of RTS dendrites can be divided into two growth parts: normal and splitting areas. So we measure the velocities v_n and v_s, the growth lengths λ_n and λ_s, and the growth periods τ_n and τ_s for the normal and splitting areas. And, then, a wave length $\lambda = \lambda_n + \lambda_s$, a period of oscillation $\tau = \tau_n + \tau_s$, and the average growth velocity $v = \lambda/\tau$ are calculated.

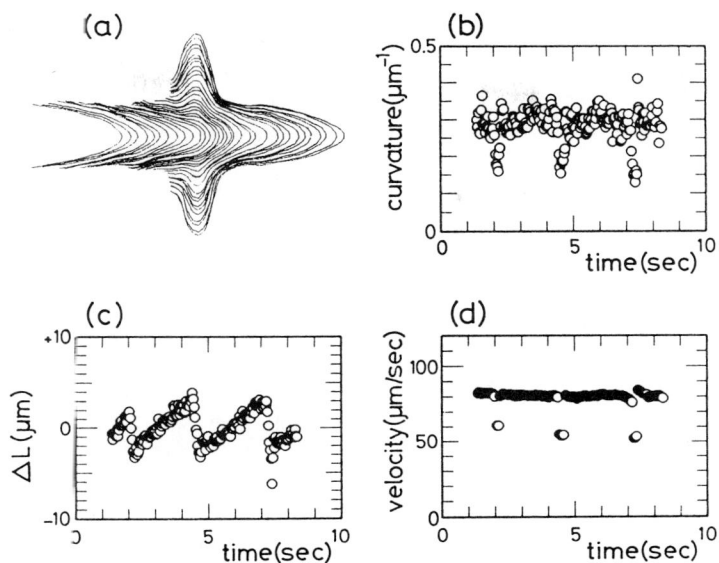

FIGURE 4 Growth features of a regular tip-splitting dendrite. Overlapped picture of tips (a), tip curvature (b), deviation ΔL of tip location from steady growth (c), and growth velocity (d).

Experimental results on the RTS dendrites are summarized in Figure 5 as the function of σ. In Figure 5(a) λ decreases rapidly compared to σ. The longest λ below 7.5% shows 270 μm, which is close to the cell thickness of 400 μm rather than l_d of 64 μm. This fact led us to construct a longer length than l_d. The plots of $1/\lambda$ seem to have a linear dependence on σ as shown in Figure 5(b). A critical supersaturation $\sigma_c = 7.4\pm0.3\%$ for the RTS phenomena can be derived by the fitting line, which is in quite good agreement with Chan et al.'s 7.5%.[2]

As indicated in Figure 5(c), no dependences on the velocities of v_n and v_s are clear. The obtained mean values are $v_n = 81\pm4$ μm/sec and $v_s = 48\pm4$ μm/sec. These results yield the peculiar binary structure of the RTS dendrites independent of σ. The agreement between v_n of RTS dendrites and v_{max} of normal dendrite is quite excellent. This feature supports the idea that the growth behavior in the normal area of RTS dendrites coincides with that in the normal dendrite not only with regard to the morphology but also the growth velocity. The growth velocity v_s in the splitting area indicates about 60% of v_n. The average velocities v of RTS dendrites are drawn in Figure 5(d). Although the experimental errors are considerably large, it is obvious that v has negative slope compared to σ. Comparing this with our previous result of $v = 43$ μm/sec in Figure 1(b) and with the result of

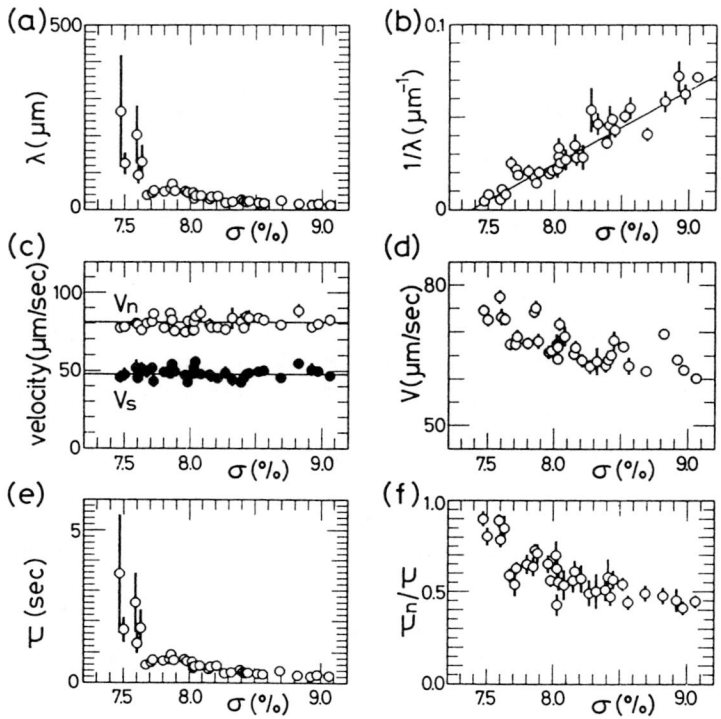

FIGURE 5   Experimental results against supersaturation σ of (a) λ, (b) $1/\lambda$, (c) v_n and v_s, (d) v, (e) τ, and (f) τ_n/τ.

$v \simeq 55$ μm/sec from Chan et al., the present results give large values of $v = 60 \sim 78$ μm/sec. Thus the cell boundary seriously the growth velocities.

The period τ and the ratio τ_n/τ, which corresponds to the duty ratio of pulse wave, are shown in Figure 5(e) and 5(f) respectively. From Figure 5(f), we understand that two periods in the lower σ show $\tau_n > \tau_s$ and that those in the higher σ show $\tau_n < \tau_s$. Therefore, the growth velocity v of RTS dendrites is decreasing as σ is increased because τ_n/τ is decreasing. Eventually, the growth processes of RTS dendrites are constituted of two constant velocities corresponding to the normal and splitting growths. Although the wave length and the period of oscillation in RTS phenomena are reduced as supersaturation is increased, the essential effects for the average velocity depends on the relative reduction of normal area. The stability on the collective tip growth of DBM fronts, as seen in Figure 3(d), is deduced from

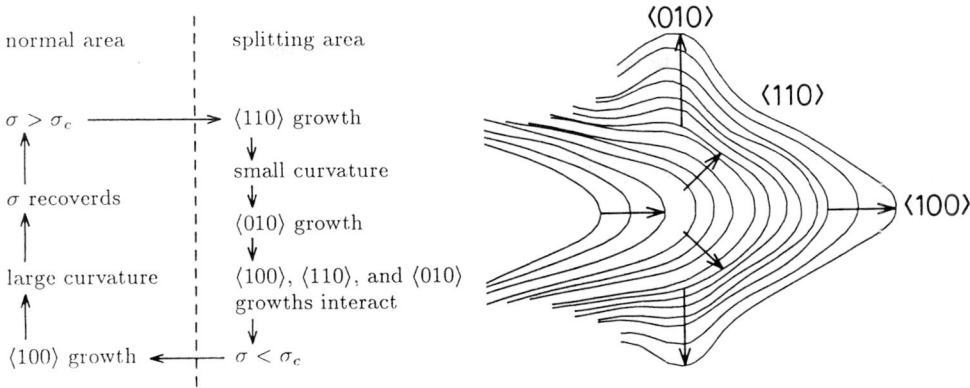

FIGURE 6 Schematic drawing of a model for a regular tip-splitting process. When local supersaturation σ near the tip is more than critical supersaturation σ_c, $\langle 110 \rangle$ growth occurs.

such a negative growth property of RTS dendrites. Forward tips compared with the DBM envelope grow slowly with frequent RTS phenomena because of the rich concentration, while backward tips grow fast with the long RTS period because of the dilute solution. By the way, in a homogeneous DBM observed in higher nonequilibrium condition than DLA, a similar negative growth property may be expected.

Dendritic pattern formation is a complex problem determined by the growth boundary and by the diffusion field. Moreover, in the RTS dendrite, transition phenomena on the growth direction are related to the formation. However, the oscillatory behavior in RTS process can be deduced from a simple model coupled with growth directions and supersaturation. Our considerations, shown in Figure 6, are as follows: The local supersaturation σ in front of a normal $\langle 100 \rangle$ tip is more than the critical supersaturation σ_c. Then $\langle 110 \rangle$ growth occurs; hence, splitting area forms. The tip becomes flat by $\langle 110 \rangle$ growth, and then $\langle 010 \rangle$ growth forms sidebranches, while the tip is growing slowly in the $\langle 100 \rangle$ direction. Three different kinds of growths interact with each other through the diffusion field. As a result, $\langle 110 \rangle$ growth is suppressed and at the same time σ is less than σ_c. After that the $\langle 100 \rangle$ tip grows, and the splitting area is switched off. In normal growth, σ recovers gradually toward macroscopic supersaturation, and then σ is more than σ_c, and so on. This model is constituted of two states of the $\langle 100 \rangle$ growth in $\sigma < \sigma_c$ and of the $\langle 110 \rangle$ growth in $\sigma > \sigma_c$. The local supersaturation σ near the tip controls switching between them, and the switching period is determined by a relaxation of σ. A theoretical approach based on this model is of interest.

Let us discuss some attractive points from the results of RTS dendrites. An outstanding characteristic of RTS dendrites is symmetrically well-developed $\langle 010 \rangle$ sidebranches, which arise from the tip-splitting process at locations close to the tip as mentioned above. Such a deterministic mechanism builds the RTS sidebranches, which is quite different from the random sidebranches of normal dendrites formed behind the tip by noise. We pointed out a relation of $\lambda \sim 1/(\sigma - \sigma_c)$ from the present experimental results, but we did not refer to the σ-dependences of τ and v. According to these data, it is possible to determine the relations for these quantities. However, these three variables should be compatible with a basic relation of $v = \lambda/\tau$. From this viewpoint, reasonable pursuits for determining these variables are still open. On the length scale of the present system, one can consider the three lengths of tip (a few μm), diffusion length (l_d=64 μm), and cell thickness (400 μm), while the intervals of RTS dendrites have been measured up to $\lambda \sim 450$ μm. In accordance with the one-dimensional diffusion equation, it is difficult to deduce a longer length than l_d for the relaxation processes of the diffusion field. Higher dimensional approaches may be necessary to understand the RTS intervals theoretically.

SUMMARY

Regular tip-splitting processes of NH_4Cl dendrites observed in the transition region between $\langle 100 \rangle$ and $\langle 110 \rangle$ growth are studied experimentally in the open space without the effects of cell boundaries. Growth morphology transfers from normal dendrite to a regular tip splitting one at the critical supersaturation of 7.4%. The periodic growth of a regular tip-splitting dendrite is clearly divided into two parts: a normal process for $\langle 100 \rangle$ growth and a splitting process for $\langle 110 \rangle$ growth with the constant growth velocities of 81 μm/sec and 48 μm/sec, respectively. The tip-splitting period decreases as supersaturation increases, and we observed a negative growth property that the average growth velocity decreases as supersaturation increases. Binary structure and periodicity of regular tip-splitting dendrites are theoretically interesting problems.

ACKNOWLEDGMENT

We would like to T. Hurubayashi for helpful contributions in experiments.

REFERENCES

1. Brener, E. A. "Effects of Surface Energy and Kinetics on the Growth of Needle-Like Dendrites." *J. Crys. Growth* **99** (1990): 165–170.
2. Chan, S. K., H. H. Reimer, and M. Kahlweit. "On the Stationary Growth Shapes of NH₄Cl Dendrites." *J. Crys. Growth* **32** (1976): 303–315.
3. Dougherty, A., P. D. Kaplan, and J. P. Gollub. "Development of Side Branching in Dendritic Crystal Growth." *Phys. Rev. Lett.* **58** (1987): 1652–1655.
4. Godrèche, C. *Solids far from Equilibrium.* Cambridge: Cambridge University Press, 1992
5. Honjo, H., S. Ohta, and Y. Sawada. "New Experimental Findings in Two-Dimensional Dendritic Crystal Growth." *Phys. Rev. Lett.* **55** (1985): 841–844.
6. Honjo, H., S. Ohta, and M. Matsushita. "Irregular Fractal-Like Crystal Growth of Ammonium Chloride." *J. Phys. Soc. Japan* **55** (1986): 2487–2490.
7. Honjo, H., S. Ohta, and M. Matsushita. "Phase Diagram of a Growing Succinonitrile Crystal in Supercooling-Anisotropy Phase Space." *Phys. Rev. A* (Rapid Comm.) **36** (1987): 4555–4558.
8. Honjo, H., and S. Ohta. "Dense-Branching Morphology of an NH₄Cl Crystal." *Phys. Rev. A* **45** (1992): R8332–R8335.
9. Honjo, H., and S. Ohta. "Regular Tip-Splitting Growth of an NH₄Cl Dendrite." *J. Crys. Growth* (1993): in press
10. Langer, J. S. "Dendrites, Viscous Fingers, and the Theory of Pattern Formation." *Science* **243** (1989): 1150–1156.
11. Matsushita, M., M. Sano, Y. Hayakawa, H. Honjo, and Y. Sawada. "Fractal Structures of Zinc Metal Leaves Growth by Electrodeposition." *Phys. Rev. Lett.* **53** (1984): 286–289.
12. Muschol, M., D. Liu, and H. Z. Cummins. "Surface-Tension-Anisotropy Measurements of Succinonitrile and Pivalic Acid: Comparison with Microscopic Solvability Theory." *Phys. Rev. A* **46** (1992): 1038–1050.
13. Ohta, S., and H. Honjo. "Growth Probability Distribution in Irregular Fractal-Like Crystal Growth of Ammonium Chloride." *Phys. Rev. Lett.* **60** (1988): 611–614.
14. Ohta, S., and H. Honjo. "Homogeneous and Self-Similar Diffusion-Limited Aggregation Including Surface-Diffusion Processes." *Phys. Rev. A* **44** (1991): 8425–8428.
15. Ohta, S. "Homogeneous Diffusion-Limited Aggregation Including Surface-Diffusion Process." In *Pattern Formation in Complex Dissipative Systems,* edited by S. Kai, 93–97. Singapore: World Scientific, 1992.
16. Pelcé, P. *Dynamics of Curved Fronts.* San Diego: Academic Press, 1988.
17. Vicsek, T. *Fractal Growth Phenomena.* Singapore: World Scientific, 1989

Yukio Saito and Kenji Shiraishi
Department of Physics, Keio University, 3-14-1 Hiyoshi, Kohoku-ku, Yokohama 223, Japan

The Effect of Noise on Dendritic Growth

The effect of noise on the dendritic growth is investigated by adding, at a point near the tip, a Gaussian white random velocity to the systematic part obtained by numerically solving a quasi-stationary evolution equation of the interface. With small noise or with strong anisotropy in the surface tension, the tip is stable and forms a regular dendrite, but the tip velocity and the tip radius oscillate. The product of the growth rate v and the tip radius ρ satisfies the Ivantsov relation. With large noise or with weak anisotropy, the tip destabilizes and splits.

1. INTRODUCTION

In nature, diverse patterns are formed from an orderless environment. Among them, the interfacial pattern formation in diffusion field has been intensively studied lately. A propagating front in the diffusion field undergoes a morphological instability, called Mullins-Sekerka instability, due to the point effect.[1,9,10] The front is unstable against the perturbation, and the structure is expected to become irregular and fine. Diffusion-limited aggregation (DLA) is a typical pattern formed by this effect with

Spatio-Temporal Patterns, Ed. P. E. Cladis and P. Palffy-Muhoray,
SFI Studies in the Sciences of Complexity, Addison-Wesley, 1995   **157**

a perpetual splitting of a dendrite tip.[15] The microscopic effects such as interface tension and interface kinetics compete with diffusional instability. A microscopic solvability theory[1,9,10] and numerical simulations,[7,13,14] however, showed that an isotropic interface tension is insufficient to stabilize the tip. In fact, in the viscous finger experiment,[11] where a fluid with low viscosity pushes another fluid with high viscosity, similar tip splitting is observed even the interface has a finite but an isotropic surface tension. The theory[1,9,10] shows that an infinitesimal anisotropy in the system is sufficient to stabilize the tip and to select the operating point. Since the crystal has an intrinsic anisotropy due to its microscopic structure, the formation of a tip-stable dendrite is a natural consequence of the theory. The stabilizing effect by the anisotropy is demonstrated in the viscous finger experiment by cutting a groove on a glass plate sandwiching two viscous fluids.[11] The fluid interface takes a profile similar to the dendritic crystal without any tip splitting.

In nature, however, not all the crystals growing under diffusion control show tip-stable dendritic form. There are various examples of tip-unstable dendrites, in electrochemical deposition or metal vapor deposition, for instance. Here we study the effect of noise as the origin of morphological crossover from tip-stable to the tip-unstable dendrite.

The effect of noise is investigated previously as an origin of the dendritic sidebranches.[12] In the microsopic solvability theory,[1,9,10] the crystal profile is expected to be parabolic without any sidebranches. The real experiments[5,2] or numerical simulation,[13,14] however, always show sidebranches in dendrites. The discrepancy is settled by the theory of convective instability, which tells that sidebranches induced by noise may be enhanced, but drift down the main stem and thus keep the dendrite tip stable to any perturbation.[1,9,10] The result may hold for small noise or random fluctuation. If the randomness is strong enough to wash out the effect of anisotropy, the tip stability will be lost and the tip splitting will result. This is realized in the solution growth experiment of NH_4Cl in the Hele-Shaw cell, where the crystal takes an irregular and fractal dendrite.[4]

The first simulational study of the noise effect on the dendritic growth is done in the boundary layer model.[12] The addition of a uniformly distributed random velocity at the tip point leads to the formation of well-developed sidebranches. Here we investigate the noise effect on the dendrite growing in a diffusion field in a quasi-stationary approximation. Noise is taken into account by a Gaussian white random velocity added to the systematic part at a point near the tip. Our main concern is the competition between the noise and an anisotropy on the stability of the tip, and thus, we explore a wide region of noise and anisotropy strengths.

2. INTERFACE EVOLUTION

The system considered is a crystal growing in the undercooled melt,[1,9,10] where the growth is controlled by the thermal diffusion process. The diffusion field u denotes the temperature measured from the value far from the interface and normalized by the temperature increase due to the latent heat production. It obeys the diffusion equation

$$\frac{\partial u}{\partial t} = D\nabla^2 u, \tag{1}$$

with D being the thermal diffusivity. When the crystal is growing with a velocity v, the latent heat produced at the solidification front should be transported by the heat flow to fulfill the energy conservation as

$$v_n = -D\vec{n} \cdot \vec{\nabla} u, \tag{2}$$

with $\vec{n}$ being the normal vector of the interface.

A rough interface can be regarded in local equilibrium such that the diffusion field u_i at the interface with a curvature κ takes the equilibrium value

$$u_i = \Delta - d\kappa. \tag{3}$$

Here Δ is the dimensionless undercooling, and d is an anisotropic capillary length with a four-fold symmetry as

$$d(\theta) = d_0(1 - d_4 \cos 4\theta). \tag{4}$$

When $d_0 = 0$, Ivantsov showed that the parabolic needle crystal grows steady with a tip radius ρ.[6] The undercooling Δ determines the product of ρ and the growth velocity v, or the Peclet number $p \equiv v\rho/2D$; for example, in two dimensions the relation is

$$\Delta = 2\sqrt{p}e^p \int_{\sqrt{p}}^{\infty} dx \exp(-x^2). \tag{5}$$

Relation (5), however, cannot determine the growth mode uniquely for a given Δ since there are infinite combinations of v and ρ possible for a given product p. Furthermore, all Ivantsov parabolas are shown unstable to small perturbations. The microscopic solvability theory[1,9,10] showed that the interface tension with a finite anisotropy stabilizes a unique Ivantsov parabola from infinitesimal perturbation, and the growth rate v satisfies a universal scaling behavior

$$vd_0/2Dp^2 = \sigma(d_4). \tag{6}$$

Here σ depends only on the anisotropy of the surface tension d_4, but is independent of the dimensionless undercooling Δ. The scaling relation (6) has been confirmed by the previous numerical simulation.[13,14]

The purpose of the present paper is to study numerically the effect of finite noise on the tip stability or the competition between the noise and anisotropy. The interface is susceptible under the thermal fluctuation and it may undergo some thermal motion with a random modulation of the velocity. We consider this effect by adding at a point $\vec{r}_I$ near the tip a random Gaussian velocity v^R to the systematic part obtained by Eq. (2). The variable v^R is related to the Gaussian white random number R with a zero mean and a unit variance as $v^R(\vec{r}_I, t) = \gamma R(t)$. Here γ characterizes the strength of the noise, and v^R has the correlation $\langle v^R(r_I, t) v^R(r_I, t') \rangle = \gamma^2 \delta(t - t')$.

3. NUMERICAL SIMULATION

Numerical simulation method we use is similar to the one used previously for the free dendritic growth from the diffusion field.[13,14] A quasi-stationarity is assumed such that the diffusion field adjusts its value very quickly compared to the front propagation. If a crystal is solidifying steadily in y-direction with a velocity v, then the field follows the time-independent diffusion equation in the comoving frame with the crystal as

$$\hat{L}u \equiv \nabla^2 u + \frac{2l^{-1}\partial u}{\partial \zeta} = \frac{D^{-1}\partial u}{\partial t} = 0 \qquad (7)$$

with $\zeta = y - vt$. The variable l is the diffusion length defined by $l = 2D/v$.

Since the main interest lies in the detailed profile of the interface Γ, which is represented as $\zeta = \zeta(x)$, we transform the stationary diffusion equation (7) to an integral equation for the interface profile Γ by applying Green's theorem as

$$-D^{-1} \int d\Gamma' g(\vec{r}, \vec{r}') v_n(\vec{r}') = \int d\Gamma' H(\vec{r}, \vec{r}') u_i(\vec{r}'). \qquad (8)$$

Here $v_n(\vec{r})$ is the local normal velocity given by Eq. (2), u_i is given by Eq. (3), and integral kernels g is the Green's function

$$g(\vec{r}, \vec{r}') = \frac{1}{2\pi} \exp\left[\frac{\zeta(x) - \zeta(x')}{l}\right] K_0\left[\frac{|\vec{r} - \vec{r}'|}{l}\right]$$

of the operator adjoint to $\hat{L}$ of Eq. (7), and $H(\vec{r}, \vec{r}') = \partial g/\partial n' - 2l^{-1}n'_z g + 1/2\delta(\vec{r} - \vec{r}')$. The kernel H satisfies the sum rule $\int d\Gamma' H(\vec{r}, \vec{r}') = -1$. Discretizing the interface Γ into many segments with maximum length d_{max}, the integral equation (8) is transformed into a matrix equation.

Due to the limitation of the calculation time and capability, we cannot study an infinitely extended profile, but restrict the study around the dendritic tip. The profile is divided into three parts: tip, transient and tail regions. The evolution equation (8) is followed only in the tip region. In the tail region, the Ivantsov parabola that fits the frame velocity is added. The transient region connects the tip and tail regions smoothly. The detail is explained in the previous paper.[14] For a given undercooling Δ, the initial velocity of the moving frame v is chosen appropriately, and the crystal shape is set to a corresponding Ivantsov parabola.[6] Then one can calculate values of integral kernels, g and H, as well as the values of the field u_i along the interface. The remaining unknown is the local velocity normal to the interface $v_n(\vec{r})$, which is obtained by solving the matrix equation, derived from Eq. (8). To this systematic velocity we add a Gaussian white velocity fluctuation v^R at a single mesh point I near the dendrite tip. In order to prevent the direct influence of the random velocity on the tip, the operating mesh point I is set at a distance of three to five times d_{max} separated from the tip. The time evolution of the solidification front is described by shifting the ith mesh point $\vec{r}_i(t)$ to $\vec{r}_i(t + \Delta t)$ determined by

$$\vec{r}_i(t + \Delta t) = \vec{r}_i(t) + [v_n(\vec{r}_i, t)\Delta t + v^R(\vec{r}_i, t)\sqrt{\Delta t}] \cdot \vec{n}_i. \tag{9}$$

Generally, the frame velocity is different from the crystal growth velocity. Therefore the frame velocity is relaxed to the velocity of the forefront velocity of the interface. By repeating this procedure of solving the local velocity, developing a new profile and adjusting the frame velocity, one may eventually reach a stable configuration, if it exists.

---

4. RESULTS

We consider a one-sided model, where the diffusion takes place only in the liquid but not in the solid. The length and time scales are so chosen that the isotropic part of the capillary length $d_0 = 10^{-3}$, and the diffusion constant in the liquid $D = 1$. The undercooling is chosen to be $\Delta = 0.25$. The Peclet number $p(\Delta)$ takes the value $p(\Delta = 0.25) = 0.0286$.

Figures 1(a) and 1(b) show the time evolution of the dendrite for a fixed anisotropy $d_4 = 0.125$ with various noise strength γ. The tip radius ρ expected from the scaling relation (6) is about 0.7, and the maximum mesh separation d_{max} is chosen to be 0.1. The tip velocity v is expected from Eq. (6) about 0.08. With a small noise $\gamma = 0.002$, the dendrite grows with a stable tip, as is shown in Figure 1(a).

The growth velocity shows some fluctuation as is shown in Figure 2(a). With a strong noise, $\gamma = 0.007$, the sidebranch amplitude of the dendrite becomes very large as is shown in Figure 1(b). The tip velocity as well as tip radius show oscillation, as is shown in Figure 2(b). The coarse-grained product of ρv is

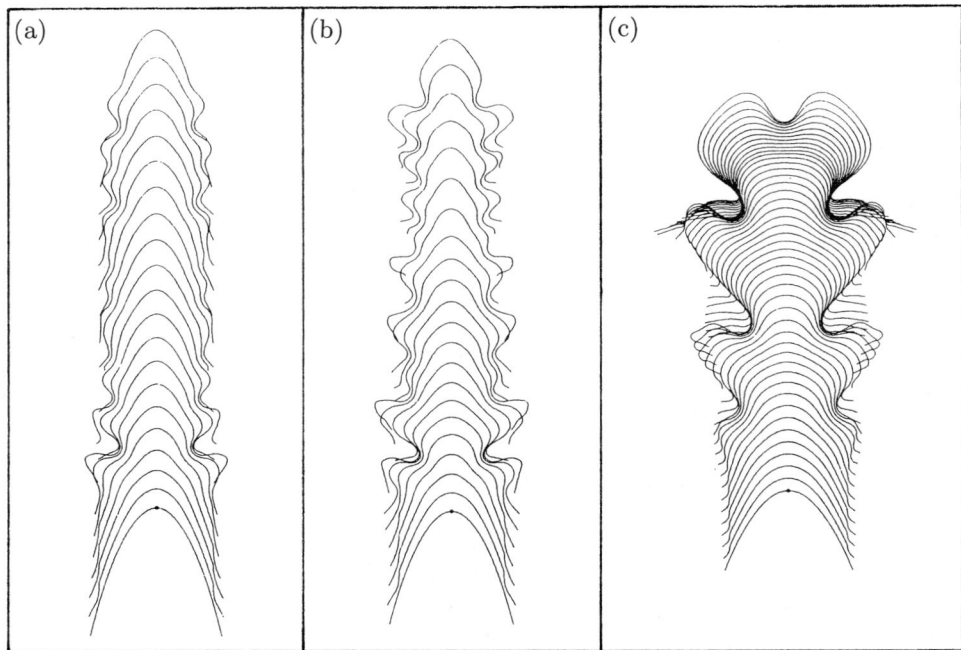

FIGURE 1 Time evolutions of the dendritic profile with various strong anisotropy d_4 and noise strengthes γ: (a) $d_4 = 0.125$, $\gamma = 0.002$, (b) $d_4 = 0.125$, $\gamma = 0.007$, and (c) $d_4 = 0.068$, $\gamma = 0.002$.

found to remain almost constant near the value given by Eq. (5), and the deviation of the dendritic profile from the Ivantsov parabola is small.

On decreasing the anisotropy strength d_4, the tip looses stability. In Figure 1(c) we show a tip-unstable dendrite at the noise level $\gamma = 0.002$ with $d_4 = 0.068$. At $d_4 = 0.070$ (not shown) the tip is stable and the growth velocity $v \approx 0.03$ agrees with the theoretical expectation $v \approx 0.035$ from the scaling relation (6). At the noise level $\gamma = 0.007$, we can also obtain the threshold of the tip stability. These results are summarized in the phase diagram of the parameter space γ and d_4, shown in Figure 3.

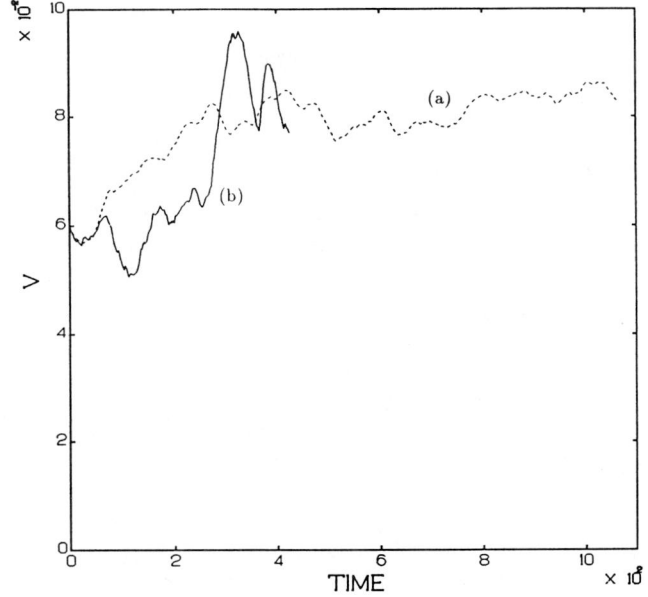

FIGURE 2 Time variation of the growth velocity v for $d_4 = 0.125$ and (a) $\gamma = 0.002$ and (b) $\gamma = 0.007$.

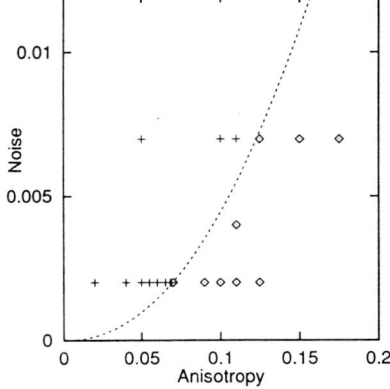

FIGURE 3 Phase diagram in a parameter space of the interface anisotropy d_4 and the noise strength γ. A diamond represents the point at which a tip-stable dendrite is realized, and a cross represents a tip-splitting event. The line is merely a guide for the eye.

5. CONCLUSIONS

By numerically solving the evolution equation of the solidification front, we found a crossover in the growth mode from a tip-stable to a tip-unstable dendrite. Without noise, we have previously found a steady dendrite growth, if a crystal has anisotropic surface tension.[13,14] With small noise or with strong anisotropy, the growth of a tip-stable dendrite is still controlled by the anisotropic interface tension. The noise influences on the amplitude of the sidebranches, and the growth velocity v and the tip radius ρ show irregular oscillation. Even with the oscillation, the Peclet number $p = v\rho/2D$ remains almost constant, as given by the Ivantsov relation.

With strong noise or with weak anisotropy, the dendrite tip looses stability and splits. Perpetual tip splitting results in an irregular aggregate.

Similar variation of the growth modes has been found in the phase field model simulation of crystal growth.[8] With strong anisotropy, a regular dendrite is found, but when the anisotropy decreases the tip oscillation starts; for very weak anisotropy, the tip becomes unstable and splits. In this phase field model simulation the system does not contain physical noise, and crossover may be induced by the numerical noise inevitable in the algorithm. An oscillating tip was also found by Hayakawa in the simulation of Laplacian growth with eight-fold interface anisotropy.[3]

ACKNOWLEDGMENTS

This work was supported by the Grant-in-Aid for Scientific Research on Priority Areas from the Ministry of Education, Science and Culture under Contract No. 04227108.

REFERENCES

1. Brener, E. A., and V. I. Mel'nikov. "Pattern Selection in Two-Dimensional Dendritic Growth." *Adv. Phys.* **40** (1991): 53.
2. Dougherty, A., and J. P. Gollub. "Steady-State Dendritic Growth of NH_4 from Solution." *Phys. Rev. A* **38** (1988): 3043.
3. Hayakawa, Y. "Shape Selection of Free Laplacian Dendrites." Preprint, 1993.
4. Honjo, H., S. Ohta, and M. Matsushita. "Irregular Fractal-Like Crystal Growth of Ammonium Chloride." *J. Phys. Soc. Japan* **55** (1986): 2487.
5. Huang, S. C., and M. E. Glicksman. "Fundamentals of Dendritic Solidification-I and II." *Acta Metall.* **29** (1982): 701, 707.

6. Ivantsov, G. P. "The Temperature Field Around a Spherical, Cylindrical or Pointed Crystal Growing in a Cooling Solution." *Dokl. Acad. Nauk. SSSR* **58** (1947): 567.

7. Kessler, D., J. Koplik, and H. Levine. "Pattern Selection in Fingered Growth Phenomena." *Adv. Phys.* **37** (1988): 255.

8. Kobayashi, R. "Simulation of Three-Dimensional Dendrite." In *Pattern Formation in Complex Dissipative Systems*, edited by S. Kai, 121. Singapore: World Scientific, 1992. Also private communication, 1992.

9. Langer, J. S. "Instabilities and Pattern Formation in Crystal Growth." *Rev. Mod. Phys.* **52** (1980): 1.

10. Langer, J. S. "Lectures in the Theory of Pattern Formation." In *Chance and Matter*, edited by J. Souletie, J. Vannimenus, and R. Stora. Amsterdam: North Holland, 1987.

11. Matsushita, M., and Y. Yamada. "Dendritic Growth of a Single Viscous Finger Under the Influence of Linear Anisotropy." *J. Crys. Growth* **99** (1990): 161.

12. Pieters, R., and J. S. Langer. "Noise-Driven Sidebranching in the Boundary-Layer Model of Dendritic Solidification." *Phys. Rev. Lett.* **56** (1986): 1948.

13. Saito, Y., G. Goldbeck-Wood, and H. Müller-Krumbhaar. "Dendritic Crystallization: Numerical Study of the One-Sided Model." *Phys. Rev. Lett.* **58** (1987): 1541.

14. Saito, Y., G. Goldbeck-Wood, and H. Müller-Krumbhaar. "Numerical Simulation of Dendritic Growth." *Phys. Rev. A* **38** (1988): 2148.

15. Witten, T. A., and L. M. Sander. "Diffusion-Limited Aggregation." *Phys. Rev. B* **27** (1983): 5686.

H. Müller-Krumbhaar, T. Ihle, and A. Bösch
Institut für Festkörperforschung, Forschungszentrum Jülich, Postfach 1913, D–5170 Jülich, Federal Republic of Germany

Fractal and Compact Morphology of Growth Patterns

A first-order phase transition is initiated by the formation of a supercritical nucleus of the new phase inside the old phase. This nucleus then grows at the expense of the old phase. It has been an open problem so far into what kind of morphology such a nucleus evolves at long times. This problem is relevant for many types of phase transformations.[12] For the cases where diffusional transport is the rate-determining step in such a growth process a semiquantitative theory was formulated recently.[2,3] Growth patterns there were classified as dendrites or seaweed with compact or fractal appearance. The most significant parameters controlling the growth morphology accordingly are supercooling as the driving force for growth and crystalline anisotropy as a symmetry-breaking quantity. Various morphology transitions and crossover behavior were predicted. The inclusion of kinetic coefficients[2] furthermore leads to a possibly more complex series of transitions, and at large driving forces[10,7] chaotic behavior is encountered.

This theory[2,3] is based on scaling relations and asymptotic matching. Because of the speculative character of certain assumptions, we have tried to perform some numerical checks of these predictions. A previously developed diffusion-transition scheme[14,15] for the numerical simulation of diffusion-controlled growth processes already gave some indication for a transition between seaweed and dendrites (see also Figures 11 and 13 in Brener et al.[3]), but was inconclusive concerning the limit of small anisotropy. We have, therefore, constructed a number of numerical methods to handle the problem of a moving boundary in a diffusion field for the fully isotropic situation in two space dimensions. Patterns which are stationary in a frame of reference moving at constant speed can be handled efficiently by a Greens-function method.[4,5] The fully time-dependent problem, however, is treated more efficiently by a direct method. The most unbiased procedure, presumably, is a generalization of our one-dimensional interface-discretization[4,5,12] to two dimensions.[8] One defines the diffusion field on a random lattice of anchor points. These points are moving around in the plane due to some short-ranged repulsive force between them. In this way an isotropic (random) lattice of approximately equal distances can be dynamically maintained, even with a moving boundary. Making the repulsive force between anchor points a function of the diffusion field or its gradient, a locally selfadjusting grid is easily constructed dynamically. Unfortunately this procedure is not efficient on vector computers since it requires complicated interpolations between the varying random lattice points. Somewhat more efficient[1] is an implementation of a phase-field model,[9,11] where every few time steps the whole field is mapped onto a lattice which was randomly rotated against the previous lattice.[1] Again this involves interpolation, but at least the lattice is regular. Obviously, there is no anisotropy left on average. A disadvantage is that the interface thickness cannot become less than about five lattice units for precision reasons. For the problem under consideration one must keep at least three length scales clearly separated: the microscopic capillary length, the "stability" length and the diffusion length.[2,3] Therefore our preferred method for the moment is the following.[6] A sharp interface like in Brener et al.[4] and Classen et al.[5,4] is moving in four copies on four different square lattices, which are rotated at a fixed angle against each other. After each time step the four interface positions are averaged and the interface on each lattice is corrected accordingly. This procedure leaves no four- and eight-fold anisotropy and was able to reproduce free dendritic growth just like the Greens-function method.[4,5,12] A 1000×1000 lattice obviously allows for a clear separation of the three mentioned length scales.

Some basic results[6] can be summarized as follows. At dimensionless supercoolings both near zero and near one and at zero anisotropy we find that the interface evolves into a wrinkled structure but apparently it can move at constant speed. This is in agreement with the predictions.[2,3] At large supercoolings $\Delta \approx 0.7$ the pattern looks like *compact seaweed*; at small supercoolings $\Delta \approx 0.35$ the pattern looks *fractal* over some range of length scales, with a fractal dimension D_f between 1.67 and 1.73. This is consistent with the fractal dimension ≈ 1.71 for Laplacian growth in continuous space.[13] The growth rate varies depending on supercooling like $V \sim \Delta^\psi$, the exponent ψ being consistent with the prediction[2,3] $\psi = 2/(2-D_f)$. So far the agreement is quite encouraging, but of course there are also some problems. The results are still not fully conclusive how the discontinuous transition between compact seaweed and dendrites takes place in time. Also the origin of the noise in the system is not yet clear. For the moment we think that a permanent source of noise is needed in order to produce the fractal structures at small driving forces, but we certainly cannot rule out yet that the intrinsic instability of the front moving into the supercooled region is sufficient to produce intrinsic chaos even at low driving forces. We are optimistic, however, to be able to elucidate at least a few more details in the near future.

REFERENCES

1. Bösch, A., H. Müller-Krumbhar, and O. Shochet. Unpublished.
2. Brener, E., H. Müller-Krumbhaar, and D. Temkin. *Europhys. Lett.* **17** (1992): 535.
3. Brener, E., K. Kassner, H. Müller-Krumbhaar, and D. Temkin. *Int. J. Mod. Phys. C* **3** (1992): 825.
4. Brener, E. H. Müller-Krumbhaar, Y. Saito, and D. Temkin. *Phys. Rev. E* **47** (1993): 1151.
5. Classen, A., C. Misbah, H. Müller-Krumbhaar, and Y. Saito. *Phys. Rev. A* **43** (1991): 6920.
6. Ihle, T., and H. Müller-Krumbhaar. *Phys. Rev. Lett.* **70** (1993): 3083.
7. Kassner, K., C. Misbah, and H. Müller-Krumbhaar. *Phys. Rev. Lett.* **67** (1991): 1551.
8. Kassner, K. "Numerical Simulation of Crystal Growth." In *Informatik-Fachberichte*, edited by D. Krönig and M. Lang, vol. 306, 259. Berlin: Springer-Verlag, 1992.
9. Kupferman, R., O. Shochet, E. Ben-Jacob, and Z. Schuss. *Phys. Rev. B* **46** (1992): 16045.
10. Misbah, C., H. Müller-Krumbhaar, and D. Temkin. *J. Phys. (France) I* **1** (1991): 585
11. Müller-Krumbhaar, H. *Phys. Rev. B* **10** (1974): 1308.
12. Müller-Krumbhaar, H., and W. Kurz. In *Phase Transformations in Materials*, edited by P. Haasen. Weinheim: VCH-Verlag, 1991; see also related articles in this volume.
13. Ossadnik, P. *Phys. Rev. A* **45** (1992): 1058.
14. Shochet, O., K. Kassner, E. Ben-Jacob, S.G. Lipson, and H. Müller-Krumbhaar. *Physica A* **181** (1992): 136.
15. Shochet, O., K. Kassner, E. Ben-Jacob, S.G. Lipson, and H. Müller-Krumbhaar. *Physica A* **187** (1992): 87.

R. Pratibha and N. V. Madhusudana
Raman Research Institute, Bangalore 560080, India

Pattern Formation in the Growth of Smectic A Liquid Crystals in Some Binary Mixtures

We report the formation of several unusual patterns in the growth of smectic A liquid crystals from the isotropic phase in some binary mixtures. If the composition is such that the interfacial tension anisotropy ($\Delta\gamma$) is weakly positive, the smectic A separates as ellipsoidal drops with flat layers. As the temperature is lowered, changes in the composition lead to a reversal of $\Delta\gamma$ and to the growth of a spherical cap at one end. Some spherical drops appear to have a variable angle between the layer normal and the interface, implying that $\Delta\gamma \simeq 0$. Flat drops which have concentric smectic A layers develop interfacial instabilities as the temperature is lowered. In some cases, a transformation of cylindrical or spherical structures to discs is observed.

1. INTRODUCTION

The smectic A liquid crystal consists of a periodic stacking of liquid layers.[3] The director is parallel to the layer normal and the medium is uniaxial. The lamellar phase which is exhibited in some concentration range of solutions of amphiphilic molecules

Spatio-Temporal Patterns, Ed. P. E. Cladis and P. Palffy-Muhoray,
SFI Studies in the Sciences of Complexity, Addison-Wesley, 1995

in water and/or alcohol also has the same symmetry. The growth of smectic liquid crystals from the isotropic phase was studied by Friedel and Grandjean long ago. They found that the A phase separates in the form of often highly decorated rodlike structures called batonnets. The structure arises because of the positive interfacial tension anisotropy in these materials, i.e., $\gamma_{\parallel} > \gamma_{\perp}$, where the subscripts refer to the mutual orientation of the interface and the layers. Such systems were studied in great detail by Fournier and Durand[4] who have also shown that the focal conic domains in the batonnets are in thermodynamic equilibrium.

Recently there have been several studies[1,2,5,6,7] of highly elongated cylindrical growths of smectic A liquid crystals that result when the interfacial tension anisotropy is negative. This type of growth occurs efficiently by absorption of molecules without nucleation of new layers. While, in the case of single component systems, these structures are metastable,[5] we have shown that in the case of two component systems, concentration gradients can actually stabilise such cylindrical structures.[6,7] There are some similarities between these cylindrical structures and the Myelene forms that are found in lyotropic lamellar systems. In the present paper we report some novel structures which occur when $\Delta\gamma$ is around zero.

2. OBSERVATIONS

We have studied in some detail the binary system consisting of octyloxy cyanobiphenyl and dodecyl alcohol. For low percentages of alcohol ($< 30\%$, say), the smectic A liquid crystal separates in the form of batonnets mentioned above. For $> 50\%$ of alcohol, we get the elongated cylindrical structures. In the following we discuss the growth of smectic A liquid crystals for intermediate compositions in which the interfacial tension anisotropy is quite small. When we cool a mixture with an alcohol concentration of $\sim 34\%$ at $0.1°C/min.$, the smectic A drops that separate are not much elongated, though they are made of focal conic domains. (See Figure 1(a).) For higher percentages ($\sim 40\%$) of alcohol, the smectic A separates usually as elliptical drops with flat layers or occasionally as spherulites (Figure 1(b)). The elliptic shape arises due to the positive interfacial tension anisotropy, but, unlike in mixtures with lower concentrations of alcohol, these are not decorated with focal conics. It is clear that in the coexistence range of smectic A and isotropic phases in this two component system, the smectic liquid crystal which initially separates from the isotropic phase will have a lower concentration of alcohol than the average and, hence, has positive interfacial tension anisotropy. But as the temperature is lowered and the smectic regions grow, the alcohol concentration in the isotropic phase increases and the interfacial tension anisotropy of the newly condensing smectic A liquid crystal progressively reduces and can even change sign, and then becomes increasingly negative. A variety of structures are formed in this regime. We describe them below.

a. The elliptic object develops a fine line all along the major axis. Then a hemispherical cap develops at one end. The orientation of the director can be easily visualised by dissolving anisotropic and dichroic dye molecules in the medium and making observations with a polarizer. (See Figure 2.) Apparently as $\Delta\gamma$ becomes increasingly negative, the layer at one end can fluctuate to develop a *hump*, which will not have an unfavourable energy. Layers now grow around this hump to form the structures shown in Figure 2(c) and 2(d). In one of the elliptical drops in Figure 2(a), a point defect is seen near one of the foci. The dark line in this case extends from this point to the farther periphery of the drop and a spherical structure develops at the latter end (Figure 2(d)) giving the appearance of a round-bottomed bottle. These observations were made between two glass plates which were treated with a polyimide solution (but not rubbed), and without any spacers. This treatment favours an orientation of the molecules parallel to the glass plates. Similar structures were found in 50 μm thick cells, but in that case the drops were not isolated as two or three of them would lie along any particular line of sight.

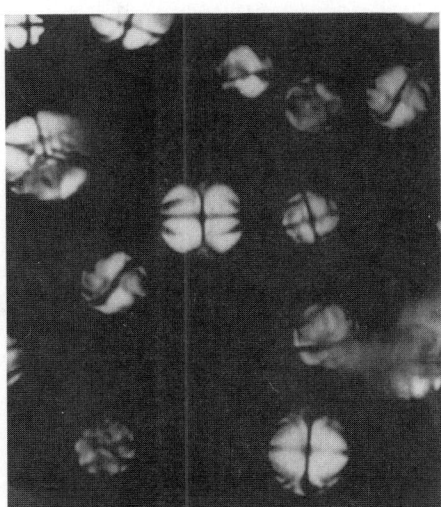 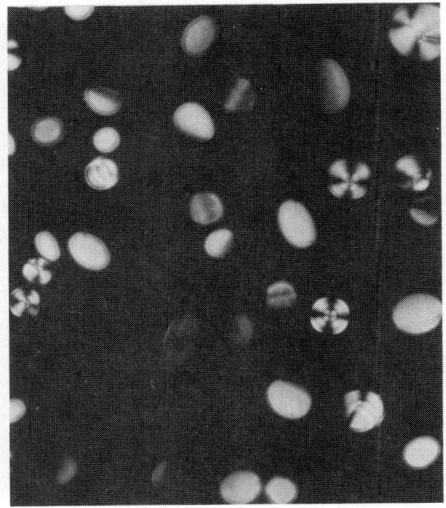

FIGURE 1 (a) (left) Smectic A drops made up of focal conic domains in a mixture with $\sim$ 34% of dodecyl alcohol. Sample thickness $\simeq$ 50 μ, crossed polarizers ($\times$400).
(b) (right) Elliptic drops of S_A with flat layers in a mixture with $\sim$ 40% of alcohol. Crossed polarizers ($\times$300).

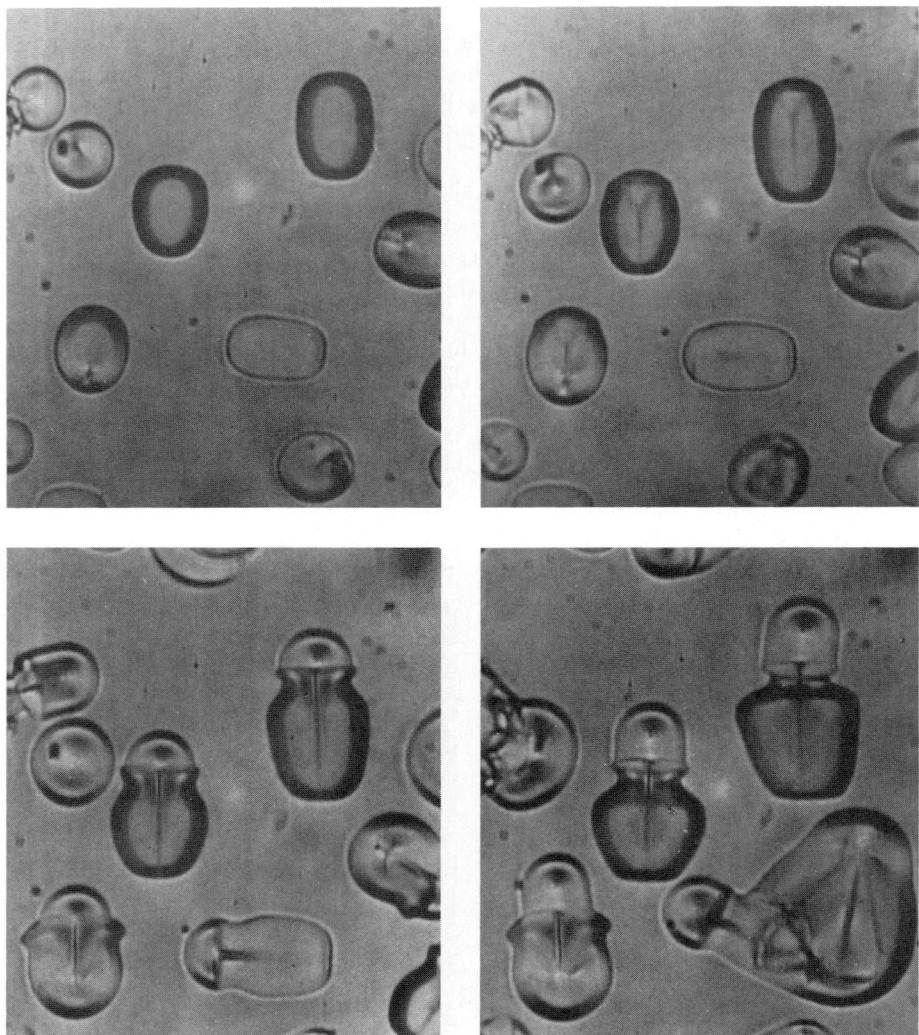

FIGURE 2 Smectic drops in a mixture with $\sim$ 40% of alcohol and a polariser set parallel to the vertical edge of the photograph ($\times 400$). (a) (upper left) sample at $44°$C; (b) (upper right) at $42.4°$C, the elliptic drops have developed fine lines; (c) (lower left) at $41.7°$C, well-developed hemispherical *caps* are seen at one end of the drops; (d) (lower right) at $40.7°$C, the caps have slightly elongated. Note the round-bottomed shape of one of the drops.

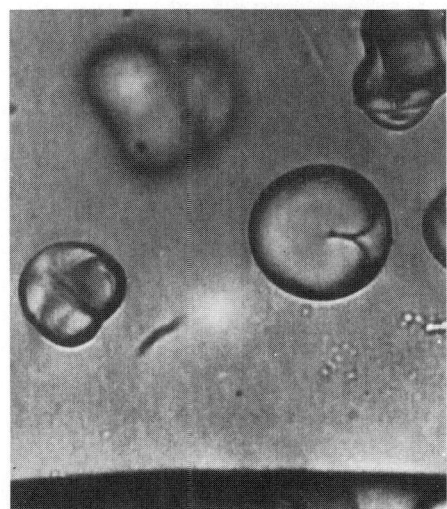

FIGURE 3 A nearly spherical drop with a radial line in a mixture with ~ 42% of alcohol. (a) (left) A polariser is set perpendicular to the radial line. (×400); (b) (right) with crossed polarisers, only two dark brushes are seen in a direction parallel to the radial line.

b. If the alcohol content in the mixture is slightly higher, ~ 42%, the smectic A drop grows with a nearly spherical shape, but with a radial line defect (Figure 3). Between crossed polarizers set parallel and perpendicular to the line, only two dark brushes are seen (Figure 3(b)). The spherical shape shows that $\Delta\gamma$ may be close to zero in this case. Further, in a few drops which have defects whose location could be monitored, we have observed that the drops continuously *rotate* as they grow in size with the lowering of temperature. In one case we could observe the rotation for more than one hour.

c. In the case of a mixture with an alcohol content of about 45% taken between two plates treated with polyimide, often the A phase grows in the form of circular discs, in which the layers are concentrically arranged implying that $\Delta\gamma < 0$. As the temperature is lowered, the drops develop interfacial instabilities which may have 3-fold and higher (up to 20-fold) symmetry (Figures 4(a) and 4(b)). The process appears to be somewhat analogous to the case (a) discussed earlier: *humps* formed on the surface giving rise to radial lines. The alcohol concentration in the newly condensing material being greater, the humps usually grow into cylindrical stubs (Figures 5(c) and 5(d)). Similar growth processes have been seen in thicker (~ 50 μm) cells.

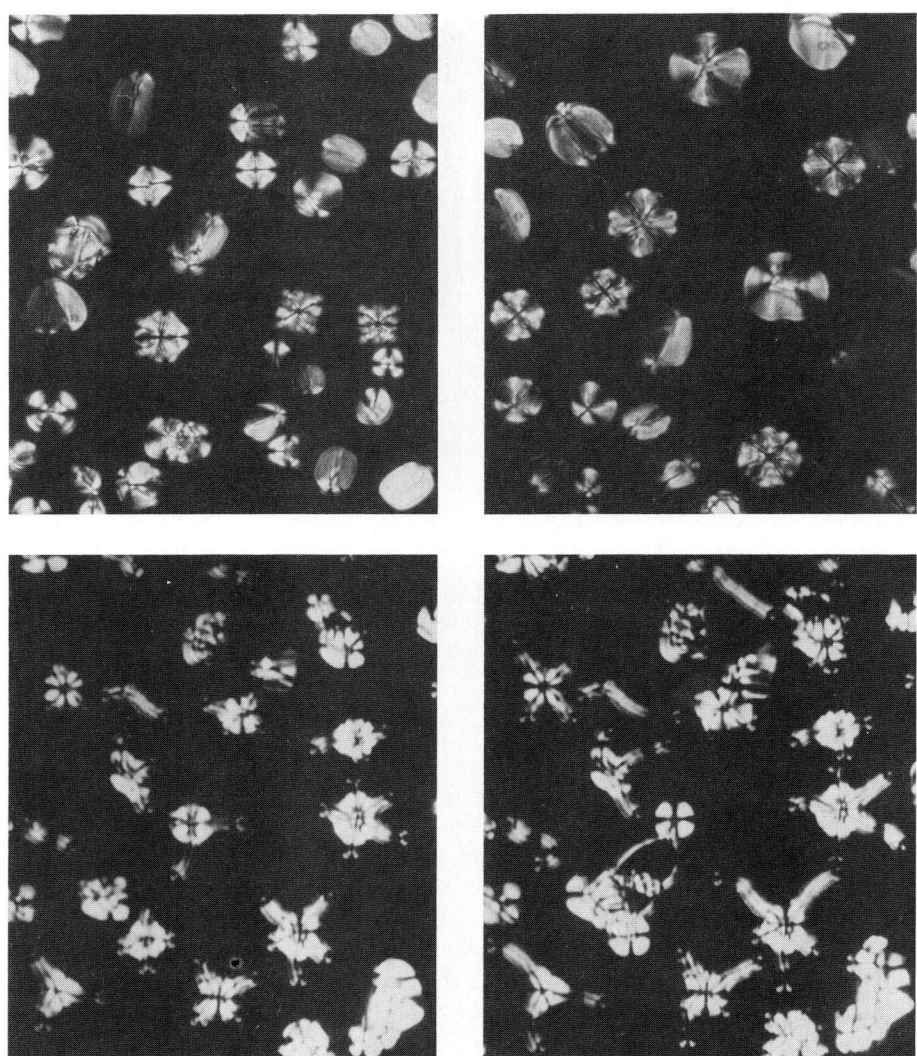

FIGURE 4   Smectic A drops that have developed multifold interfacial instabilities in a mixture with ~45% of alcohol. (a) and (b) (upper left and right) show two regions between crossed polarizers (× 300); and (c) and (d) (lower left and right) at lower temperatures, cylindrical growths are seen from such drops. In (d) a cylinder has split to form a disclike structure.

d.   Often one of the cylinders elongates considerably and at some stage *splits* to form a disclike object which encloses well-aligned smectic layers (Figure 4(d)). As this structure forms, the material of the disc obviously comes from the main drop and simultaneously all the other cylindrical stubs disappear.

e.   In a thick cell filled with a mixture having an alcohol concentration of $\sim 47\%$, spherical drops with one cylindrical tail can often be seen on maintaining the temperature for several hours. On a slight warming of the sample, the cylinder splits to form a disc whose rim becomes circular (Figure 5(a) and 5(b)). Eventually the disc shrinks to form a sphere.

f.   When the alcohol concentration is even higher, say, $\sim 50\%$, $\Delta\gamma$ will be sufficiently negative and, as we mentioned earlier, long cylindrical structures grow as the temperature is lowered. As the alcohol content in the isotropic phase progressively increases, the negative concentration gradient needed for stabilising the cylinder cannot be sustained and the cylinders collapse to form compact objects.[6] If relatively thick cylinders collapse slowly at a fixed temperature,

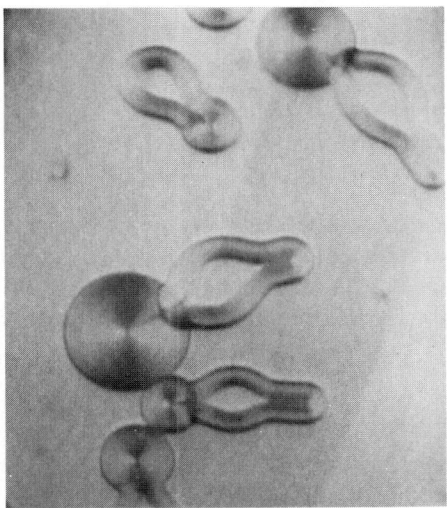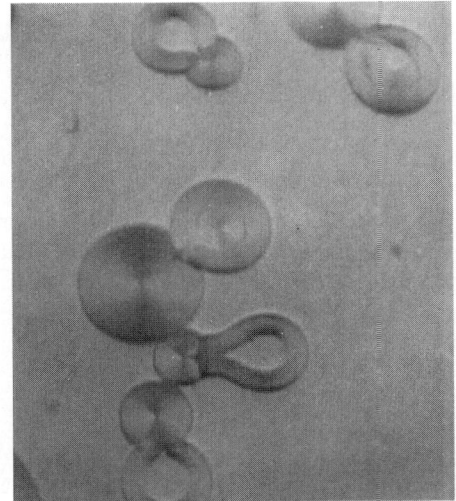

FIGURE 5   (a) (left) Spherical drops with cylindrical tails which have just split to form discs in a mixture with $\sim 47\%$ of alcohol. A polariser is set parallel to the vertical edge of the photograph ($\times 400$); (b) (right) about half a minute later, the rims of the discs have become more circular in shape.

they often form objects that consist of a large spherical object with one section chopped and, hence, forming a flat surface and a smaller spherical drop attached to it (Figure 6(a)). As time progresses, the small sphere grows at the expense of the large one whose shape changes to form a disc (Figure 6(b) and 6(c)). The disc later shrinks to form a smaller sphere which is finally swallowed up by the large sphere. In some cases, large spherical objects with one side *flattened* are also seen (Figure 6(d)).

3. DISCUSSION

The batonnets with focal conic domains studied by Friedel and Grandjean and more recently by Fournier and Durand minimise the total energy which has contributions from the curvature in the bulk and the interfacial energy with the constraint that smectic layers are orthogonal to the interface. In crystals the shapes are usually decided by dynamical processes and may not correspond to true minimisation of the interfacial energy. In smectic A which has a one-dimensional crystalline order, it is easier to attain equilibrium shapes. Fournier and Durand[4] have shown that the focal conics occurring in batonnets are indeed in an equilibrium configuration.

In our experiments, in which we add considerable quantities of a long chain alcohol, we can change the interfacial tension anisotropy. Indeed, if our starting composition is appropriate, the anisotropy can change sign during the growth of the smectic liquid crystals. This produces the variety of patterns that we have observed in the present study.

In systems that give rise to batonnets decorated with focal conic domains, the energy of the focal conic arises from curvature elasticity which increases linearly with size $[(\sim K/L^2) \cdot L^3 \propto L]$. (The presence of disclinations gives rise to only a logarithmic correction.) The interfacial energy is $\propto \bar{\gamma} L^2$. Hence, if there is anisotropy of γ ($\Delta\gamma > 0$), focal conic domains occur within the elongated structures for a size[4]

$$L > l_o \sim \frac{K}{\bar{\gamma} - \gamma_\perp} \sim \frac{K}{\Delta\gamma}. \tag{1}$$

Hence, if $\Delta\gamma$ is very small, we can have very large domains without focal conic defects as has been found in our samples (Figure 1(b)). This also probably means that both $\gamma_\parallel$ and $\gamma_\perp$ are considerably smaller in the mixture than in the pure component. Reduction of the interfacial tension by impurities is a well-known phenomenon.

As the smectic liquid crystals condense with lowering of temperature, the concentration of alcohol in the isotropic liquid increases. As a consequence, the alcohol

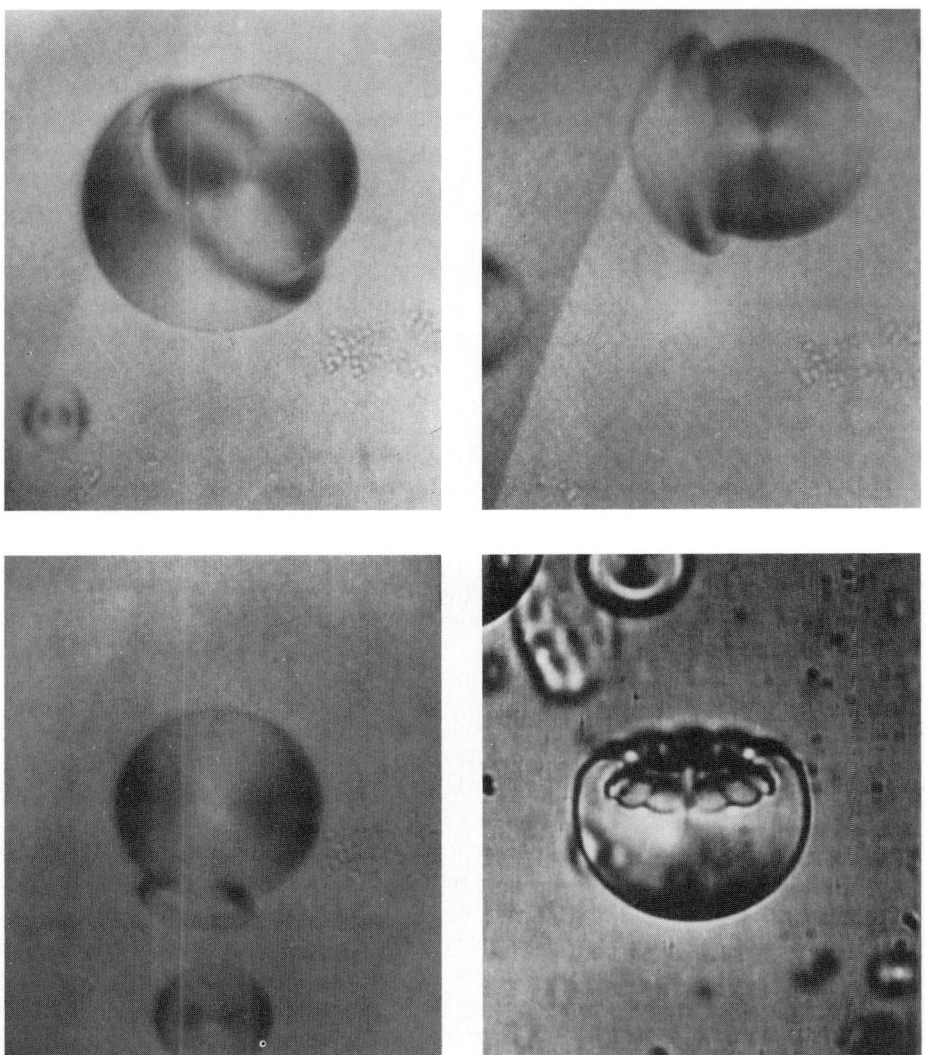

FIGURE 6   Spherical drops in samples with ∼ 50% alcohol with flattened portions.
(a) (upper left) A thick cylinder has slowly collapsed to form a pair of drops; (b) (upper right) after 35 minutes the smaller sphere has grown at the expense of the larger one; (c) (lower left) after another 15 minutes, the latter has shrunk to a disc; (d) (lower right) an isolated large drop with a flattened portion.

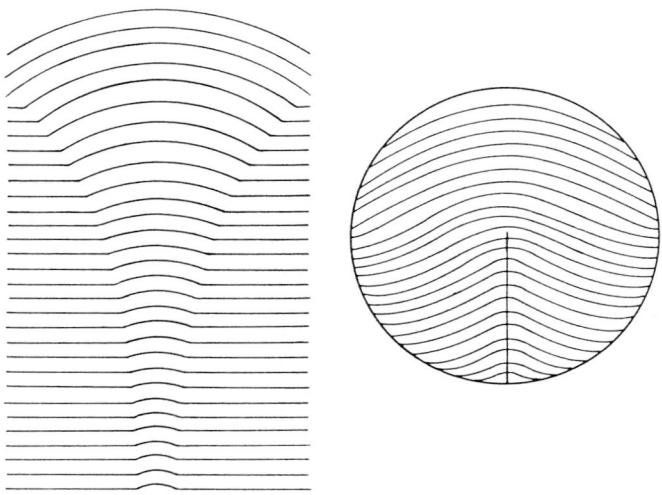

FIGURE 7   (a) (left) Schematic diagram of the arrangement of layers in smectic drops of the type shown in Figure 2(c). (b) (right) Schematic arrangement of layers in spherical smectic drops with a radial line seen in Figure 3.

content of newly condensing smectic A will also be larger and the interfacial tension anisotropy can eventually change sign. This favours the growth of smectic such that the layers are parallel to the interface. In the elongated and uniformly aligned drops, the process starts by the formation of a *hump* in the centre of a layer at one end. The strong anisotropy of elasticity of smectic phase (the compression coefficient $B \gg$ the curvature elastic constant K) results in the *hump* propagating through the layers to the other end of the drop (Figure 7(a)). The formation of the hump is, of course, facilitated by the high concentration of alcohol in the top layers, which also lead to a swelling of the layers. Indeed, X-ray measurements indicate that the layer spacing for mixtures with $\sim 40\%$ alcohol is about 10% larger than that in pure 8OCB. The new layers can now grow around the hump with spherical symmetry.

Qualitatively we can see that the formation of a hump can actually reduce the energy of the drop. For the sake of argument, assuming that the original drop consisted of a cylinder of length L and radius R of perfectly flat layers, if a hemi-spherical hump with a radius r develops such that the area of the layer is unaltered, the new lateral radius R' is given by

$$\pi R'^2 + 2\pi r^2 = \pi R^2. \tag{2}$$

The change in energy is

$$\Delta E \simeq 2\pi \gamma_\perp L (R' - R) + \frac{K}{2r^2} \cdot \pi r^2 L \tag{3}$$

which becomes negative for

$$r \geq \sqrt{\frac{K}{\gamma_\perp}} R. \tag{4}$$

For $K \sim 10^{-6}$ dyne, $\gamma_\perp \sim 10^{-2}$ dyne/cm and $R \simeq 10\mu$m, $r \simeq 3\mu$m.

Of course the energy will be lower for larger r. This leads to the formation of a *neck* near the top layers as seen in Figure 2. The final shape as seen in Figure 2(d) implies that the *flat* layers which are orthogonal to the interface grow at a faster rate than those that make smaller angles.

The other unusual object seen in our study is a spherical drop with a radial line when $\Delta\gamma \simeq 0$. The experimental observations indicate that the layers are arranged as shown in Figure 7(b). As we mentioned earlier, we see that many drops in which the line appears to be along the direction of observation rotate continuously as they grow in size. We feel that the radial line may also incorporate a screw dislocation with a relatively large Burgers vector which may be responsible for the rotation of the drop.

We have also seen that in some cases, at constant temperature some cylinders can split to form disclike objects sucking the material from the cylindrical stubs protruding from a parent spherical object. As we have argued elsewhere,[6] concentration gradients can stabilise a cylinder with respect to a sphere. As the concentration of alcohol in the surrounding liquid increases, the required type of gradient cannot be sustained. In this case the cylinder would usually be expected to collapse to a sphere. We can now compare the energy of a disc compared to that of a sphere with the same volume. A disc with lateral dimension of $2R$ and thickness $2r$ has an energy

$$F^{\text{disc}} = 2\pi\gamma_\|(R^2 + \pi rR) + (2\pi R)\frac{\pi K}{8}\ln\frac{r}{a} \tag{5}$$

where a is of the core radius of the line dislocation of strength $+1/2$ around the periphery and is of the order of a molecular dimension.

A sphere with the same volume will have a radius R_{sp} such that

$$R_{\text{sp}} = [(3/4)rR(2R + \pi r)]^{1/3}$$

and an energy

$$F^{\text{sp}} = 8\pi K R_{\text{sp}} + \gamma_\| 4\pi R_{\text{sp}}^2.$$

We can compare the energies of the two objects for the usual material parameters; viz., $K \sim 10^{-6}$ dyne, $\gamma_\| \sim 10^{-2}$ dyne/cm for different values of R and r. Calculations show that for a given value of r, $F^{\text{disc}} < F^{\text{sp}}$ for R less than some value R_o say. It is found that, for example, if $r \simeq 1$ μm, $R_o \sim 2.5$ μm. It is also seen that r/R_o decreases as r increases. If $r \simeq 4$ μm, $R_o \simeq 5.5$ μm. Hence, it would appear that thin discs, which form when a cylinder splits, become thicker by adding layers and, hence, finally become unstable with respect to spheres, as seen in the experiments.

We have also observed other structures in these systems. A more detailed paper on these structures will be published in due course.

REFERENCES

1. Adamczyk, A. "Phase Transitions in Freely Suspended Smectic Droplets. Cotton-Mouton Technique, Architecture of Droplets and Formation of Nematoids." *Mol. Cryst. Liq. Cryst.* **170** (1989): 53.
2. Arora, S. L., P. Palffy-Muhoray, and R. A. Vora. "Reentrant Phenomena in Cyano Substituted Biphenyl Esters Containing Flexible Spacers." *Liq. Cryst.* **5** (1989): 133.
3. de Gennes, P. G. *The Physics of Liquid Crystals.* Oxford: Oxford University Press, 1974.
4. Fournier, J. B., and G. Durand. "Focal Conic Faceting in Smectic-A Liquid Crystals." *J. Physique II* **1** (1991): 845.
5. Palffy-Muhoray, P., B. Bergersen, H. Lin, R. B. Meyer, and Z. Racz. "Filaments in Liquid Crystals: Structure and Dynamics." In *Pattern Formation in Complex Dissipative Systems*, edited by S. Kai, 504. Singapore: World Scientific, 1992.
6. Pratibha, R., and N. V. Madhusudana. "Cylindrical Growth of Smectic A Liquid Crystals from the Isotropic Phase in Some Binary Mixtures." *J. Physique II* **2** (1992): 383.
7. Pratibha, R., and N. V. Madhusudana. "Unusual Growth of Smectic A Liquid Crystals." *Curr. Sci.* **62** (1992): 419.

Lihong Pan and John R. de Bruyn
Department of Physics, Memorial University of Newfoundland, St. John's, Newfoundland,
Canada, A1B 3X7

Traveling Cellular Patterns and Parity Breaking at a Driven Interface

We have studied the parity-breaking bifurcation that occurs in the fluid-dynamical system known as the printer's instability. At this bifurcation, the pattern of fingers that forms at a driven fluid-air interface loses its reflection symmetry and starts to drift at constant speed. We quantify the asymmetry of the drifting fingers by analyzing video images of the pattern, and find that the drift speed is proportional to the asymmetry, in agreement with recent theoretical predictions.

I. INTRODUCTION

Many laboratory dynamical systems undergo instabilities that lead to the formation of one-dimensional patterns. Well-known examples include Rayleigh-Bénard convection in pure fluids and in binary mixtures, Taylor-Couette and Taylor-Dean flow, and directional solidification. The simplicity of these systems—relative to fully three-dimensional systems—allows meaningful comparison of experimental results with theoretical or numerical predictions, and a quantitative understanding of several one-dimensional pattern-forming systems has been achieved.

In all of these systems, the basic one-dimensional pattern itself is subject to secondary instabilities as the experimental control parameter is varied. These secondary instabilities can lead to complex and interesting, but still essentially one-dimensional, dynamical behavior. Insight into many aspects of the dynamics of one-dimensional patterns can be gained by a consideration of the patterns' symmetry properties. Coullet and Iooss[3] identified ten different symmetry-breaking instabilities that can affect a stationary, one-dimensional pattern. Of particular interest for the work presented here is the parity-breaking instability, at which the pattern's reflection, or parity, symmetry is broken. Such broken-parity patterns will propagate, with a speed proportional to the order parameter that characterizes the parity breaking.

Broken-parity states have been observed in many experimental systems. Localized regions of broken parity, that propagate through an otherwise stationary and symmetric pattern, were observed in experiments on the Mullins-Sekerka instability at the nematic-isotropic transition in liquid crystals.[7,18] Similar behavior has been reported in Taylor vortex flow,[23] the printer's instability,[13,19,20] and other systems. Extended, rather than localized, regions of broken parity have been studied in the directional solidification of succinonitrile,[8] and of laminar eutectic mixtures,[5] and in experiments on Taylor-Dean flow.[16]

Drifting patterns resulting from broken parity symmetry were first considered theoretically by Malomed and Tribelsky,[12] who found that a stationary, periodic pattern with wave number q drifts with constant velocity when the damping of the $2q$ mode is weak. This $q - 2q$ coupling has been studied by several groups.[1,6,11,21,22] Goldstein and coworkers[2,9] have also studied the parity-breaking transition, using a model that does not rely on any specific mode-coupling mechanism. In all cases the results of interest in this work are the same. If we represent our one-dimensional pattern by a function of space, y, and time, t,

$$U(y,t) = S(y,t)U_S\left(y + \phi(y,t)\right) + A(y,t)U_A\left(y + \phi(y,t)\right), \tag{1}$$

where $U_S(x)$ and $U_A(x)$ are even and odd functions of their arguments, respectively, then the amplitude A of the antisymmetric (or odd) term can be taken as the order parameter of the parity breaking. Here ϕ is a variable that gives the deviation of the pattern's phase from that of the underlying symmetric pattern. At a supercritical parity-breaking transition, the asymmetry is expected to grow continuously from zero with a square-root dependence on the relevant control parameter, μ:

$$A \propto \mu^{1/2}. \tag{2}$$

Coupling between the asymmetry amplitude A and the phase variable ϕ (or, equivalently, between modes of wave number q and $2q$) leads to a steady drift of the pattern with a phase velocity v_ϕ given by

$$v_\phi = \frac{\partial \phi}{\partial t} = \omega A + \frac{\partial^2 \phi}{\partial x^2} + \dots, \tag{3}$$

where ω is a coupling constant. From Eqs. (2) and (3) we therefore also expect the drift velocity to grow like the square root of the control parameter.

Our own work[17] concerns broken-parity waves in a fluid dynamical system known as the printer's instability. Many aspects of pattern formation in this system have been studied by Rabaud and coworkers.[4,10,13,14,15,19,20] The experimental apparatus consists of two horizontal cylinders, mounted eccentrically, one inside the other. The gap between the cylinders is narrowest at the bottom of the apparatus, and this nip region is filled with a viscous oil. The patterns form at the oil-air interface running the length of the cylinders, as it is driven by rotating one or both of the cylinders about their axes. The dynamical phase diagram for our system is shown in Figure 1, in the space defined by the two control parameters v_i and v_o, the inner and outer cylinder surface speeds, respectively. Initially the interface is straight. When one of the two cylinders rotates, there is a rotation speed above which the straight interface becomes unstable and a pattern of stationary fingers develops.[10] When both cylinders rotate, the finger pattern displays a number of interesting dynamical states.[20] When the two cylinders corotate, the pattern is spatio-temporally intermittent.[15] When they counter-rotate,[20] depending on the relative rotation velocities of the two cylinders, localized regions of propagating broken-parity fingers, moving through the stationary pattern; symmetric stationary fingers; and spatially uniform patterns of traveling fingers with broken reflection symmetry[17] can be observed.

Here we present some results from a study of the uniform traveling-finger state.[17] In our experiments, we fixed the rotation speed of the outer cylinder (v_o),

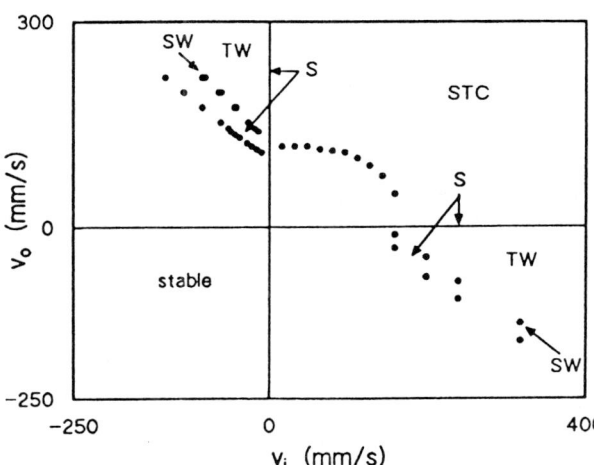

FIGURE 1 Diagram showing the dynamical states of the pattern that develops at the oil-air interface. S: stationary fingers; TW: traveling waves; STC: spatio-temporal chaos; SW: solitary waves.

and increased the speed of the inner cylinder (v_i) from zero in the opposite directon. We observe that the traveling state appears via a supercritical parity-breaking transition. We have measured the asymmetry of the traveling fingers, along with the pattern's traveling speed, and so are able to confirm the predictions of Eqs. (2) and (3) above.

II. EXPERIMENT

The two cylinders which formed the heart of our apparatus were mounted accurately parallel and horizontal, but vertically offset, such that the cross-section of the apparatus was as in Figure 2. The inner cylinder, made of Delrin, had a radius $r_1 = 50.4$ mm and length $l_1 = 202$ mm. The transparent outer cylinder, made of Plexiglas, had radius $r_2 = 66.7$ mm and length $l_2 = 210$ mm. Annular end caps on the outer cylinder contained the experimental fluid. The cylinders could be independently rotated about their axes by two computer-controlled microstepping motors, with a minimum increment in rotation speed of approximately 0.1 mm/s. In the present experiments the minimum width of the gap between the cylinders was 0.5 mm, set with micrometer screws. The stability of the stationary fingering pattern observed when only one cylinder rotated was very sensitive to the parallelism of the cylinder axes; this fact was used to optimize the cylinder alignment.

A small amount of silicone oil—enough to keep the nip region between the rotating cylinders full—was poured into the space between the cylinders. The experiments reported here were done using an oil (Aldrich Chemical Co., catalog

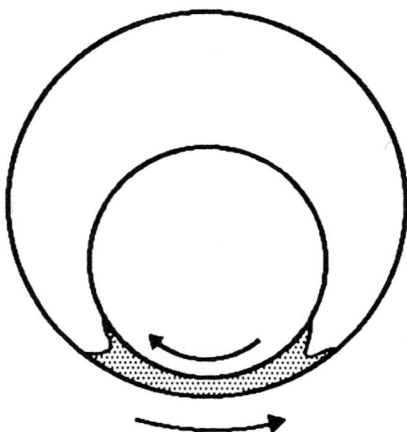

FIGURE 2   Cross-sectional view of the experimental apparatus.

no. 14,615-3) with viscosity $\mu = 0.525$ g/cm s, surface tension $\sigma = 21.8$ g/s, and density $\rho = 0.963$ g/cm^3 at room temperature. The oil-air interface at the front of the apparatus was monitored with a CCD video camera and monitor, and data were recorded on a VCR or stored on a personal computer using a video frame grabber.

The broken-parity traveling-wave state exists in the areas labelled TW in Figure 1. Experimental runs followed horizontal lines in the second quadrant of Figure 1, with v_o fixed at a selected value above the onset of the stationary fingering pattern. When that pattern had stabilized, v_i was increased in small steps and the system allowed to settle down. The fingering pattern was recorded for later analysis, and v_i further increased.

III. DATA ANALYSIS

Figure 3 shows the pattern that develops at the oil-air interface as v_i is increased, for $v_o = 139.4$ mm/s. Figure 3(a), corresponding to $v_i = 0$, shows the stationary, reflection-symmetric pattern observed when only one cylinder rotates. As v_i is increased, the pattern's reflection symmetry is broken, and the entire pattern, which now consists of fingers skewed slightly to one side, drifts with constant velocity along the length of the apparatus (Figure 3(b)). (Very close to the onset of the broken-parity state, we observe a state of many small regions of drifting fingers separated by defects.[13] At low v_o this state is a transient and eventually settles down to a uniform pattern of traveling fingers. At higher v_o, however, this disordered state appears to be stable. An instability of the broken-parity state at its onset is predicted by Caroli et al.[1] and Fauve et al.[6].) As v_i is further increased, the fingers become more and more skewed, as illustrated in Figures 3(b)–(f), and their drift velocity increases. Eventually, a value of v_i is reached at which there is a discontinuous, hysteretic transition from the traveling state back to a state of symmetric, stationary fingers, shown in Figure 3(g). As v_i is increased further, the stationary fingers become smaller, and finally the straight meniscus state returns. At higher values of v_o, the sequence of events as v_i increases is somewhat more complicated, as will be described in the next section.

The behavior of these traveling fingers is qualitatively what one would expect from the theoretical work on broken-parity states.[2,6,9] We would like, however, to make a quantitative comparison with the theoretical predictions; to do this we need a consistent way of calculating the asymmetry of the pattern. Following Goldstein et al.,[2,9] we describe the cellular pattern at the oil-air interface by a function $U(y,t)$, as in Eq. (1):

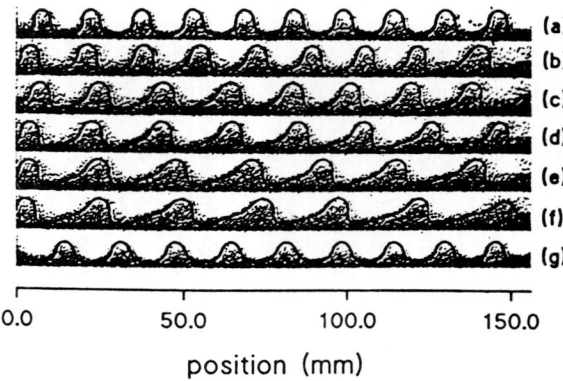

0.0   50.0   100.0   150.0

position (mm)

FIGURE 3   Patterns at the oil-air interface at $v_o = 139.4$ mm/s. (a) Stationary, symmetric fingers at $v_i = 0$; (b)–(f) asymmetric fingers moving to the left at (b) $v_i = 6.3$ mm/s; (c) $v_i = 7.9$ mm/s; (d) $v_i = 9.5$ mm/s; (e) $v_i = 11.1$ mm/s; (f) $v_i = 12.7$ mm/s; (g) stationary, symmetric fingers at $v_i = 15.8$ mm/s.

$$U(y,t) = S U_S(y + \phi(y,t)) + A U_A(y + \phi(y,t)). \tag{4}$$

Here the coefficients of the symmetric and antisymmetric terms, S and A respectively, and are assumed to be independent of y and t. $U(y)$ represents the length of the fingers at the oil-air interface, as a function of position along the apparatus.

Assuming that the interface pattern at a particular time corresponds to a single-valued function $U(y)$, as it does for the patterns shown in Figure 3, we can expand $U(y)$ as a Fourier series:

$$U(y) = \sum_{j=1}^{\infty} a_j \cos jqy + \sum_{j=1}^{\infty} b_j \sin jqy, \tag{5}$$

where q is a wave number. We are free to define the point of zero phase as we see fit. Since a pattern described by a perfect cosine wave should be perfectly symmetric, we choose our zero point so that the coefficient $b_1 = 0$, and so we consider the pattern to be of the form

$$U(y) = a_1 \cos qy + \sum_{j=2}^{\infty} a_j \cos jqy + \sum_{j=2}^{\infty} b_j \sin jqy, \tag{6}$$

i.e., a cosine wave plus higher wave number components. The total power in the Fourier spectrum is independent of the above choice of phase and is equal to $\sum(|a_j|^2 + |b_j|^2)$. We define the asymmetry parameter $\mathcal{A}$ as the square root of the fraction of the total power contained in the antisymmetric terms:

$$\mathcal{A} = \left(\frac{\sum |b_j|^2}{\sum (|a_j|^2 + |b_j|^2)} \right)^{1/2}. \tag{7}$$

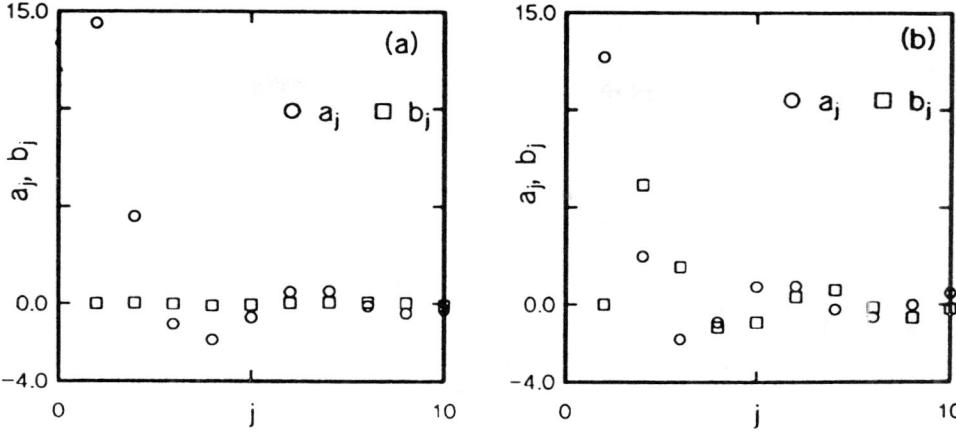

FIGURE 4 The Fourier amplitudes a_j and b_j obtained from a Fourier transform of (a) the stationary pattern shown in Figure 3(a), and (b) the traveling pattern shown in Figure 3(e).

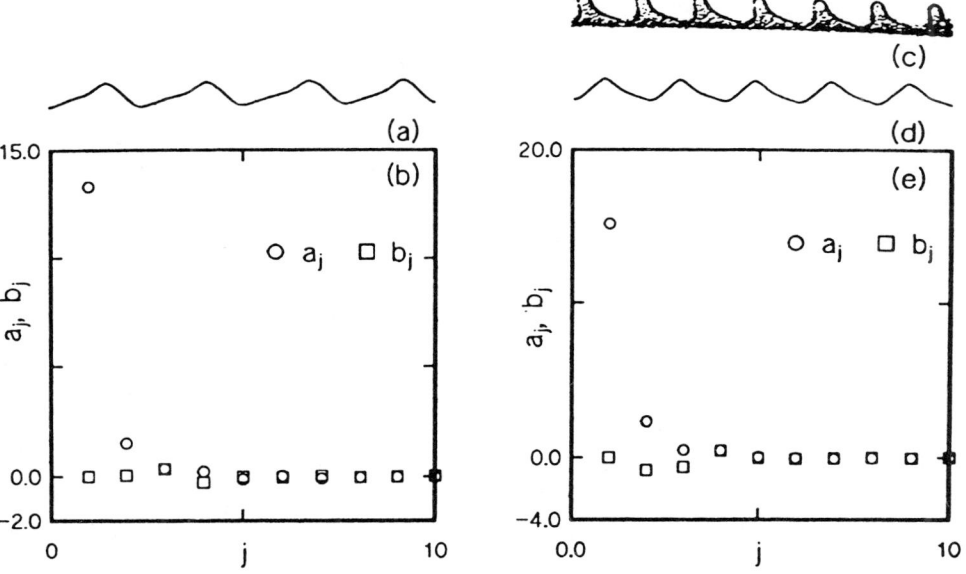

FIGURE 5 (a) The pattern of Figure 3(e) transformed into a function $U(\ell)$ of the length along the interface. (b) The Fourier amplitudes of the transformed pattern. (c) A multiple-valued interface pattern. (d) The pattern of (c) transformed as in (a). (e) The Fourier amplitudes of the transformed pattern.

This quantity is proportional to the coefficient A of the antisymmetric term in Eq. (1) above, as long as A/S (or $\mathcal{A}$) is small.

To determine $\mathcal{A}$ for a given interface pattern, we analyze the digitized video image of the pattern. A computer program is used to semi-automatically trace the location of the interface. The function $U(y)$ obtained in this way is Fourier transformed for each finger individually and the results averaged over the pattern. In Figure 4(a) we show the Fourier amplitudes determined by applying this procedure to the symmetric, stationary finger pattern of Figure 3(a). The amplitudes b_j of the asymmetric contributions are all close to zero; Eq. (7) gives $\mathcal{A} = 0.02 \pm 0.03$ for this pattern. In Figure 4(b) we show analogous results for the drifting pattern of Figure 3(e). The b_j are now clearly nonzero, with the amplitude of the second spatial harmonic, b_2, being largest; in this case $\mathcal{A} = 0.46 \pm 0.03$.

This technique works only if the function $U(y)$ is single valued. For much of the experimental range of existence of the traveling finger state, however, the fingers lean over sufficiently far that $U(y)$ is not single valued, and so cannot be directly Fourier transformed. An example is shown in Figure 5(c). We can still apply the technique described above to this data, but with an extra step in the process. Consider the length ℓ of the curve defined by the interface. In contrast to $U(y)$, the function $U(\ell)$, i.e., the length of the fingers as a function of the distance along the interface, is always single valued. We thus write

$$U(\ell) = \sum_{j=1}^{\infty} a'_j \cos jq'\ell + \sum_{j=1}^{\infty} b'_j \sin jq'\ell, \tag{8}$$

with q' being a new wave number, and as above we set the coefficient b'_1 equal to zero. We then define a modified asymmetry parameter $\mathcal{A}'$, in analogy to Eq. (7), by

$$\mathcal{A}' = \left(\frac{\sum |b'_j|^2}{\sum (|a'_j|^2 + |b'_j|^2)} \right)^{1/2}. \tag{9}$$

In Figure 5 we show the results of using this procedure on the same data as in Figure 4(b), and also on a pattern for which the function $U(y)$ is multiple valued. In the first case (Figure 5(a) and (b)) $\mathcal{A}' = 0.28 \pm 0.02$, while in the second case (Figure 5(c)–(e)), $\mathcal{A}' = 0.17 \pm 0.02$.

In Figure 6 we show the relationship between the two parameters $\mathcal{A}$ and $\mathcal{A}'$. The line in the figure is a fit to a power law; we find $\mathcal{A} = 1.01\mathcal{A}'^{0.62}$. In this chapter we present our results in terms of the parameter $\mathcal{A}$.

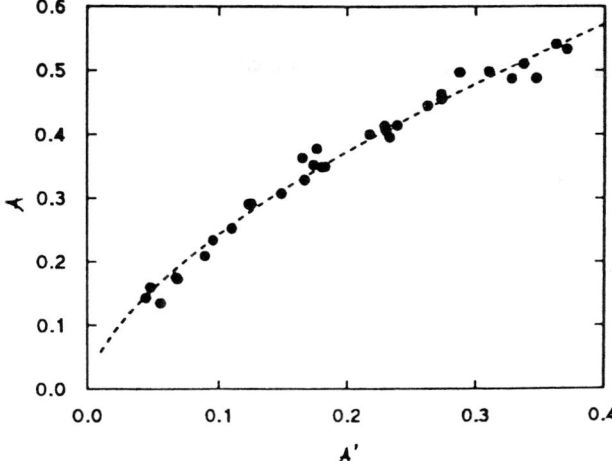

FIGURE 6 A plot of the two asymmetry parameters, $\mathcal{A}'$ vs. $\mathcal{A}$, determined from single-valued patterns at relatively low values of v_o. The line is a fit discussed in the text.

IV. RESULTS

In Figure 7(a) we show the square of the phase speed as a function of v_i, for several relatively low values of v_o. In this range of v_o, the pattern's behavior is straightforward—v_ϕ^2 increases linearly with v_i over the whole range of existence of the traveling state. In Figure 7(a) we show data for five different values of v_o between 139.4 mm/s and 156.9 mm/s; within our experimental scatter, these data are indistinguishable. A straight line fit to the v_ϕ data in Figure 7(a) indicates that v_ϕ becomes nonzero at $v_i^* = 3.9 \pm 0.6$ mm/s. In the same figure we plot the square of the asymmetry parameter for the same set of measurements. Again the different data sets overlap within our experimental scatter. $\mathcal{A}^2$ also grows linearly with v_i, and a straight line fit to this data also gives $v_i^* = 3.9 \pm 0.6$ mm/s, in agreement with the value obtained from the phase velocity data. Figure 7(b) is a plot of v_ϕ vs. $\mathcal{A}$ for the same data. The relationship is accurately linear over the entire range of existence of the traveling state.

In Figure 8(a) we show analogous data for a slightly higher value of the outside cylinder rotation speed, $v_o = 174.3$ mm/s. Here v_ϕ^2 again grows linearly from zero for low v_i. The bifurcation point is zero within our experimental resolution. At $v_i \approx 17$ mm/s, however, the pattern's drift velocity suddenly decreases, then begins to increase again. A straight line through the data after the drop in v_ϕ extrapolates to a value of v_i somewhat below the value at which the drop occurred. The drop in phase velocity is accompanied by a change in the shape of the fingers making up the pattern, and by a decrease in the pattern's wavelength of, in this case, about 30%. The change in finger shape can be seen from the asymmetry data, which are also plotted in Figure 8(a). The pattern's asymmetry drops suddenly at the same

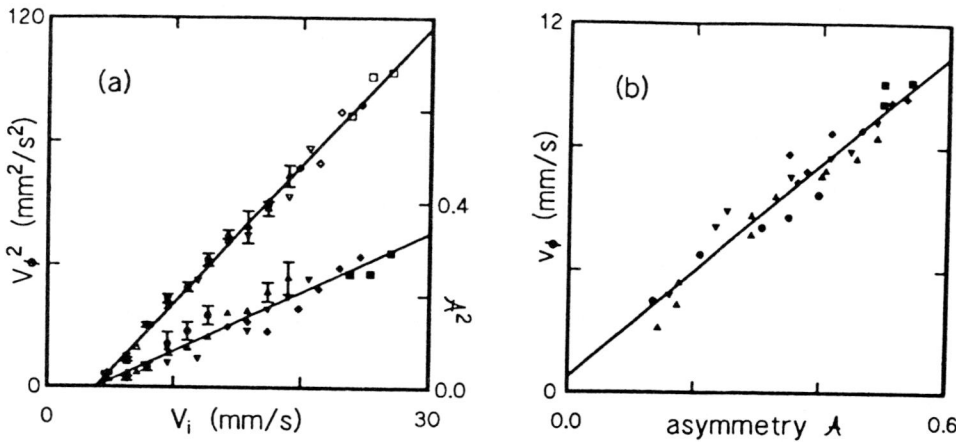

FIGURE 7   (a) The square of the pattern's phase speed, v_ϕ^2 (open symbols, left axis), and the square of the asymmetry, $\mathcal{A}^2$ (solid symbols, right axis), plotted vs. the inner cylinder rotation speed v_i. The different symbols are for different values of v_o between 139.4 mm/s and 156.9 mm/s. (b) Phase speed vs. asymmetry for the same values of v_o as in (a). The straight lines are fits to all the data shown in all cases.

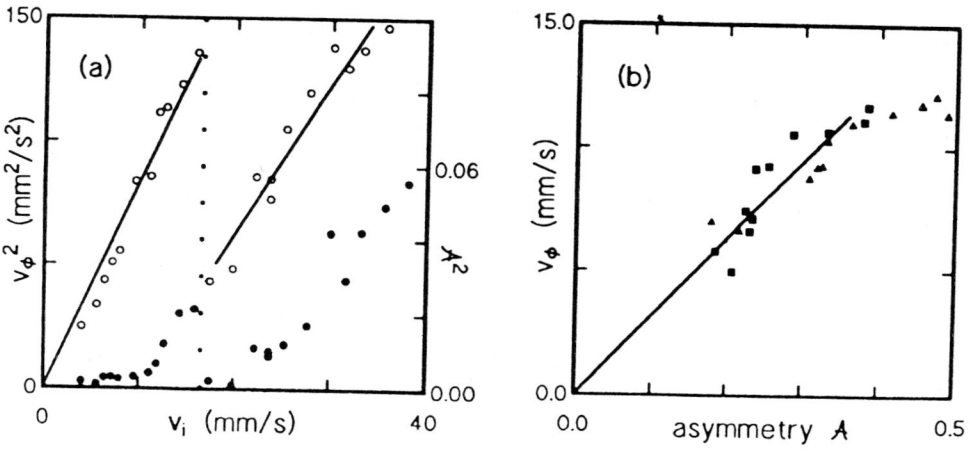

FIGURE 8   (a) v_ϕ^2 (open symbols, left axis), and $\mathcal{A}^2$ (solid symbols, right axis), vs. v_i for $v_o = 174.3$ mm/s. The straight lines are fits to the v_ϕ data before and after the wavelength-changing transition at $v_i = 17$ mm/s. (b) v_ϕ vs. $\mathcal{A}$ for the same value of v_o. The different symbols indicate data from below (squares) and above (triangles) the transition.

point at which the phase velocity drops. Thus, in a plot of v_ϕ vs. $\mathcal{A}$, (Figure 8(b)), no sign of this transition is seen—the data from above and below the jump fall on the same curve. v_ϕ is linear in $\mathcal{A}$ at low asymmetry, but in this case the linearity does not persist over the whole range of existence of the traveling state.

V. DISCUSSION

Our data confirm the theoretical predictions for the behavior of a one-dimensional pattern near a supercritical parity-breaking bifurcation.[1,2,6,9] The parity-breaking order parameter, $\mathcal{A}$, and the pattern's phase speed, v_ϕ, both increase smoothly from zero with a square-root dependence on the experimental control parameter, $v_i - v_i^*$. The phase speed is linear in the asymmetry over the whole range of existence of the traveling finger state for low values of the outer cylinder rotation speed, and for low asymmetry at higher values of v_o. At higher outer cylinder rotation speeds, the relationship between asymmetry and phase speed appears to be unaffected by the presence of a wavelength-changing transition as v_i is increased.

A more complete analysis of our results, in terms of a $q - 2q$ mode-coupling model, has yet to be carried out. Similarly, the wavelength-changing transition that occurs at higher values of the outer cylinder speed has not yet been studied in detail, although the observed behavior appears to be consistent with the series of successive bifurcations described by Cummins et al.[4] These topics will be discussed in a future publication.

ACKNOWLEDGMENTS

We are grateful to J. Gleeson and R. Goldstein for helpful discussions during the course of this work. This research was supported by a grant from the Natural Sciences and Engineering Research Council of Canada.

REFERENCES

1. Caroli, B., C. Caroli, and S. Fauve. "On the Phenomenology of Tilted Domains in Lamellar Eutectic Growth." *J. Phys. I France* **2** (1992): 281–290.
2. Coullet, P., R. E. Goldstein, and G. H. Gunaratne. "Parity-Breaking Transitions of Modulated Patterns in Hydrodynamic Systems." *Phys. Rev. Lett.* **63** (1989): 1954–1957.

3. Coullet, P., and G. Iooss. "Instabilities of One-Dimensional Cellular Patterns." *Phys. Rev. Lett.* **64** (1990): 866–869.

4. Cummins, H. Z., L. Fortune, and M. Rabaud. "Successive Bifurcations in Directional Viscous Fingering." *Phys. Rev. E* **47** (1993): 1727–1738.

5. Faivre, G., and J. Mergy. "Tilt Bifurcation and Dynamical Selection by Tilt Domains in Thin-Film Lamellar Eutectic Growth: Experimental Evidence of a Tilt Bifurcation." *Phys. Rev. A* **45** (1992): 7320–7329.

6. Fauve, S., S. Douady, and O. Thual. "Drift Instabilities of Cellular Patterns." *J. Phys. II France* **1** (1991): 311–322.

7. Flesselles, J. M., A. J. Simon, and A. J. Libchaber. "Dynamics of One-Dimensional Interfaces: An Experimentalist's View." *Adv. Phys.* **40** (1991): 1–52.

8. Gleeson, J. T., P. L. Finn, and P. E. Cladis. "Travleing-Wave States in Deep-Groove Directional Solidification." *Phys. Rev. Lett.* **66** (1991): 236–239.

9. Goldstein, R. E., G. H. Gunaratne, L. Gil, and P. Coullet. "Hydrodynamic and Interfacial Patterns with Broken Space-Time Symmetry." *Phys. Rev. A* **43** (1990): 6700–6721.

10. Hakim, V., M. Rabaud, H. Thomé, and Y. Couder. "Directional Growth in Viscous Fingering." In *New Trends in Nonlinear Dynamics and Pattern-Forming Phenomena*, edited by P. Coullet and P. Huerre, 327–337. New York: Plenum, 1990.

11. Levine, H., W.-J. Rappel, and H. Riecke. "Resonant Interactions and Traveling Solidification Cells." *Phys. Rev. A* **43** (1991): 1122–1125.

12. Malomed, B. A., and M. I. Tribelsky. "Bifurcations in Distributed Kinetic Systems with Aperiodic Instability." *Physica* **14D** (1984): 67–87.

13. Michalland, S. "Etude des Differents Régimes Dynamiques de l'Instabilité de l'Imprimeur." Ph.D. Thesis, Université de Paris VI, 1992.

14. Michalland, S., and M. Rabaud. "Localized Phenomena During Spatio-Temporal Intermittency in Directional Viscous Fingering." *Physica* **61D** (1992): 197–204.

15. Michalland, S., M. Rabaud, and Y. Couder. "Transiton to Chaos by Spatio-Temporal Intermittency in Directional Viscous Fingering." *Europhys. Lett.* **22** (1993): 17–22.

16. Mutabazi, I., and C. D. Andereck. "Drift Instability and Second Harmonic Generation in a One-Dimensional Pattern-Forming System." *Phys. Rev. Lett.* **70** (1993): 1429–1932.

17. Pan, L., and J. R. de Bruyn. "Broken-Parity Waves at a Driven Fluid-Air Interface." *Phys. Rev. Lett.* **70** (1993): 1791–1794.

18. Simon, A. J., J. Bechhoefer, and A. Libchaber. "Solitary Modes and the Eckhaus Instability in Directional Fingering." *Phys. Rev. Lett.* **61** (1988): 2574–2577.

19. Rabaud, M., Y. Couder, and S. Michalland. "Wavelength Selection and Transients in the One-Dimensional Array of Cells in the Printer's Instability." *Eur. J. Mech. B* **10** (1991): 253-260.

20. Rabaud, M., S. Michalland, and Y. Couder. "Dynamical Regimes of Directional Viscous Fingering: Spatiotemporal Chaos and Wave Propagation." *Phys. Rev. Lett.* **64** (1990): 184–187.

21. Rappel, W.-J., and H. Riecke. "Parity Breaking in Directional Solidification: Numerics versus Amplitude Equations." *Phys. Rev. A* **45** (1992): 846–859.

22. Riecke, H., and H.-G. Paap. "Parity Breaking and Hopf Bifurcation in Axisymmetric Taylor Vortex Flow." *Phys Rev. A* **45** (1992): 8605–8610.

23. Wiener, R. J., and D. F. McAlister. "Parity-Breaking and Solitary Waves in Axisymmetric Taylor Vortex Flow." *Phys. Rev. Lett.* **69** (1992): 2915–2918.

Hong Zhao† and J. V. Maher
Department of Physics and Astronomy, University of Pittsburgh, Pittsburgh, PA 15260, USA
†Present Address: Service de Physique de l'Etat Condensé, Centre d'Etude de Saclay, 91191 Gif-sur-Yvette, France

The Fracture Transition in Swollen Polymer Networks

A deep understanding of the mechanism and dynamics of the fracture of brittle materials would be a triumph for statistical physics and a result of great technological importance.[1] Recent theoretical work has shown promise for a deeper understanding.[10,14] A discussion of the current state of the theory is presented elsewhere in this conference.[15] Experiments are only just beginning to be able to follow the dynamics of a propagating crack tip, and a remarkable experiment of that type is presented elsewhere in this conference.[7,8]

Lemaire et al.[13,17] have reported fracturelike behavior when fluids were forced into viscoelastic clays. They report fracture transitions whose dynamics is rather like the results we report below. We have chosen to work with polymer solutions because the variations of their linear viscoelastic properties[3,5,9] with molecular weight, concentration and molecular architecture are generally better understood than clays, because there is much technological interest in the relation of polymer molecular properties and nonlinear pattern-formation behavior, and because we can use the softness of polymer solutions to put tip velocity for fracturelike behavior in a slow enough regime to study it easily.

Associating polymers that have more than one associating group per molecule are known to form networks in solution,[20] and these networks might be expected to fracture in much the same fashion as does a brittle solid. On the other hand, these solutions are much softer than brittle solids and the velocities of crack propagation for fracture might well fall in an experimentally more accessible regime for such polymer solutions. In this chapter, we report the results of an experiment in which we have studied the fracturelike response of an associating polymer solution to the injection of water in a confined container. Our understanding of the mechanism of the processes we observe will have to advance significantly over its present state before we will be able to say with confidence whether or not the fracture of brittle solids shares universal properties with the fracture-like response of our associating polymer solutions. However, the polymer networks are of interest of themselves, and the dynamics of the two problems are at least similar, with the polymer case showing slower, more experimentally accessible, propagation. Through the remainder of this discussion we will call the polymer phenomenon "fracture" as if we knew that the processes were the same.

While we have measured velocities of tip propagation, in this chapter, we will concentrate on the conditions for the onset of fracture, and our attempts to understand those conditions in terms of fluid dynamics and the architecture of the dissolved polymers. We have measured pattern formation when water is injected into aqueous polymer solutions in a Hele-Shaw cell.[21,22] In Figure 1, we show a schematic diagram of the experimental apparatus. A syringe pump is used to inject dyed water at a carefully controlled rate between the plates of a Hele-Shaw cell, and the resulting patterns are observed with a CCD television camera whose signal is recorded by a medical-grade video cassette recorder. The recorded images are later digitized and processed. There are five control parameters in the experiment. Two of these are hydrodynamic: Q, the water injection rate; and b, the gap between the plates of the Hele-Shaw cell. The other three involve the properties of the polymeric material in solution: M_w, the molecular weight of the polymer; c, the concentration of polymer in the solution; and z, the number of associating groups on the polymer molecule.

In Table 1, we list the polymer molecules included in this study. Three distinct architectures are involved. Most of the data reported herein was taken with end-capped associating polymers of the sort called Type 1 in the table.[1] These have a backbone of poly(oxyethylene) (PEO) with hydrophobic endcaps at each end of the PEO chain. This appears to be the simplest architecture that will form networks at low concentration. Molecules with only one associating group have been seen to form very monodisperse micelles at low concentration,[2,6] whereas measurements of light scattering from dilute concentrations of these endcapped polymers have shown very large networks at very low concentrations.[2] The Type 2 molecules listed in

[1] Union Carbide Chemicals and Plastics Company Inc., P.O. Box 8361, South Charleston, WV 25303.

[2] Fluorescence Measurements by Wang and Winnik,[19] and unpublished static and dynamic light scattering performed in our laboratory.

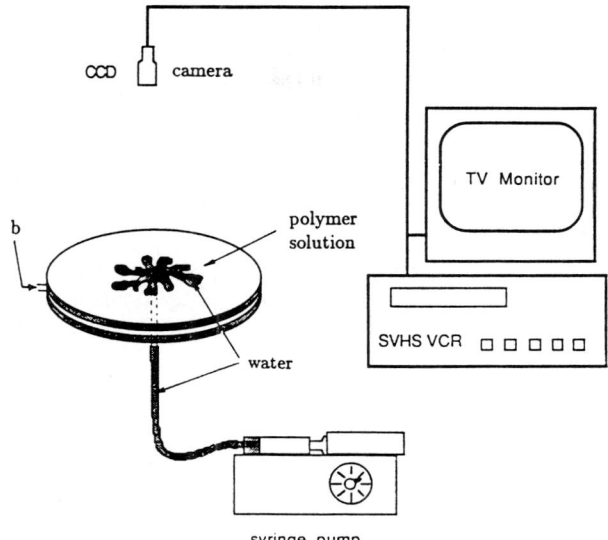

FIGURE 1 A schematic diagram of the experimental setup.

TABLE 1 Model polymers investigated

Type 1	**Endcapped associating polymers:** R–O–(DI–PEO)$_y$–DI–O–R, where PEO is polyethylene oxide, DI is isophorone diisocyanate, y is the number of repeating units of the chain backbone, and R is a hexadecyl alkyl end-group.
Type 2	**Comb associating Polymer:** $-[(-\text{DI})_w-(\text{PEO}-\text{DI})_y]_{z-1}-(\text{DI}-\text{PEO})_y-$, where w, y, z ($w=1$, $y=2$, $z=6$) indicate the number of "teeth" in a bunch of hydrophobes, the number of PEO monomers between hydrophobe bunches, and the degree of polymerization, respectively.
Type 3	**Homopolymer:** Poly(oxyethylene) (PEO)

Table 1 also have a PEO backbone, but now there are z associating hydrophobic groups along the chain, a comb with teeth of low molecular weight. Finally, Type 3 molecules are simple homopolymer PEO which will not form networks at low concentration, and whose high concentration networks should arise from entanglements, not from associating groups.

The viscoelastic properties of these molecules in solution have been studied by Jenkins et al.[11,12] Under some conditions, solutions of these molecules can be shear-thickening, but we have only used them at high concentrations where they are shear-thinning. Using the results given by Jenkins et al., we know the low shear-rate viscosity and the shear-thinning shear-rate as a function of concentration and molecular weight.

Using polymers of Type 1, the end-capped polymers, we can always observe a transition from viscous fingering to fracture. That is, once the molecular weight and concentration are set for the end-capped polymer, there is always an injection rate, Q_l, for a given setting of the cell gap b, such that, below that injection rate, the patterns formed as the injected water drives back the polymer solution add complexity only by tip-splitting, and the tip-splitting angles are always small ($< 50°$ to the initial direction of the splitting finger). There is also always an injection rate, Q_h, such that above that injection rate the patterns always look like fracture, with large branching angles (frequently near 90°) and high likelihood of side-branching. While Q_l and Q_h are never far apart, they are always separated by much more than the uncertainty in our injection control. Thus, there is always a narrow range of injection rates in that it is not possible to predict whether viscous fingering or fracture will be observed, with fingering more likely at the low end of this range and fracture more probable at the high end. There is never a pattern which is intermediate between the two possibilities, and by observing the time series of patterns for a given flow realization, there is never any ambiguity as to which class of pattern is being observed. In Figure 2, we show a set of patterns which result from driving $M_w = 50,700$ end-capped polymer of concentration c= 0.025 by weight in a cell of gap b= 0.4 mm at various driving rates, Q, as indicated. At the lowest rate, viscous fingering is still possible, and an example of this is shown in part a of the figure. In Figure 2(b), we show a fracture near threshold; i.e., at a rate essentially identical to that used in Figure 2(a). Figures 2(c) and 2(d) are fractures driven at higher rates. In Figure 3, we show two examples of patterns formed in the comb polymers, Type 2 in Table 1. For these polymers, we are not certain whether there is only one morphology transition or whether at least one more resides outside our accessible injection range. At even our lowest injection rates, the patterns show large branching angles, approaching 70 degrees in extreme cases, but never quite as large as is routinely seen in fracture. In addition, at the low injection rates there is no side-branching; all new branches seem to form by tip-splitting. While these patterns never, even at the lowest injection rates, look like the viscous fingering pattern shown in Figure 2(a), they meet our criteria for being called viscous fingering. Above an injection rate which we will again call Q_h, patterns for the Type 2 comb molecules show a dramatic transition to clearly fracture-like behavior. An example of one of these patterns is shown in Figure 3(b). The fracture cracks are dramatically narrower than those seen in Figure 2; they are, in fact, much narrower than the gap in the Hele-Shaw cell (not the case in general for Figure 2) and the problem is no longer even quasi-two-dimensional. The tip velocities are also much slower than for the flows shown in Figure 2.

In Figure 4, we show a pattern formed at extremely high injection rate in a very concentrated solution of PEO homopolymer (Type 3 of Table 1). This pattern

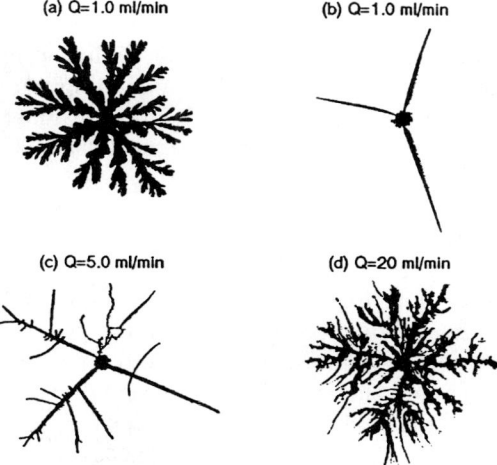

(a) Q=1.0 ml/min

(b) Q=1.0 ml/min

(c) Q=5.0 ml/min

(d) Q=20 ml/min

FIGURE 2 Patterns formed in the end-capped polymer at various driving rates.

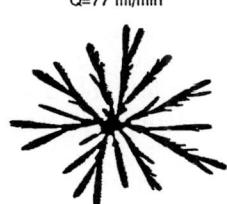

Q=77 ml/min

FIGURE 3 Patterns formed in comb polymers at two different driving rates.

(a) Q=0.06 ml/min   (b) Q=0.6 ml/min

FIGURE 4   Pattern formed in a concentrated homopolymer solution at high driving rates.

is clearly viscous fingering, as were all our homopolymer patterns at all injection rates we were able to measure. In our earlier work on homopolymer-solution pattern formation,[21] we showed that a wide variety of patterns could be formed under different injection and polymer-solution conditions, including patterns not unlike those seen for Type 3 polymers in part a of Figure 3, and including patterns with growth in the "fjords" behind the advancing fingers. However, we have never observed fracture with a homopolymer solution, no matter how highly entangled.

In looking for universal features of the fracture transition, it is natural to try to form a Deborah number; the ratio of a relaxation time for the polymer solution, t_r; and a hydrodynamic time, t_h. Unfortunately, polymer solutions have many characteristic times, so it is not obvious which of them to use. As a first place to start, we have followed DeGennes[4] and arbitrarily chosen the inverse of the shear-thinning shear rate, $\dot{\gamma}_0$ as taken from the work of Jenkins et al.[12] The hydrodynamic time is easier to define, although not rigorous. Assuming a parabolic velocity profile across the gap in the Hele-Shaw cell gives one an inverse imposed shear rate of rb^2/Q' where r is a length scale characteristic of the cell. Thus the Deborah number is

$$D_e = \frac{\tau_r}{\tau_h} \sim \frac{Q}{b^2 \dot{\gamma}_0}$$

and, since we do not know exactly what to use for the characteristic length r, we need one measurement to set our definition of D_e. We do this equivalently by observing the range of injection rates, $Q_c = [Q_l, Q_h]$, over which the fracture transition takes place for one case, and then predicting the injection rate range for the same Deborah number for all other cases. Our results are shown in Table 2. (If one assumes that the transition is centered at $D_e = 1$, then $r \simeq 30$ cm, a plausible value for our cell of diameter 20 cm.)

As can be seen from Table 2, the Deborah number as just defined works very well for all measurements with the endcapped polymers. Variations of molecular weight, solution concentration and Hele-Shaw cell-gap all show fracture setting in

TABLE 2 The threshold of the fracture transition

polymer type	Mw (g/mol)	C (%wt.)	γ_0 (1/sec)	η_0 (cP)	b (mm)	Exp. Q_c (ml/min)	Obs. Q_c (ml/min)
1	51k	2.5	20^1	12k	0.4	—	[0.6,1.0]
1	51k	2.5	20^1	12k	0.8	[2.4,4.0]	[3.0,4.5]
1	17k	2.5	7^1	43k	0.4	[0.2,0.4]	[0.3,0.5]
1	17k	2.0	10^1	32k	0.4	[0.3,0.5]	[0.2,0.6]
2	107k	1.5	0.1^1	123k	0.4	[0.003,0.005]	[0.06,0.2]
3	100k	20	>20	3k	0.4	>1.0	not obs.[2]
3	5000k	1.5	<0.1	50k	0.4	< 0.005	not obs.[2]

[1] Extracted from Jenkins et al.[12]

[2] Up to the possible accessible injection rate 77 ml/min.

at the same value of D_e. This success does not carry over when we change the molecular architecture. The comb polymer shows its transition at values of D_e roughly 30 times larger than would be expected from the endcapped results. And the homopolymer never fractures, even though we were able to drive it to values of D_e more than 10,000 times larger than that indicated by the endcapped results. Clearly the relevant polymer solution relaxation time(s) needs to be defined differently if we are to find a way to connect our results for polymers of all architectures. The success of using the inverse of the shear-thinning shear rate in the endcapped case is suggestive that there may be a single relevant time and that time is proportional to our arbitrary choice for the endcapped polymers. However, we have not succeeded in finding a useful generalization. The obvious candidate, the inverse of the shear rate at which the storage and loss moduli cross (i.e., below this shear rate viscosity dominates and above it elasticity dominates) fails miserably to produce a Deborah number that can organize our measurements.

van Damme[18] has expressed concern that all fracture observed in clays may result from plate flexing in the Hele-Shaw cell. He quite correctly points out that the normal stresses could be very large in highly viscoelastic materials like his clays and our polymer networks. While we do not know how to estimate our normal stresses, we have tested our measurements in several ways which should show us whether or not the fracture transitions we observe arise trivially from the properties of our cell. We conclude that cell flexing is highly unlikely to account for the fracture transitions we have just presented for the following reasons:

1. We have repeated our experiment using a cell with much thicker plates (22 mm rather than 13 mm) and found the fracture transition to occur at the same water injection rate.
2. Our cell (13 mm thick glass of diameter 20 cm) is much less deformable than van Damme's low deformability cell (12 mm thick glass of diameter 45 cm). Deformability changes as cell diameter to the fourth power.[16]
3. We measured the water pressure at the input to the cell and found it to be below 7×10^3 Pa at the fracture transition, much too low a pressure to bend the glass significantly.
4. The same cell, driven much nearer its limit in the homopolymer case, never showed fracturelike patterns in the homopolymer solutions.

In summary, we observe a transition to fracture for solutions of all polymers which have two or more associating groups on their PEO backbones. When these groups are only at the ends of the backbone, we are able to organize the data with a Deborah number whose polymer relaxation time is simply the inverse of the shear-thinning shear rate. When more than two associating groups are attached to the backbone, our Deborah number becomes inaccurate by a factor of 30. And even though we exceed the critical Deborah number by more than a factor of 10,000 for our entangled homopolymer solutions, we never observe them to fracture.

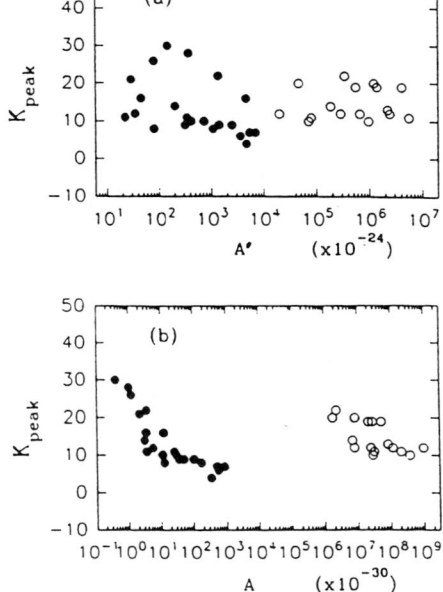

FIGURE 5 Peak Fourier mode versus the scaling numbers A and A' as defined in the text for many different flow realizations.

It is attractive to try to relate systematic changes in dynamical behavior of polymer solutions to changes in polymer architecture. The success of our Deborah number scaling in the endcapped polymer case is suggestive, but clearly we need much more work to find a viewpoint in which all three kinds of polymer will be seen to show predictable behavior. In particular, it appears that we need to define a dimensionless number that would never or almost never predict merely-entangled networks to show fracture. Other features of pattern formation in the homopolymer networks did show remarkable regularity even though a dramatic variety of patterns were achievable.[21] In particular, both the initial twig width and the dominant length scales at later times as defined by a modal analysis of the patterns averaged over many different radii collapsed on to reasonably smooth curves when the control parameters were combined into a single number $A = b^2 c^{2.2}/QM_v^{3.4}$. The number A was motivated by a Deborah number, but the approximation was made that the low-shear-rate viscosity depends only on the molecular weight and not on the concentration. When the known concentration dependence is put into A, one obtains $A' = b^2/QM_w^{3.4}c^{1.2}$. Unfortunately, A' does not collapse the data well at all. This is shown in Figure 5 where the wave number of the peak Fourier mode is shown for a large variety of flow realizations at different values of molecular weight, concentration, injection rate, and cell gap for A (top) and A' (bottom).

Thus, the results of that earlier work, attempting to use the known power law dependences of linear viscoelastic parameters for homopolymer solutions, are like the present fracture case, promising but in an unsettled state. Large amounts of very complicated data can be organized empirically with control variables which have some plausible physical motivation. But in each case it is clear that the control variables either don't work well enough (fracture) or aren't using the theoretically expected power laws (homopolymer pattern lengths) and so are in a far from finished state. Since the linear viscoelastic parameters for polymer solutions can be varied in interesting and controllable ways by changing polymer molecular weight, concentration and architecture, such solutions provide a very promising set of test systems for carrying over linear viscoelastic properties to nonlinear pattern formation problems like the present fracture study or the earlier homopolymer pattern-length study.

ACKNOWLEDGEMENTS

This work was supported by the National Science Foundation.

REFERENCES

1. Atkins, A. G., and Y. W. Mai, *Elastic and Plastic Fracture: Metals, Polymers, Ceramics, Composites, Biological Materials*. Chichester: Ellis Horwood, 1985.
2. Davidson, N. S., L. J. Fetters, W. G. Funk, W. W. Greassley, and N. Hadjichristidis. "Association Behavior in End-Functionalized Polymer. 1. Dilute Solution Properties of Polyisoprenes with Amine and Zwitterion End Groups." *Macromolecules* **21** (1988): 112.
3. deGennes, P. G. *Scaling Concepts in Polymer Physics*. Ithaca: Cornell University, 1979.
4. deGennes, P. G. "Time Effects in Viscoelastic Fingering." *Europhys. Lett.* **3** (1987): 195.
5. Doi, M., and S. F. Edwards, *The Theory of Polymer Dynamics*. New York: Oxford, 1986.
6. Fetters, L. J., W. W. Graessley, N. Hadjichristidis, A. D. Kiss, D. S. Pearson, and L. B. Younghouse. "Association Behavior of End-Functionalized Polymer. 2. Melt Rheology of Polyisoprenes with Carboxylate, Amine, and Zwitterion End Groups." *Macromolecules* **21** (1988): 1644.
7. Fineberg, J., S. P. Gross, M. Marder, and H. Swinney. "Instability in Dynamic Fracture." *Phys. Rev. Lett.* **67** (1991): 457.
8. Fineberg, J., S. P. Gross, M. Marder, and H. Swinney. "Instability in the Propagation of Fast Cracks." *Phys. Rev. B* **45** (1992): 5146.
9. Graessley, W. W. "Viscoelasticity and Flow in Polymer Melts and Concentrated Solutions." In *Physical Properties of Polymers*, edited by J. E. Mark, A. Eisenberg, W. W. Graessley, L. Mandelkern, and J. L. Koenig. Washington, DC: American Chemical Society, 1984.
10. Herrmann, H. J., and S. Roux, eds. *Statistical Models for the Fracture of Disordered Media*. Amsterdam: North-Holland, 1990.
11. Jenkins, R. D. "The Fundamental Thickening Mechanism of Associative Polymers in Latex Systems: A Rheological Study," Ph.D. Thesis, Lehigh University, 1990.
12. Jenkins, R. D., C. A. Silebi, and M. S. El-Alsser. "Steady-Shear and Linear-Viscoelastic Fracturing in Colloidal Fluids." In *Polymers as Rheology Modifiers*, edited by D. N. Schulz and J. E. Glass. ACS Symposium Series, Vol. 462. Washington, D.C.: American Chemical Society, 1991.
13. Lamaire, E., P. Levitz, G. Daccord, and H. van Damme. "From Viscous Fingering to Viscoelastic Fracturing in Colloidal Fluids." *Phys. Rev. Lett.* **67** (1991): 2009.
14. Langer, J. S. "Models of Crack Propagation." *Phys. Rev. A* **46** (1992): 3123.
15. Langer, J. S. Private communication.

16. Thomé, H., M. Raubaud, V. Hakim, and Y. Couder. "The Saffman-Taylor Instability: From the Lienar to the Circular Geometry." *Phys. Fluids A* **1** (1989): 224.
17. van Damme, H., C. Laroche, and P. Levitz, J. "Viscoelastic Effects in Fingering Between Miscible Fluids." *J. Physique* **48** (1987): 1221.
18. van Damme, H. "Flow and Interfacial Instabilities in Newtonian and Colloidal Fluids." In *The Fractal Approach to Heterogeneous Chemistry-Surfaces, Colloids, Polymers*, edited by D. Avnia, 199. New York: Wiley, 1989.
19. Wang, Y., and M. A. Winnik. "Onset of Aggregation for Water-Soluble Polymeric Associative Thickeners: A Fluorescence Study." *Langmuir* **6**, 1437 (1990)
20. Witten, T. A. "Heterogeneous Polymers and Self-Organization." *J. Phys.: Condens. Matter* **2** (1990): AS1.
21. Zhao, H., and J. V. Maher. "Viscoelastic Effects in Patterns Between Miscible Liquids." *Phys. Rev. A* **45** (1992): R8328.
22. Zhao, H., and J. V. Maher. "Associating-Polymer Effect in a Hele-Shaw Experiment." *Phys. Rev. E* **47** (1993): 4278.

Chapter 3
Hydrodynamics in Complex and Simple Systems

Helmut R. Brand
Theoretische Physik III, Universität Bayreuth, Postfach 101251, D 95440, F.R. Germany

Some Remarks About Second-Order Fluids

In this chapter I discuss the thermodynamic properties of a class of model equations thoughtsecond-order fluids to be applicable to certain types of polymer melts and polymer solutions for which the name *second-order flu-ids* is common in the literature. This term refers to the fact that an expansion, which is second order in the velocity gradients, is made for the stress tensor.

I will argue here that the terms that are quadratic in the velocity gradients in the stress tensor would have to be reversible to be compatible with general symmetry arguments and irreversible thermodynamics. I will show that the requirement of zero entropy production for these contributions to the stress tensor leads to the disappearance of the suggested coefficients in front of terms that are quadratic in symmetrized velocity gradients. This, in turn, implies that "second-order" fluids are not compatible with the second law of thermodynamics and, therefore, cannot be expected to apply to the description of experimental results.

Spatio-Temporal Patterns, Ed. P. E. Cladis and P. Palffy-Muhoray,
SFI Studies in the Sciences of Complexity, Addison-Wesley, 1995   **211**

1. INTRODUCTION AND MOTIVATION

This contribution expands on one particular point I touched upon in my review lecture "Pattern Formation in Polymers and Magnetic Liquids." To get more background about the flow behaviour of and the patterns formed in polymers, we refer the reader to Bird et al.,[2] Giesekus,[10] and Larson[15] and, for the case of pattern formation and hydrodynamic instabilities in magnetic liquids, to Bashtovoy et al.[1] and Rosensweig,[17,18] and the additional sources quoted in these books and review articles.

Here I will discuss in detail to what extent the model of second-order fluids—introduced several decades ago and used very often in the literature dealing with flows in polymer melts and solutions[2,7,13,15,19,20]—is compatible with basic symmetry requirements and the first and second law of thermodynamics. I argue that terms that enter into the stress tensor as contributions that are quadratic in the velocity gradients, would have to be reversible. Imposing the condition of zero entropy production on these terms, I show that all associated coefficients must vanish. Thus second-order fluids cannot possibly give a sensible description of experimental data, as the model might be mathematically correct, but violates basic thermodynamic constraints.

2. SYMMETRY PROPERTIES AND IRREVERSIBLE THERMODYNAMICS

To characterize the macroscopic response of a condensed matter system at long wavelengths and low frequencies, the approach of generalized hydrodynamics[9,16] based on general symmetry arguments, conservation laws, and irreversible thermodynamics[8] has turned out to be rather fruitful. The hydrodynamic regime is characterized by frequencies that are small compared to all microscopic frequencies (frequently these are collisional frequencies) and by wavelengths that are large compared to all microscopic length scales (e.g., interatomic distances). In this approach one describes the system using a small number of variables, the hydrodynamic variables, which relax infinitely slowly in the limit of vanishing wavevector. These hydrodynamic variables come in two groups[9,16]: (a) conserved variables and (b) variables associated with spontaneously broken continuous symmetries. In the former group, one has all the conserved quantities well known from simple fluids[14] including density ρ, density of linear momentum $\vec{g}$, and energy density ϵ. In magnetic systems another conserved quantity arising frequently is the magnetization density $\vec{M}$. The approach of generalized hydrodynamics has also been modified to incorporate variables that relax on a long, but finite time scale—even for vanishingly small wavevector. These variables are frequently called macroscopic variables.

To derive dynamic equations for these hydrodynamic (macroscopic) variables, one starts with the local formulation of the first law of thermodynamics, the Gibbs-Duhem relation, which connects the hydrodynamic variables density (for purposes of illustration we will discuss a simple fluid throughout this section), the density of linear momentum, and the energy density with the entropy density σ

$$T d\sigma = d\epsilon - \mu d\rho - v_i dg_i. \tag{1}$$

The so-called thermodynamic conjugate quantities or thermodynamic forces are introduced via Eq. (1). In addition, one has balance equations for all hydrodynamic (macroscopic) variables. In the case of a simple liquid these are the conservation laws, which read[9,14,16]

$$\dot{\rho} + \nabla_i g_i = 0 \tag{2}$$

$$\dot{g}_i + \nabla_j \sigma_{ij} = 0 \tag{3}$$

$$\dot{\epsilon} + \nabla_i j_i^{\epsilon} = 0 \tag{4}$$

$$\dot{\sigma} + \nabla_i j_i^{\sigma} = \frac{R}{T} \tag{5}$$

where the currents σ_{ij} (the stress tensor), j_i^{ϵ} (the energy current), and j_i^{σ} (the entropy current) are still to be determined. The current of the density ρ is the density of linear momentum $\mathbf{g}$ and Eq. (2) the continuity equation. The balance equation for the entropy density σ has, in addition to the current, a source term: the entropy production R/T with the dissipation function R. To guarantee the validity of the second law of thermodynamics, R must be positive for dissipative processes and must vanish identically for reversible processes, which are not connected with any entropy production.

When pinning down the structure of the thermodynamic forces and the currents, one must guarantee that the resulting dynamic equations satisfy all symmetry requirements and thermodynamic constraints. The former class includes the appropriate behaviour of all terms under parity and time reversal and the correct transformation behaviour under Galilei transformations and under rigid rotations.[9,14,16] To give a few examples: the density ρ and the energy density ϵ are even under parity and time reversal; i.e., they do not change sign under these operations. In contrast, the density of linear momentum $\mathbf{g}$ is odd under both parity and time reversal and, thus, changes sign under either operation. Quantities that are odd under parity, but even under time reversal, are, for example, the electric field $\mathbf{E}$ or the gradient operator ∇. For the currents it is convenient to split them into reversible and irreversible contributions. The reversible contributions have the opposite behaviour under time reversal as the variable under consideration and the dissipative parts have the same behaviour under time reversal as the variable. This can be seen very clearly for the stress tensor σ_{ij}. For a simple fluid the reversible part reads

$$\sigma_{ij}^R = p \delta_{ij} + \rho v_i v_j \tag{6}$$

while for the dissipative part it reads

$$\sigma_{ij}^D = -\eta_{ijkl}A_{kl} \tag{7}$$

where $\eta_{ijkl} = \eta(\delta_{ik}\delta_{jl} + \delta_{il}\delta_{jk} - 2/3\,\delta_{ij}\delta_{kl}) + \zeta\delta_{ij}\delta_{kl}$ and where $A_{ij} = 1/2\,(\nabla_i v_j + \nabla_j v_i)$. The viscous tensor η_{jkl} contains thus two independent viscosities, namely, shear viscosity η and bulk viscosity ζ, a result well known from textbooks on hydrodynamics.[9,14] Both contributions to Eq. (6) are even under time reversal and their structure is fixed completely by the requirements of thermodynamics and the behaviour under Galilei transformations. The two contributions in Eq. (7) are odd under time reversal; they are dissipative and, therefore, must lead to positive entropy production to be compatible with the second law of thermodynamics.

For a simple fluid the dissipation function R takes the form

$$R = \frac{1}{2}\int d\tau[\kappa(\nabla_i T)(\nabla_i T) + \eta_{ijkl}A_{ij}A_{kl}]. \tag{8}$$

From the positivity requirement for R, it follows that the thermal conductivity κ as well as the bulk and the shear viscosity must be positive.

The dissipative parts of the currents are then obtained by taking the variational derivative of the dissipation function with respect to the appropriate thermodynamic force while keeping all other forces fixed (indicated by ...). We find, for example,

$$j_i^\sigma = -\frac{\delta R}{\delta \nabla_i T}|\ldots = -\kappa\nabla_i T \tag{9}$$

and Eq. (7) follows by taking the variational derivative of the dissipation function R (given in Eq. (8)) with respect to the symmetrized velocity gradients A_{ij}.

3. APPLICATION TO SECOND-ORDER FLUIDS

Generalizing Eqs. (3), (6), and (7) to allow for a nonlinear relation between the stress tensor and velocity gradients, an approximation made frequently in polymer rheology is that of *second-order fluids*. This approximation involves an expansion of the stress tensor into terms that contain expressions quadratic in the symmetrized velocity gradients

$$\sigma_{ij} = \tilde{\eta}_{ijklmn}A_{kl}A_{mn} \tag{10}$$

where the coeffcients $\tilde{\eta}_i$ are assumed to be constants. Frequently these terms are treated in the polymer rheology literature as if they were dissipative. This is, however, not the case. Keeping in mind that the gradient operator is even under time reversal and that the velocity field is odd under time reversal, this implies that the contribution to the stress tensor listed in Eq. (10) must be even under time reversal and is thus a reversible term, which is not allowed to give rise to any entropy production; that is, R must vanish identically for this term: $R \equiv 0$. We note in passing that a simple way of seeing that these contributions must be reversible and not dissipative comes from the fact that the suggested contributions are quadratic in the velocity field just like the material derivative.

Imposing the condition of zero entropy production on the two invariants suggested for second-order fluids in the literature, we find that both coefficients must vanish identically. Thus the concept of second-order fluids with its two invariants presented in the polymer rheology literature is not compatible with basic symmetry requirements and irreversible thermodynamics.

We close this section by pointing out that, indeed, many problems are encountered when using the model of second-order fluids to describe real systems, both when it comes to the mathematical properties of the solutions (which turn out be rather unphysical quite frequently) and the ability to fit experimental data with the model (the reader is referred to Larson[15] for a detailed exposition).

4. CONCLUSIONS AND PERSPECTIVE

In the last section we have shown that one of the most frequently used models to describe the nonlinear flow behaviour of polymer solutions and polymer melts is not compatible with basic symmetry arguments and irreversible thermodynamics. This raises the question how one can interpret properly classical experiments in polymer flows such as those on the rod-climbing (Weissenberg) effect.[11,12] More generally one might ask the question whether some of the other approaches and models suggested in the literature[2,7,13,15,19,20] to describe the nonlinear flow behaviour of polymer solutions and melts also violate basic symmetry requirements and/or irreversible thermodynamics, a problem we will address in a forthcoming manuscript.[3]

Another interesting area of polymer flows, which was neglected completely in most of the literature on the subject, is the coupling of the velocity field and the strain field (originating from the transient network) to other macroscopic variables, such as electric fields or the order parameter close to a phase transition. So far only linearized considerations of such cross-coupling effects have been presented using macroscopic dynamics,[4,5,6] but a generalization to the nonlinear domain seems highly desirable.

ACKNOWLEDGMENTS

Partial support of this work through the Graduierten-Kolleg "Nichtlineare Spektroskopie und Dynamik" of the Deutsche Forschungsgemeinschaft is gratefully acknowledged.

REFERENCES

1. Bashtovoy, V. G., B. M. Berkovsky, and A. N. Vislovich. *Introduction to Thermomechanics of Magnetic Fluids.* New York: Hemisphere, 1988.
2. Bird, R. B., R. C. Armstrong, and O. Hassager. *Polymeric Liquids.* New York: Wiley & Sons, 1977.
3. Brand, H. R. "Nonlinear Constitutive Equations for Polymer Flows: Restrictions Imposed by Irreversible Thermodynamics and Basic Symmetry Requirements." Submitted manuscript.
4. Brand, H. R., H. Pleiner, and W. Renz. "Linear Macroscopic Properties of Polymeric Liquids and Melts: A New Approach." *J. Phys.* **51** (1990): 1065.
5. Brand, H. R., and K. Kawasaki. "Macroscopic Dynamics of the Isotropic-Nematic Transition in Polymeric Liquid Crystals." *J. Phys. (Paris) II* **2** (1992): 1789.
6. Brand, H. R., and H. Pleiner. "Flow Enhancement of Electrical Fluctuations in Polymer Solutions and Melts." *J. Phys. (Paris) II* **2** (1992): 1909.
7. Coleman, B. D., H. Markowitz, and W. Noll. *Viscometric Flows of Non-Newtonian Fluids.* Heidelberg: Springer, 1966.
8. De Groot, S. R., and P. Mazur. *Nonequilibrium Thermodynamics.* Amsterdam: North Holland, 1962.
9. Forster, D. *Hydrodynamic Fluctuations, Broken Symmetry and Correlation Functions.* Reading, MA: Benjamin, 1975.
10. Giesekus, H. "Flow Phenomena in Viscoelastic Fluids and Their Explanation Using Statistical Methods." *J. Non-Equilib. Thermodyn.* **11** (1986): 157.
11. Joseph, D. D., and R. L. Fosdick. "The Free Surface on a Liquid Between Cylinders Rotating at Different Speeds: Part I." *Arch. Rat. Mech. Anal.* **49** (1973): 321.
12. Joseph, D. D., G. S. Beavers, and R. L. Fosdick. "The Free Surface on a Liquid Between Cylinders Rotating at Different Speeds: Part II." *Arch. Rat. Mech. Anal.* **49** (1973): 381.
13. Joseph, D. D. *Fluid Dynamics of Viscoelastic Liquids.* Heidelberg: Springer, 1990.

14. Landau, L. D., and E. M. Lifshitz. *Hydrodynamics*. New York: Pergamon Press, 1959.

15. Larson, R. G. *Constitutive Equations for Polymer Melts and Solutions*. Boston: Butterworths, 1988.

16. Martin, P. C., O. Parodi, and P. S. Pershan. "Unified Hydrodynamic Theory for Crystals, Liquid Crystals, and Normal Fluids." *Phys. Rev.* **A6** (1972): 2401.

17. Rosensweig, R. E. "Fluid Dynamics and Science of Magnetic Liquids." In *Advances inn Electronics and Electron Physics*, edited by L. Morton, Vol. 48, 103. New York: Academic Press, 1979.

18. Rosensweig, R. E. *Ferrohydrodynamics*. Cambridge: Cambridge University Press, 1985.

19. Rivlin, R. S., and K. N. Sawyers. "Nonlinear Continuum Mechanics of Viscoelastic Fluids." *Ann. Rev. Fluid Mech.* **3** (1971): 117.

20. Truesdell, C., and W. Noll. *The Nonlinear Field Theories of Mechanics*. Heidelberg: Springer, 1965.

R. G. Larson
AT&T Bell Laboratories, Murray Hill, NJ 07976

Patterns in Sheared Polymeric Nematics

I. INTRODUCTION

Liquid crystalline polymers, either in solution or in the molten bulk state, are formed by molecules that are either intrinsically stiff, or are composed of small mesogenic units covalently bonded together, with or without intervening flexible "spacer" units. The simplest liquid crystalline polymers (LCP's), of course, are nematics. The high viscosity of nematic polymers, compared to their nonpolymeric cousins, makes them generally unsuitable for many optical applications, such as in displays, in which rapid switching between highly oriented states is required. However, although they are sluggish compared to low molecular-weight liquid crystals, polymeric liquid crystals are much more readily oriented by flow fields than are nonliquid crystalline polymers. As a result, polymeric liquid crystals can readily be drawn into high strength fibers. Other unusual properties of materials made from LCP's, such as high thermal and chemical resistance, and highly anisotropic properties, are creating a variety of possible applications for these rather new materials.[4]

When subjected to shearing flow, virtually all liquid crystalline polymers develop orientational patterns; when viewed in a polarizing microscope these appear as patterns of light intensity and coloration. The types of exotic patterns that form have been described[1,3,4,7,8,19] by the terms "bands," "stripes," "worm texture," "thread texture," and "mottled texture." The most regular of these textures are the "stripe" and "band" textures, which are quasi-one-dimensional patterns. Here, we shall distinguish "stripe" patterns from "band" patterns by the convention that the stripes are oriented parallel to the flow direction within the plane of the sample, while the bands are oriented orthogonal to the shearing flow.

Besides viscous effects, there are two sources of elasticity that might govern these patterns. (Inertia is negligible because of the high viscosity of LCP's.) One source of elasticity comes from the nematic character of the material, the other from the polymeric character. The first source of elasticity, which is also present in nonpolymeric nematics, is *Frank elasticity*. It is quantified by an *Ericksen number*,

$$ER \equiv \frac{\gamma_1 V d}{K_1}, \tag{1}$$

where γ_1 is the Leslie-Ericksen twist viscosity, V is the velocity of the moving surface that generates flow, d is the sample thickness, and K_1 is the Frank splay elastic constant. The other source of elasticity, which is present even in nonliquid crystalline polymers, is *molecular elasticity*; it is produced when the flow distorts the distribution of molecular configurations away from equilibrium. Molecular elasticity is quantified by the *Deborah number*,

$$De \equiv \lambda \dot{\gamma} \tag{2}$$

where λ is a molecular relaxation time and $\dot{\gamma} = V/d$ is the shear rate.

Among the most studied liquid crystalline polymers are solutions of poly(benzyl-L-glutamate), or PBG, in the solvent metacresol.[7,8,12,13,15] PBG molecules are polypeptides that form long, rather rigid, helices in many solvents, including metacresol. Nematic liquid crystalline phases form at concentrations of PBG in metacresol as low as 12 weight percent. At modest concentrations such as this, the Onsager theory for excluded-volume interactions and nematic ordering of long rigid rods seems to provide a reasonable description of the molecular ordering. When the Onsager expression for excluded-volume forces is coupled with expressions for Brownian and hydrodynamic forces generated in shearing flow, a Smoluchowski equation can be derived for the evolution of the molecular order in a shearing flow.[9] This Smoluchowski equation contains a molecular relaxation time, which is defined to be inversely proportional to the Brownian rotary diffusivity of the rodlike molecules. By numerical solution of the Smoluchowski equation, nonlinear effects arising from molecular elasticity can be computed; rheological properties computed in this way, such as the shear-rate dependencies of the viscosity and the normal stress differences, are in excellent agreement with experimental measurements of these properties in shearing flow.[10,15,18] Thus, the Smoluchowski equation might be useful for predicting some of the patterns to be discussed below.

From the solution of the Smoluchowski equation at low shear rates, the Leslie-Ericksen viscosities can also be computed. From these computed viscosities, one finds that the ratio of two of these viscosities, namely α_2/α_3, is *negative*. This implies that in a shearing flow such nematic polymers should be of the *tumbling* type.[6] For tumbling nematics, there is no alignment angle of the director at which shearing torques vanish; hence, large shearing torques do not drive the director toward a well-defined orientation angle. Nonpolymeric tumbling nematics have been found to have flow instabilities, including the formation of roll cells,[5,17] and the existence of an irregular state called *director turbulence* at high Ericksen number.[2,6,16] The discovery that polymeric nematic are sometimes, and perhaps frequently, of the tumbling type may account for their tendency to form complex patterns in shearing flows, as discussed below.

II. STRIPE AND BAND PATTERNS

In studies of shear-induced patterns in PBG solutions, four well-defined quasi-one-dimensional patterns have been found, tabulated below.[7,8,12,13]

A. STEADY-STATE STRIPES AT LOW SHEAR RATES

The first of these patterns, shown in Figure 1, occurs in PBG solutions at Ericksen numbers of 100–1000 during steady-state shear.[13] The optical characteristics of this pattern show that it is produced by roll cells.[13] Roll cell instabilities are both predicted and observed in nonpolymeric nematics[5,17] at similar Ericksen numbers. The linear stability analysis that predicts roll cells is simplest when the initial director orientation is uniform in the vorticity direction, that is, perpendicular to both the flow direction and the shear gradient direction. In Figure 1, and all following figures, the vorticity direction is vertical, and the How direction is horizontal. The striped pattern in liquid crystalline polymers does not seem to be qualitatively different from that for nonpolymer nematics. The main difference is that in polymeric nematics the shear rates at which the pattern can be observed are very much lower than those for nonpolymers, because the high viscosity of polymers makes the Ericksen number large even at low shear rates.[11] At shear rates of 0.1–1.0, the Ericksen number of polymeric nematics is around 10^4, which is so much above the critical condition for roll cells that regular roll cells are replaced by fine-scale turbulence, with only distorted remnants of roll cells still visible.

TABLE 1 Stripe and Band Patterns in Sheared PBG Solutions.

Ericksen No., Er	Deborah No., De	Pattern	Flow Condition
100–1000	$\ll 1$	Stripes	Steady-State Shear
> 100	$\ll 1$	Bands	Transient, During Shear
> 1000	$\lesssim 1$	Bands	Transient, After Shear
$\gg 1000$	≈ 2	Stripes	Steady-State Shear

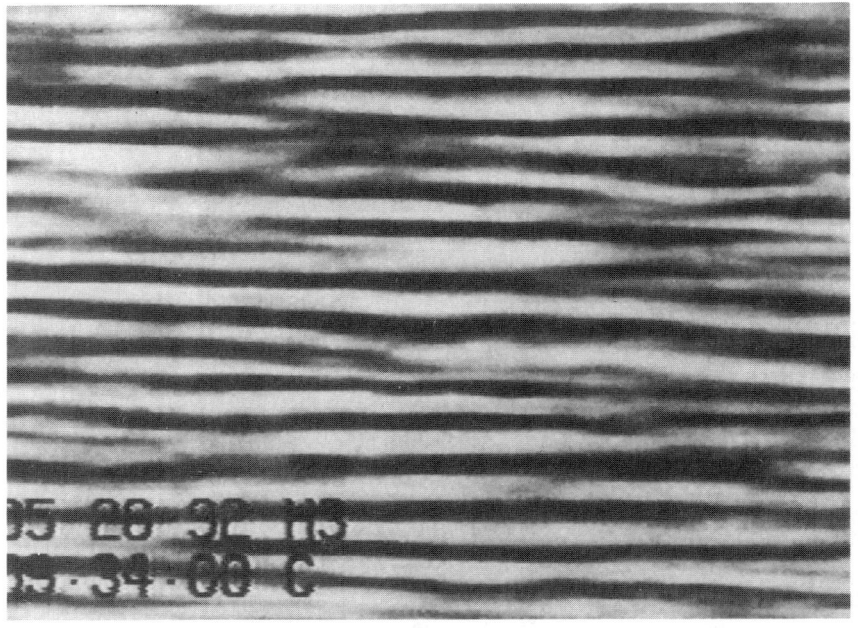

FIGURE 1 Stripe pattern formed during shearing of 15% PBG (molecular weight = 198,000) in metacresol, at a shearing velocity of 3.5μ/sec and a gap of 50μ, so that the Ericksen number is about 245. Here the initial director orientation was in the vorticity direction, and the image was obtained 7 minutes after start-up of shearing. The field of view is 890μ wide. The sample was observed between crossed $0°–90°$ polarizers with a quarter-wave plate at $45°$.

B. TRANSIENT BANDS DURING SHEAR

The roll-cell pattern just discussed is the only pattern in Table 1 that has been convincingly explained. The second pattern listed is a band structure (i.e., orthogonal to flow) that temporarily appears after the start-up of shearing (see Figure 2). This pattern only appears for high molecular-weight PBG if the director has first been oriented more or less uniformly in the flow direction.[12] If the director is initially oriented uniformly orthogonal to the sample plane—i.e., in the homeotropic orientation—then unidirectional shearing flow will rotate the director until it becomes parallel to the flow direction. Thereafter the band pattern emerges, and subsequently disappears, being replaced by stripes or by a turbulent pattern, depending on how high the Ericksen number is. The band pattern does not appear if the director is initially in the vorticity direction; instead, stripes appear. The band pattern is produced by a periodic tilting of the director with respect to the flow

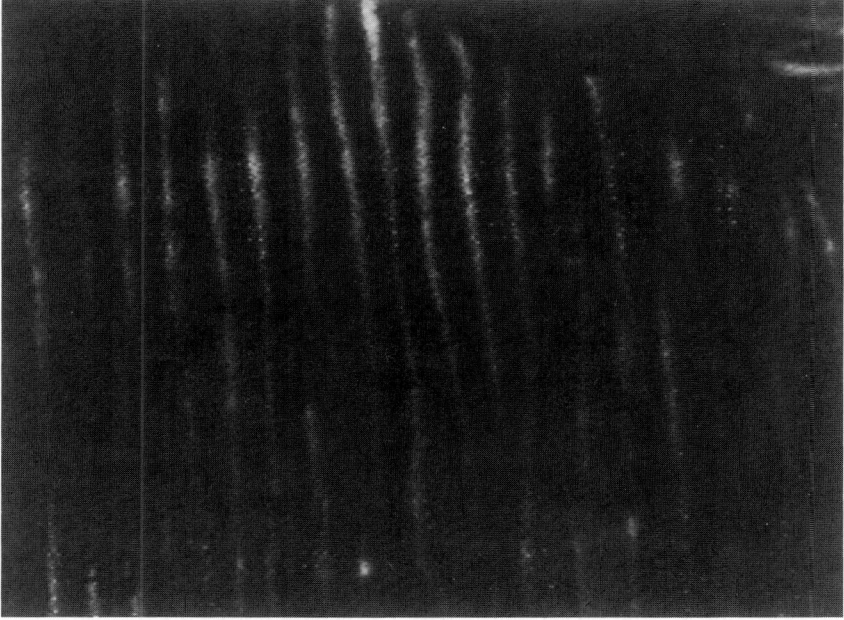

FIGURE 2   Transient band pattern formed during shearing of 15% PBG (molecular weight = 198,000) at a shearing velocity of 5μ/sec and a gap of 10μ, so that the Ericksen number is about 70. Here the initial director orientation was homeotropic, and the image was obtained 140 sec after start-up of shearing. The field of view is 890μ wide. The sample was observed between crossed $0°$–$90°$ polarizers.

direction. This tilting of the director is probably related to the tumbling character of the nematic, since a flow-aligning nematic would be expected to orient at its flow-alignment angle at high Ericksen numbers. Tumbling nematics, on the other hand, are known to be susceptible to instabilities that drive the director toward the vorticity direction. The formation of these transient bands has been observed in both PBG and in bulk, or thermotropic, LCP's. Since these transient bands form at shear rates low enough that the Leslie-Ericksen theory should apply, linear stability analysis of the Leslie-Ericksen equations for shear flow of a tumbling nematic with elastic constants and viscosities characteristic of high molecular-weight PBG solutions might predict band formation. Such a calculation has not yet been carried out.

Calculations have been carried out, however, for magnetically induced bands that are perhaps analogous to those induced by shear.[14] In a nematic sample with director oriented parallel to the plane of the sample, say in the x direction, bands perpendicular to x form when a strong magnetic field is applied perpendicular to x. Like the shear-induced bands, the bands induced by a magnetic field are transient, and eventually disappear. These transient bands form because there is a periodic secondary flow that permits more rapid reorientation of the director than would be possible if the director were to rotate without such a secondary flow.[14] Both linear and the nonlinear characteristics of the magnetically induced bands have been calculated from the Leslie-Ericksen equations, and the results compare favorably with experiments.[14,20] Perhaps a similar mechanism might explain the shear-induced transient bands.

C. TRANSIENT BANDS AFTER SHEAR

Next we come to the quasi-regular pattern that is most frequently observed, not only in PBG solutions, but also in most other liquid crystalline polymers, both thermotropic and lyotropic.[7] This is the band pattern that follows cessation of shearing (see Figure 3). After these bands have formed, they become wider, and then disappear again. The time required after shearing ceases for the bands to form is roughly inversely proportional to the shear rate. These bands, while seemingly universal in polymeric nematics, have never been reported for nonpolymeric nematics. A minimum shear rate is required for formation of these bands; for PBG, the minimum shear rate seems to correspond to a Deborah number of around 0.01. The bands do not form above a maximum shear rate corresponding to a Deborah number of around unity. While the correlation of band formation with the Deborah number suggests that molecular elasticity plays a key role, the time after shearing ceases at which the bands form can be several minutes, which is much longer than the time required for molecular-elastic effects to relax. Also, the width of the bands seems to be independent of, or at least insensitive to, the gap between the shearing surfaces, the shear rate, and the viscosity of the polymer. Thus the scaling properties

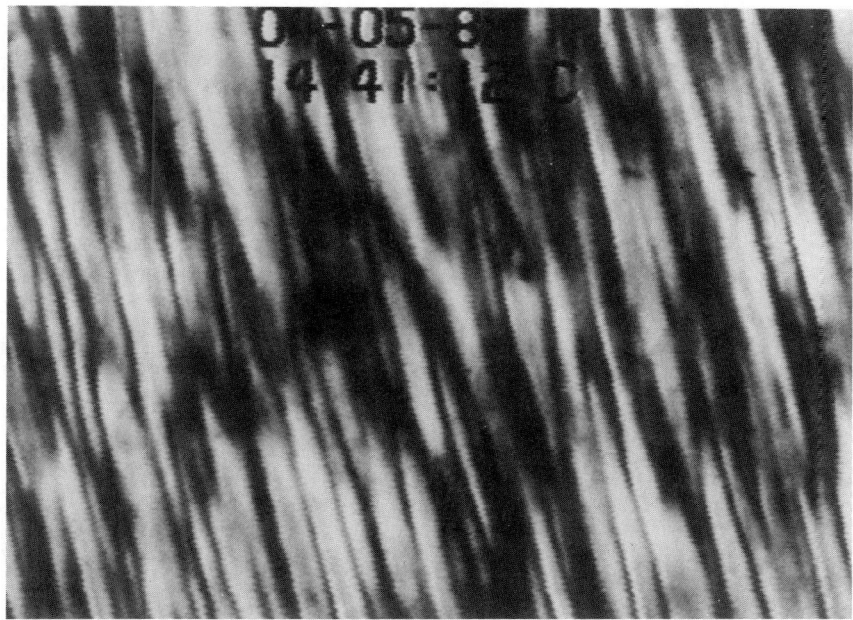

FIGURE 3 Transient band pattern formed 90 secs after cessation of shearing of 14% PBLG (molecular weight = 186,000) at a shearing velocity of 3000μ/sec and a gap of 400μ, so that the Ericksen number is about 2×10^5 and the Deborah number is 0.25. The field of view is 890μ wide. The sample was observed between crossed $0°-90°$ polarizers.

of these bands are extremely peculiar, and do not suggest any dimensionless group that might correlate their characteristics. Despite much discussion in the literature, the mechanism of formation of these bands has not yet been demonstrated.

D. STEADY-STATE STRIPES AT HIGH SHEAR RATES

Finally, we consider the steady-state stripes that form at high shear rate (see Figure 4). The extremely straight, regular, stripes are observed in PBG solutions for Deborah numbers in the range 0.3–3.[13] They are most prominent when observed between crossed polarizers oriented at 10^0 and 100^0 with respect to the flow direction, indicating that these stripes consist of a periodic tilting in the director with respect to the flow direction, with a tilt angle of about $\pm10^0$ with respect to the flow direction. The range of Deborah numbers over which the stripes form correlates with a regime of molecular dynamics called "wagging,"[10] which is predicted by the Smoluchowski equation mentioned earlier. The wagging regime occurs at shear rates above those for which simple tumbling—described by the Leslie-Ericksen theory—occurs, yet below the shear rates at which molecular-elastic effects to overcome the tendency to drive the molecules into a steady-state orientation. The wagging regime is produced by the competition between shearing torques acting on the director, which cause it to tumble, and the torques acting directly on the molecules, which cause them to align parallel to the flow. It is in this wagging regime that the stripes shown in Figure 4 are observed in PBG. These steady-state stripes are distinct from the steady-state stripes that occur at much lower shear rates (see Section II-A) in that the former are much straighter, are not surrounded by disclination lines, and have different optical characteristics, as shown by rotating the polarizers. The mechanism for formation of the high shear-rate stripes involves the Deborah number, whereas the low-shear-rate stripes are controlled by the Ericksen number.

FIGURE 4 Steady-state stripe pattern formed after shearing of 13.5% PBLG (molecular weight = 238,000) at a shearing velocity of 2500μ/sec and a gap of 100μ, so that the Ericksen number is about 3.5×10^5 and the Deborah number is 2.5. The field of view is 890μ wide. The sample was observed between crossed $10°-100°$ polarizers.

III. SUMMARY

We have described four distinct quasi-one-dimensional stripe and band patterns that form during or after steady shearing of liquid crystalline PBG solutions. Only one of these patterns has been explained, namely a steady-state stripe pattern produced by roll cells at low Deborah number, but moderately high Ericksen number. The other patterns are evidently influenced by Frank elasticity, molecular elasticity, or combinations of these, but the detailed mechanisms are still undiscovered.

REFERENCES

1. Berry, G. C. "Rheological Properties of Nematic Solutions of Rodlike Polymers." *Mol. Cryst. Liq. Cryst.* **165** (1988): 333–360.
2. Cladis, P. E., and W. van Saarloos. "Some Nonlinear Problems in Anisotropic Systems." In *Solitons in Liquid Crystals*, edited by L. Lam and J. Prost, 111–150. New York: Springer Press, 1992.
3. De'Neve, T., P. Navard, and Kléman, M. "Shear Rheology and Shear-Induced Textures of a Thermotropic Copolyesteramide." *J. Rheology* **37** (1993): 515–529.
4. Donald, A. M., and A. H. Windle. *Liquid Crystalline Polymers.* New York: Cambridge University Press, 1992.
5. Dubois-Violette, E., G. Durand, E. Guyon, P. Manneville, and P. Pieranski. "Instabilities in Nematic Liquid Crystals." *Solid State Physics (Supplement)* **14** (1978): 147–208.
6. Gähwiller, C. "Temperature Dependence of Flow Alignment in Nematic Liquid Crystals." *Phys. Rev. Lett.* **28** (1972): 1554–1556.
7. Gleeson, J. T., R. G. Larson, D. W. Mead, G. Kiss, and P. E. Cladis. "Image Analysis of Shear-Induced Textures in Liquid Crystalline Polymers." *Liq. Cryst.* **11** (1992): 341–364.
8. Kiss, G., and R. S. Porter. "Rheo-Optical Studies of Liquid Crystalline Solutions of Helical Polypeptides." *Mol. Cryst. Liq. Cryst.* **60** (1980): 267–280.
9. Kuzuu, N., and M. Doi. "Constitutive Equation for Nematic Liquid Crystals under Weak Velocity Gradient Derived from a Molecular Kinetic Equation." *J. Phys. Soc., Japan* **52** (1983): 3486–3494; **53** (1984): 1031–1038.
10. Larson, R. G. "Arrested Tumbling in Shearing Flows of Liquid Crystal Polymers." *Macromolecules* **23** (1990): 3983–3992.
11. Larson, R. G. "Roll-Cell Instabilities in Shearing Flows of Nematic Polymers." *J. Rheology* **37** (1993): 175–197.

12. Larson, R. G., and D. W. Mead. "Development of Orientation and Texture During Shearing of Liquid Crystalline Polymers." *Liq. Cryst.* **12** (1992): 751–768.

13. Larson, R. G., and D. W. Mead. "The Ericksen Number and Deborah Number Cascades in Sheared Polymeric Nematics." *Liq. Cryst.* **13** (1993): in press.

14. Lonberg, F., S. Fraden, A. J. Hurd, and R. E. Meyer. "Field-Induced Transient Periodic Structures in Nematic Liquid Crystals: The Twist-Freédericksz Transition." *Phys. Rev. Lett.* **52** (1984): 1903–1996.

15. Magda, J. J., S.-G. Baek, K. L. DeVries, and R. G. Larson. "Shear Flows of Liquid Crystal Polymers: Measurement of the Second Normal Stress Difference and the Doi Molecular Theory." *Macromolecules* **24** (1991): 4460–4468.

16. Manneville, P. "The Transition to Turbulence in Nematic Liquid Crystals." *Mol. Cryst. Liq. Cryst.* **70** (1981): 223–250.

17. Manneville, P., and E. Dubois-Violette. "Shear Flow Instability in Nematic Liquids: Theory Steady Simple Shear Flows." *J. Phys. (Paris)* **37** (1976): 285–296.

18. Marrucci, M., and P. L. Maffettone. "Description of the Liquid-Crystalline Phase of Rodlike Polymers at High Shear Rates." *Macromolecules* **22** (1989): 4076–4082.

19. Navard, P. "Formation of Band Textures in Hydroxypropylcellulose Liquid Crystals." *J. Polym. Sci.: Polym. Phys. Ed.* **24** (1986): 435–442.

20. Srajer, G., S. Fraden, and R. B. Meyer. "Field-Induced Nonequilibrium Periodic Structures in Nematic Liquid Crystals: Nonlinear Study of the Twist Frederiks Transition." *Phys. Rev. A* **39** (1989): 4828–4834.

Jian-Yang Yuan* and David M. Ronis
Department of Chemistry, McGill University, 801 Sherbrooke St. W. Montreal, Quebec H3A 2K6, Canada
*Present address: Imperical Oil Resources Limited, Research Centre, 3535 Research Road, N.W., Calgary, Alberta, T2L-2K8, Canada

New Instabilities in Colloidal Suspension Taylor-Couette Flow

We briefly review a theory of Taylor-Couette flow in colloidal crystals formed from suspensions of highly charged, poorly screened particles. We show that the usual Taylor instability is suppressed at low shear rate and two new types of instability emerge. Possible light scattering experiments are analyzed.

INTRODUCTION

Charged colloidal particles can form very soft crystalline suspensions due to electrostatic repulsion between the colloidal particles.[1] These suspended particles, although they may be very dilute, can drastically alter the dynamical properties of a fluid.[1,2,3] Our focus will be on Taylor-Couette flow (a coaxial cylindrical system with a fluid-filled gap) at relatively low shear rates generated by rotating either the inner or outer cylinder.

Spatio-Temporal Patterns, Ed. P. E. Cladis and P. Palffy-Muhoray,
SFI Studies in the Sciences of Complexity, Addison-Wesley, 1995   **229**

When these cylinders are rotated and the corresponding Taylor number reaches its critical value,[4] one would expect the Taylor instability to occur. In a crystalline colloidal suspension, one should naively expect that the critical Taylor number will increase because the resulting Taylor roll would cost extra elastic energy. However, other instability mechanisms associated with a sheared crystal or fluid may come into the picture[5,6] since the elastic moduli are so small. Furthermore, in the colloidal system the shear is transmitted primarily through the motion of the underlying liquid, which is very different from the case of molecular crystals. In this review, we summarize a theory for the long-wavelength properties of a sheared colloidal crystal in Taylor-Couette flow[7] in which we find that the critical Taylor number is increased as expected and two new instabilities occur at low shear rates when either the inner or outer cylinder is rotated. Linear stability analysis and study of nonlinear corrections show that the patterns associated with these two instabilities are characteristically different from those in the Taylor instability. Moreover, we show that the Bragg scattering pattern will be distorted by the rotation of the fluid, the shearing of the lattice due to the presence of a shear gradient in the fluid, and the distortion of the lattice associated with the instabilities.

The equations governing the perturbation of displacement fields, $\eta(\mathbf{r}, t)$, and of hydrostatic pressure, δp_h, in cylindrical coordinates can be expressed as[7]

$$D_t^2 \eta_r - 2\Omega(D_t \eta_\phi - r\Omega' \eta_r)$$
$$- \nu \left[\left(\nabla^2 - \frac{1}{r^2} \right) D_t \eta_r - \frac{2}{r^2} \frac{\partial}{\partial \phi} (D_t \eta_\phi - r\Omega' \eta_r) \right] = f_r(\mathbf{r}, t) - \frac{\partial}{\partial r} (\delta p_h) \quad (1)$$

$$D_t^2 \eta_\phi + 2\Omega D_t \eta_r$$
$$- \nu \left[\left(\nabla^2 - \frac{1}{r^2} \right) (D_t \eta_\phi - r\Omega' \eta_r) + \frac{2}{r^2} \frac{\partial}{\partial \phi} (D_t \eta_r) \right] = f_\phi(\mathbf{r}, t) - \frac{1}{r} \frac{\partial}{\partial \phi} (\delta p_h) \quad (2)$$

$$D_t^2 \eta_z - \nu \nabla^2 \eta_z = f_z(\mathbf{r}, t) - \frac{\partial}{\partial z} (\delta p_h) \quad (3)$$

where $D_t \equiv \partial/\partial t + \Omega(r)\partial/\partial \phi$, ν is the kinematic viscosity, and $\Omega(r)$ is angular velocity field of the laminar states, which are assumed to be the same as those in normal Taylor-Couette flow. The force, $\mathbf{f}(\mathbf{r}, t)$, can be written as a gradient of a stress tensor: $f_i(\mathbf{r}, t) = \partial_j \sigma_{ji}(\mathbf{r}, t)$ which is due to the inter-colloidal interaction (a lattice sum), and specifically, is a functional of the tensor $\partial_i \eta_j(\mathbf{r}, t)$. No-slip boundary conditions are imposed on the fluid at the walls, i.e., the tangential components of the fluid velocity equal the corresponding components of the surface velocity, while the normal component vanishes.

In obtaining these equations, we have applied the continuum limit and considered that (1) the drag force on colloidal particles is so high that particles flow with the fluid; (2) the gap between the two cylinders is relatively narrow so that the colloidal suspension within can be treated as a slightly bent and very soft crystal; and (3) the range of the inter-colloid forces is small compared with the Taylor cell radii so that local inversion symmetry is preserved.

The fluid is also assumed incompressible and the incompressibility condition for the velocity fields is shown to be equivalent to

$$\nabla \cdot \eta(\mathbf{r}, t) = 0 \, . \tag{4}$$

For the purposes of a linear stability analysis, the force may be further expanded to linear order in $\partial_i \eta_j(\mathbf{r}, t)$. This suggests the following model expression[6]

$$\mathbf{f}(\mathbf{r}, t) = c^2[1 + 2\varepsilon \cos(\Gamma t)]\nabla^2 \eta(r, t) \equiv c(t)^2 \nabla^2 \eta(r, t) \tag{5}$$

where Γ is an average velocity gradient in the r direction, c plays the role of the speed of sound, and $\varepsilon < 0.5$. The time-dependent term models the modulation of the local elastic constants that results from the periodic distortion of the crystal lattice caused by the laminar flow. We have ignored the contributions of higher order harmonics and have replaced the local modulation rate (i.e., the local velocity gradient) by the average gradient. This last assumption should be valid as long as the distance between the inner and outer cylinders is small compared with the radii. Furthermore, we have assumed that the system is locally isotropic, at least for the purposes of calculating the elastic moduli. Should this not be the case, the acoustic mode structure will be slightly more complicated, but the basic physical ideas will still be valid.

When the elastic forces are switched off, i.e., when $c = 0$, it is easy to show that Eqs. (1)–(3) are equivalent to the usual Navier-Stokes equations for Taylor-Couette flow,[4] and hence, Taylor roll instabilities result. On the other hand, if $c \neq 0$, i.e., when inter-colloid interactions are included, forming a Taylor roll requires that the colloid lattice be strongly deformed. It is then not surprising that the system becomes more stable to these kinds of deformations and convection related (such as Taylor) instabilities are strongly suppressed at low shear rates.

We can find stability boundaries which correspond to instabilities that are qualitatively similar to those found in the usual Taylor instability. While the mode to first go unstable is slightly different from the usual Taylor instability[7] the continuation of the mode corresponding to the usual Taylor instability will end up growing faster at high rotation rates, and should result in the Taylor roll pattern being formed. This is observed in the experiments of Weitz et al.[5] and should be expected because the convective terms which drive the Taylor instability will eventually dominate the elastic ones (remember that colloidal crystals are extremely soft) as the shear rate increases.

For reasonable values of the material parameters, the Taylor-like instabilities are not the first to occur. They are preempted by one of two resonance instabilities. Two types of these instabilities have been found numerically: one strongly depends on ε and the other is practically independent of ε.

The first class of instabilities is qualitatively similar to Mathieu instabilities[6,7] that occur when the modulation frequency of the crystal elastic properties is an integer subharmonic of twice a natural oscillation frequency of the system. Unlike the Taylor instability, the Mathieu instability only requires that the elastic property modulation rate be resonant, and this in turn only depends on the magnitude of the local shear gradient. As a result, the instability remains even when only outer cylinder is rotating. Unlike the normal Mathieu instabilities, it involves a pair of transverse acoustic modes: clockwise and counter-clockwise propagating relative to the laminar rotation. It is also shown that exchange of stability does not occur at the onset of such instabilities.[7] Furthermore, the motion associated with the first unstable mode is uniform in and pointing to z direction only, i.e.,

$$\begin{cases} \eta \cdot \nabla \mathbf{v} = 0 \\ \eta \cdot \nabla \eta = 0 \end{cases} \tag{6}$$

at the instability.

The second class of resonance-like instabilities is essentially independent of the value of ε, at least for physical values of ε, (i.e., $\varepsilon < 0.5$). As was the case in the Mathieu instabilities, the first mode to become unstable corresponds to vertically uniform motion and hence Eq. (6) is valid here as well.

The mechanism for this instability is not completely understood.[7] One observation worth mentioning is that for such an instability to occur, the tangential velocity difference across the gap between the two cylinders must be larger than twice the crystal transverse sound speed although there are *no* shock waves within the gap. It is possible to imagine the trapping of acoustic waves due to nonuniformity of the effective speed of sound under shear, although a detailed analysis remains to be performed.

The instability regions corresponding to the first instability for the two resonance mechanisms are compared in Figure 1. In the experiment of Weitz et al.,[5] vertical stripes of spacing about 1cm and 3mm were observed. The wider ones seem to be very close to what our theory has predicted. Detailed comparison of the smaller spacing ones to the acoustic instability discussed here remains to be done since an accurate value of sound speed is needed.

Nonlinear terms, as was discussed by Yuan and Ronis,[7] merely arise from the anharmonic contribution to the elastic interactions. The resulting equation is

$$D_t^2 \eta_z - \nu \nabla^2 D_t \eta_z = \nabla \cdot \left\{ \left[c(t)^2 + b|\nabla \eta_z|^2 + d|\nabla \eta_z|^4 + \ldots \right] \nabla \eta_z \right\} \tag{7}$$

It is shown[7] that (1) for $b > 0$, the system is stabilized and forming stripe patterns; (2) for $b < 0$ and $d > 0$, the system behaves differently depending on initial conditions.

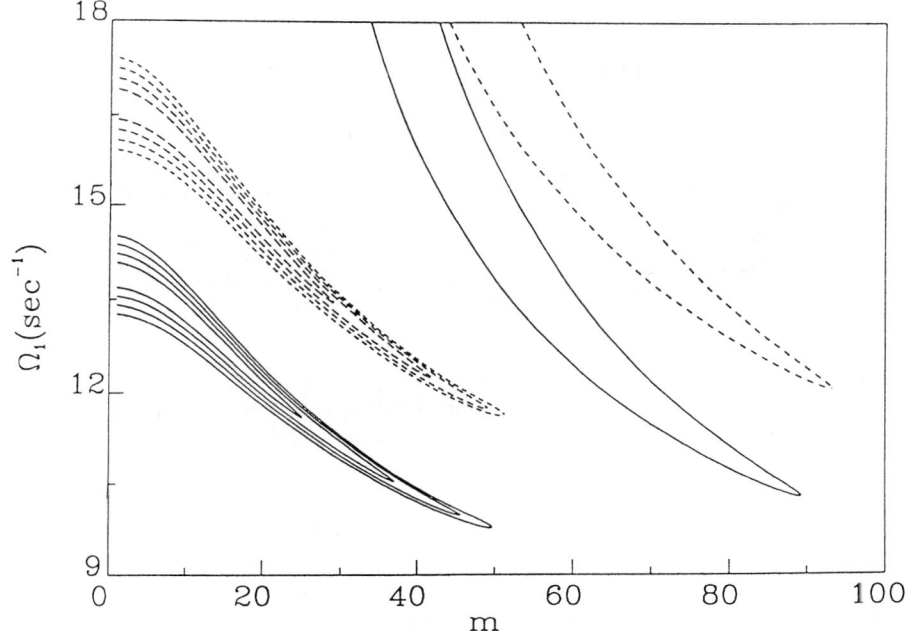

FIGURE 1 Mathieu and acoustic unstable regions for several values of ε; specifically, 0.09, 0.07, 0.05, and 0.03. Ω_1 is the angular velocity of the inner cylinder. m is the Fourier coordinate conjugate to ϕ. The radii of inner and outer cylinders are 4.52cm and 4.8375cm, respectively. Note that the acoustic unstable regions (at the upper right) are unchanged when ε is changed, while the Mathieu unstable regions (lower left) shrink when ε decreases. The acoustic instability can occur before (for $\varepsilon = 0.05$ and 0.03) or after (for $\varepsilon = 0.09$ and 0.07) the Mathieu instability. Moreover, reducing the speed of sound also makes the first instability occur earlier, and the behavior for $c = 10$cm/sec (solid) and $c = 12$cm/sec (dashed) is shown.

In an ideal light scattering experiment, it is easy to show that the momentum transfer for the scattering process, $\mathbf{q}$, and the reciprocal lattice vector, $\mathbf{K}$, is linearly related by

$$\begin{pmatrix} q_x \\ q_y \\ q_z \end{pmatrix} = \begin{pmatrix} \cos\Omega t & -\sin\Omega t - R\Omega't\cos\Omega t & -Q_r \\ \sin\Omega t & \cos\Omega t - R\Omega't\sin\Omega t & -Q_\phi \\ 0 & 0 & 1 \end{pmatrix} \begin{pmatrix} K_x \\ K_y \\ K_z \end{pmatrix} \qquad (8)$$

where

$$Q_r(\mathbf{R}, t) = \frac{\partial \eta_z}{\partial r}\big|_{\mathbf{r}=\mathbf{R}}, \text{ and } Q_\phi(\mathbf{R}, t) = \frac{1}{R}\frac{\partial \eta_z}{\partial \phi}\big|_{\mathbf{r}=\mathbf{R}} \qquad (9)$$

and **R** denotes the center of the illuminated region. Ω is the angular velocity at **R**. There are three contributions to the change in the scattering pattern, cf. Eqs. (8)–(9). The first is simply rotation about the z axis at rate Ω. The second results from the locally linear shear distortion of the lattice and appears in the terms proportional to $R\Omega' t$. Note that this results in a periodic distortion of the scattering pattern, where the period is roughly $(R\Omega')^{-1}$. Finally, there are the terms in $Q_{r(\phi)}$. These are the only ones which are nonuniform in space, and are responsible for the formation of the stripes.

Each of the foregoing contributions can be examined separately by choosing the direction of the momentum transfer. For example, the effect of the shear distortion of the lattice will not be seen if $K_x = 0$. Similarly, only the terms in the Q's will be seen if $q_x = q_y = 0$.

The observation of stripes is solely related to $Q_{r(\phi)}$. It can be shown[7] that their power spectra do not consist of a single frequency, and hence, the observed pattern is a superposition of clockwise and counter-clockwise rotating striped patterns, each rotating with angular velocity $-f/m$, where f are peak frequencies in the spectra and m is the Fourier coordinate conjugate to ϕ. These should be measurable by spectrally analyzing the scattering data. (Strictly speaking, we should compute the structure factor, and scattering intensity for a given experimental configuration in order to compare).

If the system is viewed by eye, the slowest motion will be the one most easily perceived. Of course, this conclusion is dependent on the relative magnitude of the slowest component being not too much smaller than the next slowest, etc., and this in turn depends on where the measurement is done; in general, the amplitude of the corotating component increases as **R** is moved to the inner cylinder. In reality, there are several slowly rotating components, and thus, the analysis of the experiment may be more complicated.

The theory presented here is valid only for long wavelength phenomena, and in particular we cannot preclude mechanisms which would lead to reconstruction of the lattice, either before or after the instabilities discussed here. Nonetheless, the instabilities discussed here must still occur, as long as the long wavelength properties can be discussed in terms of sheared elastic continua.

While we have used a specific model for the dynamics of the colloid and the underlying fluid, several points must be stressed. First, much of the structure of the continuum equations, cf. Eqs. (1)–(3), could have been deduced without reference to an underlying microscopic model, and the instabilities discussed here would still be obtained. An extreme example of this would arise if the shear of the colloidal lattice took place by having relatively large, but weakly aligned, domains of unsheared crystal slide over each other. The continuum equations do not include more microscopic effects like dislocations, and hence, if microscopic changes to the lattice occur, these will not be described in the current approximation. Nonetheless, the long wavelength phenomena still take place, even in the presence of dislocations, and the instabilities discussed in this work will occur. Moreover, the only adjustable parameter of the theory is ε in Eq. (5). Finally, we have ignored the space dependence of the local shear gradients in modeling the time dependent transverse sound speed. This is justified for thin samples. When this is not the case, then the modulation is not purely periodic, and the effect on the instability, and in particular on the time dependence of the stripped pattern is unclear. This point will be investigated in a future work.

ACKNOWLEDGMENTS

Portions of this work were supported by the National Sciences and Engineering Research Council of Canada and by Le Fonds pour la Formation de Chercheurs et l'Aide a la Recherche du Quebec. JYY would acknowledge IORL, Research Centre for support.

REFERENCES

1. Chaikin, P. M., J. M. di Meglio, W. D. Dozier, and H. M. Lindsay. In *Physics of Complex and SuperMolecular Fluids*, edited by S. A. Safran and N. A. Clark. New York: John Wiley and Sons, 1987.
2. Ackerson, B. J., and N. A. Clark. *Physica (Utrecht)* **118A** (1983): 221.
3. Jorand, M., E. Dubois-Violette, B. Pansu, and F. Rothen. *J. Phys. (Paris)* **49** (1988): 1119.
4. Chandrasekhar, S. *Hydrodynamic and Hydromagnetic Stability*. Dover, 1981.
5. Weitz, D. A., W. D. Dozier, and P. M. Chaikin. *J. Phys. (Paris) Colloq.* **46** (1985): C3-257.
6. Ronis, D., and S. Khan. *Phys. Rev. A* **41** (1990): 6813.
7. Yuan, J.-Y., and D. M. Ronis. *Phys. Rev. E* **48** (1993): 2880.

Jean-Marc Flesselles,[†] **Vincent Croquette,**[‡] **Stéphane Jucquois,**[‡] **and Béatrice Janiaud**[‡]

[†]INLN, UMR 129 CNRS, 1361 route des Lucioles, Sophia-Antipolis, 06560 Valbonne, France

[‡]LPS, ENS, URA D 1306 CNRS, 24 rue Lhomond, 75231 Paris Cedex 05, France

Behavior of a One-Dimensional Chain of Nonlinear Oscillators

Rayleigh-Bénard convection of argon gas in a narrow annular cell provides an experimental realization of a quasi-one-dimensional chain of nonlinear coupled oscillators, with periodic boundary conditions. Two different behaviors are observed, depending on whether the oscillators are in phase or not. If in phase, the evolution towards chaos occurs through a small system-like scenario: the period doubling of a T^2 torus. As soon as the chaotic behavior sets in, localized objects randomly propagate. If the oscillators are not in phase, the whole pattern moves through coherent global motions, which can be partially described by coupled Landau-Ginzburg equations. Similar phenomena occur in coupled map lattices.

1. INTRODUCTION

Chaotic behavior in nonlinear systems having a small number of degrees of freedom has been essentially understood during the past two decades. Spatial degrees of freedom introduce a nontrivial complexity to this kind of systems. Even for the a *priori* simplest case—systems with a single spatial dimension—no generic scenario

Spatio-Temporal Patterns, Ed. P. E. Cladis and P. Palffy-Muhoray,
SFI Studies in the Sciences of Complexity, Addison-Wesley, 1995

of transition to chaos has yet been found, though a mechanism for spreading of chaotic patches called spatiotemporal intermittency (STI) has been pointed out.[5,9] We study here a quasi-one-dimensional chain of nonlinear oscillators in a Rayleigh-Bénard convection experiment and analyze the interplay between the individual behavior of each oscillator and the global one of the entire chain.

Since, the convection fluid we use (argon gas) has a low Prandtl number; convection rolls undergo a bifurcation toward an oscillating state when driven beyond the convection threshold.[4] Hence, oscillating convection rolls provide nonlinear oscillators. Given the aspect ratio of the straight section of our cell, the convection pattern might be seen as a nearly one-dimensional chain of coupled nonlinear oscillators, with periodic boundary conditions.

Our measurements display two distinct behaviors depending on whether the oscillators oscillate in phase or not. If they do, we observe that spatial degrees of freedom are frozen as long as the system is not chaotic. Hence the transition to chaos occurs via a "small system" scenario, which is actually rarely observed in those systems, but which is theoretically well understood: a T^2 torus bifurcation. Within chaos, spatial features appear as localized wave holes.

If they do not oscillate in phase, the phase shift induces a global rotation of the pattern. Regular in the monoperiodic regime, where it can be described by coupled Landau-Ginzburg equations, this global motion loses its regularity when the Rayleigh number is increased. This behavior is compared to that observed in coupled map lattices.

---

2. EXPERIMENTAL SETUP

The setup used has already been described in other papers.[21,11] We recall here its characteristics.

2.1 CONVECTION CELL

The convection container has an annular shape, with an outer diameter of 40 mm, an inner diameter of 30 mm, and a thickness of 1.4 mm, giving an aspect ratio $\Gamma = 3.57$. It is filled with argon, pressurized around 60 atm in order to provide a high enough density to allow for visualization.

The cell is heated from below through a regulated copper block and its upper sapphire surface is maintained at fixed temperature with circulating water. The control parameter is the Rayleigh number, $Ra = g\alpha\Delta T d^3/\nu\kappa$ where g is the acceleration of gravitation, α is the thermal expansion coefficient, ΔT is the temperature difference between the top and bottom plates, ν and κ are the kinematic viscosity and the thermal diffusivity, and d is the thickness. ν, κ, and α are pressure and

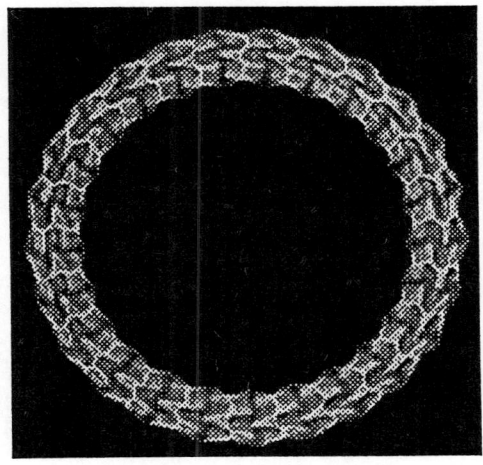

FIGURE 1 Snapshot of a pattern of 25 pairs of contrarotating oscillating rolls. White lines are due to uprising hot fluid, black lines to down-falling cold fluid. Here rolls are not oscillating in phase: diametrically opposite rolls are in phase opposition; hence, this pattern is spinning.

temperature dependent. At 60 atm, $Pr = \nu/\kappa \simeq 0.72$ for Ar. We achieve a ΔT ranging from 0 to 40 K. The onset of convection occurs at $\Delta T \simeq 2$ K (or $Ra \simeq 2000$) and the onset of oscillatory instability at $Ra \simeq 4Ra_c$. The incertitude of these measurements comes from the difficulty of appreciating the onset of the convection bifurcation, but not from the temperature stability or accuracy (± 2 mK).

Convection gives rise to a pattern of 24 to 30 pairs of contrarotating radial rolls, visualized by shadowgraphy (see Figure 1). Flows of fluid at different temperatures appear with different gray levels.

2.2 DIGITAL ACQUISITION AND TREATMENT

Raw data come from the digitization of the shadowgraphy image of the pattern. Data acquisition is provided by a video camera hooked up to a frame grabber in a computer. Special care has been taken to reduce geometric distortions by synchronizing the acquisition card with the pixel clock of the camera; this ensures that pixels are actual squares. The 512 pixels regularly spaced on a circle, which intercept every roll of the pattern, are recorded at a frequency of 25 Hz. This sampling rate is large enough for the oscillation frequency, which is typically around 1 Hz. Any other motion in the system is slower. The recorded pixels are dumped to a hard disk for later processing. Series as long as 32,768 time steps (about 22 min) have been recorded, giving a absolute precision of 0.76 mHz from 0 to 12.5 Hz. The data are spatially filtered and complexified through Fourier transforms. In order to get rid of the illumination inhomogeneities, the phase of the (complex) amplitude at the 64 points regularly spaced in the cell, extracted for each time step, provides the actual relevant signal rather than the raw grey-level signal. This quantity contains the whole spatio-temporal characteristics of the system.

3. IN PHASE OSCILLATION

At the threshold of the oscillatory instability, the oscillators can be prepared in two different states: either they all oscillate in phase or they do not. In this case they present a uniform phase shift from roll to roll such that the integrated phase is $2\pi n$, where n is found to be ± 1. We first consider the case where all oscillators are in phase $(n = 0)$.

3.1 THE DIFFERENT REGIMES

Analysis of the data consists first in calculating the Fourier power spectra at each of the 64 positions of the cell where the signal is recorded. As a first approximation, these spectra are identical. As long as the oscillators are not chaotic, they are spikes spectra, made of sharp peaks originating from a small number of frequencies and their harmonics.

In Figure 2 we show a typical example of such a spectrum. As the forcing parameter—the Rayleigh number—is varied, these spectra evolve. Instead of stacking the plots at each temperature, we have presented the results as an image where the gray level is proportional to the logarithm of the power. Results for an increasing ramp are shown in Figure 3.

Fine details show small hysteresis for a decreasing ramp.

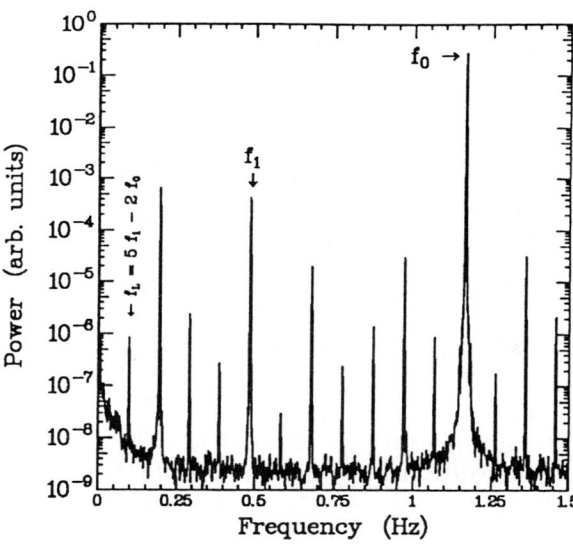

FIGURE 2 Low frequency part of time spectrum in the biperiodic regime (26 pairs of rolls, $\Delta T = 19.84$ K). Here $f_1/f_0 = 0.416757 \cdots \simeq 5/12$. All spikes can be labeled with f_0 and f_1. The lowest frequency f_L has been indicated.

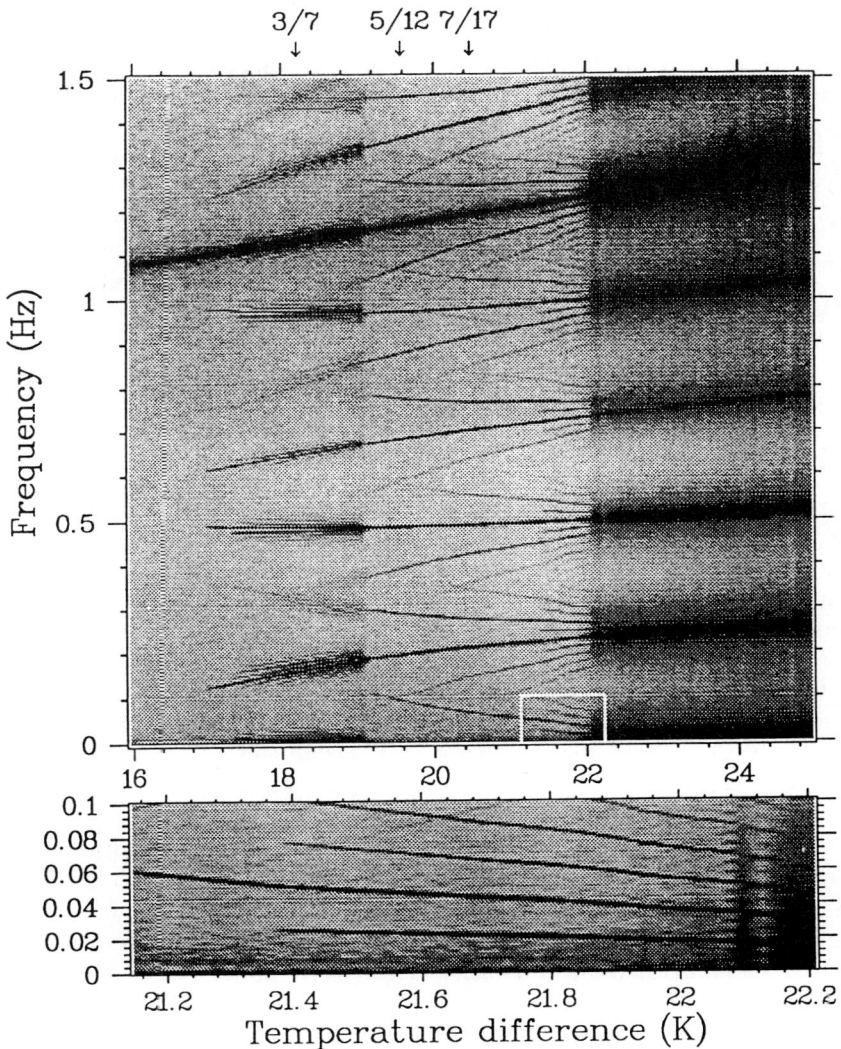

FIGURE 3 Evolution of Fourier spectra as a function of ΔT for an increasing ramp (128 temperatures) and 26 pairs of rolls. Spectra are monoperiodic for $\Delta T < 17.01$ K, and chaotic for $\Delta T > 22.08$ K. Inhomogeneous biperiodism occurs for 17.37 K $< \Delta T < 19.13$ K. The torus doubling takes place at $\Delta T = 21.38$ K. The lower part of the figure is a blowup on 64 temperatures of the white rectangle. Some rational values of f_1/f_0 have been indicated with arrows.

Three regimes appear distinctly, with different spatiotemporal behaviors. At low Ra (here $\Delta T \leq 17.01$ K), the system is monoperiodic with frequency f_0, as expected from the Hopf bifurcation to the oscillating state. At high Ra (here $\Delta T \geq 22.08$ K), the system is chaotic. In between, it is (basically) biperiodic, the secondary frequency f_1 being essentially temperature independent. Within this regime, three different sub-behaviors are observed: close to the biperiodism threshold, frequency f_1 is not constant along the annulus but the cell splits into two domains with slightly different secondary frequencies f_1. This is probably induced by cell imperfections but will not be discussed here. Above 19.13 K, these two domains disappear, f_1 recovers spatial homogeneity and biperiodism is pure. Finally, above 21.38 K we observe the doubling of the T^2 torus as the scenario for the transition towards chaos.

3.2 THE TORUS BIFURCATION

The characteristics of this transition require the description of the initial state, hence of the pure biperiodic state. There, spectra are composed of fine spikes that can be labeled using only the two fundamental frequencies f_0 and f_1, where f_0 varies linearly with Ra and f_1 is nearly constant. Note that the lowest frequency of the spectra is given by $f_L = 5f_1 - 2f_0$.

Moreover, in contrast with what is generally observed in two frequency systems,[13] no frequency locking has been detected and the ratio f_1/f_0 evolves monotonically with the Rayleigh number. Hence the whole system is essentially equivalent to two monoperiodic oscillators, with incommensurable frequencies. The dynamical trajectory explores a T^2 torus and its Poincaré section has the topology of a circle.

As Ra is increased further, we observe a bifurcation leading to the appearance of a subharmonic of the lowest frequency component $f_L/2 = 5f_1/2 - f_0$ (see blowup of Figure 3). Within experimental noise (90 dB), no component $f_0/2$ nor $f_1/2$ is detectable. Due to the absence of locking, the T^2 torus has doubled, as seen on the Poincaré section (Figure 4). Further increase of Ra induces more subharmonic bifurcations interrupted by chaotic windows before a final chaotic dynamics. Here a bifurcation leading to $f_L/6$ (see Figure 5) is followed by a chaotic window and precedes the occurrence of $f_L/4$. The bifurcation cascade ends up at this stage and chaotic dynamics becomes the rule, sometimes disrupted by small windows of regular motion. This cascade has strong hysteresis and the actual subharmonic bifurcations observed depends on the run. We suspect this dependence to be due to the spatial degrees of freedom.

The possibility of period doubling of a T^2 torus has been numerically explored by Kaneko[17] on mappings and numerically observed by Franceschini[12] on a system of differential equations arising from a seven-mode truncation of the Navier-Stokes equations. It has been studied from a mathematical point of view by Iooss and

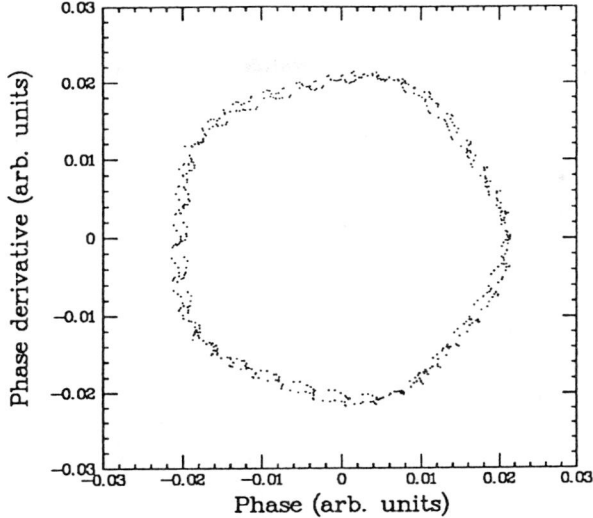

FIGURE 4 Poincaré section showing the torus doubling. $\Delta T = 21.56$ K. Radius of torus in the other direction is of order 1. Here $f_1/f_0 = [2, 2, 6, 2, 5, 2, 1, 4, \cdots]$. The fine modulations are related to the $13/32$ nearby locking.

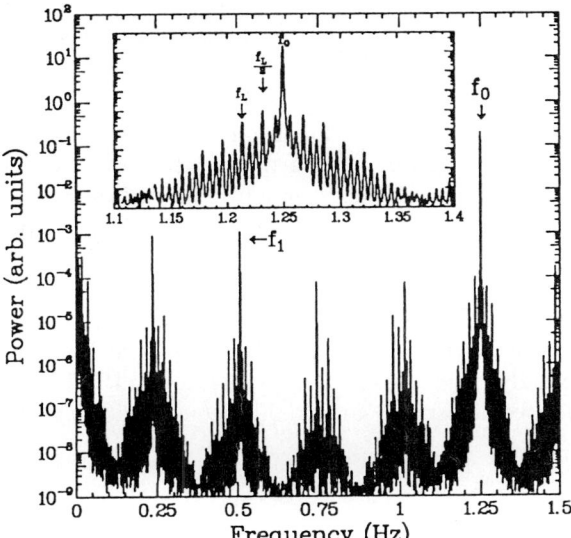

FIGURE 5 Biperiodc spectrum showing $f_L/6$. Insert is a blowup around f_0. $\Delta T = 22.08$ K. About 220 peaks are present on this spectrum.

Los,[15] and a physical approach has been used by Arneodo, Coullet, and Spiegel.[1] The scope of these studies was to determine whether a period-doubling cascade could be observed on a system having two incommensurable frequencies in a way similar to the well-known scenario occurring on a limit cycle.[10,23] The results may be summarized as follow: for most experimental conditions, period doubling of a T^2 torus is possible; however, small chaotic domains exist near each bifurcation point. It is thus expected that the cascade will abort after a few steps, leading to the destruction of the T^2 torus.

To our knowledge, experimental observations of torus doubling have been reported in three different systems: electrochemical reaction,[2] convection in molten gallium,[22] and in helium.[14] In this latter case, both oscillators seem to be essentially decoupled. In the two former cases, the first torus bifurcation is clearly seen but its unfoldings and especially the bubbles of chaos are lacking. We are thus able to observe the previously unreachable behaviors.

3.3 WAVE HOLES

In our system, as long as its dynamical behavior is regular, all rolls oscillate in phase at all frequency components of the spectra. As soon as the chain of oscillators undergoes a chaotic dynamics, bandpass filtering around frequency f_1 displays abrupt phase variations that we consider to be wave holes.

Such filtering is made because frequency peaks corresponding to the main and the secondary frequencies are still well defined close to the onset of chaos. Since f_1 is close to a $2/5$ subharmonic of f_0, and because of the chaotic enlarging of peaks, there are only five peaks in the relevant low frequency part of the spectrum, which allows for an unambiguous definition of the component of signal at f_0 or f_1.

Beside small phase fluctuations due to experimental imperfections of the cell, all oscillators remain in phase at f_0, whereas mobile localized defects exist on f_1. They are characterized by a strong amplitude dip and a phase jump localized at their cores (see Figure 6). Moreover, the faster a defect is moving, the higher its core amplitude. These defects appear and disappear spontaneously and their averaged number increases with the forcing parameter, but the large statistics, which are necessary for quantitative laws, are not yet available.

These objects are strongly reminiscent of the so-called Nozaki-Bekki holes[3] and of the related objects studied by Manneville and Chaté[6,7] on the Landau-Ginzburg equation. They have already been observed by Janiaud et al.[16] In the present experiment, this localized wave defect appears simultaneously with chaotic dynamics.

4. OUT OF PHASE OSCILLATION

We now consider the case where the temporal phase of the roll oscillation at the main frequency is not constant throughout the cell (see Figure 1). Such a state is

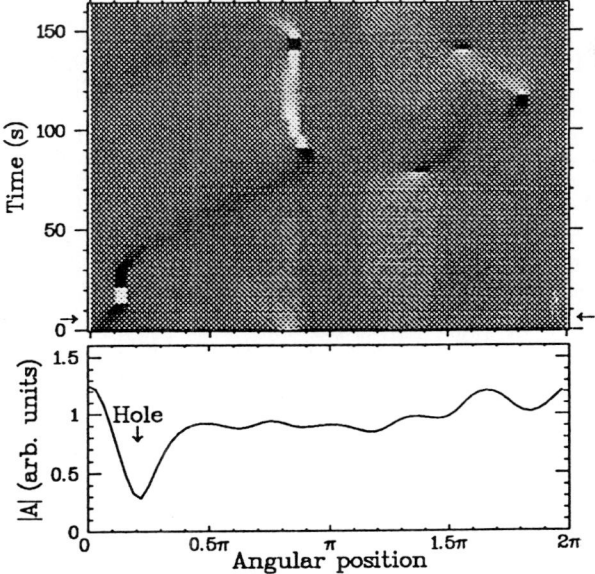

FIGURE 6
Top: Spatiotemporal evolution of holes in the f_1 component for a 26-roll pattern in the chaotic regime. Gray level is proportional to the local wavenumber. Bottom: Amplitude profile at time 6 s as indicated by arrows.

prepared by performing a small temperature jump around the oscillation threshold. Since this bifurcation is supercritical, it should develop homogeneously around the cell with a vanishing amplitude. But because of small inhomogeneities, oscillation starts at a given point in the cell and propagates until it has invaded the whole pattern. If temperature is slowly raised, phase has time to diffuse and no phase shift is obtained once all rolls are oscillating. Conversely, if the bifurcation temperature is crossed fast enough, one might get a nonzero phase shift state. Phase diffusion insures the equality of the phase shift from roll to roll. With this method, we have never been able to get an integrated phase shift other than $\pm 2\pi$.

4.1 EXPERIMENTAL RESULTS AT LOW Ra

We first describe the monoperiodic situation where the temporal phase is unambiguously defined. Then the whole pattern spins at a Ra-dependent velocity Ω. This rotation rate depends linearly on Ra and is independent of the roll number N_r, as long as the oscillation is monoperiodic. A typical order of magnitude for the rotation rate is 1 cycle per hour, whereas the oscillation frequency is of order 1 Hz. As a first approximation, the rotation is steady. However, small modulations due to the passing of the rolls are also present. Hence this modulation has a pulsation $\Omega_{mod} = N_r\Omega$. We interpret this phenomenon as a pinning of the roll pattern by cell imperfections.

Hence Fourier spectra in such a state are not truly monoperiodic: the main spike (corresponding to f_0) as well as the low-frequency continuum part corresponding to the global motion are enlarged by nonlinear combinations with the low frequency corresponding to Ω_{mod}.

4.2 AMPLITUDE EQUATIONS

This behavior can easily be explained as a result of the coupling between the spatial phase of the pattern which expresses the broken phase invariance of the pattern and the temporal phase of the roll oscillation which expresses the broken parity.

The study of a secondary Hopf bifurcation occurring on a one-dimensional periodic pattern has already been analyzed and studied numerically by Lega.[20] The initial pattern, $U_0(x)$, is periodic. Close to threshold, when the system undergoes its Hopf bifurcation, the signal reads:

$$U(x,t) = U_0(x+\phi) + \exp(i\omega_0 t)A(X,T)\zeta(x+\phi) + \text{c.c.} + \ldots, \qquad (1)$$

where ω_0 is the frequency associated to the bifurcation, and ζ its eigenvector, which is assumed to have the same period or twice that of U_0. The real phase ϕ and the complex amplitude A are functions of the slow space and time variables X and T, thereafter simply noted x and t. The phase ϕ represents the mean position of the oscillating rolls. Hence the derivative $\partial\phi/\partial t$ is the rotation rate Ω of the pattern.

According to the classification of generic forms of secondary instabilities occurring on a one-dimensional periodic pattern,[8] the dynamical equations for A and ϕ are written as:

$$\tau_A \frac{\partial A}{\partial t} = \mu A + \xi_A^2(1+i\alpha)\frac{\partial^2 A}{\partial x^2} - g(1+i\beta)|A|^2 A - \xi_\phi(\gamma+i\delta)\frac{\partial\phi}{\partial x}A, \qquad (2)$$

$$\tau_\phi \frac{\partial\phi}{\partial t} = \xi_A\gamma\frac{\partial}{\partial x}|A|^2 + i\xi_A\eta\left[\frac{\partial A}{\partial x}\overline{A} - \frac{\partial\overline{A}}{\partial x}A\right] + \xi_\phi^2\frac{\partial^2\phi}{\partial x^2}. \qquad (3)$$

We assume that ϕ and $|A|$ do not depend on space, and that space and time phase modulations of A are uncorrelated. These assumptions together with the periodic boundary conditions suggest the following form for A:

$$A_n(x,t) = a_n(t)\exp[in\frac{2\pi}{L}x + i\theta_n(t)], \qquad (4)$$

where n is a signed integer that we call the index of the pattern.

When inserted into Eqs. (2) and (3), we get:

$$a_n = \left[\frac{\left(\mu - (\xi_A n 2\pi/L)^2\right)}{g}\right]^{1/2} \qquad (5)$$

$$\frac{\partial\phi}{\partial t} = -\xi_A\eta\frac{4\pi}{L}n\left[\frac{\left(\mu - ((\xi_A 2\pi/L)n)^2\right)}{g}\right]. \qquad (6)$$

Equation (5) gives the instability threshold as a function of the mode $\mu_n = (n2\pi\xi_A/L)^2$. But we are not experimentally able to measure such a small deviation. We think that ξ_A is very small in our experiment so that Eq. (6) reduces to

$$\frac{\partial\phi}{\partial t} = -\xi_A\eta\frac{4\pi}{L}n\left(\frac{\mu}{g}\right). \tag{7}$$

This equation provides an appropriate description of the experimental behavior: first, rotation occurs for nonzero n and the direction of rotation depends on the sign of n; secondly, the rotation rate varies linearly with the control parameter μ and is independent of the wave number.

4.3 BEYOND MONOPERIODICITY

This description in terms of amplitude equations breaks down if the oscillation is not monoperiodic, where no appropriate description of the state is available. But it is experimentally still possible to measure a rotation rate. The secondary frequency, though 30 dB below the main one, tends to slow down the rotation, which then varies more slowly than linearly with Ra (see Figure 7). As forcing is increased, the rotation loses its smoothness and the rotation rate starts to fluctuate. Temperatures where such changes occur are clearly correlated with the changes of regimes observed in the in-phase case.

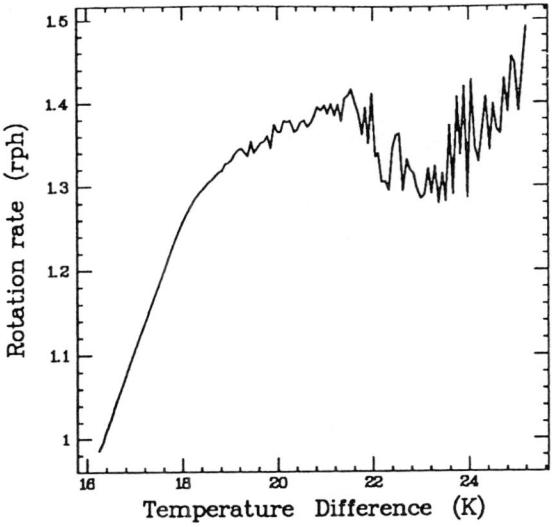

FIGURE 7 Average rotation speed for a $n = 1$ state of 26 rolls.

In the biperiodic and chaotic regimes, the pattern develops states where the secondary frequency may present transient phase shifts whereas the main one remains in phase. In such a state, small motions of the pattern are detectable. But the pinning of the pattern due to the cell's imperfections prevents us from seeing permanent rotations. At higher Ra, phase shifts appear spontaneously on the main frequency and induce global motion.

---

5. COUPLED MAP LATTICES

Such behavior is strongly reminiscent of phenomena recently observed by Kaneko[19] in coupled map lattices.

5.1 MODEL AND PREVIOUS RESULTS

Coupled map lattices (CML) consist of a one-dimensional set of N logistic maps linearly coupled to their nearest neighbors. Periodic boundary conditions at the end of the chain are assumed.

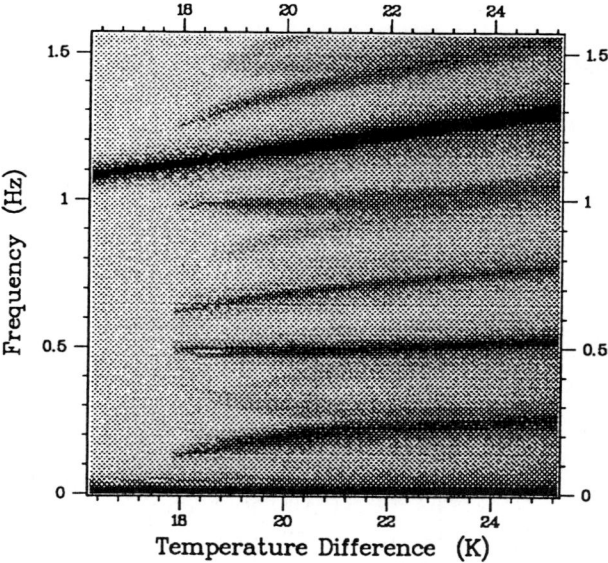

FIGURE 8   Similar to Figure 3 except here we show an $n = 1$ state of 26 rolls. Note how the spike structure disappears as soon as the secondary frequency sets in.

Using notations of Kaneko,[18] this model writes:

$$x_{n+1}(i) = (1 - \epsilon)f(x_n(i)) + \epsilon/2[f(x_n(i+1)) + [f(x_n(i-1))]] \qquad (8)$$

where $x_1(i)$ is the state of the map at site i and time n, $f(y)$ is the standard logistic map $(f(y) = 1 - ay^2)$, ϵ is a coupling parameter, and a is the control parameter. The evolution of the model is fully determined once the initial conditions have been given, i.e., the set of $x_0(i)$, $0 \leq i < N$.

We first summarize the main results of Kaneko.[18] Simulations were performed in the strong coupling regime $\epsilon = 0.5$, and with values of a ranging from 1.55 to 1.75 and the number of sites between 50 and 100. The system spontaneously generates a wavelength of approximately seven or eight lattice spacings, which defines a wave number k_0. As expected from the logistic map behavior, period 2 and 4 temporal oscillations successively appear for appropriate parameter settings as a is increased.

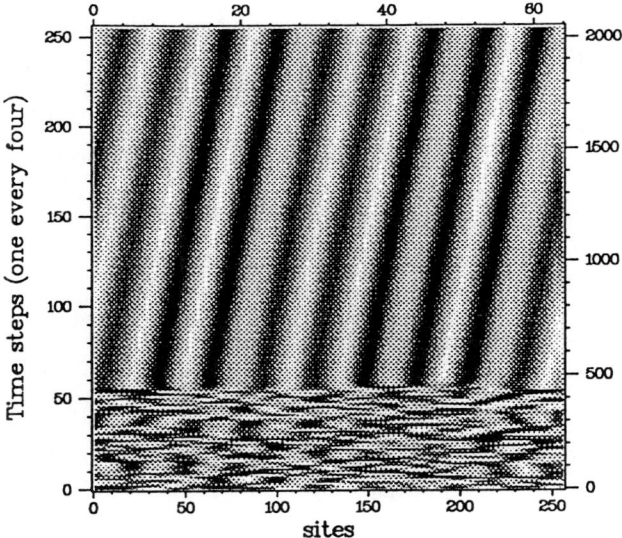

FIGURE 9   Space time evolution of a coupled map lattice, with $\epsilon = 0.5$ and $a = 1.77$. Here 64 oscillators are displayed in the horizontal direction and time runs in the vertical direction. Profiles are displayed modulo 4 so that the fast dynamics is frozen. The beginning of this sequence corresponds to a transient where temporal evolution is chaotic. At the end, the spatial pattern has reached an ordered state of index -2 which slowly drifts.

For some initial conditions, the period 4 oscillation exhibits spatially localized phase defects. In these cases the whole pattern slowly drifts with a velocity approximately proportional to the integrated phase shift.

5.2 ORIGIN OF ROTATION

In order to identify precisely the conditions for the appearance of the phase shifts and the induced rotation, we have performed simulations on systems of 64 sites. Coupling between maps stabilizes the dynamics: For an isolated (noncoupled) logistic map, the period doubling leading to period 4 occurs at $a_{1/4,\epsilon=0} = 1.232$. whereas this threshold occurs at $a_{1/4,\epsilon=5} = 1.701$ when $\epsilon = .5$. We have numerically observed that phase shifts, and thus rotation, could take place only if period 4 had settled in the system. It is actually a straightforward consequence of the fact that it is the lowest period for which a phase can be defined.

In Figure 10 we show that the drift velocity for a fixed integrated phase shift changes linearly with the control parameter, and its extrapolation to zero velocity corresponds to the onset of period 4.

Actually the discretization due to the mapping locks the phase to four preferred values distant from $\pi/2$. Performing four-time-step Fourier transforms over the N

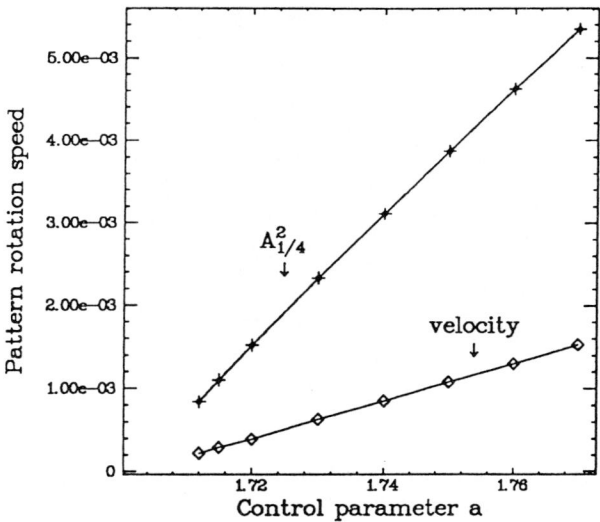

FIGURE 10 Velocity of a drifting pattern and amplitude of the period 4 mode $A_{1/4}$ versus the control parameter a. ($N = 64$, $\epsilon = 0.5$, index $= -2$.)

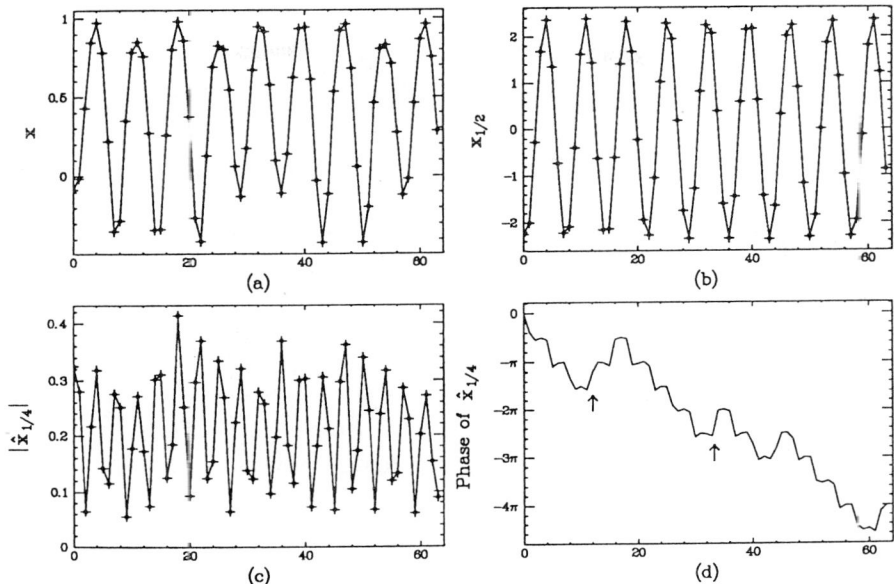

FIGURE 11 Decomposition of a traveling pattern; (a) full signal; (b) period 2 component; (c) amplitude of the period 4 component with its phase on (d). ($N = 64$, $\epsilon = 0.5$, $A = 1.77$, index $= -2$.)

sites, one obtains the spatial dependence of the (temporal) phase of the $\omega = 1/4$ mode. A typical result for an integrated phase of 4π is seen on Figure 11. In this regime, the $\omega = 0$ and $\omega = 1/2$ modes present a wave vector $k_0 = 2\pi N_0/N$, with $N_0 = 9$. The 1/4-mode amplitude is strongly modulated in space; its amplitude presents 18 minima. Its phase displays 18 quantized jumps of $\pm\pi/2$ between which the phase remains approximately constant, corresponding to the places where the amplitude is noticeable. Boundary conditions impose the total phase shift to be a multiple of 2π, hence a multiple of four jumps.

Extraction of the phase permits the preparation of initial conditions with given phase shifts by adding the right amount of phase to the different sites and doing the inverse Fourier transform. The efficiency of this procedure shows that considering the phase as a succession of discretized phase jumps is appropriate. We assume that this series of jumps is enough to characterize a pattern. Supposing that the phase jumps of $\pm\pi/2$ each time the amplitude of the 1/4 mode is depressed, it is easy to calculate the distribution of states with a given index I. The probability of having a state of index I with $2N_0$ steps is equivalent to a random walk. Thus:

$$P(2N_0; I) = 2^{1-2N_0} \begin{pmatrix} 2N_0 \\ N_0 + 2I \end{pmatrix} \tag{9}$$

$$\simeq_{(N_0 \gg 1)} \frac{2}{\sqrt{\pi N_0}} \exp(-4I^2/N_0). \tag{10}$$

This result quantitatively agrees with Kaneko's result of this probability as a function of the index for $N_0 = 28$ (see Figure 3(b) in Kaneko[19]).

6. CONCLUSION

Our experiment exemplifies the behavior of a linear chain of oscillators. It shows that the route to chaos depends strongly on whether the system has a single phase or not. In the former case, chaos occurs via an uncompleted cascade expected for a T^2 torus. It demonstrates that a chain of oscillators may transit towards spatio-temporal chaos by following a scenario typical of systems with a small number of degrees of freedom. This experiment also illustrates how the spatial degrees of freedom of the system unfreeze as soon as chaotic behavior sets in; this happens first through the hysteresis of the transition, and second via the localized defects of the Nozaki-Bekki type.

Nevertheless, why they appear on the secondary frequency and are not seen on the main one remains unexplained. If the oscillators do not oscillate in phase, the spatial extension is thus embedded in the very pattern and cannot be ruled out. The route to chaos is very different and has not yet been clarified. The pattern is subject to a global motion, regular as long as the oscillation is monoperiodic, irregular and probably chaotic at higher Rayleigh number. The origin of this motion is well understood with amplitude equations describing the coupling between the spatial phase of the pattern and the local temporal phase of the oscillation. But such a description breaks down for nonmonoperiodic situations. Similar behavior is observed on coupled map lattices, a model system for coupled nonlinear oscillators. We have shown in particular that a phase shift—localized or not—is enough to induce a drift of the pattern.

ACKNOWLEDGMENTS

We thank D. Bensimon, C. Bouchiat, A. Chiffaudel, and Y. Pomeau for discussions. We are grateful to A. Arneodo, A. Chenciner, G. Iooss, and D. Ruelle for enlightening us on mathematics of torus bifurcations.

REFERENCES

1. Arnéodo, A., P. H. Coullet, and E. Spiegel. "Cascade of Period Doublings of Tori." *Phys. Lett.* **94A(1)** (1983): 1–6.
2. Basset, M., and J. Hudson. "Experimental Evidence of Period Coupling of Tori During an Electrochemical Reaction." *Physica D* **35** (1989): 289–298.
3. Bekki, N., and K. Nozaki. "Formation of Spatial Patterns and Holes in the Generalized Ginzburg-Landau Equation." *Phys. Lett.* **110A(3)**: 133–135.
4. Busse, F. "The Oscillatory Instability of Convection Rolls in a Low Prandtl Number Fluid." *J. Fluid Mech.* **52(1)** (1972): 97–112.
5. Caponeri, M., and S. Ciliberto. "Thermodynamic Aspects of the Transition to Spatiotemporal Chaos." *Physica D* **58** (1992): 365–383.
6. Chaté, H. "Spatiotemporal Intermittency Regimes of the One-Dimensional Complex Ginzburg-Landau Equation." *Nonlinearity* **7** (1994): 185–204.
7. Chaté, H., and P. Manneville. "Stability of the Bekki-Nozaki Hole Solutions to the One-Dimensional Complex Ginzburg-Landau Equation." *Phys. Lett. A* **171** (1992): 183–188.
8. Coullet, P., and G. Iooss. "Instabilities of One-Dimensional Cellular Patterns." *Phys. Rev. Lett.* **64(8)** (1990): 866–869.
9. Daviaud, F., J. Lega, P. Bergé, P. Coullet, and M. Dubois. "Spatio-Temporal Intermittency in a 1-D Convective Pattern: Theoretical Model and Experiments." *Physica D* **55** (1992): 287.
10. Feigenbaum, M. "Quantitative Universality for a Class of Nonlinear Transformation." *J. Stat. Phys.* **19**: 25.
11. Flesselles, J.-M., V. Croquette, and S. Jucquois. "Period Doubling of a Torus in a Chain of Oscillators." *Phys. Rev. Lett.* **72(18)** (1994): 2871–2874.
12. Franceschini, V. "Bifurcations of Tori and Phase Locking in a Dissipative System of Differential Equations." *Physica* **6 D** (1983): 285–304.
13. Glazier, J. A., and A. Libchaber. "Quasi-Periodicity and Dynamical Systems: An Experimentalist's View." *IEEE Trans. Circ. Syst.* **35(7)** (1988): 790–809.
14. Haucke, H., Y. Maeno, and J. Wheatley. "Period-Doubling Two-Torus State in a Convecting ^{3}he-Superfluid ^{4}he Solution." In *LT17, Proceedings of the Seventeenth International Conference on Low-Temperature Physics*, edited by U. Eckern, A. Schmid, W. Weber, and H. Wuhl, 1123–1124. Karlsruhe, Germany: Elsevier, 1984.
15. Iooss, G., and J. Los. "Quasi-Genericity of Bifurcations to High-Dimensional Invariant Tori for Maps." *Commun. Math. Phys.* **119** (1988): 453–500.
16. Janiaud, B., S. Jucquois, J. Lega, and V. Croquette. "Experimental Evidence of Bekki Nozaki Holes" In *Pattern Formation in Complex Dissipative Systems*, edited by S. Kai, 538–550. Conference held in Kitakyushu, Japan, 18–20 September 1991. Singapore: World Scientific, 1992.
17. Kaneko, K. "Doubling of Torus." *Prog. Theor. Phys.* **69** (1983): 1806–1810.

18. Kaneko, K. "Global Traveling Wave Triggered by Local Phase Slips." *Phys. Rev. Lett.* **69(6)** (1992): 905–908.

19. Kaneko, K. "Chaotic Traveling Waves in a Coupled Map Lattice." *Physica D* **68** (1993): 299–317.

20. Lega, J. "Secondary Hopf Bifurcation of a One-Dimensional Periodic Pattern." *Eur. J. Mech. B/Fluids* **10(2)** (Suppl.) (1991): 145–150.

21. Lega, J., B. Janiaud, S. Jucquois, and V. Croquette. "Localized Phase Jumps in a Propagating Wave Train." *Phys. Rev. A* **45** (1992): 5596.

22. McKell, K., D. Broomhead, R. Jones, and D. Hurle. "Torus Doubling in Convecting Molten Gallium." *Europhys. Lett.* **12(6)** (1990): 513–518.

23. Tresser, C., and P. Coullet. "Iterations d'endomorphismes et groupe de renormalisation." *CRAS Paris* **A 287** (1978): 577.

David Raitt and Hermann Riecke

Department of Engineering Sciences and Applied Mathematics, Northwestern University, Evanston, IL 60208, USA

Domain Structures and Zig-Zag Patterns Modeled by a Fourth-Order Ginzburg-Landau Equation

1. INTRODUCTION

The formation of steady spatial structures in systems far from equilibrium has been studied in great detail over the past years, the classical examples being Rayleigh-Bénard convection and Taylor vortex flow.[9] In quasi-one-dimensional geometries, they usually share a common feature: the stable structures that arise after the decay of transients are strictly periodic in space, and if they are weakly perturbed they relax diffusively back to the periodic state. This relaxation has been investigated experimentally in various systems,[7,29] and the results agree with theoretical results based on the phase diffusion equation.[16,19,25] It describes the slow dynamics of the local phase, the gradient of which is the local wave number. The structures are generally stable over a range of wave numbers that is limited (at least) by the Eckhaus instability which is characterized by a vanishing of the diffusion coefficient. It leads, through the destruction (or creation) of one or more unit cells, i.e., a roll pair in convection or a vortex pair in Taylor vortex flow, to a new stable periodic state with a different wave number.

Recently, it has been pointed out that a vanishing of the diffusion coefficient does not necessarily invoke the Eckhaus instability.[4,5,10,26,27] Instead, under certain conditions, the system can go to a state consisting of distinct domains in which the wave number has different values and which does *not* relax to a strictly periodic structure. The domains are separated by domain walls in which the wave number changes rapidly. This situation has been treated successfully using a higher-order phase equation.[4,5,10,26,27]

Experimentally, inhomogeneous structures have been found in a variety of systems. Most of them involve at least one time-dependent structure, e.g., counter-propagating traveling waves (or spirals),[17] (turbulent) twist vortices amidst regular Taylor vortices,[1] localized traveling-wave pulses in binary-mixture convection,[2] and steady Turing patterns amidst chemical traveling waves.[20] Recently, however, domain structures involving only steady convection rolls of two different sizes have been observed in Rayleigh-Bénard convection in a very narrow channel.[12] It has been speculated that these structures may be related to the phase-diffusion mechanism discussed above.[4,5,10]

Within the framework of the phase equation, domain structures consisting of an array of domain walls are not stable due to the attractive interaction between the walls. This leads to a coarsening of the structures during which domain walls annihilate each other. This dynamics is closely related to that observed in spinodal decomposition of binary mixtures after quenches into the miscibility gap.[13] In general, the coarsening will eventually lead to periodic structures with a constant wave number. Only if the boundary conditions conserve the total phase, i.e., require the total number of convection rolls, say, to be constant, will the final state consist of a (single) pair of domain walls.

Here, we investigate the possibility of stable domain structures in the absence of phase conservation. To do so, we study systems in which two periodic patterns differing only in their wave numbers are equally likely to arise. In certain cases, the competition between the two wave numbers can be described by an extended Ginzburg-Landau equation. We study this equation with spatially ramped control parameter in order to allow the total phase to change and find that even in this general case, domain structures of varying sizes can be stable. This Ginzburg-Landau equation can also be viewed as a one-dimensional version of the Ginzburg-Landau equation for two-dimensional patterns in isotropic and anisotropic systems. The domain structures correspond then to "zig-zag" patterns. In two dimensions, phase conservation is less common than in one dimension due to the possibility of focus singularities in the pattern that often arise at the boundaries.[8] Our results may therefore also shed some additional light on the behavior of the "zig-zag" patterns studied previously.[21,3]

2. NUMERICAL SIMULATION OF FOURTH-ORDER GINZBURG-LANDAU EQUATION

A situation in which the competition between two wave numbers can be investigated with relative ease is obtained if the neutral curve that marks the stability boundary of the basic state has two almost equal minima at (slightly) different wave numbers. For values of the control parameter Σ slightly above these minima, the pattern can be described by a Ginzburg-Landau equation for the amplitude A,[22]

$$\partial_T A = D_2 \partial_X^2 A + i D_3 \partial_X^3 A - \partial_X^4 A + \Sigma A - |A|^2 A, \tag{1}$$

which gives for instance the vertical fluid velocity in convection *via*

$$v_z(x, z, t) = \epsilon e^{iq_c x} A(X, T) f(z) + h.o.t. + c.c. \tag{2}$$

For simplicity, we assume in the following that the neutral curve has reflection symmetry with respect to q_c and set $D_3 = 0$. In this case, the neutral curve has two minima if D_2 is negative. The control parameter Σ is proportional to the temperature difference across the fluid layer, say.

In order to allow the number of rolls to change freely without any pinning by the boundaries, we apply a subcritical spatial ramping to Σ. Thus, A goes to zero before the boundary of the system and the wave number selected by the ramp[14,25] corresponds to one of the minima of the neutral curve. For the numerical simulation, a Crank-Nicholson scheme with $dx \approx 0.034$ is used. Such a small grid spacing is required to reduce the pinning by the grid below the small attractive force between domain walls. For slowly varying wave numbers $Q(\tau, \xi) \equiv \partial_\xi \phi(\xi = \delta X, \tau = \delta^4 T, \delta \ll 1)$, the Ginzburg-Landau Equation (1) can be reduced to an equation for the phase of the amplitude $A = R e^{i\phi}$,

$$\partial_\tau \phi = (D + E \partial_\xi \phi + F(\partial_\xi \phi)^2)\partial_\xi^2 \phi - G \partial_\xi^4 \phi. \tag{3}$$

Domain structures arise for negative values of D. For large values of Σ wave-number gradients across domain walls are sufficiently small to be described by Eq. (3). Within this framework the interaction between domain walls is purely attractive. Thus, in the presence of a subcritical ramp, domain structures are expected to evolve to a strictly periodic pattern. This is confirmed for $\Sigma = 100$ and $D_2 = -1$, using initial conditions with three domains, i.e., a region of low wave number between two regions of high wave number.

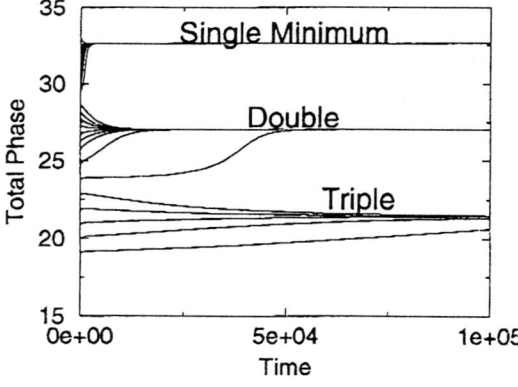

FIGURE 1 Evolution of the total phase without phase conservation. The corresponding final states are shown in Figure 2.

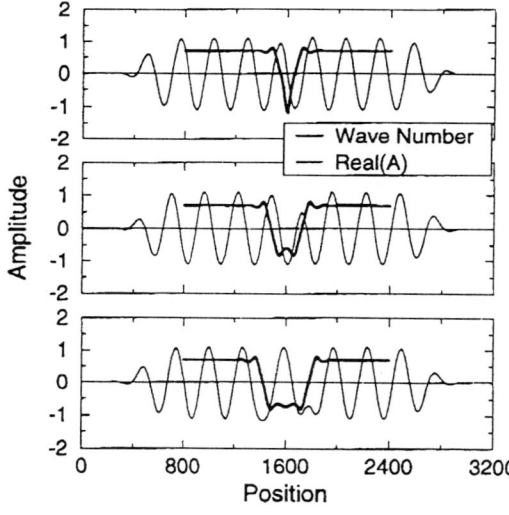

FIGURE 2 Coexisting stable final states obtained from the time evolution shown in Figure 1. The thick line gives the local wave number in the bulk ($800 < i < 2400, dx = 0.034$). Note that the amplitude goes to zero toward the boundaries due to the subcritical ramping.

For $\Sigma = 1$, on the other hand, the behavior is quite different, as shown in Figure 1. It gives the total phase $\int Q\,dX$ in the system (between the ramps) as a function of time for different initial conditions which are obtained by changing the width of the central (low wave number) domain. While for large Σ the system evolves to the same periodic state independent of the initial total phase, it converges here to one of several final states with different total phase. These states are pictured in Figure 2. They differ in the width of the central domain and, most strikingly, their local wave number exhibits oscillatory behavior in space.

The possibility for spatial oscillations in the wave number can be seen in a linear stability analysis of the periodic state. It shows that solutions which approach a periodic state for $X \to \pm\infty$ can do so in an oscillatory manner if the fourth derivative is present. The purely attractive interaction between domain walls within the phase equation, and therefore also for large Σ, is due to the monotonic behavior of the wave number across the wall. The nonmonotonic behavior found for smaller Σ therefore strongly suggests an oscillatory contribution to the interaction that would explain the stability of bound pairs of domain walls.

To study the transition between the two regimes, one could investigate the persistence of the domain structures when increasing Σ. Since the scale for Σ is set by D_2, which determines the depth of the wells in the neutral curve, increasing Σ is equivalent to making D_2 less negative. Scaling shows that the relevant parameter is D_2^2/Σ. We therefore investigate the persistence of bound pairs by changing D_2 rather than Σ since this also sheds some light on the stability of the two-dimensional patterns discussed below. The result is shown in Figure 3. It gives the total phase within the unramped region of various states as a function of D_2. For these simulations the system has been chosen much larger in order to avoid interactions between the domain walls and the ramps. The periodic state is indicated by a dashed line. The different symbols denote where the bound pairs with 1 to 5 minima in the local wave number, respectively, disappear. The states with 1 to 3 minima jump to the periodic solution above that value of D_2. The longer states are relatively hard to track due to the rapid decay of the oscillations away from the domain walls. For $D_2 < -0.8$ all six states coexist stably. In this regime, arrays of domain walls should be possible in which the widths of successive domains alternate chaotically.[6] Clearly, none of the bound pairs investigated persists all the way to $D_2 = 0$. This corresponds to the previous result that for large Σ the phase equation becomes valid. If one were to start with a chaotic array of domain walls and to increase D_2, Figure 3 suggests that domains of various widths would successively be eliminated beginning with the longest and the shortest ones until only domains with three minima remain together with a few long domains in which the domain walls are not locked into each other. However, preliminary results indicate the existence of an additional state with four minima, the stability regime of which has not yet been sufficiently established. Note that these simulations only address the interaction between two domain walls. In a general array with many domains, the regime of existence may be therefore somewhat different.

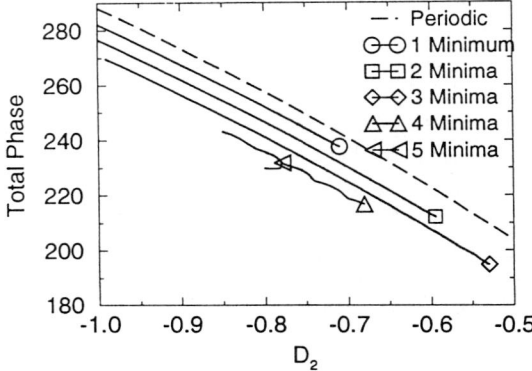

FIGURE 3   Regime of existence of bound pairs of domain walls for $\Sigma = 1$. The calculation was performed with 12800 points and a grid spacing of 0.034. The ramped part is 400 points wide on each side.

3. APPLICATION TO TWO-DIMENSIONAL SYSTEMS: ZIG-ZAGS

The above results are expected to capture also certain aspects of patterns extended in two dimensions. A particularly clear example is given by electro-hydrodynamic convection (EHC) in nematic liquid crystals. In the nematic phase, the rodlike molecules are predominantly oriented in one direction rendering the fluid anisotropic. Depending on parameters, convection arises in the form of rolls perpendicular or oblique to that preferred direction. Due to reflection symmetry, the oblique rolls can have wave number (Q, P) or $(Q, -P)$. Often, domains of oblique rolls with opposite orientation are observed to coexist, leading to "zig-zag" patterns. They are characterized by sharp transitions between the domains with the two orientations.

Quite generally, in isotropic systems, as for instance in Rayleigh-Bénard convection (RBC), straight roll patterns exhibit a secondary instability that leads to an undulatory deformation of the rolls. It arises for small wave numbers and tends to increase the local wave number. In the course of the nonlinear evolution, the undulations can grow sharper and form zig-zag-type patterns which may, in fact, be stable, as recently shown in simulations of chemical Turing patterns.[11]

Near onset, both systems can be described by suitable Ginzburg-Landau equations. In the isotropic case, one obtains[18,28]

$$\partial_T A = -(i\partial_X + \partial_Y^2)^2 A + \lambda A - |A|^2 A, \tag{4}$$

whereas the anisotropic case yields[21]

$$\partial_T A = (\partial_X^2 - iZ\partial_X\partial_Y^2 + W\partial_Y^2 - \partial_Y^4)A + \lambda A - |A|^2 A. \tag{5}$$

In the anisotropic case, the normal/oblique transition occurs at $W = -ZQ$.

Focusing on solutions that are strictly periodic in X, $A = A_1(Y,T)e^{iQX}$, one obtains Eq. (1) for A_1,

$$\partial_T A_1 = D_2 \partial_Y^2 A_1 - \partial_Y^4 A_1 + \Sigma A - |A_1|^2 A_1, \tag{6}$$

with

$$D_2 = Q, \; \Sigma = \lambda - Q^2 \text{ for Eq.}(4) \tag{7}$$
$$D_2 = W + ZQ, \; \Sigma = \lambda - Q^2 \text{ for Eq.}(5). \tag{8}$$

Note that Y and P in the two-dimensional systems correspond to X and Q in Eq. (1). Thus, the domain structures discussed in Section 2 are one-dimensional analogs of zig-zag patterns. The general stability of zig-zag patterns in anisotropic systems has been studied by Pesch and Kramer[21] and by Bodenschatz et al.[3] who find islands of stability for the zig-zag states. Their stability limits may be due to two-dimensional discussion suggests, however, identifying the islands with the discrete set of domains with different widths. Their stability could then be determined by the interaction of the domain walls separating the two kinds of oblique rolls.

Within Eq. (6), the straight (normal) roll state corresponds to a solution with constant A_1. Reducing its wave number is equivalent to decreasing D_2, which goes through zero at the onset of the zig-zag instability. It leads to the growth of undulatory deformations of the rolls which correspond to periodic variations of the local wave number in Eq. (1). Based on the simulations of that equation one would expect that D_2 has to be sufficiently negative for the resulting zig-zag structure to persist. Therefore, close to the onset of the zig-zag instability the coarsening dynamics predicted by the phase equation, Eq. (3) will prevail, and the expected final state would be a pattern of straight (oblique) rolls of either sign if the phase is not conserved. For deeper quenches into the unstable regime,[1] however, the undulations are expected to lock into each other due to the oscillatory interaction and form an array of domain walls, i.e., an array of zigs and zags, which again could be spatially chaotic.

4. CONCLUSION

The competition between patterns differing only in their wave number can lead to complex spatial patterns in which the interaction between the walls separating the domains with different wave number plays an important roll. Previously, we investigated them for conditions under which the total phase, i.e., the number of convection rolls in the system,is conserved.[23] There we found a very rich bifurcation

---

[1]Note $|D_2^2/\Sigma| \to \infty$ at the neutral curve $\Sigma = \lambda - Q^2$.

structure which originates from interactions with other modes arising from the Eckhaus instability.

In experiments, the phase is not always conserved. In the thin layers used in electro-convection of liquid crystals, for instance, inhomogeneities in the layer thickness are often strong enough that convection arises in large patches which are separated by nonconvecting regions. In isotropic two-dimensional systems, focus singularities are expected to arise in the corners of the container which also allow rolls to disappear smoothly without strong pinning.[8] Here, we therefore studied domain structures in the presence of spatial ramps. We found that they can exist even without phase conservation and we attribute this to the oscillatory behavior of the local wave number which should lead to an oscillatory interaction between adjacent domain walls. The distance between the walls is discretized accordingly. Close to the onset of the zig-zag instability in two-dimensional systems, these oscillations are small; to stabilize zig-zag patterns one therefore has to quench deeper into the unstable regime.

ACKNOWLEDGMENTS

This work has been supported by grants from NSF/AFOSR (DMS-9020289) and DOE (DE-FG02-92ER14303).

REFERENCES

1. Baxter, B. W., and C. D. Andereck. "Formation of Dynamical Domains in a Circular Couette System." *Phys. Rev. Lett.* **57** (1986): 3046.
2. Bensimon, D., P. Kolodner, and C. M. Surko. "Competing and Coexisting Dynamical States of Traveling-Wave Convection in an Annulus." *J. Fluid Mech.* **217** (1990): 441.
3. Bodenschatz, E. , M. Kaiser, L. Kramer, W. Pesch, A. Weber, and W. Zimmermann. "Patterns and Defects in Liquid Crystals." In *The Geometry of Non-Equilibrium*, edited by P. Coullet and P. Huerre, 111. New York: Plenum, 1990.
4. Brand, H. R., and R. J. Deissler. "Confined States in Phase Dynamics." *Phys. Rev. Lett.* **63** (1989): 508.
5. Brand, H. R., and R. J. Deissler. "Properties of Confined States in Phase Dynamics." *Phys. Rev. A* **41** (1990): 5478.
6. Coullet, P., C. Elphick, and D. Repaux. "Nature of Spatial Chaos." *Phys. Rev. Lett.* **58** (1987): 431.

7. Croquette V., and F. Schosseler. "Diffusive Modes in Rayleigh-Bénard Structures." *J. Phys.* **42** (1982): 1182.

8. Cross, M. C., and A. C. Newell. "Convection Patterns in Large Aspect Ratio Systems." *Physica D* **10** (1984): 299.

9. Cross, M. C., and P. C. Hohenberg. "Pattern Formation Outside of Equilibrium." *Rev. Mod. Phys.* **65** (1994): 851.

10. Deissler, R. J., Y. C. Lee, and H. R. Brand. "Confined States in Phase Dynamics: the Influence of Boundary Conditions and Transient Behavior." *Phys. Rev. A* **42** (1990): 2101.

11. Dufiet, V., and J. Boissonade. "Conventional and Unconventional Turing Patterns." *J. Chem. Phys.* **96** (1992): 664.

12. Hegseth, J. , J. M. Vince, M. Dubois, and P. Bergé. "Pattern Domains in Rayleigh-Bénard Slot Convection." *Europhys. Lett.* **17** (1992): 413.

13. Kawasaki, Y. "Spinodal Decomposition in Binary Fluid Mixtures." This volume.

14. Kramer, L., E. Ben-Jacob, H. Brand, and M. Cross. "Wavelength Selection in Systems Far From Equilibrium." *Phys. Rev. Lett.* **49** (1982): 1892.

15. Kramer, L., and W. Zimmermann. "On the Eckhaus Instability for Spatially Periodic Pattern." *Physica D* **16** (1985): 221.

16. Lücke, M., and D. Roth. "Structure and Dynamics of Taylor Vortex Flow and the Effect of Subcritical Driving Ramps." *Z. Phys. B* **78** (1990): 147.

17. Mutabazi, I., J. J. Hegseth, C. D. Andereck, and J. E. Wesfreid. "Spatiotemporal Pattern Modulations in the Taylor-Dean System." *Phys. Rev. Lett.* **64** (1990): 1729.

18. Newell, A. C., and J. A. Whitehead. "Finite Bandwidth Finite Amplitude Convection." *J. Fluid Mech.* **38** (1969): 279.

19. Paap, H.-G., and H. Riecke. "Drifting Vortices in Ramped Taylor Vortex Flow: Quantitative Results from Phase Equation." *Phys. Fluids A* **3** (1991): 1519.

20. Perraud, J.-J., E. Dulos, P. De Kepper, A. De Wit, G. Dewel, and P. Borckmans. "Turing-Hopf Localized Structures." In *Spatio-Temporal Organization in Nonequilibrium Systems*, edited by S. C. Müller and T. Plesser, 205. Dortmand: Projekt Verlag, 1992.

21. Pesch, W., and L. Kramer. "Nonlinear Analysis of Spatial Structures in Two-Dimensional Anisotropic Pattern Forming Systems." *Z. Phys.* **B63** (1986): 121.

22. Proctor, M. R. E. "Instabilities of Roll-Like Patterns for Degenerate Marginal Curves." *Phys. Fluids A* **3** (1991): 299.

23. Raitt, D., and H. Riecke. "Domain Structures: Existence and Stability in a Fourth-Order Ginzburg-Landau Equation." *Physica D*, submitted.

24. Riecke, H., and H.-G. Paap. "Stability and Wave-Vector Restriction of Axisymmetric Taylor Vortex Flow." *Phys. Rev. A* **33** (1986): 547.

25. Riecke, H., and H.-G. Paap. "Perfect Wave-Number Selection and Drifting Patterns in Ramped Taylor Vortex Flow." *Phys. Rev. Lett.* **59** (1987): 2570.

26. Riecke, H., "Stable Wave-Number Kinks in Parametrically Excited Standing Waves." *Europhys. Lett.* **11** (1990): 213.

27. Riecke, H., "On the Stability of Parametrically Excited Standing Waves." In *Nonlinear Evolution of Spatio-Temporal Structures in Dissipative Continuous Systems*, edited by F. H. Busse and L. Kramer, 437. New York: Plenum, 1990.

28. Segel, L. A. "Slow Amplitude Modulation of Cellular Convection." *J. Fluid Mech.* **38** (1969): 203.

29. Wu, M., and C. D. Andereck "Phase Modulation of Taylor Vortex Flow." *Phys. Rev. A* **43** (1991): 2074.

S. J. Linz† and S. H. Davis‡

†Institut für Physik, Universität Augsburg, W-8900 Augsburg, Germany
‡Department of Engineering Sciences and Applied Mathematics, Northwestern University, Evanston, IL 60208

Onset of Convection in Small-Aspect-Ratio Systems and Its Application to the $\vec{g}$-Jitter Problem

1. INTRODUCTION

There has been a great deal of interest in the investigation of the evolution of steady patterns in systems that are off equilibrium.[4] Rayleigh-Bénard convection has been proven to be a very useful paradigm for the study of pattern formation from both the experimental point of view and from theory. Most of the studies, however, deal with extended layers of fluid. The case of narrow or even tall cells has attracted much less attention.[7,11] This might be caused by the fact that in small systems the vertical boundaries have a strong impact on the quantitative properties of convection and exact or even approaches for the study of the onset of convection in small cells. We give simple expressions for the stability thresholds to a one-roll convection state for several types of velocity-field boundary conditions. As an important application, we also discuss changes of the onset of convection for the vertical $\vec{g}$-jitter problem[2] which is of relevance in the understanding of microgravity convection.

Spatio-Temporal Patterns, Ed. P. E. Cladis and P. Palffy-Muhoray,
SFI Studies in the Sciences of Complexity, Addison-Wesley, 1995   **265**

2. THE SYSTEM

The system that we consider is a two-dimensional convective flow in a confined cell of width L and height d with a coordinate system attached at the center of the cell. This two-dimensional model simulates a three-dimensional flow near the cell center, distant from the end walls. It is convenient to introduce an *inverse* aspect ratio $\Gamma = d/L$ and to rescale the system to a square cell by scaling the horizontal coordinate x by width L and the vertical coordinate z by height d. We scale time t by the vertical diffusion time d^2/κ (where κ is the thermal diffusivity) and the temperature T by $\kappa\nu/\alpha g d^3$ (where ν is the kinematic viscosity, α thermal expansion coefficient, and g the gravitational constant). The boundary conditions for the temperature field $T(x,z)$ are $T(x,1/2) = T_0$ and $T(x,-1/2) = T_0 + R$ for top and bottom boundaries, respectively. Here R is the Rayleigh number (the nondimensionalized temperature difference between bottom and top boundaries). The conditions on the vertical boundaries, consistent with a purely conductive state, are that either the temperature distribution at the vertical boundaries varies linearly between the top/bottom temperatures or that there is zero horizontal heat flux. We focus on the latter case. The conducive state, T_{cond}, is given by $T_{\text{cond}}(z) - T_0 = R(1/2 - z)$. For the spatio-temporal evolution of the stream function $\psi(x,z,t)$ and the deviation of the temperature field from the conductive state, $\theta(x,z,t) = T(x,z,t) - T_{\text{cond}}(z)$, one gets:

$$(\partial_t - \sigma\nabla^2)\nabla^2\psi = \Gamma J(\nabla^2\psi, \psi) + \Gamma\sigma\partial_x\theta,$$
$$(\partial_t - \nabla^2)\theta = \Gamma R\partial_x\psi + \Gamma J(\theta, \psi), \tag{1}$$

where $\nabla^2 = \Gamma^2\partial_x^2 + \partial_z^2$ and $J(f,g) = \partial_x f \partial_z g - \partial_x g \partial_z f$. Here $\sigma = \nu/\kappa$ is the Prandtl number. The thermal boundary conditions are $\theta(x, \pm 1/2, t)$ and $\partial_x\theta(\pm 1/2, z, t) = 0$. We do not specify the slip conditions for the moment; we only set, without loss of generality, $\psi = 0$ at $z = \pm 1/2$ and $x = \pm 1/2$. In what follows we restrict the inverse aspect ratio Γ to values $\Gamma \geq 0.7$, since numerical calculations[7,11] suggest that, below this value, more than a single horizontal roll is preferred. In a sense the inverse aspect ratio plays the role of the wave number in an extended system. However, in a horizontally infinite system,[3] the most unstable wave number is the one that minimizes the Rayleigh number for convection, whereas here the sidewall geometry fixes the critical Rayleigh number.

3. GENERAL STABILITY ANALYSIS FOR THE CONDUCTIVE STATE

Here we present a general discussion of the structure of the stability threshold to convection valid for all types of slip conditions on the walls. Let us suppose that we know the exact eigenfunctions of the linear problem and that they are separable (when there is no slip at *all* boundaries, this is only an approximation). Then we can write down an ansatz

$$\psi(x, z, t) = A(t)f_1(x)f_2(z), \quad \theta(x, z, t) = B(t)g_1(x)g_2(z). \tag{2}$$

Since we expect a one-roll-convective state as the marginal state, one can see that the stream function ψ should be a symmetric function of x and z. The temperature field θ should be antisymmetric in x and symmetric in z. Without loss of generality we assume that f_i and g_i $(i = 1, 2)$ are normalized to unity, i.e., $\langle f_1^2(x) \rangle = \langle f_2^2(z) \rangle = \langle g_1^2(x) \rangle = \langle g_2^2(x) \rangle = 1$. Here the brackets denote an area integration over the cell. Using the ansatz (2), we get, for the stability problem, two linear equations for the amplitudes $A(t)$ and $B(t)$ which read

$$(\Gamma^2 s_1 + s_2)\dot{A}(t) = \sigma(\Gamma^4 s_3 + \Gamma^2 s_4 + s_5)A(t) + \Gamma\sigma s_6 B(t) \tag{3}$$

$$\dot{B}(t) = (s_7\Gamma^2 + s_8)B(t) + \Gamma R s_9 A(t) \tag{4}$$

where

$$\begin{aligned} s_1 &= -\langle(\partial_x f_1(x))^2\rangle; \quad s_2 = -\langle(\partial_z f_2(z))^2\rangle; \quad s_3 = \langle f_1(x)\partial_x^4 f_1(x)\rangle; \\ s_4 &= \langle f_2(x)\partial_z^4 f_2(z)\rangle; \quad s_5 = 2\langle f_1(x)\partial_x^2 f_1(x)\rangle\langle f_2(z)\partial_z^2 f_2(z)\rangle; \\ s_6 &= \langle f_1(x)\partial_x g_1(x)\rangle\langle g_2(z)f_2(z)\rangle; \quad s_7 = -\langle(\partial_x g_1(x))^2\rangle; \\ s_8 &= -\langle(\partial_z g_2(z))^2\rangle; \quad s_9 = \langle g_1(x)\partial_x f_1(x)g_2(z)f_x(z)\rangle. \end{aligned} \tag{5}$$

Despite the generality of the approach, we get valuable information on the coefficients. Firstly, s_1, s_2, s_7, and s_8 are negative and, secondly, there are not nine independent coefficients that enter in Eqs. (3) and (4), but only seven, since $s_5 = 2s_1 s_2$, and hence positive, and $s_9 = -s_6$. The general structure of the stability threshold to convection is given by

$$R_T(\Gamma) = k_1(k_2 + 1/\Gamma^2)(\Gamma^4 + k_3\Gamma^2 + k_4), \tag{6}$$

$$k_1 = \frac{s_3 s_8}{s_6 s_9}; \quad k_2 = \frac{s_7}{s_8}; \quad k_3 = \frac{s_4}{s_3}; \quad k_4\frac{s_5}{s_3}. \tag{7}$$

Using this approach, it is possible to separate the dependence on Γ from the effect of the boundary conditions, which solely enters in numerical values of the coefficients s_i $(i = 1, \ldots, 9)$. In fact, to calculate the s_i, one has only to find the eigenfunctions of the square cell, either exactly (when possible), or using approximations, or numerical solutions. Different boundary conditions lead to different quantitative results; however, qualitatively the instabilities for different types of boundary conditions have two points in common: (i) $R_T(\Gamma)$ depends only on even powers of Γ, so that for $\Gamma \approx 10$ the limit $\Gamma \to \infty$ (tall cells) is reasonably good approximation, and (ii) $R_T(\Gamma)$ increases like Γ^4 as $\Gamma \to \infty$. Thus, the expected and well-known phenomenon, that the conductive state in taller cells is drastically stabilized, is not caused by the no-slip boundary condition at the vertical walls but is just an effect of the roll's stretching. It is energetically more expensive to keep an elongated roll alive, quite similarly to the case of Rayleigh-Bénard convection in thin layers. The effect of no slip gives an additional enhancement by a factor of 5, as we shall see below.

4. STABILITY ANALYSIS FOR DIFFERENT TYPES OF BOUNDARY CONDITIONS FOR THE VELOCITY FIELD

In this section we want to calculate the coefficients entering in $R_T(\Gamma)$, Eq. (7). Thus we have to specify the eigenfunctions of the linearized version of Eq. (1). The temperature field obeys fixed-temperature boundary conditions at the horizontal walls and zero-heat-flux boundary conditions at the vertical walls. Thus, an ansatz

$$\theta(x, z, t) \approx \sqrt{2} \sin(\pi x)\sqrt{2} \cos(\pi z) \qquad (8)$$

is appropriate. Equation (8) is the lowest order term in a Fourier expansion of the x- and z- dependence, which fulfills boundary conditions as well as the expected symmetries from Section 3. This implies that $k_2 = 1$ in Eq. (7). For the stream function $\psi(x, z, t)$, we distinguish among three cases.

4.1 FREE SLIP AT VERTICAL AND HORIZONTAL WALLS (FF)

For free-slip boundaries at all walls, the choice

$$\psi^{FF}(x, z, t) \approx \sqrt{2} \cos(\pi x)\sqrt{2} \cos(\pi z) \qquad (9)$$

for the stream function is appropriate. In fact, Eqs. (8) and (9) solve exactly the stability problem for one-roll convection. The resulting stability problem is equivalent to the Rayleigh-Bénard problem in extended systems.[3] The stability threshold can be simply obtained by replacing the wave number k by $\pi\Gamma$ since Γ corresponds to half of the wavelength $\lambda = 2\pi/k$ of the periodic pattern. The resulting stability threshold to convection is given by

$$R_T^{FF}(\Gamma) = \pi^4(1 + 1/\Gamma^2)(\Gamma^4 + 2\Gamma^2 + 1) \qquad (10)$$

and has an asymptotic limit $R_T^{FF} \simeq \pi^4\Gamma^4$ as $\Gamma \to \infty$.

4.2 NO SLIP AT HORIZONTAL AND FREE SLIP AT VERTICAL WALLS (NF)

For realistic no-slip boundary conditions at horizontal walls, but still idealized free-slip boundaries at the vertical ones, it is possible to use Rayleigh-Chandrasekhar[3,10] functions as approximations. The choice

$$\psi^{NF}(x, z, t) \approx \sqrt{2} \cos(\pi x)C_1(z) \qquad (11)$$

in combination with Eq. (8) leads to a stability threshold

$$R_T^{NF}(\Gamma) = 1.029(1 + 1/\Gamma^2)(\pi^4\Gamma^4 + 24.6\pi^2\Gamma^2 + \lambda^4) \qquad (12)$$

with $C_1(z) = [\cosh(\lambda z)/\cosh(\lambda/2)] - [\cos(\lambda z)/\cos(\lambda/2)]$ and $\lambda = 4.73$. This result can also be simply reobtained by identifying the $\pi\Gamma$ with k of stability analysis for the extended system.[3,10] We note that Eq. (12) has the asymptotic limit $R_T^{NF} \simeq 1.029\pi^4\Gamma^4$ as $\Gamma \to \infty$, which shows that in very tall cells the influence of the boundary conditions at top and bottom are almost negligible, as expected.

4.3 NO SLIP AT ALL WALLS (NN)

If one considers no-slip boundary conditions at all boundaries, the exact linear-theory eigenfunction is not separable in x and z. A simple choice,

$$\psi^{NN}(x, z, t) \approx C_1(x)C_1(z),\tag{13}$$

however, leads to surprisingly good results for the stability threshold. A comparison with numerical[7,11] calculations show that the error varies from 6.5% for $\Gamma = 1$ to 1% for $\Gamma \to \infty$. The stability threshold R_T calculated from Eqs. (7) with (8) and (13) is

$$R_T^{NN}(\Gamma) = 1.057(1 + 1/\Gamma^2)(\lambda^4\Gamma^4 + 302.58\Gamma^2 + \lambda^4)\tag{14}$$

and has the asymptotic limit $R_T^{NN} \simeq 529\,\Gamma^4$ for $\Gamma \to \infty$ being only 1% from the parallel-flow calculation by Kurzweg[7] for $\Gamma \to \infty$. The above-mentioned accuracy implies that the separability approximate in Section 3 is good.

4.4 COMPARISON

In Figure 1 we show the dependence of $\ell n R_T$ on Γ for the three difference types of boundary conditions discussed in the subsections 4.1–4.3. The no-slip results are systematically larger than the corresponding free-slip results. This is not surprising, since the walls exert extra drag which has to be overcome by buoyancy to maintain convection. For small Γ, i.e., $\Gamma \approx 1$ to 2 the NF result lies between FF and NN. For $\Gamma \geq 5$, NF and FF are almost identical which means that no slip or free slip at the top and bottom boundaries have a relatively small effect. We see here that the effect of the boundary conditions at the sidewalls is dominant. In the limit of large Γ, all stability thresholds differ just by a numerical factor and, thus, can be rescaled for $\Gamma \gg 1$ giving

$$R_T^{NN} \simeq 5.43 R_T^{FF}, \quad R_T^{NF} \simeq 1.029 R_T^{FF}.\tag{15}$$

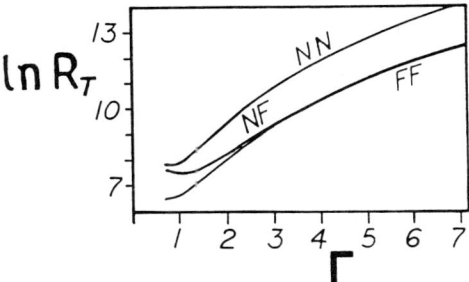

FIGURE 1 Dependence of the logarithm of the stability threshold $R_T(\Gamma)$ versus the inverse aspect ratio Γ for different sets of boundary conditions: slip at all boundaries (FF); no (free) slip at horizontal (vertical) boundaries (NF); and no slip at all boundaries (NN). The result FF is exact; NF and NN are approximate.

5. $\vec{g}$-JITTER CONVECTION

An interesting application of some technological relevance is the effect of gravitational modulation in the above system. Convection in small-aspect-ratio systems is of importance for the understanding of solidification in a microgravity environment.[1] There, however, the system is subject to external perturbations ("$\vec{g}$-jitter"), which are ever present.

Here we only consider vertical $\vec{g}$-jitter. In that case, one replaces $\vec{g}$ by $\vec{g}(t) = -g_0(1 + \Delta\xi(t))\vec{e}_z$, where g_0 is the mean gravitational acceleration, Δ the modulation amplitude, and $\xi(t)$ a time-dependent function with zero mean. The conductive temperature profile remains the same as in the nonmodulated case, only the nondimensionalized conductive pressure profile p_{cond} oscillates according to $\partial_z p_{\text{cond}} = -(1 + \Delta\xi(t))(g_0 d^3/\kappa^2 - \sigma R(1/2 - z))$. In the stability problem (3,4) one replaces the coefficient s_6 in Eq. (5) by $\tilde{s}_6 = s_6(1 + \Delta\xi(t))$, assuming that the approximate eigenfunctions (8) and (13) are still good, at least for small Δ. Equations (5) and (7) can easily be rewritten as a damped, harmonic oscillator equation with *parametric* forcing (damped Hill equation) for the marginal temperature field mode $B(t)$.

$$m\ddot{B}(t) + m\gamma\dot{B}(t) + \epsilon(t)B(t) = 0 \qquad (16)$$

where

$$\varepsilon = \frac{R(t) - R_T(\Gamma)}{R_T(\Gamma)}, \quad m = -\frac{1}{\sigma}\frac{\Gamma^2 s_1 + s_2}{(\Gamma^4 s_3 + \Gamma^2 s_4 + s_5)(\Gamma^2 s_1 + s_8)},$$

$$\gamma = \frac{\Gamma^4(s_1 s_7 + s_3\sigma) + \Gamma^2(s_1 s_8 + s_2 s_7 + s_4\sigma) + s_2 s_8 + s_5\sigma}{\Gamma^2 s_1 + s_2},$$

and $R(t) = R(1 + \Delta\xi(t))$. For Eq. (16) one can see that vertical $\vec{g}$-jitter acts like modulating the Rayleigh number. In the remainder of this section we calculate for weak perturbations Δ the value of R at which the conductive state becomes unstable.

5.1 SINUSOIDAL MODULATION

Here we discuss the special case that $\xi(t) = \cos\omega t$, where ω is the modulation frequency. For small modulation amplitudes Δ, straightforward perturbation theory can be used to determine the change of the stability threshold as well as the first-order corrections in Δ of the marginal modes. Skipping all details of the calculations, we get the following results for realistic NN-boundary conditions.

i.   The onset of convection is given by

$$R_{MT}(\Delta, \omega, \sigma, \Gamma) \sim R_T(\Gamma)[1 + s(\omega, \sigma, \Gamma)\Delta^2], \quad \Delta \to 0, \qquad (17)$$

where $R_T(\Gamma) = R_{MT}(0, \omega, \sigma, \Gamma)$ is given by Eq. (14) and

$$s(\omega, \sigma, \Gamma) = \frac{1}{2} \frac{a_1 a_2}{\omega^2 + (a_1 + a_2)^2},$$

$$a_1 = 40.69 \frac{(\Gamma^4 + 0.6\Gamma^2 + 1)}{\Gamma^2 + 1} \sigma, \quad a_2 = \pi^2(1 + \Gamma^2).$$

Let us, first of all, note that s is positive for all combinations of Γ, ω, and σ, implying that small modulation strength leads to a stabilization of the conductive state. For large Γ and $\omega/\Gamma \ll 1$, s reaches a Γ- and ω-independent value

$$s(\omega, \sigma, \Gamma) \sim \frac{1}{2} \frac{40.69\sigma}{(1 + 4.12\sigma)^2}, \quad \Gamma \to \infty. \tag{18}$$

Beyond that, we can see from Eq. (17) that s decays inversely proportional to the Prandtl number σ as $\sigma \to \infty$. This demonstrates the phenomenon, well known in extended systems,[1,5] that the inertia term in Eq. (16) is responsible for the stabilization due to modulation. The typical feature is that the stabilization is most pronounced if Γ is close to, but larger than unity.

ii. The marginal temperature field mode is only determined up to a constant factor and reads

$$Y(t) \propto \left[1 - 2\frac{a_1 a_2}{\omega} \cos(\omega t + \phi)\Delta\right]. \tag{19}$$

The phase ϕ in Eq. (19) is given by $\phi = -\arctan[(a_1 + a_2)/\omega]$ and describes the phase shift of $B(t)$ with respect to $g(t)$. We note that, for large Γ, $\phi \approx -\arctan((40.69\sigma + 9.87)\Gamma^2/\omega)$, implying that ϕ approaches $-\pi/2$ as $\Gamma \to \infty$.

5.2 STOCHASTIC MODULATION

In a seminal paper, Lücke and Schank[9] discussed the effect of small, stochastic, parametric modulation of an anharmonic oscillator. Their linearized problem has exactly the same structure as Eq. (16) and, thus, we can take advantage of their results. Assuming that $\xi(t)$ has zero mean but arbitrary dynamics and statistics with a modulation spectrum[9] $D(\omega) = \int_{-\infty}^{+\infty} dt \, e^{i\omega t} \langle \xi(t)\xi(0)\rangle$, one can easily obtain the stability threshold R_{SMT} for the stochastic case, given by

$$R_{SMT}(\Delta) \sim R_T \left[1 + \Delta^2 \int_{-\infty}^{+\infty} \frac{d\omega}{\pi} s(\omega, \sigma, \Gamma)D(\omega)\right], \Delta \to 0. \tag{20}$$

Note that $s(\omega, \sigma, \Gamma)$ and $D(\omega)$ are positive. Thus, the threshold shift $R_{SMT}(\Delta) \sim R_T$ is positive, implying that vertical g-jitter stabilizes one-roll convection as long as the modulation strength Δ is small.

6. CONCLUSION AND OUTLOOK

We have shown that simple one-mode approximations for the stream function and the temperature field lead to quite an accurate approximation of the stability threshold in small-aspect-ratio systems. Based on this experience, we have recently investigated the nonlinear convective flow structure and its hydrodynamic properties, a work that will be discussed in detail elsewhere.[8]

ACKNOWLEDGMENTS

This work was supported partially by NASA Microgravity Sciences and Applications Program. SJL also acknowledges partial support from Deutsche Forschungsgemeinschaft (Li 498/1–1&2).

REFERENCES

1. Ahlers, G., P. C. Hohenberg, and M. Lücke. "Thermal Convection Under External Modulation of the Driving Force. I. The Lorenz Model." *Phys. Rev. A* **32** (1985): 3493.
2. Alexander, J. I. D. "Low-Gravity Experiment Sensitivity to Residual Acceleration: A Review." *Microgravity Sci. Technol. III* (1990): 52.
3. Chandrasekhar, S. *Hydrodynamic and Hydromagnetic Stability.* Oxford: Oxford University Press, 1961.
4. Cross, M. C., and P. C. Hohenberg. "Pattern Formation Outside of Equilibrium." *Rev. Mod. Phys.* **65** (1993): 351.
5. Davis, S. H. "The Stability of Time-Periodic Flow." *Ann. Rev. Fluid Mech.* **8** (1976): 56.
6. Gresho, P. M., and R. L. Sani. "The Effects of Time-Periodic Flows." *J. Fluid Mech.* **40** (1970): 783.
7. Kurzweg, U. *Intl. J. Heat Mass Transfer* **8** (1965): 35.
8. Linz, S. J., and S. H. Davis. "Convection in Small-Aspect-Ratio Systems: A Minimal Model." In preparation.
9. Lücke, M., and F. Schank. "Response to Parametric Modulation Near an Instability." *Phys. Rev. Lett.* **54** (1985): 1465.
10. Niederländer, J., M. Lücke, and M. Kamps. "Weakly Nonlinear Convection: Galerkin Model, Numerical Simulation, and Amplitude Equation." *Z. Phys.* **B82** (1991): 135.
11. Samuels, M. R., and S. W. Churchill. "Stability of a Fluid in a Retangular Region Heated from Below." *A. I. Ch. E. J.* **13** (1967): 77.

Walter Zimmermann and Rainer Schmitz
IFF, Forschungszentrum Julich, KFA, D-52425 Julich, Germany

Hopf Bifurcation by Frustrated Drifts

It is demonstrated how broken symmetries induce temporal oscillating solutions behavior. For Rayleigh-Bénard convection it is shown that undulated container boundaries and spatially modulated temperature gradients lead to an oscillatory onset of convection: Hopf-bifurcation via frustrated drifts.

1. INTRODUCTION

Theoretically, a highly appreciated approach to a physical problem is to assume the simplest and most symmetric configuration that still captures the physical phenomena. The use of a simple fluid between two ideal flat and infinitely extended plates, which are heated from below, is one of the simplest system for investigations in pattern formation (see Figure 1). Rayleigh-Bénard convection in simple fluids, such as water or oil, is a systems with a low number of dynamical degrees of freedom and, in an infinitely extended layer, there is translational as well as rotational symmetry.

Spatio-Temporal Patterns, Ed. P. E. Cladis and P. Palffy-Muhoray,
SFI Studies in the Sciences of Complexity, Addison-Wesley, 1995   **273**

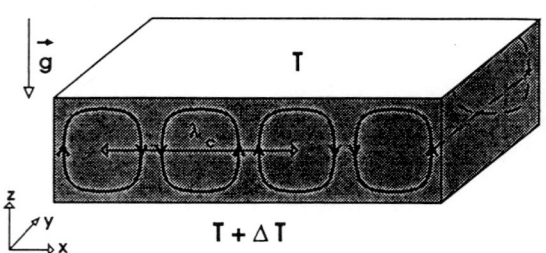

FIGURE 1 A fluid between two flat boundaries is heated from below. When the temperature difference ΔT crosses a well-defined threshold, convection sets are manifested in parallel convection rolls. The convection rolls' wave length is of the order of the fluid layer thickness d.

Above some critical difference between the temperatures at the confining bottom and top boundaries, stationary convection rolls occur. Increasing the temperature difference further, more complex spatio-temporal convection structures set in. Further dynamical degrees of freedom can be introduced when a simple fluid is replaced, for instance, by a binary fluid mixture or by a nematic liquid crystal or by a viscoelastic fluid. As a consequence, spatio-temporally richer convection structures occur much earlier. Sometimes further transitions occur immediately above the onset of convection, or the first instability itself is changed. For instance, from a stationary into an oscillatory one, when a binary fluid mixture (water-alcohol) is used. The richer bifurcation structure makes more complex materials attractive for experimental as well as theoretical investigations.

Using a simple fluid in Rayleigh-Bénard convection, however, and breaking the above-mentioned symmetries may also lead to qualitatively new and unexpected convection and pattern formation behavior. An extreme step in that direction is the choice of very small convection cells. Then the usually high numbers of possible modes in fluid dynamics are restricted to a few active ones, which may lead to well-known low-dimensional chaotic behavior such as period doubling, etc.[14] Convection in a rotating disc[8,9,19] or ramps[4,7,11] are further possibilities in changing the geometrical configuration. Another very natural generalization can be obtained by dropping the assumption of ideal (mathematical) flat container boundaries. The container boundaries of pattern-forming systems have usually a finite roughness, breaking the translational symmetry. Often the roughness is small compared to the container extensions and the related effects are beyond the experimental resolution. For such situations the usual assumption of ideal flat container boundaries is a good approximation. However, recent experiments are designed with smaller container extensions and the detection sensitivity is also under continuous improvement. Both tendencies make it more likely that, in experiments, roughness effects possibly arise, which will be puzzling within an analysis based on ideal flat container boundaries. It is therefore important to know the container dimensions and parameter ranges at which the typical roughness effects occur. Thermal convection, for instance, plays also an important role for many processes in geophysics and meteorology and several of them are investigated in the laboratory, however, with the fluids or gases in containers having rather (ideal) flat boundaries. To extrapolate back to the outdoor systems with ill-defined boundaries, it is important to know about the robustness of the phenomena under laboratory conditions against irregularities. On the other hand, irregularities break symmetries and may give rise to new effects.

While effects of a special class of statistically distributed container imperfections have been investigated recently,[18] aspects of irregularities in Rayleigh-Bénard convection can sometimes also be modeled in a first approach by periodic modulation of the container boundaries or periodic modulation of the temperature gradients. Investigations in that direction are, for instance, described by Kellery and Pal,[6,10] Coullet and Huerre,[3] and Hartung et al.[5]

Here we consider a special class of boundary undulations and temperature modulations for Rayleigh-Bénard convection, which induce the new phenomena: Hopf bifurcation via frustrated drifts (for more details, we refer the reader to a more extensive work[12]). Obtaining an oscillatory onset of convection via geometric constraints is somehow complementary to the case where an oscillatory onset of convection is achieved by using a complex fluid, such as binary fluid mixtures, between homogeneous boundaries (see Schöpf and Zimmerman[13] and references therein).

2. FRUSTRATED DRIFTS—A QUALITATIVE PICTURE

The convection rolls indicated in Figure 1 set in when the temperature difference ΔT crosses some threshold ΔT_c. In terms of the dimensionless Rayleigh number

$$R = \frac{\alpha g d^2}{\kappa \nu} \Delta T, \tag{1}$$

the critical value is $R_c = 1708$. This value is independent of the choice of the single component fluid and the respective fluid properties are contained in the material parameters such as the thermal expansion coefficient α, the thermal diffusivity κ, and the kinematic viscosity ν. (In Eq. (1) g is the gravitational constant and d the thickness of the convection layer.) When the top and the bottom boundary are parallel, then for $R < R_c$ ($\Delta T < \Delta T_c$) the heat is diffusively transported from the bottom to the top boundary and convectively beyond R_c. In the presence of an undulated boundary, as indicated in Figure 2, there is convective flow for arbitrary temperature differences ΔT. This may be weak for small temperature differences $R \ll R_c$; however, it can be already considerable for $R \sim R_c$, depending on the modulation amplitude of the boundary. This convective flow, which we call *primary flow*, has the periodicity of the modulation of the boundary as indicated in Figure 3. The presence of this primary flow has various consequences for the onset of the usual convection rolls. Those have a wave length of the order of the thickness d and we call them *secondary flow.*

When the top and the bottom plate are modulated by the same wave number and both modulations are in phase, as indicated in Figure 4(a), then the fluid layer thickness is everywhere the same and there is qualitatively no influence on the onset of the secondary convection. However, if there is a relative phase shift between both modulations, then the secondary flow sets in with a drift.[5] The drift direction depends on the sign of the phase (as indicating in Figure 4(b) and (c)), whereas the velocity depends on the modulus of the phase and the amplitude of the undulation. In the case of a finite phase shift, the thickness of the fluid layer is spatially varying and, as a consequence, the flow and temperature fields are spatially modulated, where the modulation wave length is identical with the one externally imposed.

Those drift effects are interesting by themselves. However, when they are considered to model heterogeneities at container boundaries, a homogeneous phase

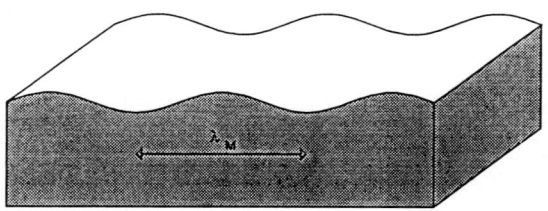

FIGURE 2 Sketch of a convection cell with modulated top boundary.

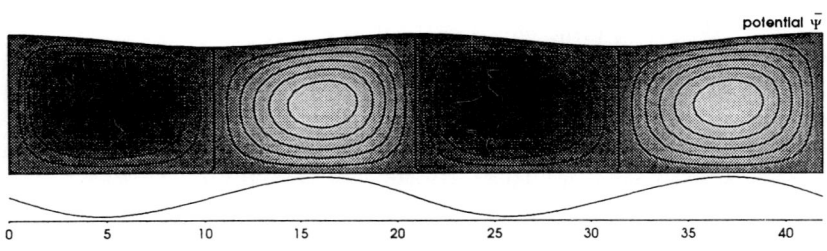

FIGURE 3 For the spatially periodic undulated top plate, we show the primary flow just at the threshold of the secondary flow. The modulation amplitude is $F_0 = 0.5$, the wave number is $k_0 = 0.3$, and the Rayleigh number is $R = 592.62$. In the lower part of the figure the velocity potential $\overline{\Psi}$ is plotted at the cell center $z = 0.5$.

a) no phaseshift

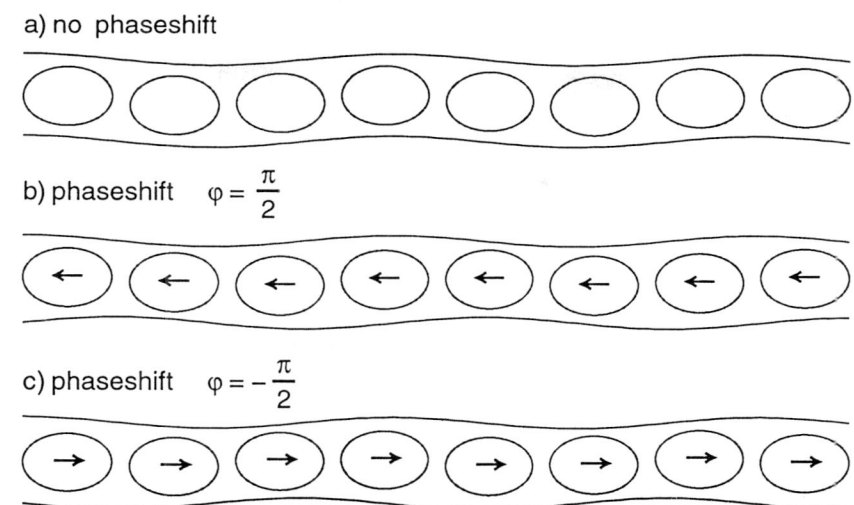

b) phaseshift $\varphi = \dfrac{\pi}{2}$

c) phaseshift $\varphi = -\dfrac{\pi}{2}$

FIGURE 4   An experimental configuration with modulated upper and lower boundaries is sketched. Periodic in phase undulation of the top and bottom boundaries leads to stationary secondary convection rolls as indicated in (a). A phase difference between the top and bottom boundary undulation leads to a drifting convection roll pattern as indicated in (b) and (c). The drift direction depends on the sign of the relative phase $\tilde{\varphi}$.

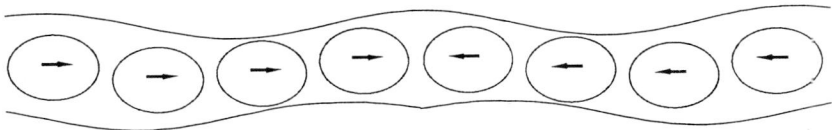

FIGURE 5   Here we propose the experimental geometry which combines the situations in Figures 4(b) and 4(c). The sign of the relative phase shift between the top and the bottom boundaries is reversed at the bottom boundary by a phase jump. The two phases of opposite sign lead to frustrated virtual drifts. As we show in our manuscript, such frustrated drifts lead to a new phenomena, the Hopf bifurcation by frustrated drifts.

shift seems rather unlikely. A geometry sketched in Figure 5 seems more appropriate to mimic some aspects of statistically distributed imperfections. Especially when the phase jumps in the undulation of bottom boundary are repeated. We will investigate the effect when similar phase jumps are periodically repeated.

What might be the consequences of periodically repeated phase jumps on the onset of the secondary flow? From Figures 4(b) and (c), one is indented in expectation of drifting secondary flow in every interval of homogeneous phase shifts. However, since they have opposite drift direction in neighboring intervals, the drifting patterns may compensate each other and the secondary convection could be stationary and nondrifting. A further possibility seems imaginable. The secondary flow drifts in every interval of homogeneous phase shift and, in the regions where the phase shift is reversed, there might be a source or a sink for the drifting waves, depending on whether drifts point towards each other or away from each other. We found a third possibility: the onset of the secondary flow is oscillatory in time.

3. HOPF BIFURCATION VIA FRUSTRATED DRIFTS IN RAYLEIGH-BÉNARD CONVECTION

A simple fluid in a gravitational field under the action of an external temperature gradient is considered. The investigations are restricted to a two-dimensional situation, with the vertical z-coordinate and the horizontal x-coordinate. We describe thermal convection of a simple fluid in Boussinesq approximation. *All length are measured in units of the layer thickness d*; therefore, the top and bottom boundaries are located at $z = 0$ and $z = 1$. Since we are investigating Rayleigh-Bénard convection in two dimensions and we are assuming an incompressible fluid, we can also introduce a velocity potential with $v_x = \partial_z \Psi$ and $v_z = -\partial_z \Phi$. For a detailed explanation of the system configurations, the calculations, and the results, overviewed below, we refer the reader to a comprehensive description by Schmitz and Zimmermann.[12]

We have investigated qualitatively equivalent configurations for Rayleigh-Bénard convection as sketched in Figure 5, which are defined by the following boundary conditions.

3.1 BOUNDARY CONDITIONS

Periodic modulations of the temperature at the top and bottom plates as well as a periodically undulated top plate are considered. The bottom plate is located at $z = 0$ and the top plate at

$$z = 1 + H_0, \qquad \text{with } H_0(x) = F_0 \cos(n_0 kx). \tag{2}$$

Ideal conductivity of the bounding plates is assumed, which leads to the following boundary conditions for the temperature field

$$T(x, z) = T_u + H_1(x) \qquad \text{at } z = 1 + H_0(x) \tag{3}$$
$$\text{and } T(x, z) = T_l - H_2(x) \qquad \text{at } z = 0, \tag{4}$$

with the modulations

$$
\begin{aligned}
H_1(x) &= F_1 \cos(n_1 kx + \varphi_1), \\
H_2(x) &= F_2 \cos(n_2 kx - \varphi_2).
\end{aligned}
\tag{5}
$$

Commensurate ratios between the wave numbers are investigated, as described below. For the velocities, we assume rigid boundary conditions which means that the components perpendicular and parallel to the boundaries vanish.

3.2 LONG WAVE LENGTH MODULATION

Wave length $\lambda_M = 2\pi/k$ of the boundary modulations that are much larger than the thickness $\lambda_M \gg d$ of the fluid layer are considered. This has several advantages mentioned below.

Without modulations the primary state is just a linear temperature profile interpolating between the top and bottom temperature without convective flow and the heat is transported from below to above by heat conduction. In our configuration with undulations in the bounding top plate or/and with modulations of the temperature at the confining plates, a primary convective flow is already induced at arbitrary mean-temperature differences $T_l - T_u$. This flow as well as the temperature field, superimposed on the linear vertical temperature gradient, have the same periodicity as the external modulations (see Figure 3). For long wave length modulations $k_i \ll q_c \sim 1/d$, the primary flow and the secondary flow live on different length scales (similar to the unmodulated case, where the wave length for the primary state—the linear temperature profile—is infinite). This has the advantage that the equations of motion for the primary state and the secondary state can be separated.

Therefore the whole solution may be split into two additive parts

$$
\Psi(x, z, t) = \overline{\Psi}(x, z) + \Phi(x, z, t),
\tag{6}
$$

$$
T(x, z, t) = \underbrace{\overline{T}(x, z)}_{P} + \underbrace{\Theta(x, z, t)}_{RB}.
\tag{7}
$$

$\overline{\Psi}(x, z)$ is the velocity potential for primary flow and $\overline{T}(x, z)$ the temperature distribution in the primary state. The respective fields of the secondary flow are $\Phi(x, z, t)$ and $\Theta(x, z, t)$.

The equations of motion (see, for instance, Schmitz and Zimmermann[12] and references therein) for thermal convection may be rewritten symbolically into the form

$$
\partial_t \vec{u} = \mathcal{L}\vec{u} + \vec{N}(\vec{u}, \vec{u})
\tag{8}
$$

with

$$
\vec{u} = \begin{pmatrix} \Psi \\ T \end{pmatrix}, \qquad \vec{u} = \vec{u}_1 + \vec{u}_2 = \begin{pmatrix} \overline{\Psi} \\ \overline{T} \end{pmatrix} + \begin{pmatrix} \Phi \\ \Theta \end{pmatrix}.
\tag{9}
$$

$\mathcal{L}$ describes the linear time-independent operators in Eqs. (8) and $\vec{N}(.,.)$ the nonlinear part. Equation (8) decomposes into two separate equations for $\vec{u}_1$ and $\vec{u}_2$, which have symbolically the following form.

$$\partial_t \vec{u}_1 = \mathcal{L}\vec{u}_1 + \vec{N}(\vec{u}_1, \vec{u}_1) \tag{10}$$

$$\partial_t \vec{u}_2 = \mathcal{L}_2 \vec{u}_2 + \vec{N}(\vec{u}_2, \vec{u}_2) \tag{11}$$

$$\text{with } \mathcal{L}_2 \vec{u}_2 = \mathcal{L}\vec{u}_2 + \vec{N}(\vec{u}_1, \vec{u}_2) + \vec{N}(\vec{u}_2, \vec{u}_1). \tag{12}$$

$\mathcal{L}_i$ are always linear operators, whereas $\mathcal{L}_2$ has periodic coefficients depending on $\vec{u}_1$. In the remainder of this section we describe the properties of the primary state obtained by solving Eq. (10) and especially its stability with respect to linear perturbations. For the stability calculation of the Floquet type, the linear part of Eq. (11) has to be solved. $\mathcal{L}_2$ includes periodic terms and, in the case of the geometrical undulation of the top plate, becomes rather involved (see Schmitz and Zimmermann[12]).

The separation of the equations of motion into those for the slowly varying primary flow $\vec{u}_1$ and for the secondary flow $\vec{u}_2$ now has several advantages: The primary flow can be calculated in a perturbation expansion with respect to small values of k; similar to the unmodulated case the bifurcation from the primary into the secondary state is again sharp; and in the case of small modulation amplitudes $F_i \sim k$, we can derive an amplitude equation for our Rayleigh-Bénard-convection configurations.[17] The linear solutions of Eq. (11) can be written in the form

$$\Psi(x, z, t) = A_0 e^{iq\xi + \lambda t} F(x, z), \tag{13}$$

where the function $F(x, z)$ has the periodicity of the external modulation. The ansatz (13) transforms the equations of motion into an eigenvalue problem. The primary state becomes unstable at the parameter set R, F_i, k, q when $Re(\lambda_i)$ becomes positive .

3.3 RESULTS FOR SPATIALLY MODULATED RAYLEIGH-BÉNARD CONVECTION

Now we describe various consequences of the introduced external modulations. In all of the following examples, we have already convective flow in the primary state below the onset of the secondary flow as indicated in Figure 3 for a top boundary modulation. The primary convective flow is periodic with the wave length of the external modulation, where the amplitudes of the flow field and the temperature modulation increases with external modulation amplitudes. In the remainder of this section, we describe explicitly only the properties of the secondary flow at its threshold.

SINGLE MODULATION. Modulating the temperature at one boundary or the top boundary itself, the eigenvalue spectrum is real. Therefore the onset of the secondary flow is stationary. However, the eigenstates are amplitude-modulated with the wave length of the external modulation.

TWO TEMPERATURE MODULATIONS. When the temperature at the top and the bottom boundary is modulated with identical wave length, then the secondary flow also is modulated and the eigenvalue spectrum is again real. When we choose, for instance, the two wave lengths of the top and bottom temperature modulation at a ratio $k_1/k_2 = 3/2$ or $2/3$ and additionally a relative phase $\phi_2 \neq 0$ ($\phi_1 = 0$), then the eigenvalue spectrum includes complex eigenvalues. Even the eigenvalue with the largest real part may have an imaginary part. In that case the linear secondary flow is drifting at onset (see Eq. 13). The drift direction depends on the sign of the relative phase ϕ_2 and the drift velocity depends continuously on ϕ_2.

TEMPERATURE AT THE BOTTOM PLATE AND UNDULATED TOP PLATE: $k_0 = k_2$. In this case the situation for the secondary flow is similar to that for two geometric boundary undulations indicated in Figure 4. When the relative phase $\phi_2 = O$ vanishes, then the secondary flow is again amplitude-modulated and stationary. At a finite phase difference $\phi_2 \neq 0$ the secondary flow drifts, where the drift direction depends on the sign of ϕ_2. The drift velocity again depends continuously on ϕ_2 and has its extrema at the values $\phi_2 = \pm 2/3 \ldots$.

MODULATED TEMPERATURE AT THE BOTTOM AND UNDULATED TOP PLATE: $k_2/k_0 = 3/2$. Here the situation as sketched in Figure 5 is modeled by modulating the temperature at the bottom plate and using a simultaneously undulated top plate. The wave numbers have the ratio $k_2/k_0 = 3/2$. When the modulation amplitudes F_0 and F_2 are small, the changes in the phase shift have similar consequences as in the case of equal wave numbers: Drifts are induced. When the phase is varied at medium values of the amplitudes, then the imaginary part of the eigenvalues with the largest real part $\max[Re(\lambda) = 0]$, changes abruptly its sign at the phase $\phi_2 = \pi/2$. At this phase there is in the eigenvalue spectrum a pair of complex conjugate eigenvalues $\lambda_{1,2} = \sigma \pm i\omega$ (with $\sigma = 0$ at onset): *a Hopf bifurcation.* This corresponds to the situation described in qualitative terms above. Since both modulation wave numbers k_0 and k_2 are different, the virtual drift direction is periodically reversed with the periodicity $2\pi/k$. When the modulation amplitudes F_0 and F_2 cross some threshold, the bifurcation from the primary to the secondary flow is of Hopf type. In Figure 6 the range for that Hopf bifurcation is shown in the $F_0 - F_2$ plane. In the unshaded region the periodically varying virtual drifts compensate each other and in the shaded region those drifts are, so to say, frustrated and lead to Hopf bifurcation. In Figure 7(a) the linear eigenfunction for the secondary flow is shown for a stationary bifurcation (unshaded region) and in (b) for the Hopf bifurcation. The eigenfunction to the Hopf bifurcation is of subharmonic nature. It is also worthwhile to say that, near the separation line in Figure 6, one has a codimension 2 bifurcation similar to that in binary fluid convection.[13]

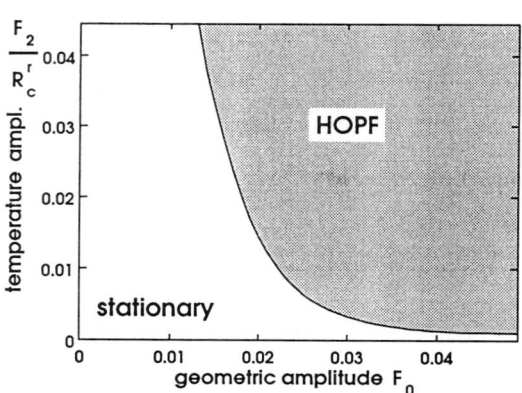

FIGURE 6 In the plane of the modulation amplitudes F_0 and F_2, the ranges are shown where the transition from the primary flow to the secondary flow occur via a stationary or a Hopf bifurcation (shaded). The ratios between the wave numbers of the geometric modulation of the top boundary and the temperature modulation at the bottom boundary is $k_0/k_2 = 2/3$ ($k = 0.25$) and the relative phase $\varphi_2 = 90°$ (rigid boundaries).

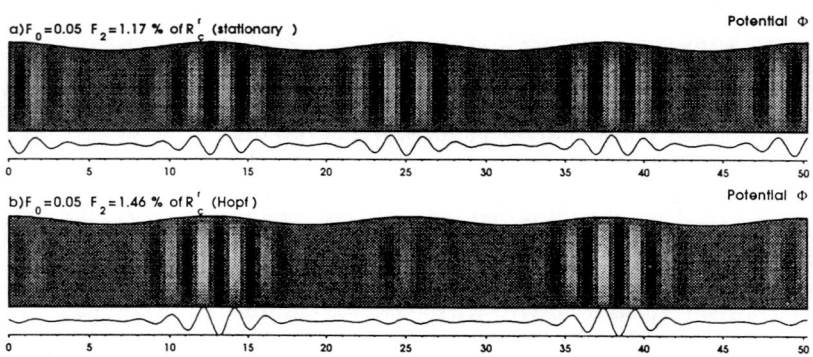

FIGURE 7 Here we show the linear secondary flow (a) in the stationary regime and (b) in the Hopf regime.

4. DISCUSSION

It has been demonstrated that broken symmetries in extended systems may lead
to a Hopf bifurcation. Beyond that Hopf bifurcation, we expect nonlinear standing
waves due to the broken translational symmetry, instead of traveling waves observed
in other systems.[13] Also in the nonlinear regime, interesting transition scenarios
from the standing wave back into stationary convection are expected similar
to those we observed in model equations. Instead of inducing frustrated drifts and,
therefore, a Hopf bifurcation via undulated boundaries in Rayleigh-Bénard convection,
one could force the same phenomena, for instance, in electroconvection in
nematic liquid crystals (see Zimmermann[15] and references therein) by introducing a
periodically varying tilt angle of the director with respect to the container boundary.
While we have discussed here the long wavelength external modulations for
two-dimensional convection, the use of short wavelength external modulations allows
in three-dimensional convection the forcing of interesting new two-dimensional
convection pattern.[16]

REFERENCES

1. Busse, F. H. "Nonlinear Properties of Convection." *Rep. Prog. Phys.* **41** (1978): 1929.
2. Busse, F. H. "Transition to Turbulence in Rayleigh-Bérnard Convection." In *Hydrodynamical Instabilities and the Transition to Turbulence*, edited by H. L. Swinney and J. P. Gollub. New York: Springer, 1981.
3. Coullet, P., and P. Huerre. "Resonance and Phase Solutions in Spatially-Forced Thermal Convection." *Physica D* (Nonlinear Phenomena) **23** (1986): 27.
4. Dominguez-Lerma, M. A., D. S. Cannell, and G. Ahlers. "Eckhaus Boundary and Wavenumber Selection Rotating Couette-Taylor Flow." *Phys. Rev. Lett.* **34** (1986): 4956.
5. Hartung, G., F. H. Busse, and I. Rehberg. "Time-Dependent Convection Induced by Broken Spatial Symmetries." *Phys. Rev. Lett.* **66** (1991): 2741.
6. Kelly, R. E., and D. Pal. "Thermal Convection with Spatially Periodic Boundary Conditions; Resonant Wavelength Excitation." *J. Fluid Mech.* **86** (1978): 433.
7. Kramer, L., E. Ben-Jacob, H. R. Brand, and M. C. Cross. "Wavelength Selection in Systems far From Equilibrium." *Phys. Rev. Lett.* **49** (1982): 1891.
8. Kuo, E. Y., and M. C. Cross. "Traveling-Wave Wall States in Rotating Rayleigh-Bérnard Convection." *Phys. Rev. E* **47** (1993): 2245.

9. Ning, L., and R. Ecke. "Küppers-Lortz Transition at High Dimensionless Rotation Rates in Rotating Rayleigh-Bérnard Convection." *Phys. Rev. E* **47** (1993): 2991.

10. Pal, D., and R. E. Kelly. "Thermal Convection with Spatially Periodic Nonuniform Heating: Nonresonant Wavelength Excitation." In *Proceedings 6th International Heat Trans. Conference, Toronto*, Vol. 2, 1978.

11. Rehberg, I., E. Bodenschatz, B. Winkler, and F. H. Busse. "Forced Phase Diffusion in a Convection Experiment." *Phys. Rev. Lett.* **59** (1987): 282.

12. Schmitz, R., and W. Zimmermann. "Rayleigh-Bérnard Convection in a Spatially Periodic Environment." To be published, 1994.

13. Schöpf, W., and W. Zimmermann. "Convection in Binary Fluids: Amplitude Equations, Codimension-2 Bifurcation, and Thermal Fluctuations." *Phys. Rev. E* **47** (1993): 1739.

14. Stavans, J., F. Heslot, and A. Libchaber. "Fixed Winding Number and the Quasiperiodic Route to Chaos in a Convective Fluid." *Phys. Rev. Lett.* **55** (1985): 596.

15. Zhong, F., R. E. Ecke, and V. Steinberg. "Rotating RBC: The Küppers-Lortz Transition." *Phys. Rev. Lett.* **67** (1991): 2473.

16. Zimmermann, W. "Pattern Formation in Electrohydrodynamic Convection." *Mat. Res. Bulletin* **16** (1991): 46.

17. Zimmermann, W., A. Ogawa, S. Kai, K. Kawasaki, and T. Kawakatsu. "Wavelength Competition in Convection Systems." *Europhys. Lett.* **24** (1993): 217.

18. Zimmermann, W., and R. Schmitz. "Hopf Bifurcation by Frustrated Drifts." *Phys. Rev. Lett.* submitted.

19. Zimmermann, W., M. Sesselberg, and F. Petruccione. "Effects of Disorder in Pattern Formation." *Phys. Rev. E* **48** (1993): 2699.

H. F. Goldstein and E. Knobloch
Department of Physics, University of California, Berkeley, CA 94720, USA

Convection in a Rotating Cylinder in the Low Prandtl Number Limit

1. INTRODUCTION

Rayleigh-Bénard convection in a pure fluid has been studied in the past as a particularly simple system exhibiting spontaneous pattern formation when the Rayleigh number R exceeds a critical value (hereafter referred to as R_c). Substantial progress has been made in understanding the process of pattern selection at small amplitudes, i.e., for $|R - R_c| \ll R_c$. Most theoretical treatments (see, e.g., Chandrasekhar[3]) assume that the system is unbounded in the horizontal plane. This assumption facilitates the solution of the linear stability problem and, hence, the location of R_c. In a nonrotating system it is known that the resulting instability is a steady-state one, and leads to either a pattern of rolls or of hexagons. Rolls are found when the boundary conditions at top and bottom are identical and non-Boussinesq effects are absent. If either of these requirements is not fulfilled, the initial transition is hysteretic and gives rise to hexagons.[2,11] The primary difference between these two cases is the absence of a reflection symmetry in the midplane of the layer in the latter case. The plane layer has another important reflection symmetry as well. This is a reflection in any *vertical* plane. When this symmetry is broken by rotating the layer about the vertical, the initial instability may set in as overstability. Chandrasekhar shows that in an unbounded layer this occurs only

Spatio-Temporal Patterns, Ed. P. E. Cladis and P. Palffy-Muhoray,
SFI Studies in the Sciences of Complexity, Addison-Wesley, 1995

for Prandtl numbers $\sigma < 0.68$; for water ($\sigma = 6.7$) the instability continues to be a steady state instability.[3] The broken reflection symmetry in vertical planes has, however, a profound effect on the *stability* of the resulting rolls. As the rotation rate increases, the initial rolls lose stability to rolls oriented at an angle α with respect to the initial pattern in the direction of rotation, and these are in turn unstable to a roll pattern at α relative to them, and so on.[9] Thus, for large enough rotation rates, no stable rolls are present near onset, even in systems with a midplane reflection symmetry. The angle α and the critical rotation rate for the instability depend on the Prandtl number, as discussed in detail by Clune and Knobloch.[4] The situation becomes quite different when the translation invariance of the plane layer is broken as well, e.g., by considering convection in a *finite* container. To preserve rotational invariance we consider here convection in cylindrical containers. In the nonrotating case the onset of convection continues to be a steady-state one, although the pattern that forms near onset may take a form quite different from the roll pattern characteristic of the unbounded system. This pattern, described by Buell and Catton,[1] is a reflection symmetric spokelike pattern with a nonzero azimuthal wavenumber m. Recent experiments on convection in water revealed, however, that in a rotating cylinder the corresponding pattern is no longer reflection symmetric (in fact, it is an m-fold *spiral*) and that it precesses in the rotating frame.[5,13]

As pointed out by Ecke et al.,[5] the observation of precessing patterns implies that the onset of convection is now a Hopf bifurcation. In fact, in such a system a Hopf bifurcation is to be expected whenever the azimuthal wavenumber m is nonzero, regardless of the Prandtl number. The argument goes as follows. Consider, say, the temperature perturbation Θ from the conduction state. In a cylindrical container, it follows that

$$\Theta(r, \phi, z, t) = \Re\{a(t)e^{im\phi}f_m(r, z)\} + \cdots, \tag{1}$$

where (r, ϕ, z) are cylindrical coordinates, $f_m(r, z)$ is the eigenfunction of the mode m, and $a(t)$ is its complex amplitude. Here, the omitted terms ($\cdots$) involve spatial harmonics of the fundamental generated by nonlinear terms; these modes are "slaved" to the evolution of a. We assume that $m \neq 0$ so that the instability *breaks* azimuthal symmetry. When the cylinder is nonrotating and the boundary conditions are homogeneous in ϕ, the equation satisfied by a must commute with the following symmetries:

$$\text{rotations} \quad \phi \to \phi + \theta : a \to ae^{im\theta} \tag{2}$$

$$\text{reflections} \quad \phi \to -\phi : a \to \overline{a}. \tag{3}$$

It follows that for $\epsilon \equiv (R - R_c)/R_c \ll 1$ the amplitude a satisfies an equation of the form

$$\dot{a} = g(|a|^2, \epsilon)a, \tag{4}$$

where the function g is forced by Eq. (3) to be real. Since a is small, we may expand g in a Taylor series,

$$\dot{a} = \epsilon a + \alpha |a|^2 a + \cdots, \tag{5}$$

obtaining

$$\dot{A} = \epsilon A + \alpha A^3 + \cdots, \quad \dot{\Phi} = 0, \tag{6}$$

where $a = Ae^{i\Phi}$ and A and Φ are real. The equation $\dot{\Phi} = 0$ is forced by the reflection symmetry of the system and embodies the requirement that the resulting pattern be neutrally stable with respect to rotations. When the cylinder is rotated about the vertical with angular velocity Ω, the reflection symmetry (3) is broken. Consequently the function g acquires an imaginary part, and Eq. (5) becomes

$$\dot{a} = (\epsilon + i\Omega\delta)a + (\alpha + i\Omega\beta)|a|^2 a + \cdots, \tag{7}$$

where now ϵ, δ, α, and β are all functions of Ω^2. In terms of the real variables A, Φ we now have

$$\dot{A} = \epsilon A + \alpha A^3 + \cdots, \quad \dot{\Phi} = \Omega(\delta + \beta A^2 + \cdots). \tag{8}$$

Consequently the broken reflection symmetry turns the steady state bifurcation in the nonrotating system into a Hopf bifurcation in the rotating system. Moreover, since $\dot{\Phi}$ is the rate of change of the azimuthal phase, it is to be identified with the precession frequency ω_p in the rotating frame. In particular, Θ takes the form

$$\Theta = \Re\{Ae^{i(m\phi + \omega_p t)} f_m(r, z)\} + \cdots. \tag{9}$$

The bifurcation is thus to a *rotating wave*.[10] Equation (8) shows that the precession frequency of a stationary pattern is given by

$$\omega_p = \Omega \left(\delta - \frac{\beta}{\alpha}\epsilon \right) + O(\epsilon^2). \tag{10}$$

We denote the precession frequency at onset ($\epsilon = 0$) by $\omega_c (\equiv \Omega\delta)$. The dependence of ω_p on Ω and ϵ suggested by the above theory was tested experimentally by Ecke et al.,[5] and the quantities R_c and ω_c, as well as the selected wavenumber m, were computed by Goldstein et al.[6] for two types of boundary conditions: (a) stress-free boundaries on top and bottom, and no-slip on the sidewall, and (b) no-slip everywhere. The solution in the latter case with fixed temperature at top and bottom and insulating sidewall was found to be in excellent agreement with the experimental data of Zhong et al.[14]; Goldstein et al.[7] extended these calculations to sufficiently low Prandtl numbers to produce overstability in an unbounded layer. In the circular container there are now two distinct sources of oscillation: overstability, present for all azimuthal wavenumbers m, and precession, present for all nonaxisymmetric patterns ($m \neq 0$). In particular, for nonaxisymmetric patterns these two sources of oscillations compete, and it is therefore of interest to determine the oscillation spectrum and behavior of the corresponding eigenfunctions.

2. THE EQUATIONS

We consider Boussinesq convection in a right circular cylinder of radius d and height h, filled with a pure fluid and rotating with constant and uniform angular velocity Ω about a vertical axis. We denote by Γ its aspect ratio d/h. The linearized, nondimensionalized equations of motion take the form[3]

$$\frac{1}{\sigma}\partial_t \mathbf{u} = -\nabla p + \nabla^2 \mathbf{u} + R\Theta\hat{\mathbf{z}} + \mathcal{T}\mathbf{u}\times\hat{\mathbf{z}}, \qquad (11)$$

$$\partial_t \Theta = w + \nabla^2\Theta, \qquad (12)$$

$$\nabla\cdot\mathbf{u} = 0, \qquad (13)$$

where $\mathbf{u} = u\hat{\mathbf{r}} + v\hat{\phi} + w\hat{\mathbf{z}}$ is the velocity field, Θ and p are the departures of the temperature and pressure from their conduction profiles, and $\hat{\mathbf{z}}$ is the unit vector in the vertical direction. The quantities $\mathcal{T} \equiv 2\Omega h^2/\nu$, $R \equiv g\alpha\Delta T h^3/\kappa\nu$, and $\sigma \equiv \nu/\kappa$ denote, respectively, the square root of the Taylor number, the Rayleigh number, and the Prandtl number. In these equations, length is in units of the layer thickness h, and time is in units of the vertical thermal diffusion time h^2/κ. Note that in writing Eq. (11) we have assumed that the centrifugal acceleration is sufficiently small so that the effective gravity remains vertical. This assumption is justified for the rotation rates used by Zhong et al.[13] for which $d\Omega^2/g < 0.01$ ($\mathcal{T} < 8548$).

For each value of the azimuthal wavenumber m, the conduction solution $u = v = w = \Theta = 0$ is stable to small perturbations below some critical value of the Rayleigh number $R_c^{(m)}$ which depends, in general, on the aspect ratio, the Taylor number and the Prandtl number, as well as the boundary conditions. In other words, if we write the time dependence of a solution to the linear problem as $e^{s_m t}$, then $\Re(s_m) < 0$ for all solutions when $R < R_c^{(m)}$. At $R = R_c^{(m)}$ there is for the first time a neutrally stable solution to the linear problem with wavenumber m, i.e., $\Re(s_m) = 0$. If $\Im(s_m) = 0$, the bifurcation is steady state, and if $\Im(s_m) = \omega_c^{(m)} \neq 0$, we have a Hopf bifurcation with Hopf frequency $\omega_c^{(m)}$. The critical Rayleigh number is given by $R_c = \min_m R_c^{(m)}$. The above problem has been solved for a $\Gamma = 1$ cylinder using both types of boundary conditions. The results for the boundary conditions (a), given by

$$\partial_z u = \partial_z v = w = \Theta = 0 \quad \text{on} \quad z = 0, 1, \qquad (14)$$

$$u = v = w = \partial_r\Theta = 0 \quad \text{on} \quad r = \Gamma, \qquad (15)$$

are shown in Figures 1 and 2 as dashed lines, and were obtained for $\sigma = 0.025$, the Prandtl number of liquid mercury, and $\Gamma = 1$.

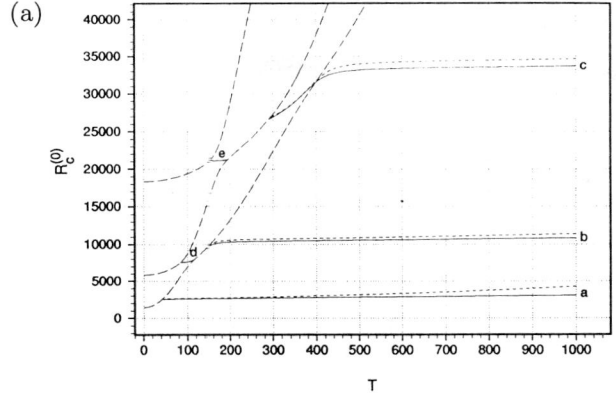

(a)

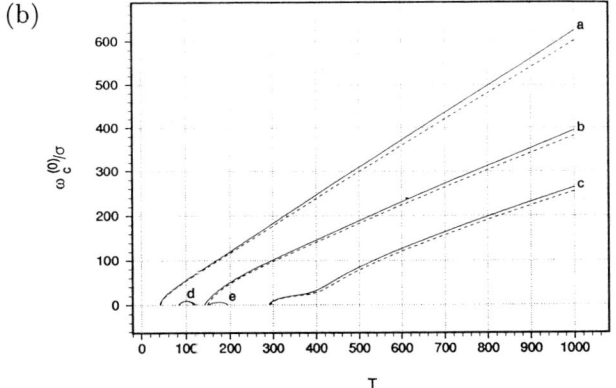

(b)

FIGURE 1 The first few neutrally stable axisymmetric modes. (a) $R_c^{(0)}$ vs. $\mathcal{T}$ and (b) the corresponding $\omega_c^{(0)}/\sigma$. Solid lines are the zero Prandtl number limit; dashed lines are for $\sigma = 0.025$. The long dashed lines in the Rayleigh number plot indicate steady state modes, which are independent of the Prandtl number. Only the zero Prandtl number modes are labeled.

In the following we discuss the Taylor number dependence of the results for $m = 0$ and $m = 1$. The former correspond to axisymmetric patterns; in the latter case the pattern breaks the circular symmetry. For $\sigma < 1$ the Taylor number must exceed some minimum value that depends on σ (the Takens-Bogdanov (TB) point) before overstable axisymmetric modes become possible. There are two types of such TB points, those that give rise to oscillations whose frequency increases monotonically with $\mathcal{T}$, and those that connect back to another TB point. The former correspond to the oscillations one expects at larger rotation rates by analogy with the unbounded problem. The latter are associated with Hopf curves that connect different steady modes. Notice the deformation of the steady state neutral curves in the vicinity of these connections. The behavior seen here differs significantly from the unbounded system where exactly one TB point exists on each steady mode. The complete

(a)

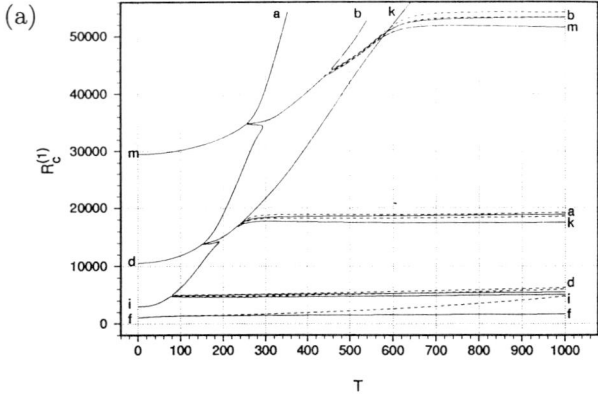

(b)

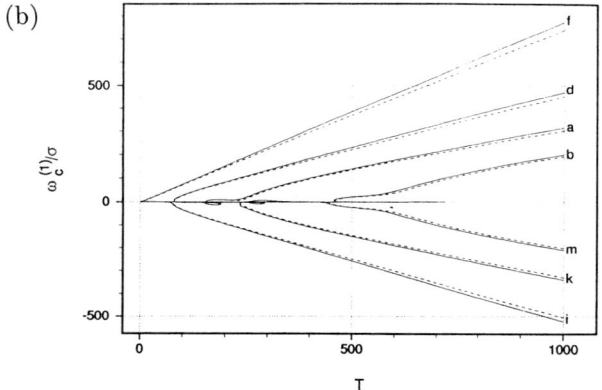

FIGURE 2   As for Figure 1, but for $m = 1$; in this case all modes are oscillatory.

spectrum of oscillation frequencies is obtained by reflecting Figure 1(b) in $\omega = 0$. When this is done, the Hopf curves connecting steady state neutral curves produce disconnected ovals. The deformation of $\omega_c^{(0)}$ for mode c seen in Figure 1(b) suggests that these ovals are created by a pinching off process as parameters are varied. This process allows a large T oscillatory mode to change the steady state mode on which it terminates, and leaves behind a short segment of oscillations connecting the two steady state modes involved (cf. Figure 1(a)). In Figure 2(a) we show the corresponding results for $m = 1$. For such nonaxisymmetric modes the rotation of the container splits the neutral stability curves for all the oscillatory $m = 0$ modes. This rotational splitting is expected on general grounds (cf. Knobloch,[8]) and arises because prograde ($\omega < 0$) and retrograde ($\omega > 0$) modes no longer couple to the rotation in the same way. As a result, one type of mode comes in first. In the present case the *prograde* mode always has the lower critical Rayleigh number at large T. The difference between the prograde and retrograde modes can also be seen in the asymmetry of the frequency spectrum with respect to reflection in $\omega = 0$ (Figure 2(b)). Thus, it is the rotational splitting of the modes a, b, and c in Figure 1(a) that is responsible for the modes (d, i), (a, k), and (b, m) in Figure 2(a), respectively. In Figure 2(a) we reveal, however, the presence of an extra mode at larger T, labeled f. This mode does not originate through rotational splitting; it has no (nearly) symmetric counterpart in Figure 2(b) and its precession frequency decreases linearly to zero with decreasing T. It is also the preferred mode in the range of T shown. The existence of such a mode is in accord with general considerations.[5] Since this mode does not appear to have an axisymmetric counterpart, we identify it as a Poincaré mode of the rotating cylinder, in contrast to the other modes (d, i, a, k, b, m) which are nonaxisymmetric counterparts of the usual oscillatory modes. Note that this identification is supported both by the relatively high precession frequency of the mode (see Figure 2(b)), and by the fact that this precession is counter to the direction of rotation ($\omega > 0$).

We discuss next the effect of nonaxisymmetry on the transition between steady and oscillatory modes. Recall first that generically the bifurcation to a nonaxisymmetric mode cannot be stationary. Consequently, all the curves in Figure 2(a) correspond to Hopf bifurcations. This is clearly seen in Figure 2(a) at small T where the neutral stability curves for steady state bifurcations shown in Figure 1(a) are seen to have unambiguous Hopf counterparts. In the following we characterize the modes as prograde or retrograde depending on whether $\omega < 0$ or $\omega > 0$ at large ω. With increasing T, one sees two types of interaction between the Hopf curves that correspond to the $m = 0$ steady and oscillatory neutral stability curves. One type of interaction is organized around the Takens-Bogdanov points, and results in what we call *mode repulsion*, as in the interaction between modes (i, d) near $T = 80$, between modes (k, a) near $T = 240$ and between modes (m, b) near $T = 450$. Note that the interaction near $T = 240$ is of the same type as the other two, except for the fact that it is the mode k, i.e., the prograde mode, that turns around, and not the retrograde mode. This type of interaction comes about because of the rotational splitting of the Hopf curves for nonaxisymmetric modes. The other type of interaction comes about through the rotational splitting of the Hopf curves connecting the steady state neutral curves in the axisymmetric case, and results in either a *crossing* or an *avoided crossing* of the corresponding nonaxisymmetric Hopf curves. Examples of the former are found near $T = 255$, and of the latter near $T = 154$ (see Figure 2(a)). Both interactions are of the same type except for the fact that in the latter case the interaction takes place between two retrograde modes (modes a and d both precess counter to the direction of rotation at large T), while in the former it takes place between a retrograde mode (mode a) and a prograde one (mode m). Additional insight into these interactions can be found from their Prandtl number dependence.[7]

3. THE ZERO PRANDTL NUMBER LIMIT

In order to assist in the interpretation of the small Prandtl number results, we examined the zero Prandtl number limit of Eqs. (11–13). As pointed out by Thual,[12] there are several ways of taking this limit. This is because the Prandtl number can be small either because the viscosity is small or because the thermal diffusivity is large. In the former case the motion is dominated by the constraint arising from the Taylor–Proudman theorem. This is the limit that is of interest in geophysical applications, and corresponds to very large Taylor numbers. In laboratory experiments, the Taylor numbers do not reach high values, however, and the more appropriate limit is that obtained by assuming that $\mathcal{T}$ remains of order one. In this case the zero Prandtl number limit corresponds to large thermal diffusivity. The resulting critical Rayleigh number will be of order one, though the temperature difference required to drive convection will have to be large. The correct scaling is to scale $\mathbf{u}$ with σ and the time with σ^{-1}, while the temperature fluctuation Θ scales with σ. This procedure implies that the relevant time scale of motion is now the viscous time scale, the thermal time scale being so short that temperature fluctuations equilibriate essentially instantaneously. Consequently, the allowed temperature fluctuations are very small. In the limit the resulting linear equations are

$$\partial_t \mathbf{u} = -\nabla p + \nabla^2 \mathbf{u} + R\Theta\hat{\mathbf{z}} + \mathcal{T}\mathbf{u} \times \hat{\mathbf{z}}, \tag{16}$$

$$0 = w + \nabla^2\Theta, \tag{17}$$

$$\nabla \cdot \mathbf{u} = 0. \tag{18}$$

It follows that the precession velocity of nonaxisymmetric patterns will be $O(\sigma)$ in the scaling leading to Eqs. (11–13), or $O(1)$ in the above scaling. In dimensional terms the frequency is $O(\nu/h^2)$. This scaling is the appropriate one if one is interested in precessing patterns in slowly rotating containers in the limit of small Prandtl numbers. The solutions to Eqs. (16–18) with the boundary conditions (14–15) are shown as solid lines in Figures 1 and 2 and for low rotation rates are very close (i.e., $O(\sigma)$ close) to those of Eqs. (11–13) with $\sigma = 0.025$, the error increasing typically linearly with the rotation rate. This behavior is expected from the scaling leading to Eqs. (16–18).

The "other" low Prandtl number limit corresponds in the scaling of Eqs. (11–13) to the large rotation limit. In this limit the dimensional precession frequency is $O(2\Omega)$, where Ω is the dimensional rotation rate of the cylinder. In the scaling of equations (11–13) one expects the dimensionless frequencies to scale as $\sigma\mathcal{T}$. It is easy, therefore, to pick out from Figures 1 and 2 those modes that persist in this limit.

ACKNOWLEDGMENTS

This work was done as part of a joint project with I. Mercader and M. Net, and was supported by an INCOR grant from Los Alamos National Laboratory.

REFERENCES

1. Buell, J. C., and I. Catton. "The Effect of Wall Conduction on the Stability of Fluid in a Right Circular Cylinder Heated from Below." *J. Heat Transfer* **105** (1983):255–260.
2. Busse, F. H. "Non-linear Properties of Thermal Convection." *Rep. Prog. Phys.* **41** (1978):1929–1967.
3. Chandrasekhar, S. *Hydrodynamic and Hydromagnetic Stability.* New York: Dover, 1961.
4. Clune, T., and E. Knobloch. "Pattern Selection in Rotating Convection with Experimental Boundary Conditions." *Phys. Rev. E* **47** (1993):2536–2550.
5. Ecke, R. E., F. Zhong, and E. Knobloch. "Hopf Bifurcation with Broken Reflection Symmetry in Rotating Rayleigh-Bénard Convection." *Europhys. Lett.* **19** (1992):177–182.
6. Goldstein, H. F., E. Knobloch, I. Mercader, and M. Net. "Convection in a Rotating Cylinder. Part 1. Linear Theory for Moderate Prandtl Numbers." *J. Fluid Mech.* **248** (1993):583–604.
7. Goldstein, H. F., E. Knobloch, I. Mercader, and M. Net. "Convection in a Rotating Cylinder. Part 2. Linear Theory for Low Prandtl Numbers." *J. Fluid Mech.* **262** (1994): 293–324.
8. Knobloch, E. "Bifurcations in Rotating Systems." In *Theory of Solar and Planetary Dynamos: Introductory Lectures*, edited by M. R. E. Proctor and A. D. Gilbert. Cambridge: Cambridge University Press, in press.
9. Küppers, G., and D. Lortz. "Transition from Laminar Convection to Thermal Turbulence in a Rotating Fluid Layer." *J. Fluid Mech.* **35** (1969):609–620.
10. Rand, D. "Dynamics and Symmetry. Predictions for Modulated Waves in Rotating Fluids." *Arch. Rat. Mech. Anal.* **79** (1982):1–37.
11. Schlüter, A., D. Lortz, and F. H. Busse. "On the Stability of Steady Finite Amplitude Convection." *J. Fluid Mech.* **23** (1965):129–144.
12. Thual, O. "Zero-Prandtl Number Convection." *J. Fluid Mech.* **240** (1992): 229–258.
13. Zhong, F., R. E. Ecke, and V. Steinberg. "Asymmetric Modes and the Transition to Vortex Structures in Rotating Rayleigh-Bénard Convection." *Phys. Rev. Lett.* **67** (1991):2473–2476.

14. Zhong, F., R. E. Ecke and V. Steinberg. "Rotating Rayleigh–Bénard Convection: Asymmetric Modes and Vortex States." *J. Fluid Mech.* **249** (1993):135–159.

L. S. Tsimring† and M. I. Rabinovich††
†Institute for Nonlinear Science, University of California, San Diego, CA 92093-0402, USA
‡Permanent address: Institute of Applied Physics, Russian Academy of Sciences, 46 Ulyanov str., 603600 Nizhni Novgorod, Russia.

Dislocations in Hexagonal Patterns

We discuss the dynamics of dislocations of roll systems forming hexagonal patterns. The analysis and numerical simulations are carried out within a model of three resonantly coupled Newell-Whitehead-Segel equations for complex amplitudes of the roll structures. It is shown that an individual dislocation of one roll system is driven away as a result of phase synchronization among roll patterns. Two dislocations with opposite topological charges, belonging to different roll systems, are attracted to each other and form a "penta-hepta" defect on the background of the perfect hexagonal pattern, which remains stable.

INTRODUCTION

Hexagonal patterns appear in many different physical situations when a large two-dimensional rotationally invariant system undergoes an instability with characteristic nonzero wavenumber. The key physical mechanism leading to the formation of hexagonal structures is the three-wave interaction between spectral components with different wavevectors. This interaction is described by the second term in the right-hand side of the general equation for the complex amplitude $a_{\mathbf{kk}}(t)$ of the spectral component with wavevector $\mathbf{k}$ (the real field is its Fourier transform $u(\mathbf{r},t) = \int a_{\mathbf{kk}} \exp(i\mathbf{kr})d\mathbf{k} + c.c.$):

$$
\frac{da_{\mathbf{kk}}}{dt} = \mu(\mathbf{k})a_{\mathbf{kk}} + \int \alpha_{\mathbf{kk}_1\mathbf{k}_2} a_{\mathbf{k}_1} a_{\mathbf{k}_2} \delta(k - k_1 - k_2) dk_1 dk_2
$$
$$
- \int \gamma_{\mathbf{kk}_1\mathbf{k}_2\mathbf{k}_3} a_{\mathbf{k}_1} a_{\mathbf{k}_2} a_{\mathbf{k}_3} \delta(k - k_1 - k_2 - k_3) dk_1 dk_2 dk_3 + O(a_{\mathbf{k}}^4). \tag{1}
$$

The growth rate of the instability $\mu(|\mathbf{k}|) = \mu - \mu_2(|\mathbf{k}| - q_0)^2$ reaches the maximum value μ on the circle $|\mathbf{k}| = q_0$. The Delta function in the quadratic term picks triplets of wavevectors which in case of small supercriticality $(\mu/\mu_2)^{1/2} \ll q_0$ form equilateral triangle, and the corresponding three waves form a hexagonal pattern. Such patterns are observed in Rayleigh-Bénard convection in non-Bussinesq fluids,[1,2,3,4] in Bénard-Marangoni convection,[5] auto-catalytic reactions,[6] etc. (see also the review by Cross and Hohenberg[7]). However, perfect hexagonal patterns are rather difficult to observe in large aspect-ratio systems. Typically, different line or point defects emerge on the background of a hexagonal pattern. Line defects usually take the form of grain boundaries between hexagons with different orientation, or grain boundaries separating hexagons and rolls, the latter case was investigated in detail by Malomed et al.[8] Among point defects most typical are so-called "penta-hepta" defects, or pairs of cells with 5 and 7 ridges. These defects, once having been created, remain very stable. It was shown recently by Ciliberto et al.[3] and by Bodenschatz[9] the direct demodulation of the optical images obtained for non-Boussinesq convection, that a penta-hepta defect consists in fact of two dislocation of two of the three underlying roll systems. An interesting question of how these "bound states" of dislocations are created has been left unanswered. In this chapter we consider the dynamics of dislocations of individual roll systems within the framework of amplitude equations. We will show that an individual dislocatons distorts strongly the phase synchronization of the hexagonal pattern and therefore cannot be stationary. The Peach-Köhler-type force drives the dislocation away or towards another dislocation depending on the initial phase distribution in the pattern. In the latter case, if the two dislocations belong to different roll patterns, they stick together and form a stationary penta-hepta defect.

As we have already mentioned before, the perfect hexagon pattern is a result of the resonant interaction of three roll systems $\{A_i \exp[i\mathbf{k}_i\mathbf{r}], i = 1, 2, 3\}$ with wavevectors $\mathbf{k}_{1,2,3}$ satisfying the three-wave resonant condition

$$
\mathbf{k}_1 + \mathbf{k}_2 + \mathbf{k}_3 = 0.
$$

(In the following we will assume that the wavenumbers of all three waves correspond to the most unstable wavenumber $|\mathbf{k}_1| = |\mathbf{k}_2| = |\mathbf{k}_3| = q_0$ which we set to unity in dimensionless variables.) For convection in a horizontal layer of fluid three-wave interaction appears only if the "up-down" symmetry $A \rightarrow -A$ of the system is broken. For example, in Rayleigh-Bénard convection this symmetry breaking is caused by temperature dependence of viscosity, in Bénard-Marangoni convection—by temperature dependence of the surface tension, etc. As a result of the nonlinear interaction, systems of three rolls oriented at $120°$ to each other grow simultaneously, their amplitudes become equal, and the sum of their phases approaches zero. This process can be described by the following set of three amplitude equations for the complex amplitudes of three roll systems which can be easily deduced from Eq. (1):

$$\partial_t A_1 = \mu A_1 + \alpha A_2^* A_3^* - (|A_1|^2 + \gamma |A_2|^2 + \gamma |A_3|^2) A_1 \,,$$
$$\partial_t A_2 = \mu A_2 + \alpha A_1^* A_3^* - (|A_2|^2 + \gamma |A_1|^2 + \gamma |A_3|^2) A_2 \,, \qquad (2)$$
$$\partial_t A_3 = \mu A_3 + \alpha A_1^* A_2^* - (|A_3|^2 + \gamma |A_1|^2 + \gamma |A_2|^2) A_3 \,.$$

Here μ is the supercriticality parameter, α is the coefficient of quadratic nonlinearity, describing non-Boussinesq effects, γ is the ratio of the coefficient of cubic interaction of rolls of different orientation to the coefficient of cubic self-interaction, and we can always make appropriate scaling such that μ, α, γ, and A are all $O(1)$. Note, that in nonscaled variables $\bar{\mu}$ is $O(\epsilon)$, $\bar{A}$ and $\bar{\alpha}$ are $O(\epsilon^{1/2})$ and $\bar{\gamma}$ is $O(1)$, where ϵ is a small number characterizing deviation of the control parameter (e.g., Rayleigh number) R from the threshold value R_c, $\epsilon \sim (R - R_c)/R_c$. Linear stability analysis shows (Busse,[10] see also Malomed and Tribelsky[11] and Ciliberto et al.[3]) that if $\alpha \neq 0$ a stable hexagon pattern appears as a result of subcritical bifurcation when $-\alpha^2/4(1 + 2\gamma) < \mu < \alpha^2(\gamma + 2)/(\gamma - 1)^2$. When the defects are present, the amplitudes of individual roll systems become functions of space as well as time. The structure of possible envelope equations was discussed by Haken[12] and Malomed et al.,[8] who introduced transversal diffusion terms in Eq. (2) (see also Brand[13]). Following them, we restrict ourselves with the most natural assumption that, besides quadratic nonlinearity, the envelopes of individual roll systems are governed by the Newell-Whitehead-Segel equations[14,15]

$$\partial_t A_1 = \mu A_1 + \alpha A_2^* A_3^* - (|A_1|^2 + \gamma |A_2|^2 + \gamma |A_3|^2) A_1 + \hat{D}_1^2 A_1 \,,$$
$$\partial_t A_2 = \mu A_2 + \alpha A_1^* A_3^* - (|A_2|^2 + \gamma |A_1|^2 + \gamma |A_3|^2) A_2 + \hat{D}_2^2 A_2 \,, \qquad (3)$$
$$\partial_t A_3 = \mu A_3 + \alpha A_1^* A_2^* - (|A_3|^2 + \gamma |A_1|^2 + \gamma |A_2|^2) A_3 + \hat{D}_3^2 A_3 \,.$$

These differ from Eq. (2) by the linear terms $\hat{D}_i^2 A_i$ in the right-hand sides. Here, $\hat{D}_i = \partial_{X_i} - i\epsilon^{1/2}/2 \partial_{Y_i}^2$, $X_i = \epsilon^{1/2} x_i/\xi_0$, and $Y_i = \epsilon^{1/2} y_i/\xi_0$ are pairs of rescaled dimensionless Cartesian coordinates orthogonal and parallel to rolls axes, respectively, and $\xi_0 (\equiv q_0^{-1})$ is the dimensional length scale of the system. It is important to note that in description of hexagons we have to assume the same scaling for both x_i and y_i coordinates. Since the defects in hexagons presumably have the same characteristic size in x and y directions, parallel diffusion (the terms proportional to $\partial_{Y_i}^2$ in $\hat{D}_i$) is usually much weaker than transversal diffusion. We keep those terms here in order to have a unified set of equations for both roll and hexagon pattern formation depending on values of parameters μ, α, and γ.

The set of Eq. (3) can be represented in a variational form

$$\partial_t A_i = -\frac{\delta \mathcal{F}}{\delta A_i^*},$$

with the free energy functional

$$\mathcal{F} = \int dxdy\{-\mu(|A_1|^2 + |A_2|^2 + |A_3|^2) - \alpha(A_1^* A_2^* A_3^* + c.c.)$$
$$+ \frac{1}{2}(|A_1|^4 + |A_2|^4 + |A_3|^4)$$
$$+ \gamma(|A_1|^2|A_2|^2 + |A_1|^2|A_3|^2 + |A_2|^2|A_3|^2) + (|\hat{D}_1 A_1|^2 + |\hat{D}_2 A_2|^2 + |\hat{D}_3 A_3|^2)\},$$
$$(4)$$

so only static patterns can be expected as $t \to \infty$.

In our numerical simulations of Eq. (3), we employed a pseudo-spectral split-step method on a square mesh with the system size 64×64 and zero-gradient boundary conditions. We used time step 0.1, mesh size 0.5, and parameters $\epsilon = 10^{-2}$, $\mu = 0.5$, $\alpha = 1$, and $\gamma = 2$. Here we are concerned with the evolution and interaction of dislocations of individual roll systems forming a hexagon pattern.

Let us first consider as initial conditions sets of three rolls of small inital amplitude, and one of sets contains a dislocation which has appeared due to, say, initial fluctuations. Specifically, we used the following complex amplitudes $A_i = R_i \exp i\phi_i$:

$$R_1 = 0.011 \tanh\left(0.2\sqrt{(X_1 - X_1^0)^2 + (Y_1 - Y_1^0)^2}\right),$$
$$\phi_1 = \arctan\frac{Y_1 - Y_1^0}{X_1 - X_1^0} + \phi_0, \qquad (5)$$
$$R_2 = 0.01, \ \phi_2 = 0.0, \ R_3 = 0.01, \ \phi_3 = 0.0.$$

If there were no quadratic nonlinearity ($\alpha = 0$), one of three roll system (for our initial conditions this is the roll set 1 with a dislocation) would suppress the other two and the dislocation itself would remain motionless since the period of rolls we take, corresponds to the optimal wavenumber. On the contrary, within the range of parameters favoring hexagons ($\alpha = 1.0$), amplitudes of all three roll sets grow, and their phases get synchronized to form hexagonal structures. However, because of the dislocation, synchronization process is quite peculiar. In what follows, we use the parameter $F(x, y, t) = -\cos(\phi_1 + \phi_2 + \phi_3)$ as a measure of synchronization.

In Figures 1 and 2 the sequences of spatial distributions of $F(x, y)$ and the corresponding real-field structures reconstructed from the complex amplitude distributions, are plotted for different moments of time. Initially, F is a smooth function of (x, y) everywhere except the core of dislocation where it takes all values from -1 to 1 (see Figure 1(a), 2(a)). At $t > 0$, F decays towards $F_0 = -1$ everywhere except the line where it initially was equal to 1. After a rather short transient ($t \simeq 10.0$) $F \simeq -1$ almost everywhere but some narrow corridor starting from the core of dislocation, where it reaches the value of 1 (see Figure 1(b), 2(b)). The axis of this corridor coincides with the line $F(x, y, 0) = 1$. After that the dislocation starts moving quickly along this corridor (see Figure 1(c), 2(c)), and soon leaves the region of integration, so almost perfect hexagon pattern with $F = -1$ everywhere eventually establishes (see Figure 1(d), 2(d)).

In the second series of numerical experiments we began with two dislocations which belong to two different roll structures:

$$R_1 = 0.01 \tanh \left(0.2\sqrt{(X_1 - X_1^0)^2 + (Y_1 - Y_1^0)^2} \right), \quad \phi_1 = \arctan \frac{Y_1 - Y_1^0}{X_1 - X_1^0} + \phi_0,$$

$$R_2 = 0.01 \tanh \left(0.2\sqrt{(X_2 - X_2^0)^2 + (Y_2 - Y_2^0)^2} \right), \quad \phi_2 = -\arctan \frac{Y_2 - Y_2^0}{X_2 - X_2^0},$$

$$R_3 = 0.1, \quad \phi_3 = 0.0.$$

(6)

Here $\{X_1^0, Y_1^0\}$ and $\{X_2^0, Y_2^0\}$ are the coordinates of cores of dislocations. Note the sign "$-$" in ϕ_2 which indicates different topological charges (opposite directions of phase rotation) of the two dislocations. In Figures 3 and 4 we illustrate the evolution of the synchronyzation parameter F and the real field for $\phi_0 = \pi/4$. Again,

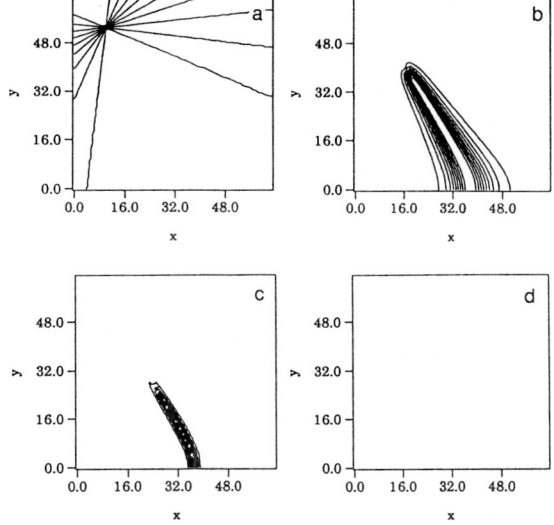

FIGURE 1   Profile of the synchronization parameter $F = -\cos \Phi$ for one dislocation in the hexagonal pattern at different times: (a) $t = 0$; (b) $t = 10$; (c) $t = 13$; and (d) $t = 20$.

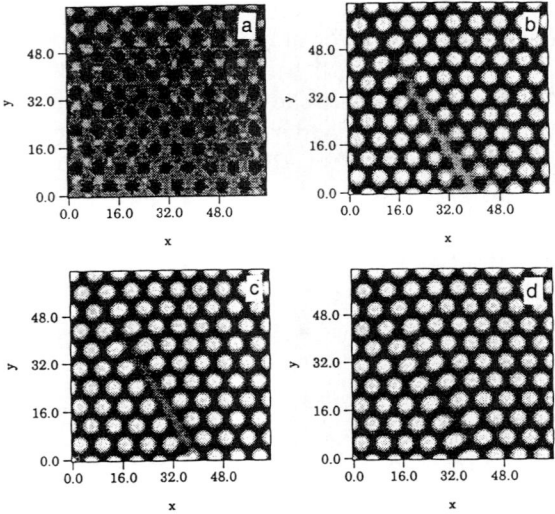

FIGURE 2   Structure of hexagonal pattern reconstructed from the amplitude distributions when one dislocation is present, at different times: (a) $t = 0$; (b) $t = 10$; (c) $t = 13$; and (d) $t = 20$.

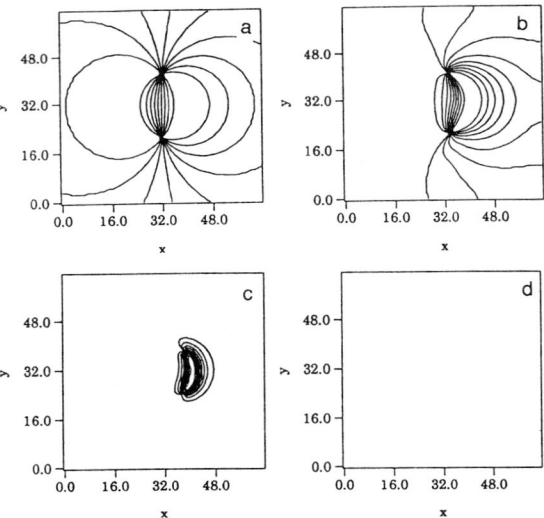

FIGURE 3   Profile of the synchronization parameter $F = -\cos\Phi$ for two dislocations in the hexagonal pattern with $\phi_0 = \pi/4$ at different times: (a) $t = 0$; (b) $t = 8$; (c) $t = 10$; and (d) $t = 20$.

at $t = 0$ the synchronization parameter is distributed smoothly within the range $(-1, 1)$ (with the given initial phase distribution the isolines of F are circular, see Figure 3(a)). Now, instead of a straight corridor we get a curved (circular) corridor connecting both defects (see Figure 3(b)). After some transient period dislocations begin to move toward each other along the corridor (see Figure 3(c)) and eventually they stick together and form a penta-hepta defect (see Figure 3(d)). Corresponding plots of the real field are presented in Figure 4(a)–(d). The trajectories of the dislocations motion and the position of the penta-hepta pair depend strongly on the initial phase distribution (see Figure 5). It is worth mentioning that the structure of the penta-hepta defect (see Figure 6) resulting from the merging of two dislocations, is remarkably similar to one found in the laboratory experiment.[3]

Qualitatively, the behavior described above can be easily understood within the *phase approximation* (see, e.g., Cross and Hohenberg[7]). After initial transients all three magnitudes R_i become equal and adiabatically follow the phase dynamics. This approximation is valid outside cores of dislocations. The summation of the real parts of Eq. (3) then gives the equation for R:

$$(1 + 2\gamma)R^2 - \alpha \cos \Phi R - \mu = 0, \qquad (7)$$

where we neglected time and space derivatives and denote $\Phi \equiv \phi_1 + \phi_2 + \phi_3$. The imaginary parts of Eq. (3) describe the phase variations

$$\partial_t \phi_i = -3\alpha R \sin \Phi + \partial^2_{X_i} \phi_i, \ i = 1, 2, 3. \qquad (8)$$

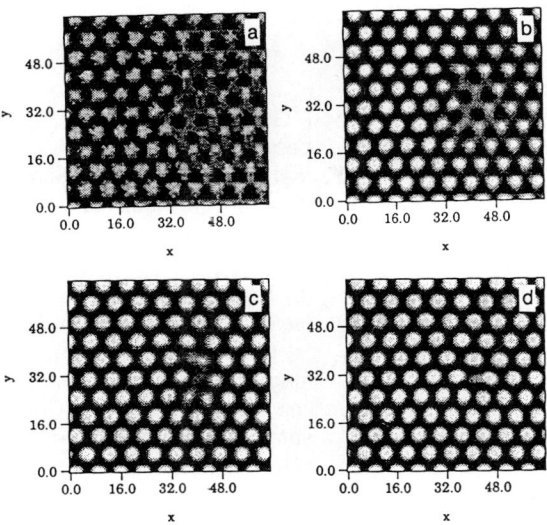

FIGURE 4 Structure of hexagonal pattern when two dislocation of two underlying roll structures are present: (a) $t = 0$; (b) $t = 8$; (c) $t = 10$; and (d) $t = 20$.

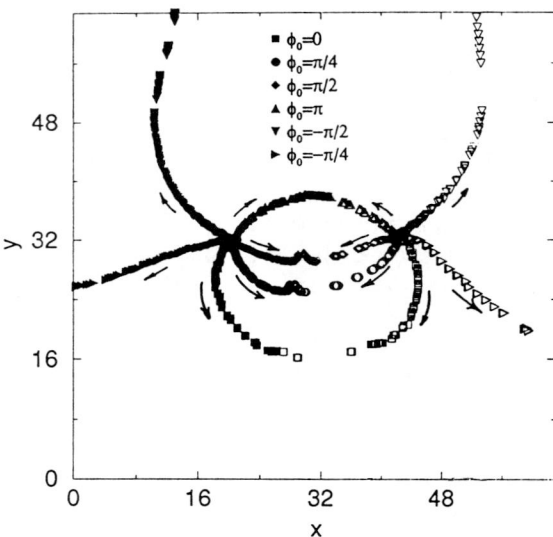

FIGURE 5   Trajectories of the cores of dislocations in the hexagonal pattern for several different phase shifts between roll systems ϕ_0.

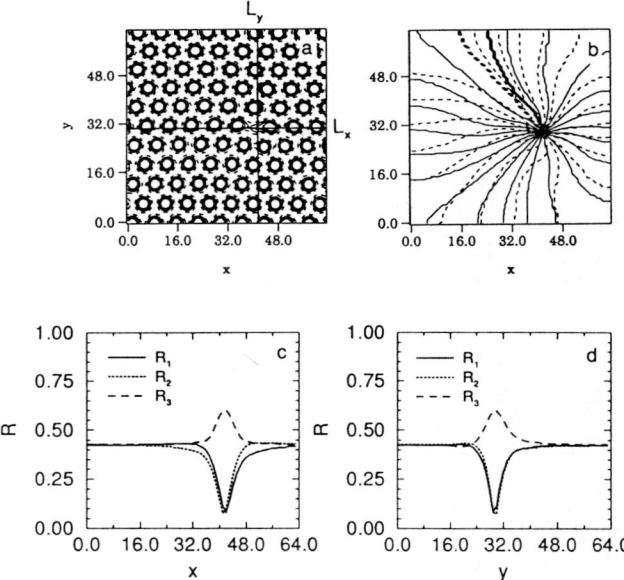

FIGURE 6   "Penta-hepta" pair formed by merging of two dislocations ($t = 40$): (a) isolines of the "temperature" field $T = \sum_{i=1}^{3} A_i \exp(ix_i)$; (b) Equiphase lines of ϕ_1 (solid) and ϕ_2 (dashed); (c) Cross sections of the magnitudes R_i along the line L_x in (a); and (d) The same as in (c) along the line L_y in (a).

Equation (8) with R taken from Eq. (7) also form a variational model with the free energy functional

$$
\mathcal{F}_\phi = \int dx dy \{ -\frac{\alpha}{4(1+2\gamma)}
$$
$$
\times \left[\cos\Phi(\cos\Phi + \sqrt{\cos^2\Phi + B}) + B\ln(\cos\Phi + \sqrt{\cos^2\Phi + B}) \right] \tag{9}
$$
$$
+ \frac{1}{2}[(\partial_{X_1}\phi_1)^2 + (\partial_{X_2}\phi_2)^2 + (\partial_{X_3}\phi_3)^2] \}
$$

where $B = 4\mu(1+2\gamma)/\alpha^2$. It is easy to see now that as long as the phase variations are smooth and one can neglect space derivatives in Eq. (9), the density of the free-energy functional is a monotonous function of the parameter F. Within the same assumptions the equation for Φ is simply

$$
\partial_t \Phi = -\frac{\alpha}{2(1+2\gamma)} \left[\cos\Phi + \sqrt{\cos^2\Phi + B} \right] \sin\Phi, \tag{10}
$$

so Φ goes to 0 ($F \to -1$) everywhere except on the lines where $\Phi = \pi$ ($F = 1$), and therefore corridors of large F are formed. The only points where those lines can originate and/or end (except infinity) are the dislocations. Therefore, the dislocations start to move (due to a Peach-Köhler-type force, see Sigga and Zippelius[16]) as to decrease optimally the free energy of the system, i.e., along the lines of maximum $F = 1$ and either go to infinity or meet each other and stick, thus creating a "penta-hepta" pair. Unfortunately, this simple local phase approximation fails not only in the cores of defects, but eventually within the corridors as well. Indeed, in accordance with Eq. (10) the corridors get more and more narrow, and at large times the assumption of smoothness of phase dynamics is also violated in their vicinities and the diffusion terms in Eq. (8) become important. In the long run some stationary profiles of corridors are established due to phase diffusion (these stationary corridors are analogous to the kink solutions of the sine-Gordon equation). That is why the above arguments hold only qualitatively. Nevertheless, they explain correctly the main features of behavior of individual roll dislocations and the penta-hepta pair formation.

ACKNOWLEDGMENTS

Authors are grateful to H. Abarbanel and E. Bodenschatz for useful discussions. This work was supported by the U.S. Department of Energy under contract DE-FG03-90ER14138 and by the Office of Naval Research under contract N00014-D-0142 DO#15.

REFERENCES

1. Walden, R. W., and G. Ahlers. "Non-Boussinesq and Penetrative Convection in a Cylindrical Cell." *J. Fluid. Mech.* **109** (1981): 89.
2. Ciliberto, S., E. Pampaloni, and C. Perez-Garcia. "Competition Between Different Symmetries in Convective Patterns." *Phys. Rev. Lett.* **61** (1988): 1198–1201.
3. Ciliberto, S., P. Coullet, J. Lega, E. Pampaloni, and C. Perez-Garcia. "Defects in Roll-Hexagon Competition." *Phys. Rev. Lett.* **65** (1990): 2370–2373.
4. Bodenschatz, E., J. R. DeBruyn, G. Ahlers, and D. S. Cannell. "Transition Between Patterns in Thermal Convection." *Phys. Rev. Lett.* **67** (1991): 3078–3081.
5. Cerisier, P., C. Perez-Garcia, C. Jamond, and J. Pantaloni. "Wavelength Selection in Bénard-Marangoni Convection." *Phys. Rev. A* **35** (1987): 1949–1952.
6. Ouyang, Q., and H. L. Swinney. "Transition From a Uniform State to Hexagonal and Striped Turing Patterns." *Nature* **352** (1991): 610.
7. Cross, M., and P. Hohenberg. "Pattern Formation Out of Equilibrium." *Rev. Modern Phys.* **65** (1993): 851.
8. Malomed, B. A., A. A. Nepomnyashchy, and M. I. Tribelsky. "Domain Boundaries in Convection Patterns." *Phys. Rev. A* **42** (1990): 7244–7263.
9. Bodenschatz, E. Private communication.
10. Busse, F. H. "The Stability of Finite Amplitude Cellular Convection and Its Relation to an Extremum Principle." *J. Fluid Mech.* **30** (1967): 626.
11. Malomed, B. A., and M. I. Tribelsky. "Stability of Stationary Periodic Structures for Weakly Supercritical Convection and in Related Problems." *Sov. Phys.—JETP* **65** (1987): 305–310.
12. Haken, H. *Advanced Synergetics.* New York: Springer-Verlag, 1983.
13. Brand, H. R. "Envelope Equations Near the Onset of a Hexagonal Pattern." *Progr. Theor. Phys. Suppl.* **99** (1989): 442–449.
14. Newell, A. C., and J. A.Whitehead. "Finite Bandwidth, Finite Amplitude Convection." *J. Fluid Mech.* **38** (1969): 279.
15. Segel, L. A. "Distant Side-Walls cause Slow Amplitude Modulation of Cellular Convection." *J. Fluid Mech.* **38** (1969): 203.
16. Siggia, E. D., and A. Zippelius. "Dynamics of Defects in Rayleigh-Bénard Convection." *Phys. Rev. A* **24** (1981): 1036–1049.

Chapter 4
Electrohydrodynamic Convection in Liquid Crystals

Stephen W. Morris,* John R. de Bruyn,† and A. D. May‡
*Department of Physics, University of California, Santa Barbara, California, U.S.A. 93106
Current address: Department of Physics and Erindale College, University of Toronto,
Toronto, Ontario, Canada M5S 1A7
†Department of Physics, Memorial University of Newfoundland, St. John's, Newfoundland,
Canada A1B 3X7
‡Department of Physics, University of Toronto and Ontario Laser and Lightwave Research
Centre, 60 St. George Street, Toronto, Ontario, Canada M5S 1A7

Electroconvective Patterns in Freely Suspended Liquid Crystal Films

We briefly review some results from our experiments on electrically driven
convective flow in thin, freely suspended films of smectic-A liquid crystal.

INTRODUCTION

We have studied the convective flow pattern that develops when an electric field is
applied in the plane of a thin, freely suspended film of smectic-A liquid crystal.[3,4,5]
Interactions between the applied field and ions in the film material lead to the onset
of an organized convective flow above a critical voltage. The layered structure of the
smectic phase[2] constrains the film to behave like a two-dimensional isotropic fluid,
making this an interesting system in which to study nonlinear hydrodynamics. In
this chapter we briefly review the main results of our earlier work. More details on
the research described here, as well as a more complete list of references, can be
found in Morris et al.[3,4,5]

Spatio-Temporal Patterns, Ed. P. E. Cladis and P. Palffy-Muhoray,
SFI Studies in the Sciences of Complexity, Addison-Wesley, 1995   **307**

The smectic-A phase is a layered phase in which the molecules are oriented normal to the plane of the layers.[2] Smectic liquid crystals can form robust freely suspended films an integer number of layers thick. Within the layers, the smectic-A liquid crystal flows like an ordinary fluid, with no flow-induced reorientation of the molecular axis. Flow perpendicular to the plane of the layers is, in contrast, very difficult. Thus, a freely suspended film of smectic-A will behave like an isotropic, two-dimensional fluid.

The films we studied were less than one micron in thickness, and so exhibited bright interference colors when illuminated with white light. Films of nonuniform thickness display many different colors corresponding to different thicknesses, and one can pick out by eye the change in color due to a thickness change of a single layer. In these nonuniform films, the electroconvection that results when a large enough electric field is applied in the plane of the film can easily be visualized through the beautiful, swirling patterns of color produced as the regions of different thickness flow. In Color Plate 4 we show such a pattern.

The patterns observed in films of nonuniform thickness are aesthetically pleasing, but not particularly useful in terms of a quantitative understanding of the system. We carried out quantitative measurements on uniformly thick films, that displayed a single color over their entire extent. We studied electroconvection in films of 8CB (4,4'-n-octylcyanobiphenyl) doped with 7.5 ± 0.2 mMol/l of TCNQ (tetracyanoquinodimethane) to control the concentration and species of ionic impurities in the liquid crystal. The films used were between 20 and a few hundred molecular layers (0.06 to 1 μm) thick, and typically 2 mm wide by 20 mm long. The experimental setup is shown schematically in Figure 1. The dc electric field that drove the electroconvection was produced by applying a constant potential difference across the 15 μm diameter wires which supported the film along its long sides. In the uniform films, the flow was visualized by following the motion of dust particles intentionally introduced onto the film.

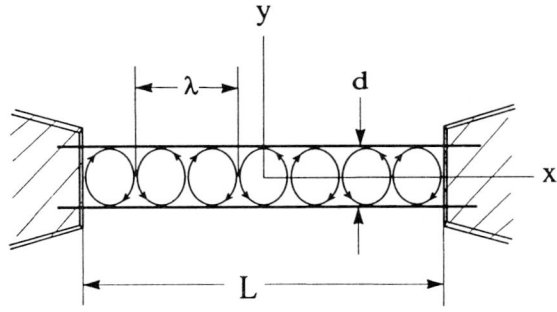

FIGURE 1 A schematic illustration of the film holder, showing the wavelength of the convection pattern.

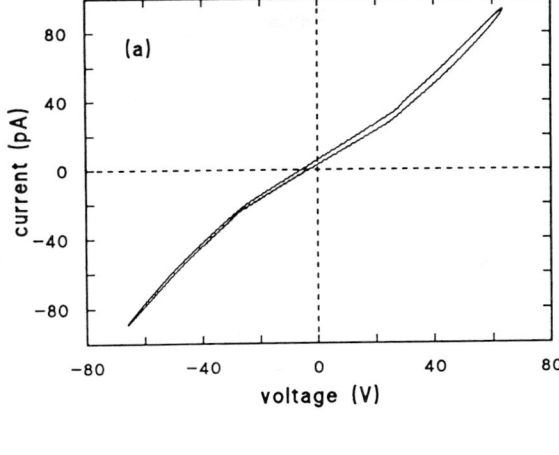

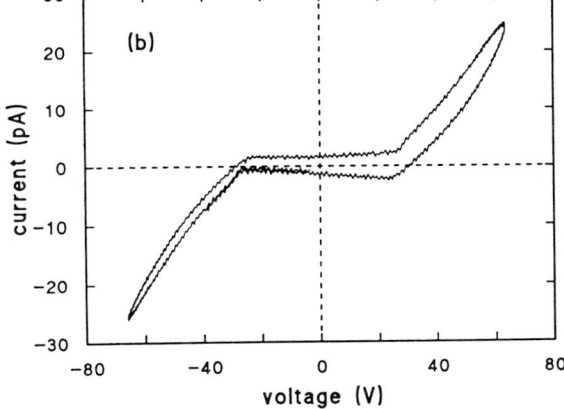

FIGURE 2　(a) The current-voltage curve of a typical film. The small hysteresis loop is due to capacitive effects. The kink in the curve at ±25 V signals the onset of electroconvection. (b) The same data with the slope below onset subtracted, to highlight the enhanced charge transport due to convection above onset.

　　As the electric field across the film was increased, there was a well-defined critical voltage above which the film began to flow in a vortexlike pattern with a wavelength equal to $(1.30 \pm 0.05)d$, where d is the film width.[4] Measurements of the current through the film (see Figure 2) indicate an increase in the effective conductance of the film at the onset of electroconvection, as charge begins to be transported convectively as well as by electrical conduction.[5] This increase in charge transport is analogous to the increase in heat transport that accompanies the onset of Rayleigh-Bénard convection in simple fluids. The continuous increase in current as the field is increased, as well as the measurements of the flow velocity discussed below, indicate that the electroconvection appears via a supercritical bifurcation.

By following the motion of dust particles on the film, we were able to map out the flow velocity field. The velocity components v_x and v_y, in the x and y directions, respectively (as defined in Figure 1), were well described by an expansion in terms of sines and cosines in the x direction, and in terms of Chandrasekhar functions[1] in the y direction.[5] The terms of this expansion satisfy the appropriate boundary conditions at the edges of the film.[5] The expressions for v_x and v_y are

$$v_x(x, y) = \sum_{i,j} B_{ij} M_i(y) \frac{\sin\left[(2j-1)p(x-x_0)\right]}{(2j-1)p} \tag{1}$$

and

$$v_y(x, y) = \sum_{i,j} B_{ij} C_i(y) \cos\left[(2j-1)p(x-x_0)\right], \tag{2}$$

where p is the dimensionless pattern wavenumber, and x_0 is an arbitrary origin. C_i is a Chandrasekhar function, and M_i, which is defined by Morris et al.,[5] is a linear combination of the Chandrasekhar functions S_i. The coefficient B_{ij} is the amplitude of the i,jth mode.

Close to the onset of flow, only the B_{11} mode was needed to describe the measured velocity field accurately. Further above onset, where the flow is more fully developed, a satisfactory fit to the measured field required two modes in the y direction and three in the x direction. The variation in the mode amplitudes B_{1j} as a function of the dimensionless control parameter $\epsilon = (V/V_c)^2 - 1$, where V is the applied voltage and V_c its critical value, is shown in Figure 3. We fitted these results to expressions of the form

$$B_{1j} = a_j \epsilon^{j/2}(1 + b_j \epsilon); \tag{3}$$

the results of the fit are shown in Figure 3 and describe the data well.

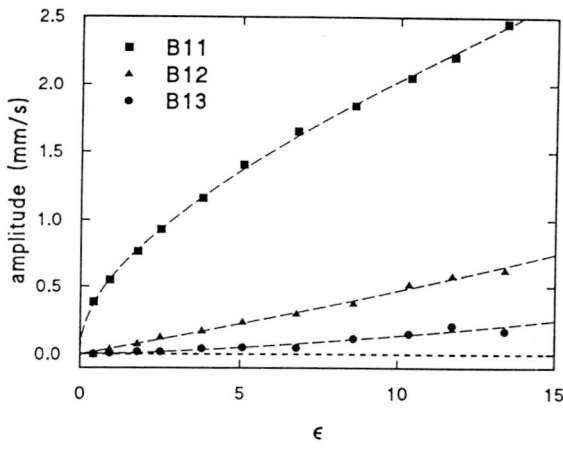

FIGURE 3 The mode amplitudes B_{1j} of the convective velocity field, as a function of the dimensionless control parameter $\epsilon = (V/V_c)^2 - 1$. The dashed lines are fits of the data to expressions like Eq. (3).

Our work to date on this system has been primarily concerned with the behavior of the steady convecting state close to the onset of electroconvection.[3,4,5] There remain many interesting questions to be addressed. Far above onset (i.e., above $\epsilon \approx 15$), the ordered, steady convective flow we studied becomes unstable, and an apparently turbulent—but still two-dimensional—flow state appears. A study of this state, and of the stability boundaries of the basic convecting state, is currently underway. Finally, in Color Plate 5 we show an interesting variant of this experiment, in which the film (of nonuniform thickness in this case) is suspended between concentric circular electrodes, so forming an annulus. In this arrangement, the driving field is oriented radially.

ACKNOWLEDGMENTS

This work was supported by research grants from the Natural Science and Engineeering Research Council of Canada, and from the Province of Ontario through the Ontario Laser and Lightwave Research Centre.

REFERENCES

1. Chandrasekhar, S. *Hydrodynamic and Hydromagnetic Stability.* Oxford: Clarendon, 1961.
2. de Gennes, P. G. *The Physics of Liquid Crystals.* Oxford: Clarendon, 1979.
3. Morris, S. W., J. R. de Bruyn, and A. D. May. "Electroconvection and Pattern Formation in a Suspended Smectic Film." *Phys. Rev. Lett.* **65** (1990): 2378–2381.
4. Morris, S. W., J. R. de Bruyn, and A. D. May. "Patterns at the Onset of Electroconvection in Freely Suspended Smectic Films." *J. Stat. Phys.* **64** (1991): 1025–1043.
5. Morris, S. W., J. R. de Bruyn, and A. D. May. "Velocity and Current Measurements in Electroconvecting Smectic Films." *Phys. Rev. A* **44** (1991): 8146–8157.

Shoichi Kai,[†] **Yukiko Adachi,**[†] **and Satoru Nasuno**[‡]

†Department of Electrical Engineering, Kyushu Institute of Technology, Kitakyushu 804, Japan

‡Department of Applied Physics, Kyushu University, Fukuoka 812 Japan

Stability Diagram, Defect Turbulence, and New Patterns in Electroconvection in Nematics

The pattern formation in electrohydrodynamic convection on nematic liquid crystals is described. Current topics include defect mediated structures, both periodic and nonperiodic. Structures and phase diagram of the electroconvection roll are given, which agree with recent theoretical results. The route to defect turbulence is then studied in detail. New patterns constructed by defects, called defect-mediated periodic structures, are also shown. The defect lattice pattern is observed above a roll convection state at certain frequency ranges of applied field, in which defects are regularly oriented. The phase waves in the oscillating grid pattern, which is a three-dimensional cellular convection, show macroscopic spiral and target patterns. Several other new patterns are introduced as well as those in homeotropically oriented nematics. The formation mechanisms for some patterns are briefly discussed.

Spatio-Temporal Patterns, Ed. P. E. Cladis and P. Palffy-Muhoray,
SFI Studies in the Sciences of Complexity, Addison-Wesley, 1995

1. INTRODUCTION

Nematic liquid crystals (NLCs) are long organic molecules which show, in some temperature range, a nematic phase.[6,7,8,9,10,11,13,14] A nematic phase shows dielectric and conductive anisotropies due to orientational order of the molecules defined by a director which represents the mean direction of the molecules. The electrohydrodynamic (EHD) convection in NLCs occurs when an electric voltage V higher than a critical voltage V_c is externally applied to a thin layer of NLC, with typical thickness $d = 50\mu$m.[6,7,8,9,10,11,13,14]

The horizontal geometry of the cell on EHD convection in NLCs is rectangular.[14,18] The director aligns parallel to the cell surface along one fixed direction x, and due to elastic forces the same direction is obtained in the whole sample in the absence of any external force. For $V > V_c$, a periodic pattern of convection rolls appears with periodic distortions of the director, which results from the competition between elastic and electrostatic forces.[6,7] Then there are two frequency regimes separated by a critical frequency f_c which depends on two relaxation processes, i.e., space charge and director deformation. Lower frequency regime ($f < f_c$) is called the "conduction regime" where space charge plays a main role for the instability.[8] Higher frequency regime ($f > f_c$) is called the "dielectric regime" where dielectric and elastic effects play major roles. Thus two different routes to turbulence can be observed.[11,14,19,40]

In the "conduction" regime above a threshold V_c, typically several voltages, one has two-dimensional convection rolls, called the Williams domain (WD).[6,7,8] The basic mechanism of WD in dc fields was explained by Helfrich in the late '60s.[10] At a higher voltage one finds a bifurcation to the fluctuating WD (FWD), which is defect turbulence.[11,18] Defects are continuously created and annihilated: the defect number in a unit area fluctuates and the defects never become stationary.[11,21] Increasing the voltage further, one observes a transition to the grid pattern (GP) first reported by Kai et al.[20] GP can be stationary or oscillatory. In oscillatory GP, spiral or target phase waves of these oscillation can be observed similar with ones in the Belousov-Zhabotinsky chemical reaction system.[42,43] At even higher voltages one observes a transition to a fully developed turbulent state. The turbulent state in EHD convection is called the dynamic scattering mode (DSM) and this mode, which occurs immediately above GP, we call DSM1.[16,20,39,40,41] Increasing the voltage further a transition from DSM1 to another dynamic scattering mode (DSM2) happens.[16,19,23,39,40,41]

Increasing the voltage at higher frequencies but still below f_c, at a threshold often convection sets in via travelling waves.[11,12,13,16,18,34] This has been originally reported by Hirakawa and Kai.[11] In the "dielectric" regime immediately above threshold, one observes often a short wavelength periodic stripe pattern.[11] Here the wavelength is rather independent of the thickness. At a second threshold, one observes a transition to a quasi-periodic pattern, the so-called chevrons.[6,7,11] The chevron pattern observed in the dielectric regime can be regarded, in a sense, as a kind of periodic defect orientation due to modurated potential for defects.[19] Sometimes chevrons are already visible immediately above V_c.[11] Then finally it becomes turbulent with an increase of V.

Now some theoretical trends on pattern dynamics will be described below. The Orsay group[8] extended the Carr-Helfrich mechanism[10] to an ac-voltage and it was recognized within the one-dimensional approximation which explained two different frequency regimes, i.e., the conduction and the dielectric regimes. In the conduction regime, convection rolls have a wavelength of the order of the thickness of the fluid layer and the director is mainly stationary in time with small modulations, whereas in the dielectric regime the director is mainly oscillating with the external frequency, and the wavelength is much smaller than the layer thickness. These could be roughly understood in the framework of the above one-dimensional model calculation.[44,45,46]

For further discussions of nonlinear states in EHD convection near V_c, we have to deal with a nonlinear equation like the Newell-Whitehead equation and the Swift-Hohenberg equation in isotropic Rayleigh-Bérnard (R-B) convection.[5,24,25,28,45] This has been, however, very complicated and more effort is needed to derive these equations because of non-Newtonian and anisotropic fluid. Taking only slowly, varying modes in space and in time and considering a symmetry argument, one can obtain a two-dimensional nonlinear amplitude equation.

$$\partial_t A = \epsilon A + [\partial_x^2 - ik_2\partial_x\partial_y^2 + w\partial_y^2 - \partial_y^4]A - |A|^2 A, \qquad (1)$$

which was originally derived by the Bayreuth group.[1,24,25,26,33,45] Here an equation with $k_2 = 2$ and $w = 0$ for isotropic R-B convection is called the Swift-Hohenberg equation.[28] Equation (1) successfully describes the obtained experimental facts, such as oblique. zig-zag, and defect motions.[1,26,33,37,38] In order to describe these instabilities in more detail, however, phase equations are needed which are derived from reducing variables further.[26,37,38] The phase waves in the GP region could be treated along these approaches.[30,31,36] Defect turbulence called FWD, on the other hand, slightly above V_c, is mainly characterized by finite defect density and complex defect dynamics.[9,22] However, the properties and the formation route are not well understood yet.

We will describe these pattern formation phenomena and introduce patterns newly observed by several authors. Finally, we will give some prospects.

2. STABILITY DIAGRAM AND PATTERN SELECTION

The stability diagram of WD rolls in EHD convection is shown in Figure 1, which is experimentally obtained in a large aspect ration $\Gamma(= 1000)$.[32] The experiment has been done using frequency-voltage jump methods.[2,3,29] The liquid crystal used here was N-p-methoxybenzilidene-p-butylaniline (MBBA) doped with 0.05 w% of tetra-n-butylammonium bromide (TBAB) to obtain a suitable cutoff frequency f_c by controlling its concentration. Three different types of long-wavelength instabilities—Eckhaus, zig-zag, and skewed varicose[4,5,28]—are found in the stability diagram. These appear during the period when the system relaxes from an initial state to a final state.

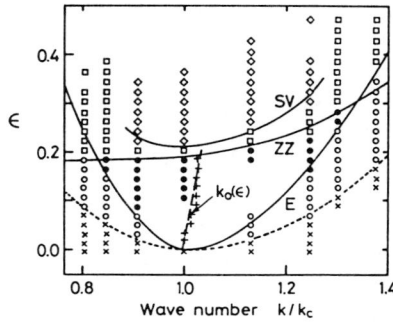

FIGURE 1   Stability diagram experimentally obtained in EHD convection.[29] E: Eckhaus, ZZ: zigzag, SV: skewed varicose, dotted line: marginal stability line.

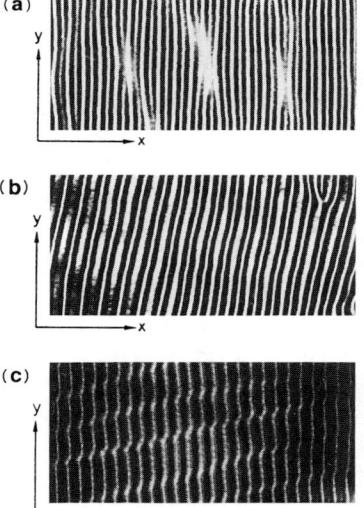

FIGURE 2   Transient deformations due to secondary instabilities. (a) Eckhaus, (b) zig-zag, and (c) skewed varicose instabilities.

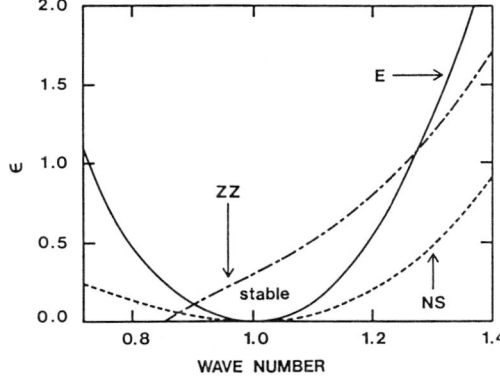

FIGURE 3 Stability diagram theoretically obtained in EHD convection.[37,38] The symbols are the same as those in Figure 4.2. In the "stable" region, simple decays of defects and patterns take place. NS: neutral stability line.

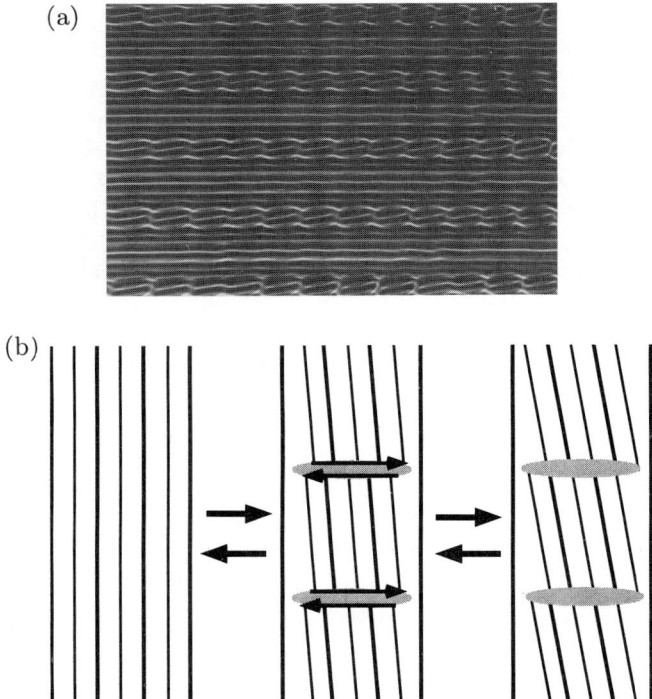

FIGURE 4 Defect lattice. Transition state from WD to Defect lattice (DL) (a) and schematic illustration of its oscillation (b).[32] (a) Each fourth roll-pair has a periodic defect street. (b) Temporal oscillation takes place from left to right, then back to the left again.

Including these three long-wavelength instabilities, pattern evolutions can be classified into three behaviors in the stability diagram; these transient behaviors are experimentally observed as follows:

1. simple decay (marked by cross in Figure 1): The initial pattern disappears very quickly leading to uniform state for $\varepsilon < 0$ or to new periodic patterns for $\varepsilon > 0$.
2. long mode instabilities: Transient pattern deformations can be observed due to long-wavelength instabilities through three different types of characteristic deformations: Eckhaus, zig-zag, and skewed varicose. Typical examples of such deformations are shown in Figure 2. All deformation patterns described here only appear as transient patterns. (However, in some cases, transient modes are frozen as the most stable modes, for example, as a zig-zag pattern frequently observed in EHD.[14,35]) Finally, the stable pattern appears different from both initial and transient modes.
3. evolution without excitation of unstable mode: For points marked by a solid circle in Figure 1, a pattern either keeps the initial one or evolves toward the optimal one on the line $k_0(\varepsilon)$. Often the evolution is quite slow, mainly due to the defect motion.

In Figure 1 the dotted curve denotes the neutral stability curve (NS) for the onset of convection. The solid line represents stability boundaries for Eckhaus (E), zig-zag (ZZ), and skewed varicose (SV) instabilities. The present stability diagram shows clear asymmetry with respect to k_c. The asymmetry of ZZ instability curve qualitatively agrees with the stability diagram (see Figure 3) theoretically obtained based on an anisotropic model (1) with mean-flow effect.[37,38]

The final state is either steady or sustained time-dependent one, depending on ε. In the range $0 < \varepsilon < 0.18$ of Figure 1, the final steady state is stationary periodic roll. For $\varepsilon > 0.18$, the final state is weakly turbulent, called FWD. The optimal branch becomes unstable at $\varepsilon_c = 0.19$ with respect to zig-zag instability, and defect turbulence (FWD) occurs. The mean density of defects increases with very steep ε-dependence in the vicinity of a certain critical value ε_c.[15,22]

Thus, we found three different routes to defect turbulence (FWD)[32]: (1) from WD through zig-zag, (2) directly from WD, and (3) from WD through defect lattice, with an increase in an applied field frequency that is still lower than f_c. These scenarios are due to the major role of the corresponding instabilities: (1) zig-zag instability, (2) the combination of skewed varicose and zigzag instability, and (3) the skewed varicose instability. The details have been described.[32]

3. NEW PATTERNS

3.1 DEFECT LATTICE IN WD AND PHASE WAVE IN GP

A very interesting and curious pattern is shown in Figure 4. Namely defects form regular latticelike structures above a second threshold in a certain frequency range. Originally this type of pattern, observed by Yamazaki et al.,[39,40,41] was named the Bamboo structure but recently has been more suitably renamed the defect lattice (DL).[32] This structure appears via a skewed varicose instability and characterized by the two-dimensional defect lattice as shown in Figure 4.[32] Increasing a voltage, we usually have perfect DL; i.e., each roll has oriented defects. The defect pair in DL is collectively oscillating in time (see Figure 4). The maximum displacement in the oscillation is about one roll width. The lattice wave numbers, k_{x1} and k_{y1}, increase with an increase in V as $(V - V_d)^{0.5}$ where V_d is the threshold of DL as shown in Figure 5. Thus, transition from WD to DL seems to be a supercritical bifurcation. No qualitative explanation on the mechanism for these patterns is given, and no theoretical study and prediction has been done yet.

Further increasing voltages from DL and FWD states, the three-dimensional convective patterns (GP) is observed.[20] This pattern is often oscillatory, and the phase of the oscillation is incollective in space in common cells but it is collective in well-prepared cells. The collective motions of oscillation shows clear phase waves such as spiral and target patterns. An example is shown in Figure 6. This type

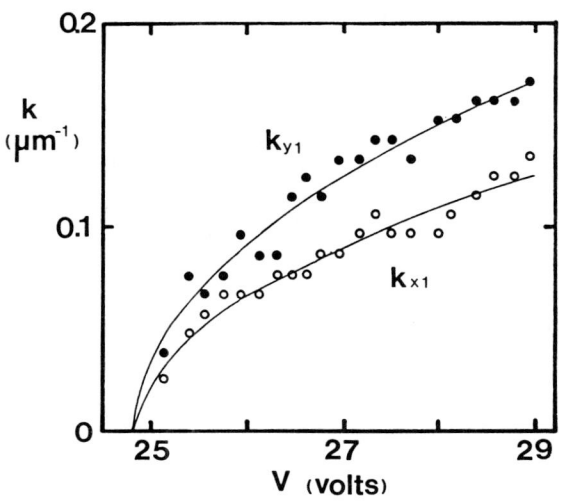

FIGURE 5 Voltage dependences of lattice wave numbers k_{x1} and k_{y1} for the defect lattice ($V_c = 22.4V$ for WD, $f = 850Hz$, and $d = 25\mu m$). The solid lines indicate respectively $k_{x1} = 0.062(V - 24.8)^{0.5}$ and $k_{y1} = 0.084(V - 24.8)^{0.5}$; $k_{x10} = 0.062, k_{y10} = 0.084$ and its ratio $r = 0.74$.

of pattern has been discovered and investigated by Nasuno et al.[30,31] The velocity of these phase waves is very fast: 50 to 100 lattices per second (about 200–1000 μm/s). The shape is eliptical with the ratio of 1:2 for long and short axes after they are normalized by the respective scales of a GP lattice. This is probably related to the anistropy of liquid crystals. The short axis lies in the rubbing direction. When two phase waves collide, both disappear and never pass through. This is typical property of dissipative waves. These target patterns have a rather short lifetime and always repeat their collapse and formation randomly in space. Increasing the voltage further, the lifetime and the size of these patterns become respectively shorter and smaller, and finally turbulent structure appears. These waves could be explained by regarding them as nonlinear oscillators. These oscillators couple with each other just as coupled nonlinear oscillators and show collective motions. To understand better, however, we need more progress in the theoretical work on phase equations, and in EHD convection. Some experimental and theoretical works by Sano et al. on this regime are in progress and further details are expected.[36]

3.2 CHEVRON PATTERN: PERIODIC DEFECT ORIENTATION ABOVE f_c

In the dielectric regime at threshold, one often observes the oscillatory behavior of the director in transmitted light signals as well as the appearance of a herringbone structure, the so-called chevrons. The pictures in Figure 7 give an example of the transition to chevrons from narrow WD. At the first bifurcation point, say, $V_c = 81$ V, the propagating WD (PWD) can be observed whose wavelength w is about 5 μm with the propagating velocity $v_p \simeq 1\mu$m. Such a propagating mode in EHD was first observed in the late '70s and has been called PWD.[11] Immediately above the threshold, one observes very narrow WD and a finite defect density similar to FWD as shown in Figure 7(a). Those are the skin layer convection.[39,41] By increasing further the applied voltage, the total defect number increases and the defect density becomes locally higher on straight lines as shown in Figure 7(b). Then finally at the second bifurcation point V_c^* defects become stationary and well ordered (see Figure 7(c)). The inside narrower rolls, then, are inclined with respect to the original WD and propagate, and a large number of defect pairs regularly aligns along one dimension (see, for example, Figures 7(d) and 8). Thus the chevron pattern can be said to be a periodic defect-orientation pattern. By further increasing the field, these periodic defect orientations produce again a large-scale defect as marked by arrows in Figure 7(d). This is a hierarchical pattern in the route to turbulence (DSM) in the dielectric regime.

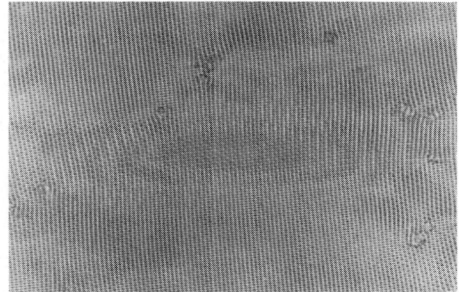

FIGURE 6 Typical phase wave (target pattern) in oscillatory GP. The clear target pattern can be observed at the center.

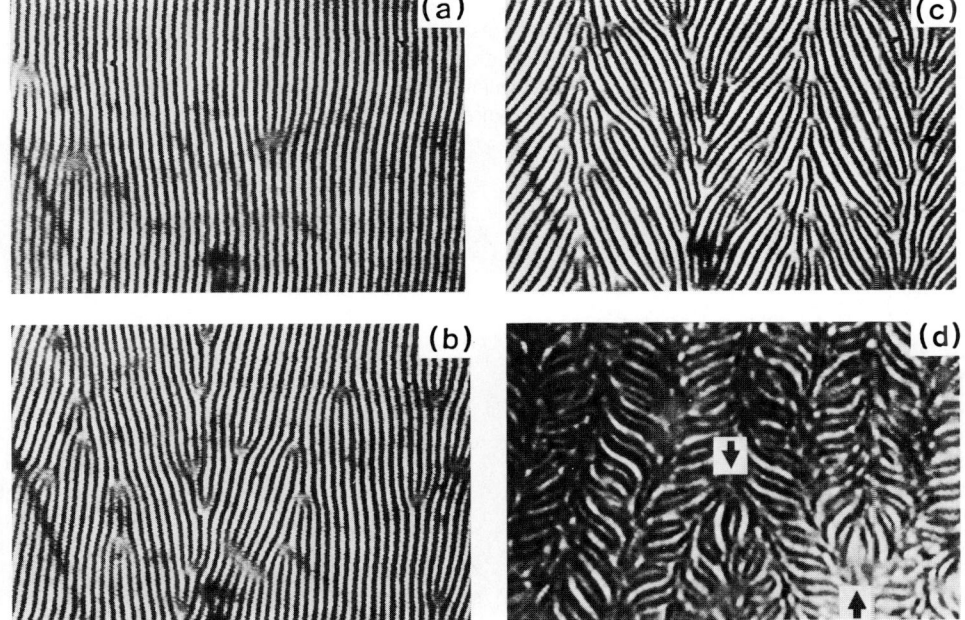

FIGURE 7 Formation process of the chevron pattern.[19] Applied voltage of $f > f_c$ is increased from (a) to (d). The arrows indicate the macroscopic defects in new scale.

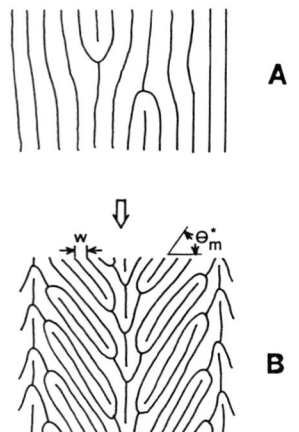

A

B

FIGURE 8 Schematic drawings of chevron structure and definition of short wavelength w, long wavelength L, inclination angle of rolls θ_m^*, and propagation velocity v_p. A: defects in for example FWD. B: chevron structure (periodic defect orientation).

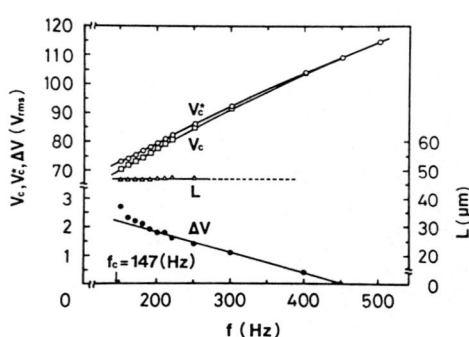

FIGURE 9 The difference between two bifurcation points as a function of f. V_c and V_c^* are defined from anomalies.

   Three parameters characterizing chevron patterns—i.e., a maximum inclination angle θ_m^*, the long wavelength L of defect orientation line, and the propagating velocities v_p—are schematically shown in Figure 8. These variables are characteristic parameters to define bifurcations of patterns. The details of these have been already reported elsewhere.[19,23] The difference between V_c and V_c^* becomes smaller with an increase of an applied voltage (see Figure 9). Finally both thresholds come together and no difference is observed at another cut off frequency $f = f_c^*$. This is a codimension two bifurcation point. No such prediction can be given in a theoretical result.[19] The chevron threshold comes down to the first threshold at approximately $f_c^* \sim 4f_c$. The threshold difference $\Delta V = V_c^* - V_c$ shows roughly a linear decrease to f_c^*. The chevrons thus could be explained as a result of the nonlinear interaction of two linear modes.[19,45,46]

The picture for the formation of the chevrons was shown in a previous publication.[19] It can be briefly explained as follows. Immediately above the first threshold, defects are produced. When the voltage is further increasing the coexistence of the two modes becomes possible and, due to the beating between them, the defects are trapped on the line at the minimum of the potential. Because the defects lead to the zig-zag deformation, the sketched pattern in Figure 8 is realized.

3.3 COMPLEX SPIRAL PATTERNS IN A SAMPLE WITH HOMEOTROPIC ORIENTATION

An additional aspect of patterns in dielectric regime will be given in the homeotropic orientation sample. The previous discussion has been concentrated to the planar orientation of nematics. The phase diagram for homeotropically oriented samples is shown in Figure 10. In this simple, the first Fréedericksz transition takes place at much lower voltage than that of the onset of electroconvection. Increasing the voltage, then, leads to EHD. The EHD pattern in the dielectric regime are also quite different from that for planar case. An example is shown in Figure 11.[17] The typical pattern is a spiral with sinusoidal waves which travel and wind into the center of

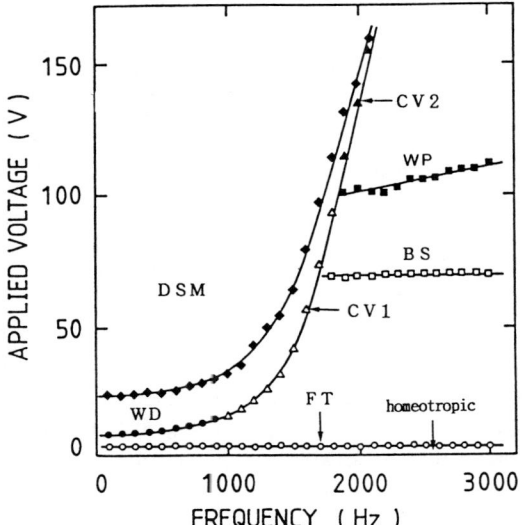

FIGURE 10 Phase diagram for homeotropic samples ($\Gamma = 400 \times 400, d = 50\mu m, T = 30°C$, a doping material TBAB of 0.01 wt% is mixed.) FT: Fréedericksz transition point WD: Williams domain convection BS: bend-stripe pattern (we previously named this WP1 in Figure 1 of another paper[17].) WP: wavy pattern.

FIGURE 11 Complex spiral patterns in homeotropic EHD convection in the dielectric regime $f = 1\text{kHz} > f_c, d = 30\mu\text{m}$, a different sample from that for the phase diagram in Figure 10.[17]

the spiral. An example of the voltage dependence of the traveling velocity v and the wavelength λ of sinusoidal waves is shown in Figure 12. Obviously two different phases can be distinguished, i.e., the wavy pattern (WP: the slow traveling velocity pattern) and the propagating wavy pattern (PWP: the fast traveling pattern). In WP, the velocity v is almost constant, independent of an applied voltage, but in PWP it steeply increases with increases of voltages.[17] The bend-stripe (BS) pattern is similar to WP but it is not wavy but stripes. There is a big difference in the width of spiral rolls between PWP1 and PWP2 states. Transition to turbulence (T in Figure 12) from these patterns occurs through speeding-up the traveling velocity v of sinusoidal waves and also through narrowing both wavelengths of spiral and sinusoidal waves with an increase of applied voltages, as shown in Figure 11.

4. SUMMARY AND CONCLUSION

The transient behavior from steady WD to another steady WD can perhaps be understood by comparison with transient behavior in R-B convection. Our results from the stability diagram shows three different secondary instability lines. These closely connected to the formation of defect turbulence, FWD, and spatio-temporally complex phenomena. We have been already reported that there are three

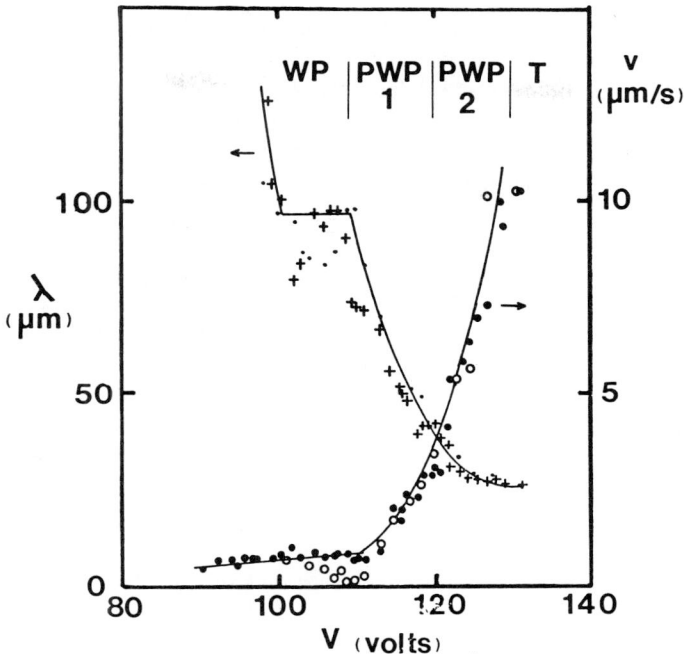

FIGURE 12 Voltage dependences of traveling velocity v and wavelength of sinusoidal patterns λ in spiral. WP: wavy pattern, PWP: propagating wavy pattern, T: turbulence. $\cdot$, o: for increasing, $+$, •: for decreasing.

different routes to FWD related to these three instabilities. Many theoretical works in general frameworks attempt to understand such complicated aspects by use of lines are complex time-dependent Ginzburg Landau (TDGL) equations. The quantitative description on defect turbulence by these equations has not been fully successful. For example, complex TDGL can create a pair of defects but cannot describe characteristic defect interaction often observed in FWD.

This situation will be the same for the formation mechanism of DL, consists of periodic alignment of real defects. Defects can be described as when both the imaginary part and the real part of convective amplitude become zero, i.e., a real singular point. This seems to be impossible when using a simple Swift-Hohenberg equation. Neighboring defects in DL are oscillatory and steadily combining and parting. This is very similar to the Biot-Savart's electromagnetic interaction between two wires flowing alternating currents. Also DL wavelength, both x- and y-directions, seems to start from the zero value at its threshold, indicating supercritical bifurcation. This hints at a theoretical description for DL. We will report elsewhere about the formation mechanism of DL in the near future.

We showed that the chevron pattern was formed by the periodic defect orientation via a second bifurcation point. Two different linear modes can coexist and make periodic potential wells where defects are trapped. The threshold difference between these two modes goes to zero with an increase of f and the co-dimension two bifurcation point appears at f_c^*. There is no quantitative theoretical work in this region, except linear theory done by the Bayreuth group. No great progress has been done yet in dielectic regime. Most will be in the future.

Several new patterns are also introduced in this article, for instance, spirals in a homeotropic sample and phase waves in GP. Both show similar patterns in the Belousov-Zhabotinsky (B-Z) reaction. The spirals in a homeotropic orientation are much more complicated and we found two different phases, distinguished by their dynamical behavior. They might be described by the phase dynamic approach similarly to B-Z reaction pattern. However, there is also a big difference. In the B-Z reaction, spatial degrees of freedom are not essential and play rather a secondary role because it has no spatial structure in its background. In EHD it has a spatial structure in its background, for example, GP. Then the oscillating units are clearly distinguishable unlikely in B-Z reaction. This may lead to the possibility of quite different pattern properties. We do not know, however, those details at the moment. Here we stress a rich variety of pattern formation and the new trends on the higher instabilities in electroconvection as well as universality of pattern formation. For WP and PWP appearing in the high-frequency regime in a homeotropically oriented sample, there is no theoretical work and, up to now, very little experimental activity has been done. Very recently, however, some theoretical study has begun.[27]

ACKNOWLEDGMENTS

This work has been in part supported by the Grant in Aid for Scientific Research from the Ministry of Education, Science and Culture of Japan (No. 04640372).

REFERENCES

1. Bodenshatz, E., W. Zimmermann, and L. Kramer. "On Electrically Driven Pattern-Forming Instabilities in Planar Nematics." *J. Phys. (Paris)* **49** (1988): 1875.
2. Braun, E., S. Rasenat, and V. Steinberg. "Mechanism of Transition to a Weak Turbulence in Extended Anisotropic Systems." *Europhys. Lett.* **15** (1991): 597.
3. Braun, E., S. Rasenat, and V. Steinberg. "Eckhaus Instability and Defect Nucleation in Two-Dimensional Anisotropic Systems." *Phys. Rev. A* **43** (1991): 5728.
4. Busse, F. H. "The Oscillatory Instability of Convection Rolls in a Low Prandtl Number Fluid." *J. Fluid Mech.* **52** (1972): 97.
5. Busse, F. H. "Nonlinear Properties of Thermal Convection." *Rep. Prog. Phys.* **41** (1978): 1929.
6. Chandrasekhar, S. "Hydrodynamic Instabilities of Nematic Liquid Crystals Under AC Electric Fields." In *Liquid Crystals*. Cambridge: Cambridge University Press. 1977.
7. de Gennes, P. G. *The Physics of Liquid*, 3rd ed. Oxford: Clarendon Press, 1982.
8. Dubois-Violette, E., P. G. de Gennes, and O. Parodi. *J. Phys. (Paris)* (1971): 305.
9. Gil, L., J. Lega, and J. L. Meunier. "Statistical Properties of Defect-Mediated Turbulence." *Phys. Rev. A* **41** (1990): 1138.
10. Helfrich, W. "Conduction-Induced Alignment of Nematic Liquid Crystals: Basic Model and Stability Considerations." *J. Chem. Phys.* **51** (1969): 4092.
11. Hirakawa, K., and S. Kai. "Analogy Between Hydrodynamic Instabilities in Nematic Liquid Crystal and Classical Fluid." *Mol. Cryst. Liq. Crust.* **40** (1977): 261.
12. Joets, A., and R. Ribotta. "Hydrodynamic Transitions to Chaos in the Convection of an Anisotropic Fluid." *J. Physique* **47** (1986): 1595.
13. Joet, A., and R. Ribota. "Localized, Time-Dependent State in the Convection of a Nematic Liquid Crystals." *Phys. Rev. Lett.* **60** (1988): 2164.
14. Kai, S. "Electrohdrodynamic Instability of Nematic Liquid Crystal." In *Noise in Nonlinear Dynamical Systems*, edited by F. Moss and P. V. E. McClintock, Vol. 3, Ch. 2, 22–76. Cambridge: Cambridge University Press, 1989.
15. Kai, S. "Stability Diagram, New Patterns and Turbulence of Electroconvection in Nematics." *Forma* **7** (1992): 189.
16. Kai, S., and K. Hirakawa. "Anolalies Near the Electrohydrodynamical Instability Points in the Nematic Liquid Crystal MBBA." *Solid State Comm.* **18** (1976): 1579.
17. Kai, S., and K. Hirakawa. "Phase Diagram of Dissipative Structures in the Nematic Liquid Crystal Under AC field." *Solid State Comm.* **18** (1976): 1573.

18. Kai, S., and K. Hirakawa. "Successive Transitions in Electrohydrodynamic Instabilities of Nematics." *Prog. Theor. Phys. Supp.* **64** (1978): 212.
19. Kai, S., and W. Zimmermann. "Pattern Dynamics in the Electrohydrodynamics of Nematic Liquid Crystals." *Prog. Theor. Phys.* (Supplement) **99** (1989): 458.
20. Kai, S., K. Yamaguchi, K. Hirakawa. "Observation of Flow Figures in Nematics Liquid Crystal MBBA." *Jpn. J. Appl. Phys.* **14** (1975): 1653.
21. Kai, S., M. Kohno, and N. Chizumi. "Spatial and Temporal Behavior on Pattern Formation and Defect Motions in the Electrohydrodynamic Instability in Nematic Liquid Crystals." *Phys. Rev. A* **40** (1989): 6554.
22. Kai, S., M. Andoh, M. Kohno, M. Imasaki, and W. Zimmermann. "Defect Dynamics and Statistics of Defect Chaos in Electrohydrodynamic Convection in Nematics." *Mol. Cryst. Liq. Cryst.* **198** (1990): 247.
23. Kai, S., W. Zimmermann, M. Andoh, and N. Chizumi. "Local Transition to Turbulence in Electrohydrodynamic Convection." *Phys. Rev. Lett.* **64** (1990): 1111.
24. Kaiser, M., P. Pesch, and E. Bodenshatz. "Mean Flow Effects in the Electrohydrodynamic Convection in Nematic Liquid Crystals." *Physica D* **59** (1992): 320.
25. Kaiser, M., P. Pesch, and E. Bodenshatz. "Amplitude Equations for the Electrohydrodynamic Instability in Nematics." *Phys. Rev. E.* **48** (1993): 4510.
26. Kramer, L., Bodenshatz, E., W. Pesch, W. Thom, and W. Zimmermann. "New Results on the Electrohydrodynamic Instability in Nematics." *Liq. Cryst.* **5** (1989): 699.
27. Kramer, L., A. Hertrich, and W. Pesch. "Electrohydrodynamic Convection in Nematics: The Homeotropic Case." Preprint, 1993.
28. Manneville, P. *Dissipative Structures and Weak Turbulence.* Boston: Academic Press, 1990.
29. Nasuno, S., and S. Kai. "Instabilities and Transition to Defect Turbulence in Electrohydrodynamic Convection of Nematics." *Europhys. Lett.* **14** (1991): 779.
30. Nasuno, S., and Y. Sawada. "A New Scheme of Spatio-Temporal Chaos Created by the Interaction Between the Phase Waves and the Topological Defects." *Prog. Theor. Phys.* (Supplement) **99** (1989): 450.
31. Nasuno, S., M. Sano, and Y. Sawada. "Phase Wave Propagation in the Rectangular Convective Structure of Nematic Liquid Crystal." *J. Phys. Soc. Japan* **58** (1989): 1875.
32. Nasuno, S., O. Sasaki, S. Kai, and W. Zimmermann. "Secondary Instabilities in Electroconvection in Nematic Liquid Crystals." *Phys. Rev. A* **46** (1992): 4954.
33. Pesch, W., and L. Kramer. "Nonlinear Analysis of Spatial Structures in Two-Dimensional Anisotropic Pattern Forming Systems." *Z. Phys.* **B63** (1986): 121.

Color Plates

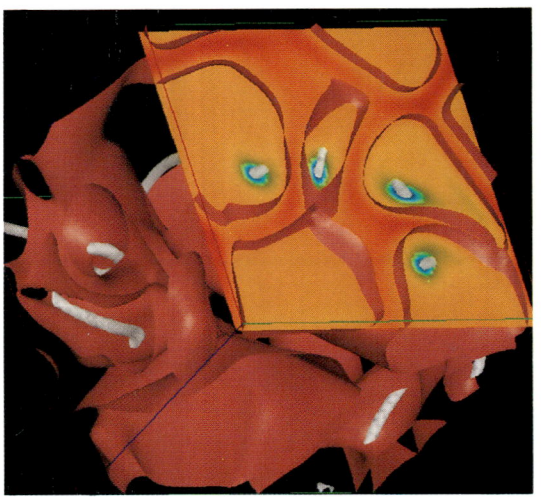

Color Plate 1:
Amplitude isosurfaces in a three-dimensional CGL simulation. Tubes are low-amplitude surfaces around vortex filaments; sheets are high-amplitude surfaces bordering domain walls. A two-dimensional slice with amplitude shading is also shown. [Huber]

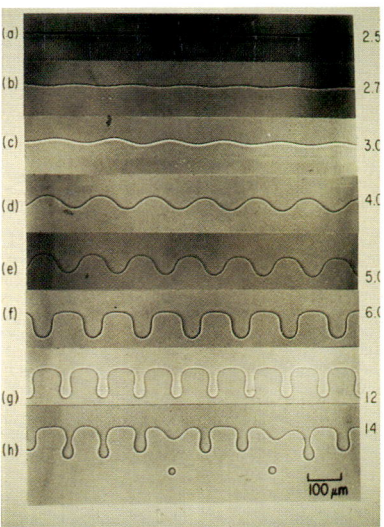

Color Plate 2 (left):
A moving nematic-isotropic interface goes unstable as the velocity is increased. The interface is on average horizontal, with the isotropic phase on the top and the nematic phase on the bottom. For $v < v^* = 2.5\ \mu m/$sec, the front is flat. For $v > v^*$, the front is wavy. The bifurcation is supercritical. [Bechhoefer]

Color Plate 3 (right):
Front splitting in a moving nematic-isotropic interface. The isotropic phase (phase 0) is on top. The homeotropically oriented nematic (phase 2) is on the bottom. A region of metastable planar nematic (phase 1) is present inside the bright triangle. The simultaneous presence of 20 and 10 interfaces indicates that the splitting transition here is hysteretic. [Bechhoefer]

Color Plate 4:
Electroconvective flow in a nonuniformly thick, freely suspended regtangular film of smectic-A liquid crystal. Regions of different thickness appear as different colors. [Morris]

Color Plate 5 (left):
Convection in a nonuniformly thick film of smectic-A liquid crystal in an annular geometry; the applied field is oriented radially. [Morris]

Color Plate 6 (right):
Reconstructions of the spatial patterns using the first eight modes. The images correspond to the first four images in Figure 1. [Herzel]

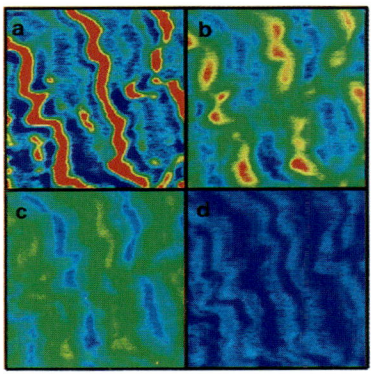

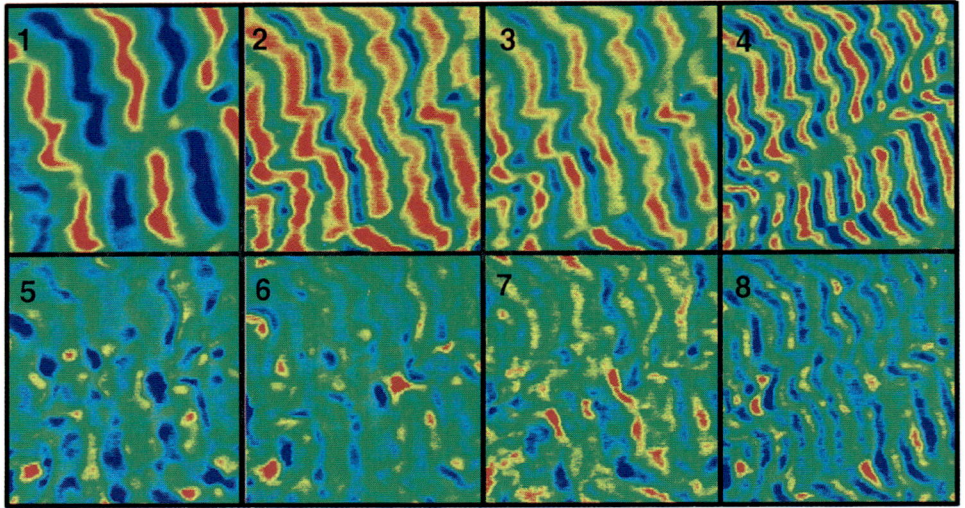

Color Plate 7:
First eight empirical orthogonal functions (EOF) corresponding to the pattern shown in Figure 1. The EOFs were calculated using 750 snapshots over approximately 25 seconds. The first four and the eighth mode are mainly composed of wavy stripes whereas in modes 5, 6, and 7 irregularly distributed spots of different sizes are the dominant features in modes 5, 6, and 7. [Herzel]

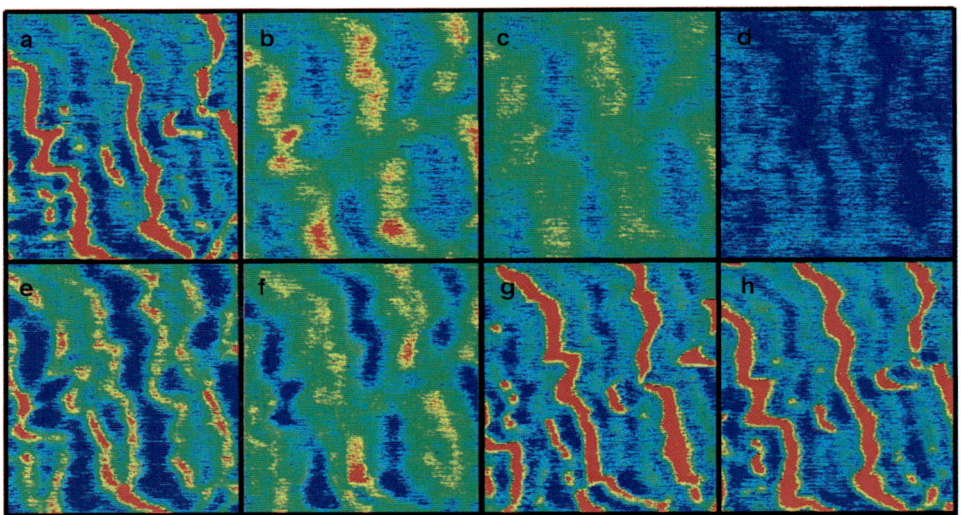

Color Plate 8:
Mode amplitudes for normal (a) and chaotic (b) vocal fold vibrations. In (a), mode 1 is illustrated as a solid line, and mode 2 as a dashed line. In (b), the first three mode amplitudes are displayed. [Herzel]

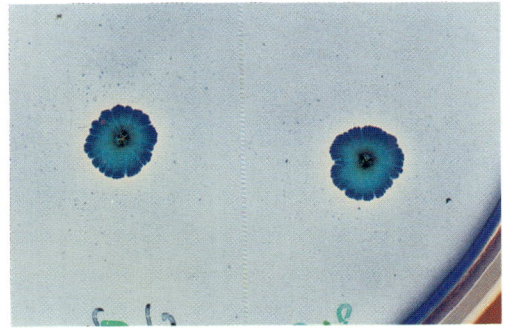

Color Plate 9:
Growth of *Bacillus subtilis 168* on low pepton level—the colony is compact. [Ben-Jacob]

(a)

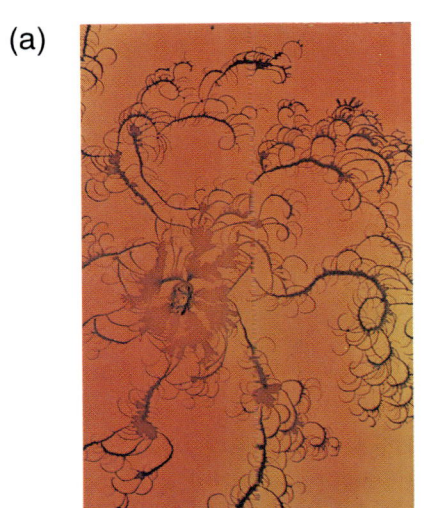

(b)

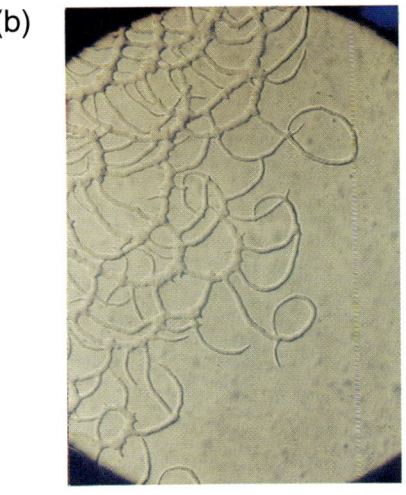

Color Plate 10:
Example of the chiral nature of phase *C*. (a) Demonstrates the complexity of this phase. (b) Shows the three-dimensional structure using Numarsky prism. [Ben-Jacob]

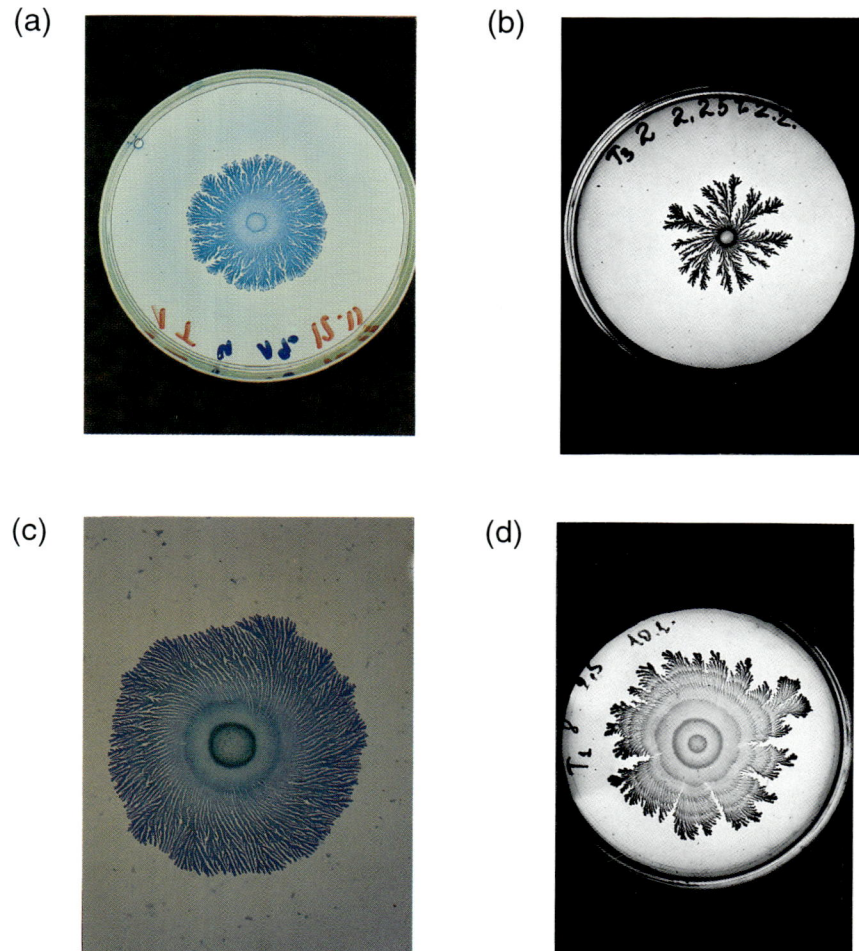

Color Plate 11:
Growth patterns of phase $\mathcal{T}$. (a) Tip-splitting growth (1% agar and 4 g/l pepton). (b) DLA-like growth (2.25% agar and 2 g/l pepton). (c) Tip-splitting growth with a global twist (2.5% agara and 5 g/l pepton). (d) Rings are observed on top of growing tip-splitting (2.5% agar and 8 g/l pepton). [Ben-Jacob]

Color Plate 12:
Demonstration of the coexistence of phases $\mathcal{T}$ and $\mathcal{C}$. [Ben-Jacob]

(a)

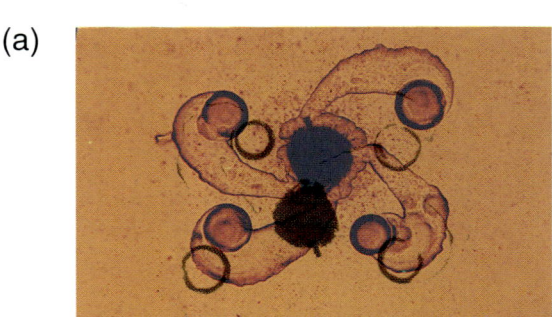

Color Plate 13:
The Vortex mode. (a) Initial stage showing the overall twist and the vortices—the darker droplets at the leading tips. (b) Late stage growth. Each of the branches has a vortex at the tip. [Ben-Jacob]

(b)

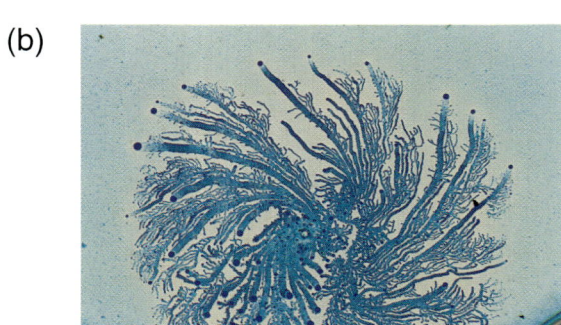

34. Rehberg, I., S. Rasenat, J. Fineberg, M. de la Torre Juarez, and V. Steinberg. "Temporal Modulation of Traveling Waves." *Phys. Rev. Lett.* **61** (1988): 2449.

35. Ribbota, R., A. Joet, and L. Lei. "Oblique Roll Instability in an Electroconvective Anisotropic Fluid." *Phys. Rev. Lett.* **56** (1986): 595.

36. Sano, M., K. Sato, S. Nasuno, and H. Kokubo. "Complex Dynamics of a Localized Target Pattern in Electrohydrodynamic Convection." *Phys. Rev. A* **46** (1992): 3540.

37. Sasa, S. "A Model for Defect Chaos in Electrohydrodynamic Convection of Nematic Liquid Crystals." *Prog. Theor. Phys.* **83** (1990): 824.

38. Sasa, S. "The Dynamics Near ZigZag Instability." *Prog. Theor. Phys.* **84** (1990): 1009.

39. Yamazaki, H., K. Hirakawa, and S. Kai. "Spatial Fourier Analysis of Convective Cells in the Region From WD to the GP." *J. Phys. Soc. Japan* **52** (1983): 1878.

40. Yamazaki, H., K. Hirakawa, and S. Kai. "Two Kind of Turbulent Flows in DSM of EHD Instabilities." *Mol. Cryst. Liq. Cryst.* **122** (1985): 41.

41. Yamazaki, H., K. Hirakawa, and S. Kai. "Convective Pattern in the Dielectric Regime of Electrohydrodynamic Instability." *J. Phys. Soc. Jpn.* **56** (1987): 502.

42. Zhabotinsky, A. M., and A. N. Zaikin. "Concentration Propagation in Two-Dimensional Liquid-Phase Self-Oscillating System." *Nature* **225(5253)** (1970): 535.

43. Zhabotinsky, A. M., and A. N. Zaikin. "Aotowave Process in a Distributed Chemical System." *J. Theor. Biol.* **40** (1973): 45.

44. Zimmermann, W., and L. Kramer. "Oblique-Roll Electrohydrodynamic Instability in Nematics." *Phys. Rev. Lett.* **55** (1985): 402.

45. Zimmerman, W. "Pattern Formation in Electrohydrodynamic Convection." *MRS Bull.* **16** (1991): 46.

46. Zimmerman, W., W. Thom, and V. A. Raghunathan. "Broken Symmetries in the Electrohydrodynamic Instability in Nematics." Preprint, 1992.

Shin-ichi Sasa,† Tsuyoshi Mizuguchi,‡ and Masaki Sano*
†Department of Physics, Kyoto University, Kyoto 606, Japan
‡Yukawa Institute for Theoretical Physics, Kyoto University, Kyoto 606, Japan
*Research Institute of Electrical Communication, Tohoku University, Sendai 980, Japan

Phase Jump Lines in Two-Dimensional Patterns

We introduce exciting phenomena arising in the electrohydrodynamic convection of nematics. The phenomena are related to "phase jump lines"
which may have endpoints at the dislocations of periodic patterns and
which move in various forms. We clarify the topological nature of the phase
jump lines and investigate complex motion by introducing a suitable model
equation.

1. INTRODUCTION

When an A.C. electric field is applied to nematic liquid crystals, a rich variety
of spatio-temporal convective patterns arise upon altering the field strength and
frequency.[2] For a certain range of field strength, a grid pattern forms via a complicated route and, when the voltage increases further, the grid pattern begins to
oscillate as the system experiences a Hopf bifurcation. In our experiments, we parametrically forced the oscillating grid pattern by driving the oscillations voltage of
the electric field at a frequency close to twice that of the grid oscillation. Since the
phase of the grid oscillation is locked at two points, a domain wall connecting the

two locked states forms. Such a domain wall is called a phase jump line, phase slip line, or simply a string.

Experimental results exhibit two exciting phenomena. First, it has been found that phase jump lines may have endpoints at the dislocations of the grid patterns. These are called "open phase jump lines." (In contrast, phase jump lines without endpoints are called "closed phase jump lines.") In Figure 1 we show open and closed phase jump lines arising in a convective pattern obtained experimentally. Second, phase jump lines have been observed to move in various forms. Roughly speaking, when a frequency misfit is sufficiently small, closed phase jump lines vanish, and open phase jump lines tend to stationary straight lines. Increasing a frequency misfit, two different phenomena occur depending on the sign of the frequency misfit: For one sign, phase jump lines form a periodic pattern, while for the opposite sign, they move actively. We have found two types of characteristic motion the "drifting ring" and "swaying string." These are illustrated in Figure 2.

From the theoretical point of view, we wish to solve two problems: (1) Why can dislocations in the grid pattern become endpoints of phase jump lines? (2) How are the complex motions of phase jump lines explained? We will try to give answers to these problems in the following sections.

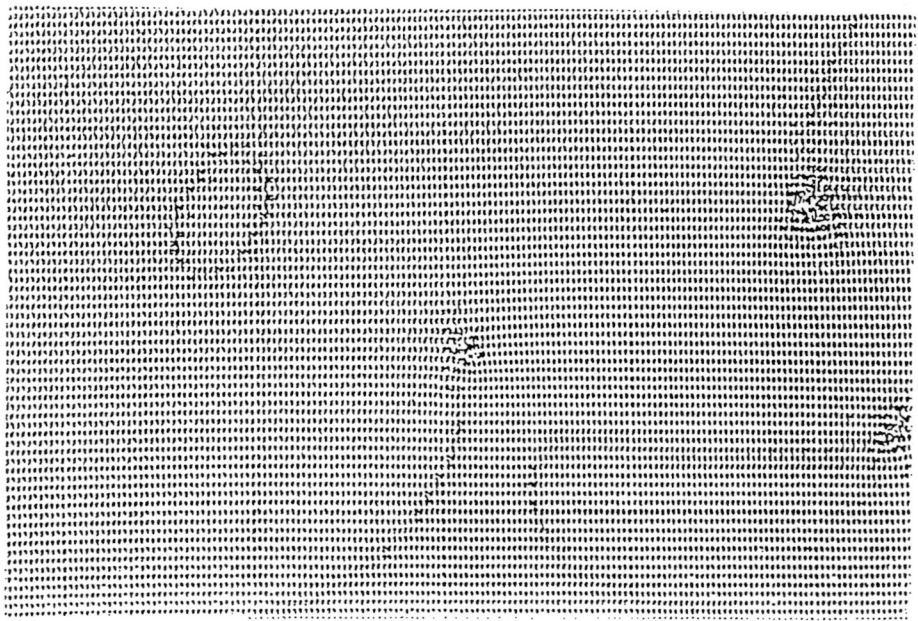

FIGURE 1   Snapshot of a convective pattern obtained experimentally.

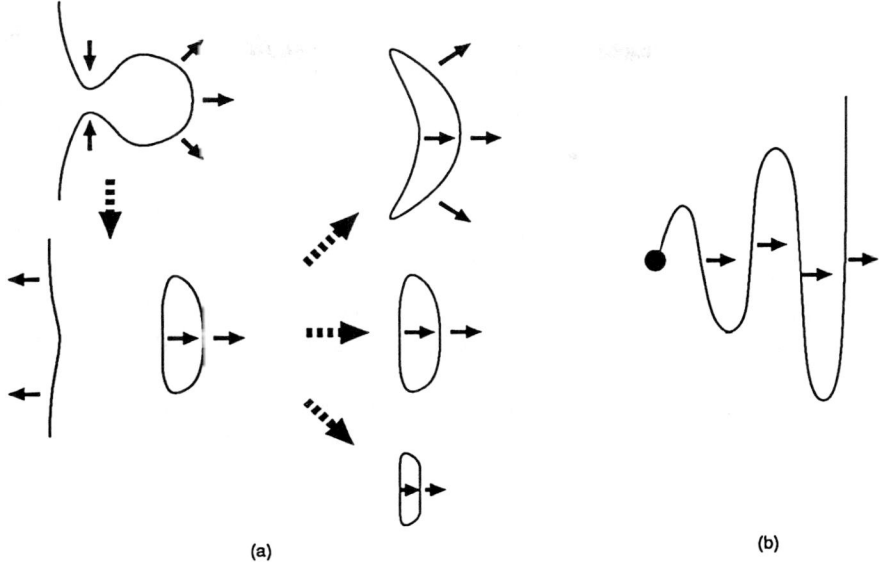

FIGURE 2 Schematic figures of (a) "drifting ring" and (b) "swaying string."

2. TOPOLOGICAL ARGUMENTS

The grid oscillation is regarded as a standing wave with the form

$$
\begin{aligned}
U(x,y) = {}& G_1 \cos(k_1 x + \phi_1) \cos(k_2 y + \phi_2) \\
& + G_2 \sin(k_1 x + \phi_1) \sin(2k_2 y + 2\phi_2) \cos(\omega t + \psi),
\end{aligned} \tag{1}
$$

where ϕ_1, ϕ_2, and ψ are arbitrary constants. (See Figure 3.) A family of stable standing waves is then obtained by applying the transformation

$$
(\phi_1, \phi_2, \psi) \rightarrow (\phi_1 + \delta_1, \phi_2 + \delta_2, \psi + \delta_t) \tag{2}
$$

that acts on the pattern expressed by Eq. (1). Vortex-type defects, therefore, can form with topological charges

$$
\kappa_x = \frac{1}{2\pi} \oint d\vec{s} \cdot \vec{\nabla}\phi_1, \quad \kappa_y = \frac{1}{2\pi} \oint d\vec{s} \cdot \vec{\nabla}\phi_2, \quad \text{and} \quad \kappa_t = \frac{1}{2\pi} \oint d\vec{s} \cdot \vec{\nabla}\psi. \tag{3}
$$

After some calculation, we obtain the equality:

$$
(\kappa_x, \kappa_y, \kappa_t) = \left(\frac{m+n}{2}, \frac{m-n}{2}, \frac{-3m+n}{2} + l \right), \tag{4}
$$

where m, n, and l are integers. When we drive the system periodically with frequency ω_e at sufficiently large amplitude, the grid oscillation will be locked to the external force with the frequency of a fractional number of ω_e. Since we consider the case that ω_e is close to twice the frequency of the grid oscillation, the temporal phase ψ is locked to ψ_* or $\psi_* + \pi$. The symmetry of a family of patterns is then broken as

$$(\phi_1, \phi_2) \rightarrow (\phi_1 + \delta_1, \phi_2 + \delta_2) \tag{5}$$

and

$$\psi \rightarrow \psi + \pi. \tag{6}$$

The first symmetry property leads to the formation of vortex-type defects with the topological charge $(\kappa_x, \kappa_y) = ((m+n)/2, (m-n)/2)$, and the second discrete symmetry property leads to formation of interfaces connecting the two locked states. We should note here that the two symmetry properties are not independent. Therefore, the two types of topological defects can fuse as shown in Figure 4. Let us consider Figure 4(b) to see how the configuration is consistent. Suppose that the field (ϕ_1, ϕ_2, ψ) has a value $(\phi_1^{(0)}, \phi_2^{(0)}, \psi^{(0)})$ at a point P near the interface. Then, at a point Q across the interface, the field has a value $\phi_1^{(0)}, \phi_2^{(0)}, \psi^{(0)} + \pi)$. On the other hand, the value of the field (ϕ_1, ϕ_2, ψ) can vary smoothly to $(\phi_1^{(0)} + \pi, \phi_2^{(0)} + \pi, \psi^{(0)})$ moving counterclockwise around the vortex with topological charge $(1/2, 1/2)$. Noting that the pattern with $\phi_1^{(0)}, \phi_2^{(0)}, \psi^{(0)} + \pi)$ is identical to that with $(\phi_1^{(0)} + \pi, \phi_2^{(0)} + \pi, \psi^{(0)})$, we see that the interface must end at a vortex.

This discussion was generalized so as to be applied to many systems.[6] As a result, we can show that phase jump lines appear even for the case that ω_e is close to ω_0. This result may be somewhat mysterious because the temporal phase ψ is locked at one point. Experimental results, however, support our result. As another example exhibiting open phase jump lines, we considered systems in which the

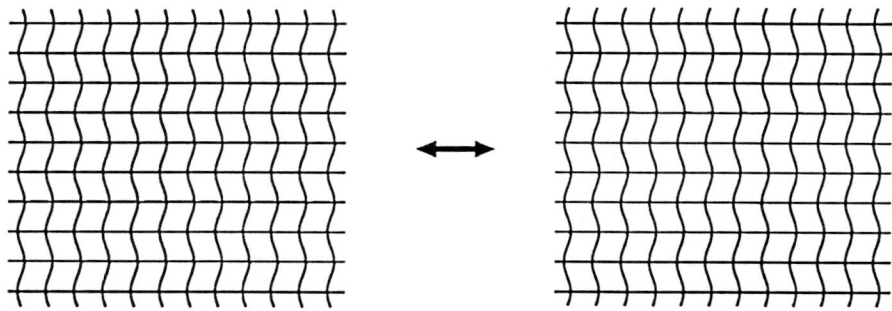

FIGURE 3   Schematic figure of a regular oscillating grid pattern.

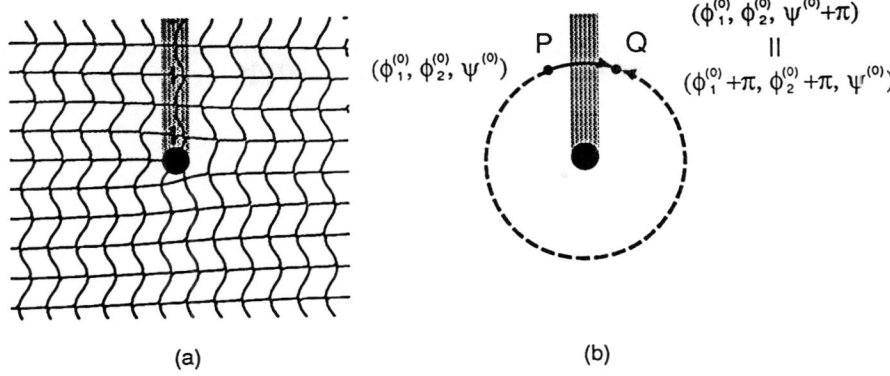

(a)

(b)

$(\phi_1^{(0)}, \phi_2^{(0)}, \psi^{(0)})$ P Q

$(\phi_1^{(0)}, \phi_2^{(0)}, \psi^{(0)}+\pi)$

||

$(\phi_1^{(0)}+\pi, \phi_2^{(0)}+\pi, \psi^{(0)})$

FIGURE 4 Open phase jump line. (a) Snapshot of a convective pattern. (b) Configuration near the topological defect.

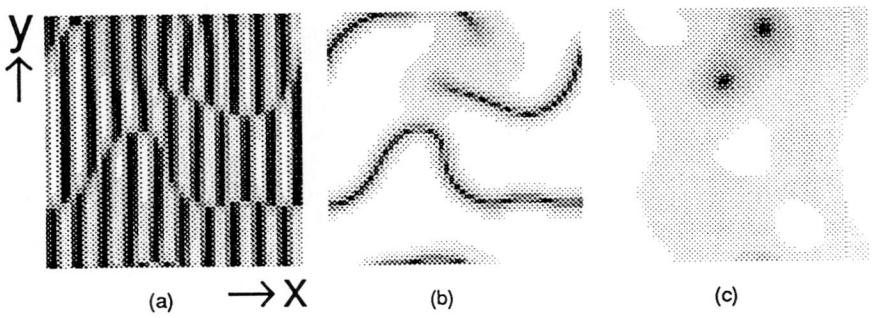

y

(a) → X

(b)

(c)

FIGURE 5 Numerical simulation of Eq. (8). Black and white regions correspond to the minima and maxima of (a) pattern U, (b) $|W_1|$, and (c) $|W_2|$.

resonant interaction between two modes with wave numbers k and $2k$ is relevant. In such a system, a pattern U is expressed by

$$U = W_1 e^{ikx} + W_2 e^{2ikx} + c.c. \tag{7}$$

The dynamics of this pattern are described by the time evolution of the two amplitudes W_1 and W_2. The following equation is used[6]:

$$\partial_t W_1 = (r_1 - g_1|W_1|^2 - g_3|W_2|^2)W_1 + \alpha W_2 W_1^* - d_1\Delta W_1,$$
$$\partial_t W_2 = (r_2 - g_2|W_1|^2 - g_4|W_1|^2)W_2 + \beta W_1^2 - d_2\Delta W_2. \tag{8}$$

This is the simplest model equation which exhibits open phase jump lines. Indeed, open phase jump lines arise in ordering processes from random initial conditions as shown in Figure 5. From the viewpoint of symmetry breaking, we have also showed that the topological nature of these phase jump lines are similar to cosmic strings connected by monopoles.[7]

3. MOTIONS OF PHASE JUMP LINES

In this section, we discuss the motion of phase jump lines. A pattern tends to relax to a uniform stable state in variational systems. This is the general principle by which we can understand the motion of dislocations (i.e., endpoints) as they invade phase jump lines. Also, the simplest model equation introduced above can not exhibit any complex behaviour even if it is nonvariational. We thus return to the experiments.

Let us recall that the regular convective pattern $U(x, y)$ is expressed as

$$U(x, y) = U_0(x, y) + W\Phi_c(x, y)e^{i(\omega_0 + \Delta)t} + c.c., \tag{9}$$

where ω_0 corresponds to a natural frequency observed without the external forcing, and Δ denotes a frequency misfit. The basic pattern U_0 and the critical mode Φ_c are given as

$$U_0(x, y) = G_1 \cos(k_1 x + \phi_1) \cos(k_2 y + \phi_2), \quad \text{and}$$
$$\Phi_c(x, y) = \sin(k_1 x - \phi_1) \sin(2k_2 y + 2\phi_2). \tag{10}$$

Under the assumption that the grid pattern is regular and stationary, the dynamics of two phases are irrelevant and the complex amplitude W obeys the complex Ginzburg-Landau equation supplemented with a forcing term

$$\dot{W} = (1 + ic_0)W - (1 - ic_2)|W|^2 W + (1 + ic_1)\Delta W + \gamma W^*, \tag{11}$$

where c_0, c_1, and c_2 are real and γ is positive. The variable W^* denotes the complex conjugate of W and γ measures the amplitude of the external force. The frequency misfit Δ is now given by $\Delta = c_2 - c_0$. This equation, of course, can describe only closed phase jump lines. We will now discuss the motion described by this equation.

We first note that the equation is invariant under the transformation

$$W \to We^{i\pi}, \tag{12}$$

and that phase locking corresponds to a symmetry breaking of this transformation. In the one-dimensional system, the domain wall connecting the two locked states is called an Ising-type or Bloch-type depending on whether or not the pattern is invariant under the chiral transformation

$$W \to We^{i\pi},$$
$$x \to -x.$$

(13)

Recently, Coullet et al. showed that a Bloch wall moves with a constant velocity proportional to a real-order parameter defined near the Ising-Bloch transition point in a one-dimensional system.[1] We further found numerically that various types of oscillating interfaces appear[3] and also that spatio-temporal intermittency occurs due to the "splitting of the Bloch walls." (See Figure 6.)

In two-dimensional systems, domain walls are one-dimensional, and Ising and Bloch walls are defined by introducing a local coordinate system (ξ, η), where ξ is a coordinate along the normal direction of the wall, and η is orthogonal to the basis ξ. (See Figure 7(a).) A domain wall is called Ising type when it is invariant under the transformation

$$W \to We^{i\pi}, \quad \xi \to -\xi.$$

(14)

Otherwise, it is called Bloch type. In analogy to the one-dimensional case, Bloch walls in two-dimensional systems move in the direction normal to the wall. Here, we should note that the classification is defined locally. A wall, therefore, can have a singular point at which two Bloch walls with opposite chiralities are connected. One may interpret that such a singular point as a domain wall connecting the two Bloch walls.

Let us discuss the motion of two Bloch walls connected at a singular point as shown in Figure 7(a). If the singular point is stationary, the two Bloch walls move in opposite directions. Eventually, the wall forms a spiral pattern. We stress here that there are some cases in which the singular point moves and that the motion of the singular point is related to the symmetry of the pattern under the transformation

$$W \to We^{i\pi}, \quad (\xi, \eta) \to -(\xi, \eta).$$

(15)

If the pattern near the singular point is symmetric for the transformation, the point is called Ising type and remains fixed; otherwise, the point is called Bloch type and will move.

We now explain the phenomenon we have named "drifting ring." A drifting ring is composed of two types of Bloch walls which are connected at two Bloch-type singular points (Figure 7(b)). If the direction in which the singular point moves is the same as that of the Bloch walls, the ring moves, shrinking or expanding as shown in Figure 8. The deformation of the ring depends on the ratio of the two velocities of the Bloch wall and the singular points. Numerical simulations suggest that the ratio seems to be related to the frequency misfit, but this point is not yet clear. We have also found that, for another set of parameter values, phase jump

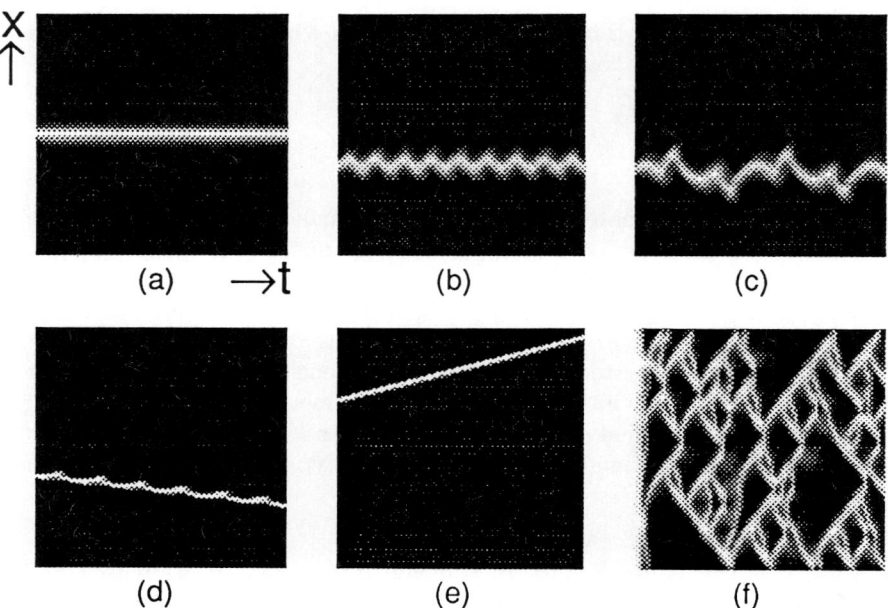

FIGURE 6 One-dimensional pattern evolutions by numerical simulation of Eq. (11) with parameters $(c_1, c_2, -0.5, 1.0, 0)$ and (a) $\gamma = 0.28$, (b) 0.24, (c) 0.20, (d) 0.18, (e) 0.06, and (f) $(c_1, c_2, -0.5, 1.0, -0.3, 0.25)$. (a) stationary (Ising wall), (b) sinusoidal oscillation, (c) "double step" oscillation, (d) oscillatory drift, (e) steady drift (Bloch wall), and (f) spatio-temporal intermittency. Black and white regions correspond to the maxima and minima of $|W|^2$.

lines form periodic patterns and "sucker patterns," and exhibit "string turbulence" as shown in Figure 9.

Finally, we discuss the motion of open phase jump lines. Our goal is to explain the "swaying string" phenomenon. This behaviour may occur if the connecting point is emitted from an endpoint. Unfortunately, however, we have not yet confirmed this statement, because we have no suitable model to analyze. To construct a model corresponding to experiments, at least six complex fields are needed as relevant dynamical variables, and necessarily many parameters must be included in the model. It may be quite difficult to find suitable parameters for which "swaying string" behaviour appears. We have studied two models exhibiting "open Bloch strings." One model consists of a set of amplitude equations which describe the resonant interaction of two frequencies ω and 2ω. The other describes a standing wave coupled with a nonoscillatory higher mode. In any case, there are still many parameters in the model equations, and we must study them further, keeping in mind the physical mechanism of the "swaying string."

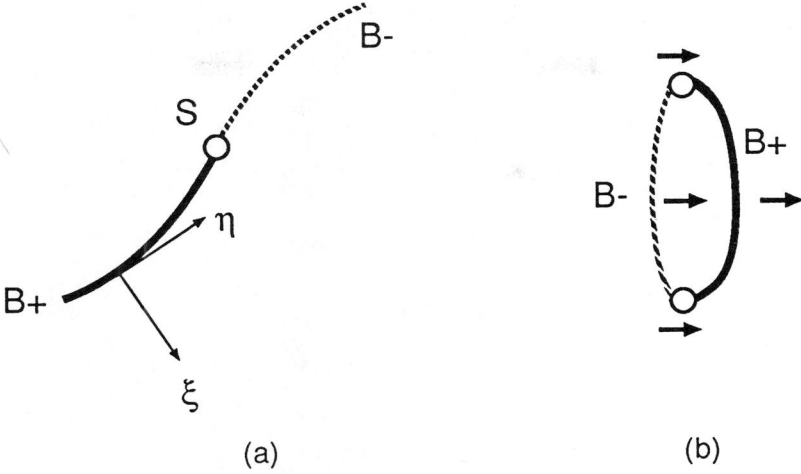

FIGURE 7   Schematic figures of (a) a Bloch wall with the singularity and (b) two Bloch walls connected at two points.

4. CONCLUSION

Let us summarize our paper. First, we introduced open and closed phase jump lines arising in electrohydrodynamic convection of nematics. Second, we developed a topological theory for open phase jump lines. These are topological defects called hybrid type and should be observed in many experimental systems. Third, based on the complex Ginzburg-Landau equation with a forcing term, we explained the motion of closed phase jump lines called "drifting rings." The key mechanism of a "drifting ring" is related to the symmetry breaking of a singular point of ring walls. Explaining the motion of "swaying string" remains an open question.

ACKNOWLEDGMENTS

The authors thank Y. Kuramoto and B. Janiaud for stimulating discussions. They also thank G. C. Paquette for a critical reading of the manuscript.

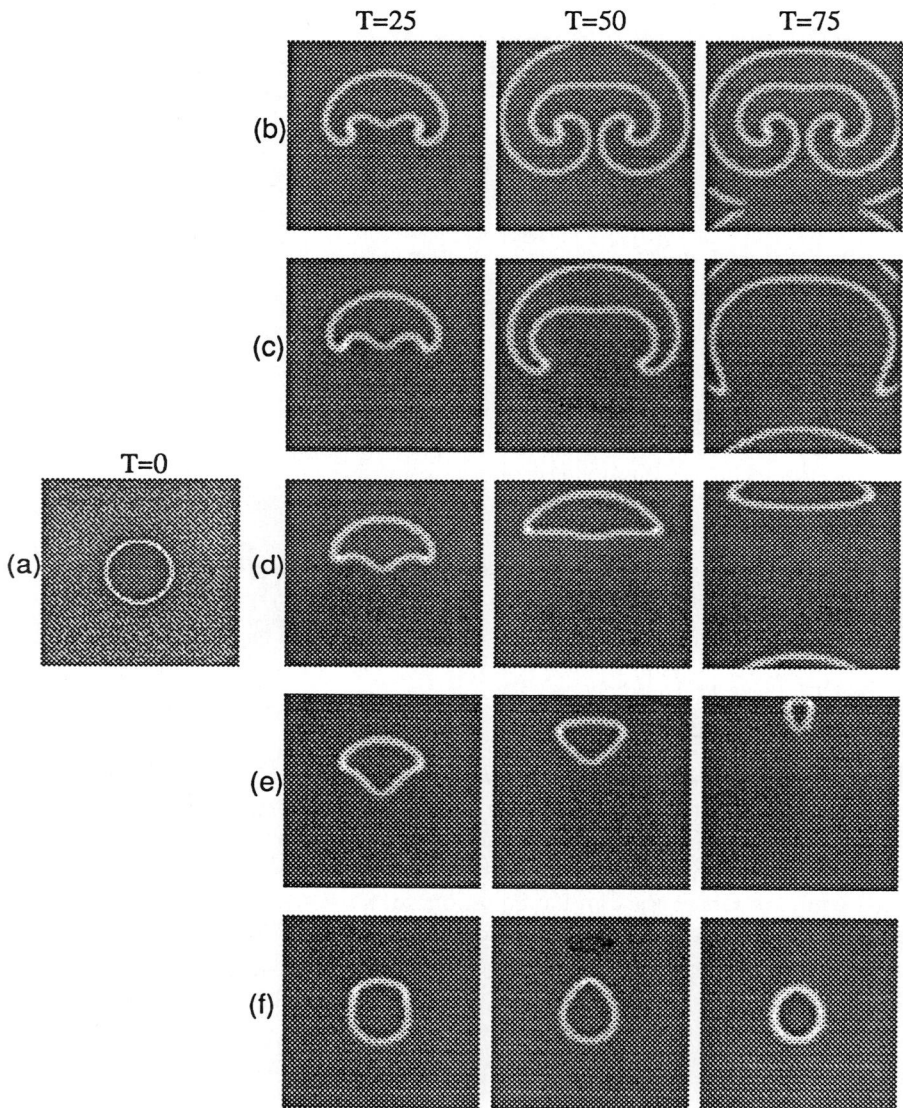

FIGURE 8   Drifting rings with parameters $(c_1, c_2, y) = (1.0, 0.0, 0.2)$ and (b) $\Delta = 0.15$, (c) 0.125, (d) 0.1, (e) 0.05, and (f) -0.15, starting from (a), the initial condition. Black and white regions correspond to the maxima and minima of $|W|^2$.

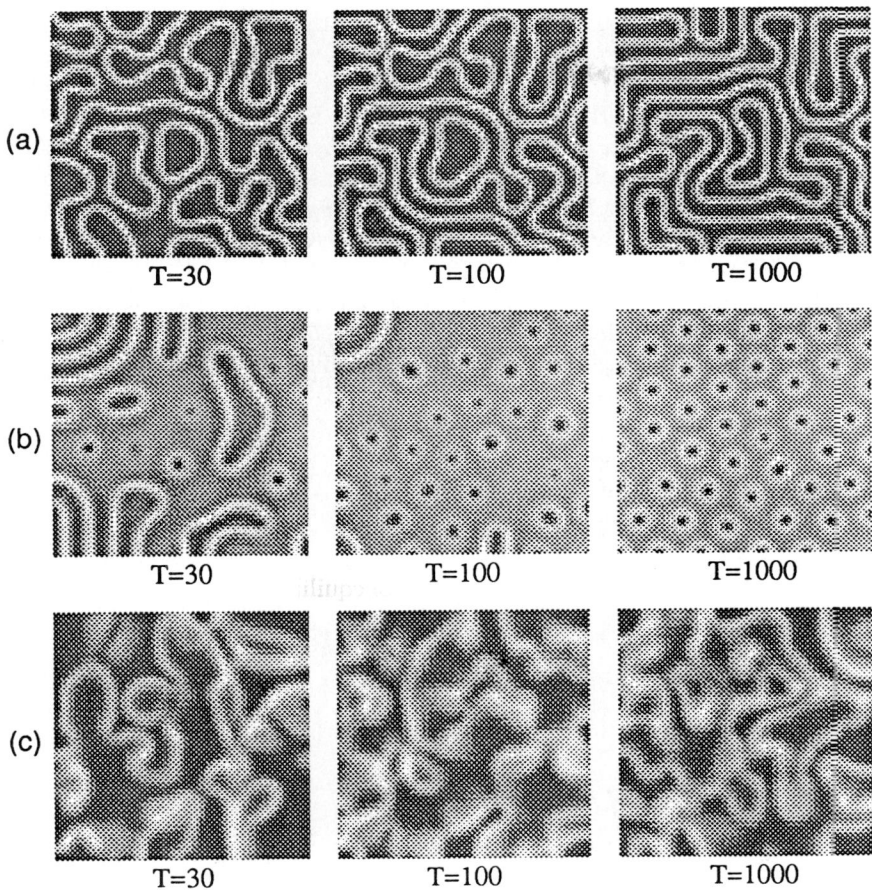

FIGURE 9 Two-dimensional pattern evolutions by numerical simulation of Eq. (11). (a) periodic pattern, (b) sucker pattern, and (c) string turbulence. A set of parameters $(c_1, c_2, \Delta, \gamma)$ is chosen as (a) $(-0.5, 0., 0.79, 0.9)$, (b) $(-0.5, 1.0, 0.7, 0.5)$, and (c) $(-0.5, 1.0, -0.5, 0.4)$. Black and white regions correspond to the maxima and minima of $|W|^2$.

REFERENCES

1. Coullet, P., J. Lega, B. Houchmanzadeh, and J. Lajzerowicz. "Breaking Chirality in Nonequilibrium Systems." *Phys. Rev. Lett.* **65** (1990): 1352–1355.
2. Kai, S., and K. Hirakawa. "Successive Transitions in Electrohydrodynamic Instabilities of Nematics." *Prog. Theor. Phys. Suppl.* **64** (1978): 212–243.
3. Mizuguchi, T., and S. Sasa. "Oscillating Interfaces in Parametrically Forced Systems." *Prog. Theor. Phys.* **89** (1993): 599–605.
4. Sasa, S., T. Mizuguchi, and M. Sano. "Open and Closed Phase Jump Lines in Two-Dimensional Cellular Patterns." *Europhys. Lett.* **19** (1992): 593–597.
5. Sasa, S. "Hybrid Topological Defects in Nonequilibrium Systems." *Phys. Rev.* **A46** (1992): 5268–5270.

Helga Richter,† Agnes Buka,†‡ and Ingo Rehberg†

†Physikalisches Institut, Postfach 101251, Universität Bayreuth, Fax: 0921-552999; e-mail: helga.richter@uni-bayreuth.de

‡Permanent address: Research Institute for Solid State Physics, H-1525 Budapest, P.O. Box 49, Hungary

The Electrohydrodynamic Instability in a Homeotropically Aligned Nematic: Experimental Results

Visual observations and measurements of the onset of electroconvection in a homeotropically aligned nematic liquid crystal MBBA are presented. Because of its negative dielectric anisotropy, the first instability to cccur when increasing the voltage across the layer is the electrically driven bend Fréedericksz transition. Further increase of the voltage leads to electrcconvection. This paper characterizes the convection pattern close to the onset of convection in order to allow a comparison with a linear stability analysis of the Fréedericksz state.

1. INTRODUCTION

In a recent interesting paper, Hertrich et al.[2] pointed out the possibility of complicated spatio-temporal behavior close to the onset of convection, when a nematic liquid crystal with negative dielectric anisotropy in a homeotropically aligned configuration is used. The reason for this expectation is the fact that, because of the rotational symmetry of the boundary conditions, no orientation of the expected roll pattern is preferred. Because of the negative dielectric anisotropy and the homeotropic alignment, the first instability to occur is the Fréedericksz transition. This is not a pattern-forming instability, but a spontaneous break in the rotational symmetry. Under idealized conditions there are no forces fixing the orientation of the director in the plane of the layer; thus, this state is expected to be metastable. If convection sets in and if the rolls are normal to the director projection in the plane of the layer, there is no reason to believe that this metastability would be distorted. If, on the other hand, the rolls are oblique with respect to this director projection, this additionally broken symmetry gives rise to the expectation that the orientation of the rolls would not be metastable any more: the rolls might constantly reorient their position. As such a reorientation cannot be expected to occur simultaneously over the entire cell, one might have a complicated, time-dependent pattern directly at threshold. The necessary condition for this interesting scenario is the existence of oblique rolls. The Hertrich paper[2] did indeed predict oblique rolls for small frequencies of the driving electric field, and normal rolls for larger frequencies. The goal of the work presented here is a first attempt to check this prediction experimentally. The most interesting result is that we do not find oblique rolls, but rather normal rolls within the whole experimentally accessible frequency regime. At high frequencies, however, the normal rolls break the left-right symmetry in a peculiar way which is not yet understood.

2. EXPERIMENTAL SETUP

The nematic liquid crystal MBBA is sandwiched between two transparent electrodes with a spacing of 24 μm. In order to achieve homeotropic alignment, these electrodes are sputtered with chrome[1]. An attractive feature of this novel procedure seems to be a good stability of the sample—within half a year there were no obvious signs of deterioration. A disadvantage of this coating is a tremendous reduction of the light intensity by the chrome layer, which has a thickness of about 200 Å. Using an exposure time of 0.1 s, we get an image measured by means of a 512×512 square pixel, 14-bit slow-scan CCD-camera. The Fréedericksz transition and the electroconvection are driven by applying an ac-coupled sinusoidal current to the cell. In order to allow an easy access to the cell, we did not use any temperature control of the sample. Thus, most of the measurements shown here have been performed at (24 ± 2) °C, unless otherwise stated.

3. RESULTS

Figure 1 is the phase diagram of our sample. We measured the voltage across the sample for the Fréedericksz transition and the onset of electroconvection for different frequencies of the driving field. The Fréedericksz transition occurs at an rms voltage of $V_F = 3.2$ V. Within our resolution and in the range of frequencies shown here, we found no sign for a frequency dependence of this threshold (such an effect is not expected, either); thus, the threshold $V_F(f)$ is just a straight line. The curved line shows the onset of electroconvection. It has been determined by visual inspection of the sample, a method which gives reasonable results within a resolution of 1%. In order to give an idea of the reproducibility of this measurement, open squares are shown representing the measured onset of convection when increasing the frequency of the ac-field, and solid circles correspond to the measurement when decreasing the frequency. The visual deviation of these two points at 3000 Hz might reflect the temperature influence in this sensitive part of the threshold curve. In order to avoid deterioration of the sample, we did not perform measurements above 25 V. From a comparison of this threshold curve with the one presented by Hertrich et al.[2] (see Figure 5 therein), we would expect the predicted Lifshitz-point, where the crossover from normal to oblique rolls takes place, to be located around a driving frequency of 1600 Hz.

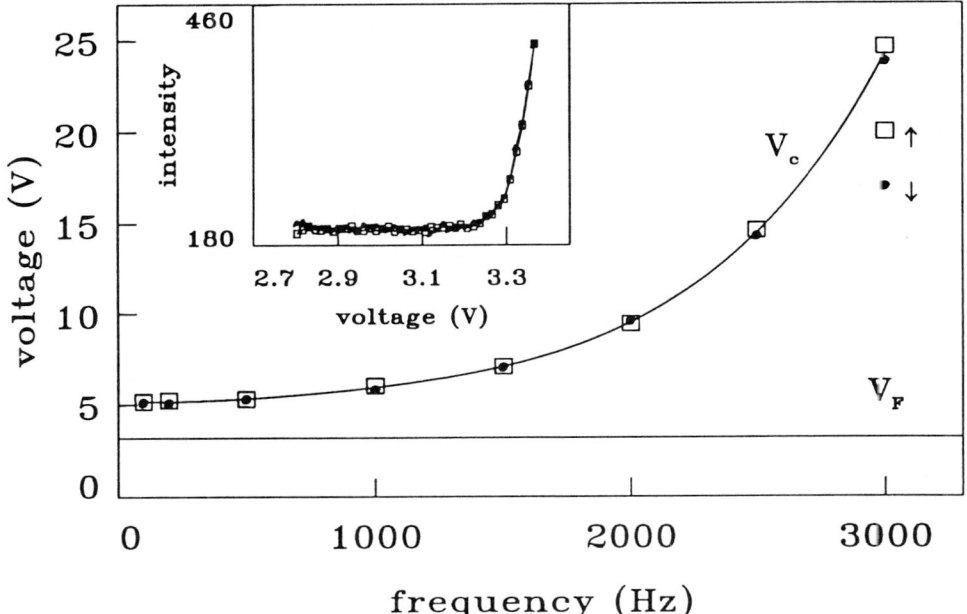

FIGURE 1   Phase diagram showing the Fréedericksz transition and the onset of convection. The inset is a detailed measurement of the Fréedericksz transition.

It is important to notice that the Fréedericksz transition is not perfect; it is rather slightly rounded as indicated by the onset measurement shown in the inset of Figure 1. Here, we crossed the polarizer and the analyzer and measured the intensity of one single pixel of the CCD-array. Due to the absorption of the polarizer and the analyzer, an exposure time of 1 s has to be used in this case. The underlying idea of this procedure is the fact that the CCD-array, being dc-coupled to the 14-bit AD-converter, is capable of measuring absolute light intensities. We measured these light intensities for different voltages of the driving electric field. The open squares correspond to increasing voltage, and the solid circles were obtained by decreasing the voltage. We would like to mention that there is a specific difficulty with this method for the homeotropic alignment. If the orientation of either the polarizer or the analyzer is parallel to the director, the image remains black above threshold. For any other orientation, an increase in the intensity occurs once the director starts to reorient. Fortunately, it turns out that the director orients always in the same direction, if the transition is performed sufficiently slow. Thus it is possible to orient the polarizer and analyzer in the appropriate direction in order to be sensitive to the bend Fréedericksz transition.

We believe that the rounding of the measurement curve shown in the inset of Figure 1 indicates that the homeotropic alignment is not geometrically perfect, although there is no obvious distortion of this alignment when observing the cell under crossed polarizers below threshold. This imperfection presumably also manifests itself in the fact that the orientation of the director is reproducible, which is considered as a great help performing the measurements of the convection onset.

The measurements were made in a frequency range from 15 Hz up to 3 kHz. We observed two different kinds of convection patterns with a crossover taking place at about 100 Hz. Below this frequency we believe to get normal rolls as demonstrated below, and for higher frequencies a peculiar pattern. We choose two frequencies, 30 Hz and 1000 Hz, to demonstrate the two different regions.

The orientation of the director is measured by the procedure indicated in Figure 2. Here we rotated the crossed polarizer and analyzer, and measured the resulting intensity within a small fraction of the observation field (150×150 pixels), where the orientation of the director was fairly homogeneous. The rotation angle $\alpha = 0$ of the polarizer is chosen arbitrarily. In Figure 2 (a) we represent the result of this procedure above the Fréedericksz transition and below the onset of convection. The data are taken at a frequency of 30 Hz and a voltage of 4.02 V. From this curve we conclude that the mean orientation of the director φ_d is either $0°$ or $90°$. The lower part of the figure shows the result obtained above the onset of convection (30 Hz, 5.76 V). The solid line, which is supposed to be a guide for the eye, is a fit to harmonical functions. Our interpretation of these two figures is that the convection did not change the mean orientation of the director, at least within the resolution given by this procedure (about $10°$).

In Figure 3 we show the same comparison for a higher frequency of 1000 Hz. In Figure 3, view(a) is taken at 4.31 V, and (b) at 6.64 V. The conclusion is

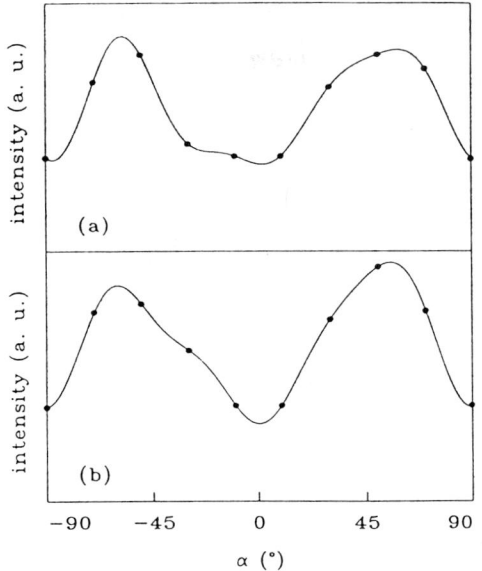

FIGURE 2   Determination of the director orientation at 30 Hz. (a) Below the onset of convection, 4.02 V, (b) with convection, 5.76 V. The solid line is a guide for the eye.

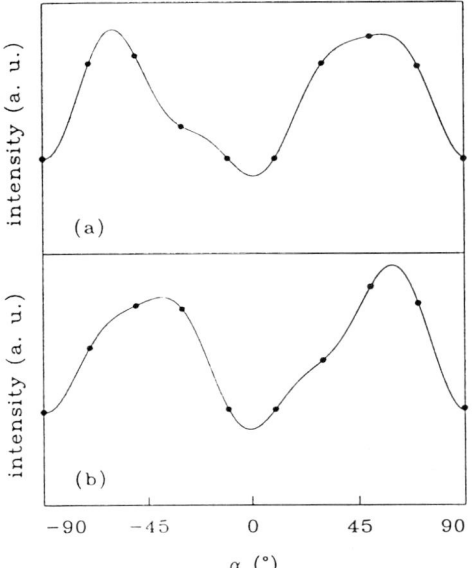

FIGURE 3   Determination of the director orientation at 1000 Hz. (a) Below the onset of convection, 4.31 V, (b) with convection, 6.64 V.

the same as in Figure 2, namely that the mean orientation of the director field is not much changed by the onset of convection. This statement seems to apply for all frequencies. The angle for the orientation of the director is measured as $\varphi_d = 0° \pm 10°$.

FIGURE 4   Image of the convection pattern of an area of 1000 μm $\times$ 1000 μm at 30 Hz, $\epsilon = (V/V_c)^2 - 1 \approx 0.1$ (28 °C).

FIGURE 5   Image of the convection pattern of an area of 1000 μm $\times$ 1000 μm at 1000 Hz, $\epsilon = (V/V_c)^2 - 1 \approx 0.1$ (28 °C).

The convection patterns above threshold for these two frequencies of 30 Hz and 1000 Hz are shown in Figures 4 and 5. They indicate that there is clearly a preferred direction for the rolls which varies slightly with the position in the sample, presumably caused by the change of the preferred orientation of the director. There is no obvious change in the orientation of the rolls as a function of the frequency. Thus, by means of visual observations like these, together with the determination of the orientation of the director as indicated above, and the contrast measurements described below, we draw the conclusion that the rolls are normal to the director field for any frequency within the range shown in Figure 1. The theory, on the other hand, would predict oblique rolls for small frequencies below 1600 Hz or so. It is worth mentioning that from the measurements of Figures 2 and 3 it cannot be excluded that the roll orientation is parallel to the director. This would, however, be very improbable from a theoretical point of view. Moreover, it would make the interpretation of the following contrast measurements very difficult.

There is something peculiar about the convection rolls measured above about 100 Hz. At first sight this effect is best described as a somewhat unsatisfactory optical contrast of the images. In order to quantify this statement, we measured the contrast of an image as a function of the orientation of the polarizer—with no analyzer used in this case. The contrast is defined as A_0/I_0, where A_0 is the amplitude of the first harmonic of the spatially periodic light intensity modulation, and I_0 is the mean value of the light intensity. This method is well understood for planarly aligned nematics in the normal roll case. There, the contrast is high when the director is aligned parallel to the polarizer, and almost zero when the director is aligned perpendicular to the polarizer, as shown in Figure 6. For normal rolls in the homeotropic case, one would expect similar behavior.[4] As indicated by Figure 7(a), which was measured at 30 Hz and at 5.76 V, this seems to be the case: The contrast is indeed low when the polarizer is parallel to the roll axis. Thus these rolls can consistently be interpreted as normal rolls, although this fact seems to be in disagreement with the theoretical calculations. If the rolls were oriented parallel to the director, they would not be expected to change its direction and, thus, they would be invisible.

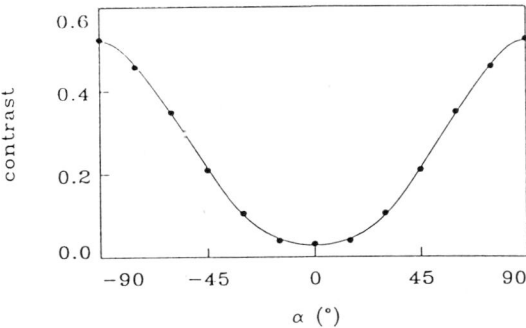

FIGURE 6   Optical contrast of normal rolls in planar alignment.

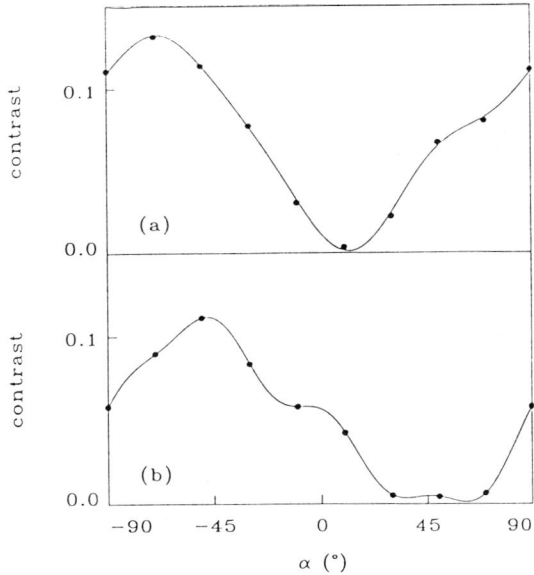

FIGURE 7   Optical contrast of convection rolls in homeotropic alignment, (a) 30 Hz, 5.76 V, (b) 1000 Hz, 6.64 V.

The strangest behavior observed so far is shown in Figure 7(b): These rolls do not behave like normal rolls. They have a minimal contrast when the polarizer has an angle of approximately 45° with the roll axis, and a maximum when this angle is about −45°. Thus, these rolls are not oblique rolls in a usual sense, because they appear to be aligned normal to the mean orientation of the director. They are not regular normal rolls, either, because they have obviously broken the symmetry of the normal rolls. We thus call them abnormal rolls.

There is no ideas about the geometrical shape of the abnormal rolls. They do not seem to agree with the prediction from the linear stability theory. As mentioned by Hertrich et al.,[2] however, the linear stability theory in the case of oblique rolls might be of limited use, because the ensuing mode is unstable, roughly speaking with respect to rotations of the roll. Thus, the abnormal rolls might already be some manifestation of this instability of the linear solution. In order to make closer contact to the theory, we tried to measure the most unstable mode by analyzing subcritical thermal fluctuations below the onset of electroconvection, a method which has been successfully applied in the planar case.[6] The result is shown in Figure 8 and must be considered as negative: No obvious pattern can be observed. The reason for this difficulty in observing the unstable mode is unclear: The optical sensitivity is expected to be as twice as high compared to the planarly aligned sample.[4]

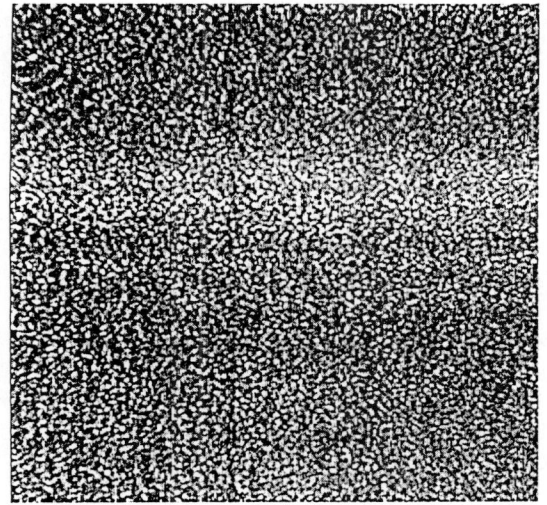

FIGURE 8 Attempt to observe subcritical convection patterns, $\epsilon \approx -0.1$, 30 Hz (28°C).

In summary, there seems to be a qualitative disagreement between the theoretical predictions and experimental observation in this interesting system. One speculation we would like to investigate in the near future is the question of the possible time dependence of these patterns: Since they appear on an already tilted director orientation, the electric ac-field might induce oscillations of this field phase-locked with the driving frequency. This effect is also present in planarly aligned normal rolls, where it is enhanced when approaching the cut-off frequency,[5] but it might be stronger in the present case where the convection is an instability originating from a less trivial ground state. The natural way to explore the time dependence would be stroboscopic illumination, a method that is harder to implement here because of the large light absorption due to the chrome coating.[1]

Finally, we would like to mention that spiral convection patterns are observed in our cell as well (two images have been presented by Pesch et al.[3]). Before exploring their behavior in detail, a clarification of the nature of the most unstable mode at the onset of convection seems to be the most pressing task.

ACKNOWLEDGMENTS

We would like to thank W. Decker, A. Hertrich, L. Kramer, and W. Pesch for calling our interest to this fascinating problem and for many stimulating discussions. This work was supported by Deutsche Forschungsgemeinschaft through Graduiertenkolleg Nichtlineare Spektroskopie und Dynamik, and SFB 213, Bayreuth, the NATO

CRG 910112, and the Hungarian Academy of Sciences (OTKA 2976). A. B. wishes to thank the University of Bayreuth for its hospitality.

REFERENCES

1. Chuvyrov, A. N., O. A. Scaldin, and V. A. Delev. "Auto-Waves in Liquid Crystals. II. Uniform Fast Oscillating Flows." *Mol. Crys. & Liquid Crys.* **215** (1992): 187–198.
2. Hertrich, A., W. Decker, W. Pesch, and L. Kramer. "The Electrohydrodynamic Instability in Homeotropic Nematic Layers." *J. Phys. II* France **2** (1992): 1915–1930.
3. Pesch, W., A. Hertrich, and L. Kramer. "Electrohydrodynamic Convection in Nematics: The Homeotropic Case." In *Spatio-Temporal Organization in Nonequilibrium Systems*, edited by S. C. Müller and Th. Plesser, 211–213. Dortmund: Projekt-Verlag, 1992.
4. Richter, H., S. Rasenat, and I. Rehberg. "The Shadowgraph Method at the Fréedericksz Transition." *Mol. Crys. & Liquid Crys.* **222** (1992): 219–228.
5. Schneider, U., M. de la Torre Juárez, W. Zimmermann, and I. Rehberg. "Phase Shift of Dielectric Rolls in Electroconvection." *Phys. Rev. A* **46** (1992): 1009–1013.
6. Winkler, B. L., W. Decker, H. Richter, and I. Rehberg. "Measuring the Growth Rate of Electroconvection by Means of Thermal Noise." *Physica D* **61** (1992): 284–288.

Michael Dennin,† Guenter Ahlers, and David S. Cannell
Department of Physics and Center for Nonlinear Science, University of California, Santa Barbara, CA 93106; †email: mike@tweedledee.ucsb.edu.

Measurement of Material Parameters of the Nematic Liquid Crystal I52

We report measurements of the anisotropy in the dielectric constant $\Delta\epsilon$ and the ratio of the bend elastic constant K_{33} to $\Delta\epsilon$ for the nematic liquid crystal 4-ethyl-2-fluoro-4'-[2-(*trans*-4-*n*-pentylcyclohexyl)-ethyl]-biphenyl (I52). We used the electric-field-induced bend Fréedericksz transition. I52 is a nonpolar compound recommended recently as a liquid-crystalline benchmark material. The liquid-crystal cell consisted of two parallel glass plates coated with transparent conductors. The liquid crystal was aligned with its director perpendicular to the plates, and an AC electric field was applied between the plates. Above a critical voltage, there is a nonhysteretic transition (the Fréedericksz transition) to a state where the director is no longer normal to the plates. From measurements of the onset voltage and capacitance, we deduce $K_{33}/\Delta\epsilon$ and $\Delta\epsilon$, respectively. We also report early measurements on the electro-hydrodynamic instability in this system when the director is parallel to the plates.

Spatio-Temporal Patterns, Ed. P. E. Cladis and P. Palffy-Muhoray,
SFI Studies in the Sciences of Complexity, Addison-Wesley, 1995
353

1. INTRODUCTION

A nematic liquid crystal (NLC) consists of long rod-like molecules that possess orientational order but do not have positional order, and their average direction of alignment is referred to as the director.[1,4] Cells consisting of parallel glass plates and containing the NLC can be treated so that the director aligns in a uniform direction parallel to the plates. This is known as planar alignment.[3] When an AC voltage is applied across the plates of such a cell containing a NLC with a negative anisotropy of the dielectric constant, there is a critical value of the applied voltage, V_c, at which hydrodynamic flows occur and convection rolls develop.[2,9,10] When the wavevector of the rolls is at an angle with respect to the direction of the director alignment, the rolls are referred to as oblique rolls. Bodenschatz et al.[2] have calculated both the onset voltage and the angle at onset. The calculation depends on thirteen parameters. Recently, Finkenzeller et al.[7] have proposed the use of 4-ethyl-2-fluoro-4'-[2-(*trans*-4-*n*-pentylcyclohexyl)-ethyl]-biphenyl (I52) as a liquid-crystalline benchmark material. I52 is available in high purity and is chemically stable. Our electro-convection experiments have shown that this material has a strong tendency to form oblique rolls and squares (see Figure 1). Determinations of the material parameters of I52 thus are clearly of interest. Using the bend Fréedericksz transition,[6,11,13,14] we have measured two combinations of these parameters: the anisotropy in the dielectric constant, $\Delta\epsilon = \epsilon_\parallel - \epsilon_\perp$ where $\epsilon_\parallel$ and $\epsilon_\perp$ are respectively the dielectric constant parallel and perpendicular to the director, and the ratio $K_{33}/\Delta\epsilon$ where K_{33} is the bend elastic constant.

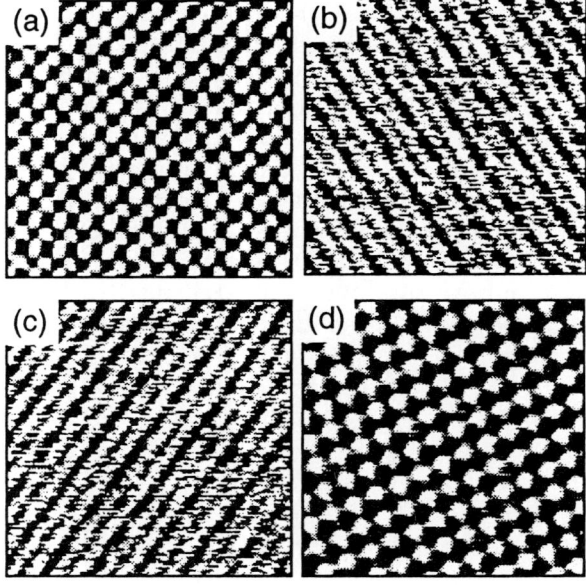

FIGURE 1   Four examples of electro-convection patterns observed in I52. Image (a) is at $T = 45.0\ °C$, $f = 100$ Hz, $\sigma = 6.4 \times 10^{-8}\ \Omega^{-1}m^{-1}$, and $V = 14.0$ V. Images (b), (c), and (d) are at $T = 65.0\ °C$, $\sigma = 1.7 \times 10^{-7}\ \Omega^{-1}m^{-1}$, and $f = 200$ Hz with $V = 10.67$ V, 10.69 V and 12.33 V, respectively. In all of these pictures, the director is aligned along the horizontal axis and the wavelength is on the order of $1.4d$ where $d = 10\ \mu m$ is the thickness of the cell.

For the bend Fréedericksz transition, the liquid crystal is aligned with the director perpendicular to the plates. When an electric field is applied and $\Delta\epsilon < 0$, there is a continuous transition at $V = V_c^F$ to a state where the director is no longer normal to the plates. Data for V_c^F and for the capacitance of the liquid-crystal cell allow determination of $\Delta\epsilon$ and $K_{33}/\Delta\epsilon$. Obviously, the measurements can be used to obtain K_{33} and $\Delta\epsilon$ separately. As a check on our bend Fréedericksz measurements, we performed the same experiments using N-(4-methoxybenzylidene)-4'-n-butylaniline (MBBA), a liquid crystal for which the material parameters are known,[2,5,8] and obtained good agreement with the accepted values. The values of $\Delta\epsilon$ and K_{33} we obtain for I52 are consistent with the observed behavior in the electro-convection experiments.

2. EXPERIMENTAL APPARATUS

We used two types of liquid-crystal cells. For the electro-convection experiments, commercial cells were used.[1] They were 10 μm thick with the spacing set by glass beads. The surfaces of the glass plates had been treated with a rubbed polyimide to obtain planar alignment. The cells were 1 cm^2 in area, and the central 0.5 cm^2 was coated with a transparent conductor made of indium-tin-oxide (ITO). The I52 used in the convection experiments was doped with 1% I_2 by weight.

For the Fréedericksz experiments, the entire area of the cell was coated with ITO, and the surfaces of the glass plates were treated with lecithin[3] to obtain perpendicular alignment. For the experiments with I52, undoped I52 was used, and the height was set at (28 ± 6) μm by mylar spacers. The MBBA used was doped with 0.01% tetrabutylammonium bromide by weight and placed in cells with a height of (75 ± 6) μm set by mica spacers. The bend Fréedericksz onset voltage does not depend on cell thickness, but the thicker cell enabled a more accurate height determination.

Both the electro-convection and the Fréedericksz transition experiments were performed in the same apparatus which consisted of three parts: a shadowgraph apparatus for visualization, a temperature control stage, and electronics for applying the voltage and measuring the capacitance and resistance of the cells. This is shown schematically in Figure 2(a). The cell was located within the temperature control stage which was mounted on a translation stage. The shadowgraph apparatus consisted of two parts: a light source mounted below the translation stage, and a lens and camera system mounted above the temperature control stage. The light source consisted of a point source at the focal point of a lens. The point source was made

[1] Cells and the information concerning their characteristics were obtained from Display Tech Inc., 2200 Central Ave, Boulder, CO 80301.

by mounting a 40 μm pinhole in front of a high-power red LED. The plastic lens

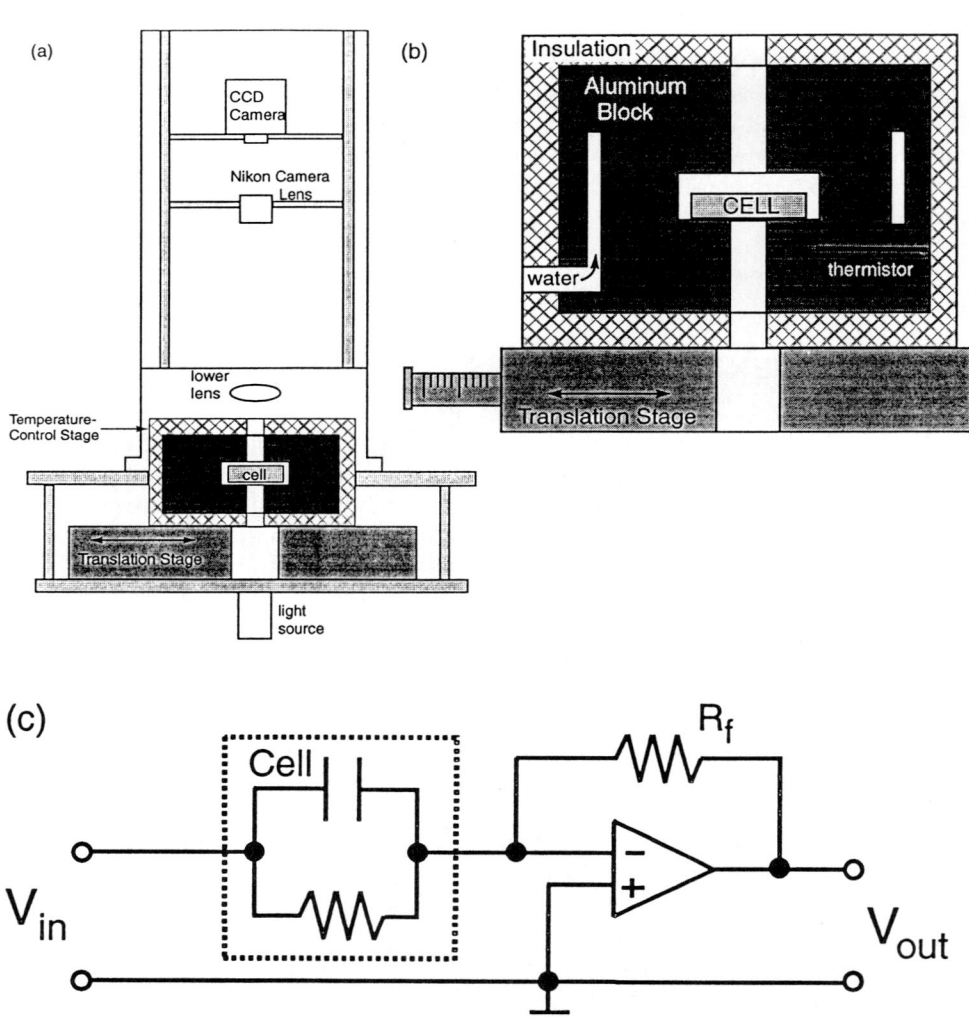

FIGURE 2 (a) Schematic diagram, not to scale, of the apparatus showing the optical system and temperature control stage. (b) Schematic diagram, not to scale, showing a more detailed view of the temperature control stage. (c) Schematic diagram of the circuit used to measure the resistance and capacitance of the cell. V_{out} is measured with an A/D converter, and the cell is modeled as a capacitor and resistor in parallel.

of the LED had been machined away so that the pinhole was located approximately 250 μm from the active element of the LED. The light was converted into a parallel beam by a 14 mm diameter achromatic lens with a 21 mm focal length. A dichroic sheet polarizer was placed between the light source and the liquid crystal cell and could be rotated with respect to the cell. For the Fréedericksz transition experiments, another dichroic sheet polarizer was placed between the cell and the lens system. This was used as an analyzer and was crossed with respect to the lower polarizer. When the director is perpendicular to the glass plates, the polarization of the light passing through the cell is not affected by the liquid crystal, so the intensity of the light passing through the analyzer is a minimum. As the director begins to tilt with respect to the plates, the polarization is rotated, so the intensity of the light passing through the analyzer is nonzero.

The lens system consisted of two lenses and a CCD camera which were mounted in a 1.2 m high aluminum tube.[2] The camera used a 1.27 cm charge-coupled device (CCD) with 510 × 492 picture elements. The image was digitized in the computer using an 8-bit gray scale.[3] The lower lens was a 20 mm diameter achromat with a 53 mm focal length. The lens was fixed in place 6.63 cm above the cell. The second lens was a Nikon 50 mm $f/1.4$ camera lens. The Nikon lens and CCD camera were mounted on separate movable carriages. The design of the carriages allowed the relative position of the Nikon lens and CCD camera and the position of the Nikon lens and camera as a unit to be adjusted independently. With this arrangement the magnification of the Nikon lens-camera system was adjustable from 1x to 30x, and the plane imaged by the Nikon lens-camera system could be adjusted. The flexibility of the system was important for imaging the electro-convection experiments where the image is strongly dependent on the plane imaged.[12]

The temperature-control stage is shown schematically in Figure 2(b) and consisted of a cylinder of aluminum 6.78 cm high with a diameter of 9.78 cm wrapped with 0.64 cm of insulating foam. The aluminum cylinder contained the sample cell and was mounted on an x-y translation stage which was equipped with micrometers accurate to 1 μm. To calibrate the magnification of the shadowgraph system, a reference object was imaged in two measured lateral positions. We illuminated the cell from below and viewed it from above through a 1.40 cm diameter hole which ran along the axis of the aluminum cylinder. The hole was closed at the top and bottom with glass windows. A 0.318 cm wide and 2.54 cm high circular channel with an inner radius of 3.56 cm was located with its midplane at the midplane of the aluminum cylinder and surrounded the center of the cylinder. Water which was temperature-controlled to better than ±1 mK flowed through the channel maintaining the temperature of the aluminum cylinder. The temperature of the cylinder was measured with a thermistor that had been calibrated against a mercury thermometer and which was embedded in the aluminum.

[2]Sony HVM-200 CCD camera.
[3]PCEYE Video Capture System manufactured by Chorus Data Systems, Inc., P.O. Box 370, 6 Continental Blvd., Merrimack, NH 03054.

The waveform for the AC voltage which was applied to the cell was generated using a computer-controlled synthesizer card capable of generating arbitrary waveforms.[4] For these experiments, only sinusoidal waveforms were used. For finer resolution of steps in the amplitude of the applied waveform, a 12-bit multiplying D/A chip was used. The signal was amplified with a power amplifier.[5] The frequencies used for the electro-convection experiments ranged from 25 Hz to 2000 Hz and the voltages used ranged from 0 V to 85 V (all voltages quoted are root mean square values). The Fréedericksz experiments were performed using the same range of voltages and a frequency of 2000 Hz. When determining V_c^F of the Fréedericksz transition, the voltage was increased in steps of 0.05 V every 5 minutes, and then decreased at the same rate to check for hysteresis. For determination of V_c for electro-convection, the voltage was increased/decreased in steps of 0.01 V every 15 minutes.

The resistance R and the capacitance C of the cell were measured using the circuit shown in Figure 2(c). The cell was modeled as a resistor and capacitor in parallel, and measuring the phase and amplitude of output voltage relative to that of the applied voltage yielded the resistance and capacitance of the cell. The applied voltage was

$$V_i(\omega) = V_{\text{in}} \cos(\omega t).$$

The output voltage was digitized in the computer with an A/D converter,[6] and the phase and amplitude of the output voltage were extracted by fitting the digitized data to

$$V_o(\omega) = V_{\text{out}} \cos(\omega t + \phi).$$

The amplitude and phase of the output voltage are related to the resistance and capacitance of the cell by

$$V_{\text{out}} = \frac{R_f \sqrt{1 + \omega^2 R^2 C^2}}{R} V_{\text{in}},$$
$$\phi = \pi + tan^{-1}(\omega RC).$$

The accuracy of our measurements was checked by measuring the resistance and capacitance of the parallel combination of a known resistor and capacitor similar in value to those of a filled cell.

[4] WSB - 10 arbitrary waveform synthesizer manufactured by Qua Tech, Inc., 478 East Exchange St., Akron, OH 44304.

[5] Hewlett-Packard 6827A bipolar power supply/amplifier.

[6] Labmaster DMA manufactured by Scientific Solutions, Inc., 6225 Cochran Rd., Solon, OH 44139-3377.

3. EXPERIMENTAL RESULTS

It has been reported that $\Delta\epsilon$ for I52 is temperature dependent, changing from negative to positive at 75.2°C as temperature is increased.[7] To check this behavior, we measured the onset voltage V_c^F for the Fréedericksz transition as a function of temperature. The theoretical value of the onset voltage is given by

$$V_c^F = \pi \left(\frac{K_{33}}{\epsilon_0 \Delta\epsilon} \right)^{1/2}.$$

As $\Delta\epsilon$ goes through zero, the onset voltage should diverge.

To determine V_c^F, the intensity of the light transmitted through the NLC cell was measured as a function of voltage. In Figure 3, we show a typical transmission curve for I52. The initial rounding is due to a slight inhomogeneity in the cell, and V_c^F is determined by a linear extrapolation of the sharp rise in intensity to a value of zero. For this case, $V_c^F = 25.16$ V. In Figure 4, we show the onset voltage as a function of temperature, and the curve is diverging as expected. In Figure 5, we plot $(V_c^F)^{-2}$ versus temperature. Extrapolating this curve to zero determines $\Delta\epsilon = 0$. The results are fit quite well by a straight line and give $\Delta\epsilon = 0$ at $T = 58.50°C$. The nematic range[7] is from 13°C to 103.4°C, so it is reasonable to assume that K_{33} is roughly constant in the range of temperatures used and that $\Delta\epsilon$ varies linearly with temperature. However, it is possible that the linear dependance is due to any variation in K_{33} being compensated by variations in $\Delta\epsilon$.

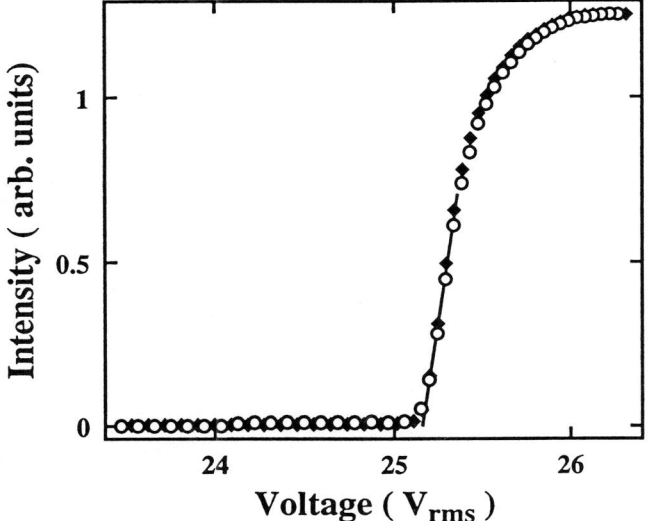

FIGURE 3
Transmission curve used to determine the onset of the Fréedericksz transition in I52 at 30.9°C. Shown here is both the intensity of light transmitted through the cell as a function of voltage while stepping up the voltage (diamonds) and while stepping down the voltage (circles). The solid line shows the linear extrapolation to zero intensity used to determine V_c^F.

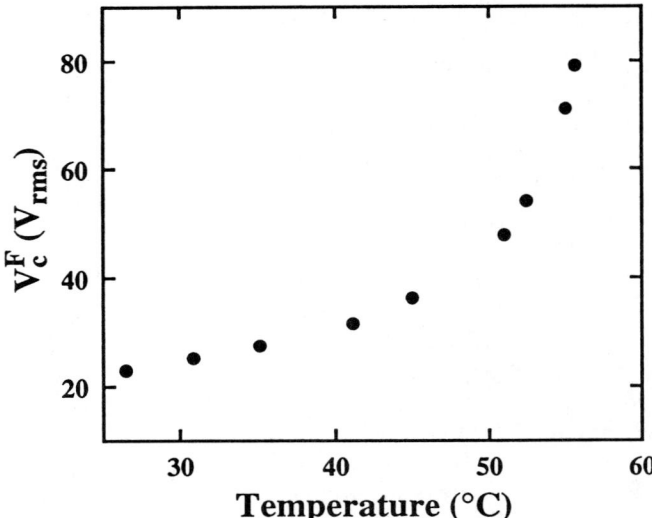

FIGURE 4   Results for the critical voltage V_c^F of the bend Fréedericksz transition in pure I52 as a function of temperature.

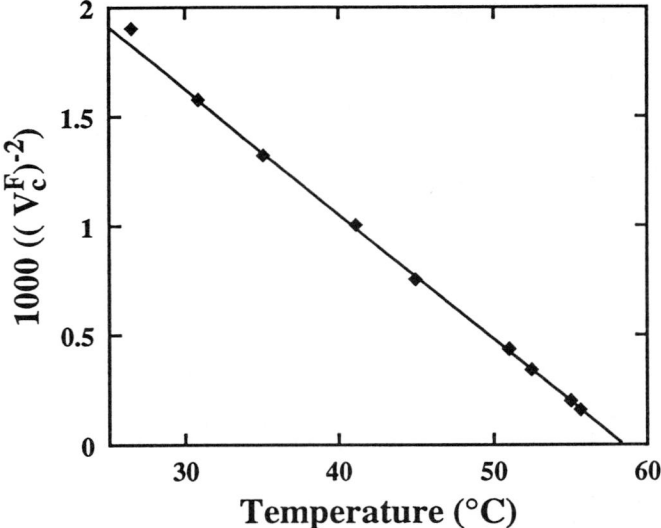

FIGURE 5   Data from Figure 4 plotted as $(V_c^F)^{-2}$ vs. T. The solid line is a linear fit to the data with the point at 26.5°C excluded. The fit gives $\Delta\epsilon = 0$ at 58.50°C.

To obtain $\Delta\epsilon$ separately, we measured the capacitance of the liquid-crystal cell both above and below the Fréedericksz transition and computed

$$\Delta\epsilon = \frac{(C_\| - C_\perp)d}{\epsilon_o A},$$

where $C_\|$ and $C_\perp$ are the capacitance of the cell with the director parallel and perpendicular to the field respectively, A is the area of the cell containing liquid crystal, and d is the thickness of the cell. Measuring $\Delta\epsilon$ by this method automatically excludes the contribution of the spacers to the capacitance of the cell. Below the transition, one measures $C_\|$ directly because the director is aligned parallel to the electric field. Above the transition, the director forms an angle with respect to the field, and the high voltage limit of the capacitance is given by[6,11,13]

$$C = C_\perp + \frac{S}{V},$$

where the slope S is a function of the material parameters and independent of A/d.[11,13] Extrapolating to $1/V = 0$ as shown in Figure 6 for $T = 30.9°C$, one obtains $C_\perp$. The main sources of error in computing $\Delta\epsilon$ are the uncertainty in the height and area of the cell. The theoretical formula for the onset voltage is used to determine K_{33}. This formula assumes infinite anchoring strength of the director at the boundaries. In practice, measured values of V_c^F are usually lower than the theoretical ones. Performing the experiment with MBBA gives a rough idea of the applicability of the theoretical formula; however, there is no guarantee that MBBA and I52 have the same anchoring strengths when lecithin is used to align the director.

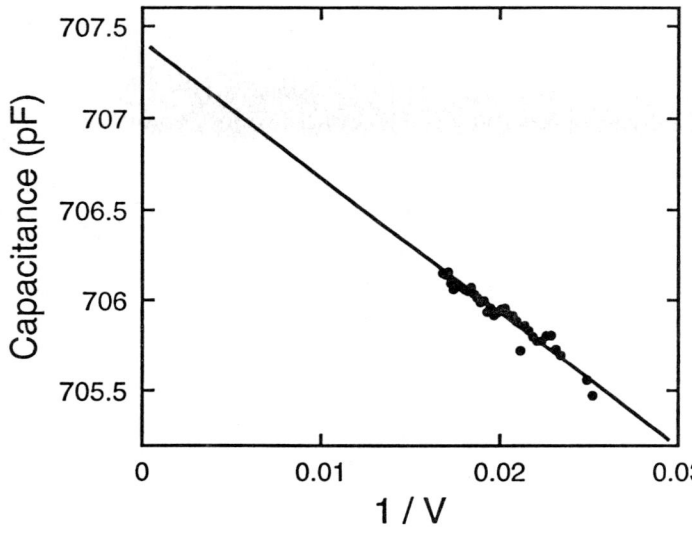

FIGURE 6
Capacitance (pF) versus $1/V$ for I52 at 30.9°C. The solid line is a linear fit to the data extrapolated to $1/V = 0$.

For I52 at $T = 30.9°C$, we obtain $\Delta\epsilon = -0.051 \pm 0.014$ and $K_{33} = (28 \pm 8) \times 10^{-12}$ N. For MBBA at $T = 25.0°C$, we measure $V_c^F = 3.95$ V. The value computed using the previous measured values[2,5,8] $\Delta\epsilon = -0.53$ and $K_{33} = 8.61 \times 10^{-12}$ N is $V_c^F = 4.26$ V. Using our measured value of V_c^F and our capacitance measurements for MBBA, we find $\Delta\epsilon = -0.6 \pm 0.1$ and $K_{33} = (8 \pm 1) \times 10^{-12}$ N.

4. SUMMARY

We have measured the temperature dependence of $K_{33}/\Delta\epsilon$ for I52 and found that $\Delta\epsilon = 0$ at $T = 58.50°C$. We have also measured $\Delta\epsilon$ which allows us to compute K_{33}. We have checked our measurements by repeating them for MBBA, and we found reasonable agreement between our measured values of $\Delta\epsilon$ and K_{33} and previous measurements of these constants.

When the measured values of K_{33} and $\Delta\epsilon$ are used to compute the onset voltage and angle of the rolls at onset for electro-convection, reasonable agreement is found with our preliminary measurements.[7] In Figure 7, we show the measured onset voltage for electro-convection in I52 (squares) and MBBA (triangles). For comparison, the solid line and dashed line are the free-boundary-condition theory[2] computed for I52 and MBBA respectively.[8] The diamonds in Figure 8 are the results for the angle between the director and the wavevector of the pattern at onset for I52 at higher frequencies. For MBBA, the wavevector of the pattern was always parallel to the director. The solid line is the theoretical calculation using the same parameters that were used to compute the onset voltage curve for I52. For both of these curves, there are 13 parameters[2] in the theory, six of which remain unknown for I52. Reasonable values based on those for other similar liquid crystals were used in computing the theoretical curves.[6] One source of the increasing discrepancy between theory and experiment for the onset voltage of I52 at higher frequencies is a voltage dependence of the conductivity of I52. As shown in Figure 9, the conductivity of I52 increases with voltage, which in the theory lowers V_c. The theory curve plotted in Figure 7

---

[7] See Bodenschatz, et al.[2] The values of the material parameters used for I52 were as follows. The elastic constants, in units of 10^{-12} N, were $K_{11} = 30$, $K_{22} = 20$, and $K_{33} = 30$. The viscosities, in units of 10^{-3} kgm^{-1}s^{-1}, were $\alpha_1 = 6$, $\alpha_2 = -110$, $\alpha_3 = -5$, $\alpha_4 = 21.3$, $\alpha_5 = 110$, and $\alpha_6 = -5$. The other required parameters were $\epsilon_\parallel = 2.9$, $\epsilon_\perp = 2.927$, $\sigma_\parallel/\sigma_\perp = 1.38$ and $\sigma_\perp = 6.0 \times 10^{-8}$ Ω^{-1}m^{-1}.

[8] The values of the material parameters used for MBBA were as follows. The elastic constants, in units of 10^{-12} N, were $K_{11} = 6.66$, $K_{22} = 4.2$, and $K_{33} = 8.61$. The viscosities, in units of 10^{-3} kgm^{-1}s^{-1}, were $\alpha_1 = -18.1$, $\alpha_2 = -110.4$, $\alpha_3 = -1.1$, $\alpha_4 = 82.6$, $\alpha_5 = 77.9$, and $\alpha_6 = -33.6$. The other required parameters were $\epsilon_\parallel = 4.72$, $\epsilon_\perp = 5.25$, $\sigma_\parallel/\sigma_\perp = 1.5$, and $\sigma_\perp = 1.4 \times 10^{-7}$ Ω^{-1}m^{-1}. The perpendicular conductivity $\sigma_\perp$ was measured by us, and the other values were taken from Bodenschatz et al.[2]

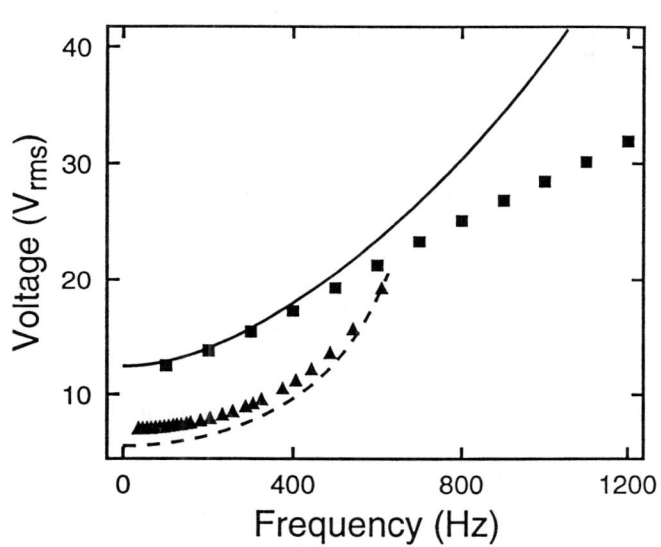

FIGURE 7 Onset voltage versus frequency for I52 at $45.0°C$ with a conductivity of $6.4 \times 10^{-8} \ \Omega^{-1}m^{-1}$ (squares) and for MBBA at $25°C$ with a conductivity of $1.4 \times 10^{-7} \ \Omega^{-1}m^{-1}$ (triangles). The dashed line is the calculated boundary using the measured parameter values for MBBA. The solid line is the calculated boundary using the measured parameter values for I52 with reasonable values used for the unknown I52 parameters. Free boundary conditions were used for both calculations.

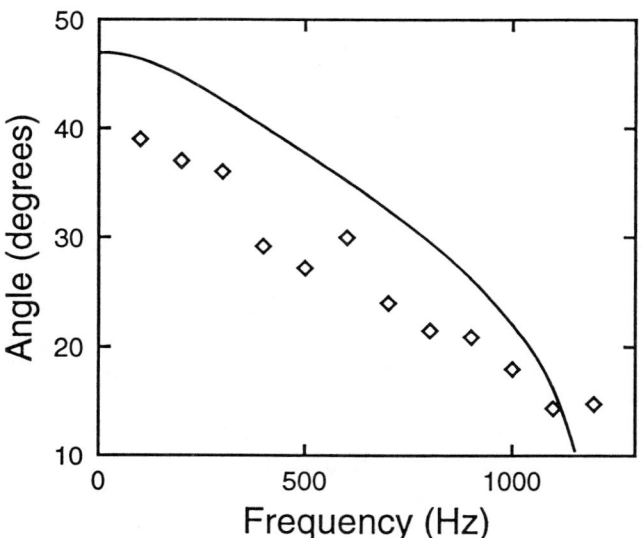

FIGURE 8 The angle between the wavevector of the rolls and the director at onset for I52 under the same conditions as Figure 7. The solid line is the result of the theoretical calculation using the same parameters as in Figure 7.

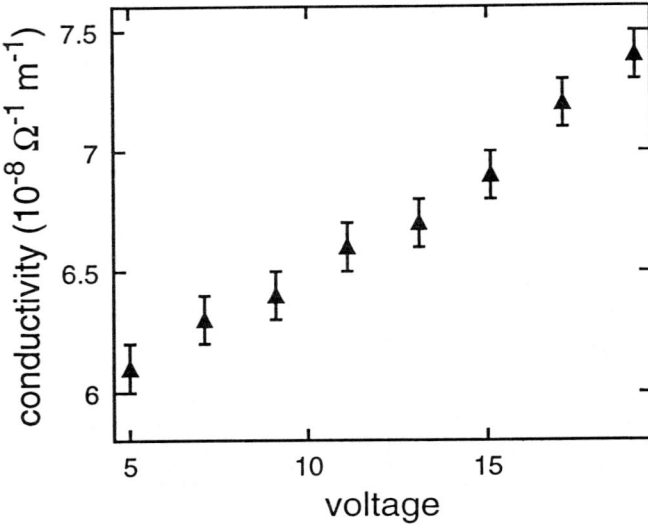

FIGURE 9 Conductivity measured at $f = 1000$ Hz as a function of voltage for I52 doped with 1% I_2 at $45.0°C$.

was computed using a fixed conductivity of 6×10^{-8} $\Omega^{-1}m^{-1}$. The initial reasonable agreement for the onset voltage and angle is promising, but further work is needed to resolve the discrepancies between theory and experiment for I52.

ACKNOWLEDGMENTS

We thank I. Rehberg, P. Mather, S. Morris, and B. Frisken for helpful insights on the preparation of liquid crystal cells. Conversations with I. Rehberg were especially beneficial with regards to the subtleties of electro-convection, and S. Morris was of particular assistance with the Fréedericksz experiments. We thank F. Wudl and F. Klavetter for many useful discussions on the doping of liquid crystals. We also thank U. Finkenzeller of E. Merck for providing us with information on liquid crystals and samples of the liquid-crystal I52. This work was supported by NATO through travel grant CRG910112 and the National Science Foundation through grant DMR91-17428. M. Dennin acknowledges support through an Office of Naval Research National Defense Science and Engineering Graduate Fellowship.

REFERENCES

1. Blinov, L. M. *Electrooptical and Magnetooptical Properties of Liquid Crystals.* New York: John Wiley, 1983.
2. Bodenschatz, E., W. Zimmermann, and L. Kramer. "On Electrically Driven Pattern-forming Instabilities in Planar Nematics." *J. Phys. (France)* **49** (1988): 1875–1899.
3. Cognard, J. *Alignment of Nematic Liquid Crystals and Their Mixtures. Mol. Cryst. Liq. Cryst. Sup. 1.* New York: Gordan and Breach, 1982.
4. de Gennes, P. G. *The Physics of Liquid Crystals.* Oxford: Clarendon Press, 1975.
5. De Jeu, H. W., W. Claassen, and A. Spruyit. "The Determination of the Elastic Constants of Nematic Liquid Crystals." *Mol. Cryst. Liq. Cryst.* **37** (1976): 269–280.
6. Deuling, H. J. "Deformation of Nematic Liquid Crystals in A an Electric Field." *Mol. Cryst. Liq. Cryst.* **19** (1972): 123–131.
7. Finkenzeller, U., T. Geelhaar, G. Weber, and L. Pohl. "Liquid-Crystalline Reference Compounds." *Liquid Crystals* **5** (1989): 313–321.
8. Kneppe, H., F. Schneider, and N. K. Sharma. "Rotational Viscosity γ_1 of Nematic Liquid Crystals." *J. Chem. Phys.* **77** (1982): 3203–3208.
9. Kramer, L., E. Bodenschatz, W. Pesch, W. Thom, and W. Zimmermann. "New Results on the Electrohydrodynamic Instability in Nematics." *Liquid Crystals* **5** (1989): 699–715.
10. Madhusudana, N. V., V. A. Raghunathan, and K. R. Sumathy. "Flexoelectric Origin of Oblique-Roll Electrohydrodynamic Instability in Nematics." *Pramana J. Phys.* **28** (1987): L311–L316.
11. Morris, S. W., P. Palffy-Muhoray, and D. A. Balzarini. "Measurements of the Bend and Splay Elastic Constants of Octylcyanobiphenyl." *Mol. Cryst. Liq. Cryst.* **139** (1986): 263–280.
12. Rasenat, S., G. Hartung, B. L. Winkler, and I. Rehberg. "The Shadowgraph Method in Convection Experiments." *Experiments in Fluids* **7** (1989): 412–420.
13. Uchida, T., and Y. Takahashi. "New Method to Determine Elastic Constants of Nematic Liquid Crystal from C-V Curve." *Mol. Cryst. Liq. Cryst. Lett.* **72** (1981): 133–137.
14. Winkler, B. L., H. Richter, I. Rehberg, W. Zimmermann, L. Kramer, and A. Buka. "Nonequilibrium Patterns in the Electric-Field-Induced Splay Fréedericksz Transition." *Phys. Rev. A* **43** (1991): 1940–1951.

J. P. McClymer, E. F. Carr, and H. Shehadeh
Department of Physics and Astronomy, University of Maine, Orono, ME 04469-5709

Dynamic Scattering Modes of Nematic Liquid Crystals in a Magnetic Field

In the late 1960s, there was considerable interest in the dynamic scattering modes of nematic liquid crystals for potential display applications.[8] With the advent of better technology as exemplified by the twisted nematic cell, interest in the dynamic scattering modes waned. Renewed interest in these highly scattering states occurred in light of new views of pattern formation and transitions to soft and hard turbulence.[12]

This chapter is a preliminary report of the dynamic scattering regimes in the presence of a strong magnetic field and how the field leads to qualitatively different dynamic scattering modes than previously reported. In particular, our results are understood in terms of viscous interactions between the convective flow cells due to the electric field and surface walls, that result from the magnetic field. A more detailed paper will appear elsewhere.

We first present a very brief review of electrohydrodynamic convection in nematic liquid crystals before describing our experimental setup. We then compare and contrast the dynamic scattering modes that are observed with and without a magnetic field. Finally, we present our model of surface walls and evidence in support of it.

Nematic liquid crystals typically consist of elongated organic molecules in which the long axis of the molecules align in a preferred direction.[5,7,18] This direction is loosely referred to as the director. These "soft solids" have small elastic constants and are easily deformed by electric and magnetic fields as well as surface forces.

Hydrodynamic convection can be induced in films of nematic liquid crystal with negative dielectric anisotropy by applying either DC or AC electric fields. The convection is usually observed through transparent conducting electrodes. A low-frequency regime, called the conduction regime, is observed in which the long axis of the molecules (the director) anomalously align with a component parallel to the electric field. This behavior is explained by the Carr-Helfrich-Orsay mechanism.[1,9,14] In this model, the conductivity anisotropy leads to charge separation and build-up. The localized ions are accelerated by the electric field and drag the fluid along. This fluid motion, in turn, reorients the director orientation that reinforces the charge separation, setting up counter rotating flow cells. At sufficiently small voltages above the threshold value, the director pattern undulates smoothly and forms what is commonly referred to as Williams domains.[19] Application of larger voltages aligns the director at the flow alignment angle and forms walls separating the flow cells[2,10,17] due to the strong shear flow. A schematic of such a defect wall and the flow cells are shown in Figure 1 with the charges accumulating at the walls.

A higher frequency regime, called the dielectric regime, also exists. At these higher frequencies the ionic diffusivity is too slow to allow charge separation to occur so that the charge density is stationary and the director field follows the resultant oscillating electric field. The frequency at which the system behavior changes is called the cutoff frequency.

At high voltages within the low-frequency conduction regime a transition to two highly scattering regimes have been observed. The lower voltage state is called

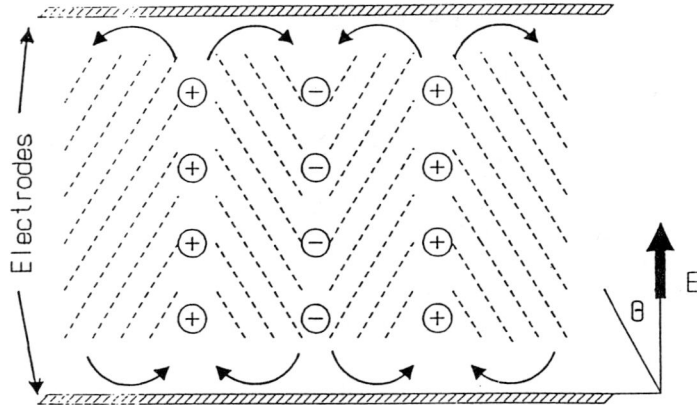

FIGURE 1   A model for molecular alignment (the dashed lines) and material flow in response to an external electric field showing charges accumulated at the defect walls.

primary or dynamic scattering mode 1 (DSM 1)[6,13] while the higher voltage state is referred to as secondary or dynamic scattering mode 2 (DSM 2).[16]

EXPERIMENTAL

The nematic liquid crystal N-(p-meth-oxybenzylidine)-p-butylaniline (MBBA), sandwiched between two conducting glass electrodes separated by 135 microns, is placed between the pole faces of an electromagnet. The sample is aligned such that the width of the sample cell, containing a region of liquid crystal approximately 1 cm wide by 3 cm long, is parallel to the magnetic field, causing the molecules to line up parallel with the magnetic field. An AC electric field, well below the cutoff frequency, causes hydrodynamic convection in the nematic liquid crystal which induces molecular alignment along the thickness of the sample in competition with the magnetic field alignment. The temperature is maintained to better than $0.1°C$, well below the isotopic transition temperature.

Our results are very reproducible providing care is taken in two areas:

1. Fine temperature control is not necessary for our observations but we have been unable to find traveling wave states at lower temperatures near the crystal-nematic transition temperature.
2. We sometimes, but not often, find it necessary to leave the sample in the field for hours before observing traveling waves or oriented dynamic scattering modes. We assume that the arraignment of surface walls, which we will discuss later, may take some time to form. Once the surface walls are formed the state persists indefinitely, even in the absence of a magnetic field.

One consequence of the nonzero magnetic field is that all the threshold voltages are increased. For example, the William's domains are normally observed at about 8 volts while we observe them at about 40 volts in the presence of an 8-kG field. Observations and photographs are made with a low-power microscope equipped with rotatable polarizers. The microscope itself can be rotated about the magnetic field direction thus allowing views away from the surface normal of the electrode.

DYNAMIC SCATTERING MODES IN THE ABSENCE OF A MAGNETIC FIELD

In the absence of a magnetic field the dynamic scattering modes are considered a transition to turbulence.[1] On the basis of the spatial power spectrum of transmitted light, DSM 1 is considered anisotropic turbulence while DSM 2 is an example of

[1]See, for instance, Kai et al.[12] and references contained therein.

isotropic turbulence.[20] If the system is subjected to a voltage jump in which the preferred state is DSM 2, the DSM 2 is observed to inhomogeneously nucleate out of DSM 1. When the electric field is removed, the disclination density of the DSM 1 is seen to be significantly less and to decay away much more rapidly than that of DSM 2.

DYNAMIC SCATTERING MODES IN THE PRESENCE OF A MAGNETIC FIELD

With a strong magnetic field (0.8 T) applied perpendicular to the electric field, and in the plane of the sample, the nature of DSM 1 and 2 is changed substantially. This change can clearly be seen in Figure 2, a transmitted light photograph showing the coexistence of both DSM 1 and DSM 2. This photograph was taken approximately 4 seconds after applying a voltage jump from 0 to 140 volts using a sine wave at 50 Hz. The lighter region corresponds to DSM 1 while the darker regions are the more stable DSM 2 which is shown nucleating near the center of the picture. DSM 2 has also nucleated outside the field of view and is growing in towards the center.

As can clearly be seen, the flow cells are visible in both dynamic scattering regimes. Both modes are anisotropic and neither mode is turbulent, although the DSM 2 state does scatter more light. An obvious question to ask is if these states we observe with the magnetic field are indeed analogous to the DSM states observed in the absence of a field? In addition to the qualitative similarity in scattering power, the DSM 2 state is observed to nucleate and grow at the expense of DSM 1 on similar time scales. More importantly, the structure of the scattering modes is similar as can be seen by switching off the electric field and observing the decay of the disclination loops. The director field in DSM 2 appears significantly more entangled than in DSM 1 and requires more time to relax to a homogeneous state.

DSM 1 retains its anisotropic appearance but the convective rolls retain the symmetry imposed by the strong magnetic field unlike the case of zero magnetic field in which the roll orientation is controlled by weaker surface interactions allowing the orientation of the convecting rolls to freely move about.[11]

The similarly anisotropic, but more turbid, DSM 2 phase requires a more subtle explanation. We propose a mechanism for the formation of DSM 2 in terms of surface bend-splay Helfrich walls as shown in Figure 3. We earlier postulated this structure to explain traveling waves observed in this same system in the presence of a magnetic field.[3,4]

FIGURE 2   Coexistence of DSM 1 and DSM 2 showing convective flow cells 4 seconds after a voltage jump from 0 to 140 volts.

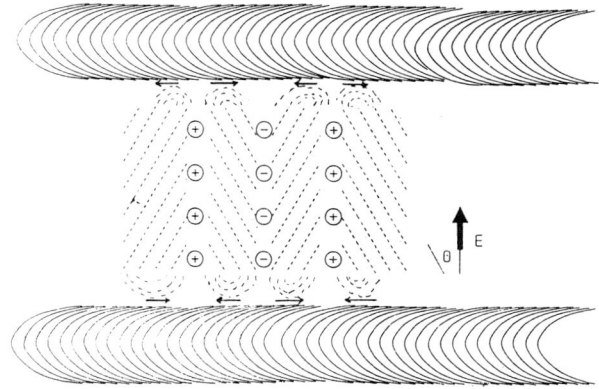

FIGURE 3   A schematic of the interaction between the convective flow cells and the surface walls (solid lines). The flow directions are indicated by arrows at the flow cell surface wall interface.

When the surface walls are oriented in the same direction on both surfaces, as in Figure 3, the counter-rotating flow cells see different director orientations near the surface. The resulting viscous interaction is not symmetric, resulting in a net viscous force driving the flow cells along the magnetic field direction. If the surface walls are oriented in the opposite sense on both surfaces, each flow cell sees the same director configuration and there is no net force on the flow cell.

In the traveling wave state, in which there is fluid flow transverse to the electric field, the surface walls in a particular region bend parallel to the magnetic field on both surfaces while in adjacent regions they bend antiparallel to the field. The two regimes are separated by a Bloch line (a twist disclination line) between the two-bend Neel (bend-splay Helfrich) walls discussed by Ranganath[15] (see Figure 4) which we term a Ranganath defect.

These well-organized regions of uniform surface wall orientation may be destroyed at higher voltages leading to smaller, irregular fluctuating patches which are separated from other patches by Ranganath defects. If we assume that an interaction between the flow cells and the surface structure exists, then the variation in surface structure can lead to either transverse forces parallel or antiparallel to the magnetic field or no force at all. The rapidly changing surface structure can then cause increased scattering as the convective flow cells are subjected to rapidly varying forces.

The existence of a coherent surface structure may be shown by observing the sample in polarized light at some oblique angle from the surface normal. In this way light traversing the sample will see a twist in the director field near the surfaces due to the surface walls. The light will adiabatically follow the twist and its plane of

(a)

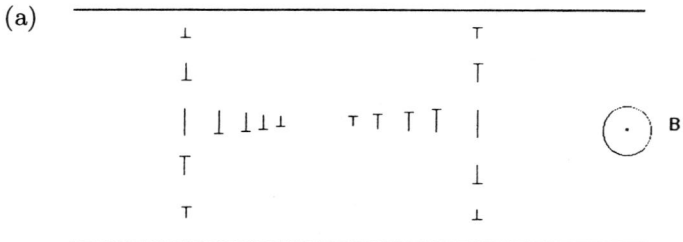

(b)

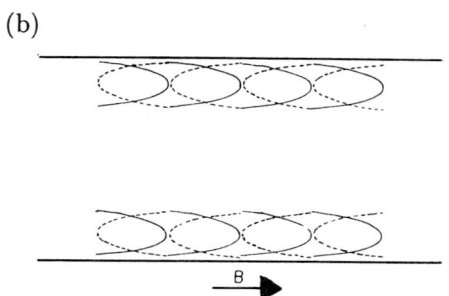

FIGURE 4 Two orthogonal views of the Ranganath defect using the "nail" view of the director to indicate that the director is tilted with respect to the plane of the paper. (a) Bloch (twist disclination) line between two Neel (bend-splay) walls. (b) The orientation of the bend-splay walls with the solid line indicating the structure in the foreground and the dashed line indicating structures in the background. A Bloch line separates the two domains.

(a)

(b)
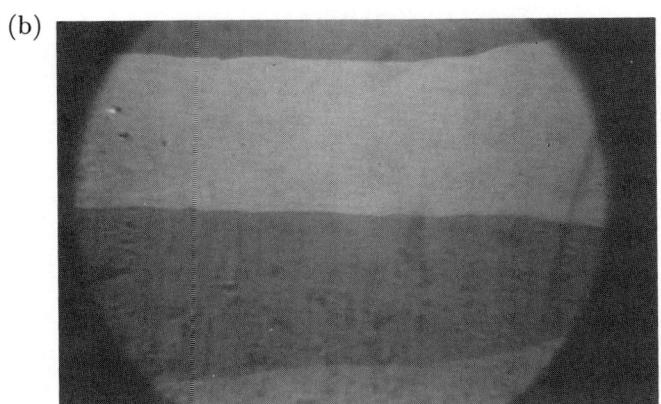

(c)

FIGURE 5
(a) Regions exhibiting transverse flow separated by lines of intensive scattering. The flow within each region is in opposite directions.
(b) Photograph taken after turning off the electric field while retaining the magnetic field. The polarizer was parallel to the magnetic field and the analyzer set 10 from the crossed position.
(c) Same as in (b) except the analyzer has been rotated to −10.

(a)

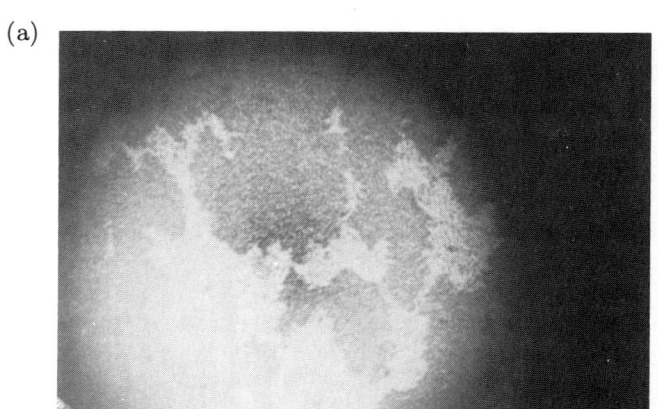

(b)

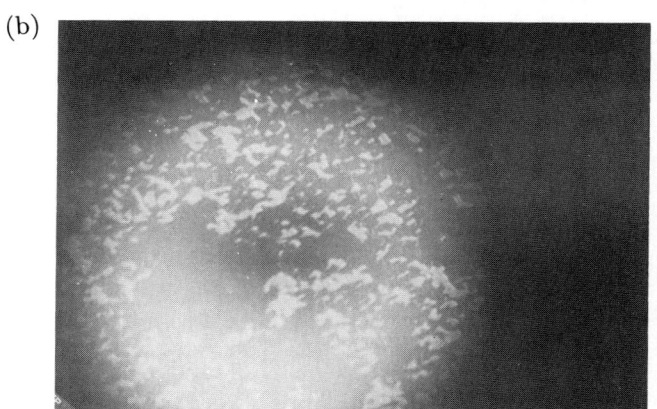

(c)

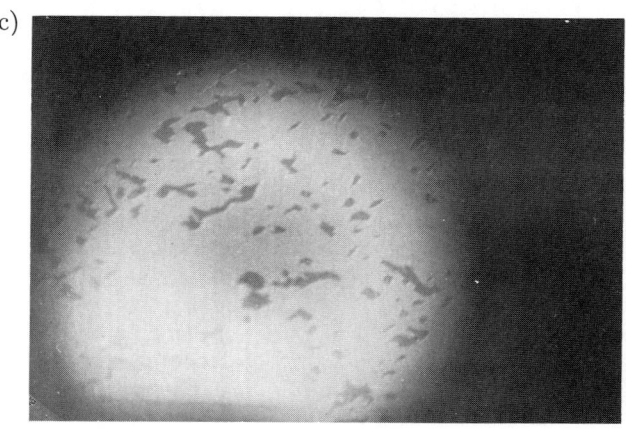

FIGURE 6 (a) Sample showing both DSM 1 and DSM 2 shortly after a voltage jump. (b) Photograph taken seconds after removing the electric field with same polarizer orientation as in Figure 5(b). (c) Effect of rotating analyzer in the opposite direction.

polarization will be rotated. The amount of rotation can be measured by rotating the analyzer away from its crossed position to extinguish the light.

In Figure 5, we show this effect when the sample is undergoing transverse flow. The dark center region of Figure 5(a) is moving to the right while the region above and below it (some what out of focus) are moving in the opposite direction. Removal of the electric field, when viewed along the surface normal with crossed polarizers, shows a homogeneous sample in which the different flow regions are separated by two disclination lines, one each at the top and bottom surface. In Figure 5(b), we show what happens when the sample is viewed at an angle of 10 degrees from the surface normal and the analyzer is rotated 10 degrees(ccw) to extinguish part of the sample. When the analyzer is rotated 10 degrees in the other direction or the direction of observation is changed to −10 degrees, the light and dark regions interchange as shown in Figure 5(c).

The results of a similar experiment on a sample showing both dynamic scattering modes is shown in Figure 6. The dark regions in Figure 6(a) corresponds to DSM 2 while the light regions correspond to DSM1. Removal of the electric field and rotation of the analyzer extinguishes part of the sample as shown in Figure 6(b) and (c). The region that originally showed only DSM 1 shows a uniform color indicating a homogeneous alignment of the surface walls. In contrast, the region that exhibited DSM 2 consists of patches of bright and dark regions indicating that the surface walls reverse at the boundary through Ranganath defects. The matchup is not perfect as the system continued to evolve during and between the photographs. These time varying irregular patches cause the surface walls on the two surfaces to be oriented in the same sense, either parallel or antiparallel to the magnetic field, or oriented in opposite senses. The former states result in net transverse forces while the latter results in no net force. The strong local shearing motion nucleates disclinations that are transported by the convective flow and that are responsible for the strong scattering.

In summary, we have shown that there are two high-voltage scattering modes in the presence of a strong magnetic field. These dynamic scattering modes are analogous to the modes seen in the absence of a field. They differ from the latter in that they are not turbulent sates and coexist with well-defined flow cells. The DSM 2 state is seen to be associated with inhomogeneous surface walls that interact with the flow cells to produce strong fluctuations in the director field. A more detailed paper involving more quantitative measurements is in preparation.

ACKNOWLEDGMENTS

J.P.M. would like to thank the sponsors of the Petroleum Research Fund of the American Chemical Society for partial support of this work. The authors also thank Geetha Seshaaiyar for her careful reading of this manuscript.

REFERENCES

1. Carr, E. F. "Influence of Electric Fields on the Molecular Alignment in the Liquid Crystal P-(Anisalamino)phenyl Acetate." *Mol. Cryst. Liq. Cryst.* **7** (1969): 253.
2. Carr, E. F. "Domains Due to Electric and Magnetic Fields in Bulk Sample sof Liquid Crystals." In *Liquid Crystals and Ordered Fluids III,* edited by J. F. Johnson and R. S. Porter, 165. New York: Plenum, 1978
3. Carr, E. F., and J. P. McClymer. "Transverse Flow in Electric Fields and Defects in a Nematic Liquid Crystal." *Mol. Cryst. Liq. Cryst.* **182B** (1990): 245.
4. Carr, E. F. "Transverse Flow in an Electric Field Using a Nematic LC Exhibiting Homeotropic Texture." *Mol. Cryst. Liq. Cryst. Lett.* **8** (1992): 117.
5. Chandrasekhar, S. *Liquid Crystals.* Cambridge: Cambridge University Press, 1992.
6. Elliot, G., and J. G. Gibson, "Domain Structure in Liquid Crystals, Induced by Electric Fields." *Nature* **205** (1965): 995.
7. de Gennes, P. G. *The Physics of Liquid Crystals.* Oxford: Clarendon Press, 1974.
8. Heilmeir, G. H., L. A. Zanoni and L. A. Barton. "Dynamic Scattering: A New Electrooptic Effect in Certain Classes of Nematic Liquid Crystals." *Proc. IEEE* **56** (1968): 1162.
9. Helfrich, W. "Conduction-Induced Alignment of Nematic Liquid Crystals: Basic Model and Stability Considerations." *J. Chem. Phys.* **51** (1969): 4092.
10. Igner, D., and J. H. Freed. "Transverse Viscous Forces in Carr Walls and Possible Dynamic Consequences." *Mol. Cryst. Liq. Cryst.* **101** (1983): 301.
11. Joets, A., and R. Ribotta. "Hydrodynamic Transitions to Chaos in the Convection of an Anisotropic Fluid." *J. Physique* **47** (1986): 595.
12. Kai, S., W. Zimmerman, M. Andoh, and N. Chizumi. "Local Transitions to Turbulence in Electrohydrodynamic Convection." *Phys. Rev. Lett.* **64** (1990): 1111.
13. Kapustin, A. P., and L. K. Vistin. "Ferroelectric Properties of Liquid Crystals." *Kristallografiya* **10** (1965): 118.
14. Orsay Liquid Crystal Group. "Quasielastic Rayleigh Scattering in Nematic Liquid Crystals." *Phys. Rev. Lett.* **22** (1969): 1361.
15. Ranganath, G. S. "Defect States in a Nematic Liquid Crystal in a Magnetic Field." *Mol. Cryst. Liq. Cryst.* **154** (1988): 43.
16. Sussmann, A. "Secondary Hydrodynamic Structures in Dynamic Scattering." *Appl. Phys. Lett.* **21** (1972): 269.
17. Tarr, C. E., and E. F. Carr. "NMR Evidence for an Ionic Conduction-Induced Flow Alignment in Nematic Liquid Crystals." *Solid State Comm.* **33** (1980): 459.

18. Vertogen, G., and W. H. de Jeu. *Thermotropic Liquid Crystals, Fundamentals.* New York: Springer-Verlag 1988
19. Williams, R. "Domain Walls in Liquid Crystals." *J. Chem. Phys.* **39** (1963): 384.
20. Yamazaki, H., S. Kai, and K. Hirakawa. "Two Kinds of Turbulent Flows in DSM of EHD Instabilities." *Mol. Cryst. Liq. Cryst.* **122** (1985): 41.

Chapter 5
Phase Separation in Complex Systems

Kyozi Kawasaki,† Tsuyoshi Koga,† Mikihito Takenaka,‡ and Takeji Hashimoto‡

†Department of Physics, Faculty of Science, Kyushu University 33, Fukuoka 812, Japan
‡Department of Polymer Chemistry, Kyoto University, Kyoto 606, Japan

Spinodal Decomposition in Binary Fluid Mixtures

This chapter summarizes our recent work on computer simulation of the extended Cahn-Hilliard equation for spinodal decomposition of a binary fluid mixture and on the detailed comparison with the experiment on a polymer blend. We found that not only the growth law but also the scaling function favor the fluid model rather than the solid model.

INTRODUCTION AND MODEL

The dynamics of ordering processes in fluid systems quenched from a disordered state to the two-phase coexistence region has been studied using various systems such as critical fluids and polymer blends.[12] Latest extensive experimental studies on spinodal decomposition in such fluids[1,4,13,14,21] provide us with quantitative details especially on the late stage dynamics.

On the other hand, analytical approaches to the problem are still on a qualitative or semiquantitative level due to the nonlinearity of this phenomenon.[6,7,8,20]

Recently, computer simulations using a discretized version of the time-dependent Ginzburg-Landau (TDGL) equation[15,18] have been extended to the studies of the dynamics of spinodal decomposition in fluids and quantitative results on the late stage dynamics have become available.[9,10] These new developments enable us to attempt quantitative comparisons between simulation and experimental data.

In this chapter, we present such detailed comparisons of the results on the late stage dynamics obtained by the computer simulation of the TDGL model with the hydrodynamic interaction[10,11] and the experiments of a critical mixture of polybutadiene (PB) and polyisoprene (PI).[4,21]

Here we consider binary fluids and assume that the fluid is incompressible. In order to describe the spinodal decomposition of binary fluids, we introduce a local order parameter $S(\mathbf{r})$, e.g., the local concentration, and the transverse component of the local velocity $\mathbf{v}(\mathbf{r})$. When the relaxation of the local velocity is more rapid, compared to that of the local order parameter, we can eliminate the local velocity by assuming that the local velocity instantaneously follows the changes in the local order parameter. As a result, we obtain the following TDGL-type equation for binary fluids[5] :

$$\frac{\partial}{\partial t}S(\mathbf{r}, t) = L\nabla^2\mu(\mathbf{r}) - \nabla S(\mathbf{r}) \cdot \int \mathbf{T}(\mathbf{r} - \mathbf{r}') \cdot \nabla' S(\mathbf{r}')\mu(\mathbf{r}')d\mathbf{r}', \qquad (1)$$

where $\mu(\mathbf{r}) \equiv \delta H\{S\}/\delta S(\mathbf{r})$ is the chemical potential, $H\{S\}$ the Ginzburg-Landau type free-energy functional, L the Onsager kinetic coefficient, and $\mathbf{T}(\mathbf{r})$ the Oseen tensor given by

$$\mathbf{T}(\mathbf{r}) = \frac{1}{8\pi\eta}\left(\frac{\mathbf{1}}{|\mathbf{r}|} + \frac{\mathbf{r}\mathbf{r}}{|\mathbf{r}|^3}\right). \qquad (2)$$

Here η is the shear viscosity and $\mathbf{1}$ the unit tensor. The second term on the right-hand side of Eq. (1) express the change of $S(\mathbf{r})$ due to convective flow. The latter is driven by the osmotic body force $\mathbf{f}(\mathbf{r}) = \mu(\mathbf{r})\nabla S(\mathbf{r})$. We have ignored the thermal noise in Eq. (1) to focus on the dynamics in the very late stage of spinodal decomposition. In order to elucidate the meaning of the unfamiliar osmotic body force, we have related it to the surface tension force of phase boundary in the appendix.

To integrate Eq. (1) numerically, we employ the cell-dynamic method.[15,18] Details of the numerical method have been reported elsewhere.[9,10] The system used is a three-dimensional cubic lattice of size 128^3 with periodic boundary conditions.

RESULTS

GROWTH LAW

In the computer simulation, we calculate the spherically averaged structure factor $I_k(t)$ of the order parameter. The normalization condition for $I_k(t)$ is $\int_0^\pi k^2 I_k(t)dk = 1$, which is the same as that used in analyzing the experiments.[4,21] We use the peak position of $I_k(t)$ denoted as $k_m(t)$ as the characteristic wave number. In the present study, we employ the following reduced variables for wave number and time which are the same as those used in the experimental works[4,21]: $K_m(t) \equiv k_m(t)/k_m(0)$ and $\tau \equiv tD_{app}k_m(0)^2$. Here $k_m(0)$ is a wave number at which the growth rate of the scattering intensity at the early stage is maximum and D_{app} is the collective diffusion constant determined in the early stage.

In Figure 1, we show the time dependence of $K_m(\tau)$ obtained by the simulation and the experiment on a double-logarithmic scale. From this figure, we find that $K_m(\tau)$ decreases linearly with time in the late time region in both the simulation and the experiment. This indicates that the domain growth in the late stage proceeds by the flow induced by the osmotic body force created by the inhomogeneous order parameter.[8,20]

On the other hand, we also find that there are some differences of the time dependence of $K_m(\tau)$ in the intermediate stage. They are probably due to the fact that the conditions used in the simulation, such as the strength of the hydrodynamic interaction, the initial state and the strength of the thermal noise, are not the same as those in the experiment. It is, however, important to note that we can expect that the quantities obtained in the late stage, e.g., the coefficient of the linear growth law of the characteristic length and the scaling function, are not expected to be affected by such effects, since these describe universal aspects of the phenomenon.

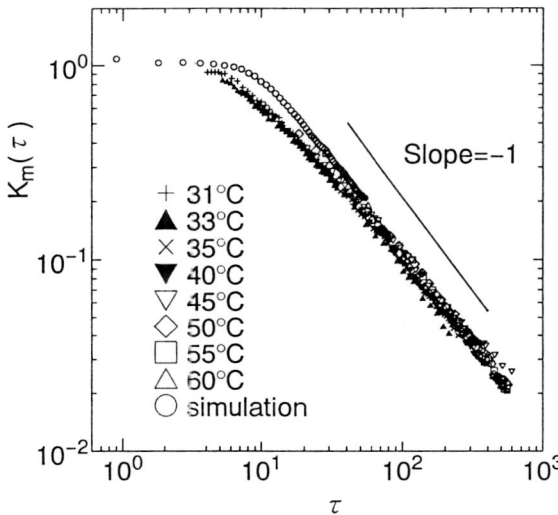

FIGURE 1 Comparison of $K_m(\tau)$ obtained from the computer simulation and the experiments.[11]

DYNAMICAL SCALING LAW

We study the scaling law for $I_k(t)$ in the hydrodynamic domain growth region. When the dynamical scaling law holds, $I_k(t)$ is scaled as

$$I_k(t) = k_m(t)^{-3}F(x), \tag{3}$$

where $F(x)$ is the scaling function and $x \equiv k/k_m(t)$ is the scaled wavenumber. The scaling plots for $I_k(t)$ obtained by the simulation and the experiments are presented in Figures 2 and 3, respectively. From these figures, we find that the scaling function for $x < 3$ obtained by the simulation is very similar to the experimental one. For large $x > 3$, the experimental scaling function obeys the Porod's law[16] x^{-4} at the x region covered in the experiment, while the simulation data do not satisfy the Porod's law and are still time dependent. This behavior at large x is due to the fact that the ratio of the thickness of interfaces to the characteristic length of domains is not small enough. When interfaces have a finite thickness, Porod's law is modified and the tail of the structure factor is given by[17]

$$I_k^{(t)}(t) = \frac{\Sigma(t)}{\pi\phi(1-\phi)}k^{-4}\exp(-\sigma_I^2 k^2), \tag{4}$$

where σ_I is a parameter related to the thickness of interfaces and superscript (t) on the lhs denotes the tail of $I_k(t)$. The method for estimating σ_I in the computer simulation and the experiment was presented by Koga et al.[11] and Hashimoto et al.,[4] respectively.

Using $x \equiv k/k_m(t)$, Eq.(4) is written as follows :

$$\hat{F}^{(t)}(x,t) \equiv k_m(t)^3 I_k^{(t)}(t)$$
$$= \frac{1}{\pi\phi(1-\phi)}\frac{\Sigma(t)}{k_m(t)}x^{-4}\exp(-\sigma_r(t)^2 x^2), \tag{5}$$

where $\sigma_r(t) \equiv \sigma_I k_m(t)$ is the relative interfacial thickness. $\hat{F}^{(t)}(x,t)$ satisfies Porod's law only for $\sigma_r(t)x \ll 1$. Since $\sigma_r(t)$ decreases with time as coarsening proceeds, the range of validity of Porod's law increases with time. Note that since the dimensionless ratio $\Sigma(t)/k_m(t)$ is independent of time in the late stage, that has been confirmed in the experiments[21] and in the simulation,[11] $\hat{F}^{(t)}(x,t)$ for $\sigma_r(t)x \ll 1$ is independent of time in the late stage.

The value of $\sigma_r(t)$ for the experimental data in Figure 3 is about 0.05, while $\sigma_r(t) \simeq 0.18$ for the simulation data in Figure 2, which is about 4 times larger than the experimental value in Figure 3. This is the reason why the simulation data deviates from Porod's law at the relatively small x values. In order to remove such effects, we use the hardened data obtained by the transformation[18]: $S(\mathbf{r}) \rightarrow S_e\text{sgn}(S(\mathbf{r}))$, where S_e is the equilibrium value of $|S|$. The hardened simulation data

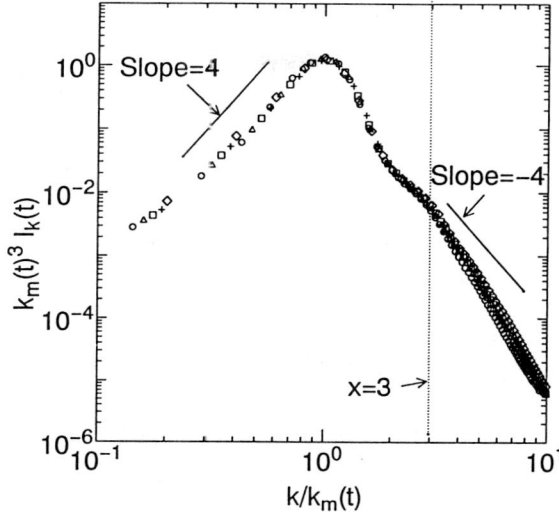

FIGURE 2 Scaled structure factor obtained from the computer simulation. The symbols $\bigcirc$, $\triangle$, $\square$, $+$, and $\diamond$ correspond to $\tau = 36.1, 40.6, 45.1, 49.6,$ and 54.1, respectively.[11]

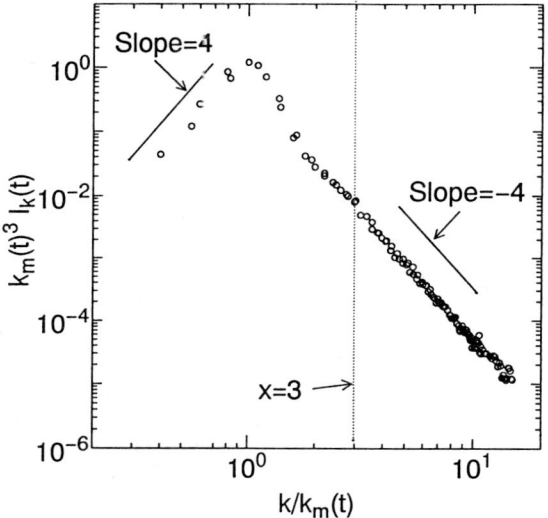

FIGURE 3 Scaled structure factor obtained from the experiment at $\tau = 223 - 289$.[11]

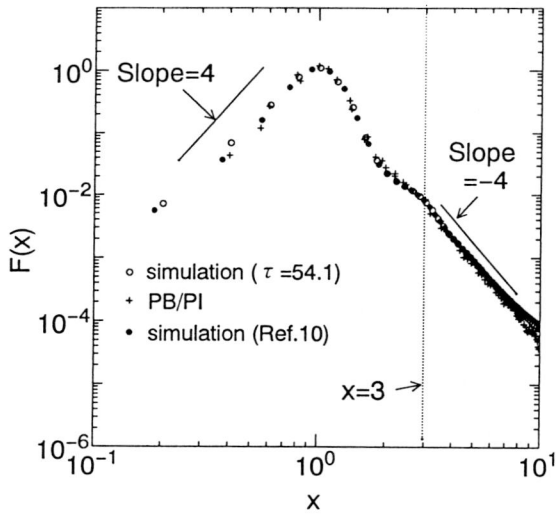

FIGURE 4 Comparison of the hardened scaling function obtained from the simulation ($\tau = 54.1$) with the experimental scaling function ($\tau = 223 - 289$).[11]

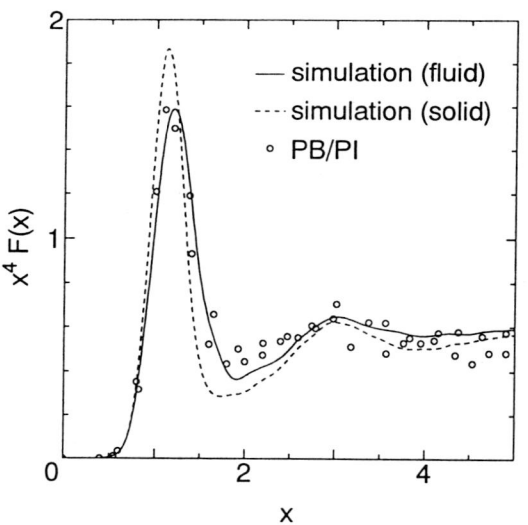

FIGURE 5 Porod plots of the hardened scaling function for fluid and solid and the experimental scaling function ($\tau = 223 - 289$).[11]

together with the experimental data are presented in Figure 4. From this figure, we see that the hardened data show Porod's law and are in good agreement with the experimental data for $x < 5$. The upward deviation of the hardened scaling function from Porod's law for $x > 5$ is due to the discreteness of the lattice. In this figure, we also present the scaling function obtained very recently by the similar simulation method using the different initial condition,[10] where the temporally linear domain growth law has been clearly observed. This scaling function also agrees with the experimental data and we do not find the difference between the scaling functions due to different initial conditions.

In Koga and Kawasaki's work,[10] the scaling function for the conserved system without the hydrodynamic interaction, which is called "solid," was also obtained. Porod plots of the scaling functions for fluid and solid are presented in Figure 5 together with the experimental results. In this figure, the scaling functions for fluid and solid are obtained by an interpolation using a spline function from the data given by Koga and Kawasaki.[10] We find from this figure that the experimental scaling function, which is obtained in the hydrodynamic growth region, is closer to the simulation results for fluid than that for solid.

CONCLUSION

In this chapter, we compared the results on the late stage dynamics of spinodal decomposition obtained from the computer simulation using the extended TDGL model and the experiments on polymer mixture of PB and PI in detail. We found that the domain growth in the late stage in both the simulation and the experiment obeys the well-known temporally linear growth law, which indicates that the hydrodynamic interaction dominates the domain growth. We compared the scaling functions obtained in such hydrodynamic domain growth region in detail and showed that the scaling functions for the two systems are in good agreement. We note that recently Shinozaki and Oono[19] performed a computer simulation similar to ours. Elsewhere[10] we also obtained the coefficient of the linear growth law of $k_m(t)^{-1}$ and found the excellent agreement with the experimental result for a small molecule binary fluid mixture.[2,3]

ACKNOWLEDGMENTS

The work was supported by Grant-in-Aid for Scientific Research on Priority Areas "Computational Physics as a New Frontier in Condensed Matter Research," from the Ministry of Education, Science and Culture, Japan as well as by the Joint Program of the National Institute for Fusion Science.

APPENDIX

When the chemical potential $\mu(\mathbf{r})$ takes the usual form $\mu(\mathbf{r}) = -K\nabla^2 S(\mathbf{r}) + h'(S(\mathbf{r}))$, the osmotic body force $\mathbf{f}(\mathbf{r})$ is written as

$$
\begin{aligned}
\mathbf{f}(\mathbf{r}) &= \mu(\mathbf{r})\nabla S(\mathbf{r}) \\
&= -K\nabla^2 S(\mathbf{r})\nabla S(\mathbf{r}) + h(S(\mathbf{r}))\nabla S(\mathbf{r}) \\
&= -K\nabla^2 S(\mathbf{r})\nabla S(\mathbf{r}) + \nabla h(S(\mathbf{r})).
\end{aligned} \tag{A1}
$$

The force acting on the unit area of interface f_σ is given by

$$
\begin{aligned}
f_\sigma &\equiv \int_{-\infty}^{\infty} \mathbf{f}(\mathbf{r}) \cdot \mathbf{n}\, dn \\
&= \kappa \left[K \int_{-\infty}^{\infty} (\partial_n S)^2 dn \right] + \int_{-\infty}^{\infty} dn \frac{\partial}{\partial n} \left[-\frac{K}{2}(\partial_n S)^2 + h \right] \\
&= \sigma\kappa,
\end{aligned} \tag{A2}
$$

where $\mathbf{n}$ and n are the unit normal to the interface and the coordinate along $\mathbf{n}$, respectively, and κ the mean curvature and we have used

$$
\nabla^2 S = \partial_n^2 S - \kappa\partial_n S. \tag{A3}
$$

Here

$$
\sigma \equiv K \int_{-\infty}^{\infty} (\partial_n S)^2 dn, \tag{A4}
$$

is the surface tension.

REFERENCES

1. Bates, F. S., and P. Wiltzius. "Spinodal Decomposition of a Symmetric Critical Mixture of Deuterated and Protonated Polymer." *J. Chem. Phys.* **91** (1989): 3258–3274.
2. Beysens, D., P. Guenoun, and F. Perrot. "Phase Separation of Critical Binary Fluids under Microgravity: Comparison with Matched-Density Conditions." *Phys. Rev. A* **38** (1988): 4173–4185.
3. Guenoun, P., R. Gastaud, F. Perrot, and D. Beysens. "Spinodal Decomposition Patterns in an Isodensity Critical Binary Fluid: Direct-Visualization and Light-Scattering Analyses." *Phys. Rev. A* **36** (1987): 4876–4890.

4. Hashimoto, T., M. Takenaka, and H. Jinnai. "Scattering Studies of Self-Assembling Processes of Polymer Blends in Spinodal Decomposition." *J. Appl. Cryst.* **24** (1991): 457–466.

5. Kawasaki, K. "New Method in Non-Equilibrium Statistical Mechanics of Cooperative Systems." In *Synergetics*, edited by H. Haken, 35–44. Stuttgart: Teubner, 1973.

6. Kawasaki, K. "Theory of Early Stage Spinodal Decomposition in Fluids Near the Critical Point I." *Prog. Theor. Phys.* **57** (1977): 826–839.

7. Kawasaki, K., and T. Ohta. "Theory of Early Stage Spinodal Decomposition in Fluids Near the Critical Point II." *Prog. Theor. Phys.* **59** (1978): 362–374.

8. Kawasaki, K., and T. Ohta. "Kinetics of Fluctuations for Systems Undergoing Phase Transitions Interfacial Approach." *Physica* **118A** (1983): 175–190.

9. Koga, T., and K. Kawasaki. "Spinodal Decomposition in Binary Fluids: Effects of Hydrodynamic Interactions." *Phys. Rev. A* **44** (1991): R817–R820.

10. Koga, T., and K. Kawasaki. "Late Stage Dynamics of Spinodal Decomposition in Binary Fluid Mixtures." *Physica* **196A** (1993): 389–415.

11. Koga, T., K. Kawasaki, M. Takenaka, and T. Hashimoto. "Late Stage Spinodal Decomposition in Binary Fluids: Comparison Between Computer Simulation and Experimental Results." *Physica* **198A** (1993): 473–492.

12. Komura, S., and H. Furukawa, eds. *Dynamics of Ordering Processes in Condensed Matter.* New York: Plenum, 1988.

13. Kubota, K., N. Kuwahara, H. Eda, and M. Sakazume, "Spinodal Decomposition in a Critical Isobutyric Acid and Water Mixture." *Phys. Rev. A* **45** (1992): R3377–R3379.

14. Kubota, K., N. Kuwahara, H. Eda, M. Sakazume. and K. Takiwaki. "Dynamic Scaling Behavior of Spinodal Decomposition in a Critical Mixture of 2,5-hexanediol and Benzene." Department of Biological and Chemical Engineering, Gunma University, 1992.

15. Oono, Y., and S. Puri. "Study of Phase-Separation Dynamics by use of Cell Dynamical Systems I: Modeling." *Phys. Rev. A* **38** (1988): 434–453.

16. Porod, G. "General Theory." In *Small Angle X-Ray Scattering*, edited by O. Glatter and O. Kratky, 17–51. New York: Academic, 1982.

17. Ruland, W. "Small-Angle Scattering of Two-Phase Systems : Determination and Significance of Systematic Deviations from Porod's Law." *J. Appl. Cryst.* **4** (1971): 70–73.

18. Shinozaki, A., and Y. Oono. "Asymptotic Form Factor for Spinodal Decomposition in Three-Space." *Phys. Rev. Lett.* **66** (1991): 173–176.

19. Shinozaki, A., and Y. Oono. "Spinodal Decomposition in 3-Space." *Phys. REv. E* **48** (1993): 2622–2654.

20. Siggia, E. D. "Late Stages of Spinodal Decomposition in Binary Mixtures." *Phys. Rev. A* **20** (1979): 595–605.

21. Takenaka, M., and T. Hashimoto. "Scattering Studies of Self-Assembling Processes of Polymer Blends in Spinodal Decomposition II: Temperature Dependence." *J. Chem. Phys.* **96** (1992): 6177–6190.

P. Palffy-Muhoray,† J. Y. Kim,‡ M. Mustafa,* and T. Kyu*
†Liquid Crystal Institute, Kent State University, Kent, OH 44242
‡Department of Physics, Kent State University, Kent, OH 44242
*Institute of Polymer Engineering, University of Akron, Akron, OH 44325

Polymerization-Driven Phase Separation in Polymer–Liquid Crystal Systems

We have used light scattering to study the kinetics of polymerization-induced phase separation in liquid crystal-polymer mixture. The evolution of the structure factor is compared with scaling predictions for thermally quenched systems. We have also observed a cascading phenomenon where phase-separated domains become unstable and undergo phase separation for a second time.

Polymer-dispersed liquid crystals (PDLC) materials,[5] consisting of liquid crystal droplets dispersed in a polymer matrix, have received considerable attention recently, both for reasons of fundamental scientific interest and because of their potential for display applications. They are formed by phase separation of the initially homogeneous mixture of the constituents; the phase separation may be due to thermal quench, solvent evaporation, or polymerization. A great deal is known about the kinetics of phase separation induced by a thermal quench in liquid mixtures, alloys, and polymer blends; however, little work has been done to date on the kinetics of phase separation due to polymerization.[11,12,16,17] The system under study consists of the commercial nematic liquid crystal mixture E7, the epoxy Epon 828, and the curing agent Capcure[1] 3-800. The molecular structures of Epon 828 and Capcure 3-800 are shown in Figure 1. E7 is a eutectic liquid crystal mixture of cyanobiphenyls and terphenyls, Epon is a bifunctional epoxide of bisphenol, and Capcure is a trifunctional mercaptan. We studied a ternary mixture consisting of 0.45 weight fraction E7, 0.275 weight fraction Epon, and 0.275 weight fraction Capcure. The components were combined and stirred to form a homogeneous mixture which was centrifuged to remove dissolved air. The sample was contained between parallel glass plates separated by a 30-μm mylar spacer. The sample temperature was held at $62 \pm 0.5°$C, $4°$C above the nematic-isotropic transition temperature of the pure liquid crystal. Static light scattering measurements were carried out using a He-Ne laser.

The scattered intensity $I(q, t)$ is shown in Figure 2 as a function of the scattering vector $q = (4\pi/\lambda)\sin(\theta/2)$, where $\lambda = 6328$Å and θ is the scattering angle. Although the polymerization process commences on mixing, no evidence of phase separation is seen for ~ 37 min. After this induction period, there is an abrupt increase in the intensity of scattered light. The maximum scattered intensity increases with time, while q_{max}, the magnitude scattering vector where the scattered intensity is a maximum, decreases. After ~ 42 minutes no further change in the angular distribution of scattered intensity profile can be detected.

The phase separation takes place at constant temperature, and is driven by polymerization instead of the usual thermal quench. As the condensation reaction proceeds, the average degree of polymerization increases, and eventually the system becomes unstable against phase separation. If the prepolymer and the curing agent in the PDLC-forming mixture are considered as one species with molecules having the average degree of polymerization, then the system may be regarded as a pseudo-binary mixture. The structure factor $S(q, t)$ is the Fourier transform of the spatial correlation function of the dielectric constant at time t, and is proportional to the intensity $I(q, t)$ of the scattered light.

In the theory of Cahn and Hilliard[2,3,4] for the early stages of spinodal decomposition (SD), the time dependence of the structure factor is given by $S(q, t) = S(q, 0)e^{2R(q)t}$, where

$$R(q) = Mq^2\left[-\left(\frac{\partial^2 f}{\partial \phi^2}\right) - 2\kappa q^2\right]. \tag{1}$$

Here $R(q)$ is the growth rate, M is the mobility, f is the free energy density, and ϕ is the concentration. In a polymerizing system, M and $\partial^2 f/\partial \phi^2$ are both expected to vary with the degree of polymerization, and time. Hence, unlike in thermally quenched binary mixtures, q_{max} is expected to change with time even in the early stage.

[1] from Miller-Stephenson Chemical Co.

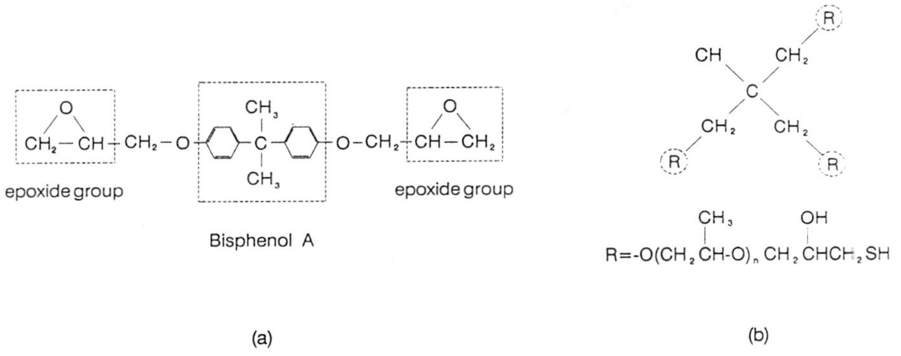

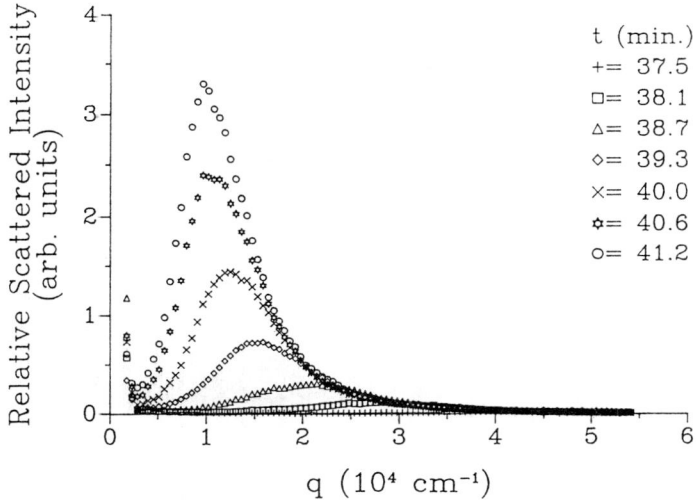

FIGURE 1 Molecular structure of Epon 828 (a) and Capcure 3-800 (b).

FIGURE 2 Scattered intensity $I(q,t)$ vs. the scattering vector q.

The liquid crystal–polymer system under study is a multicomponent mixture, whose components, in addition to the liquid crystal, are the n-mers whose distribution changes with time. A simple stability argument is as follows. An approximate expression for the free energy density of a multicomponent liquid mixture is,[10] in the Van der Waals approximation,

$$f = -\left(\sum_{i=0} \alpha_i \phi_i\right)^2 + kT \sum_{i=0} \frac{\phi_i}{v_i} \ln \frac{\phi_i}{v_i}, \tag{2}$$

where α_i is a coupling constant (independent of molecular volume) and ϕ_i and v_i are the volume fraction and the molecular volume, respectively, of species i. The stability of the system is determined by the eigenvalues of the Hessian,

$$\frac{\partial f}{\partial \phi_i \partial j} = -2(\alpha_0 - \alpha_i)(\alpha_0 - \alpha_j) + kT \left(\frac{1}{v_0 \phi_0} + \frac{d_{ij}}{v_i \phi_i}\right). \tag{3}$$

If $i = 0$ refers to the liquid crystal and, assuming that for the polymer $\alpha_i = \alpha_p$ for $i \geq 1$, the eigenvalues are the zeroes[9] of $f(\lambda) = (1 + \epsilon \sum 1/(\lambda - b_i))$, where $b_i = (kT/v_i \phi_i)\delta_{ij}$, δ_{ij} is the Kroenecker delta, and $\epsilon = 2(\alpha_0 - \alpha_p)^2 + (kT/v_0 \phi_0)$. The expression $f(\lambda)$ is shown in Figure 3; the smallest eigenvalue, $\lambda_{\min} = \lambda_1$, is

$$\lambda_{\min} \propto -2(\alpha_0 - \alpha_p)^2 + kT \left(\frac{1}{v_0 \phi_0}\right) + \left(\frac{1}{\sum v_i \phi_i}\right). \tag{4}$$

Since v_i is proportional to i, as the polymerization proceeds, $\sum v_i \phi_i$ increases with time, and $\lambda_{\min}$ decreases. When $\lambda_{\min} < 0$, the system becomes unstable and phase separation begins.

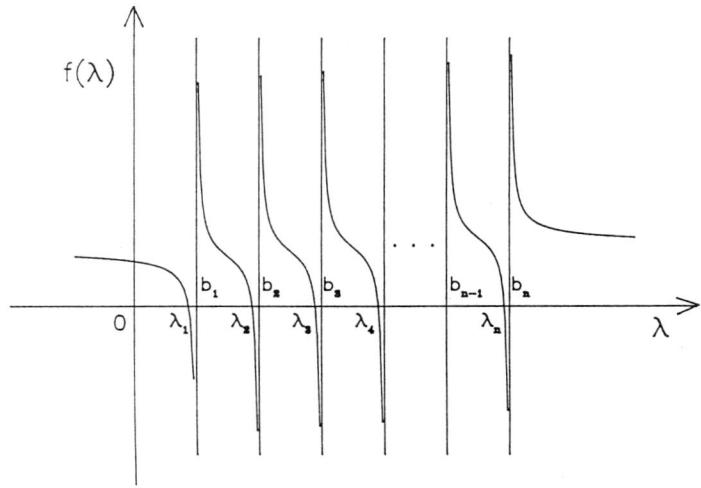

FIGURE 3 $f(\lambda) = 1 + \epsilon \sum(1/(\lambda - b_i))$; the zeros of $f(\lambda)$ are the eigenvalues of $\partial^2 f/\partial \phi_i \partial \phi_j$.

In the late stage of phase separation, the scaling ansatz[14] for the structure factor is

$$\tilde{S}(q,t) = q_{max}^{-3}(t)F(q/q_{max}). \tag{5}$$

Since $\tilde{S}(q,t) = I_{nor}(q,t)$, $F(q/q_{max}) = I_{nor}q_{max}^3$. In Figure 4 we show the normalized scaled structure factor $F(q/q_{max})$ at different times. Some deviation from the scaling prediction is observed near the wings and for the late time data; nonetheless, there is reasonable overall agreement. For the late stage of SD, Langer et al.,[13] Binder and Stauffer,[1] and Siggia[15] have considered, for thermally quenched binary mixtures, a power-law dependence of q_{max} on t,

$$q_{max} \sim t^{-\nu}. \tag{6}$$

Langer et al. predicted $\nu = 0.21$. Binder and Stauffer considered cluster dynamics with the result that $\nu = 1/3$. Siggia predicted that, for mixtures with the critical composition, the initial growth is diffusional, and $\nu = 1/3$, and that at long times the domain growth proceeds by coalescence due to surface tension and $\nu = 1$. In Figure 5 we show the dependence of q_{max} on the time $t - t_0$ where t_0 is the induction time. For small $t - t_0$ our results are consistent with an exponent of $-1/3$. However, in this regime, the exponent sensitively depends on the value of t_0, and thus more data is required for its unambiguous determination. At later times, the exponent is clearly -1.

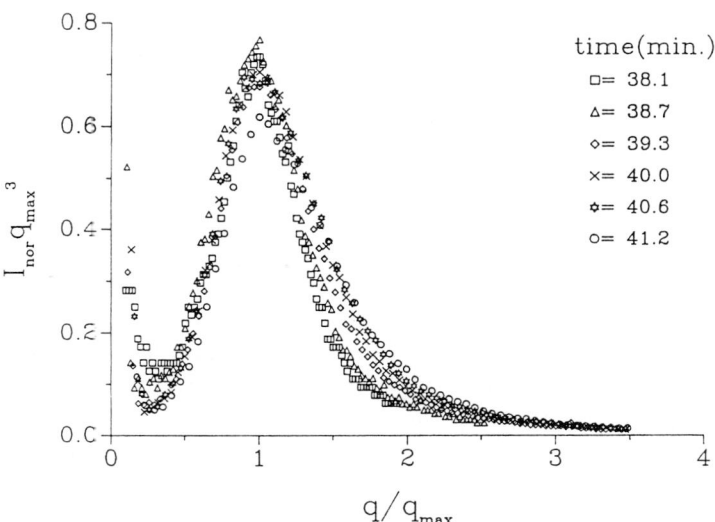

FIGURE 4 The scaled structure factor $F(q/q_{max}) = I_{nor}q_{max}^3$ vs. the scaled wave vector q/q_{max}.

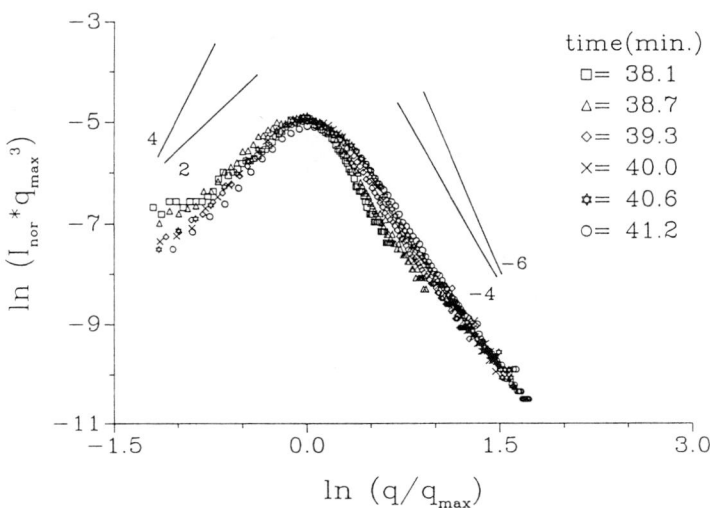

FIGURE 5 $\ln q_{max}$ vs. $\ln(t - t_0)$. The slope is $-\nu$.

Furukawa[6] proposed that the scaled structure factor during the late stage of SD in d dimensions has the form

$$F(x) = \frac{(1 + \gamma/2)x^2}{\gamma/2 + x^{2+\gamma}}, \tag{7}$$

where $x = q/q_{max}$, and $\gamma = d+1$ for an off-critical mixture and $\gamma = 2d$ for a critical mixture. For growth in three dimensions, $F(x) \sim x^2$ for $x < 1$, and $x > 1$, for off-critical quenches $F(x) \sim x^{-4}$, while for a critical mixture $F(x) \sim x^{-6}$. Furukawa subsequently also proposed[7,8] $F(x) \sim x^3/(2 + x^9)$ and $F(x) \sim x^4/(3/2 + x^{10})$ for the late stage of phase separation in three dimensions.

The exponent of the scaled scattering vector x for $x < 1$ in Figure 6 takes on values between 2 and 4, in agreement with the prediction of Furukawa. The exponent for $x > 1$ is -6 near $x \simeq 1$, and -4 for $x \gg 1$. The exponent of -6 is typical for a thermally quenched critical binary mixture. In our system it may be expected to arise if the system is predominantly unstable, rather than metastable, during the phase separation process.

Phase separation was also observed using a microscope while the sample was curing at 45°C. In Figure 7, two separated regions were observed during the early time of the phase separation. The domains grow both by direct coalescence, and by the Lifshitz-Slyozov evaporation-condensation mechanism. A striking phenomenon observed at a later time is that, as polymerization continues, the material in both liquid-crystal-rich and polymer-rich phase-separated domains again becomes unstable, and undergoes phase separation again, for a second time, as shown in Figure 7.

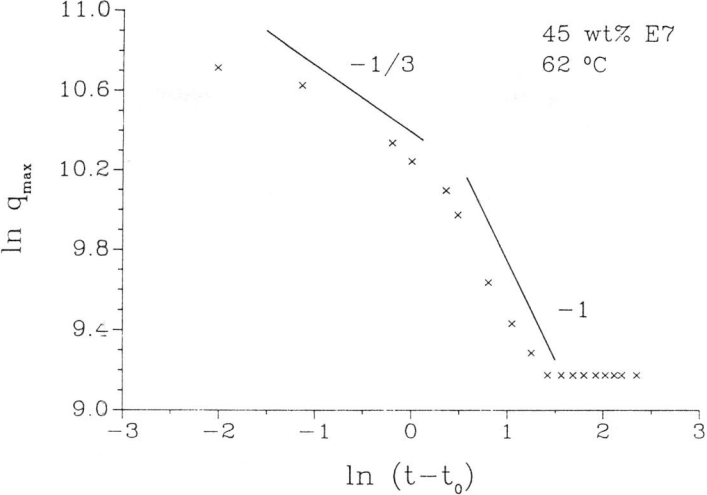

FIGURE 6 $\ln(I_{nor}q_{max}^3)$ vs. $\ln(q/q_{max})$.

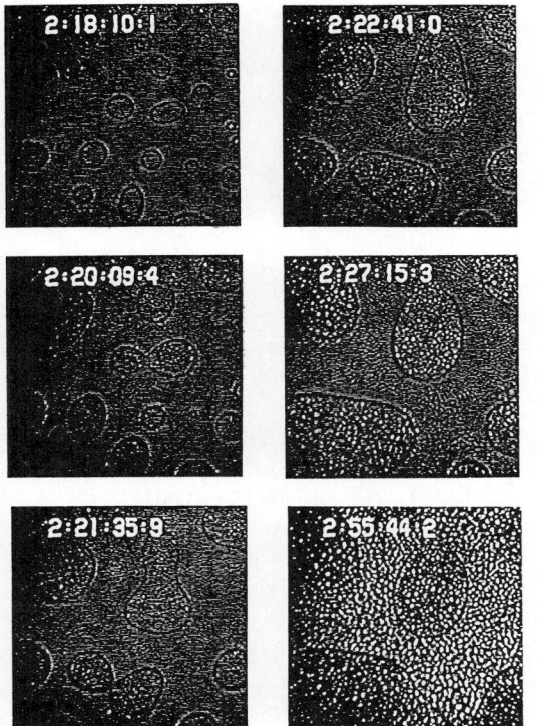

FIGURE 7 Micrographs of the cascade phenomena where phase separation occurs for a second time.

In summary, we have studied phase separation in a liquid crystal and epoxy mixture where the phase separation is induced by polymerization. The scaling relations for quenched binary systems appear to hold for our system. We have observed the crossover of q_{max} from $t^{-1/3}$ to t^{-1} as predicted by Siggia; the scaling behavior of the structure factor is in reasonable agreement with Furukawa's predictions. We have also observed a novel cascade phenomenon where phase separation takes place for a second time in already phase-separated domains.

ACKNOWLEDGMENTS

The authors acknowledge a support from the National Science Foundation under the Science and Technology Center grant no. DMR 89-20147. We thank R. Petscheck, E. C. Gartland, and L. Reichel for helpful discussions.

REFERENCES

1. Binder, K., and D. Stauffer. *Phys. Rev. Lett.* **33** (1974): 1006.
2. Cahn, J. W., and J. E. Hilliard. *J. Chem. Phys.* **28.** (1958): 258.
3. Cahn, J. W. *J. Chem. Phys.* **30** (1959): 1121.
4. Cahn, J. W., and J. E. Hilliard. *J. Chem. Phys.* **31** (1959): 258.
5. Doane, J. W., N. A. Vaz, B.-G. Wu, and S. Zumer. *Appl. Phys. Lett.* **48** (1986): 269.
6. Furukawa, H. *Physica* **123 A** (1984): 497.
7. Furukawa, H. *Phys. Rev. B* (Rapid Commun.) **33** (1986): 638.
8. Furukawa, H. *J. Phys. Soc. Jpn.* **58** (1989): 216.
9. Golub, G. H., and M. V. Van Loan. *Matrix Computations*, 2nd ed. Johns Hopkins University Press, 1989. (Thm. 8.6.2 in §8.6 "A Divide and Conquer Method").
10. Kehlen, H., M. T. Ratzsch, and J. Bergmann. *Z. Chem.* **26(1)** (1986): 1–6.
11. Kim, J. Y., and P. Palffy-Muhoray. *Mol. Cryst. Liq. Cryst.* **203** (1991): 93.
12. Kim, Do Hyun, and Sung Chul Kim. *Polymer Eng. & Sci.* **31** (1991): 289.
13. Langer, J. S., M. Bar-on and D. Miller. *Phys. Rev. A* **11** (1975): 1417.
14. Marro, J., J. L. Lebowitz, and M. H. Kalos. *Phys. Rev. Lett.* **43** (1979): 282.
15. Siggia, E. *Phys. Rev.* **20 A** (1979): 595.
16. Smith, G. W., and N. A. Vaz. *Liq. Cryst.* **3(5)** (1988): 543.
17. Visconti, S., and R. H. Marchessault. *Macromolecules* **7** (1974): 913.

Hiroshi Furukawa
Faculty of Education, Yamaguchi University, Yamaguchi 753, Japan

Pattern Formation in Ordering Dynamics Under Inhomogeneous External Conditions

Morphologies of the ordering domain in the first-order phase transition are investigated under inhomogeneous external conditions. Domain morphology found by the homogeneous quenching is isotropic, i.e., irregular. When the quenching is inhomogeneous, domain morphologies are more or less regular. However, the regular morphology is not a simple reflection of the inhomogeneous external condition, but the morphology may change discontinuously. Several example are shown numerically using solid and fluid models. The present study is significant from not only the fundamental physics but also the technology to make ordered materials efficiently.

1. INTRODUCTION

When a binary solid or fluid mixture is suddenly and homogeneously quenched from a one-phase state to an unstable state, the phase separation takes place. Initially small and irregular precipitates appear in homogeneous single phase. Precipitates soon start to coarsen. In almost all cases coarsening obeys simple scaling

(a)   (b)

(c)   (d)

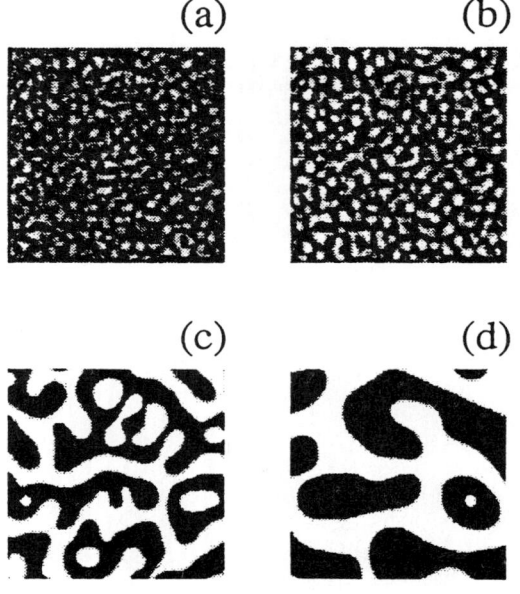

FIGURE 1   (a,b) Early and (c,d) late time patterns of the two-dimensional phase-separating fluid mixture given by numerical simulation. The quench is done at critical composition. The (aging) time of (a) is half that of (b), and the time of (c) is half that of (d).

law (Figure 1) where the pattern in the phase-separating mixture becomes statistically stationary under a suitable space-time rescaling (for reviews see Gunton et al.,[8] Furukawa,[4] and Komura and Furukawa[10]). The phase separation is a typical example of the ordering dynamics. In nature, however, such orderings may be inhomogeneous, because there are temporal delays in coarsening from position to position. The inhomogeneous ordering may yield morphologies completely different from that of homogeneous quenching. In this paper we shall discuss morphological transitions in ordering dynamics using numerical simulations. In Section 2 two models are presented. In Section 3 numerical result for nonhydrodynamic models are presented. In Section 4 numerical result for hydrodynamic model is presented. In Section 5 short remarks are presented.

2. MODELS

1.1 CONSERVED CASE

The equation of motion is described by the set of following equation (see, for instance, Farrel and Valls[1,2,3]):

$$\frac{d\psi}{dt} = -\vec{\nabla} \cdot \vec{j}, \qquad \frac{d\vec{v}}{dt} = \nu\nabla^2\vec{v} + \psi\vec{\nabla}\mu, \qquad (1)$$

$$\vec{j} \equiv M\vec{\nabla}\mu + \gamma\vec{v}\psi, \qquad \mu \equiv K\nabla^2\psi + \mu_0, \qquad (2)$$

where ν, γ, K, and M are constants, ψ is the order parameter, $\vec{j}$ is the current where in Eq. (2) the first term on the right-hand side is due to the diffusion and the second term is due to the drift, and μ and μ_0 are the nonlocal and local chemical potentials. The first term on the right-hand side of Eq. (1) is the dissipation term of the Navier-Stokes equation, the second term is the force term, and the left-hand side is the inertial term. The nonlinear term of the velocity $\vec{v}$, which is also the inertial term, is neglected. The set of equations (1) and (2) are solved on the square lattice by a simple discretization. The lattice spacing is chosen to be unity. The partial set (1) and (2) with $\gamma = 0$ also describes the diffusional motion of solid mixture (alloy).

1.2 NONCONSERVED CASE

The second model is a lattice gas model with a simple cubic or square lattice with a nonconserved order parameter.[6,7] The attribute of each atom is represented by a spin variable σ with infinite number of spin state $\sigma = 1, 2, 3, \ldots, \infty$. The interaction between atoms is simplified so that the Hamiltonian H may be given by that of ferromagnetic Potts model:

$$H = -\frac{J}{2} \sum_{i,j} \delta_{\sigma_i, \sigma_j}, \qquad \sigma = 1, 2, 3, \ldots, \infty, \tag{3}$$

where J (> 0) is the coupling constant, and the summation is taken among neighboring sites. We take the interaction up to the second nearest-neighbor sites for square lattice and upto the third nearest-neighbor sites for the simple cubic lattice. The thermal fluctuation is taken only as the initial random configuration. Monte Carlo samplings are done under the condition that the state of any spin is changeable only into the same spin state of one of neighboring interacting spins. If all neighboring spins including central spin are the same, further Monte Carlo samplings are skipped. In this dynamical scheme the infinite number of spin states is not an obstacle to the numerical simulation.

Two examples of the boundary condition with a stable interface between unstable and stable phases are shown in Figure 2. The interface between unstable and stable phase is assumed to move as a function of time t as $T(z, t) = f(z - ut)$ with $u > 0$. Here T represents a state parameter like temperature. It is assumed that the thermal flow is not responsible for the interfacial instability. For instance, when the molten metal in a crucible solidifies, the heat flows from inner to outer. In this case no interfacial instability is possible. Atoms or molecules near the interfaces are updated by shearing or otherwise by their own rapid motions in disordered phases.

Let the z-axis be perpendicular to the quench boundary, and let the position of the quench boundary be $z = z_b$. Inside the quench boundary ($z \leq z_b$) ψ or σ evolves

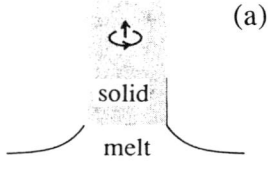

(a)

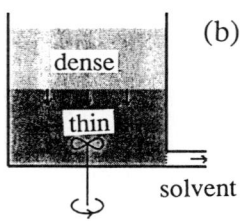

(b)

solvent

FIGURE 2   Examples of inhomogeneous boundary condition with stable interfaces. (a) Nonconserved case. A piece of material is being drawn-up in contact with molten materials and the drawn-up part is solidifying by the cooling. This method is widely used to make a large perfect crystal. (b) A suggested experiment for polymer or polymer blend solutions. The ratio of the solvent is decreasing, by evaporation, by extraction, or by other method. Then under the gravitational field there occurs the gradient in the density of the solvent. This causes the inhomogeneous boundary condition.

according to the equation shown above, whereas outside the quench boundary ($z > z_b$) ψ or σ is set random. The location of the quench boundary is shifted once in each set of n iterations. Therefore the shift velocity of the quench boundary is

$$u = 1/n \qquad \text{(per iteration)} \tag{4}$$

The inhomogeneous quench starts from the position near $z = 0$, say, at $z = z_0 + 1$. Here z_0 indicates a length about a few lattice spacing.

2. NUMERICAL SIMULATIONS (NONHYDRODYNAMIC CASE)

2.1 CONSERVED ORDER PARAMETER[5]

Let us first consider the nonhydrodynamic situation ($\gamma = 0$). For the numerical simulation of inhomogeneous quench with conserved order parameter, we use the chemical potential[11,12,13]:

$$\mu_0 \equiv \frac{\tanh \psi}{\tanh 1} - 1. \tag{5}$$

We have chosen

$$M\Delta t = 1, \qquad K = \frac{1}{2}. \tag{6}$$

In the present boundary condition the order parameter is not conserved at the quench boundary. In an actual experiment, such a violation of the conservation law corresponds to the rapid change in the order parameter outside the quench boundary. In the case of the asymmetric order parameter the simulation starts with the symmetric order parameter $\psi_0 = 0$, with the same fluctuation strength $\delta\psi_0$, and then the ψ_0 is switched to a nonzero value $\psi_0 \neq 0$. Three- (Figure 3) and two-dimensional numerical simulations exhibit similar morphological changes against the quench boundary velocity u. In three dimensions, columnar domains are percolated in two-dimensional plane parallel to the quench boundary when the quench is deep; i.e., the composition is near 0.5.

In the asymmetric order parameter case ($\langle\psi\rangle \neq 0$) the morphological change against the quench boundary velocity u is somehow unexpected (Figure 4). Before quenching the system had positive average density $\langle\psi\rangle > 0$, but after the quench the system may have a negative average density $\langle\psi\rangle < 0$. This change in the average density before and after the quench is due to the violation of the conservation of order parameter at the quench boundary. However, the reversal of the average density cannot readily be understood. A plausible explanation of this density reversal is due to the nucleationlike process in an unstable phase. At the quench boundary columnar domains contact with unstable phase, which has a small positive

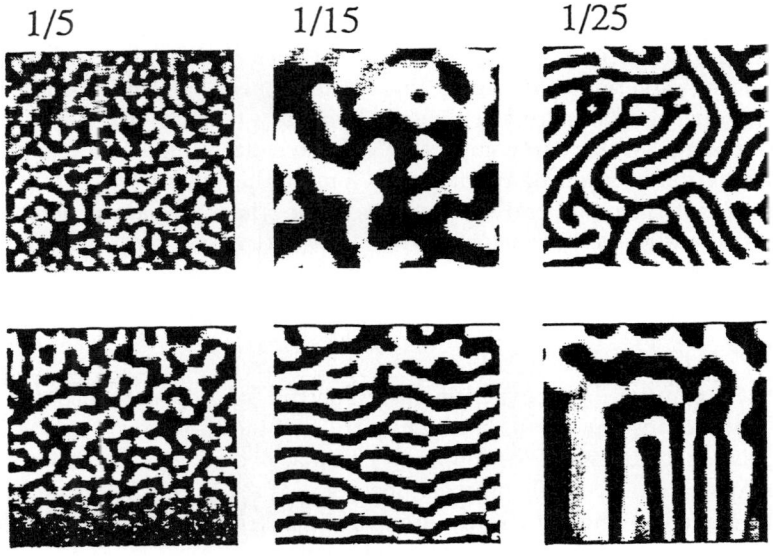

FIGURE 3 Domain morphology for three-dimensional system with various quench boundary velocities $u = 1/5, 1/15, 1/25$. The system size is $80 \times 80 \times 80$. For each value of u the upper figure represents the sectional view perpendicular to the quench boundary, at the 1/4 position from the bottom (quench boundary). The lower figure represents the side view. Initially the order parameter is randomly distributed between $-0.05 \leq \psi \leq 0.05$.

density. In such a phase a droplet with a negative density is unstable and tends to grow larger. In this way the columnar domain with a negative density tends to grow in contact with unstable phase with small positive density. This type of instability is stronger in the case of droplet with a negative (positive) density in the unstable media with positive (negative) density, than in the case of droplet with positive (negative) density in the unstable media with positive (negative) density. This must be the origin of the density reversal before and after the quench and the process is the same as the nucleation in the bulk system.

2.2 NONCONSERVED ORDER PARAMETER[6,7]

We observed several grain morphologies. When the quench starts from a single-ordered surface (Figure 5(a)), we find three typical grain patterns. When the shift velocity is large, grains are compact except near the initial surface: The large single-ordered phase extends slowly near the initial surface. For small shift velocities there is no grain any more but the inside of the quench boundary is occupied by a single-ordered phase. The appearance of the third grain morphology depends on the initial condition. If the surface of the single crystal is covered by a thin disordered phase initially, then for intermediate velocities grains become longer in the extending direction of ordering phase, except near the single-ordered surface. When the quench starts from a random boundary (Figure 5(b)), no large single grain is created. Grains are compact or columnar according to the shift velocity. This observation of grain morphologies is almost the same both in two and three dimensions, but the quantitative properties of the transitions are somehow different between two and three dimensions.

For the hydrodynamic conserved case we have chosen

$$\Delta t = 0.5 \qquad (M = 1), \tag{7}$$

instead of 1.0 in nonhydrodynamic case. For the sake of simplicity we have set $\nu = 1$, and various values are tried for γ to examine the hydrodynamic effect on the morphological change. The following boundary conditions for the velocity field $\vec{v}$ is set. At $z = 0$, i.e., at the starting position, we set $\vec{v} = 0$. The order parameter outside the quench boundary is frozen in the initial random value, and the velocity field is set zero before the quench boundary comes.

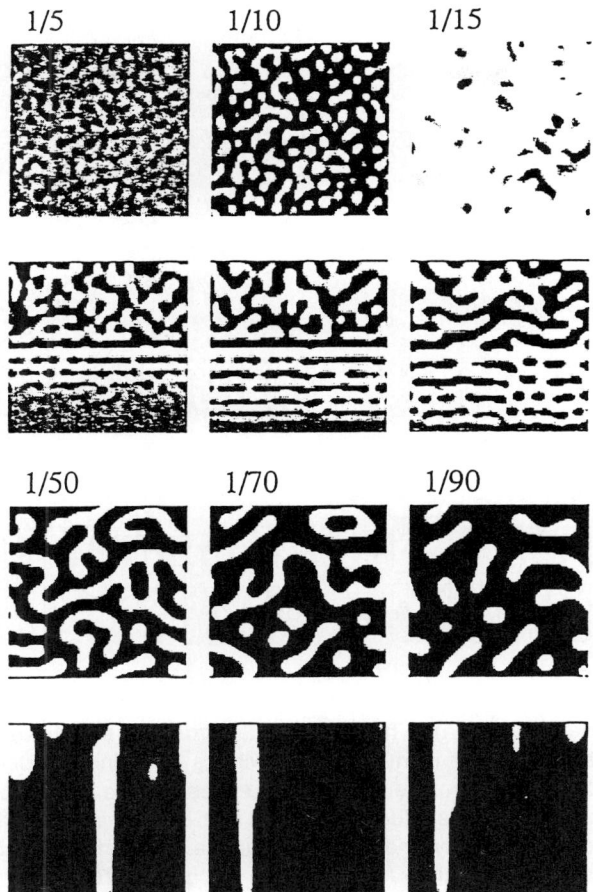

FIGURE 4 Domain morphology for three-dimensional system with various quench boundary velocities $u = 1/5 - 1/90$. The system size is $80 \times 80 \times 80$. For each value of u the upper figure represents the sectional view perpendicular to the quench boundary, at the 1/4 position from the bottom (quench boundary). The lower figure represents the side view. Before the quench boundary reaches the middle of the vertical system size ($z < 40$), order parameter is randomly distributed between $-0.1 \leq \delta\psi \leq 0.1$. After the quench boundary reaches, the order parameter is randomly distributed between $0.1 \leq \delta\psi \leq 0.3$. White and black correspond to the positive and negative order parameters.

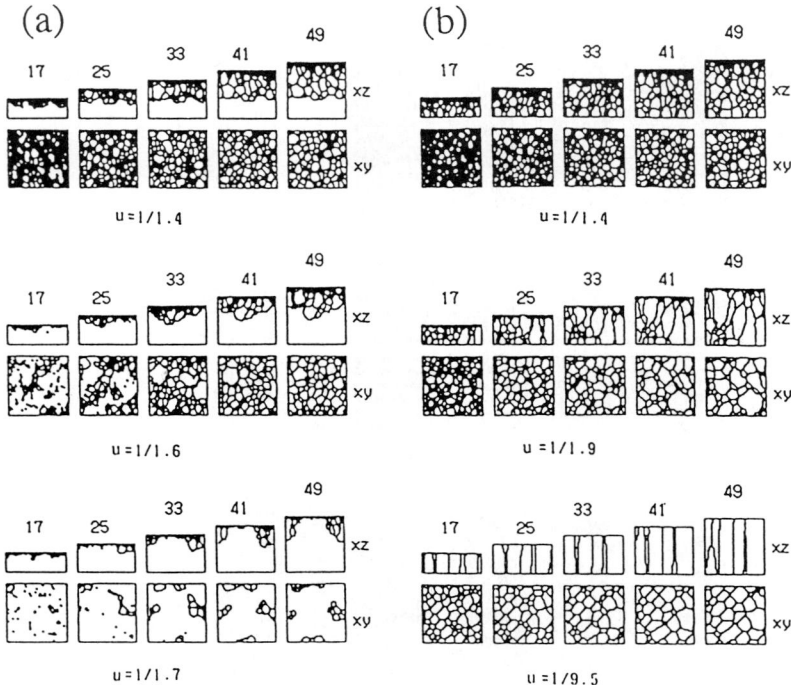

FIGURE 5 Grain morphologies (a) in the case of start from the single-ordered phase and (b) in case of starting from random boundary, with $z_0 = 0$. In each set of figures upper ones indicate the side views. Here the upper edges are the locations of quench boundary which is shifting upwards. Numbers shown with figures indicate the distance (L_z) between the quench boundary and the initial surface. Lower figures indicate sections of the system at the distance $(1/3)L_z$ from the quench boundary. The variable u indicates the shift velocity of the quench boundary (lattice space/MCS). The system size is $50 \times 50 \times L_z$ $(L_z < 50)$.

Here we see the morphological change according to the fluidity parameter γ. The hydrodynamic effect on the domain morphology depends on the dimensionality. For the inhomogeneous quenching of the two-dimensional hydrodynamic system, the lamella-type domain becomes unstable as the fluidity increases (Figure 6(a)). On the other hand the hydrodynamic effect on the column type domain is weak (Figure 6(b)). The disappearance of the lamella-type domains by the hydrodynamic motion is not a linear effect, because there is no linear hydrodynamic term in the equation of motion (1) and (2). In three dimensions, the domain morphology corresponding to the column-type also disappear as the fluidity increases (Figure 7).

This is because columnar domains percolate and therefore interfaces are curved. The hydrodynamic motion quickly disturb such domains.

3. REMARKS

By the inhomogeneous quenching ordered materials are generated easily and efficiently. For instance, in the case of nonconserved order parameter the average grain size grows in proportion to $t^{1/2}$. But the volume of the single-ordered phase with inhomogeneous quenching is proportional to t. This is an important characteristics to the technology of producing an ordered material. Besides the technological importance, the inhomogeneous ordering is interesting problem from the fundamental nonequilibrium statistical mechanics.

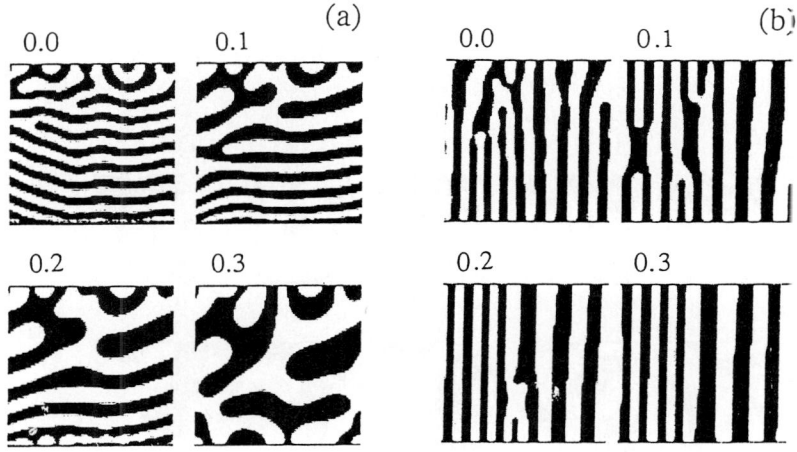

FIGURE 6   Domain pattern in two-dimensional inhomogeneous quench in hydrodynamic system. (a) Quench boundary velocity $u = 1/20$, (b) $u = 1/50$. Numbers shown with figures indicate values of γ, which measures the fluidity. Other numerical parameters are $\nu = M = a = g = 1$, $K = 1/2$, and $\Delta t = 0.5$. The initial conditions are $\vec{v} = \vec{0}$, $|\psi| \leq 0.05$. The lower edge of each figure corresponds to the quench boundary; these boundaries are shifting downwards. The system size is 80×80.

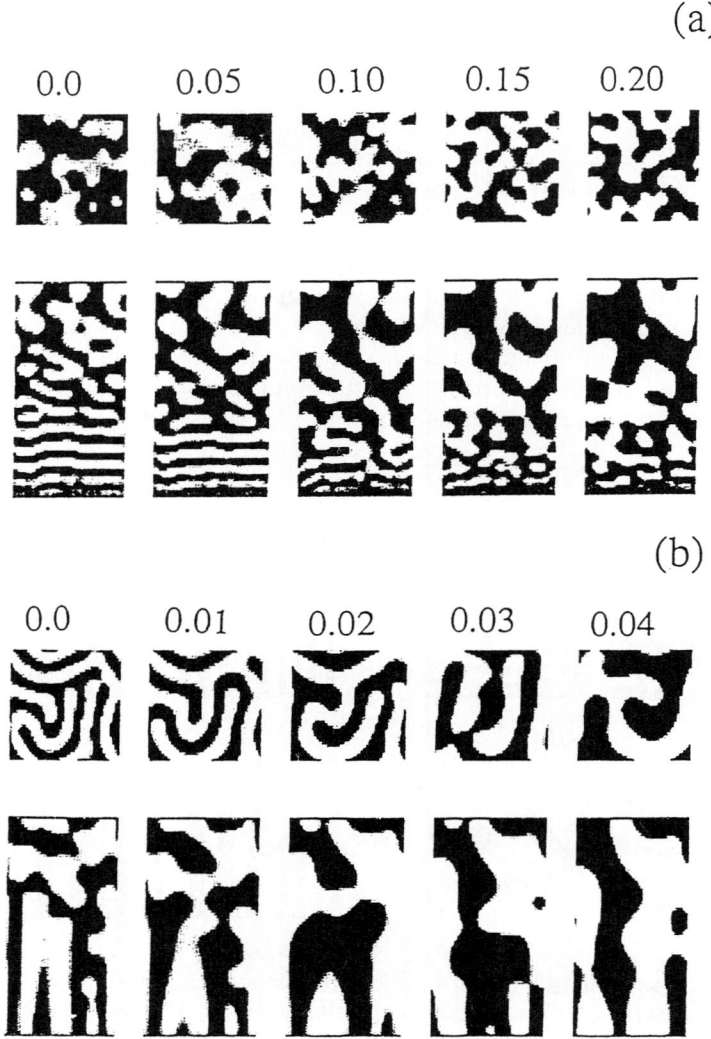

FIGURE 7 Domain pattern in three-dimensional inhomogeneous quench in hydrodynamic system. (a) Quench boundary velocity $u = 1/20$, (b) $u = 1/50$. Numbers shown with figures indicate values of γ, which measures the fluidity. Notice that values of γ in (b) are very small. Other numerical parameters are the same as in Figure 6. The initial conditions are $\vec{v} = \vec{0}$, $|\psi| \leq 0.1$. Upper figures are sectional views at 1/4 position from the quench boundary (lower edges of lower figures). Quench boundaries are shifting downwards. The system size is $40 \times 40 \times 80$.

REFERENCES

1. Farrel, J. E., and O. T. Valls. "Spinodal Decomposition in a Two-Dimensional Fluid Model." *Phys. Rev.* **B 40** (1989): 7027–7039.
2. Farrel, J. E., and O. T. Valls. "Spinodal Decomposition in a Two-Dimensional Fluid Model: Heat, Sound, and Universality." *Phys. Rev.* **B 42** (1990): 2353–2362.
3. Farrel, J. E., and O. T. Valls. "Growth Kinetics and Domain Morphology After Off-Critical Quenches in a Two-Dimensional Fluid Model." *Phys. Rev.* **B 43** (1991): 630–640.
4. Furukawa, H. "Dynamical Scaling for Phase Separation." *Adv. Phys.* **34** (1984): 703–750.
5. Furukawa, H. "Phase Separation by Directional Quenching and Morphological Transition." *Physica* **A 180** (1992): 128–155.
6. Furukawa, H. "Ordering Dynamics with Moving Plane Boundary Between Ordered and Disordered Phases." *Prog. Theor. Phys.* **87** (1992): 871–878.
7. Furukawa, H. "Grain Growth by Inhomogeneous Quenching: Numerical Simulation in Three Dimensions." *Prog. Theor. Phys.* **88** (1992): 857–863.
8. Gunton, J. D., M. San Miguel, and P. S. Sahni. "The Dynamics of First-Order Phase Transition." In *Phase Transitions and Critical Phenomena,* edited by C. Domb and J. L. Lebowitz, vol. 8, 267–460. London: Academic, 1983.
9. Kawasaki, K. "Kinetic Equations and Time Correlation Functions of Critical Fluctuations." *Ann. Phys.* **61** (1970): 1-56.
10. Komura, S., and H. Furukawa, eds. *Dynamics of Ordering Processes in Condensed Matter.* New York: Plenum, 1988.
11. Oono, Y., and S. Puri. "Computationally Efficient Modeling of Ordering of Quenched Phase." *Phys. Rev. Lett.* **58** (1987): 836–839.
12. Oono, Y., and S. Puri. "Study of Phase-Separation Dynamics by Use of Cell Dynamical Systems. I. Modeling." *Phys. Rev.* **A 38** (1988): 434–453.
13. Puri, S., and Y. Oono. "Study of Phase-Separation Dynamics by Use of Cell Dynamical Systems. II. Two-Dimensional Demonstrations." *Phys. Rev.* **A 38** (1988): 1542–1565.

Karl Amundson

AT&T Bell Laboratories, 600 Mountain Ave., Murray Hill, NJ 07974

Alignment of Lamellar Block Copolymer Microstructure in an Electric Field

1. INTRODUCTION

Block copolymers are polymer molecules made up of chemically distinct blocks of monomers. Among them, the diblock copolymer exhibits the simplest architecture; each polymer molecule consists of two blocks of monomers (Figure 1(a)). For many diblock copolymers, at high temperature, the two halves of the polymer chain intermingle and the material behaves much like a homopolymer. Upon cooling (or perhaps solvent removal in the case of a solvated block copolymer), the material undergoes a weakly first-order phase transition to an *ordered*, microphase separated state.[6,9,16] In the ordered state, the two halves of the polymer chain do their best to get away from each other while maintaining their covalent bond; that is, they microphase separate. The ordered state is characterized by a quasi-static, composition pattern of molecular scale period (Figure 1(b)). Several ordered states distinguished by their symmetries have been identified; the selection depends primarily upon the polymer composition.[6,9,16] Some common patterns are lamellar sheets, hexagonally packed cylindrical arrays, and bcc spherical microstructures. In this report, we are interested in the lamellar microstructure.

Spatio-Temporal Patterns, Ed. P. E. Cladis and P. Palffy-Muhoray,
SFI Studies in the Sciences of Complexity, Addison-Wesley, 1995   **411**

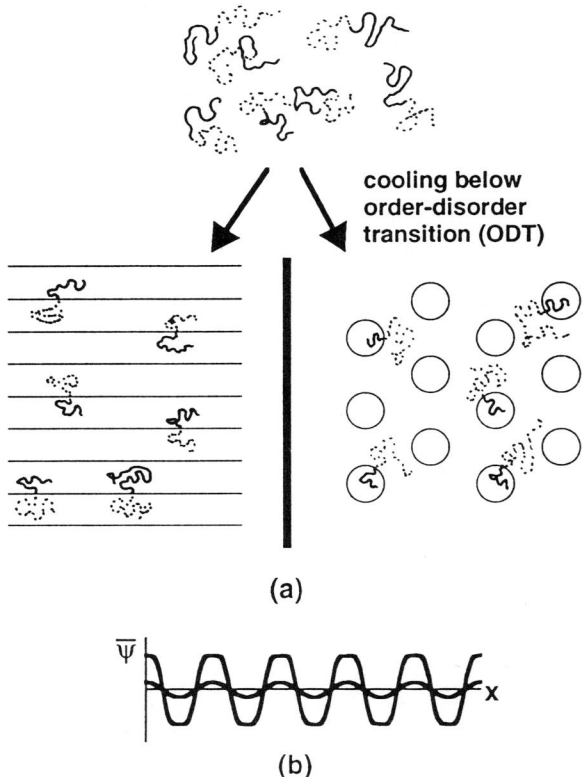

(a)

(b)

FIGURE 1   (a) Diblock copolymer material in the disordered phase (above) and in ordered phases (below). Chemically distinct portions of the polymer chain are denoted by the solid and dotted lines. The quasi-static composition pattern is suggested by the periodic shaded patterns in the ordered states. In (b) the quasi-static composition pattern is shown for the lamellar ordered state as a function of a position variable. The low-amplitude sine wave suggests the composition pattern near the order-disorder transition, and the larger-amplitude function represents the pattern further away from the transition.

The long-range order of the ordered state is interrupted by defects, much like a polycrystalline or liquid crystalline material. However, global alignment of block copolymer microstructure can be achieved by deformation or flow.[11,13,15,17,18] Various mechanisms have been proposed to explain the alignment process.[11,13,15,17,18] Recently, we have studied the alignment of lamellar block copolymer microstructure in the relatively weak aligning force induced by an electric field.[1,4,5] Although the alignment force provided by an electric field is relatively weak, it is educational to study the alignment process in the simple electric-field-induced body force. Also, interesting morphology results from the interplay between field-induced forces and inter defect interactions. Electric field alignment may have special applications as well, since various aligned states can be achieved that would be impossible to achieve by flow or deformation.

A block copolymer with a lamellar (or cylindrical) microstructure is an example of an anisotropic fluid, much like many liquid crystal phases. In an electric field, the electrostatic contribution to the free energy density of a block copolymer with lamellar microstructure contains an anisotropic contribution[4]

$$f - f_0 = \frac{1}{8\pi} \frac{\beta^2}{\varepsilon_D} |\mathbf{E}_0|^2 \langle \overline{\psi}^2 \rangle \, (\hat{\mathbf{e}}_q \cdot \hat{\mathbf{e}}_Z)^2 \tag{1}$$

which is the driving force for alignment. { is the free energy density and f_0 the free energy density in the absence of an electric field. ε_D is the dielectric constant in the limit of vanishing stationary composition pattern, β is the sensitivity of the dielectric constant to composition: $\beta = d\varepsilon/d\psi$. $\overline{\psi}(\mathbf{r})$ is the compositional order parameter associated with the stationary composition pattern, and is the local time-averaged volume fraction of one component minus its mean value. $\langle \overline{\psi}^2 \rangle$ is the space-averaged square of $\overline{\psi}(\mathbf{r})$. $\hat{\mathbf{e}}_q$ is the unit wave vector of the lamellae and $\hat{\mathbf{e}}_Z$ the unit vector in the direction of the applied field, $\mathbf{E}_0$. The microstructural orientations with lowest free energy are those for which $\hat{\mathbf{e}}_q$ is orthogonal to the applied field.

2. EXPERIMENTAL

The block copolymer of this study is a symmetric polystyrene-poly(methyl methacrylate) diblock copolymer (PS-PMMA) of molecular weight 37,000, and an order-disorder transition in the range of 251 to 256°C.[2] A sample was placed in a heated chamber between electrodes.[4,5] It was heated to 247°C (several degrees below the order-disorder transition), at which point an 18 kV/cm electric field was applied for approximately 10 min, then cooled 3–4°C/min to room temperature. The field was applied until the sample cooled to below the glass transition. The small angle X-ray scattering (SAXS) pattern of the sample after alignment is shown in Figure 2. The X-ray beam direction was orthogonal to the direction the electric field was previously applied. The SAXS pattern reveals strong alignment of lamellae so that their normal vectors are orthogonal to the electric field direction. This is the orientation of minimum free energy (see Eq. (1)).

Microtome slices were taken from the aligned sample (far from any surfaces), stained with ruthenium tetroxide and viewed with an electron microscope. Some slice planes were perpendicular to the electric field, and are referred to as "perpendicular slices." Other slice planes contained the electric field direction and are referred to as "parallel slices" as shown in Figure 3.

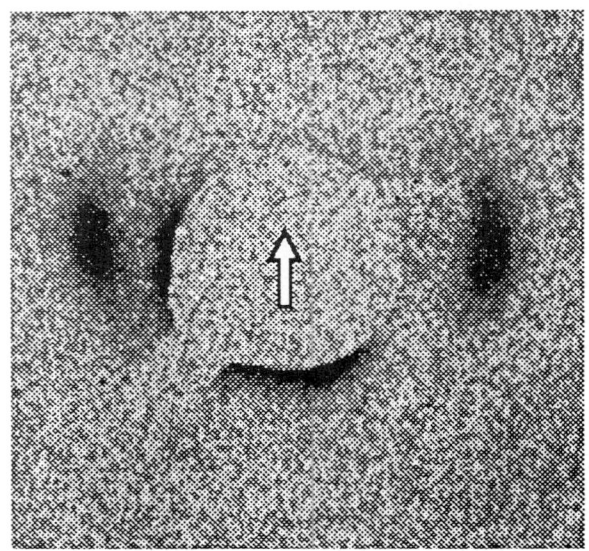

FIGURE 2   Small-angle X-ray scattering pattern from the block copolymer sample after alignment in the electric field. The X-ray beam was orthogonal to the applied field direction (indicated by the arrow). The two peaks result from scattering off the lamellar microstructure. Thin arcs near the beamstop outline are artifacts from imperfect beam alignment.

"perpendicular slice": slice plane $\perp$ E

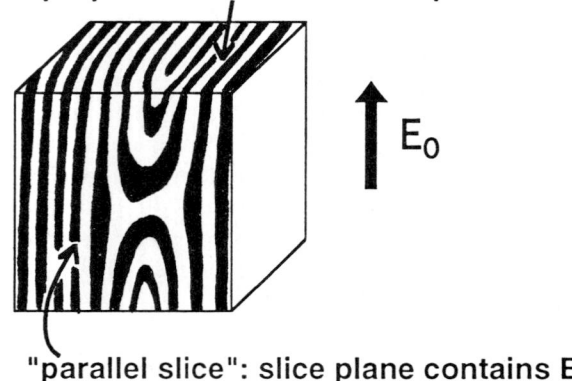

E_0

"parallel slice": slice plane contains E

FIGURE 3   Orientation of slice planes with respect to the direction of the applied field.

3. MORPHOLOGY OF ALIGNED BLOCK COPOLYMER

In Figure 4 we show portions of typical transmission electron micrographs (TEMs) of "perpendicular" and "parallel" slices. There are distinct differences between the micrographs of the two types of slices; both are consistent with extensive alignment of the lamellar microstructure by the electric field. In addition, four classes of defect structures are identifiable. $+1/2$ disclination lines and defect walls are most prevalent. $-1/2$ disclination lines and isolated edge dislocations are also observed, but are rarer. These defect structures as seen in smectic-A systems are discussed in books by de Gennes[10] and Kleman[14] and in numerous reports.[7,8] Disclination lines are lines of singularity in the $\hat{e}_q$ field and are identified by a $\pm\pi$ rotation of the lamellar wave vector upon counterclockwise travel along a loop that circumscribes a type $\pm1/2$ disclination line. Wall defects are analogous to grain boundaries in polycrystalline materials. Intersection points and lines of some of these defects with the slice plane are indicated in Figure 4(a).

Micrographs of "parallel" sections show predominately large regions of uniform striations along the electric field direction, as in the lower right and upper left of Figure 4(b). In addition, "disturbances" such as the one running through the center of Figure 4(b) are common. Typically, these regions are elongated in the direction of the electric field. In gross view, they are regions of coarse striation,

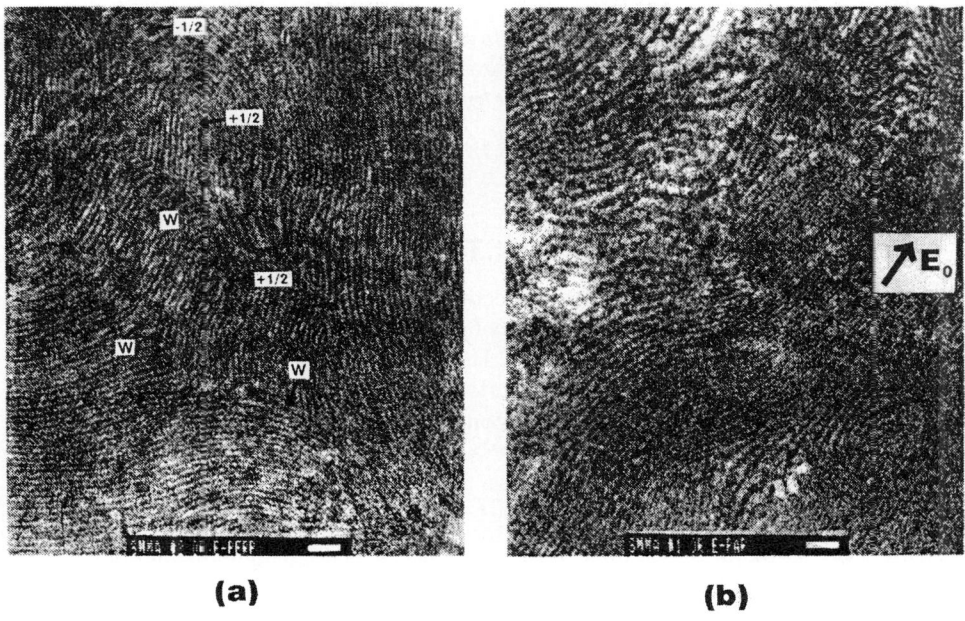

(a)   **(b)**

FIGURE 4   TEMs of microtome slices from an aligned PS-PMMA block copolymer sample. The contrast was achieved by staining the slices with Ruthenium Tetroxide. In (a), the slice plane was perpendicular to the electric field direction, and in (b) the slice plane contained the electric field direction, and the view shows a region of high defect density. Intersection of the slice plane with line and wall defects are indicated in (a).

indicating shallow intersection of the slice plane with the lamellae. It is likely that these features reflect a disclination line with a trajectory approximately in the slice plane, but offset from the slice plane (as illustrated in Figure 3). A more detailed analysis of these features reveals additional complexity; localized regions of high curvature of the lamellae and sharp jumps in lamellar orientation indicate a high density of line and wall defects. The form of these regions is likely due to defect-defect interactions and defect pinning, as will be discussed in the next section.

From the two types of micrographs, it is apparent that the defect structures are arranged anisotropically in the field-aligned sample. Disclination lines run predominately in the direction of the applied field. The anisotropy of wall defects is best understood by first defining a rotation axis associated with the wall defect. A rotation plus a translation is required to map the lamellar pattern on one side of a wall defect to that on the other. In the field aligned sample, the rotation axes of the wall defects are mainly aligned along the direction of the applied field, $\hat{\mathbf{e}}_Z$.

The anisotropic defect arrangement in the field-aligned sample is strong evidence that alignment of the lamellar microstructure is accomplished through movement of the defect structures, and selective annihilation of some of them. An alternative explanation would be rotation of some disclination lines and wall defects. However, such a movement is rather formidable; it would involve very long-range motion and associated large-scale reformation of the lamellar pattern, and seems unlikely. In the next section, forces on defect structures are considered.

---

4. FORCES ON DEFECT STRUCTURES

A full description of the forces on the defects is complicated because of defect-defect interactions and the many arrangements of defects manifested in the material. It is instructive to consider some simple cases. Here, we consider two scenarios: a general wall defect and a pair of parallel disclination lines of opposite sign. First, we focus on field-induced forces and next interactions between defects are considered.

4.1 ELECTRIC FIELD-INDUCED FORCES ON A WALL DEFECT

Because the alignment of the microstructure with respect to an external field generally differs on the two sides of a wall defect, there is a free energy density difference across a wall defect which is proportional to the square of the electric field strength. This gives a stress on the wall

$$\sigma_{\approx \text{ elec}} = \frac{1}{8\pi} \mathbf{E}_0^2 \frac{\beta^2}{\varepsilon_D} \overline{\langle\psi\rangle^2} \left[(\hat{\mathbf{e}}_{q,1} \cdot \hat{\mathbf{e}}_Z)^2 - (\hat{\mathbf{e}}_{q,2} \cdot \hat{\mathbf{e}}_Z)^2 \right] \hat{\mathbf{n}}\hat{\mathbf{n}} \qquad (2)$$

where $\hat{e}_{q,1}$ and $\hat{e}_{q,2}$ are the unit vectors in the direction of the lamellar pattern wave vectors on the two sides of the wall, and $\hat{n}$ is a unit normal vector to the wall. The stress drives the wall to move toward the side for which $(\hat{e}_q \cdot \hat{e}_Z)^2$ is largest.

The only walls that do not experience a field-induced force are those for which on both sides the alignment of lamellae with respect to the electric field is equally good (i.e., $(\hat{e}_{q,1} \cdot \hat{e}_Z)^2 = (\hat{e}_{q,2} \cdot \hat{e}_Z)^2$). This condition is equivalent to the condition in which the rotation axis associated with the wall defect is in the direction of the electric field. These are precisely the wall defects that remain after the field alignment.

4.2 FORCES ON DISCLINATION LINES

Disclination lines have associated with them an extended distortion field. The total free energy involves an integration of the free energy density due to distortion[3,5,12]

$$f = \frac{1}{2}\overline{B}\gamma^2 + \frac{1}{2}K_1(\nabla \cdot \hat{e}_q)^2 + \frac{1}{8\pi}\frac{\beta^2}{\varepsilon_D}\langle\overline{\psi}^2\rangle(\hat{e}_q \cdot \mathbf{E_O})^2. \tag{3}$$

The first term expresses the energy of compression or expansion of the layers. γ is the compression or expansion layer-strain and $\overline{B}$ is the layer compressibility. The second term represents the energy associated with a splay distortion (the molecules undergo splay, the layer planes bend), and K_1 is the splay elastic coefficient. The third term is the second-order anisotropic electrostatic energy from Eq. (1).

Here we consider a pair of straight, parallel disclination lines of strengths $+1/2$ and $-1/2$ in an otherwise homogeneous block copolymer sample, as shown in Figure 5(a). The distortion field around the pair of disclination lines contributes both mechanical and electrical energy to the sample according to the formula:

$$W = W_{\text{core}} + \frac{1}{2}K_1 L \ln\left(\frac{b/2}{r_{\text{core}}}\right) + \frac{1}{128}\frac{\beta^2}{\varepsilon_D}\langle\overline{\psi}^2\rangle b^2 L(E_1^2 - E_3^2) \tag{4}$$

where b is the Burgers vector for the disclination line pair (equal to twice the distance between the lines), r_{core} is the radius of the $+1/2$ disclination line core (also the lower limit of integration of the energy density), and E_1 and E_3 are the components of the external field in the x_1 and x_3 directions, respectively (defined in Figure 5 (a)). L is the length of the disclination lines. W_{core} is the core energy of the disclination lines. In calculating Eq. (4) only the distortion within the semicylindrical region centered on the $+1/2$ disclination and between the two disclination lines is considered. The far-field distortion is not changed by movement of the disclination lines toward each other, but only through the approach of disclinations of opposite Burgers vector.[5]

(a)

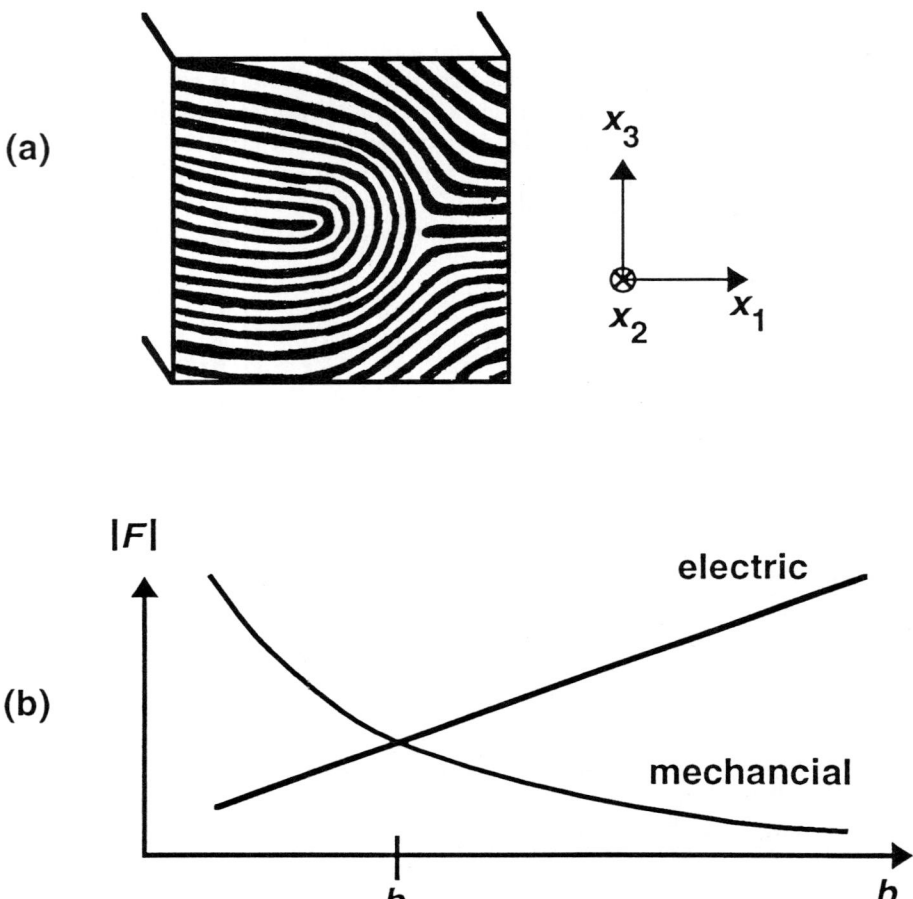

(b)

FIGURE 5   (a) Cross-sectional cut through a block copolymer material that is well ordered except for a pair of straight, parallel $+1/2$ and $-1/2$ disclination lines. A local coordinate system (x_1, x_2, x_3) is defined by the inter-line vector (along x_1) and the direction of translational symmetry (x_2), as shown. In (b) the magnitude of the electronic field-induced force and the mechanical force are shown as a function of the Burgers vector (twice the separation between disclination lines).

The magnitudes of the forces between the disclinations as predicted by this equation are shown qualitatively in Figure 5(b). The field-induced force is attractive if, in far field, the pattern is aligned in a preferred direction with respect to the electric field $(E_1 > E_3)$, or repulsive if the pattern, in far field, is aligned contrary

to the preferred direction far from the disclinations ($E_3 > E_1$). A general principle
is that the electric field drives disclination lines to move in directions that improve
the overall alignment of the microstructure with respect to the applied field.

For small separations the force due to mechanical distortion dominates, while
for large separations the electrical force dominates. The crossover distance at which
the two are equal is at the Burgers vector

$$b_0 = \left(\frac{32K_1}{\frac{\beta^2}{\varepsilon_D} \langle \overline{\psi}^2 \rangle |E_1^2 - E_3^2|} \right)^{1/2} \tag{5}$$

which for the experiment of this report is on the order of a micron (when E_1 or E_3
equals E_0).

There is one case where the disclination line pair does not experience an electric
field-induced force: when the lines run in the direction of the applied field (so
$E_1 = 0 = E_3$). These are precisely the types of disclination lines that remain in the
field-aligned sample.

4.3 DEFECT-DEFECT INTERACTIONS

Both line and wall defects interact with each other either through direct contact
or through the distortion field emanating from disclination lines. The interplay between field-induced forces and defect interactions leads to several interesting phenomena which will be discussed in this section.

A couple of ways a wall defect can interact with other defects are suggested
in Figure 6(a). Disclination lines must either form complete loops or end either at
the edge of the sample or, more likely, at a wall defect. Wall defects may share
common borders with other wall defects. In either of these cases, the total length
or area of defects must be changed in order for the wall defects to translate, and the
wall defect loses translation invariance. When a wall defect loses this translation
invariance through interaction with other defects, its mobility can be reduced, or
if the interaction is overpowering, the wall defect may be pinned and may not
respond to the electric field-induced stress. An important point is that inter-defect
interactions will increase and electric field-induced forces decrease (because of the
reduced size of defects) when the density of defects increases. Thus, the mobility of
defects should be a decreasing function of the local defect concentration.

Line defects interact with one another through their distortion fields, as was
discussed in Section 4.2. Line defects can also interact with wall defects as suggested
in Figure 6(c). Here, depending upon how the wall energy changes with the angle
of the lamellar pattern just below the wall, the two defects will either be attracted
toward or repelled from each other. Line defects that form complete loops such
as in focal conic structures, will be self-interacting. For all of these scenarios, the

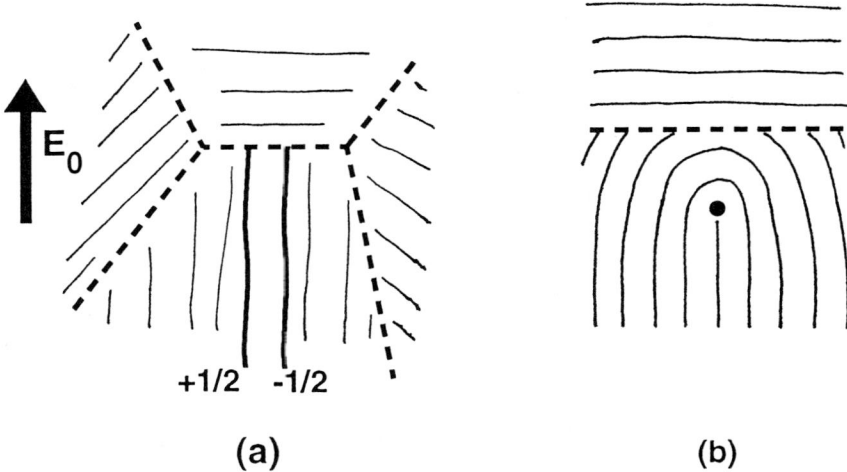

(a)   **(b)**

FIGURE 6   Defect-defect interactions are illustrated. The thin lines represent the lamellar pattern, the thick dashed lines indicate wall defects, and thick solid lines disclination lines. In (a), wall defects are in contact with other walls and disclination lines. In (b) the distortion field from a line defect extends to a wall defect.

magnitude of forces on the defects as a function of the distance between defects (or radius of a disclination loop) has the character shown in Figure 5(b), with a crossover point between the two forces on the order of an electric length scale

$$\xi = \left(\frac{K_1}{\frac{1}{4\pi} \frac{\beta^2}{\varepsilon_D} \langle \overline{\psi}^2 \rangle E^2} \right)^{1/2} \tag{6}$$

where E_2 is an appropriate square field strength.

Thus, just as with the wall defects, as the density of defects increases, the mean distance between them decreases, and the inter-defect forces become more significant compared to the electric field-induced forces. Once again, as the defect density increases, the mobility of defects decreases, and ultimately defects may become pinned.

The interplay between electric field-induced forces and defect interactions results in several observable phenomena. For example, the rate of alignment was found to be much slower in a block copolymer that was cooled quickly from the disordered phase than for a sample that was cooled slowly.[4] The rapidly cooled sample had a much higher density of defects. Also, the clumping of defects in the aligned sample discussed in Section 3 is likely due to an interplay between electric field-induced and inter-defect forces. Consider, for instance, a defect that migrates in response to an electric field-induced force. As it enters a region of higher defect density, its mobility will be reduced and it may become pinned. The result is that defects will linger in regions where the density of defects is higher, and the condition of uniform defect density will be unstable. Perhaps the interplay between electric field-induced forces and inter-defect forces is responsible for the clumps of defects in the field aligned sample (as shown in Figure 4(b)). It is noteworthy that many of the regions of high defect density seen in "parallel" microtome slices (cf. Figure 4(b)) have a width transverse to the applied field that is on the order of the electric length scale of Eq. (6).

ACKNOWLEDGMENTS

This work was a collaboration with Eugene Helfand and Xina Quan at AT&T Bell Laboratories and Steven D. Smith at the Procter and Gamble Co. We thank Frank Padden and Paul Szajowski for help with electron microscopy and Don D. Davis for small-angle X-ray scattering work.

REFERENCES

1. Amundson, K., E. Helfand, D. D. Davis, X. Quan, S. S. Patel, and S. D. Smith. "Effect of an Electric Field on Block Copolymer Microstructure." *Macromolecules* **24** (1991): 6546.
2. Amundson, K., E. Helfand, X. Quan, S. S. Patel, and S. D. Smith. "Optical Characterization of Ordering and Disordering of Block Copolymer Microstructure." *Macromolecules* **25** (1992): 1935.
3. Amundson, K. R., and E. Helfand. "Quasi-Static Mechanical Properties of Lamellar Block Copolymer Microstructure in an Electric Field. I. Alignment Kinetics." *Macromolecules* **26** (1993): 1324.
4. Amundson, K., E. Helfand, X. Quan, and S. D. Smith. "Alignment of Block Copolymer Microstructure in an Electric Field. II. Alignment Mechanism." *Macromolecules* **26** (1993): 2698.
5. Amundson, K., E. Helfand, X. Quan, S. D. Hudson, and S. D. Smith. "Alignment of Block Copolymer Microstructure in an Electric Field. II. Alignment Mechanism." *Macromolecules* (1994): submitted.
6. Bates, F. S., and G. H. Fredrickson. "Block Copolymer Thermodynamics: Theory and Experiment." *Ann. Rev. Phys. Chem.* **41** (1990): 525.

7. Bouligand, Y. "Recherches Sur Les Textures Des États Mésomorphes 3. Les Plages a Éventails dans Les Cholestériques." *J. Phys.* (Paris) **34** (1973): 603.

8. Chandrasekhar, S., and G. S. Ranganath. "The Structure and Energetics of Defects in Liquid Crystals." *Adv. Phys.* **35** (1986): 507.

9. Fredrickson, G. H., and E. Helfand. "Fluctuation Effects in the Theory of Microphase Separation in Block Copolymers." *J. Chem. Phys.* **87** (1987): 697.

10. de Gennes, P. G. *The Physics of Liquid Crystals.* Oxford: Clarendon Press, 1974.

11. Hadziioannou, G., A. Mathis, and A. Skoulios. *Colloid Polymer Sci.* **257** (1979): 136.

12. Kawasaki, K., and T. Ohta. "Phase Hamiltonian in Periodically Modulated Systems." *Physica* **139A** (1986): 223.

13. Keller, A., E. Pedemonte, F. M. Willmouth, "Macro Lattice From Segregated Amorphous Phases of a Three Block Copolymer." *Kolloid-Z. u. Z. Polymere* **238** (1970): 385.

14. Kléman, M. *Points, Lines and Walls.* New York: Wiley, 1983.

15. Koppi, K. A., M. Tirrell, F. S. Bates, K. Almdal, and R. H. Colby. "Lamellae Orientation in Dynamically Sheared Diblock Copolymer Melts." *J. Phys.* (Paris) (1994): submitted.

16. Leibler, L. "Theory of Microphase Separation in Block Copolymers." *Macromolecules* **13** (1980): 1602.

17. Morrison, F., G. L. Bourvellec, and H. H. Winter. "Flow-Induced Structure and Rheology of a Triblock Copolymer." *J. App. Polym. Sci.* **33** (1987): 1585.

18. Patel, S. S., R. G. Larson, K. I. Winey, and H. Watanabe. "Rheology of Polystyrene-Poly(methyl methacrylate) Block Copolymers." *Macromolecules* (1994): submitted.

19. Winey, K. I., S. S. Patel, R. G. Larson, and H. Watanabe. "Interdependence of Shear Deformations and Block Copolymer Morphology." *Macromolecules* **26** (1993): 2542.

20. Winey, K. I., S. S. Patel, R. G. Larson, and H. Watanabe. "Morphology of a Lamellar Diblock Copolymer Aligned Perpendicular to the Sample Plane: Transmission Electorn Microscopy and Small-Angle X-Ray Scattering." *Macromolecules* **26** (1993): 4373.

Chapter 6
Chemical and Materials
Instabilities

D. Walgraef

Center for Nonlinear Phenomena and Complex Systems, Free University of Brussels, CP 231, B-1050 Brussels, Belgium;
Director of Research at the Belgian National Fund for Scientific Research

Turing Structures in Chemical and Materials Instabilities

Despite almost forty years of theoretical modeling of pattern formation in reaction-diffusion systems, only recently has the experimental observation of spatial structures induced by the mechanism proposed by Turing in 1952 occurred. Besides the generic aspects of pattern formation, selection, and stability in such systems, we briefly discuss some specific properties of two- and three-dimensional structures in chemical and materials instabilities, which are related to experimental constraints (spatial gradients, time dependence of the bifurcation parameter, etc.).

1. INTRODUCTION

One of the most interesting aspects of the complex dynamics that govern natural phenomena is, perhaps, their ability to induce instabilities that trigger the formation of coherent spatio-temporal structures on macroscopic scales.

The study of the origin of this spatio-temporal order in open systems far from thermal equilibrium, and the selection mechanisms for the structures and their symmetries, has become an important research issue, both experimentally and theoretically.[6,16,24] Such instabilities, which are known from the beginning of the century (consider the Rayleigh-Bénard, Bénard-Marangoni, and Taylor-Couette instabilities), also appear in many other physical, chemical, or biological systems.

In particular, the possibility of pattern formation in driven chemical systems has long been a puzzling phenomenon both theoretically and experimentally. In 1952 Turing proposed that the interplay between the diffusion of reacting species and their nonlinear interactions could lead to the formation of spatial patterns that should correspond to spatial modulations of their concentration.[19] Several kinetic models have been proposed to justify the existence of spatial instabilities inducing such structures,[14,15] but experimental problems have long prohibited their observation.[2] Only since 1990 have genuine Turing structures been observed experimentally[7,17] in the chlorite-iodide-malonic acid reaction. The experimental observations confirm several theoretical predictions such as for example the fact that, contrary to the case of hydrodynamic structures, the wavelength of these chemical patterns only depends on kinetic rates and diffusion coefficients and not on the geometry of the reactor. On the other hand, these structures may be tridimensional and the analysis of their selection and stability properties leads to a true dissipative crystallography. Indeed, their description in the framework of amplitude equations and numerical analysis based on kinetic models predict the existence of stable bcc lattices, hexagonal prisms, or equidistant walls in tridimensional geometries, and of hexagonal or striped patterns in two-dimensional geometries. Furthermore, these different structures may be simultaneously stable in certain parameter ranges.

I will review here some generic properties of chemical patterns but also discuss some specific aspects related either to the peculiarities of reaction-diffusion systems or due to experimental requirements. I will also show why Turing instabilities may be relevant to the formation of microstructures in materials.

2. CHEMICAL PATTERNS IN TWO AND THREE DIMENSIONS

2.1 AMPLITUDE EQUATIONS

When the evolution of nonlinear chemical systems can be described by a reaction-diffusion dynamics of the type:

$$\partial_t C = f(C) + D\nabla^2 C \tag{1}$$

(where $C(\vec{r}, t)$ is the vector representing the state of the system and containing the concentrations of the various active species, f represents the kinetic rates, and D is the diffusion tensor), it has been possible to show, on the basis of general arguments, confirmed by the analysis of particular models, how, on varying some control parameters, a uniform steady state may become unstable compared to inhomogeneous perturbations having well-defined wavevectors. This control parameter, or bifurcation parameter, may be, for example, the concentration of some species, the temperature, etc. At the bifurcation point, the linear stability analysis shows that the growth rate of all perturbations of wavevector $\vec{k}$ such that $\vec{k} = |k_c|$ (where k_c is the critical wavenumber) becomes positive. One may thus expect a transition from the uniform steady state to spatially ordered states and one has to know the behavior of the system beyond the instability threshold to be able to determine which structures will be selected and what will be their stability domain. In isotropic media, due to the orientational degeneracy, the competition between the various excited modes is studied through mode combinations defined by :

$$C(\vec{r}, t) = C_0 + \Sigma_{i=1}^{m}(A_i e^{i\vec{k}_i \vec{r}} + \bar{A}_i e^{-i\vec{k}_i \vec{r}}) \qquad (2)$$

where C_0 is the uniform steady state and $A_i = R_i \exp i\phi_i$ is the amplitude of the unstable mode of wavevector k_i. Each structure is thus characterized by m pairs of wavevectors.

The standard techniques of bifurcation analysis lead to the following equations for uniform amplitudes A_i:

$$\partial_t A_i = \epsilon A_i + v\Sigma_j\Sigma_k \bar{A}_j \bar{A}_k \delta(\vec{k}_i + \vec{k}_j + \vec{k}_k) - A_i|A_i|^2$$
$$- \Sigma_j g_{ij} A_i |A_j|^2 - \Sigma_j\Sigma_k\Sigma_l \bar{A}_j \bar{A}_k \bar{A}_l \delta(\vec{k}_i + \vec{k}_j + \vec{k}_k + k_l). \qquad (3)$$

The variable ϵ represents the reduced distance to the bifurcation point ($\epsilon = (b - b_c)/b_c$ where b is the bifurcation parameter and b_c its value at the instability threshold). The coefficients g_{ij} are functions of the angles α_{ij} between the wavevectors k_i and k_j. These equations are valid in the vicinity of the bifurcation and their structure is universal. Only the coefficients depend on the kinetic model under investigation and need to be calculated explicitly.

When the coefficients g_{ij} are such that $g_{ij}(i \neq j) > g_{ii}$, the only stable structures that appear through a supercritical bifurcation at $\epsilon = 0$ correspond to unidirectional modulations (defined by one pair of wavevectors only, i.e., $m = 1$).[11] These structures correspond to bands, lamellae, or walls and are the analogs of the convective rolls that appear in the Rayleigh-Bénard instability. The $m > 1$ structures may nevertheless be stable if they contribute to the quadratic term of the amplitude equations and appear subcritically.

Hence, on increasing the value of the bifurcation parameter, the following sequence should be observed in three-dimensional systems[21]: a bcc lattice ($m = 6$) is the first structure to appear, followed by hexagonal prisms ($m = 3$) and finally by a wall structure ($m = 1$). In two-dimensional geometries, the succession occurs between hexagons ($m = 3$, subcritical) and stripes ($m = 1$, supercritical).

One sees, in the three-dimensional bifurcation diagram sketched in Figure 1, that the different structures may coexist for a well-defined range of values of the bifurcation parameter, and that the system may present tristability. On the other hand, sufficiently far from the bifurcation point, higher-order nonlinearities and higher-order harmonics may become important. They may modify the nature of the stable structures and induce, for example, quasi-periodic structures of pentagonal or icosahedral symmetry as well as patterned domain walls.[22]

Let's return to the case of patterns with hexagonal symmetry. On writing $A_i = R_i \exp i\phi_i$, one obtains, from Eq. (3), the following equation for the sum of the phases, Φ:

$$\partial_t \Phi = -v \frac{R_2^2 R_3^2 + R_1^2 R_3^2 + R_2^2 R_1^2}{R_1 R_2 R_3} \sin \Phi. \tag{4}$$

The variable Φ thus relaxes monotonically to 0 (respectively π) when $v > 0$ (respectively $v < 0$). Hence, as is the case in the Brusselator model, v changes sign when the bifurcation parameter is increased, and there is thus an exchange of stability between the state $\Phi = 0$ and $\Phi = \pi$ at this point (the maxima of the first state, or H_0 hexagons, correspond to the minima of the second state, or H_π hexagons). This effect has been studied in detail by Verdasca et al.[20] In agreement with the numerical simulations, the following sequence of patterns emerge when the value of the bifurcation parameter is increased: the H_π structure is followed by stripes, and then another hexagonal pattern; the H_0 structure may reappear at still higher values of ϵ. Reversing the variation of the control parameter, one backtracks through the various structures undergoing hysteresis loops at each transition. A similar scenario has also been observed in the Schnackenberg[18] model and may be shown to have the same origin.[13]

As in other pattern-forming instabilities, the Turing instability described here induces the spontaneous breaking of translational and rotational symetries. As a result, the behavior of the phase of the resulting structures should be diffusive. Furthermore, the nucleation of topological defects (dislocations, grain boundaries, etc.) may be expected in extended systems. While the phase dynamics of Swift-

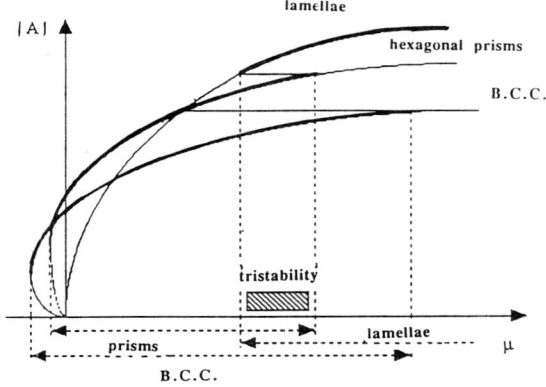

FIGURE 1 Schematic bifurcation diagram for the Brusselator model in tri-dimensional media. Heavy lines represent stable states. Note a domain where the three structures are simultaneously stable.

Hohenberg roll patterns has been studied in detail, and the reduction of the roll stability domain from the marginal one to the domain limited by the Eckhaus and zigzag instabilities has been thoroughly analyzed,[12] less attention has been paid to the phase stability of hexagonal platforms. It is why we recently computed the phase dynamics of a perfect hexagonal structure resulting from the following amplitude equations:

$$\partial_t A = \epsilon A + [(\vec{q_1} + \vec{\nabla})^2 - q_c^2]^2 A + vB^*C^* - A|A|^2 - gA(|B|^2 + |C|^2)$$
$$\partial_t B = \epsilon B + [(\vec{q_2} + \vec{\nabla})^2 - q_c^2]^2 B + vA^*C^* - B|B|^2 - gB(|A|^2 + |C|^2)$$
$$\partial_t C = \epsilon C + [(\vec{q_3} + \vec{\nabla})^2 - q_c^2]^2 A + vB^*A^* - C|C|^2 - gC(|B|^2 + |A|^2)$$
$$(|\vec{q_1}| = |\vec{q_2}| = |\vec{q_3}| = q = q_c + k)$$

(5)

After the separation of amplitude and phase variables and the adiabatic elimination of the relaxing amplitude modes, one obtains the following dynamics for the two-component phase vector[10]:

$$\partial_t \vec{\phi} = D_\perp \nabla^2 \vec{\phi} + (D_\| - D_\perp)\text{grad.div}\vec{\phi} \tag{6}$$

with

$$D_\perp = q_c^2 + 6q_c k - \frac{16q_c^4 k^2}{W_\Delta} \tag{7}$$

and

$$D_\| = 3q_c^2 + 10q_c k - 16q_c^4 k^2 \left(\frac{1}{W_\Delta} + \frac{2}{W_\Sigma}\right) \tag{8}$$

with

$$W_\Delta = 2(1 - g)R^2 + 4vR \tag{9}$$

and

$$W_\Sigma = 2(1 + 2g)R^2 - 2vR \tag{10}$$

with

$$R = \frac{2}{1 + 2g} \left[v + \sqrt{v^2 + (1 + 2g)(\epsilon - 4q_c^2 k^2)}\right]. \tag{11}$$

It is remarkable that these equations have the same form as the equations of elasticity of an isotropic bidimensional solid (except that the evolution here is diffusive and not propagative), the coefficients D_1 and D_2 playing the role of the Lamé coefficients. Furthermore, the stability analysis shows that the hexagonal patterns are stable in the domain defined by $D_1 > 0$ and $D_2 > 0$, as shown on the phase diagram in Figure 2.

We hope that this result will allow the interpretation of White's observations[25] on the stability of hexagonal convective planforms in non-Boussinesq fluids. It furthermore suggests that the destabilization of roll patterns via a zigzag instability may lead to the formation of hexagonal structures. The numerical checks of these results will be presented elsewhere.

The above discussion and results concern systems with spatially uniform control parameters. However, the experimental constraints imposed by the lateral feeding of the system trigger spatial variations of the control parameters within the reactive layer. These variations and their associated gradients may induce the coexistence of structures of different symmetry as shown by Borckmans et al.,[3] but also determine the orientation of the structures. Such gradient effects have been studied in detail in the case of specific reaction-diffusion models,[4] and an illustrative example is shown in Figure 3.

2.2 NUMERICAL ANALYSIS OF A KINETIC MODEL

Chemical pattern formation has been studied in great detail in the framework of various reaction-diffusion models,[1] such as the "Brusselator" model. This kinetic model proposed by Lefever, Nicolis, and Prigogine[15] may be considered as a paradigm for the study of the spatio-temporal organization of nonlinear chemical systems. Due to the universality of the behavior of reaction-diffusion systems close to instability points, one may hope to obtain from this model a fair description of pattern

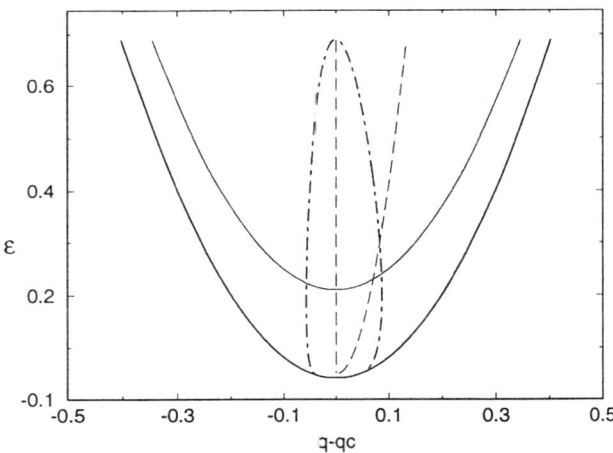

q-q_c

—— marginal stability(hexagons)
—— Amplitude roll stability
— — · Phase roll stability
— · — Phase hexagons stability

FIGURE 2 Stability diagram of roll and hexagonal patterns described by Eq. (3) for $g = 2$ and $v = 0.2$.

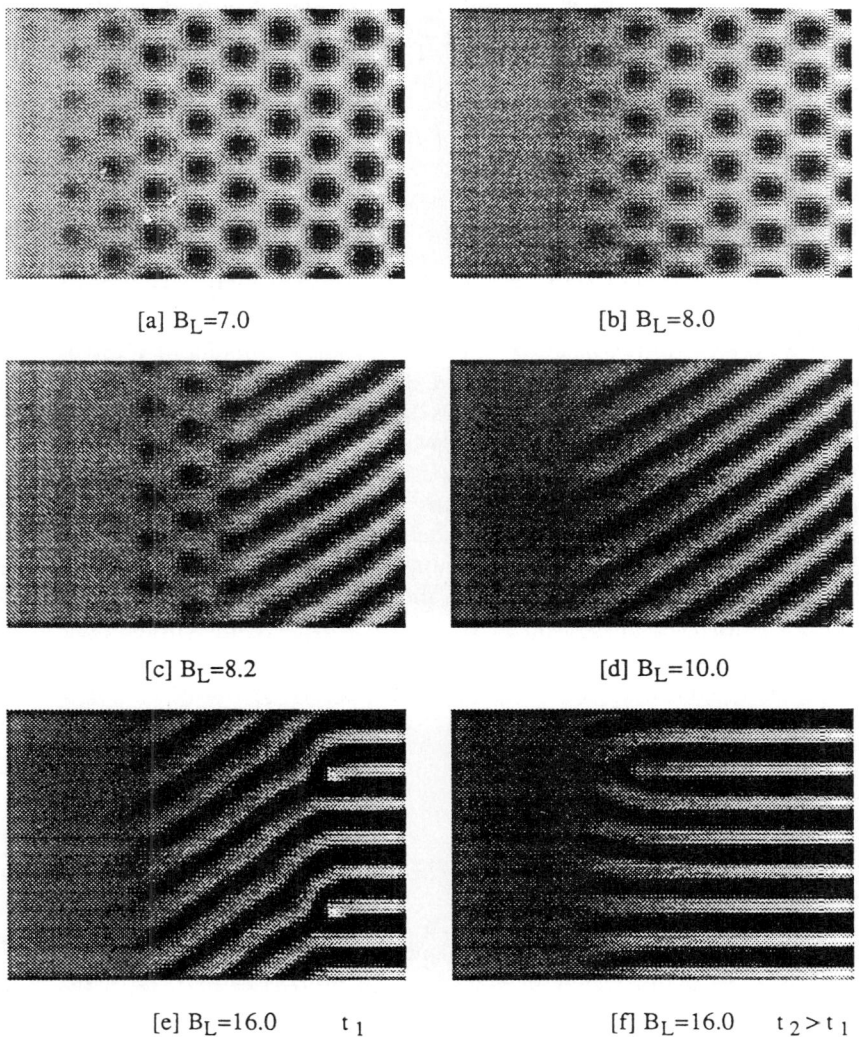

[a] B$_L$=7.0

[b] B$_L$=8.0

[c] B$_L$=8.2

[d] B$_L$=10.0

[e] B$_L$=16.0   t$_1$

[f] B$_L$=16.0   t$_2$>t$_1$

FIGURE 3   Succession of patterns obtained in the numerical analysis of the Brusselator in the presence of a linear ramp from the left to the right[1] ($B = Bc = 6.71$ is imposed at grid points $(30, y)$, boundary conditions are no-flux to the left and right and periodic at top and bottom, BL is increased quasi-statically from [a] $BL = 7$ to [b] $BL = 8-$ [c] $BL = 8.2-$ [d] $BL = 10-$ [e] and [f] $BL = 16$ at two successive times (courtesy of A. De Wit).

formation in chemical systems. Let us recall that the dynamics of this model is given by the following kinetic equations :

$$\partial_t X = A - (B+1)X + X^2 Y + D_x \nabla^2 X$$
$$\partial_t Y = BX - X^2 Y + D_y \nabla^2 Y \tag{12}$$

where A and B are control pool species with constant concentrations, which may thus be considered as control parameters; D_x and D_y are diffusion coefficients and D the Laplacian operator. The concentration of B is traditionally used as bifurcation parameter. This system presents a pattern-forming instability at $B = B_c = (1 + A\sqrt{D_x/D_y})^2$ with $k_c^2 = A/\sqrt{D_x D_y}$, where k_c is the critical wavenumber (note that, contrary to hydrodynamic structures, this wavenumber is intrinsic since it only depends on kinetic constants and diffusion rates and not on geometrical factors). At this bifurcation point, the growth rate of all perturbations of wavevector k such that $|k| = k_c$ becomes positive. Hence one may expect a transition between a spatially uniform state and a spatially structured state. One has thus to know the behavior of the system beyond the instability to be able to determine which structures are be selected and what are their stability domains.

The results of the weakly nonlinear analysis described above have been tested via a numerical analysis of this model[5] and the results are in complete analogy with the experimental observations. In a two-dimensional medium, one obtains band structures with grain boundaries and disclinations, or hexagonal structures as experimentally observed, as well as the exchange of stability between H_0 and H_π hexagons[20] (see Figure 4).

3. AUTO-ORGANIZATION OF DEFECT POPULATIONS IN MATERIALS INSTABILITIES

Many driven or damaged materials present various self-organization phenomena which are able to modify their physicochemical properties and have thus a very practical importance.[24] Most of these phenomena correspond to the spatial organization of defect populations (e.g., deformation bands and dislocation patterns in plastic instabilities, layering of corroded or oxidized metals, defect microstructures in metals and alloys under irradiation, etc.), and the corresponding spatial structures result from instabilities that are of the Turing type. Effectively, these instabilities result from the interplay between the nonlinear defect interactions and the defect mobilities which are very different according to the nature of the defect motion (e.g., point defect diffusion, dislocation glide or climb, etc.).

An illustrative example may be found in irradiated metals and alloys. Effectively, these materials present several types of microstructures which correspond to

(a) (b)

(c) (d)

FIGURE 4 Typical two-dimensional patterns for the concentration of species X in the Brusselator. The integrations were performed by A. De Wit on a square grid of size 64×64 with periodic boundary conditions.[20] The grays scale corresponds to the concentrations between the absolute minimum (black) and maximum (white); on increasing B one obtains: [a] H_π, [b] stripes, [c] H_0, and [d] zig-zag structures.

the spatial organization of defect populations. Well-known examples include void and bubble lattices, precipitate ordering, defect walls and vacancy loop ordering. In particular, the spatial ordering of vacancy dislocation loops occurs frequently in metals and alloys irradiated at moderate doses and high temperatures.[9] The uniform distribution of loops that are created by the collapse of cascades becomes unstable beyond some threshold that is determined by various material and irradiation parameters (e.g., the irradiation dose, damage rate, bias in the migration of point defects to loops and network dislocations, and temperature). We analyzed this phenomenon in the framework of a dynamical model which considers the two major elements of irradiated microstructures, namely, vacancy and interstitial clusters.[23] In this regard, we model the effects of irradiation on materials in the form of dynamical equations for two mobile atomic-size species (vacancies and interstitial atoms), and two basic immobile elements of the microstructure (vacancy and interstitial clusters). These equations are based on the basic elements of defect dynamics, namely, point defect creation, recombination, and migration to microstructures, and are written as :

$$\partial_t c_i = K(1 - \epsilon_i) - \alpha c_i c_v + D_i \nabla^2 c_i - D_i c_i (Z_{iN}\rho_N + Z_{iV}\rho_v + Z_{iI}\rho_I)$$

$$\partial_t c_v = K(1 - \epsilon_v) - \alpha c_i c_v + D_v \nabla^2 c_v - D_v(Z_{vN}(c_v - \bar{c}_{vN})\rho_N$$
$$+ Z_{vV}(c_v - \bar{c}_{vV})\rho_v + Z_{vI}(c_v - \bar{c}_{vI})\rho_I)$$

$$\partial_t \rho_I = \left(\frac{2\pi N}{|\vec{b}|}\right)\left(\epsilon_i K + D_i Z_{iI} c_i - D_v Z_{vI}(c_v - \bar{c}_{vI})\right)$$

$$\partial_t \rho_v = \frac{1}{|\vec{b}|r_V^0}[\epsilon_v K - \rho_v(D_i Z_{iV} c_i - D_v Z_{vV}(c_v - \bar{c}_{vV}))]$$

$$(13)$$

where c_v corresponds to vacancies and c_i to interstitials. The expression ρ_N is the network dislocation density, ρ_v the vacancy loop and ρ_I the interstitial loop density. Variable K is the displacement damage rate, $\epsilon_i K$ the interstitial loops production rate, and ϵ_v the cascade collapse efficiency; α is the recombination coefficient, $\vec{b}$ the Burgers vector, r_V^0 the mean vacancy loop radius and $Z_{.,.}$ are the bias factors which will be approximated by $Z_{iN} = 1 + B$, $Z_{iI} \cong Z_{iV} = 1 + B'$ and $Z_{vI} = Z_{vN} = Z_{vV} = 1$. Variable B is the excess network bias and B' is the excess loop bias $(B' > B)$. Expressions $\bar{c}_{vN}$, $\bar{c}_{vV}$, and $\bar{c}_{vI}$ are the concentrations of thermally emitted vacancies from network dislocations, vacancy loops, and interstitial loops, respectively.

We showed that this system undergoes a pattern-forming instability at sufficiently high irradiation dose and derived the corresponding slow mode dynamics and amplitude equations which turn out to be of the Swift-Hohenberg type.[23] The problem here is that, during irradiation, the microstructure is continuously evolving, and we were only able to analyze their formation in a quasi-static approximation which is nevertheless in good agreement with the experimental observations. In this description, the effective bifurcation parameter is continuously increasing in time. As a result, the following sequence of vacancy loops patterns are observed during the irradiation process: starting from a uniform distribution of loops, the system develops first three-dimensional cellular structures of cubic symmetry. Then these structures become unstable and a transition to wall structures occurs, this pattern being the final one. This scenario has been obtained in various parameter ranges corresponding to specific experimental situations, and the comparison with the experimental results has been discussed by Ghoniem and Walgraef.[8]

ACKNOWLEDGMENTS

Fruitful discussions with P. Borckmans, G. Dewel, and A. De Wit are gratefully acknowledged. This work has been supported by a grant of the Commission of the European Communities (CI1-CT92-0006).

REFERENCES

1. Baras, F., and D. Walgraef, eds. Special Issue: Nonequilibrium Chemical Dynamics: From Experiment to Microscopic Simulation. *Physica* **A 188** (1992).
2. Borckmans, P., G. Dewel, D. Walgraef, and K. Katayama. "The Search for Turing Structures." *J. Stat. Phys.* **48** (1987): 1031–1044.
3. Borckmans, P., G. Dewel, and A. De Wit, and D. Walgraef.. "Instabilités et Structures Spatiales en Chimie de non Équilibre." *Entropie* **164–165** (1991): 83–90.
4. Borckmans, P., A. De Wit, and G. Dewel. "Competition in Ramped Turing Structures." *Physica* **A 188** (1992): 137.
5. Borckmans, P., G. Dewel, A. De Wit, and D. Walgraef. "Three-Dimensional Dissipative Structures in Reaction-Diffusion Systems." *Physica* **D61** (1992): 289.
6. Busse, F., and L. Kramer, eds. *Nonlinear Evolution of Spatio-Temporal Structures in Dissipative Continuous Systems*. New York: Plenum, 1990.
7. Castets, V., E. Dulos, J. Boissonade, and P. De Kepper. "Experimental Evidence of a Sustained Standing Turing-Type Nonequilibrium Chemical Pattern." *Phys. Rev. Lett.* **64** (1990): 2953–2956.
8. Ghoniem, N. M., and D. Walgraef. "Evolution Dynamics of 3D Periodic Microstructures in Irradiated Materials." *J. Model. & Simul. in Mat. Sci. & Engr.* **1** (1993): 569–590.
9. Jäger, W., P. Ehrhart, and W. Schilling. "Dislocation Patterning Under Irradiation." In *Nonlinear Phenomena in Materials Science*, edited by L. Kubin and G. Martin, 279. Switzerland: Trans Tech Publications, 1988.
10. Lauzeral, J., S. Metens, and D. Walgraef. "On the Phase Dynamics of Hexagonal Patterns." *Europhys. Lett.* **24** (1993): 707–712.

11. Malomed, B. A., and M. I. Tribel'skii. "Stability of Stationary Periodic Structures for Weakly Supercritical Convection and Related Problems." *Sov. Phys. JETP* **65** (1987): 3013–3016.

12. Manneville, P. *Dissipative Structures and Weak Turbulence.* Boston: Academic Press, 1990.

13. Metens, S., G. Dewel, and P. Borckmans. Private communication, 1993.

14. Murray, J. "Parameter Space for Turing Instability in Reaction-Diffusion Mechanmism: A Comparison of Models." *J. Theor. Biol.* **98** (1979): 143–164.

15. Nicolis, G., and I. Prigogine. *Self-Organization in Nonequilibrium Systems.* New York: Wiley, 1977.

16. Nicolis, G., and I. Prigogine. *Exploring Complexity.* Munich: Piper and New York: Freeman, 1987.

17. Ouyang, Q., and H. L. Swinney. "Transition From a Uniform State to Hexagonal and Striped Turing Patterns." *Nature* **352** (1991): 610.

18. Schnackenberg, J. "Chemical Reaction Systems with Limit Cycle Behavior." *J. Theor. Biol.* **81** (1979): 389.

19. Turing, A. "The Chemical Basis of Morphogenesis." *Phil. Trans. Roy. Soc. London* **237B** (1952): 37–72.

20. Verdasca, J., A. De Wit, G. Dewel, and P. Borckmans. "Reentrant Hexagonal Turing Structures." *Phys. Lett.* **A 168** (1992): 194.

21. Walgraef, D., P. Borckmans, and G. Dewel. "Nonequilibrium Phase Transitions and Chemical Instabilities." *Adv. Chem. Phys.* **49** (1982): 311–355.

22. Walgraef, D., G. Dewel, and P. Borckmans. "Quasiperiodic Order in Dissipative Systems." *Nature* **318** (1985): 606.

23. Walgraef, D., and N. M. Ghoniem. "Spatial Instabilities and Dislocation-Loop Ordering in Irradiated Materials." *Phys.Rev.* **B39** (1989): 8867.

24. Walgraef, D., and N. M. Ghoniem, eds. *Patterns, Defects and Materials Instabilities.* Dordrecht: Kluwer Academic Publishers, 1990.

25. White, D. B. "The Planforms and Onset of Convection with a Temperature-Dependent Viscosity." *J. Fluid Mech.* **191** (1988): 247.

K. I. Agladze,[†‡] O. Steinbock,[†] A. Warda,[†] and S. C. Müller[†]
[†]Max-Plank-Institut für molekulare Physiologie, Rheinlanddamm 201, 44139 Dortmund, FRG
[‡]Institute of Theoretical and Experimental Biophysics Russian Academy of Science, Pushchino, 142292 Moscow region, Russia

Birth and Death of Spiral Waves

INTRODUCTION

One of the most interesting examples of self-organization in active excitable media is the appearance of rotating spiral waves.[4,7,13,16] Previous studies show that spirals can act like independent pacemakers, they can drift,[5,8] interact and collide,[5,11] or combine to more complicated structures. The motion of the spiral wave tip can be also of rather complex nature.[9] Spiral waves create a "population" in the active medium with some special rules of life.

Using recently developed methods of operation with a chemical active medium on the basis of the oscillating Belousov-Zhabotinsky (BZ) reaction (such as gel techniques,[18] control of a photosensitive version of the reaction by a laser beam[13]), we investigate experimentally the processes of birth and death of these vortices. It is known that spiral waves evolve when a propagating wave front is perturbed.[16] In the simplest case a so-called "wave break" appears due to this perturbation which curls up to form a rotating spiral.[15,16] We study the critical size of this perturbation, created by a laser beam, for spirals to be born. On the other hand, the problem of spiral birth is closely associated with the problem of keeping them alive. We show that spirals will stay alive only if they have at least enough space; otherwise, they disappear.

SPIRAL BIRTH

For perturbing the front of a travelling wave we performed experiments in which silica gel with fixed catalyst—$Ru(bpy)_3^{2+}$—was used as a support for BZ reaction. The concentration of catalyst was 4 mM. Disregarding the bromination of malonic acid the reactant concentrations were as follows: 0.20 M $NaBrO_3$, 0.17 M malonic acid, 0.09 M NaBr, and 0.2 to 0.45 M H_2SO_4. Illuminated spots were made with the aid of an argon laser (514 nm-line) and projected onto the surface of the gel.

A thin laser beam produces a small cut in the front of a propagating wave. Since under strong illumination the ruthenium-catalyzed version of BZ reaction loses its excitability,[1,6] the wave front disappears in the illuminated spot. After switching off the light, the excitability of the medium is restored and fronts propagate again into the initially inhibited the illuminated spot.

We found that there exists a critical diameter d_{krit} of the spot size, below which spirals are not created. In Figure 1 we show the result of cutting the wave front for the case when the size of the spot is smaller than the critical value. After switching off the light, the free wave ends start to propagate towards each other. After collision they merge and the wave front is fully restored. In Figure 2 we demonstrate the process of birth of a pair of spiral waves after application of a

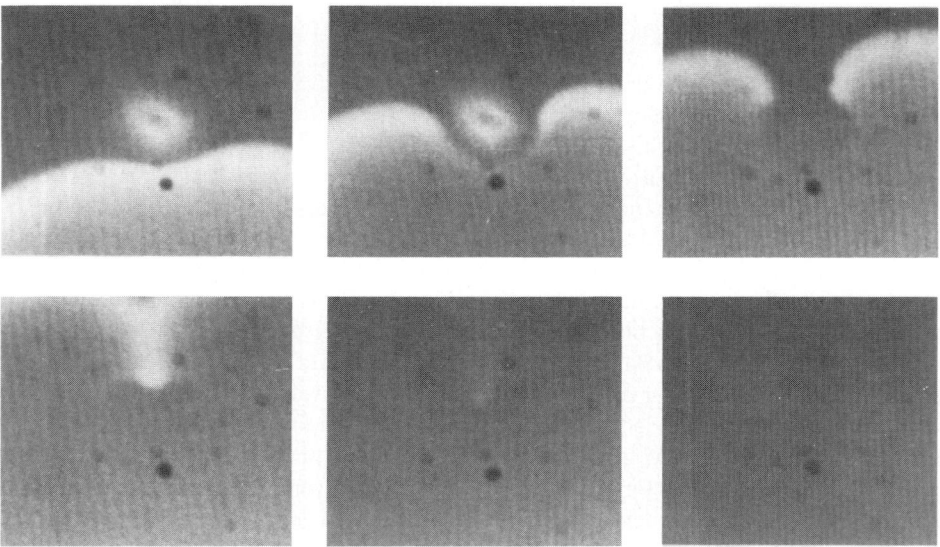

FIGURE 1   A cut of the small piece of the wave front does not lead to the spiral origination.

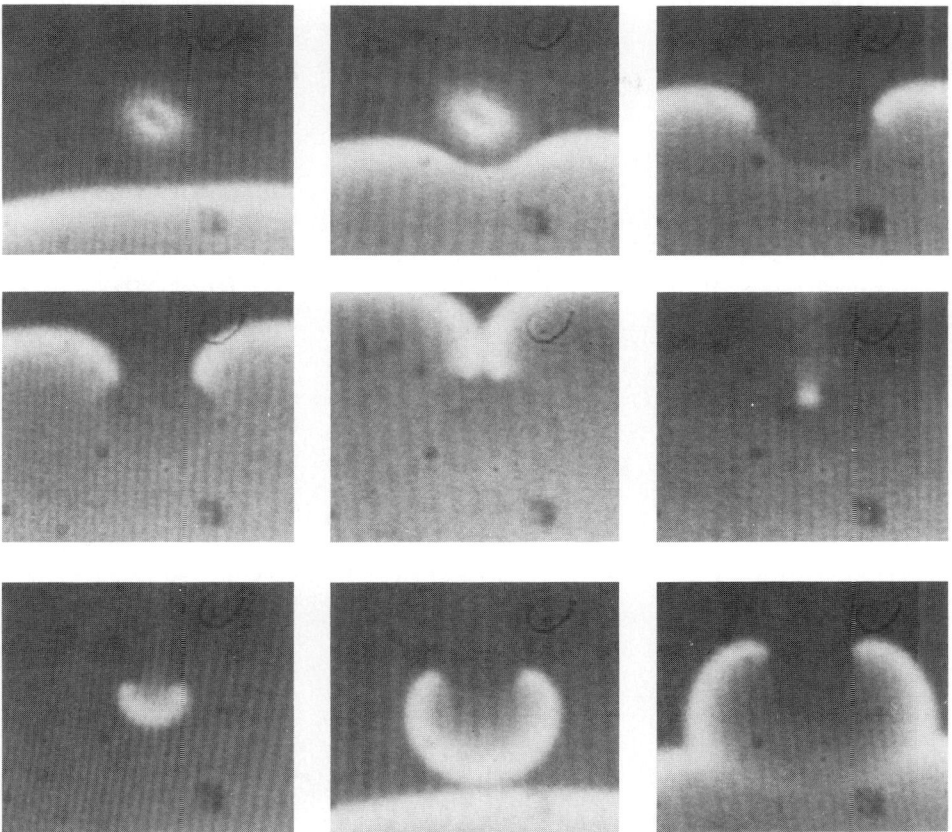

FIGURE 2   Birth of spiral waves pair because of the application of the critical perturbation.

supercritical perturbation (spot size larger than the critical one). As in the previous case, the broken wave ends move towards each other but, because of the larger initial distance, they have time to curl up sufficiently. Thus, after wave collision, they do not coalesce but continue their independent motion—a pair of spiral waves is born. In some cases we observed that a just created pair died after one or two cycles of rotation. As discussed below, this is related to the most important condition for spirals to survive—the problem of minimal space.

The critical size d_{krit} of the cut of a wave front depends on the wavelength of the spiral wave which changes, for instance, with the acidity of the medium. In Figure 3 we illustrate that with increase of the acidity the critical size decreases.

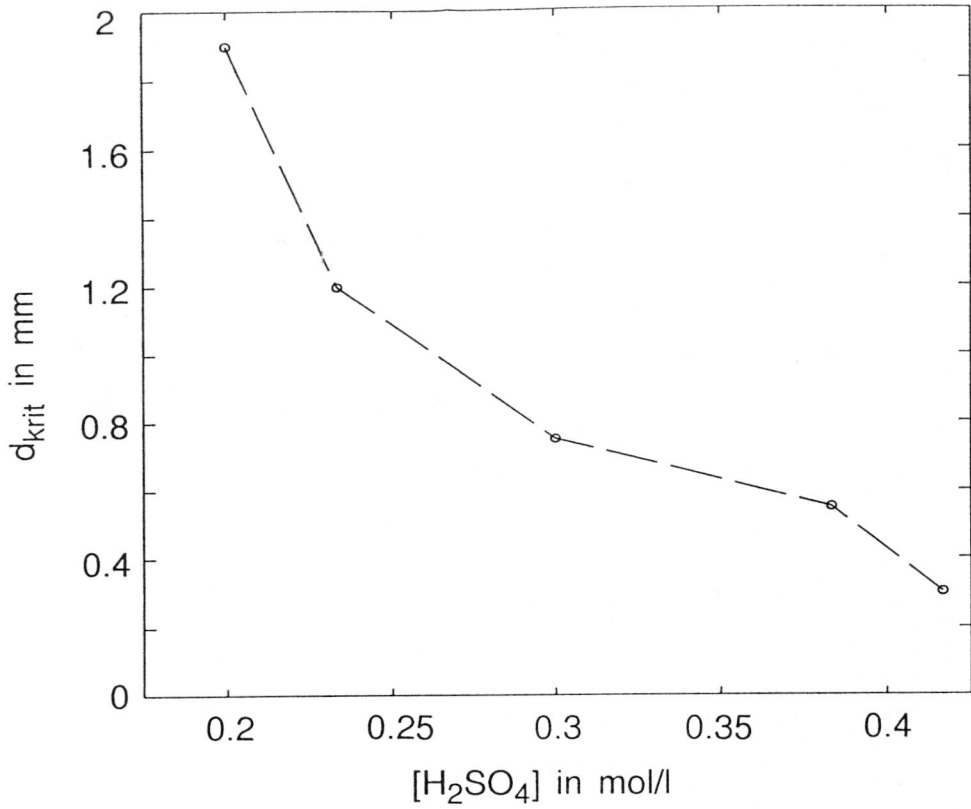

FIGURE 3 Dependence of the critical size d_{krit} on the H_2SO_4-concentration of the system.

The increase of acidity corresponds to a decrease of the spiral wavelength.[2,10] So as the wavelength decreases, a smaller perturbation should be applied for the wave front to create spirals. In fact, the wavelength of the spiral provides a characteristic spatial scale of the system. Of course, this is true only if it exceeds by far the characteristic diffusion distance. This condition is usually met in the BZ reaction, where the wavelength is of the order of a few millimeters and the characteristic diffusion length is of the order of 100 micrometers.

SPIRAL DEATH

The problem of minimal space for the spiral to live was studied experimentally in a gel system by placing spiral wave close to the boundary of the medium. In these experiments agarose gel was used because its mechanical properties make it easier to cut it. Disregarding the bromination of malonic acid the reactant concentrations were as follows: 0.33 M $NaBrO_3$, 0.24 M malonic acid, 0.06 M NaBr and 0.37 M H_2; influence on the reaction (which also inhibits the reaction) was prevented by covering the gel layer by a 2-mm layer of transparent, chemically inert silicon oil.

We found that there is a critical distance between the core of the spiral and the boundary: if we place the spiral close to the periphery, it dies because of the collision with the boundary. The sequence of images in Figure 4 illustrates the process of interaction, collision and death of the spiral when it was placed too close to the boundary of gel. The measured critical distance was 0.5 mm. This value corresponds to the critical size of the pair of spirals divided by a factor of 2. This is not a simple coincidence. Mathematically the condition of minimal size of the pair of spirals (minimal possible distance between two spiral cores) is the same as the distance between the core of a single spiral and a nonflux boundary of the medium. A nonflux boundary for which von Neuman boundary conditions hold can be regarded as a virtual mirror[17] and the medium can be extended by its "mirror reflection." In this case it is clear that a single spiral close to the border will interact with the border as if it is interacting with the "mirror reflection" of itself.

CONCLUSION

Recent developments in the field of oscillating reactions allow to study much more complicated regimes and wave patterns in extended systems. It becomes now possible for the experimentalist to be not only a passive observer of the life of a spiral, but to govern the growth and decay of a spiral population by choosing sets of external parameters and the proper geometry of the medium.

Gel techniques have become a most important tool for the experimentalist for creating of special geometries of an active medium. By using the light sensitive version of the BZ reaction one can apply well-controlled perturbations to the system and precisely change parameters in the desired direction. These advantages underline the role of the BZ reaction as an extremely valuable instrument for the modelling of self-organization processes.

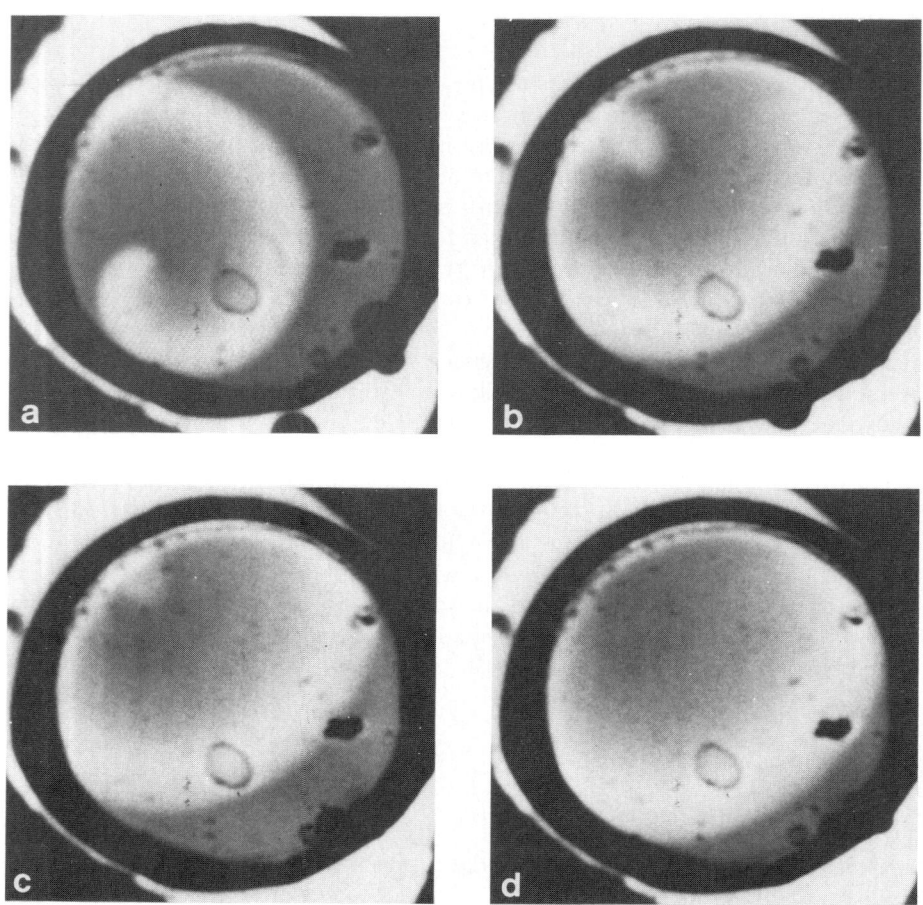

FIGURE 4 Interaction with boundary and death of the spiral wave, placed in the distance less than critical one.

REFERENCES

1. Agladze, K. I., and V. I. Krinsky. "Multi-Armed Vortices in an Active Chemical Medium." *Nature* **292** (1982): 424–426.
2. Agladze, K. I. *Proc. Biological Center of Academy of Science of USSR*, Pushchino (1983).
3. Agladze, K. I., V. A. Davydov, and A. S. Mikhailov. "Observation of a Helical-Wave Resonance in an Excitable Distributed Medium." *JETP Lett.* **45** (1987): 767-768.
4. Davidenko, J. M, A. M. Pertsov, R. Salomontsz, W. Baxter, and J. Jalife. "Stationary and Drifting Spiral Waves of Excitation in Isolated Cardiac Muscle." *Nature* **355** (1992): 349–351.
5. Krinsky, V. I., and K. I. Agladze, "Interaction of Rotating Waves in an Active Chemical Medium." *Physica D* **8** (1983): 50–56.
6. Kuhnert, L. "Photochemische Manipulation von Chemischen Wellen." *Naturwissenschaften* **73** (1986): 96–97.
7. Lechleiter, J., S. Girard, E. Peralta, and D. Clapham. "Spiral Calcium Wave Propagation and Annihilation in *Xenopus laevis* Oocytes." *Science* **252** (1991): 123–126.
8. Markus, M., Zs. Nagy-Ungvarai, and B. Hess. "Phototaxis of Spiral Waves." *Science* **257** (1992): 225–227.
9. Nagy-Ungvarai, Zs., J. Ungvarai, and S. C. Müller. "Complexity in Spiral Wave Dynamics." *Chaos* **3** (1993): 15–19.
10. Plesser, T., S. C. Müller, and B. Hess. "Spiral Wave Dynamics as a Function of Proton Concentration in the Ferroin-Catalyzed Belousov-Zhabotinskii Reaction." *J. Phys. Chem.* **94** (1990): 7501–7507.
11. Schütze, J., O. Steinbock, and S. C. Müller. "Forced Vortex Interaction and Annihilation in an Active Medium." *Nature* **356** (1992): 45–47.
12. Siegert, F., and C. J. Weijer. "Three-Dimensional Scroll Waves Organize Dictyostelium Slugs." *PNAS* **89** (1992): 6433–6437.
13. Steinbock, O., and S. C. Müller. "Chemical Spiral Rotation is Controlled by Light-Induced Artificial Cores." *Physica A* **188** (1992): 61–67.
14. Steinbock, O., J. Schütze, and S. C. Müller. "Electric-Field-Induced Drift and Deformation of Spiral Waves in an Excitable Medium." *Phys. Rev. Lett.* **68** (1992): 248–251.
15. Wiener, N., and A. Rosenblueth. *Arch. Inst. Cardiol. Mex.* **16** (1946): 205.
16. Winfree, A. T. "Spiral Waves of Chemical Activity." *Science* **175** (1972): 634–636.
17. Winfree, A. T. "Rotating Chemical Solutions." *Theor. Chem.* **1** (1978): 1–54.
18. Yamaguchi, T., L. Kuhnert, Zs. Nagy-Ungvarai, S. C. Müller, and B. Hess. "Gel Systems for the Belousov-Zhabotinskii Reaction." *J. Phys. Chem.* **95** (1991): 5831–5837.

Shoichi Kai
Department of Applied Physics, Kyushu University, Fukuoka 812 804, Japan

Gradient-Free Structures in Liesegang Phenomena

Pattern formation in precipitation processes in the gel-containing systems of two electrolytes, which can react with each other to produce insoluble products, is described. Such a precipitation is a kind of multidispersed system consisting of many small crystals. The multidispersion of crystals leads to different kinetics due to the ripening process among crystals from that of one single crystal, where certain connection between reaction and diffusion plays an important role depending on the crystal size. A recent theory containing the ripening kinetics is discussed to describe some details such as subring formations, interlayer structures, and gradient-free structures in the precipitating pattern formation.

1. INTRODUCTION

It is well known that a crystal growth phenomenon shows beautiful patterns, either periodic or fragmental.[34] In multidispersed precipitation systems, it is well known that periodic bands, the so-called Liesegang bands (rings; see Figure 1),[36] are often observed. The periodic law obtained here is the spacing law $X_{N+1} = pX_N$ where

Spatio-Temporal Patterns, Ed. P. E. Cladis and P. Palffy-Muhoray,
SFI Studies in the Sciences of Complexity, Addison-Wesley, 1995   **445**

X_N is the distance of Nth ring location from an original junction and p is a constant coefficient. Very similar patterns to those are frequently observed in various systems,[16,17,18,19,20,33] not only in precipitation systems but also, for example, in geological systems,[39,41,46] in biological systems,[1,13,14,60] and in our solar system (Titius-Bode's law).[54] As a limit case for those systems, fractal and fragmental patterns can be observed (see Figure 2).[13,14,40,61] Therefore, it is considered that there must be some universal mechanisms from physicochemical point of view in these phenomena.

In Liesegang phenomena as shown in Figure 1, two different spatial scales and dynamics are observed such as crystal growth dynamics and precipitation pattern dynamics. In this standard Liesegang experiment, a solution of one soluble electrolyte, for instance, lead nitrate ($Pb(NO_3)_2$), at relatively low concentration (of the order of 0.1 M), is placed in a container such as a test tube to which a gel-forming material is added. After a gel is formed, a second electrolyte solution, such as the potasium iodide (KI), normally at substantially higher concentrations (of

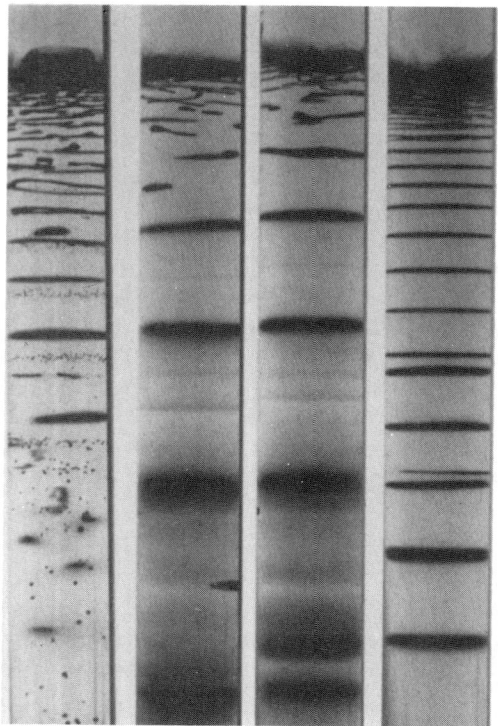

FIGURE 1 Periodic precipitation of PbI_2 in the presence of concentration gradients (Liesegang band for various concentration combinations of $pb(NO_3)_2$ and KI which change the spacings of bands.

FIGURE 2   Bacteria colony pattern (courtesy of Dr. Fujikawa). The periodic growth pattern of $B.$ $subtilis$ on agar plates is observed. The bacteria is inoculated on agar plates (15 g/l) containing 1 g/l of peptone (fractal dimension $D_f = 1.706$).

the order of 1 M), is poured on top of the gel containing $Pb(NO_3)_2$. The iodine-ions (I^-) diffuse into the gel and react with lead ions (Pb^{2+}) to form lead iodide (PbI_2) which is almost insoluble. After an interval of minutes, yellow bands (rings) appear, sequentially in a period of hours, which are parallel to the surface of the diffusion front. Similar periodic patterns have been reported for many other slightly soluble electrolytes and even gases.[16,17,18,19,20,33]

The basic formation process of Liesegang rings is periodic in the presence of gradients such as gradients of temperature, concentration, and so on. Macroscopic structures may also arise from an electrolyte solution in the absence of any gradients; that is, initially the solution of PbI_2 is prepared homogeneously. For instance, as shown in Figure 3, the precipitate is irregularly arranged in the gradient-free case, instead of having periodic Liesegang rings in gradient fields.

Such macroscopic structures are quite different from macroscopic structures of single crystal formation, such as a snow crystal. In precipitation, structures are formed from multidispersed small crystals while a snow pattern is composed of a single crystal. In the later case its formation dynamics can be described by the microscopic kinetics of crystal growth. However, in the former case, the interactions among multidispersed crystals (colloids) must be taken into account for their growth dynamics which is called the ripening process (Ostwald ripening). This happens

(a)

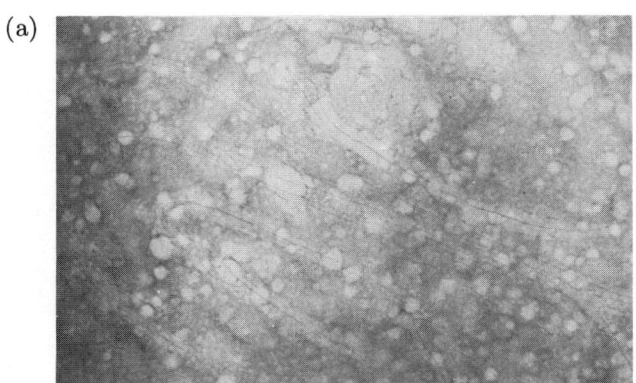

(b)

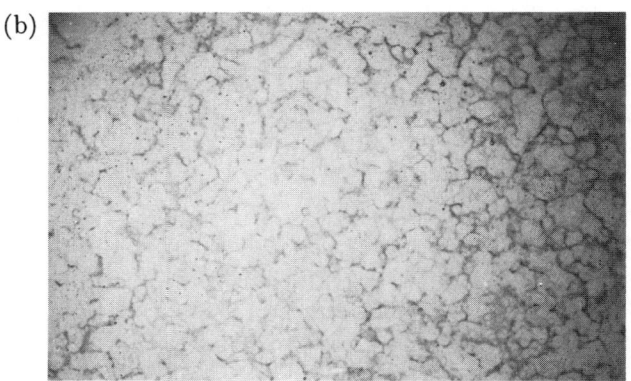

FIGURE 3   Nonperiodic patterns ((a) Swiss cheese type (S = 3.33) and (b) networks type (S = 3.72)) of PbI_2 in precipitation process in the absence of gradients.

randomly although dynamics for all crystals is deterministic, because their sizes are randomly distributed. In this sense, Ostwald ripening is an example of nonthermal fluctuation. In this article, the net dynamics of precipitation pattern formation is used to describe how both reaction process and Ostwald ripening work.

2. THEORY OF LIESEGANG RING

2.1 SUPERSATURATION THEORY

Theories for the Liesegang ring can be classified into two categories, pre-nucleation and post-nucleation theories, which will be explained below briefly. Pre-nucleation theories include: (1) supersaturation theory by Wi. Ostwald,[21,32,47,49,51,53,59] (2) diffusion wave theory by Wo. Ostwald,[48,50] (3) adsorption theory,[2,3] (4) simple diffusion theory,[12] and (5) membrane theory.[10,55] On the other hand, postnucleation theories are proposed rather recently and include: (1s) coagulation theory,[7,15,52] (2) instability theory,[8,9,11,24,27,38,45] and (3) competitive growth theory.[4,56,57,58]

In prenucleation theories, the supersaturation theory is still supported by many researchers and possibly explains some experimental facts. The idea of the supersaturation theory follows. The outer electrolyte ions A diffuse into another gel solution which contains the inner electrolyte ions B. Both A and B react each other and the reaction product C is produced. Then C forms precipitates through supersaturated state. This can be formulated as follows.[32]

$$A + B \rightarrow C, \tag{1}$$

$$C \rightarrow D, \tag{2}$$

$$r = k_+ a^{\nu_A} b^{\nu_B} - k_- c^{\nu_C}, \tag{3}$$

and then

$$\frac{\partial a}{\partial t} = D_A \frac{\partial^2 a}{\partial x^2} - \nu_A r, \tag{4a}$$

$$\frac{\partial b}{\partial t} = D_B \frac{\partial^2 b}{\partial x^2} - \nu_B r, \tag{4b}$$

$$\frac{\partial c}{\partial t} = D_C \frac{\partial^2 c}{\partial x^2} + \nu_C r - P, \tag{4c}$$

$$\frac{\partial d}{\partial t} = P. \tag{4d}$$

Here a, b, D_A, D_B, and t are the ionic concentrations of A and B, and the diffusion constants of A, B, and time, respectively. Here also C, D, $k_\pm$, $\nu_{A,B,C}$, d, and P are the monomer, the precipitate, the reaction rate constants for forward $(+)$ and backward $(-)$ reactions, the stoichiometric numbers of, respectively, A, B, and C (the order of reaction), precipitation concentration, and precipitation rate, respectively. They assumed that $P(c, d) = O$ if $c < c^*$ and $d = 0$, and $P(c, d) = q_p(c - c^s)$ if $c > c^s$ or $d > 0$, where c^* and c^s are some supersaturated and saturated concentrations of the monomer C, respectively. The important point is that precipitates do not start immediately after the reaction. Maintaining a supersaturation state for a while, it starts to form precipitates following the precipitation rate P. Therefore, the characteristic spatial scale is determined from the combination of diffusion and the precipitation rate P. Giving initial conditions, the results for a and b are familiar as error functions. After reaction at the original junction, the diffusion profiles for both $a(x,t)$ and $b(x,t)$ are determined. Then the concentration product is given by $K(x,t) = a^{\nu_A} b^{\nu_B}$ ($\nu_A = 1$ for I$^-$ and $\nu_B = 2$ for Pb^{2+}), which informs us the next precipitating location exceeding a critical saturation value K_c. Thus, at the new location, the same processes as those at the original junction start and new diffusion profiles are again obtained under the newly given boundary conditions. The repetition is made to form a sequence of precipitation bands. Unsolved problems for the results of this theory follow.

1. A precipitation location is a sink and, therefore, no width of a precipitation band can be given by the theory.
2. The spiral pattern and wide colloid distributions cannot be described.[24,42]
3. It cannot describe secondary structures, spatial bifurcation, and gradient-free patterns.[24]

Very recently, Falkovich and Keller proposed a new theory which belongs to the category of post-nucleation theory using the Cahn-Hilliard-type equation in Eq. (4c), $\partial c/\partial t = \nabla[D(c)\nabla c] - D'\nabla^4 c - P$, which is familiar in spinodal decomposition processes.[8] Then due to negative diffusion, macroscopic patterns are formed. The similar pattern formation due to negative diffusion was also obtained by Nishiyama.[46] In his theory multidiffusion processes and the Gibbs-Duhem relation were introduced to get uphill diffusion.

2.2 INSTABILITY AND COMPETITIVE PARTICLE GROWTH THEORY

The instability theory is based on the thermodynamical instability of colloid particles which shows an autocatalytic property, the so-called Ostwald ripening by a combination of diffusion process and the Gibbs-Kelvin (Thomson) effect, indicating the size-dependent solubility $c_{eq}(R)$. This ripening can be described by the following equations[23]:

$$\frac{\partial R}{\partial t} = \frac{M}{R^\mu}(\bar{c} - c_{eq}(R)), \tag{5a}$$

$$c_{eq}(R) = c_{eq}(\infty)\left(1 + \frac{\alpha}{R}\right). \tag{5b}$$

where R, $c_{eq}(R)$, $c_{eq}(\infty)$, M, μ, and $\bar{c}$ are the size of a crystal, the equilibrium (saturation) concentration of a crystal with size R, the equilibrium concentration of a crystal with $R = \infty$, the growth rate, the value depending on the types of limited growth processes ($\mu = 1$ for diffusion limited growth (DLG) and $\mu = O$ for interfacial reaction limited growth (IRLG)) and the average concentration of a solution, respectively. The rate M is given as $M = D_c V$ for IRLG. The value α is the capillary constant; $\alpha = (2\sigma_\infty/k_B T)V$ where σ_∞, k_B, T, and V are the surface free energy of the corresponding flat surface, the Boltzmann constant, temperature, and the molar volume of a crystal, respectively.

The instability due to this ripening can be simply described as follows (see Figure 4). In the solution of multidispersed crystals, the group of relatively large crystals can grow by dissolving their neighbor groups of small crystals. Spatial fluctuation modes of concentrations, which are unstable, can grow through this mechanism and form macroscopic patterns. In competitive growth theory, competitive growth among crystals with different sizes due to Ostwald ripening plays an important role to form structures. That is, initial concentration profiles determine the first growing particles which suppress the growth of other lately nucleated particles. The instability, therefore, need not occur in order for patterns to form.

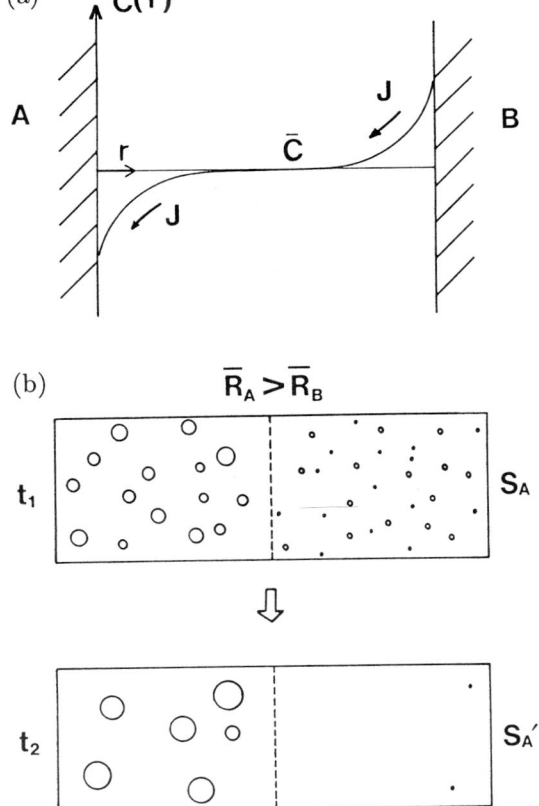

FIGURE 4   Competitive growth (a) and Ostwald ripening (b). (a) Two crystals A and B stay in the supersaturation solution with the concentration $\bar{c}$. The surface layers are very thin (compare their radii). The concentration gradient $c(r)$ for both crystals and diffusion fields are shown in figure by the solid line and the mass flow, therefore, occurs from B to A as shown by J. This means that A can grow by taking the materials of B off. This idea could be extended into size distribution functions as shown in (b). (b) At time t_l two area with supersaturation rate S_A where each mean radius is $\overline{R}_A$ and $\overline{R}_B$ respectively $(\overline{R}_A > \overline{R}_B)$ contact each other. At the later time, t_2, particles in the region A can grow and make dense precipitates by dissociation of particles in the region.

In other recent theories the very sharp concentration dependence of nucleation rate has been assumed to be important and taken into account.[6,35] Then for discontinuous change of nucleation rate $\partial N_t(x,t)/\partial t$ ($= K_n(S(x,t) - S^*)^E$) with the power law of the higher order of supersaturation rate $S = c/c_{eq}(\infty)$, the clear zone without precipitates could be obtained. On the other hand, for continuous change of $\partial N_t(x,t)/\partial t$, $\partial N_t(x,t)/\partial t = K_n S(x,t)^E \times \exp[-u/(\ln(S(x,t) + 1))^2]$, the calculated result showed a vague "clear" zone, where K_n and u are the constants. Some precipitation pattern formations may be understood based on this idea. For the length scale, E has been experimentally obtained as a factor greater than 5.[44] The detail of some of those theories will be described in a later section.

PATTERNS IN THE PRESENCE OF GRADIENTS. The main feature of Liesegang ring formation can be described by the supersaturation theory of Eq. (3). However, a recent theory has been extended possibly into describing "sub" phenomena such as secondary structures, spirals, and subring formations.[42] Here we take the precipitation rate P described in Eq. (4) as follows:

$$P = \frac{n}{V}\frac{\partial}{\partial t}\left(\frac{4\pi R^3}{3}\right). \tag{6}$$

That is, Ostwald ripening is introduced in a set of equations. Here n is the number of colloidal particles in precipitates whose shape is assumed to be sphere.[27] We have made following assumption; fast reaction and very slow diffusion, that is, local equilibrium ($r = 0$), interfacial reaction limited growth, the one-dimensional system in space, and a second-order reaction ($\nu_A = \nu_B = \nu_C = 1$). The simulation has been done using Lorenz method.[27,37]

Summarizing results obtained from the simulation, we have following facts[27,29,31]:

1. The simulation describes well the temporal difference between formations of rings and colloids that were experimentally observed.
2. The space becomes wider and Ostwald ripening becomes stronger, as the diffusion coefficient of monomers becomes larger.
3. As Ostwald ripening becomes stronger, subrings form more easily and the width between bands becomes narrower.
4. As the concentration product $K = ab$ increases, the spacing decreases and the number of rings that form increases.
5. The spacing law can be well described (see Figure 5).
6. As the reaction rate k_+ becomes smaller, the space between two rings becomes wider and the spacing coefficient increases.

Thus the simulations for the model agree well with many experimental observations. However, the width law cannot be explained yet. The reason for this is probably that no distribution of particle size has been taken into account. If the concentration gradients of reactans are decreased, periodicity among bands becomes more stochastic.[25] The transition from periodic to nonperiodic structures occurs continuously.

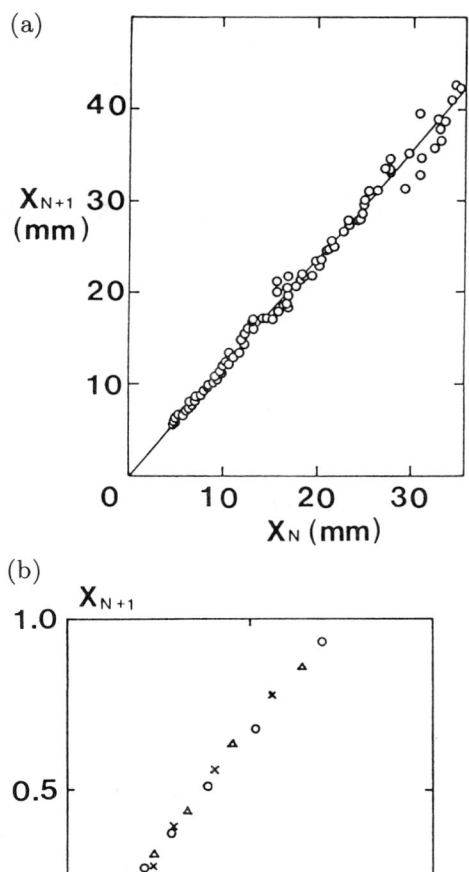

(a)

(b)

FIGURE 5   Spacing law in an experiment for PbI_2 ($Pb(NO_3)_2 = 3 \times 10^{-3}$ M, KI = 0.3 M) (a) and computer simulation (b) due to Eq. (4) using Eq. (6).

GRADIENT-FREE STRUCTURES. As ion concentrations decrease continuously, Liesegang rings finally appear stochastically in the absence of gradients.[25,43] When initially no gradient-free system is prepared, completely nonperiodic structures form.[25,26,44,45,57] This can be formalized by taking $\nabla^2 a = \nabla^2 b = 0$ and also $r = 0$ in Eq. (4),[29,45,58] i.e.,

$$\frac{\partial c}{\partial t} = D_c \frac{\partial^2 c}{\partial x^2} - P, \tag{7a}$$

$$\frac{\partial d}{\partial t} = P. \tag{7b}$$

Because colloidal particles are multidispersed, a particle distribution function $f(R, x, c, t)$ must be defined. Then the total particle number $n(x, c, t)$ is given by the equation

$$n(x, c, t) = \int_0^\infty f(x, R, c, t) dR. \tag{8}$$

Therefore, we substitute it into the precipitation rate P in Eq. (6):

$$P = \frac{4\pi}{V} \int_0^\infty dR R^2 R f(x, R, c, t). \tag{9}$$

The distribution function $f(x, R, c, t)$ follows the equation of continuity,

$$\frac{\partial f(x, R, c, t)}{\partial t} + \frac{\partial}{\partial R} \frac{dR}{dt} f(x, R, c, t) = 0. \tag{10}$$

Accordingly, equations cf motion in the gradient-free case are given by Eqs. (7)–(10). For stability and pattern formation, the important factors here will be: (1) the growth law of colloid particles, (2) the magnitude of surface energy, (3) the structure of distribution functions, and (4) the diffusion constant of monomers.

In real gradient-free experiments, total monomer concentration c is conserved as $\overline{c_0}$, because there is no supply from outside. Therefore, the average supersaturation rate $S(t)$ in a bulk solution is given by the equation

$$S(t) = S_0 - \frac{4\pi}{3 V c_{eq}(\infty)} \int_{R^*}^\infty dR R^3 f(R, t). \tag{11}$$

Here $S_0 = \overline{c_0}/c_{eq}(\infty)$ and $\overline{c_0}$ is the initial monomer concentration immediately after the system is quenched and before nucleation starts. When the monomer concentration is of small average size, $f(x, R, C, t)$ suppresses growth, but later, as the concentration size grows larger, its effect dissolves due to the conservation of monomers. Thus the macroscopic Ostwald ripening process among particle distribution functions $f(x, R, c, t)$ can be considered (see Figure 4).

For two-dimensional calculations for IRLG, we use Eqs. (7)–(11) as follows[29,31]:

$$\frac{\partial S(x,y,t)}{\partial t} = \frac{D_c}{L^2}\left(\frac{\partial^2}{\partial x^2}+\frac{\partial^2}{\partial y^2}\right)S - \frac{P}{c_e(\infty)}, \tag{12a}$$

$$\frac{P}{c_e(\infty)} = \frac{4\pi n\alpha^3}{c_e(\infty)V}R^2\frac{\partial R}{\partial t} = \overline{\alpha}R^2\frac{\partial R}{\partial t}, \tag{12b}$$

$$\frac{\partial R}{\partial t} = \frac{1}{q}\left(S - 1 - \frac{1}{R}\right). \tag{12c}$$

Here L is the total length (size) of a system (here $L = 10$ cm), n is the number of particles at a lattice point, $x = x'/L$, $y = y'/L$, $R' = \alpha R$, and $q = \alpha^2/c_e M$ which represents the inverse of the strength of Ostwald ripening through the capillarity length (original variables, x, y, and R, are dashed here). As initial conditions, we set an initial particle radius $R_0 = 0.5$ at all mesh points (100×100). Spatially homogeneous and noncorrelated random noise is added to the initial concentration S_0 as the initial fluctuation. The critical radius is determined as $1/(S_0 - 1)$. Other values to be used for calculations are as follows: $c_e = 1.4 \times 10^{18}$ cm^{-3}, $V = 1.24 \times 10^{-22}$ cm^3, $\alpha = 6.1 \times 10^{-7}$ cm, $D_c = 10^{-5}$ cm^2/s, and $n = 10^9$ cm^{-3}. As described previously, at each mesh point the monodispersed particle distribution is assumed. Therefore Ostwald ripening works only among different mesh points and the total size distribution for all mesh points shows the typical Lifshiz-Slyozov-Wagner (LSW) distribution.[22,23] This distribution becomes transiently wide, and then sharp and asymmetric.

Typical simulation results are shown in Figures 6–8. In Figure 6 we show a Swiss-cheese-type (SW) pattern (a) which appears in the weak Ostwald ripening, and a big hole (BH) pattern (b) in strong Ostwald ripening in simulation. The darkness corresponds to the size of particles; i.e., the darker is the larger. When the broadening of particle size distributions starts, Ostwald ripening works exordinarily well and the macroscopic pattern starts to form. The transition between BH and SW patterns is rather clear and seems to show different stability regimes. In Figure 7 we show different type of BH pattern from that in Figure 6(b), which appears when we use the conditions of small q^{-1} (weak Ostwald ripening) and of large monomer diffusion D_c. In that case, Ostwald ripening works with a long characteristic length through D_c; that is, it is not strong in magnitude but long in space. This shows usually a large-scale structure. To summarize, one can say the following. When S is large and ripening is weak, precipitation patterns have many small holes, called Swiss cheese (SW) patterns. When S is rather small and ripening is strong, BH (experimentally often observed as a networks-type (NW)) pattern is observed. These statements agree well with experimental observations. For inhomogeneous colloidal distributions, the pattern similar to Liesegang ring is obtained in simulation (see Figure 8). Here S is set to be inversely proportional to the distance from the origin at which the initial disturbances are added. However, the spacing law in this case does not hold and the space is almost constant.

A similar pattern observed in experiments is also shown in Figure 9, where the homogeneous solution is initially prepared and quenched. However, it is probable that temperature gradients occur. We discuss here a set of equations potentially

(a)   (b)

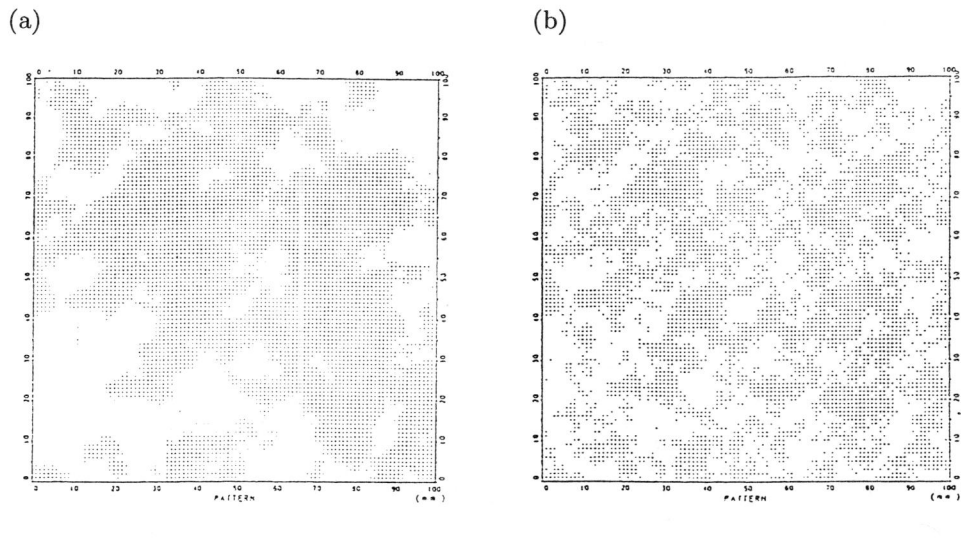

(c)

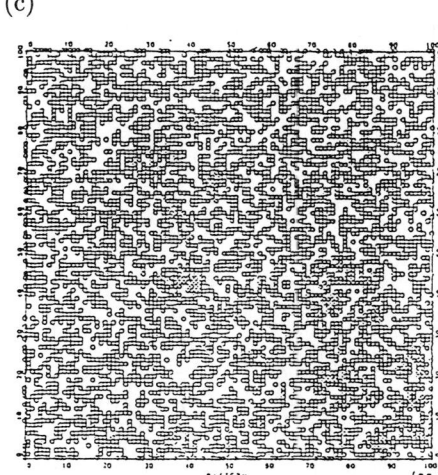

FIGURE 6   Simulation in two dimensions by Eq. (12). (a) $q^{-1} = 0.23 \times 10^{-2}$ s^{-1}, $S_0 = 1.1$, $t = 5 \times 10^5$ s; (b) $q^{-1} = 0.23 \times 10^{-4}$ s^{-1}, $S_0 = 5.0$, $t = 5 \times 10^5$ s; and (c) $q^{-1} = 0.23 \times 10^{-2}$ s^{-1}, $S_0 = 3.0$, $t = 5 \times 10^3$ s.

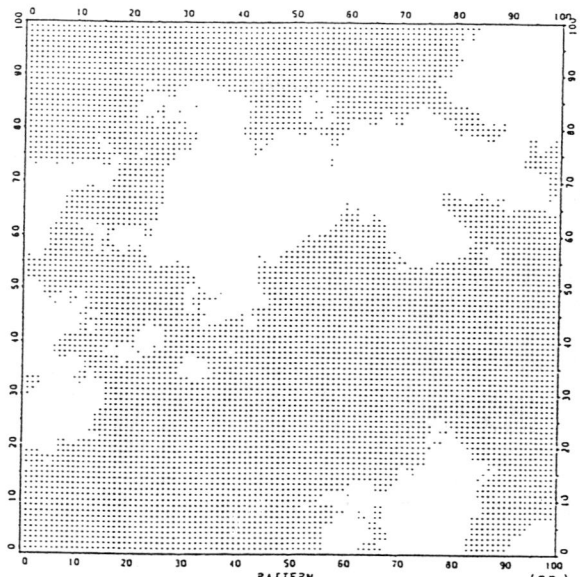

FIGURE 7 An example of a macroscopic pattern with long characteristic length. $D_c = 10^{-4}$ cm^2/s, $q^{-1} = 0.23 \times 10^{-4} s^{-1}$, $S_0 = 3.0$, $t = 4.5 \times 10^5$ s.

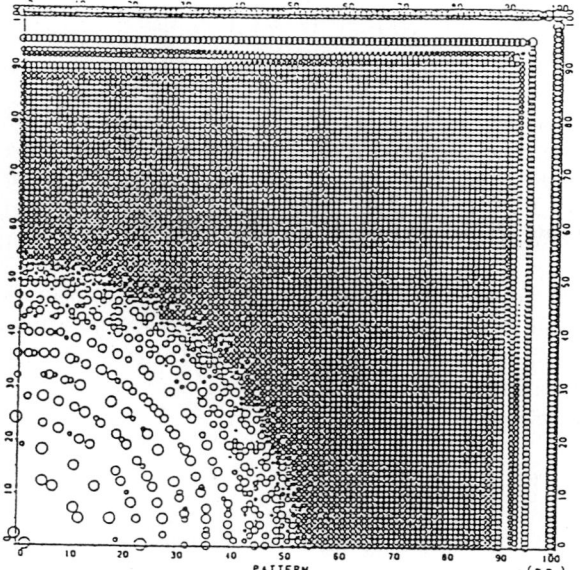

FIGURE 8 An example similar to a Liesegang ring by Eq. (12). Here the monomer concentration is taken as $c(r) \propto r^{-1}$. For $c(r) \propto r^{-1}$ and $\exp(-r)$, no qualitative difference of patterns is obtained.

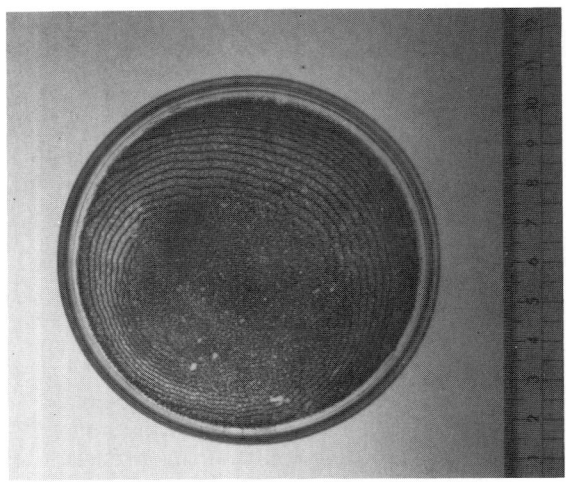

FIGURE 9 An example, similar to that in Figure 8, experimentally obtained from initially prepared homogeneous solution of PbI_2 ($S = 3.0$). An arbitrary temperature gradient exists. The bottom scale is indicated in cm.

containing instability, which comes out when no macroscopic gradients exist. These equations are limited when macroscopic gradients exist and a periodic precipitation forms. Thus they can simultaneously describe two different structures, both in the direction of gradients (i.e., Liesegang ring) and in the direction of free of gradients (i.e., interlayer structures), or in the plane perpendicular to the diffusion direction where gradients are always absent.

SPACING LAW AND LOGNORMAL DISTRIBUTION. The regular spacing law $X_{N+1}/X_N = 1 + p'_N = p_N$ is due to directional diffusion. Here p_N is constant (experimentally it becomes constant for large N). We can rewrite the spacing law as follows:

$$X_N = p_0^N X_0, \tag{13}$$

where X_0 and p_0 are some constant values which depend on only initial conditions. Let us suppose that heterogeneous nucleation randomly distributed in space occurs. These nuclei can induce inhomogeneous concentration gradients in the surrounding areas which are also randomly distributed. Such nondirectional perturbations of diffusion may lead to a random spacing law:

$$X_N(x_i) = p_0^{N_i} X_0, \tag{14}$$

which is spatially dependent; that is, N_i is a random variable depending on space x_i. Taking the logarithm of the above variables and summing over whole space, then we have

$$\log X_N = N_i \log p_0 + \log X_0, \tag{15a}$$

$$\sum_{x_i}^{x_n} \log X_N = \sum_{x_i}^{x_n} N(x_i) \log p_0 + \sum_{x_i}^{x_n} \log X_0. \tag{15b}$$

Since N_i is a random variable, its distribution must be normal; that is, $\log X_N$ must be of normal distribution. Therefore, a set of X_N becomes the lognormal distribution:

$$P(x) = \frac{P_0 x_{av}}{x} \exp\left[-\frac{(\log x/x_{av})^2}{2\sigma_x^2}\right], \tag{16}$$

with $x = X_N$, the average x_{av}, and the variance σ_x. This relation is often observed under some specific kinetics, such as cascade bifurcations and cascade coagulations.[28] This argument gives some idea about the size distribution of many holes (spaces among precipitates) in gradient-free systems. Equations (14)–(16) suggest that the distribution of hole size may be lognormal if the spacing law holds in gradient-free systems (although it is random) as well as for Liesegang ring. We have reported clear evidence of this in previous experimental results.[26,30,31]

3. EXPERIMENT IN GRADIENT-FREE SYSTEMS

3.1 PHASE DIAGRAM

As shown in Figure 10, each hole, which is an element constructing network patterns, usually has a big particle at its center. Such a big particle nucleates and grows at very early stage and works as an initiator of the gradient field of concentrations by absorbing materials surrounding it. By suppressing the growth of particles that nucleate later in the neighborhood by Ostwald ripening or by absorption, the big particle makes a clear zone between itself and the surrounding precipitate groups with many small particles. (we call it a *trigger particle* hereafter.) The experimental result shown in Figure 11 supports this idea where the linear relationship between the size of a trigger particle and the width of a clear zone is observed. That is, a bigger particle results in a bigger clear zone. The size distribution of this hole, therefore, is closely related to the lognormal distribution (see the previous section). The slope of this depends on the initial supersaturation rate S_0 and total excess concentration $\nabla C(T_s) = C(T_i) - C(T_s)$ where $C(T_i)$ is a saturation concentration prepared at T_i and $C(T_s)$ at a settle-down temperature T_s. These also influence the type of observed patterns (see Figure 12). For instance, in the region close to ΔC_H-line in Figure 12(a), which separates quasi-stable states (below the line) from absolutely unstable (above the line: homogeneous nucleation) states, the network pattern (NWP: flocculation pattern) can be observed. On the other hand, in the region close to ΔC_L-line, which separates stable states from quasi-stable (heterogeneous nucleation) states, the Swiss cheese pattern (SWP: gelation pattern) is observed. In such pattern formations, Ostwald ripening plays an important role as follows.

In a solution once supersaturated, a critical size $R^*(t)$ changes with time: $R^*(\infty)/\ln S(t)$. At every instantaneous time, only particles larger than $R^*(t)$ can grow, and smaller ones are dissolved. Accordingly, whether particles can grow or shrink is determined by competition between the growth velocity of particles and the temporal change of R^*.

In the state close to absolutely unstable region with medium values of ΔC and S_0 (initial supersaturation rate), many small particles are nucleated initially because almost-homogeneous nucleation is realized. Therefore, essentially no extra-big particles can form and each particle can grow independently without strong Ostwald ripening. Thus holes are small and SWP could be formed. However, when many particles nucleate at the beginning for the conditions with very large S_0 but small ΔC, S becomes very small immediately and R^* becomes extremely large.

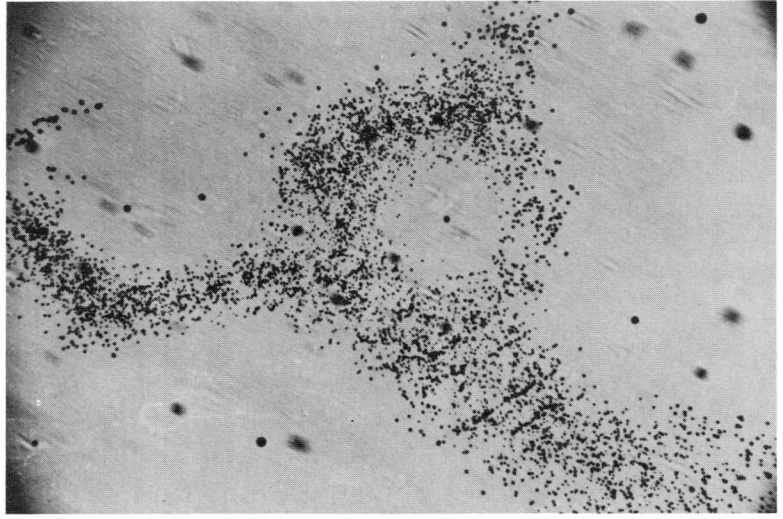

FIGURE 10   Typical networks pattern in gradient-free system ($c_i = 5.2$ mM 1% agar, $T_s = 25°$C). Closed-up pattern (a hole). Here the clear trigger particle is observed at the center of a hole.

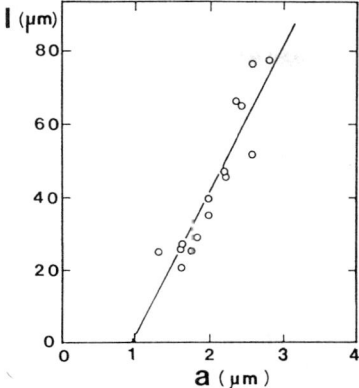

FIGURE 11 Relationship between sizes of the trigger particle diameter $a = 2R$ and the hole l. $C_i = 5.21$ mM, $T_s = 22°$C, $S = 3.04$.

This leads to the result that a very small number of big particles in particle size distributions can grow and a large number of other small particles are dissolved by strong Ostwald ripening. In this case, big holes form, some holes can connect each other, and, as a result, NW patterns are formed. By similar consideration, the following conclusions may hold: (1) At initially large S and large ΔC, Ostwald ripening is small and SWP appears. (2) At initially large S and small ΔC, Ostwald ripening is large and NWP appears. (3) At initially small S and small ΔC, as a small number of particles nucleates and Ostwald ripening is strong, partial precipitates or very faint precipitate could appear. (4) At initially small S and large ΔC, a small number of nucleation occurs and Ostwald ripening could be strong. Then NW patterns are formed. Finally the pattern formation diagram can be summarized on the $S - \Delta C$ plane as shown in Figure 12(b). Here ST, CAS, ACS, and HN denote the regime where the supersaturation theory holds, the regime where competition between ST and ACS occurs, the regime where autocatalytic structures occur, and the regime of heterogeneous nucleation, respectively.[53]

3.2 LOCAL COOLING AND RIPENING

One may expect that the locally dense precipitate will appear when a supersaturation system is further cooled down, partially in space. In Figure 13, we show the result when the center part (~ 1 cm in diameter) of the supersaturated solution of PbI_2 system with $S = 3.4$ is partially superquenched to $T = 0°$C for some fixed interval t_i. In contrast to trivial expectation, in some cases the precipitates are less at the quenched location, which generally depend on t_i.[29,31] For instance, for $t_i < 8$ s, there are rather less precipitates at the strongly quenched part than at surrounding parts. It is probable that many small particles first nucleate and then are dissolved, because S becomes smaller than the equilibrium value later due to the heating-up caused by heat conduction from other areas (see curve a in Figure 14).

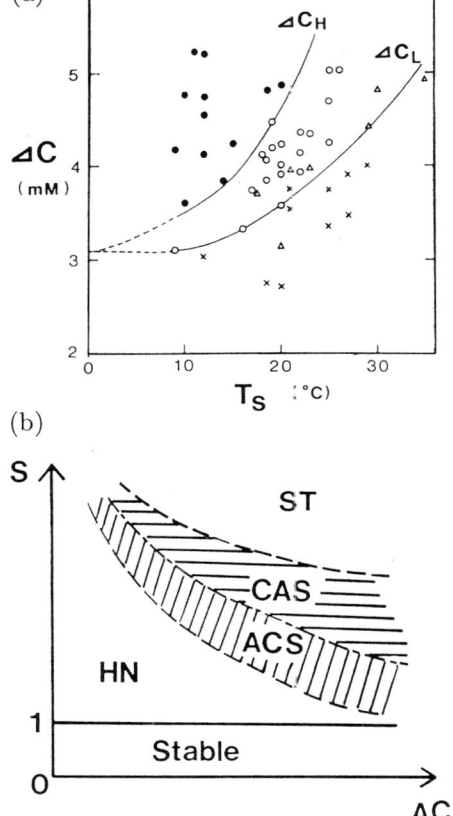

FIGURE 12   Pattern diagram
on the amount of precipitation
ΔC and settle-down temperature
T_s (a) and possible diagram for
the formation mechanism (b).
●: homogeneous and uniform
precipitation (fractal dimension by box
counting method; $D_f = 2.0$ (see Kai et
al.[29]).
o: nonperiodic structures ($1.2 < Df <
1.8$).
△: faint and inhomogeneous
precipitation (fractal-like).
×: no precipitation and structure.
ΔC_H: upper limit of structure formation.
ΔC_L: lower limit (b).
HN: heterogeneous nucleation.
ST: supersaturation theory regime.
CAS: competition between autocatalytic
process and supersaturation theory.
ACS: autocatalytic process.
Stable: no precipitation.

However, as t_i increases (see Figure 13), the quenched location has denser precipitates than the surrounding area by more precipitation and also by ripening. This aspect can be explained by the idea described in the above section, i.e., the dependence on the temporal change of S and ΔC. Namely, when the central region is superquenched ($T = 0 < T_s$), a larger number of small particles nucleate at that location. Immediately after nucleation, S becomes very small and materials start to be supplied from the surrounding area. However, if the diffusion process is relatively slow, new particles can nucleate in the surrounding solution whose sizes are relatively larger because of lower supersaturation. This leads to the growth of the surrounding particles by dissolution of the center particle. The supersaturation rate at the superquenched center area, therefore, depends on the size of superquenched area, the quenched period, and the diffusion velocity. Therefore, if the long period of the superquenched state is given at the center (curve b in Figure 14), dense precipitates could be possible because temperature at the center continuously decreases with time, i.e., maintaining S. Thus the structures are determined by the temperature change of ripening competition between the center and the surrounding particles (see Figure 14).

In relation to this, the structure formations for SWP and NWP are more complicated as shown in Figure 15. The initial size distribution $f(R)$ has an averaged size $\overline{R}$ smaller than a corresponding critical size R^* to S of the solution after nucleation (see Figure 15(a)). Then only small number of large particles can grow dissolving many small particles by competitive growth. As a result, locally dense precipitation called NWP can be realized when total precipitating materials ΔC are sufficient. On the other hand, if $\overline{R}$ is larger than R^*, many crystals can grow, and immediately the supersaturation ratio decreases (see Figure 15(b)). Therefore, rather homogeneous precipitates called SWP are realized. Thus precipitation structures depend on the relationship between a critical size and a size distribution profile. A more detailed discussion has been given in a previous publication.[31]

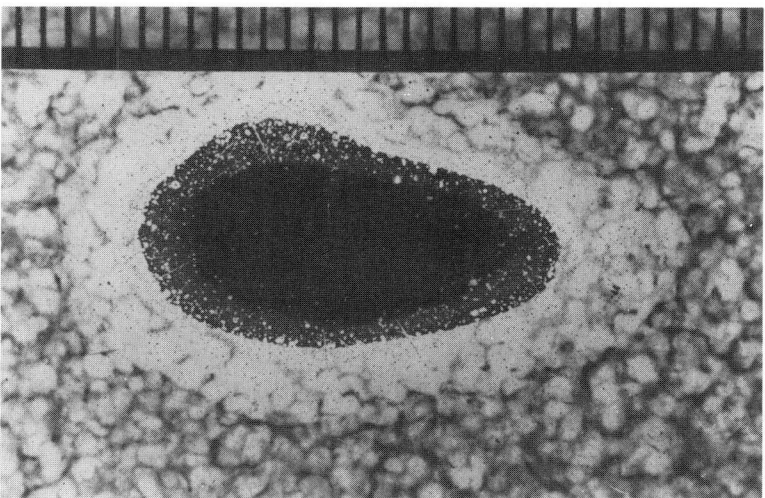

FIGURE 13 Pattern at local superquenching (the central location was cooled down further to $T_s = 0°C$ during period $t_i = 20$ s. ($S = 3.4$, PbI$_2$ system). The scale is indicated in mm.

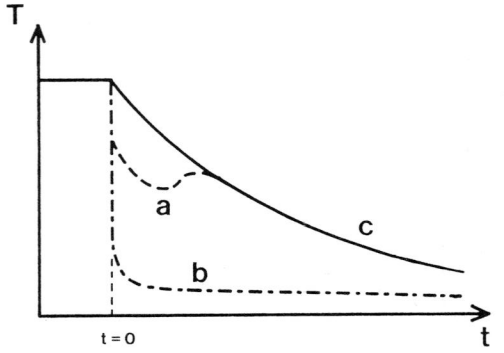

FIGURE 14 Schematic drawing of temporal change of the growth process at local cooling (a and b) (short-period (a) and long-period (b) local cooling). The curve c indicates a temporal change in global cooling at relatively high temperature, for example, at room temperature. The curves a and b show the temperature change at local cooling areas.

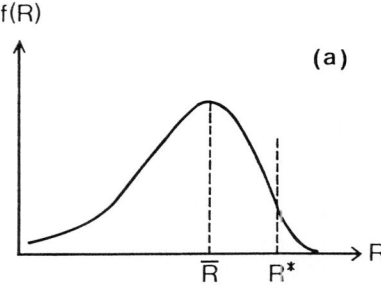

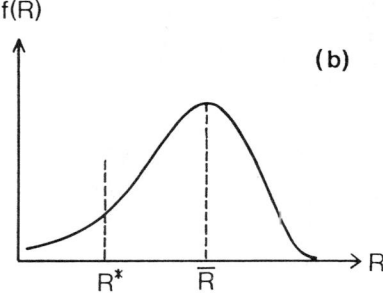

FIGURE 15 Distribution $f(R)$ of particle size at nucleation and critical size R^* at corresponding supersaturation immediately after nucleation.

It is also due to the strong dependence of nucleation rate on S (see Levan and Ross[35]) and it is said that three states in the phase diagram in Figure 12 may be simultaneously realized, i.e., absolutely unstable, quasi-stable, and stable states in space because of the existence of a local temperature gradient.

4. SUMMARY AND CONCLUSION

We have described various theories which have been proposed to describe precipitating pattern formations. The supersaturation theory described well the essential periodicity of the Liesegang ring at large concentration gradients. Therefore, it may be suitable for describing the main feature of precipitation pattern formation in the presense of macroscopic gradients. However, in order to describe the detailed structures, such as subrings and spatial bifurcation, the Ostwald ripening process must be possibly taken into account.

The "sub" phenomena, such as secondary structures, interlayer structure, and subrings, are often observed, for instance, in the lead iodide system but not in the magnesium hydroxicide system. There, in the gradient-free system of $Mg(OH)_2$, only homogeneous and uniform precipitation is observed and no pattern forms experimentally, unlike in the case of PbI_2.[27] As was shown in the text, Eqs. (4) for Liesegang ring formation are continuously connected to Eq. (7) for the gradient-free case by taking $\nabla^2 = 0$. Speaking from a theoretical point of view related to these equations, therefore, it can be said that the equation for $Mg(OH)_2$ system is stable and that for PbI_2 it is unstable for disturbances. This means the contribution of Ostwald ripening in the $Mg(OH)2$ system would be largely different from that in the PbI_2 system. Namely, if P in Eq. (7) is extremely large, homogeneous precipitates form and no macroscopic structure appears in gradient-free case. Liesegang rings form due to the existence of large macroscopic gradients of the concentration but no subring forms in the $Mg(OH)_2$ system. The instability and competitive growth theory, therefore, may be useful for gradient-free case of PbI_2.

In precipitation systems, we observe fractal-like patterns which indicate no characteristic length in the formation mechanism. Only diffusion is important in the scale of the order of centimeters. (The effect of the reaction is of smaller scale. This determines the minimum cutoff in logarithmic plot, shown in the fractal measurement by Kai et al.[30]) This suggests that Ostwald ripening is important in the scale of the order of less than cm (the mesoscopic scale). In future theoretical treatments, gelation of agar could be taken into account in order to provide a more sophisticated theory.

REFERENCES

1. Ben-Jacob, E., H. Scmueli, O. Shochet, and A. Tenenbaum. "Adaptive Self-Organization During Growth of Bacteria Colonies." *Physica* **A187** (1992): 378.
2. Bradford, S. C. "Adsorptive Stratification in Gel." *Biochem. J.* **10** (1916): 369.

3. Bradford, S. C. "Adsorption Theory of Straited Precipitates." *Kolloid-Z.* **30** (1922): 364.

4. Chernevsky, D. S., A. A. Polezhaev, and S. C. Muller. "A Model of Pattern Formation by Precipitation." *Physica* **D54** (1991): 160.

5. Dee, G. T. "The Patterns Produced by Precipitation at a Moving Reaction Front." *Phys. Rev. Lett.* **57** (1986): 275.

6. Dee, G. T. "Patterns Produced by Precipitation at a Moving Reaction Front." *Physica* **23D** (1986): 340.

7. Dhar, N. R., and A. C. Chatterji. "Theories of Liesegang Ring Formation." *Kolloid-Z.* **37** (1925): 2.

8. Falkovitz, M. S., and J. B. Keller. "Precipitation Pattern Formation." *J. Chem. Phys.* **88** (1988): 416.

9. Feinn, D., P. Ortoleva, W. Scalf, S. Schmidt, and M. Wolff. "Spontaneous Pattern Formation in Precipitating Systems." *J. Chem. Phys.* **69** (1978): 27.

10. Fischer, M. H., and G. D. McLaughlin. "Comments on the Theory of Liesegang Rings." *Kolloid-Z.* **30** (1922): 13.

11. Flicker, M., and J. Ross. "Mechanism of Chemical Instability for Periodic Precipitation Phenomena." *J. Chem. Phys.* **60** (1974): 3458.

12. Fricke, R. "Theory of Liesegang Rings." *Z. Phys. Chem.* **107** (1923): 41.

13. Fujikawa, H., and M. Matsushita. "Bacterial Fractal Growth in the Concentration Field of Nutrient." *J. Phys. Soc. Japan* **60** (1991): 88.

14. Fujikawa, H. "Periodic Growth of *Bacillus subtilis* Colonies on Agar Plates." *Physica* **A189** (1992): 15.

15. Gnanam, F. D., S. Krishnan, P. Ramasamy, and G. S. Laddha. "Periodic Precipitation of Calcium Carbonate in Agar Gel." *J. Colloid. Inter. Sci.* **73** (1980): 193.

16. Hedges, E. S., and J. E. Myers. *The Problem of Physico-Chemical Periodicity.* New York: Longmans Green, 1926.

17. Hedges, E. S., and R. V. Henley. "The Formation of Liesegang Rings as a Periodic Coagulation Phenomenon." *J. Chem. Soc.* (1928): 2714.

18. Hedges, E. S. "Periodic Structures From Interacting Gases." *J. Chem. Soc.* (1929): 1848.

19. Hedges, E. D. *Liesegang Ring and Other Periodic Precipitation.* London: Chapman and Hall, 1932.

20. Henisch, H. *Crystals in Gels and Liesegang Rings.* New York: Cambridge University Press, 1986.

21. Kahlweit, M. "Kinetics of Phase Formation in Condensed Systems. V. Periodic Precipitation and a Model Experiment on the Internal Oxidation of Metal Alloys." *Z. Phys. Chem. Neue Folge* **32** (1962): 1.

22. Kahlweit, M. "Precipitation and Aging." In *Physical Chemistry*, edited by H. Eyring, D. Henderson, and W. Jost, 719–759, Vol. 10. New York: Academic Press, 1970.

23. Kahlweit, M. "Ostwald Ripening of Precipitates." *Adv. Colloid Inter. Sci.* **5** (1975): 1.

24. Kai, S., S. C. Muller, and J. Ross. "Measurements of Temporal and Spatial Sequences of Events in Periodic Precipitation Processes." *J. Chem. Phys.* **76** (1982): 1392.

25. Kai, S., S. C. Muller, and J. Ross. "Periodic Precipitation Patterns in the Presence of Concentration Gradients II." *J. Phys. Chem.* **87** (1983): 806.

26. Kai, S., and S. C. Muller. "Spatial and Temporal Macroscopic Structures in Chemical Reaction Systems: Precipitation Patterns and Interfacial Motion." *Sci. Form* **1** (1985): 9.

27. Kai, S., S. Higaki, H. Yamazaki, and T. Yamada. "Morphogenesis in Precipitation Processes—Importance of Ostwald Ripening." *Trans. IEE Jpn.* **107c** (1987): 1011. (Japanese)

28. Kai, S., S. Higaki, M. Imasaki, and H. Furukawa. "1/f Noise, Lognormal Distribution, and Cascade Process in Electrical Networks." *Phys. Rev.* **A35** (1987): 374.

29. Kai, S., T. Yamada, and H. Yamazaki. "Pattern Formation Kinetics and Fractality in Precipitating Phenomena." *J. Miner. Petrol. Econ. Geol.* (special issue) **4** (1988): 27–36. (Japanese)

30. Kai, S., T. Yamada, S. Ikuta, and S. C. Muller. "Fractal Geometry of Precipitation Patterns." *J. Phys. Soc. Japan* **58** (1989): 3445.

31. Kai, S. "Pattern Formation in Precipitation." In *Dynamics, Statistics and Formation of Patterns*, edited by K. Kawasaki and M. Suzuki. Singapore: World Scientific, 1993.

32. Keller, J. B., and S. I. Rubinow. "Recurrent Precipitation and Liesegang Rings." *J. Chem. Phys.* **74** (1981): 5000.

33. Kirkaldy, J. S. "Spontaneous Evolution of Spatiotemporal Patterns in Materials." *Rep. Prog. Phys.* **55** (1992): 723–95.

34. Langer, J. S. "Instabilities and Pattern Formation in Crystal Growth." *Rev. Mod. Phys.* **52** (1980): 1.

35. Levan, M. E., and J. Ross. "Measurements and Hypothesis on Periodic Precipitation Processes." *J. Chem. Phys.* **91** (1987): 6300.

36. Liesegang, R. E. "Uber Einige Eigenschaften von Gallerten." *Naturwiss. Wochenschr.* **11** (1896): 353.

37. Lorenz, E. N. "Deterministic Nonperiodic Flow." *J. Atmos. Sci.* **20** (1963): 130.

38. Lovett, R., P. Ortoleva, and J. Ross. "Kinetic Instabilities in First Order Phase Transitions." *J. Chem. Phys.* *69* (1978): 947.

39. McBirney, A. R., and R. M. Noyes. "Crystallization and Layering of the Skaergaard Intrusion." *J. Petrol.* **20** (1979): 487.

40. Meakin, P. "Diffusion-Controlled Cluster Formation in 2-6-Dimensional Space." *Phys. Rev.* **A27** (1983): 1495.

41. Merino, E. "Survey of Geochemical Self-Patterning Phenomena." In *NATO ASI Ser. C (Chem. Inst.)* **20** (1984): 305.

42. Muller, S. C., S. Kai, and J. Ross. "Curiosities in Periodic Precipitation Patterns." *Science* **216** (1982): 635.

43. Muller, S. C., S. Kai, and J. Ross. "Periodic Precipitation Patterns in the Presence of Concentration Gradients I." *J. Phys. Chem.* **86** (1982): 4078.
44. Muller, S. C., S. Kai, and J. Ross. "Mesoscopic Structure of Pattern Formation in Initially Uniform Colloid." *J. Phys. Chem.* **86** (1982): 4294.
45. Muller, S. C., and G. Venzl. *Modeling of Patterns in Space and Time.* Lecture Notes in Biomathematics, edited W. Jaeger. Berlin: Springer, 1983.
46. Nishiyama, T. "CDS Modeling of Chemical Layering due to Uphill Diffusion in a Mineral-Fluid System." *Proc. 29th Interl. Geol. Congr., Part A* **7** (1994): 7–16
47. Ostwald, Wi. "M." *Z. Phys. Chem.* **27** (1897): 265.
48. Ostwald, Wo. "Theory of Liesegang Ring." *Kolloid-Z.* **36** (1925): 380.
49. Prager, S. "Periodic Precipitation." *J. Chem. Phys.* **25** (1956): 279.
50. Raman, C. V., and K. S. Ramaiah. "On the Wave-Like Character of Periodic Precipitates." *Proc. Ind. Acad. Sci.* **9A** (1939): 455.
51. Rayleigh, L. "Periodic Precipitates." *Phil. Mag.* **38** (1919): 738.
52. Shinohara, S. "A Theory of One Dimensional Liesegang Phenomena." *J. Phys. Soc. Japan* **29** (1970): 1073.
53. Smith, D. A. "On Ostwald's Supersaturation Theory of Rhythmic Precipitation (Liesegang's Rings)." *J. Chem. Phys.* **81** (1984): 3102.
54. Toramaru, A., K. Ito, and S. Kai. "The Liesegang Model of the Titus-Bode's Law." *Forma* **6** (1991): 247.
55. van Oss, C. J., and P. Hirsch-Ayalon. "An Explanation of the Liesegang Phenomenon." *Science* **129** (1959): 1365.
56. Venzl, G., and J. Ross. "Nucleation and Colloidal Growth in Concentration Gradients (Liesegangs)." *J. Chem. Phys.* **77** (1982): 1302.
57. Venzl, G. "Pattern Formation in Precipitation Processes. I. The Theory of Competitive Coarsening." *J. Chem. Phys.* **85** (1986): 1996.
58. Venzl, G. "Pattern Formation in Precipitation Processes. II. A Postnucleation Theory of Liesegang Bands." *J. Chem. Phys.* **85** (1986): 2006.
59. Wagner, C. "Mathematical Analysis of the Formation of Periodic Precipitations." *J. Colloid Sci.* **5** (1950): 85.
60. Wimpenny, J. W. T. "Spatial Order in Microbial Ecosystems." *Biol. Rev.* **56** (1981): 295.
61. Witten, T. A., and L. M. Sander. "Diffusion-Limited Aggregation." *Phys. Rev.* **B27** (1983): 5686.

ADDITIONAL REFERENCES

For students who want to know the complete history of the Lisegang phenomenon: The author recommends an old review article written by K. H. Stern. "The Liesegang Phenomenon." *Chem. Revs.* **54** (1954): 79. As well as the above references 20, 31, and 33, and a bibliography *A Bibliography of Liesegang Rings*, edited by K. H. Stern. Washington, D.C.: U. S. Government Printing Office, 1967.

V. O. Pannbacker,† **O. Jensen,**† **G. Dewel,**‡ **P. Borckmans,**‡ **and E. Mosekilde**†
†Physics Department, The Technical University of Denmark, 2800 Lyngby, Denmark
‡Service de Chimie-Physique and Centre for Nonlinear Phenomena and Complex Systems
C.P. 231, Université Libre de Bruxelles, 1050 Brussels, Belgium

Localized Structures in the Chlorine Dioxide-Iodide-Malonic Acid System

1. INTRODUCTION

In his 1952 paper "The Chemical Basis of Morphogenesis," Turing[15] described how diffusion driven instabilities can lead to stationary space-periodic-concentration patterns with an intrinsic wavelength. Such patterns are usually referred to as Turing structures. They can arise only if the diffusion coefficients of the major species are sufficiently different.

In 1990, three-dimensional Turing patterns were finally obtained[4] using the Chlorite-Iodide-Malonic Acid (CIMA) reaction. Subsequently quasi-two-dimensional structures were also produced[10] using the same reaction in a similar experimental set-up. To prevent convective disturbances, the experiments are implemented in a reactor filled with a gel implanted with starch, that serves as color indicator by complexing the iodine species. Because of its size, this complex is immobilized in the gel and its formation reduces the effective diffusion coefficient of iodide.[9]

Spatio-Temporal Patterns, Ed. P. E. Cladis and P. Palffy-Muhoray,
SFI Studies in the Sciences of Complexity, Addison-Wesley, 1995   **469**

A kinetic model of the chlorine-dioxide-iodide-malonic acid (CDIMA) reaction, a subsidiary reaction of CIMA that also gives rise to experimental Turing structures,[6] has been developed[8] taking into account the complexing effect of the starch. In terms of the dimensionless iodide $I^- = u$ and chlorite $[ClO_2^-] = v$ concentrations the corresponding reaction-diffusion equations take the form:

$$\frac{\partial u}{\partial t} = a - u - \frac{4uv}{1 + u^2} + \nabla^2 u$$

$$\frac{\partial v}{\partial t} = \delta \left[b[u - \frac{uv}{1 + u^2}] + c\nabla^2 v \right]$$

where a and b are parameters related to feed concentrations of other reactants and $c = D_{ClO_2^-}/D_{I^-}$ represents the ratio of the molecular diffusion coefficients of the inhibitor (v) and activator (u). δ, that is proportional to the starch concentration, accounts for the complex formation.

We have studied the one- and two-dimensional bifurcation diagrams of this model in a region where subcriticality is important. Because of multistability and pinning effects a host of localized structures may be generated.[7] The results were obtained by integrating the model with the following values (unless otherwise noted) of the parameters: $a = 30, c = 1.5, \delta = 8$ whereas b is used as bifurcation parameter. In the one-dimensional case use was made of the semi-implicit Crank-Nicolson scheme with either periodic or no-flux boundary conditions. In two dimensions, we used alternating direction implicit or odd-even hopscotch methods respectively for no-flux or periodic boundary conditions.

2. BIFURCATION DIAGRAM

Linear stability analysis[8] of the model shows that for realistic values of the parameters, the unique homogeneous steady state (HSS), $u_s = a/5$ and $v_s = 1 + a^2/25$, becomes diffusionally unstable to infinitesimal perturbations with a nonzero wavenumber k_c as the control parameter b is decreased below a characteristic value b_T. The system may also undergo a Hopf bifurcation at b_H where a pair of complex conjugate eigenvalues of the linearized matrix crosses the imaginary axis. However, when starch is present in sufficient amount $(\delta > 1)$, the Hopf bifurcation is pushed beyond the Turing threshold $(b_T < b_H)$ (see Figure 1) even when the molecular diffusion coefficients of the two species are of the same order of magnitude $(c = 1.5)$.

As shown in Figure 2 the bifurcation to space-periodic structures is strongly subcritical. The Hopf bifurcation at $b = b_H$ also presents a weakly inverted character. As a result, in the subcritical region $(b > b_T)$, the model exhibits bistability between stripes and the homogeneous reference steady state (HSS) for one-dimensional systems. In two dimensions, even a region of tristability is observed between the HSS, stripes and hexagons (see Figure 2).

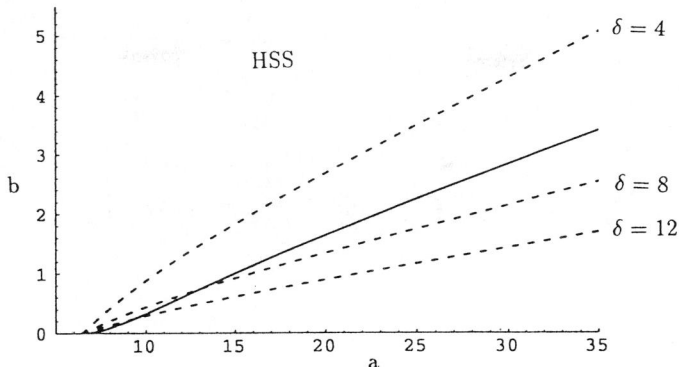

FIGURE 1   Linear stability analysis of the reference homogeneous steady state (HSS). The dashed line corresponds to the Hopf bifurcation for different values of δ. The solid line stands for the Turing bifurcation.

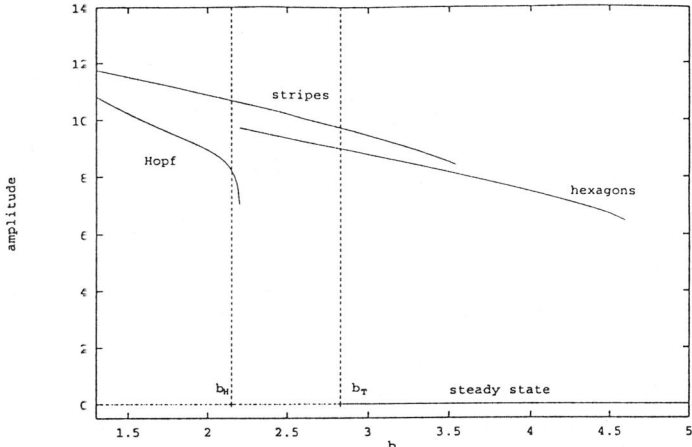

FIGURE 2   Bifurcation diagram in two dimensions for $a = 30, \delta = 8$, and $c = 1.5$.

The structures shown in Figure 3 exhibit hexagons (with two different phase relations[3]) and stripes, as also found experimentally. They are obtained by numerical integration of the reaction-diffusion system considered, starting from the HSS with addition of a low amplitude noise.

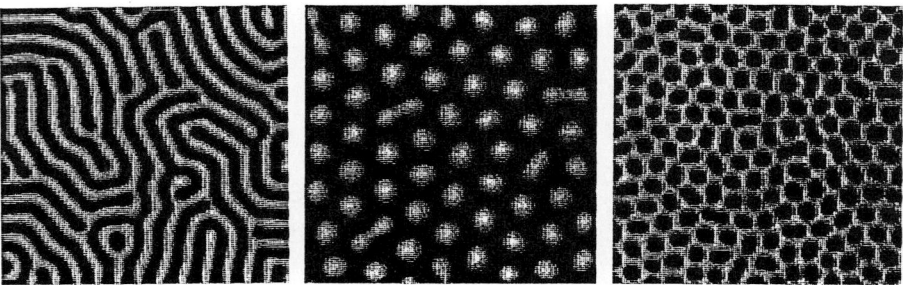

FIGURE 3 Structures obtained for the concentration of $u =$[I$^-$] for the CDIMA reaction model: stripes and hexagons with two different phase relations. $c = 1.5$ and the other parameters (a, b, δ) are from left to right: (15, 0.52, 8), (10, 0.08, 11), and (30, 2.5, 8). The gray scales correspond to the concentration variations between minimum (black) and maximum (white).

In large systems these patterns contain many defects. Patches of differently oriented hexagons nucleate at various points. In the growth process when these regions meet, grain boundaries form. Because of the frustration induced by these defects, distorted hexagons with characteristic angles slightly different from $\pi/3$ may then appear spontaneously. The boundary conditions also play a role in the mechanism of pattern selection. As in convective instabilities, no flux boundary conditions tend to align the stripes perpendicularly to the borders. This effect also induces the creation of defects (dislocations/disclinations) in the pattern.

3. STATIONARY LOCALIZED STRUCTURES

As we have discovered in the preceding section, the Turing bifurcations to hexagons (two dimensions) or stripes (one and two dimensions) are strongly subcritical for the values of the parameters we have considered. The subcritical domain thus consists of regions of multistability where a large number of localized structures may be supported as our numerical simulations have shown.[7]

In one dimension, the simplest consists of a front linking the HSS to stripes. As the corresponding bistable region is wide, the interface is sharp and its pinning[13] results from its interaction with the underlying periodic structure over a finite band of values of bifurcation parameter b. Outside this locking band, depinning takes place and, for instance on the lower edge of the band, the periodic structure becomes dominant and invades the territory of the HSS (see Figure 4). The reverse occurs on the upper edge. In the process the front moves with a nonuniform velocity[2]: it proceeds by rapid jumps over a distance of one wavelength of the growing pattern

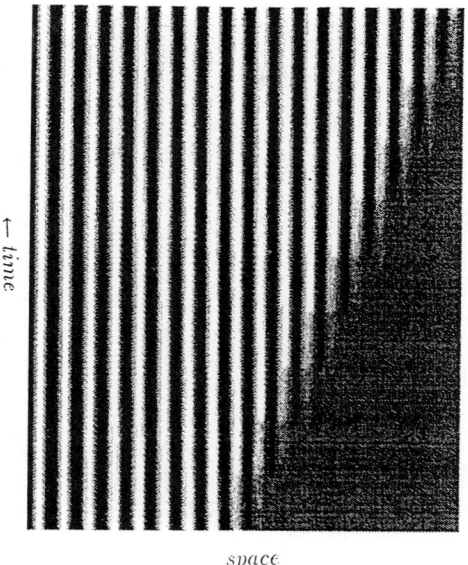

FIGURE 4 A one-dimensional periodic structure invades the homogeneous steady state (space-time plot: time runs upward).

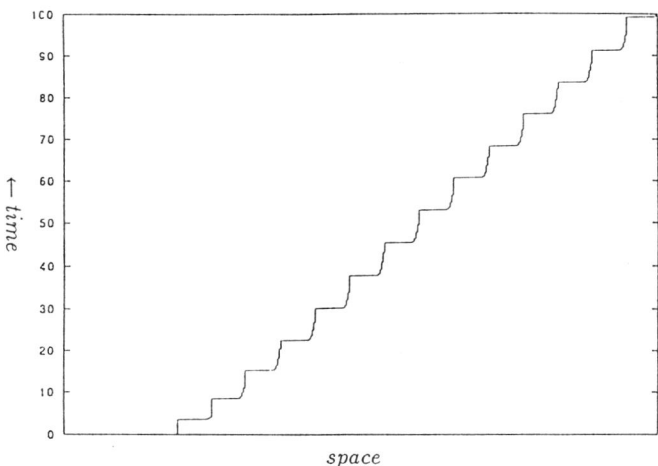

FIGURE 5 Position of the front of the space periodic structure (compare Figure 4), as a function of time when $b(= 3)$ falls in the region $b_T < b < p_{low}$, outside the pinning band.

← time

space

FIGURE 6   Pinned front connecting a Turing pattern (right) to a wavetrain with a very small wavenumber.

followed by a latency period (see Figure 5). This period increases as one comes close to the edge of the locking band where this period diverges. This behavior is characteristic of so-called nonadiabatic dynamical contributions, not contained in the weakly nonlinear amplitude equations.

Building on the pinned front structure, other coherent structures may be formed: droplets of one state embedded in the other. For instance, depending on the initial conditions, a Turing droplet may contain a variable number of wavelengths for the same value of the parameters as a result of the same pinning effects.

In two dimensions, the situation is even richer as the system may exhibit tristability. Fronts and droplets bringing into contact structures of different symmetries, for instance a droplet of an hexagonal Turing structure in the background of the HSS, are then possible. Their stability again results from the existence of pinning effects as may be shown by the characteristics of the depinning transition. Nonvariational effects[14] arising from the nongradient character of the dynamics of the CDIMA model do not come into play near the bifurcation point.

4. SPIRALS AND VORTICES IN ONE-DIMENSIONAL AND TWO-DIMENSIONAL

Front solutions connecting a striped Turing pattern domain to a train of plane waves may also be formed in the region of parameter space where these two global

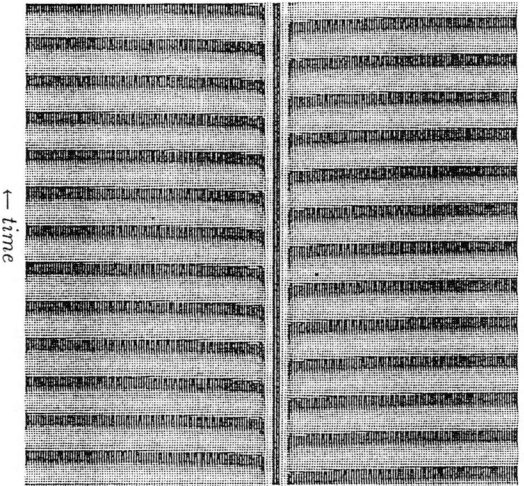

space

FIGURE 7 Embedded droplet of Turing structure in a background of homogeneous oscillations.

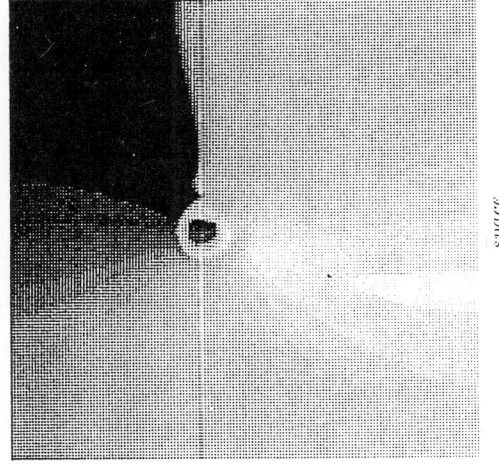

space

FIGURE 8 Two-dimensional vortex with a Turing induced core (pacemaker).

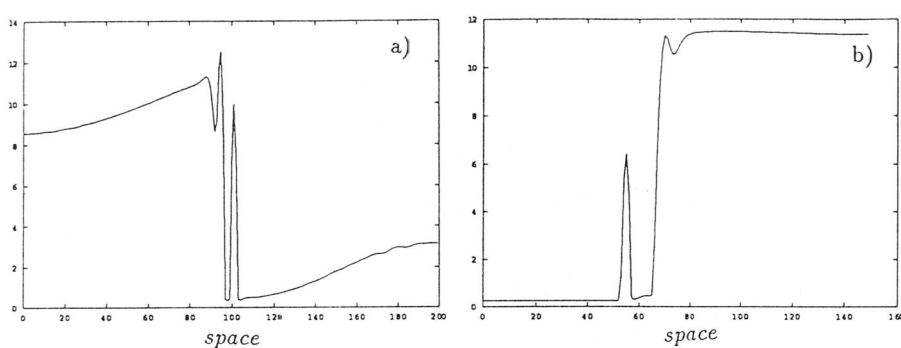

FIGURE 9 A section through the core for (a) one and (b) two dimensions (South-West to North-East).

structures are simultaneously stable and where the mixed Turing-Hopf mode is unstable. As the result of a subtle compensation between the spatial and the non-linear dispersion effects, the wavenumber of the waves generated by the front is very small and therefore the corresponding phase velocity is very large as shown in the space-time plot of Figure 6.

Because of the pinning effects described above, stationary fronts may be obtained for a finite range of values of the bifurcation parameter. By varying the control parameter b outside the pinning band, we have studied numerically the dynamics of these fronts. Close to the Hopf bifurcation, the Turing state is dominant and tends to occupy the territory of the waves, the reverse situation occuring near the lower edge of the band. In this process the front velocity exhibits characteristic oscillations as already discussed in the preceding section. Here, however, one wavelength is added (or substracted) during some multiple of the wave period.

These fronts may again serve to construct droplets, here localized Turing-Hopf structures. An exemple is shown in Figure 7 where a one-dimensional Turing structure truncated to a few wavelengths is embedded in a background of homogeneous oscillations. The amplitude of the waves goes to zero in the core where the Turing amplitude is maximum. Similar asynchronous wave sources have been obtained in the numerical simulations of the Brusselator.[12] They have also been observed[11,12] in experiments with the CIMA reaction ("Chemical Flip-Flop").

In two dimensions, the same conditions yield a spiral with a Turing spot in the core, playing the role of a pacemaker (see Figure 8). A section along a radial line passing through the core reveals the same profile as in one dimension (see Figure 9). Here again the wavelength of the emitted waves is large. For some values of the parameters vortices[1] may be obtained that exhibit radial isophases: u or v varies as $\exp i(\omega t \pm \theta)(r, \theta$: polar coordinates). The long range interactions between these vortices may give rise to nontrivial dynamical behaviors[5] that could be observed in tha CDIMA reaction.

ACKNOWLEDGMENTS

G. D. and P. B. are Senior Research Associates with the F.N.R.S. (Belgium). Discussions with A. De Wit and S. Metens are acknowledged. This work was sponsored by Twinning contract # SC1*-CT91-0706 of the EC Science Program.

REFERENCES

1. Aranson, I. S., L. Kramer, and A. Weber. "On the Interaction of Spiral Waves in Nonequilibrium Media." *Physica D* **53** (1991): 376–384.
2. Bensimon, D., B. I. Shraiman, and V. Croquette. "Nonadiabatic Effects in Convection." *Phys. Rev. A* **38** (1988): 5461–5464.
3. Borckmans, P., A. De Wit, and G. Dewel. "Competition in Ramped Turing Structures." *Physica A* **188** (1992): 137–157.
4. Castets, V., E. Dulos, J. Boissonade, and P. De Kepper. "Experimental Evidence of a Sustained Turing-Type Nonequilibrium Chemical Pattern." *Phys. Rev. Lett.* **64** (1990): 2953–2956.
5. Elphick, C., and E. Meron. "Dynamics of Phase Singularities in Two-Dimensional Oscillating Systems." *Physica D* **53** (1991): 385–399.
6. Epstein, I., I. Lengyel, S. Kadar, M. Kagan, and M. Yokoyama . "New Systems for Pattern Formation Studies." *Physica A* **188** (1992): 26–33.
7. Jensen, O., V. O. Pannbacker, G. Dewel, and P. Borckmans. "Subcritical Transitions to Turing Structures." *Phys. Lett. A* **79** (1993): 91–96.
8. Lengyel, I., and I. Epstein. "Modeling of Turing Structures in the Chlorite-Iodide-Malonic Acid-Starch Reaction System." *Science* **251** (1991): 650–652.
9. Lengyel, I., and I. Epstein. "A Chemical Approach to Designing Turing Patterns in Reaction-Diffusion Systems." *Proc. Natl. Acad. Sci. (USA)* **89** (1992): 3977–3979.
10. Ouyang, Q., and H. L. Swinney. "Transition From a Uniform State to Hexagonal and Striped Turing Patterns." *Nature* **352** (1991): 610–612.
11. Perraud, J. J., A. Agladze, E. Dulos, and P. De Kepper. "Stationary Turing Patterns Versus Time-Dependent Structures in the Chlorite-Iodide-Malonic Acid Reaction." *Physica A* **188** (1992): 1–16.
12. Perraud, J. J., E. Dulos, P. De Kepper, A. De Wit, G. Dewel, and P. Borckmans. "Turing-Hopf Localized Structures." In *Spatio-Temporal Organization in Nonequilibrium Systems*, edited by S. C. Müller and T. Plesser, 205–210. Projekt Verlag, 1992.
13. Pomeau, Y. "Front Motion, Metastability and Subcritical Bifurcations in Hydrodynamics." *Physica D* **23** (1986): 3–11.

14. Thual, O., and S. Fauve. "Localized Structures Generated by Subcritical Instabilities." *J. Phys. (France)* **49** (1988): 1829–1834.
15. Turing, A. "The Chemical Basis of Morphogenesis." *Phil. Trans. R. Soc. Lond.* **237B** (1952): 37–72.

Igor Aranson,[†] **Lorenz Kramer,**[‡] **and Andreas Weber**[‡]

†Department of Physics, Bar-Ilan University, Ramat-Gan, Israel 52900
‡Physikalisches Institut der Universität Bayreuth, 8580 Bayreuth, Postfach 101251, Germany

The Theory of Motion of Spiral Waves in Oscillatory Media

For the complex Ginzburg-Landau equation, symmetric bound spiral pairs exist in the parameter range where the perturbations behave asymptotically in an oscillatory manner. In the like-charged case one has multiple bound states which may be interpreted as multiple-armed spirals. Outside the oscillatory range, bound states do not appear to exist and well-separated spirals are found to repel each other regardless of topological charge. Bound spiral pairs appear to be unstable with respect to spontaneous symmetry breaking. The existence and the stability of symmetric and asymmetric lattices of spirals is discussed. Finally the interaction of a spiral with a localized axisymmetric inhomogeneity is investigated. It is shown that in the oscillatory range stable rotation of the spiral around the inhomogeneity (trapping) is possible. The radius of the equilibrium orbit is calculated analytically.

Spatio-Temporal Patterns, Ed. P. E. Cladis and P. Palffy-Muhoray,
SFI Studies in the Sciences of Complexity, Addison-Wesley, 1995   **479**

1. INTRODUCTION

Spatially extended oscillatory media can be found in physics, chemistry, and biology. Examples of oscillatory media are chemical oscillations like the famous Belousov-Zhabotinskii reaction[23,29] or some cases of catalytic surface reactions,[16,18] systems sustaining (nonlinear) traveling waves as observed and/or predicted in thermal convection in binary fluids,[22,26,32] homeotropically oriented nematic liquid crystals,[11,21] and electroconvection in planarly aligned nematic liquid crystals[27] (for a general review see Hohenberg et al.[14]). Transversely extended lasers are other systems where the oscillatory instability occurs.[5,6,10]

The simplest description of such media is provided by the complex Ginzburg-Landau equation (*CGLE*)

$$\frac{\partial a}{\partial t} = (1 + \nu(\mathbf{r}))a + (1 + ib)\Delta a - (1 + ic)|a|^2 a. \tag{1}$$

where the complex field a describes the amplitude and phase of the modulations of the pattern.[19,20,24] A weak inhomogeneity of the medium is included by allowing for a function $\nu(\mathbf{r})$. This *CGLE* plays the role of a normal form in the vicinity of a supercritical transition to an oscillatory state in spatially extended systems and is thus very general. Many of the results obtained from the CGLE carry over qualitatively to situations where a more complicated description, e.g., by reaction-diffusion models, is more appropriate.

The simple zeros of a represent topologically stable point defects in two dimensions (and line defects in three dimensions). One has topological quantum numbers (charges) ± 1 related to the phase change of $\pm 2\pi$ when going around the defect. When $b - c \neq 0$, the defects are sources of spiral waves whose constant phase lines behave like an Archimedean spiral, except in the immediate neighborhood of the core. When $b - c$ tends to zero, the emitted wavenumber goes to zero and the spirals go over into vortices which rotate if $b = c \neq 0$.

The asymptotic interaction is very different for the two cases: For $b - c = 0$, it is long-range decaying essentially like r^{-1} with corrections due to dynamic screening[8,17] whereas, for $b - c \neq 0$, it is short-range decaying that is essentially exponential.[2,7,25] Interaction manifests itself in a motion of each spiral. The resulting velocity has a radial (along the line connecting the spiral cores) and a tangential component.

In Aranson et al.[2] we presented detailed simulations of Eq. (1) for symmetric spiral pairs showing that a stable bound state exists if $|b-c|$ is not too small, i.e., for $(|b - c|/(1 + bc)| > c_{cr} = 0.845...)$.

There is numerical evidence that symmetric bound states after a sufficiently long evolution spontaneously break the symmetry and one spiral begins to dominate, pushing away other spirals. In fact, at least in the convectively unstable, but absolutely stable range, a symmetry-broken state is produced directly from random initial conditions.[1,33] To understand the symmetry-breaking instability[3] one may consider the perturbation of the frequency ω of the waves emitted by each spiral, caused by the interaction with the other spiral. Indeed, from the analysis of $CGLE$, it is known that the shock (or sink) where two waves with different frequencies ω_i collide moves in the direction of smaller frequency, which means that after sufficiently long time only the larger frequency (or equivalently the larger wavenumber because of the dispersion relation $\omega = -c(1 - k^2) - bk^2$) dominates in a bounded system. Therefore, if due to the interaction the frequencies of rotation of the spirals become different, one can expect a drastic breaking of the symmetry of the system.

Another interesting problem is the drift of a spiral in the gradient created by a localized inhomogeneity.[1] For a weak inhomogeneity the velocity of the motion can be found perturbatively. The spiral may be trapped at some distance from the inhomogeneity resulting in stationary rotation around the inhomogeneity.[5,6]

The problems discussed in this paper are treated within a general perturbative framework described in Sections 2 and 3. The results of Sections 3–4 have been published before by Aranson et al.[2,3]

2. THE PERTURBATION METHOD

The (one-armed or singly charged) isolated spiral solution of Eq. (1) is of the form

$$\tilde{a}(r, \theta) = F(r) \exp\left(i(\omega t \pm \theta + \psi(r) + \varphi)\right) \tag{2}$$

and satisfies the following equations for the functions $F(r)$ and $\psi(r)$:

$$\Delta_r F - \frac{1}{r^2}F - (\psi')^2 F - b[(\Delta_r \psi)F + 2\psi'F'] + F - F^3 = 0$$

$$b\left[\Delta_r F - \frac{1}{r^2}F - (\psi')^2 F\right] + (\Delta_r \psi)F + 2\psi'F' - \omega F - cF^3 = 0 \tag{3}$$

where (r, θ) are polar coordinates, $\Delta_r = \partial_r^2 + (1/r)\partial_r$. Primes denote derivatives with respect to r, and $\varphi = const$ is an arbitrary phase. The functions F and ψ have the following asymptotic behaviour

$$F(r) \to \sqrt{1 - k^2} \ , \psi'(r) \to k \qquad r \to \infty$$

$$F(r) \sim r \ , \psi'(r) \sim r \qquad r \to 0 \tag{4}$$

and $\omega = -bk^2 - c(1 - k^2)$. The constant k is the asymptotic wavenumber of the waves emitted by the spiral which is determined uniquely for given b, c. In general k has to be determined numerically (see, e.g., Hagan[13]).

[1]Some simulations of $CGLE$ with inhomogeneity were done also by K. Staliunas[30] and Gil et al.[12]

Due to the interaction with other spirals, a boundary, or an inhomogeneity, the spiral core moves with some velocity $\mathbf{v}$ which is to be calculated. We associate with each spiral a region determined by the wave emitted by the spiral. Thus in an infinite system the boundaries are given by the shocks that build up where the waves of neighboring spirals collide. Inside each region the perturbed spiral solution is written in the form

$$a(r,\theta) = \big(F(r) + W(r,\theta,t)\big)\exp\big(i(\omega t + \theta + \psi(r) + \varphi(t))\big) \tag{5}$$

where W is the (complex) correction to the unperturbed spiral solution, $\partial_t\varphi$ describes the correction to the frequency of the spiral, and r,θ are now the coordinates co-moving with the velocity $\mathbf{v}$.

For definiteness of the analysis we have chosen the $+$ sign in Eq. (2). If the spirals are well separated and the inhomogeneity is weak, one can expect the correction function W to be small except maybe in the region of the shock between spirals. Hence, the correction can be determined in principle from the linear approximation.

From now on we restrict ourselves to the case $b = 0$ to avoid lengthly expressions. The generalization to the case of arbitrary b is discussed in Section 5. Now, separating the real and the imaginary parts of $W = A + iB$ and representing the solution in the form of a Fourier series

$$\binom{A}{B} = \sum_{n=-\infty}^{\infty} \binom{A_n(r)}{B_n(r)}\exp(in\theta), \tag{6}$$

we arrive at the system

$$
\begin{aligned}
\Delta_r A_n - \frac{n^2}{r^2}A_n - 2F^2 A_n &- \frac{\Delta_r F}{F}A_n - 2\left(\psi' F \frac{\partial}{\partial r}\left(\frac{B_n}{F}\right) + \frac{in}{r^2}B_n\right) \\
&= \frac{vn}{2i}F'\delta_{\pm 1,n}\exp(in\eta) - \nu_n(r)F, \\
\Delta_r B_n - \frac{n^2}{r^2}B_n - 2cF^2 A_n &- \frac{\Delta_r F}{F}B_n + 2\left(\psi' F \frac{\partial}{\partial r}\left(\frac{A_n}{F}\right) + \frac{in}{r^2}A_n\right) \\
&= \frac{v}{2}(-in\psi' F + \frac{F}{r})\delta_{\pm 1,n}\exp(in\eta) + \partial_t\varphi\delta_{0,n},
\end{aligned}
\tag{7}
$$

where $\mathbf{v} = (v_x, v_y)$ is the velocity, $\eta = \arctan(v_x/v_y)$ is the angle of the drift, and

$$\nu_n(r) = \frac{1}{2\pi}\int_0^{2\pi} \nu(\mathbf{r} - \mathbf{r}')\exp(-in\theta)d\theta. \tag{8}$$

First we consider the interaction of two spirals in a homogeneous medium ($\nu = 0$). For simplicity the spirals are positioned on the x-axis at $\pm X$. The problem of the interaction of two symmetric, oppositely charged or like-charged spirals is equivalent to the problem of the interaction of one spiral with a plane boundary with different boundary conditions. Asymmetric states are considered later on. In the case of oppositely charged spirals the symmetry of the problem is $a(x,y) = c(-x,y)$. Numerically obtained bound states of oppositely charged spirals indeed possess such a symmetry (see Aranson et al.,[2] Figures 1 and 2). Here the term "bound state" means that the distance between the spiral cores is in stable equilibrium, but there is drift in the y-direction. For this case, therefore, the boundary conditions at $x = 0$ are

$$\partial a / \partial x = 0. \tag{9}$$

Then the velocities of the spirals obey the relation $v_{1x} = -v_{2x}$ and $v_{1y} = v_{2y}$. The case of like-charged spirals is more complicated because the symmetry of the problem is $a(x,y) = a(-x,-y)$ and the velocities obey $v_{1x} = -v_{2x}, v_{1y} = -v_{2y}$. Later on we show that this case is in some approximation also similar to the oppositely charged spirals. Then the spirals rotate around each other.

We suppose that the velocity of the spiral drift v is small which is the case when the distance $2X$ between the spirals is large. The condition $vX \ll 1$ will also be needed and it turns out to be satisfied for well-separated spirals. The terms $\sim vW$ can be omitted in the general case $b \neq c$ because it involves two small quantities, but they become important for $b = c$. The velocity enters only into the equations for $n = \pm 1$, and the correction to the frequency comes from the zero harmonic. We consider separately the behaviour of the solutions of Eqs. (7) for $r \to 0$ and for $r \to \infty$. Equations (7), together with appropriate boundary conditions, are solvable only for a distinct velocity and frequency depending on the spiral separation.

3. ASYMPTOTIC BEHAVIORS

Consider the behaviour of the homogeneous solutions ($v = 0$) of Eqs. (7) for $r \to \infty$. Then we can neglect the terms $\sim \Delta_r F, 1/r$ and $1/r^2$. Also we can replace ψ' by k and F by $\sqrt{1 - k^2}$. Then one has for arbitrary n the simplified system

$$A_n'' - 2(1 - k^2)A_n - 2k\partial_r B_n = 0, \qquad B_n'' - 2c(1 - k^2)A_n + 2k\partial_r A_n = 0 \tag{10}$$

describing perturbations of the asymptotic plane waves emitted by the spiral. Substituting the solution in the form $A_n, B_n \sim \exp(pr)$, we have the following characteristic equation:

$$p[p^3 + p(4k^2 - 2(1 - k^2)) - 4ck(1 - k^2)] = 0. \tag{11}$$

Actually Eq. (11) can be obtained directly from Eq. (3) by considering the spatial behaviour of stationary perturbations of a plane-wave state with wavenumber k. One can thus also obtain Eq. (11) from the usual dispersion relation $\sigma(k; p)$ of (side band) perturbations of the plane-wave state by setting $\sigma = 0$. We remind that stability of the wave number band is obtained from the condition $Re\sigma < 0$ for all imaginary p (Eckhaus stability) or of the saddle point $Re(d\sigma/dp) = 0$ (absolute stability).[1,33] The root $p_0 = 0$ corresponds to the translation mode. Using the numerically defined asymptotic spiral wavenumber k, it turns out that for $c > c_{cr} \approx 0.845$ the equation possesses one real negative root $p_3 < 0$ and a pair of complex-conjugated roots $p_{1,2} = \alpha \pm i\beta$ with $\alpha > 0$. In contrast for $c < c_{cr}$ all the roots are real. It will be seen later that the value $1/\beta$ sets the scale for the distance between the spirals in the bound state. For $c < c_{cr}$ there is no bound state.

A more detailed analysis of Eqs. (7), including $O(r^{-1})$ corrections, show that the asymptotic solutions are given by

$$\begin{pmatrix} A_n \\ B_n \end{pmatrix} = r^\mu \begin{pmatrix} 1 \\ \gamma \end{pmatrix} \exp(pr) \tag{12}$$

with

$$\gamma = \frac{p^2 - 2(1 - k^2)}{2pk}, \qquad \mu = \frac{4k^2 - 2p^2 + 6kp/c}{3p^2 + 6k^2 - 2}. \tag{13}$$

Thus for $c > c_{cr}$ (we consider first this case) the outer solution of Eqs. (7) is of the form

$$\begin{pmatrix} A \\ B \end{pmatrix} = \sum_{n=-\infty}^{\infty} \left[\begin{pmatrix} 1 \\ \gamma \end{pmatrix} C_{1n} r^\mu \exp(pr) + \begin{pmatrix} 1 \\ \gamma^* \end{pmatrix} C_{2n} (r^\mu)^* \exp(p^*r) \right] \exp(in\theta) \tag{14}$$

with $C_{1n} = C_{2,-n}^*$ and $p = p_1$. The homogeneous solutions dominate the asymptotic behaviour of Eqs. (7), so Eq. (14) can be used also for $v \neq 0$.

The coefficients C_{1n}, C_{2n} are to be determined from the boundary conditions for $x = 0$. From Eq. (9) (oppositely charged spirals) we have

$$\partial a(r, \theta)/\partial x = \left(F_x + W_x + i(F + W)(\psi_x + \theta_x) \right) \exp\left(i(\theta + \psi + \omega t) \right) = 0 \tag{15}$$

For $r \gg 1$ one can neglect F_x, θ_x, and using $\psi_x \approx k\cos\theta$, $W_x \approx W'\cos\theta$ and $F \approx \sqrt{1 - k^2}$ one arrives at $(W = A + iB)$

$$A' - kB = 0, \qquad B' + kA = -k\sqrt{1 - k^2}. \tag{16}$$

Now, substituting Eq. (14) into Eq. (16) and using the fact that on the boundary the radius r depends on the angle θ as

$$r(\theta) = \frac{X}{\cos\theta} \qquad -\frac{\pi}{2} < \theta < \frac{\pi}{2}$$

$$r(\theta) = \infty \qquad |\theta| > \frac{\pi}{2} \tag{17}$$

one obtains the following equations

$$\sum_{n=-\infty}^{\infty} C_{1n} \exp(in\theta) = -k\frac{\sqrt{1-k^2}}{\delta}\exp(-pr)r^{-\mu} \qquad (18)$$

where $\delta = p\gamma + k - (p^*\gamma^* + k)(p - k\gamma)/(p^* - k\gamma^*)$. For $x = 0$ we can expand $\exp(-pr)r^{-\mu}$ in the series

$$\exp(-pr(\theta))r(\theta)^{-\mu} = \sum_{n=0}^{\infty} Z_n \cos(n\theta)$$

$$Z_n = \frac{1}{\pi}\int_{-\pi}^{\pi}\exp\left(-p\frac{X}{\cos(\theta)}\right)\cos(n\theta)\left(\frac{X}{\cos\theta}\right)^{-\mu} d\theta. \qquad (19)$$

From Eqs. (18) and (19) one has $C_{1,-n} = C_{1,n}$ from which follows $C_{1,n} = C_{2,n}^*$. For $X \gg 1$ the coefficients Z_n can be estimated easily because the main contribution to the integrals comes from the region of small θ. For $\theta \ll 1$ we can replace in the exponent $X/\cos(\theta) \approx X(1 + \theta^2/2)$. After simple algebra we have

$$C_{1n} = C_{2n}^* = -\frac{k\sqrt{1-k^2}\exp(-pX)}{\delta\sqrt{2\pi pX}}X^{-\mu}. \qquad (20)$$

Consider the solutions of Eqs. (7) for $r \to 0$ and $n = 1$. We shall analyze the homogeneous solutions ($v = 0$) but again the results are applicable to the case $v \neq 0$. Because $F \sim r$, $\psi \sim r$, and $\Delta F/F \to 1/r^2$ for $r \to 0$, Eqs. (7) reduce as follows

$$A'' + \frac{1}{r}A' - \frac{2}{r^2}A - \frac{2i}{r^2}B = 0, \qquad B'' + \frac{1}{r}B' - \frac{2}{r^2}B + \frac{2i}{r^2}A = 0. \qquad (21)$$

We can look for the solution of Eq. (21) in the form

$$\begin{pmatrix} A \\ B \end{pmatrix} \sim \begin{pmatrix} a_m \\ b_m \end{pmatrix} r^m \qquad (22)$$

and this leads to $m = \pm 2$ or $m = 0$. In the case $m = 0$ also the behaviour $\sim \ln r$ is admitted. The solution of Eq. (22) with $m = 0$ and no logarithmic term is the translation mode. For $r \to \infty$ it is bounded and, therefore, not essential for our analysis. The solutions of Eqs. (7) with the other allowed behaviour for $r \to 0$ are expected to behave for $r \to \infty$ according to Eq. (14).

4. DETERMINATION OF THE VELOCITIES

If the velocity $\mathbf{v}$ is chosen zero, then it is impossible to satisfy the boundary conditions (20) for $r \to \infty$ and regularity for $r \to 0$ simultaneously. Only the proper choice of the velocity $\mathbf{v}$ allows to satisfy both conditions.

The following procedure can be employed. One determines (numerically) a particular inhomogeneous solution of Eqs. (7) for $n = 1$ with zero boundary conditions for $r = 0$ leading to a solution which is regular at $r = 0$. From the linearity of Eqs. (7) in $\mathbf{v}$, one sees that a solution for arbitrary value of $\mathbf{v}$ can be obtained by superposition of solutions with $v_x = 1, v_y = 0$ and $v_x = 0, v_y = 1$, which is equivalent to $v = 1$ with $\eta = 0$ and $\eta = \pi/2$, respectively. For $r \to \infty$ the solutions behave according to Eq. (14) and we obtain the constants C_{1x}, C_{2x} (for $v_x = 1, v_y = 0$) and C_{1y}, C_{2y} (for $v_x = 0, v_y = 1$). In general the relations $C_{1x} = C_{2x}^*$ and $C_{1y} = C_{2y}^*$ are not satisfied (first boundary condition (20)). To correct this it is useful to determine (numerically) a homogeneous solution of Eqs. (7) that behaves as Eq. (22) with $m = 2$ for $r \to 0$ and compute the asymptotic constants C_1, C_2. Then the superposition satisfying the first boundary condition (20) are determined (mixing factors ξ_x, ξ_y) leading to

$$
\begin{aligned}
C_x &= C_{1x} + \xi_x C_1 = C_{2x}^* + \xi_x^* C_2^*, \\
C_y &= C_{1y} + \xi_y C_1 = C_{2y}^* + \xi_y^* C_2^*.
\end{aligned}
\tag{23}
$$

Thus for each value of the parameter c, the constants C_x, C_y are uniquely determined. Then v_x and v_y may be calculated as a function of distance X by satisfying the second condition Eq. (20). One easily sees from Eq. (20)

$$
\begin{aligned}
v_x &= Im\left(\frac{-k\sqrt{1-k^2}\exp(-pX)}{\delta C_y\sqrt{2\pi pX}}X^{-\mu}\right) / Im(C_x/C_y), \\
v_y &= Re\left(\frac{-k\sqrt{1-k^2}\exp(-pX)}{\delta C_y\sqrt{2\pi pX}}X^{-\mu}\right) - v_x Re(C_x/C_y).
\end{aligned}
\tag{24}
$$

The bound states of the $CGLE$ correspond to the case of $v_x = 0$. Therefore the equilibrium distance $2X_e$ can be found from the equation

$$
Im[pX_e + \mu \ln X_e] = -\phi + \pi l
\tag{25}
$$

where $l = 1, 2, 3...$ and $\phi = -\arg[1/(\delta C_y p^{1/2})]$. One thus needs only one solution of Eqs. (7) in order to determine the velocities for all distances.

To check the results we performed full two-dimensional simulations of $CGLE$. The full simulations of Eq. (1) was carried out on a CRAY-YMP Supercomputer. We used a second-order quasi-spectral method based on FFT. The typical values of

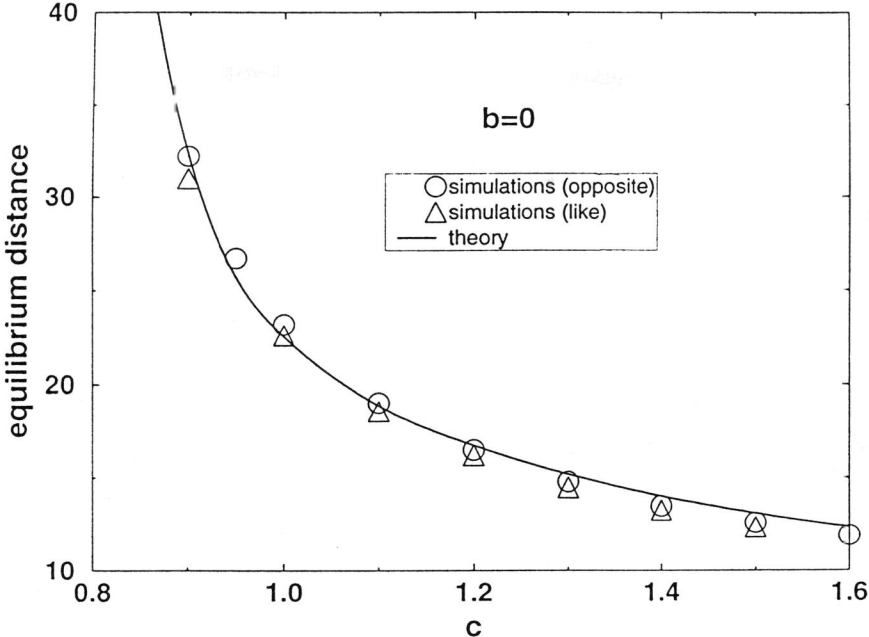

FIGURE 1   Equilibrium distance $2X_e$ given by Eq. (25) as function of c for $b = 0$.

the parameters of the code were: the number of harmonics in each directions was 256, the system size was 100×200 dimensionless units, and the time step $\Delta \tau \sim 0.1$. The results were checked with better space discretization and smaller time steps.

In Figure 1 the dependence of the velocities on the spiral separation X is plotted for $b = 0, c = 1$ and compared with results from full numerical simulations. There is reasonable agreement, particularly for the radial velocity v_x. The first zero of v_x at $2X_e \approx 11.5$ corresponds to $l = 1$ in Eq. (25). The next zero at $2X_e \approx 22.8$ corresponds to a stable bound state. The equilibrium distance obtained from the theory ($l = 2$ in Eq. (25)) is also in very good agreement with the results of the simulations of the full $CGLE$ (see Aranson et al.,[2] Figure 3).

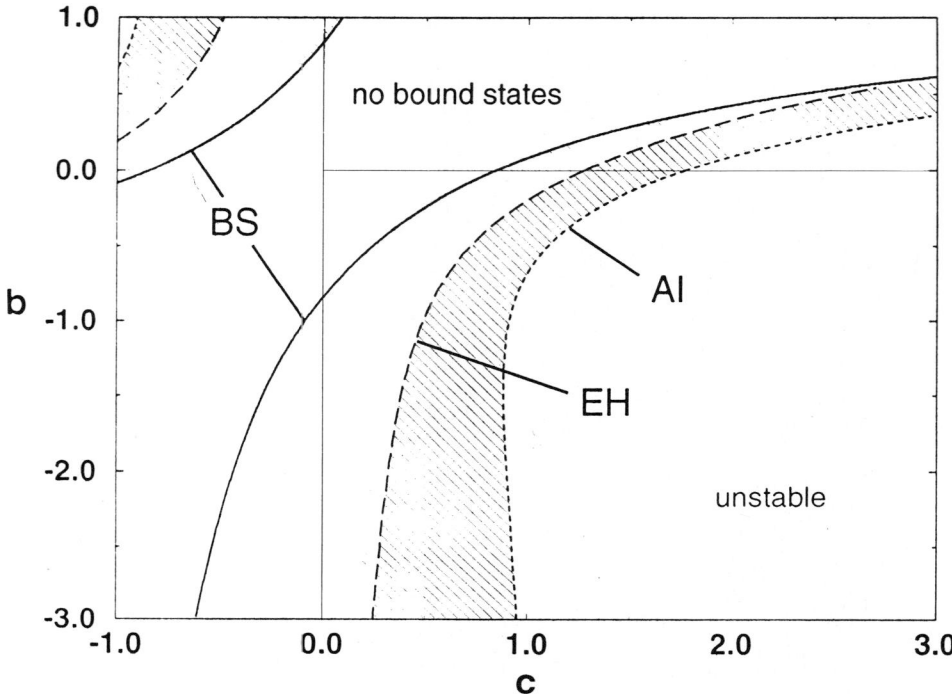

FIGURE 2   Border of the oscillatory range as given by the expression (28) (BS). The boundary of convective (EH) and absolute instability (AI) for the waves emitted by spiral are also shown (for explanation see Aranson et al.[1,33]).

5. GENERALIZATIONS

5.1 $c \leq c_{cr}$

In the case $0 < c < c_{cr}$ we have two real positive roots of Eq. (11), for definiteness $0 < p_1 < p_2$. Applying the analysis as in the previous sections, we can derive in principle from the boundary conditions the expressions for the coefficients $C_{1/2,n}$, but for small c it is technically very difficult in this form. In the general case $0 < c < c_{cr}$, the velocities $v_{x,y}$ depend on both exponents $\exp(-p_1 X), \exp(-p_2 X)$.

The results can be simplified considerably for the case $c \to 0$ and $|ck|X \gg 1$. Then one can neglect the coefficients C_{2n} because, for $c \to 0$, one has $0 < p_1 \approx -2ck \ll 1, p_2 \approx \sqrt{2}, \mu_1 \to 0$ and $\delta_1 \to -1/(2k^2)$. Therefore we have from the boundary conditions

$$C_{1n} = \frac{k^2 \exp(-|2ck|X)}{\sqrt{\pi|ck|X}}, \qquad C_{2n} = 0. \tag{26}$$

After the matching with the solution for $r \to 0$, we will obtain the result $v_{x,y} \approx C_{x,y}^{-1} \exp(-2|ck|X)/\sqrt{X}$. This coincides with an earlier analysis using a phase diffusion equation.[2,25,7] From the numerical simulations for $b = 0$ and $c = 0.5$, one finds $C_x > 0$ leading to asymptotic repulsion. Our results appear to be in contradiction with the work.[25]

5.2 THE CASE OF b $\neq$ 0

It is well known that the homogeneous Eqs. (7) admit a similarity transformation from the case $b \neq 0$ to the case $b = 0$.[13] Indeed, dividing the equation by $(1 + ib)$ and performing the following transformations of the variables

$$\tilde{c} = \frac{c - b}{1 + bc}, \qquad \tilde{\omega} = \frac{\omega + b}{1 - \omega b}, \qquad \tilde{F} = F\sqrt{\frac{1 + bc}{1 - \omega b}}$$

$$\tilde{k} = \frac{k}{\sqrt{(1 - \omega b)/(1 + b^2)}}, \qquad \tilde{r} = r\sqrt{(1 - \omega b)/(1 + b^2)} \tag{27}$$

leads to the case $b = 0$ but with a new value of c. Therefore, all of the results concerning the structure of the eigenfunctions and the characteristic roots remain the same. Then the critical value c_{cr} corresponds to a curve in the b–c plane

$$\frac{c - b}{1 + bc} = c_{cr} \approx 0.845. \tag{28}$$

This curve defines the border of existence of bound states for oppositely charged spirals. It is plotted in Figure 2 (curve BS). Also included in Figure 2 is the Eckhaus stability boundary (curve EH) and the boundary of absolute stability (curve AI) for the waves emitted by the spirals. The Eckhaus instability signalizes the onset of convective instability.[1,33] The bound states may exist only within the "oscillatory" range (between the curves BS and AI).

Since the full Eqs. (7) are not invariant under the similarity transformation, the velocities and the equilibrium distances cannot be determined in this way. Then a calculation along the lines of Sections 3 and 4, including b from the beginning on, is necessary.

5.3 INTERACTION OF LIKE-CHARGED SPIRALS

For large separation X the interaction of like-charged spirals is similar to the interaction of oppositely charged ones. The only difference is that in the like-charged case the spirals rotate around the common center of the symmetry. To see the similarity we note that, for small x and $|y| \ll X$, one has approximately $a(x, y) = a(x, -y)$. Thus the symmetry $a(x, y) = a(-x, -y)$ (like-charged) and $a(x, y) = a(-x, y)$ (oppositely charged) have nearly the same effect on the boundary conditions. This result is in good agreement with simulations. For $|b - c|$ below the critical value, one has repulsion at large distance (as in the oppositely charged case) and at small distance. So it is quite clear that the interaction is repulsive everywhere.

Like-charged spirals may form more complicated bound states or aggregates (see Aranson et al.,[2] Figure 7) In contrast to the two-spiral bound states, which are simply rotating with constant velocity, each spiral in the aggregate performs a more complicated motion (possibly nonperiodic) on the background of a steady-state rotation. Such states were observed experimentally by Müller.[31]

6. SPONTANEOUS BREAKING OF THE SYMMETRY

To construct a slightly asymmetric two spiral state, we have to suppose that the phases $\varphi_{1,2}$ are different (the other quantities except W remain the same for the two spirals). Then the position X_0 of the shock between the spirals will depend on the phase difference $\varphi_1 - \varphi_2$ and can be determined from the condition of continuity of the field $a_1 = a_2$ for $x = X_0, y = 0$. X_0 can be determined easily in the limit of large separation. Substituting Eq. (5) into the continuity conditions, we obtain at leading order of expansion in $1/X$:

$$k(X - X_0) + \varphi_1 = k(X + X_0) + \varphi_2 \tag{29}$$

and therefore $X_0 = \varphi/2k$, where $\varphi = \varphi_1 - \varphi_2$. Matching of the solutions at the position of the schock requires $\partial a_{1,2}/\partial x = 0$ for $x = X_0, y = 0$. (In a first approximation we can neglect the curvature of the shock line).

Allowing for nonzero values of φ from the beginning, within the "oscillatory" range of the parameters of $CGLE$ the analysis, along the lines of Sections 3 and 4, gives for oppositely charged spirals in the limit $X \gg \varphi/(2k)$ the following equations of motion

$$v_{1x} = Im\left(\frac{-k\sqrt{1-k^2}\exp(-p(X - \varphi/(2k)))}{\delta C_y\sqrt{2\pi pX}}X^{-\mu}\right)\Big/Im\left(\frac{C_x}{C_y}\right),$$

$$v_{1y} = Re\left(\frac{-k\sqrt{1-k^2}\exp(-p(X - \varphi/(2k)))}{\delta C_y\sqrt{2\pi pX}}X^{-\mu}\right) - v_{1x}Re\left(\frac{C_x}{C_y}\right),$$

$$\partial_t\varphi = 2Im\left(\frac{-k\sqrt{1-k^2}\exp(-pX)}{\delta C_0\sqrt{2\pi pX}}X^{-\mu}\sinh(\frac{p\varphi}{2k})\right)\Big/Im\left(\frac{C_{10}}{C_0}\right) \approx \zeta\varphi \tag{30}$$

(the last approximation is valid for $p\varphi/k \ll 1$). Analogous equations hold for the velocity of the second spiral. Here $C_{x,y}, C_{0,1}$ are constant obtained from numerical solution of the linearized problem (Eqs. (7)) for the first and zero harmonic of the

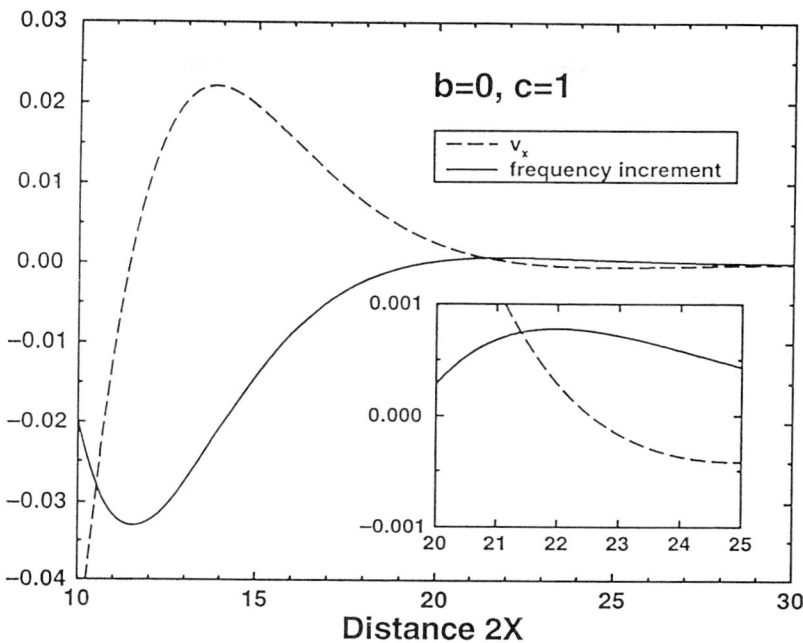

FIGURE 3 The dependence of radial velocity (v_x) and the frequency increment $\zeta = d(\partial_t \varphi)/d\varphi$ versus spiral separation $2X$ for $b = 0, c = 1$.

perturbation and μ, δ are functions of k, p obtained from Eqs. (13). The symmetric bound state corresponds to the case $\varphi = 0, v_x = 0$ and $v_{1y} = v_{2y}$.

The constant ζ determined numerically turns out to be positive for the (first) symmetric bound state (see Figure 3), so that the state is unstable with respect to growth of φ. Hence the frequencies of rotation (and, therefore, the underlying wavenumbers) become different, which means breaking of the symmetry of the two-spiral solution. However, in fact, ζ is very small, so the symmetric bound states can be rather long lived. The symmetry breaking was also observed for other parameters, including $b \neq 0$, inside the oscillatory range. As a result of the symmetry breaking in a bounded system, only one "free" spiral will remain, whereas the other spiral is pushed away to the boundary (this might not be true for very large cells; see "Conclusion"). Depending on the boundary conditions, the second spiral will finally either annihilate at the boundary (nonflux boundary conditions, i.e., zero normal derivative on the boundary), or, with periodic boundary conditions, the defect will persist for topological reasons, but will be reduced to its core and will be enslaved in the corner of the schock structure of the free spiral.[3] The last case leads to an asymmetric lattice of topological defects, which appears to be stable in the oscillatory case. We have tested the stability of such lattices in simulations of up to 4×4 cells.

The analysis can be simplified considerably for the case $|c - b| \to 0$ and $|(c - b)k|X \gg 1$. The matching with the outer solution can then be done analytically (see, e.g., Pismen and Nepomnyashchy[25] and Biktashev[7]) using a phase diffusion equation. From the analysis the equations for the frequency can be inferred in explicit form for $|(c - b)k|X \gg 1$ as

$$\partial_t \varphi = -2(1 + b^2)k^2 \sqrt{\left|\frac{\pi c'}{kX}\right|} \exp(-2|c'kX|) \sinh(|c'|\varphi) \qquad (31)$$

where $c' = (c - b)/(1 + bc)$. One sees that the phase difference always tends to zero and the symmetry of the state is restored. The stability of symmetric states together with the repulsion of the spirals shows the possibility for the existence of stable symmetric ("antiferromagnetic") lattices of spirals, i.e., lattices made up of free spirals with alternating topological charge.[2] Moreover, we have verified that starting from strongly asymmetric initial conditions leads eventually, in the non-oscillatory case, to restortion of the symmetry. The results also carry over to like-charged spiral states. For like-charged spirals, one has the same mechanism of symmetry breaking as for an oppositely charged pair.

7. INTERACTION OF SPIRAL WITH A LOCALIZED INHOMOGENEITY

For simplicity we consider an axisymmetric inhomogeneity located at the point X on the x-axis, and the spiral located at $(0,0)$. For the case of a localized inhomogeneity ($\nu \to 0$ for $r \to \infty$), the boundary conditions for the function W will be simply $W \to 0$ for $r \to \infty$. To calculate the velocities the procedure described in Section 4 can be employed. In addition, we have to calculate the inhomogeneous solution of Eqs. (7) corresponding to $v_y = v_x = 0$ but $\nu_1 \neq 0$. Thus we obtain six constants: C_{1x}, C_{2x} (for $v_x = 1, v_y = 0, \nu = 0$), C_{1y}, C_{2y} (for $v_x = 0, v_y = 1, \nu = 0$), and C_{01}, C_{02} ($v_x = v_y = 0, \nu \neq 0$). Since the superposition of these solutions do not satisfy the boundary conditions ($A_1, B_1 \to 0$ for $r \to \infty$), it is useful to determine (numerically) a homogeneous solution of Eqs. (7) that behaves as Eq. (22) with $m = 2$ for $r \to 0$ and compute the asymptotic constants C_1, C_2.

Then the superposition satisfying the boundary conditions give rise to the following expressions for the velocities:

$$v_y = Re\frac{C_{01}(x)/C_1 - C_{02}(x)/C_2}{C_{1y}/C_1 - C_{2y}/C_2}, \quad v_x = Im\frac{C_{01}(x)/C_1 - C_{02}(x)/C_2}{C_{1y}/C_1 - C_{2y}/C_2}. \qquad (32)$$

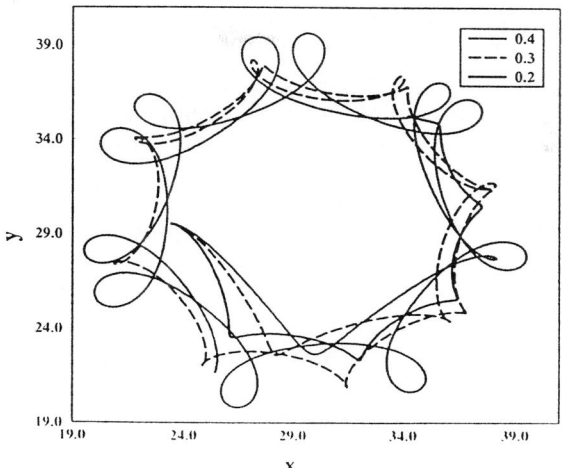

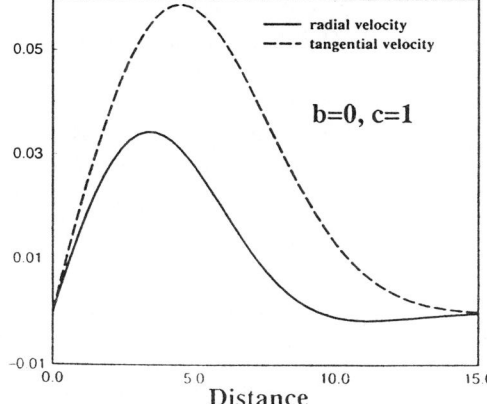

FIGURE 4 (a) The tangential and radial velocities of the spiral versus the distance X from the inhomogeneity for $b = 0, c = 1$ and $b_0 = 0.3, \sigma = 20$. From numerical simulations one has the radius of the first stationary orbit $r_0 \approx 9.8$. (b) The limiting orbits of the spiral core for different values of inhomogeneity "strength" $b_0 = 0.1; 0.2; 0.3$ for $\varepsilon = 1/b = 0.2, c = 1, \sigma = 20$.

For the particular form of inhomogeneity $\nu(r) = b_0 \exp(-(r^2)/\sigma)$ the functions $\nu_n(r, X)$ are given by:

$$\nu_n(r, X) = (-1)^n b_0 \exp\left(-\frac{r^2 + X^2}{\sigma}\right) I_n\left(\frac{2rX}{\sigma}\right) \qquad (33)$$

where I_n is a modified Bessel function. Determining the velocities as functions of X, we found that there exists a discrete spectrum of stationary orbits ($v_x = 0$). However, only the first stationary orbit is important here because the tangential velocity is rapidly vanishing as X increases. The discrete set of the orbits exists although the inhomogeneity function $\nu(r)$ is monotonic. The comparison with the numerical simulations gives very good agreement with the above theory (see Figure 4(a)). Note that in the range of validity of the above treatments the velocities are simply proportional to the prefactor b_0 of the inhomogeneity. Thus the radii of the stationary orbits are independent of b_0. Changing the sign of b_0, however, interchanges the stable with the unstable orbits. For the parameters of Figure 4(a) (b=0, c=1), the central position $X = 0$ becomes stable for the case $b_0 < 0$.

For $b \to \infty$ (which is the typical case for nonlinear optics, see, e.g., Coullet et al.[10]), from our numerical simulations, we observed that the round stationary orbits exist only if the inequality $b < b_c$ holds. In the opposite case, the simulations exhibit more complicated orbits (see Figure 4(b)) which cannot be described by Eqs. (32). In order to describe the numerical simulations, one has to include the effective acceleration term into the equation of motion:

$$\frac{d\mathbf{v}}{dt} + \varepsilon \hat{K}\mathbf{v} = \mathbf{G} \qquad (34)$$

where $\hat{K}$ is the friction tensor (or mobility tensor), and $\varepsilon = 1/b$. The appearence of the acceleration can be explained by the intrinsic Galilean invariance of the $CGLE$ for $b \to \infty$. The calculation of the friction tensor can be performed along the lines of the analysis presented above and will be reported elsewhere. Moreover, as it was shown by Aranson et al.,[4] for $\varepsilon \to 0$ one always has negative diagonal elements of $\hat{K}$, which indicates core instability. This instability is stabilized only above some critical value of ε.

8. CONCLUSION

Our work shows that the interaction of spirals is for sufficiently large separation either oscillatory, leading to stable symmetric bound states or else presumably repulsive (this was verified numerically only for one set of parameters). Therefore, in the parameter range where isolated spirals are stable (above curve AI in Figure 2), one may expect that, in an ensemble of well-separated spirals, annihilation does not occur, although it is not obvious that our results carry over to ensembles. Long-time simulations in very large cells with random initial conditions indicate that the system eventually settles down in a spatially disordered stationary state with a low density of spirals. This state appears to be similar to the "vortex glass" found by Huber,[15] although, in that work, time dependence appeared to persist in the shock structures separating the spirals. In view of the nonmonotonic interaction the existence of disordered stationary states seems quite plausible. Analogous simulations in very large cells in the nonoscillatory region (above curve BS in Figure 2) indicate that the system typically remains in a disordered dynamical state.

Regarding the interaction of a spiral with a localized inhomogeneity, it is interesting to note that in the oscillatory region one can have trapping of the spirals for either sign of the inhomogeneity. Whereas for one sign the central position is stable (pinning), there exist only orbits for the other sign. As pointed out there are cases where velocity and acceleration are needed to describe the dynamic state of a spiral. As a result one can have meandering, especially in situations where the interaction with boundaries is important (see, e.g., Sepulchre and Babloyantz[28]), and presumably random walks in a slowly varying background.

ACKNOWLEDGMENTS

We wish to thank S. Popp, K. Staliunas, and O. Stiller for useful discussions. I. A. wishes to thank the Alexander-von-Humboldt foundation for a research fellowship and the University of Bayreuth for its hospitality. Support by the German-Israeli Foundation (GIF) and Deutsche Forschungsgemeinschaft (SFB 213, Bayreuth) is gratefully acknowledged.

REFERENCES

1. Aranson, I., L. Aranson, L. Kramer and A. Weber. "Stability Limits of Spirals and Traveling Waves in Nonequilibrium Media." *Phys. Rev. A* **46** (1992): 2992.
2. Aranson, I. S., L. Kramer, and A. Weber. "The Theory of Interaction and Bound States of Spiral Waves in Oscillatory Media." *Phys. Rev. E* **47** (1993): 3221.
3. Aranson, I. S., L. Kramer, and A. Weber. "The Formation of Asymmetric States of Spiral Waves in Oscillatory Media." *Phys. Rev. E* **48** (1993): R9.
4. Aranson, I. S., L. Kramer, and A. Weber. "Core Instability and Spatiotemporal Intermittency of Spiral Waves in Oscillatory Media." *Phys. Rev. Lett.* **72** (1994): 2316.
5. Arecchi, F. T., G. Giacomelli, P. L. Ramazza, and S. Residori. "Experimental Evidence of Chaotic Itinerancy and Spatiotemporal Chaos in Optics." *Phys. Rev. Lett.* **65** (1990): 2531
6. Arecchi, F. T., G. Giacomelli, P. L. Ramazza, and S. Residori. "Vortices and Defect Statistics in Two-Dimensional Optical Chaos." *Phys. Rev. Lett.* **67** (1991): 3749.
7. Biktashev, V. N. "Drift of a Reverberator in an Active Medium Due to the Interaction with Boundaries." In *Nonlinear Waves II*, edited by A. V.

Gaponov-Grekhov and M.I. Rabinovich. Research Reports in Physics, p. 87. Heidelberg: Springer, 1989.

8. Bodenschatz, E., L. Kramer, and W. Pesch. "Structure and Dynamics and Dislocations in Anysotropic Pattern-Forming Media." *Physica D* **32** (1988): 135.

9. Coullet, P., L. Gil, and J. Lega. "Defect-Mediated Turbulence." *Phys. Rev. Lett.* **62** (1989): 1619.

10. Coullet, P., L. Gil, and F. Rocca. "Optical Vortices." *Optics Communications* **73** (1989): 403.

11. Feng, Q., W. Decker, W. Pesch, and L. Kramer. "On the Theory of Rayleigh-Bérnard Convection in Homeotropic Nematic Liquid Crystals." *J. Phys., France II* **2** (1992): 1303.

12. Gil, L., K. Emilsson, and G. L. Oppo. "Dynamics of Spiral Waves in a Spatially Inhomogeneous Hopf Bifurcation." *Phys. Rev. A* **45** (1992): 567.

13. Hagan, P. S. "Spiral Waves in Reaction Diffusion Equations." *SIAM J. Appl. Math* **42** (1982): 762.

14. Hohenberg, P. C., and M. Cross. "Pattern Formation Outside of Equilibrium." *Rev. Mod. Phys.*: in press.

15. Huber, G., P. Almstrom, and Th. Bohr. "Nucleation and Transients at the Onset of Vortex Turbulence." *Phys. Rev. Lett.* **69** (1992): 2380.

16. Jakubith, S., H. H. Rotermund, W. Engel, A. von Oertzen, and G. Ertl. "Spatiotemporal Concentration Patterns in a Surface Reaction: Propagation and Standing Waves, Rotating Spirals, and Turbulence." *Phys. Rev. Lett.* **65** (1990): 3013.

17. Kramer, L., and A. Weber. "Interaction and Dynamics of Defects in Convective Roll Patterns of Anisotropic Fluids." *J. Stat. Phys.* **64** (1991): 1007.

18. Krischer, K., M. Eiswirth, and G. Ertl. "Oscillatory CO Oxidation on Pt(110): Modelling of Tempral Self-Organization." *Surf. Sci.* **251/252** (1991): 900.

19. Kuramoto, Y., and S. Koga. "Turbulized Rotating Chemical Waves." *Prog. Theor. Phys.* **66** (1981): 1081.

20. Kuramoto, Y. *Chemical Oscillations, Waves and Turbulence.* Springer Series in Synergetics. Berlin: Springer-Verlag, 1984.

21. Lekkerkerker, H. "Oscillatory Convective Instabilities in Nematic Liquid Crystals." *J. Phys., France, Lett.* **38** (1977): 277.

22. Moses, E., and V. Steinberg. "Flow Patterns and Nonlinear Behavior of Travelling Waves in Convective Binary Mixtures." *Phys. Rev. A* **34**, (1986): 693.

23. Müller, S. C., T. Plesser, and B. Hess. "Two-Dimensional Spectrophotometry of Spirl Waves Propagation in the Belousov-Zhabotinskii Reaction I. Experiments and Digital Data Representation. II. Geometric and Kinematic Parameters." *Physica D* **24** (1987): 71.

24. Newell, A. C. "Envelope Equations." *Lect. Appl. Math.* **15** (1974): 157.

25. Pismen, L., and A. A. Nepomnyashchy. "On Interaction of Spiral Waves." *Physica D* **54** (1992): 183.
26. Rehberg, I., and G. Ahlers. "Experimental Observation of a Codimension Two Bifurcation in a Binary Fluid Mixture." *Phys. Rev. Lett.* **55** (1985): 500.
27. Rehberg, I., S. Rasenat, and V. Steinberg. "Travelling Waves and Defect-Initiated Turbulence in Electroconvecting Nematics." *Phys. Rev. Lett.* **62** (1989): 756.
28. Sepulchre, J. A., and A. Babloyantz. "Motion of Spiral Waves in Oscillatory Media and in the Presence of Obstacles." *Phys. Rev. E* **48** (1993): 187.
29. Skinner, G. S., and H. L. Swinney. "Periodic to Quasiperiodic Transition of Chemical Spiral Rotation." *Physica D* **48** (1991): 1.
30. Staliunas, K. "Dynamics of Optical Vortices in a Laser Beam." *Optics Comm.* **90** (1992): 123.
31. Steinbock, O., and S. Müller. "Multi-Armed Spirals in a Light Controlled Excitable Reaction." *Intl. J. Bif. & Chaos* (1993): in press.
32. Walden, R. W., P. Kolodner, A. Passner, and C. M. Surko. "Travelling Waves and Chaos in Convection in Binary Fluid Mixtures." *Phys. Rev. Lett.* **55** (1985): 496.
33. Weber, A., L. Kramer, I. S. Aranson and L. B. Aranson. "Stability Limits of Travelling Waves and the Transitionto Spatiotemporal Chaos in the Complex Ginzburg-Landau Equation." *Physica D* **61** (1992): 279.

Charles R. Doering† and Werner Horsthemke‡

†Clarkson Institute for Statistical Physics, Department of Physics, Clarkson University, Potsdam, NY 13699-5820

‡Center for Nonequilibrium Structures, Department of Chemistry, Southern Methodist University, Dallas, TX 75275-0314

Chemistry-Flow Interactions: Stability Analysis of a Reaction-Diffusion-Convection System

The onset of spatial pattern formation in chemical reaction-diffusion systems can be studied via linearized stability analysis—the Turing instability in multispecies systems with varying diffusion coefficients is a specific well-known example. Less attention has been given to stability analyses in the presence of an imposed flow field. We study, via a combination of analytical and numerical methods, stability characteristics of a reaction-diffusion-convection model for the particular case of constant shear rate flow. When the various chemical species have equal diffusion coefficients, this background flow generally serves to stabilize the homogeneous steady state. The convection manifests itself qualitatively in the spectrum of the linearized evolution operator: the imposed flow leads to complex eigenvalues and so to flow-induced oscillatory time scales.

Spatio-Temporal Patterns, Ed. P. E. Cladis and P. Palffy-Muhoray,
SFI Studies in the Sciences of Complexity, Addison-Wesley, 1995   **499**

INTRODUCTION

Ever since the pioneering work of Turing,[6] instabilities in diffusion-reaction systems have played a central role in our understanding of spatial pattern-forming mechanisms. The effect of hydrodynamic flows on the stability properties of reacting and diffusing systems has received relatively little attention, though, and here we present the results of a recent investigation[2] of the simplest case of linear shear flow. A multiple perturbation approach to this problem was studied by Spiegel and Zaleski nearly a decade ago.[4] They considered small-flow velocities close to a stationary bifurcation of the homogeneous system, neglected spatial variations in the direction transverse to the flow, and expanded about zero wavenumber in the direction of the flow. The linear stability analysis of the reaction-diffusion-convection equation reported in this chapter—including both analytical and numerical results—was not subject to these limitations. In contrast to the analyses of Turing and Spiegel and Zaleski, however, we restricted our attention to the case of equal diffusion coefficients for all species.

The details of the problem we considered are as follows. For a two-dimensional flow in the x-direction with linear shear rate α in the y-direction, the reaction-diffusion-convection problem in nondimensional units is

$$\partial_t c = \Delta c - Pey\partial_x c + f(c), \tag{1}$$

where c is a vector of concentrations. The Péclet number based on the width in the y-direction L, the shear velocity scale αL, and the diffusion coefficient D, is

$$Pe = \frac{aL^2}{D}. \tag{2}$$

The function $f(c)$ describes the chemical kinetics. With length units chosen so that the system extends from -1 to $+1$ in the y-direction, the boundary conditions on the concentration were taken to be impermeable:

$$\partial_y c(x, y, t)\bigg|_{\pm 1} = 0. \tag{3}$$

The system was considered to be unbounded in the x-direction.

We performed a linear stability analysis of a stationary state of the homogeneous system, C, satisfying

$$f(C) = 0. \tag{4}$$

Writing

$$c(x, y, t) = C + \delta c(x, y, t), \tag{5}$$

infinitesimal perturbations δc are seen to obey

$$\partial_t \delta c = \Delta \delta c - Pey \partial_x \delta c + J \delta c, \tag{6}$$

where

$$J = \frac{\partial f}{\partial c}\bigg|_{c=C} \tag{7}$$

is the Jacobian matrix of the kinetic term evaluated at the steady state. To obtain the spectrum of eigenvalues determining the stability of the homogeneous steady state, ones needs to find the eigenvalues κ for

$$(\partial y^2 - ikPey)\Phi(y) = \kappa\,\Phi(y), \tag{8}$$

with boundary conditions $\Phi'(-1) = 0 = \Phi'(+1)$, as well as the eigenvalues $\tilde{\lambda}$ of the homogeneous system, defined by

$$\det[J - \tilde{\lambda}1] = 0. \tag{9}$$

Then the *ansatz*

$$\delta c(x, y, t) = e^{\lambda t} e^{ikx} \Phi(y) \begin{pmatrix} \delta c_1 \\ \vdots \\ \delta c_n \end{pmatrix}, \tag{10}$$

leads to the characteristic equation

$$\det[(-\lambda - k^2 + \kappa)1 + J) = 0. \tag{11}$$

Hence,

$$\lambda = \tilde{\lambda} - k^2 + \kappa, \tag{12}$$

and the eigenvalues of the full problem are seen to be the sum of the eigenvalues for the chemistry ($\tilde{\lambda}$), diffusion in the streamwise direction ($-k^2$), and diffusion-convection in the transverse direction (κ). This simple decomposition is the result of the fact that—with equal diffusion coefficients for all species—the problem separates.

The spectral problem in Eq. (8) was solved by the independent variable substitution

$$z = (ikPe)^{\frac{1}{3}} y + (ikPe)^{-2/3} \kappa, \tag{13}$$

leading to Airy's equation for the eigenfunctions:

$$\Phi''(z) - z\Phi(z) = 0. \tag{14}$$

The solutions are linear combinations of Airy's functions[1] $Ai(z)$ and $Bi(z)$,

$$\Phi(z) = \beta Ai(z) + \gamma Bi(z), \tag{15}$$

where the coefficients β and γ are to be determined. Denoting

$$z_\pm = \pm(i\,k\,Pe)^{1/3} + (ikPe)^{-2/3}\kappa, \tag{16}$$

imposition of the boundary conditions leads to the characteristic equation for κ,

$$\det \begin{pmatrix} Ai'(z_-) & Bi'(z_-) \\ Ai'(z_+) & Bi'(z_+) \end{pmatrix} = 0. \tag{17}$$

Equation (17) determines κ as a function of kPe but, because an analytical expression for the spectrum cannot be obtained, it had to be solved numerically. Our numerical evaluation of the eigenvalues[2] is summarized in Figure 1.

Apart from the purely numerical results, though, we also derived some general properties of the spectrum. First, we found that the real part of the eigenvalue κ is not positive. In fact, for a nonzero value of kPe, the real part must actually be strictly negative:

$$Re\kappa = -\frac{\int_{-1}^{+1} |\partial_y \Phi|^2 \, dy}{\int_{-1}^{+1} |\Phi|^2 \, dy} < 0 \ (\text{when } kPe \neq 0). \tag{18}$$

(This is true because the only way that the integral of $|\partial_y \Phi|^2$ can vanish is if $\Phi(y)$ is a constant, but $\Phi \equiv$ constant is not a solution if $kPe \neq 0$.) Hence, one

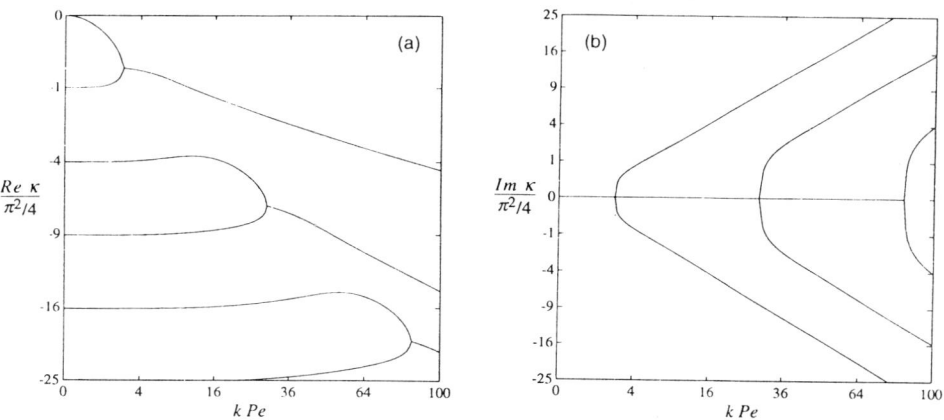

FIGURE 1   Spectra for the diffusion-convection process: (a) $Re\kappa$ vs. kPe, and (b) $Im\kappa$ vs. kPe. When two real eigenvalues merge, they become a complex conjugate pair. The first six eigenvalues, ordered according to the magnitude of $Re\kappa$, are plotted.

general result is that linear shear flow has an overall stabilizing effect beyond that of diffusion alone—at least when all the diffusion coefficients are equal. Second, and also in distinction from the effect of diffusion alone, we found that the spectrum can be complex (the operator in Eq. (8) is not self-adjoint). Thus uniform shear introduces characteristic oscillatory time scales into the reaction-diffusion-convection system. Moreover, the complex eigenvalues must appear in conjugate pairs, because for every eigenfunction-eigenvalue solution $(\Phi(y), k)$ to Eq. (8), another solution is $(\Phi(-y)^*, \kappa^*)$. The numerical results in Figure 1 illustrate these results.

If we order the eigenvalues according to increasing magnitude of the real part, then as kPe increases beyond successive threshold values, successive pairs of the real eigenvalues coalesce and form a complex conjugate pair of eigenvalues, as seen in Figure 2(b). The ith pair of eigenvalues is real for kPe smaller than $(kPe)_c^i$, and complex above this value. The first four threshold values, which we found numerically, are

$$
\begin{aligned}
(kPe)_c^1 &= 2.25813\ldots, \\
(kPe)_c^2 &= 28.6687\ldots, \\
(kPe)_c^3 &= 83.9558\ldots, \\
(kPe)_c^4 &= 168.092\ldots.
\end{aligned}
\tag{19}
$$

Because the streamwise wavenumber k runs to infinity, short wavelength perturbations of the homogeneous steady state have complex eigenvalues for arbitrarily small shear rates.

To summarize, we included a uniform shear flow in the linear stability analysis of convection-diffusion-reaction systems and proved that if all species have the same diffusion coefficient, then the shear does not have a destabilizing effect. Furthermore, when the product of the streamwise wavenumber and the shear rate are high enough (on the scale of the diffusion coefficient and the spanwise length), we discovered that the shear introduces an imaginary part to eigenvalues resulting in "underdamped" modes and flow-induced oscillatory time scales.

The fact that a uniform shear does not induce instability might not be surprising in light of the enhanced diffusion typically accompanying such flow fields; i.e., Taylor diffusion.[5,7] However, the possibility of instability when different species have different diffusion coefficients, as is the case for a Turing bifurcation, remains open—apart from Spiegel and Zaleski's perturbative analysis. The complication introduced by unequal diffusion coefficients is that the ansatz in Eq. (10) is no longer valid because the spatial structure of modes for species with different diffusion coefficients varies. This leads to higher-order eigenvalue problems reminiscent of classical hydrodynamic stability theory.[3]

ACKNOWLEDGMENTS

The research reported here was supported in part by NSF Grants PHY–8907755, PHY-8958506, and PHY-9214715. We acknowledge helpful comments from E. Titi.

REFERENCES

1. Abramowitz, M., and I. A. Stegun. *Handbook of Mathematical Functions.* New York: Dover, 1972.
2. Doering, C. R., and W. Horsthemke. "Stability of Reaction-Diffusion Systems in a Linear Shear Flow." *Phys. Lett. A* **182** (1993): 227–231.
3. Drazin, P. G., and W. H. Reid. *Hydrodynamic Stability.* New York: Cambridge University Press, 1981.
4. Spiegel, E. A., and S. Zaleski. "Reaction-Diffusion Instabilities in a Sheared Medium." *Phys. Lett. A* **106** (1984): 335–338.
5. Taylor, G. I. "Dispersion of Soluble Matter in Solvent Flowing Slowly Through a Tube." *Proc. Roy. Soc. A* **219** (1953): 196–203.
6. Turing, A. M. "The Chemical Basis of Morphogenesis." *Phil. Trans. Roy. Soc. London B* **327** (1952): 37–72.
7. Van den Broeck, C. "Taylor Dispersion Revisited." *Physica A* **168** (1990): 677–696.

Hanspeter Herzel,† Katharina Krischer,‡ David Berry* and Ingo Titze*

†Institute of Thecretical Physics, Humboldt University, Invalidenstr. 42, D-10115 Berlin, Germany
‡Fritz Haber Institute, Max Planck Society, Faradayweg 4-6, D-14195 Berlin, Germany
*National Center for Voice and Speech, The University of Iowa, Iowa City, IA 52242, USA

Analysis of Spatio-Temporal Patterns by Means of Empirical Orthogonal Functions

A classical, statistical method is described to identify coherent spatial structures from spatio-temporal patterns arising in experiments or extensive simulations of partial differential equations (PDE's). The technique is based on the diagonalization of the covariance or two-point correlation matrix.

The applicability of the method is demonstrated using spatio-temporal experimental data from heterogeneous catalysis and numerical simulations of vocal fold vibrations. In both cases complex spatio-temporal patterns can be successfully described by a few dominant spatial modes.

It is argued that the technique of empirical orthogonal functions (EOF's) provides information on the underlying mechanisms of pattern generation, allows effective data compression and can be exploited to reduce noise. Moreover, low-dimensional systems of ordinary differential equations (ODE's) can be derived using EOF's.

1. INTRODUCTION

The chapters of this volume demonstrate clearly that spatio-temporal patterns appear in a large variety of systems ranging from hydrodynamics, liquid crystals, reaction diffusion systems to bacteria populations. There are, on one hand, impressive experimental studies and, on the other hand, extensive numerical simulations of the corresponding PDE's. In both cases the qualitative features as, e.g., stripes, hexagonal patterns or spiral waves are often visualized by video movies.

In this chapter we discuss the classical method of "empiral orthogonal functions" (EOF's) which can be used to identify large-scale spatially coherent structures in such "movies." [1,3,6,9,12,17,18,20,22,23] The technique provides insight into the underlying mechanisms of pattern formation and allows data compression and noise reduction and has been applied for decades in various fields. But still, the method has not become a standard tool in connection with the spatio-temporal patterns discussed during this conference.

The method of EOF's can be embedded into the general framework of mode analysis. In linear (or linearized) systems the concept of eigenvalues and eigenfunctions is fundamental. However, the analytic determination of eigenfunctions is restricted to simple geometries. For example, small vibrations of a rectangular (or circular) membrane are simply superpositions of trigonometric (or Bessel) functions. It is shown by Breuer and Sirovich[6] that the method of EOF's allows the effective calculation of eigenfunctions in complex geometries. They can also be obtained even when the governing equations are unknown. Thus, the procedure extends down to linear systems.

However, most applications of the method are associated with nonlinear and chaotic behavior. Beyond the onset of instabilities, the occuring patterns can be described in principle by means of Galerkin expansions.[8] A prominent example where such an expansion really leads to a low-dimensional description is Lorenz's treatment of Rayleigh-Benard convection using a projection of the PDE's onto three mode-amplitudes of double-trigonometric functions.[19]

Contrarily, empirical orthogonal functions are not calculated from the equations of motion but from the statistical correlations of the spatio-temporal output variables. Hence, EOF's can be applied even to nonlinear systems in complicated geometries in which the governing equations are unknown.

In the next section, the technique is introduced in detail. Then, two novel applications of EOF's are given: experimental spatio-temporal data from CO oxidation on a Pt(110) crystal surface and PDE simulations of vibrating vocal folds. In both cases the decomposition of the data into spatially coherent modes reveals otherwise hidden features of the dynamics.

2. CALCULATION OF EMPIRICAL ORTHOGONAL FUNCTIONS

EOF's are a natural method to study spatial and temporal correlations. Hence, related techniques have appeared in many fields under various names such as singular value decomposition,[24] principal component analysis,[1] singular spectrum analysis,[29] proper orthogonal decomposition,[20] Karhunen-Loeve expansion,[12] or bi-orthogonal decomposition.[3] In this paper, we prefer the term EOF originating from the context of meteorology,[18] because this name emphasizes that merely empirical data are required.

A "movie" $U(\vec{x}, t)$ is the starting point of the analysis of spatio-temporal patterns. In hydrodynamics $U(\vec{x}, t)$ is often the velocity field.[9,20] In our applications, $U(\vec{x}, t)$ refers to grey levels of video images or to the position coordinates of mesh points in the computer simulation of vocal fold vibrations.

As a first step we separate the data into mean and time-varying part:

$$U(\vec{x}, t) = U_0(\vec{x}) + u(\vec{x}, t). \tag{1}$$

If the data are known at N discrete points in space and if M snapshots of the movie are given, the covariance matrix is obtained as

$$\mathbf{R}_{ij} = \frac{1}{M} \sum_{m=1}^{M} u(x_i, t_m) u(x_j, t_m) \quad (i, j = 1, 2, ..., N). \tag{2}$$

The normalized eigenvectors $\phi(\vec{x})$ of the symmetric $N \times N$ matrix $\mathbf{R}$ form a complete orthogonal set of EOF's. The corresponding eigenvalues λ_i tell how much of the total variance of the movie $u(\vec{x}, t)$ can be explained by the corresponding empirical mode $\phi_i(\vec{x})$.

In this way, the space-time signal can be decomposed into

$$u(\vec{x}, t) = \sum_{i=1}^{N} c_i(t) \phi_i(\vec{x}) \nu. \tag{3}$$

The mode-amplitudes $c_i(t)$ are obtained simply by projecting the data onto the EOF's:

$$c_i(t_m) = \sum_{j=1}^{N} u(x_j, t_m) \phi_i(x_j). \tag{4}$$

The coefficient vectors $c_i(t)$ are sometimes termed "chronos" (as counterpart of the spatial eigenvectors called "topos") and are orthogonal as well.

The described decomposition into orthogonal EOF's looks similar to other expansions such as Fourier analysis or wavelet transforms. However, the empirical functions $\phi(\vec{x})$ are obtained in such a way that they capture the dynamics optimally. For example, the described calculation of EOF's implies that

$$\lambda_1 = \sum_{m=1}^{M} c_1^2(t_m) = \sum_{m=1}^{M} (u, \phi_1)^2 \tag{5}$$

is maximal with the side constraint that $(\phi_1, \phi_1) = 1$. The parenthesis refer to the spatial scalar product. Generally, EOF's exhibit the property that a maximum capture of the variance in a minimum number of modes is achieved.

Consequently, the first EOF's reflect the most intense coherent spatial structures of the movie. Contrarily to eigenfunctions of linear systems, EOF's do not describe all possible modes but they show the modes which are actually excited.

If there is a predominance of a few EOF's, i.e., if the cumulated variance of the first modes approaches the total variance, then the essential features of the movie are represented by just a few spatial modes and their time-varying coefficients. In such a case, data compression, noise reduction and derivation of low-order models become possible.

3. HETEROGENEOUS CATALYSIS

The first example we want to discuss is the decomposition of experimental data that stems from the oxidation of CO on a Pt(110) single crystal surface under low pressures. The data were provided by Ertl, Rotermund, and coworkers from the Fritz Haber Institute in Berlin. This system is known to exhibit an enormous variety of different spatio-temporal patterns.[10,15] Based on a detailed picture of the reaction kinetics[16] it was possible to model some of the spatio-temporal phenomena accurately.[4] The origin and the nature of the dynamics in other parameter regimes, however, is still an open and challenging question.

An example of the latter type of spatio-temporal behavior is reproduced in Color Plate 7. The pseudo-colors correspond to different coverages of the surface: mainly CO-covered regions are shown in red, regions with a high O-coverage appear dark blue; light blue, green and yellow areas indicate a decreasing oxygen and an increasing CO coverage.

The characteristic features of this type of pattern are irregularly bent stripes together with smaller islands of different shape and size in between the stripes.

With progressing time the predominantly CO-covered stripes and islands in Color Plate 7(a) react with oxygen, and the surface is transformed in a more homogeneous state (see Color Plates 7(b)–1(d)) before CO-enriched areas emerge again (see Color Plates 7(e)–1(g)). Color Plate 7(g) seems to be a translation of Color Plate 7(a): the stripes appear now just half way between the stripes of Color Plate 7(a); the same is true for the islands. The process happens again, and some time later (see Color Plate 7(h)) we are back at a situation that is very similar to the one shown in Color Plate 7(a). Note, however, that there are differences in the patterns of Color Plate 7(a) and 7(h), and hence the dynamics are not perfectly periodic.

For the determination of the empirical eigenfunctions we used 750 images over approximately 25 seconds, each image consisting of 256×256 pixels. The direct calculation of the spatial two-point covariance matrix is obviously not possible for images of this size; it would involve the diagonalization of a matrix with $O(10^9)$ elements.

The eigenvectors of the spatial correlation matrix can, however, be represented by a superposition of the snapshots themselves:

$$\phi_j(\vec{x}) = \sum_{m=1}^{M} b_j(t_m) u(\vec{x}, t_m) \tag{6}$$

whereby the coefficients b are the normalized eigenvectors of the temporal two-point correlation matrix $\mathbf{A}$

$$\mathbf{A}_{mn} = \frac{1}{M} \sum_{i=1}^{N} u(x_i, t_m) u(x_i, t_n) \qquad (m, n = 1, 2, ..., M), \tag{7}$$

$$\mathbf{A}\vec{b} = \lambda \vec{b}. \tag{8}$$

Provided that a moderate number of snapshots gives a good representation of the data, Eqs. 6 and 7 allow the determination of EOF's in cases where the size of the spatial domain of interest prohibits their direct calculation. This procedure has been termed the "method of snapshots."[22]

In Table 1, we show the 10 largest normalized eigenvalues together with their cumulative contribution to the total variance. There are three dominant modes which capture more than 75% of the total variance. Each of the next four modes carries more than 1% of the variance, whereas the contribution of the higher modes becomes very small.

The first eight EOF's are shown in Color Plate 6. The first four modes as well as the eighth one seem to be composed of wavy stripes. The number of these stripes in the individual modes is different. The second and third mode consist of approximately twice as many stripes as the first one, and the fourth mode of three times as many stripes. This is reminiscent of a Fourier series. However, it would never be possible to describe the wavy character of the stripes with just a few Fourier modes.

TABLE 1 Individual and cumulative normalized variance captured by the first EOF's.

EOF No.	Normalized Variance	Cumulative Normalized Variance
1	0.394	0.394
2	0.257	0.651
3	0.138	0.789
4	0.047	0.836
5	0.016	0.852
6	0.014	0.866
7	0.012	0.878
8	0.008	0.886
9	0.007	0.893
10	0.0056	0.899

The prominent attributes of the fifth, sixth, and seventh mode are not stripelike features, but seemingly irregular distributed "spots." The different characteristics of the modes 1–4 and 8, on one hand, and 5–7, on the other hand, are reflected in the corresponding time series shown in Figure 1, and both, chronos and topos together, reveal interesting and unexpected information about the underlying spatio-temporal dynamics: the first four coefficients as well as the eighth one seem to have a defined period, whereas the fourth and eighth coefficient are somewhat noisy. The time-series of the 5th, 6th and 7th coefficient, however, just cannot be viewed as being a noisy version of a periodic motion; we rather have to conclude that they represent irregular dynamics. This, in turn, leads to the presumption that the full spatio-temporal dynamics is a superposition of a time-periodic oscillation of a stripe-like pattern and a temporally irregular motion of spatially disordered spots. Never would have such an insight into the dynamics been possible without the decomposition of the data into EOF's.

Moreover, as demonstrated in Color Plate 8, the decompostion of the data into EOF's not only leads to a better understanding of the dynamical process, but also results in a drastic data reduction. The images shown in Color Plate 8 were reconstructed according to Eq. 3 using the first eight modes. A comparison with Color Plate 7 shows a remarkable agreement of the two representations. In addition, the images appear to be more clear; i.e., some noise reduction has been achieved. Indeed, a very extensive application of EOF's is related to pattern recognition.[12]

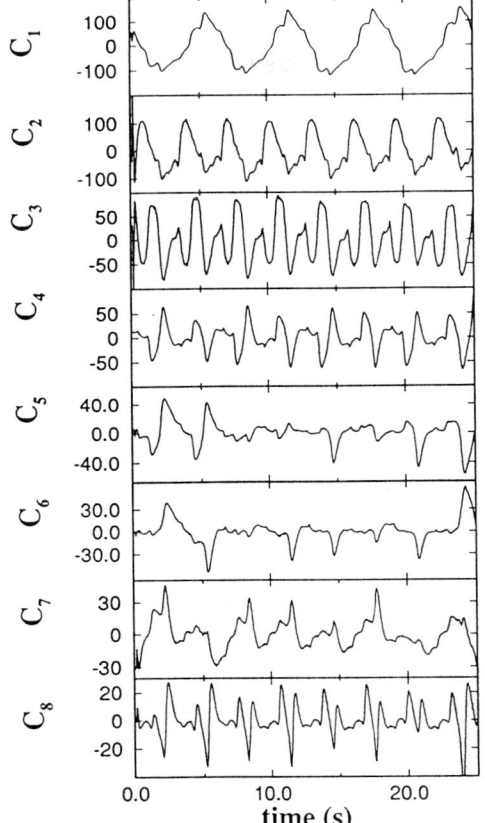

FIGURE 1 Time series of the coefficients of the first eight modes. C_1 to C_4 and C_8 are approximately periodic in time, whereas the temporal behavior of C_5 to C_7 is irregular. Note that the irregular time series correspond to spatial modes that are dominated by irregularly distributed "spots."

Of course, the truncation of the series after eight modes is somewhat arbitrary. The proper choice of the truncation threshold depends on the purpose of the decomposition. It cannot be excluded generally, that rather weak EOF's might have considerable importance for some purposes.

4. SIMULATION OF VOCAL FOLD VIBRATIONS

According to the myoelastic theory of phonation, the vocal folds are set into vibration by the combined effects of aerodynamic forces and the elastic properties of the folds. Normal phonation is characterized by nearly periodic oscillations with a period of a few milliseconds. The analysis of rough sounding voices (e.g., creaky voices, vocal disorders, and also newborn infant cries) reveals an intimate relationship between voice mechanics and bifurcations and chaos.[13,14,21,27]

In this chapter, the output of sophisticated biomechanical simulations of vocal fold vibrations are studied by means of EOF's (also termed modes in the remainder of the paper). The governing viscoelastic partial differential equations incorporate: the complex three-dimensional shape of the vocal folds, the layered tissue scheme of the folds, tissue viscosity, incompressibility, and nonlinear stress-strain curves of vocal fold tissue.[2]

The equations are solved using triangular finite elements. There are 207 nodes per fold which are free to oscillate in lateral and vertical directions. Details of the simulations have been presented elsewhere.[28] In Figure 2, we show a coronal view of the middle section together with the discretization into finite elements (left) and chaotic oscillations of the nodes (right). During simulations of normal phonation, the nodes follow nearly elliptical orbits.[5,27]

In this chapter, we focus on the statistical analysis of representative vibratory patterns. All runs employed left-right symmetry and, hence, we have 414 degrees of freedom, i.e., the spatial vector $\vec{x}$ in Eqs. 1–4 has 414 components.

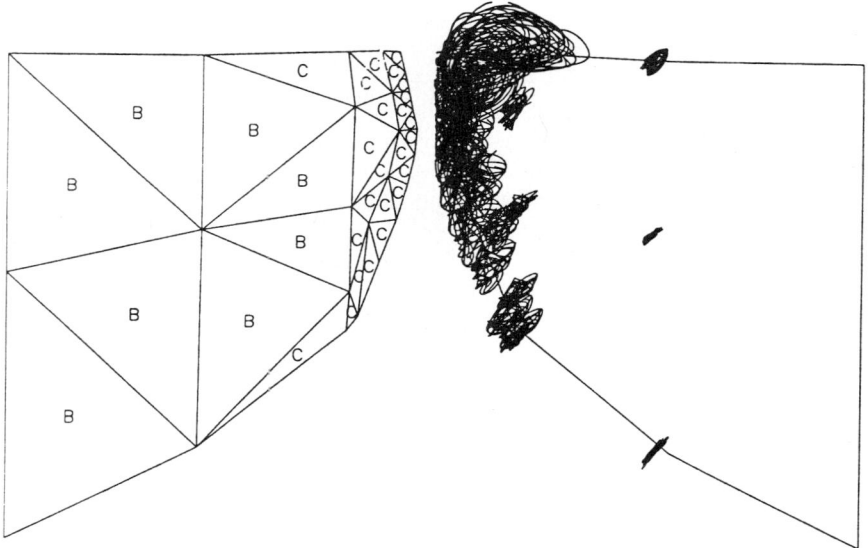

FIGURE 2 A coronal view of a longitudinal section of simulated vocal folds. On the left side, the 32 finite elements are displayed. Their viscoelastic properties are chosen to simulate the layered structure of vocal folds with body ("B") and cover ("C"). On the right side, trajectories of nodal points are shown for chaotic vibrations.

TABLE 2 Cumulative normalized variance of the first EOF's for normal phonation (left), chaotic vibrations for high pressure (right).

Mode Number	$P_s = 8$ cm H_2O		$P_s = 13$ cm H_2O	
	λ_i (%)	Cumulative sum of λ_i (%)	λ_i (%)	Cumulative Sum of λ_i (%)
1	71.2	71.2	65.1	65.1
2	25.0	96.2	16.8	81.9
3	2.4	98.6	8.0	89.9
4	0.9	99.5	2.1	92.0
5	0.2	99.7	1.8	93.8

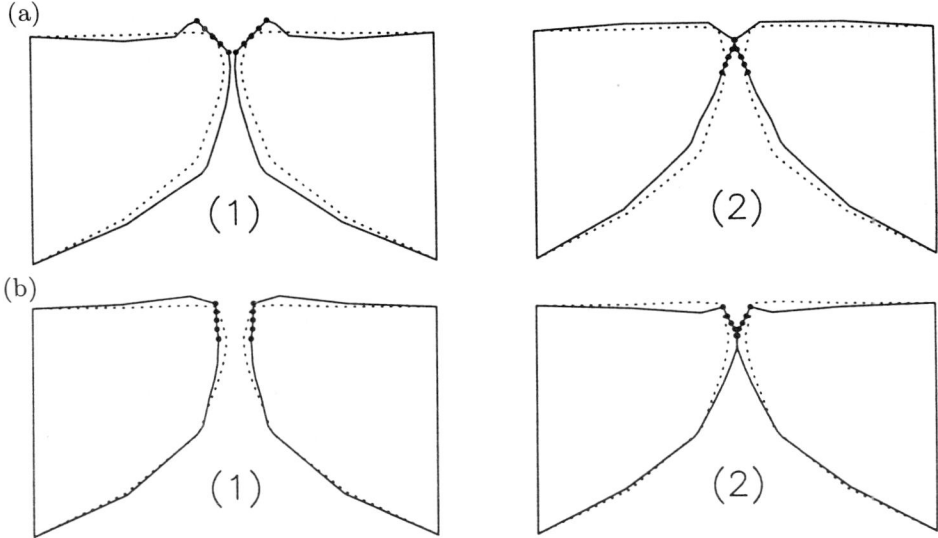

FIGURE 3 A coronal view of the two strongest EOF's for normal phonation. In both cases, frame 1 shows a maximum in the corresponding mode-amplitudes, while frame 2 corrosponds to minima. Five nodal points are bolded to emphasize the out-of-phase and in-phase motion of the upper parts of the folds.

(a)

(b)

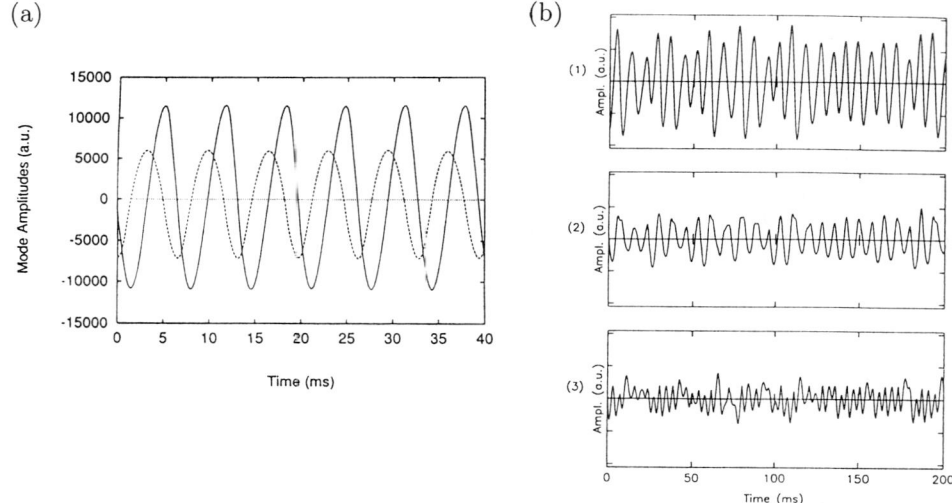

FIGURE 4   Mode amplitudes for normal (a) and chaotic (b) vocal fold vibrations. In (a), mode 1 is illustrated as a solid line, and mode 2 as a dashed line. In (b), the first 3 mode amplitudes are displayed.

In this section, periodic and chaotic vocal fold oscillations are analyzed, corresponding to normal and rough voices, respectively. In Table 2, we show that for normal phonation (taken to be the condition where the subglottal pressure P_s equals 8 cm of H_2O) two modes explain about 96% of the variance. This finding is consistent with earlier approaches claiming that chest voice can be characterized as the superposition of a uniform opening of the glottis (top to bottom) and an alternating divergent-convergent glottis.[25,26]

Indeed, the snapshots of the two dominant modes shown in Figure 3 exhibit these features. The first mode is largely responsible for the out of phase motion of the lower and higher portions. Positive amplitudes lead to a divergent shape (frame 1), whereas negative amplitudes give a convergent glottis (frame 2). Moreover, some vertical displacement is induced due to the incompressibility of the tissue. The coronal views of the second mode visualize the lateral movement of the folds near the top of the glottal air passage. The phase shift of these modes (see Figure 4(a)) is responsible for the energy transfer from the airflow to the tissue: the convergent shape of the air passage during opening allows only a weak flow, i.e. nearly the full subglottal pressure pushes the folds apart. The divergent shape during closure leads to a substantial airflow and, therefore, the static pressure is reduced by the Bernoulli force. In this way, dissipative losses can be overcome and oscillations are sustained.

Thus it is possible with the aid of EOF's to reduce hundreds of trajectories to essentially two modes of vibration, allowing an interpretation of the mechanism of self-oscillation.

Chaotic oscillations, related to rough voice quality, have been found in the simulations for decreasing stiffness of the cover and for increasing subglottal pressure. It has been suggested earlier that the origin of chaos in vocal fold vibrations can be traced back to the desynchronization of a few oscillatory modes.[27,14] Indeed, in Table 2, we show that even for complex vibratory patterns associated with relatively high subglottal pressures (see Figure 2 right), a few modes capture most of the variance. Figure 4 displays the mode-amplitudes of the dominant modes during normal and chaotic phonation.

It turns out, moreover, that during the bifurcation cascades the dominant modes remain essentially the same, i.e., dot products of the first three normalized empirical modes give about 90% agreement.[5] Hence, appropriate low-order models can be designed to describe the observed bifurcations. The analysis of ODE's which are derived by projection of the original PDE's onto the EOF's will be discussed in a forthcoming paper.

The data compression with EOF's can be exemplified by considering the output of simulations of normal phonation. The original movie of 3 seconds, sampled at 20 kHz, results in 414×60000 real numbers (about 100 Mb). Since the first two modes contain 96% of the variance, the superposition of these EOF's yield a virtually indistinguishable movie using only two spatial eigenfunctions and their time-varying amplitudes (together about 484 kb).

For rectangular compressible folds with linear stress-strain curves the eigenfunctions are analytically known [25,26]. As mentioned in the introduction, EOF's are applicable also in linear systems. Hence, empirical functions can be used in a future study for analyzing how the original eigenfunctions of the linear system are distorted due to nonlinearities, compressibility, and complex geometry.

5. DISCUSSION

The above applications have demonstrated that rather different space-time signals can be analyzed successfully with empirical orthogonal functions.

However, we do not claim that the method works as well for any pattern. The main limitations result from the requirement that the covariance matrix has to be estimated with sufficient accuracy. For periodic or quasiperiodic signals a movie of a few cycles might be sufficient for the calculation of the covariance matrix.[29] Our analysis revealed that 300 ms (about 50 cycles) lead to reasonable estimations of the EOF's for chaotic fold vibrations. However, unstationary data such as $1/f$ noise would be difficult to analyze.

A consistent estimation of the covariance matrix allows the decomposition into EOF's. However, the advantages of this specific orthogonal basis (data compression, noise reduction,...) apply only if there is a dominance of far less then N modes. Consequently, for highly irregular data with short correlations in space and time many eigenvalues have comparable magnitude and, hence, the projection onto EOF's gives no advantage.

The possibility of deriving low-order models from PDE's or experimental movies using EOF's has already been demonstrated by Lorenz.[18] For our applications, this problem will be discussed explicitly in forthcoming papers. For other data, there are already convincing results in this direction.[23,9] For example, Deane et al. derived ODE's for flow simulations in a grooved channel and for a flow around a cylinder.[9] In this way, they were able to identify the bifurcations that correspond to qualitative changes in the full simulations.

In Krischer et al.,[17] EOF's from similar chemical data as discussed in section 3 were used to train an artificial neural network (ANN). In this way, an implicit model of the spatio-temporal pattern was constructed. It has been shown that the ANN reproduces essential dynamical features of the original data.

Finally, we mention that EOF's have been applied successfully to scalar time-series[7,11] as well. In this case, a pseudo-phase-space is constructed in which the EOF decomposition is performed.

Summarizing, we emphasize that the analysis of spatio-temporal patterns with EOF's is an algorithmically rather simple technique which can give insight into a variety of complex patterns. Even though it is a linear method based upon two-point correlations in space and time, it turned out to be enlightening in a number of highly nonlinear systems.

This was demonstrated with two examples: In the case of the CO-oxidation the full dynamics is the result of a superposition of stripe-like spatial modes that oscillate periodically in time, and spatial modes, that are dominated by irregular "spots" and show an aperiodic temporal behavior.

The decomposition of simulations of vocal fold oscillations facilitated a physical interpretation of the energy transfer mechanism in vocal fold dynamics. Moreover, the EOF analysis made it possible to interpret the appearance of chaotic oscillations of rough voice in terms of a desynchronization of a few low-order modes.

ACKNOWLEDGMENTS

Partial funding for this research was provided by Grant No. P60 DC00976 from the National Institutes on Deafness and Other Communication Disorders. The authors are also grateful to G. Ertl, H. H. Rotermund and S. Jakubith for providing the experimental data of the CO-oxidation and to I. G. Kevrekidis for introducing them to the method of EOF's and for hospitality in Princeton.

REFERENCES

1. Ahmed, N., and M. H. Goldstein. *Orthogonal Transforms for Digital Signal Processing.* New York: Springer-Verlag, 1975.
2. Alipour-Haghighi, F., and I R. Titze. "Elastic Models of Vocal Fold Tissues." *J. Acoust. Soc. Am.* **90** (1991): 1326–1331.
3. Aubry, N. S., R. Guyonnet, and R. Lima. "Spatial-Temporal Analysis of Complex Signals: Theory and Applications." *J. Stat. Phys.* **64** (1991): 683–739.
4. Baer, M., M. Falcke, and M. Eiswirth. "Dispersion Relation and Spiral Rotation in an Excitable Surface Reaction." *Physica A* **188** (1992): 78–88.
5. Berry, D. A., H. Herzel, I. R. Titze, and K. Krischer. "Interpretation of Biomechanical Simulations of Normal and Chaotic Vocal Fold Oscillations with Empirical Eigenfunctions." *J. Acoust. Soc. Am.*: to appear (June, 1994).
6. Breuer, K. S., and Sirovich, L. "The Use of the Karhunen-Loeve Procedure for the Calculation of Linear Eigenfunctions." *J. Comp. Phys.* **96** (1991): 277–296.
7. Broomhead, D. S., and G. P. King. "Extracting Qualitative Dynamics from Experimental Data." *Physica D* **20** (1986): 217–236.
8. Canuto, C., M. Hussaini, A. Quarteroni, and T. A. Zang. *Spectral Methods in Fluid Dynamics.* New York: Springer-Verlag, 1988.
9. Deane, A. E., I. G. Kevrekidis, G. E. Karniadakis, and S. A. Orszag. "Low-Dimensional Models for Complex Geometry Flows: Application to Grooved Channels and Circular Cylinders." *Phys. Fluids A* **3** (1991): 2337–2354.
10. Ertl, G. "Oscillatory Kinetics and Spatio-Temporal Self-Organization in Reactions at Solid Surfaces." *Science* **254** (1991): 1750–1755.
11. Fraedrich, K. "Estimating the Dimension of Weather and Climate Attractors." *J. Atmos. Sci.* **43** (1986): 419–432.
12. Fukunaga, K. *Introduction to Statistical Pattern Recognition.* Boston: Academic Press, 1990.
13. Herzel, H., and J. Wendler. "Evidence of Chaos in Phonatory Samples." In *Proceedings EUROSPEECH*, 263–266. Genova: ESCA, 1991.
14. Herzel, H., D. A. Berry, I. R. Titze, and M. Saleh. "Analysis of Vocal Disorders with Methods from Nonlinear Dynamics." *J. Speech Hear. Res.*: to appear (August, 1994).
15. Jakubith, S., H. H. Rotermund, W. Engel, A. von Oertzen, and G. Ertl. "Spatiotemporal Concentration Patterns in a Surface Reaction: Propagating and Standing Waves, Rotating Spirals, and Turbulence." *Phys. Rev. Lett.* **65** (1990): 3013–3016.
16. Krischer, K., M. Eiswirth, and G. Ertl. "Oscillatory CO Oxidation on Pt(110): Modeling of Temporal Self-Organization." *J. Chem. Phys.* **96** (1992): 9161–9172.

17. Krischer, K., R. Rico-Martinez, I. G. Kevrekidis, H. H. Rotermund, G. Ertl and H. L. Hudson. "Model Identification of a Spatiotemporally Varying Catalytic Reaction." *AIChE J.* **39** (1993): 89–98.

18. Lorenz, E. N. "Empirical Orthogonal Functions and Statistical Weather Prediction." Scientific Report No.1, Dept. of Meteorology, Massachusetts Institute of Technology, 1956.

19. Lorenz, E. N. "Deterministic Nonperiodic Flow." *J. Atmos. Sci.* **20** (1963): 130–141.

20. Lumley, J. L. *Stochastic Tools in Turbulence.* New York: Academic Press, 1970.

21. Mende, W., H. Herzel, and K. Wermke. "Bifurcations and Chaos in Newborn Cries." *Phys. Lett.* A **145** (1990): 418–424.

22. Sirovich, L. "Turbulence and the Dynamics of Coherent Structures: I, II and III." *Quart. Appl. Math.* **XLV** (1987): 561–590.

23. Sirovich, L., and J. D. Rodriguez. "Coherent Structures and Chaos: A Model Problem." *Physics Letters A* **120** (1987): 211–214.

24. Strang, G. *Linear Algebra and Its Applications.* New York: Academic Press, 1988.

25. Titze, I. R., and W.J. Strong. "Normal Modes in Vocal Fold Tissue." *J. Acoust. Soc. Am.* **57** (1975): 736–744.

26. Titze, I. R. "On the Mechanics of Vocal-Fold Vibration." *J. Acoust. Soc. Am.* **60** (1976): 1366–1380.

27. Titze, I. R., R. Baken, and H. Herzel. "Evidence of Chaos in Vocal Fold Vibration." In: *Vocal Fold Physiology: Frontiers in Basic Science,* edited by I. R. Titze, 143–188. San Diego: Singular Publishing Group, 1993.

28. Titze, I. R., and F. Alipour-Haghighi. *Myoelastic Aerodynamic Theory of Phonation.* Forthcoming.

29. Vautard, R., P. Yiou, and M. Ghil. "Singular-Spectrum Analysis: A Toolkit for Short, Noisy Chaotic Signals." *Physica D* **58** (1992): 95–126.

Chapter 7
Noise, Turbulence, Chaos, and Intermittency

A. Hernández-Machado
Departament d'Estructura i Constituents de la Matèria, Universitat de Barcelona, Diagonal 647, E-08028 Barcelona, Spain.

The Effect of Noise on Spatio-Temporal Patterns

INTRODUCTION

The study of fluctuations on spatio-temporal patterns far from equilibrium constitutes an active field of research. Such effects have been observed experimentally in the generation of cells in Rayleigh-Bénard convection,[4,20,21] Taylor vortices in unstable Couette-Taylor flow,[33] electroconvection in nematic liquid crystals[3,37,29] and sidebranching in dendritic growth.[2,8,9,26] Some analytical work and computer simulations have been carried out to explain such phenomena.[4,35,32] However, many fundamental aspects remain to be clarified, for instance, the origin of the experimental noise or the correct modeling of the stochastic sources. Noise could have an internal or external origin, and it could appear in the dynamic equations or in the boundary conditions. One aim of the theoretical studies would be the development of a unified theory of stochastic processes in spatially extended systems by means of functional Langevin and Fokker-Planck formalism, where the noise terms take into account stochastic perturbations of internal (thermal) and/or external origin.

Spatio-Temporal Patterns, Ed. P. E. Cladis and P. Palffy-Muhoray,
SFI Studies in the Sciences of Complexity, Addison-Wesley, 1995   **521**

The simplest way to consider fluctuations is by adding a noise to the macroscopic equations. However, this is not always a correct procedure because, in general, both internal and external noise could appear in a multiplicative way. In some liquid crystals experiments,[3,37] his situation has been studied by deliberately superimposing a noise to the external parameter, the AC voltage, that appears in the deterministic equations multiplying the relevant variables. In this situation, the effects of the noise depend on the state of the system. The experimental results[3,37] imply a strong effect on the response of the system, like changes in the threshold of the instability points.[14] Furthermore, internal noise could also appear multiplicatively in situations for which the transport coefficients depend on the field variables characterizing the system. This possibility has been discussed in different context, such as, phase-separation dynamics,[12,17,18] pattern formation,[15,16,22] and polymers.[7] Then, the role of multiplicative noise in spatially extended systems has to be clarified, both for internal and external noise.

We start with amplitude equations for a scalar variable with a spatial and temporal dependence and that include additive and multiplicative noise sources. The advantage of this procedure is that amplitude equations are relatively simple to be treated analytically, specially in the aspects related to the noise contributions. Furthermore, they have substantial advantage from the numerical point of view compared with other descriptions, such as integro-differential equations. We write our stochastic amplitude equation in the following generic form:

$$\frac{\partial \psi}{\partial t} = V(\psi, \nabla\psi) + G(\psi, \nabla\psi)\xi(\vec{r}, t) + \eta(\vec{r}, t) \tag{1}$$

where $\psi(\vec{r}, t)$ is the variable and ξ and η are the multiplicative and additive noises. Both are Gaussian and white with zero-mean value and intensities D_M and D_A, respectively. The noises are assumed to be independent of each other and the multiplicative noise term will be interpreted according to the Stratonovich calculus.[31]

In Sections 2 and 3, we study some effects of external noise on models given by Eq. (1). In these cases, the multiplicative noise is modeling the external noise and the additive noise is internal (thermal). In Section 2, we consider the Swift-Hohenberg equation. We have performed a linear stability analysis and numerical simulations.[11] The results show that multiplicative noise induces convective structures (noise-induced structures) in a regime in which a deterministic analysis predicts a homogeneous solution. In Section 3, we apply a mean-field theory to model A of phase separation dynamics (i.e., with nonconserved order parameter).[34] Our analytical results predict that the transition point is also shifted by the multiplicative noise and they are corroborated by numerical simulations of the exact model. In Section 4, we consider a different situation than the ones of the previous sections.[28,27] In this case, the multiplicative noise appearing in Eq. (1) is of internal origin and there is no additive noise. This model could be obtained by a coarsed-grained procedure from a more microscopic description based in a master equation. As an example, we discuss domain growth in model B of phase separation dynamics (i.e., with conserved order parameter). The multiplicative noise gives new contributions to the Cahn-Hilliard theory and, in particular, a delay in the domain growth dynamics.

SWIFT-HOHENBERG EQUATION

The Swift-Hohenberg equation has been extensively used to study Rayleigh-Bénard convection.[4] In the deterministic situation for some value of the external control parameter, the Rayleigh number, the fluid evolves from a homogeneous situation to the generation of rolls. Fluctuations of internal origin were introduced by means of an additive noise.[10,32,36] In this case, disordered states appear for large noise intensity.[10] The effects of deterministic perturbations, like temporal ramps or sinusoidals, on the external parameter have also been studied.[13,30,32] In order to achieve better understanding of the effects of noise, it is interesting to apply stochastic perturbations deliberately to the external parameter, that could be well controlled. Here, we study a model which contains such fluctuations of external origin. This model could describe fluctuations in the temperature gradient that is externally applied to the system, giving rise to a new nonequilibrium situation. Our model is given by Eq. (1) with:

$$V(\psi) = \left[\Gamma - \left(1 + \nabla^2\right)^2\right]\psi - \psi^3 \qquad G(\psi) = \psi. \tag{2}$$

The fluctuations of the control parameter around a mean value Γ are taking into account by the term proportional to $\xi(\vec{r}, t)$ of Eq. (1). In this case, the intensity D_A is proportional to temperature and the intensity D_M is assumed to be a second independent control parameter. For $D_M = 0$, a fixed small D_A and $\Gamma < \Gamma_c = 0$, the system is in a homogeneous state of $\langle\psi\rangle = 0$ with small random fluctuations induced by the additive noise. For $\Gamma > 0$ the system evolves from the homogeneous state to a convective ordered state composed of rolls. By increasing D_A the state becomes more disordered and could even disappear.

From Eqs. (1) and (2), we write the equation for the structure function, defined as $S(\vec{q}, t) = 1/V\langle\psi(\vec{q}, t)\psi(-\vec{q}, t)\rangle$, where the average is calculated over the different realizations of the noises present in Eq. (1). In the linear approximation we obtain:

$$\frac{\partial}{\partial t}S(\vec{q}, t)$$
$$-2\left[-\Gamma - \widetilde{D_M} + \left(1 - 2q^2 + q^4\right)\right]S(\vec{q}, t) + 2D_I + 2D_M\left(\frac{1}{2\pi}\right)^d \int S(\vec{q}, t)d\vec{q}, \tag{3}$$

where we have considered $q\Delta x < 1$ and $\widetilde{D_M} = \frac{D_M}{\Delta x^2}$. The contribution of the additive noise does not change the deterministic stability condition of the linear analysis.

However, due to the multiplicative character of the external noise,[23] we find nonzero contributions proportional to D_M. Thus, the presence of a multiplicative noise leads to the existence of an effective noise-dependent control parameter, $\Gamma + \widetilde{D_M}$ and the linear analysis predicts that the system will leave the homogeneous state in situations for which $\Gamma + \widetilde{D_M} > 0$. The nonlinearity will then stabilize the system in an ordered state. This only happens when $D_M \neq 0$ and, therefore, we talk about a noise-induced convective structure.[11]

We have also performed numerical simulations of Eqs. (1) and (2) in a two-dimensional lattice.[11] Additive and multiplicative noise were introduced into the algorithm by means of a standard procedure.[28,27] In García-Ojalvo et al.[11] we have studied the cases of ordered ($\Gamma > 0$) and disordered ($\Gamma < 0$) states when only internal noise is considered ($D_M = 0$). In the first case, the characteristic roll pattern associated to the convective state are obtained, as expected. Furthermore, we have also studied the same situation with $D_M \neq 0$. In this case, instead of the homogeneous pattern associated to subcritical values of Γ ($\Gamma < 0$), we obtain a roll pattern very similar to that induced by a supercritical value ($\Gamma > 0$). We have also studied analytically and numerically other quantities,[11] like the perpendicular transmitted flux $J(t)$, defined as $J(t) = 1/V \int \psi^2(\vec{r},t)d\vec{r}$,[20,21] and the stationary flux, J_{stat}. We have obtained that $J(t)$ is zero for the homogeneous case and reaches a nonzero steady state value for a not very large intensity of the external noise. Furthermore, for small intensities of the additive noise, J_{stat} has a linear dependence versus Γ and it presents a clear positive shift proportional to the intensity of the multiplicative noise. The theoretical and numerical results of García-Ojalvo et al.,.[11] are in good qualitative and quantitative agreement.

MEAN FIELD THEORY

In this section we consider the dynamic evolution of a nonconserved order parameter, ψ, like the magnetization of a ferromagnet. The model is given by Eq. (1) with:

$$V(\psi) = \left[\Gamma + \beta\nabla^2\right]\psi - \psi^3 \qquad G(\psi) = \psi. \qquad (4)$$

Like in the previous section the multiplicative noise of Eq. (1) takes into account the fluctuations in the external control parameter Γ and the additive noise is of internal origin. To apply the mean field theory, we discretize the model, Eqs.(1,4). The parameter β is a measure of the interaction length between cells. The mean field theory decouples the behavior of one cell from the others in such way that the spatial coupling of cell i to its neighbors is replaced by a coupling to the average value $\langle\psi\rangle$:

$$\dot{\psi_i} = \Gamma\psi_i - \psi_i^3 + \beta(\langle\psi\rangle - \psi_i) + \xi_i + \eta_i \qquad (5)$$

and the mean value $\langle \psi \rangle$ has to be calculated self-consistently imposing $\langle \psi \rangle = \langle \psi_i \rangle$. The advantage of the mean-field ansatz is that Eq. (5) is now closed in the variable ψ_i. Moreover, the self-consistent equation for $\langle \psi \rangle$ is nonlinear, opening the possibility for multiple solutions which are to be expected in the case of symmetry breaking associated to a transition. In the Langevin Eq. (5), we can drop the subscript i since now we have the same equation for every site. At this point, it is revealing to give a simple argument valid in the limit of a very large spatial coupling $\beta \to \infty$. From Eq. (5) one easily obtains the exact evolution equation for $\langle \psi(t) \rangle$.[23] In the limit $\beta \to \infty$, the fluctuations of the variable ψ around its mean value $\langle \psi \rangle$ are expected to vanish since the coupling term $\beta(\langle \psi \rangle - \psi)$ in Eq. (5) prevents such fluctuations. Consequently the steady state equation becomes $(\Gamma + D_M) \langle \psi \rangle - \langle \psi \rangle^3 = 0$. This result predicts that multiple solutions arise at the value $\Gamma = -D_M < 0$ whereas no transition is possible in the absence of multiplicative noise for $\Gamma < 0$. In other words, the transition to bistability is advanced. Note that the location of the transition point coincides in this limit with the criterion of the breakdown of linear stability: the transition is predicted at the point at which the coefficient of the linear term in the equation for $\langle \psi \rangle$ vanishes. This criterion was used in the previous section in the study of the Swift-Hohenberg equation[11] to locate the transition to rolls and good agreement with simulations was found. This point has been recently discussed by Kramer.[1] In the simulations of García-Ojalvo et al.,[11] the internal noise intensity is small, which is tantamount to having a strong spatial coupling. The observed agreement with the linear instability criterion is thus in accordance with our mean-field analysis.

From the Fokker-Planck equation associated to Eq. (5) one could write the following self-consistant equation for the mean value $\langle \psi \rangle$:

$$\langle \psi \rangle = \frac{\int dz\, z \exp\left[-2 \int_0^z dy - \Gamma y + y^3 - \beta(\langle \psi \rangle - y) + D_M y/2 D_M y^2 + 1\right]}{\int dz\, exp\left[-2 \int_0^z dy - \Gamma y + y^3 - \beta(\langle \psi \rangle - y) + D_M y/2 D_M y^2 + 1\right]}. \tag{6}$$

This equation possesses the trivial solution $\langle \psi \rangle = 0$. Furthermore, a pair of new solutions symmetric around $\langle \psi \rangle = 0$ appear at the values of the parameters for which $\langle \psi \rangle' 3_{\langle \psi \rangle = 0} = 1$. At this point, we would like to stress the difference of this situation with what is happening in the well-known zero-dimensional case. Indeed consider Eq. (5) with $\beta = 0$. The stationary probability density can easily be obtained and is found to be an even function of ψ. Consequently, the corresponding stationary state value of the average of ψ is identically zero and this happens whether or not the system is linearly unstable. However, in the presence of a spatial coupling $\beta \neq 0$ our result indicates that there is a symmetry breaking and $\langle \psi \rangle \neq 0$. We have also performed[34] numerical simulations of the exact model Eq. (1) and (4) to compare with the previous theory and we obtain good qualitative agreement. In particular, the simulations of the exact model give a higher critical value of Γ than the mean-field model. This is a usual feature of mean-field theories and it also happens in the absence of multiplicative noise.

A MULTIPLICATIVE INTERNAL NOISE

Here, we consider a situation in which an internal noise appears in a multiplicative way. An example in phase separation is given by a system of two components, like a binary liquid or alloy.[12,18,17] The system is suddenly quenched from a one-phase region inside its coexistence region. Then, the homogeneous region becomes unstable and domains of the new stable phases start growing. This mechanism is called spinodal decomposition. The deterministic evolution of such a system has been studied when a field-dependent diffusion coefficient is taken into account.[17] This assumption has been considered to model deep quenching[18,12] or to take into account the presence of an external field, like gravity.[16,15] We have found[27,28] that the assumption of field-dependent transport coefficients implies multiplicative thermal fluctuations. In Ramirez,[27,28] we have obtained Langevin equations with multiplicative noise from a mesoscopic derivation using a coarse-grained procedure. In general, these equations could be written by means of Eq. (1) with:

$$V(\psi) = \nabla M \left[\nabla \left(-1 + \nabla^2 \right) \psi + \psi^3 \right] - \frac{\beta^{-1}}{2} \nabla \left(\nabla \frac{\delta}{\delta c} \right) M; \quad G(\psi) = \nabla^i [m] \quad (7)$$

where $M(\psi) = (1 - a\psi^2)$ is the field-dependent diffusion coefficient, $M(\psi) = m^2(\psi)$ and the ∇ operator in $G(\psi)$, Eq. (7), acts on the noise ξ of Eq. (1). Furthermore, $D_M = \beta^{-1}$ and $D_A = 0$. For $a = 0$ we obtain the usual model B of phase separation dynamics with additive noise (the multiplicative noise reduces to additive in this case). However, when M depends on the field variable, apart from the multiplicative noise term, we obtain a spurious term of stochastic origin, the second one in Eq. (7), which ensures the evolution of the system to the correct equilibrium solution. A linear stability analysis could be applied to the Langevin equation, Eqs. (1) and (7).[23] In this way, we have obtained new contributions to the Cahn-Hilliard theory. Our results indicate that only modes with $k < k_c = 1 - 4a\beta^{-1}/R^2$ are unstable, and they will grow in the early stages of the evolution. On the contrary, modes with $k > k_c$ remains stable during the linear regime. R is a mesoscopic length that gives the sizes of the region containing the cell points involved in the diffusion of matter at each time step.[27,28] So, one can expect that, for initial times, with a smaller number of modes growing, the domain growth would be slower than in the additive noise case $a = 0$. Regarding the numerical simulations of the stochastic Eqs. (1) and (7), the essential problem is introduced by the conservation law acting on the multiplicative noise term. In Ramirez,[27,28] we have derived numerical algorithms to simulate these equations and simulations of Eqs. (1) and (7) have been performed, starting from an homogeneous initial state $\psi = 0$. The numerical results show that the dynamics is not only slower by increasing a,[17] but also the multiplicative noise induces a delay in the short-time behavior of the domain growth. The result corroborate quantitatively the new contribution of the Cahn-Hilliard theory.

ACKNOWLEDGMENTS

I acknowledge the collaboration in the work presented here of L. Ramírez-Piscina, J. García-Ojalvo, J. M. Sancho, C. Van den Broeck, J. M. R. Parrondo, J. Armero, A. Lacasta, and J. L. Mozos. I also thank L. Kramer, P. Hohenberg, J. S. Langer, and J. Casademunt for stimulating suggestions. I am grateful for the support of the Dirección General de Investigación Científica y Técnica (Spain) Pro. No. PB90-0030.

REFERENCES

1. Becker, A., and L. Kramer. Preprint, 1993.
2. Bouissou, P., A. Chiffaudel, B. Perrin, and P. Tabeling. *Europhys. Lett.* **13** (1990): 89.
3. Brand, H. R., S. Kai, and S. Wakabayashi. *Phys. Rev. Lett.* **54** (1985): 555.
4. Cross, M. C., and P. C. Hohenberg. *Rev. Mod. Phys.* (1993).
5. Deissler, R. J. *J. Stat. Phys.* **54** (1989): 1459.
6. Deissler, R. J., A. Oron, and Y. C. Lee. *Phys. Rev. A* **43** (1991): 4558.
7. Doi, M., and S. F. Edwards. *The Theory of Polymer Dynamics.* International Series of Monographs on Physics, Vol. 73. Cambridge: Oxford University Press, 1989.
8. Dougherty, A., P. D. Kaplan, and J. P. Gollub. *Phys. Rev. Lett.* **58** (1987): 1652.
9. Dougherty, A., and J. P. Gollub. *Phys. Rev. A* **38** (1988): 3043.
10. Elder, K. R., J. Viñals, and M. Grant. *Phys. Rev. Lett.* **68** (1992): 3024.
11. García-Ojalvo, J., A. Hernández-Machado, and J. M. Sancho. *Phys. Rev. Lett.* **71** (1993): 1542.
12. Gunton, J. D., and M. Droz. *Introduction to the Theory of Metastable and Unstable States.* Lecture Notes in Physics, Vol. 183. Berlin: Springer-Verlag.
13. Hohenberg, P. C., and J. B. Swift. *Phys. Rev. A* **46** (1992): 4773.
14. Horsthemke, W., and R. Lefever. *Noise-Induced Phase Transitions.* Berlin: Springer-Verlag, 1984.
15. Jasnow, D. In *Far from Equilibrium*, edited by L. Garrido. Lectures Notes in Physics, Vol. 319. Berlin: Springer-Verlag, 1988.
16. Kitahara, K., Y. Oono, and D. Jasnow. *Mod. Phys. Lett. B* **2** (1988): 765.
17. Lacasta, A., A. Hernández-Machado, J. M. Sancho, and R. Toral. *Phys. Rev. B* **45** (1992): 5276.
18. Langer, J. S., M. Bar-on, and H. D. Miller. *Phys. Rev. A* **11** (1975): 1417.
19. Langer, J. S. *Phys. Rev. A* **36** (1987): 3350.
20. Meyer, C. W., G. Ahlers, and D. S. Canell. *Phys. Rev. Lett.* **59** (1987): 1577.

21. Meyer, C. W., G. Ahlers, and D. S. Canell. *Phys. Rev. A* **44** (1991): 2514.
22. Mozos, J. L., and A. Hernández-Machado. *J. Stat. Phys.* (January, 1994).
23. Novikov, E. A. *Soviet Phys. JETP* **20** (1965): 1290.
24. Pieters, R., and J. S. Langer. *Phys. Rev. Lett.* **56** (1986): 1948.
25. Pieters, R. *Phys. Rev. A* **37** (1988): 3126.
26. Qian, X. W., and H. Z. Cummins. *Phys. Rev. Lett.* **64** (1990): 3038.
27. Ramírez-Piscina, L., A. Hernández-Machado, and J. M. Sancho. *Phys. Rev. B* **48** (1993): 125.
28. Ramírez-Piscina, L., A. Hernández-Machado, and J. M. Sancho. *Phys. Rev. B* **48** (1993): 119.
29. Rehberg, I., S. Rasenat, M. de la Torre Juarez, W. Schopf, F. Horner, G. Ahlers, and H. R. Brand. *Phys. Rev. Lett.* **67** (1991): 596.
30. Stiller, D., A. Becker, and L. Kramer. *Phys. Rev. Lett.* **25** (1992): 3670.
31. Stratonovich, R. L. *Introduction to the Theory of Random Noise.* Gordon & Breach, 1963.
32. Swift, J. B., and P. C. Hohenberg. *Phys. Rev. Lett.* **60** (1988): 75.
33. Tsameret, A., and V. Steinberg. *Phys. Rev. Lett.* **67** (1991): 3392.
34. Van den Broeck, C., J. M. R. Parrondo, J. Armero, and A. Hernández-Machado. Preprint, 1993.
35. Viñals, J., E. Hernández-García, M. San Miguel, and R. Toral. *Phys. Rev. A* **44** (1991): 1123.
36. Viñals, J., H. Xi, and J. D. Gunton. *Phys. Rev. A* **46** (1992): 918.
37. Wu, M., and C. D. Andereck. *Phys. Rev. Lett.* **67** (1991): 596.

Michael Frey† and Emil Simiu‡
†Department of Mathematics, Bucknell University, Lewisburg, PA 17837
‡Structures Division, Building and Fire Research Laboratory, National Institute of Standards
and Technology, Gaithersburg, MD 20899

Noise-Induced Transitions to Chaos

INTRODUCTION

Multistable systems can exhibit irregular (i.e., neither periodic nor quasiperiodic) motion with jumps. Such motion is referred to as basin-hopping or stochastic chaos when induced by noise,[1] and deterministic chaos in the absence of noise. Deterministic and stochastic chaos have hitherto been viewed as distinct and have been analyzed from different, indeed contrasting, points of view.

For a wide class of systems, stochastic and deterministic chaos can be not only indistinguishable phenomenologically but also closely related mathematically. We show this for one-degree-of-freedom multistable systems whose unperturbed counterparts have homoclinic and heteroclinic orbits. (Extensions of the theory to higher-degree-of-freedom and spatially extended systems are underway.) When perturbed by weak damping and deterministic periodic forcing, the dynamics of these systems are periodic or quasiperiodic over certain regions of the system parameter

Spatio-Temporal Patterns, Ed. P. E. Cladis and P. Palffy-Muhoray,
SFI Studies in the Sciences of Complexity, Addison-Wesley, 1995   **529**

space. Over other regions of parameter space the dynamics may be sensitively dependent upon initial conditions; i.e., exhibit a topological equivalence to the Smale horseshoe map. We show that a transition from periodic or quasiperiodic motion to chaotic motion with sensitive dependence upon initial conditions is possible through the introduction of noise.

We develop computable expressions providing: (1) necessary conditions for the occurrence of stochastic chaos with jumps, and (2) measures of chaotic transport characterizing the "intensity" of the chaos. These expressions depend on the distribution and, in particular, the mean-square spectrum of the noise. We obtain these expressions using: (1) the Melnikov transform and its attendant notion of phase space flux, and (2) tail-limited noise including uniformly bounded path approximations of Gaussian noise and shot noise.

The remainder of the chapter is divided into five sections. The first section presents results for systems perturbed by weak additive noise. These results are used in the following section to treat Duffing oscillators with weak near-Gaussian noise. In particular, we describe results for the Duffing oscillator with attracting homoclinic orbits which admit comparison with results based on the Fokker-Planck equation. In the third section, we present results for multiplicatively perturbed systems. These results are used in the fourth section with a recently introduced model of shot noise to treat the Duffing oscillator with shot noise-like dissipation. The last section contains summary comments.

SYSTEMS WITH ADDITIVE EXCITATION

We consider the integrable, two-dimensional, one-degree-of-freedom dynamical system with energy potential V governed by the equation of motion

$$\ddot{x} = -V'(x), \qquad x \in \mathcal{R}. \tag{1}$$

The system governed by Eq. (1) is assumed to have two hyperbolic fixed points connected by a heteroclinic orbit $\vec{x}_s = (x_s(t), \dot{x}_s(t))$. If the two hyperbolic fixed points coincide, then $\vec{x}_s$ is homoclinic. A perturbative component is introduced into the system governed by Eq. (1), giving

$$\ddot{x} = -V'(x) + \varepsilon w(x, \dot{x}, t). \tag{2}$$

The perturbative function $w : \mathcal{R}^2 \times \mathcal{R} \to \mathcal{R}$ is assumed to satisfy the Meyer-Sell uniform continuity conditions[15] and only the near-integrable case, $0 < \varepsilon \ll 1$, is considered. In this section, we restrict our attention to the case of additive excitation and linear damping. For this case,

$$w(x, \dot{x}, t) = \gamma g(t) + \rho G(t) - \kappa \dot{x} \tag{3}$$

and system (2) takes the form

$$\ddot{x} = -V'(x) + \varepsilon[\gamma g(t) + \rho G(t) - \kappa \dot{x}]. \tag{4}$$

Here g and G represent deterministic and stochastic forcing functions, respectively. g is assumed to be bounded, $|g(t)| \leq 1$, and uniformly continuous (UC). The parameters ρ, γ and κ are nonnegative and fix the relative amounts of damping and external forcing in the model.

Consider the random forcing

$$G(t) = \sqrt{\frac{2}{N}} \sum_{n=1}^{N} \frac{\sigma}{S(\nu_n)} \cos(\nu_n t + \varphi_n), \tag{5}$$

where $\{\nu_n, \phi_n; n = 1, 2, \ldots, N\}$ are independent random variables defined on a probability space $(\Omega, \mathcal{B}, P)$, $\{\nu_n; n = 1, 2, \ldots, N\}$ are nonnegative with common distribution Ψ_o, $\{\phi_n; n = 1, 2, \ldots, N\}$ are identically uniformly distributed over the interval $[0, 2\pi]$ and N is a fixed parameter of the model. S and σ in Eq (5) are defined below. The process G is a randomly weighted modification of the Shinozuka noise model.[18,19]

Let $\mathcal{F}$ denote the linear filter with impulse response $h(t) = \dot{x}_s(-t)$ where $\dot{x}_s(t)$ is the velocity component of the orbit $\vec{x}_s$ of the system governed by Eq. (1). $\mathcal{F}$ is called the system orbit filter and its output is $\mathcal{F}[u] = u * h$ where $u = u(t)$ is the filter input and $u * h$ is the convolution of u and h. S in Eq. (5) is then defined to be modulus $S(\nu) = |H(\nu)|$ of the orbit filter transfer function

$$H(\nu) = \int_{-\infty}^{\infty} h(t) e^{-j\nu t} dt \tag{6}$$

and σ in Eq. (5) is

$$\sigma^2 = \int_0^{\infty} S^2(\nu) \Psi(d\nu).$$

Let the distribution Ψ_o of the angular frequencies ν_n in Eq. (5) have the form

$$\Psi_o(A) = \frac{1}{\sigma^2} \int_A S^2(\nu) \Psi(d\nu), \tag{7}$$

where A is any Borel subset of $\mathcal{R}$. S is assumed to be bounded away from zero on the support of Ψ, $S(\nu) > S_m > 0$ a.e. Ψ. Under this condition S is said to be Ψ-admissible. If S is Ψ-admissible, then it is also bounded away from zero on the support of Ψ_o and $1/S(\nu_n) < 1/S_m$ a.s. Ψ_o. We have the following results for G and its filtered counterpart $\mathcal{F}[G]$.

Fact G1. The processes G and $\mathcal{F}[G]$ are each zero-mean and stationary.

Fact G2. If S is Ψ-admissible then G is uniformly bounded with $|G(t, \omega)| \leq \sqrt{2N/S_m}$ for all $t \in \mathcal{R}$ and $\omega \in \Omega$.

Fact G3. The marginal distribution of $\mathcal{F}[G]$ is that of the sum

$$\sigma \sqrt{\frac{2}{N}} \sum_{n=1}^{N} \cos U_n$$

where $\{U_n; n = 1, \ldots, N\}$ are independent random variables uniformly distributed on the interval $[0, 2\pi]$.

Fact G4. The processes G and $\mathcal{F}[G]$ are each asymptotically Gaussian in the limit as $N \to \infty$. In particular, the random variables $G(t)$ and $\mathcal{F}[G](t)$ are, for each t, asymptotically Gaussian.

Fact G5. The spectrum of G is $2\pi\Psi$ and G has unit variance.

Fact G6. The spectrum of $\mathcal{F}[G]$ is $2\pi\Psi_o$ and its variance is σ^2.

Fact G7. Let the spectrum Ψ of G be continuous. Then $\mathcal{F}[G]$ is ergodic.

Proof of the first six of these results can be found in Frey and Simiu.[7] *Fact G7* is related to the fact that Gaussian processes with continuous spectra are ergodic.[10,14] It follows from *Fact G5* that a modified Shinozuka noise process G as in Eq. (5) can be constructed for any given spectrum.

Five realizations of G with bandlimited spectrum are shown for comparison in Figure 1 together with five realizations of Gaussian noise with the same spectrum. $S(\nu) = \mathrm{sech}\,\nu$ is used in this example.

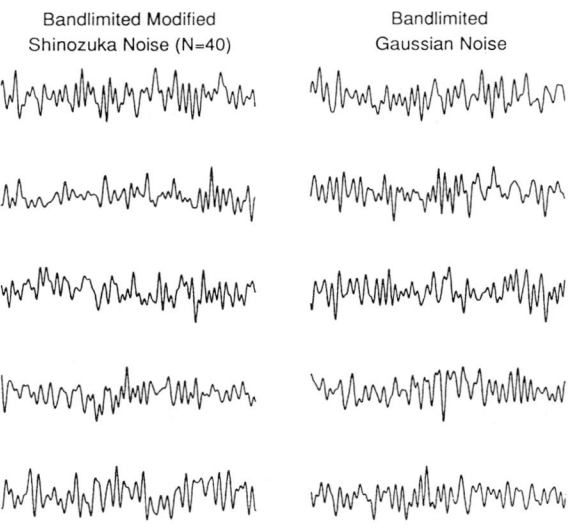

Bandlimited Modified
Shinozuka Noise (N=40)

Bandlimited
Gaussian Noise

FIGURE 1   Realizations of modified Shinozuka and Gaussian noise processes with identical bandlimited spectra and $S(\nu) = \mathrm{sech}\,\nu$.

Let us now consider the effect of the perturbation $\varepsilon w(x, \dot{x}, t)$ on the global geometry of Eq. (1). For sufficiently small perturbations, the hyperbolic fixed points of Eq. (1) are perturbed to a nearby invariant manifold and the stable and unstable manifolds associated with the homoclinic or heteroclinic orbit of Eq. (1) separate.[2] The distance between the separated manifolds is expressible as an asymptotic expansion $\varepsilon M + O(\varepsilon^2)$ where M is a computable quantity called the Melnikov function. The separated manifolds may intersect transversely and, if such intersections occur, they are infinite in number and define lobes marking the transport of phase space[24]. The amount of phase space transported, the phase space flux, is a measure of the chaoticity of the dynamics.[3] The lobes defined by the intersecting manifolds generally have twisted, convoluted shapes whose areas are difficult to determine, making analytical calculation of the flux difficult, if not impossible. For the case of small perturbations, however, the phase space flux can be expressed in terms of the Melnikov function. The average phase space flux has the asymptotic expansion $\varepsilon \Phi + O(\varepsilon^2)$ where Φ, here called the flux factor, is a time average of the Melnikov function:

$$\Phi = \lim_{T \to \infty} \frac{1}{2T} \int_{-T}^{T} M^+(\theta_1 - t, \theta_2 - t) dt, \tag{8}$$

where M^+ is the maximum of 0 and M.[3,24]

To apply Melnikov theory to a deterministic excitation g, g must be bounded and UC. In the case of random perturbations G, the theory requires that G be uniformly bounded and uniformly continuous across both time and ensemble. The noise model G in Eq. (5) is uniformly bounded as noted in *Fact G2*. However, G does not necessarily have the needed degree of continuity.

We define a stochastic process X to be ensemble uniformly continuous (EUC) if, given any $\delta_1 > 0$, there exists $\delta_2 > 0$ such that if $t_1, t_2 \in \mathcal{R}$ and $|t_1 - t_2| < \delta_2$, then $|X_{t_1}(\omega) - X_{t_2}(\omega)| < \delta_1$ for all $\omega \in \Omega$. A stochastic process can have UC paths and fail to be EUC. G is EUC if it is bandlimited.

Conditions on the perturbation function w in Eq. (3) sufficient for the Melnikov function to exist are that g be UC and that G be EUC. The Melnikov function for the system governed by Eq. (4) is then given by the Melnikov transform $\mathcal{M}[g, G]$ of g and G:

$$\begin{aligned}
M(t_1, t_2) =& \mathcal{M}[g, G] \\
=& -\kappa \int_{-\infty}^{\infty} \dot{x}_s^2(t) dt + \gamma \int_{-\infty}^{\infty} \dot{x}_s(t) g(t + t_1) dt \\
& + \rho \int_{-\infty}^{\infty} \dot{x}_s(t) G(t + t_2) dt.
\end{aligned} \tag{9}$$

Since $h(t) = \dot{x}_s(-t)$, denoting the integral of $\dot{x}_s^2$ by I, we obtain

$$M(t_1, t_2) = -I\kappa + \gamma \mathcal{F}[g](t_1) + \rho \mathcal{F}[G](t_2). \tag{10}$$

The expectation and variance of $M(t_1, t_2)$ are, respectively,

$$E[M(t_1, t_2)] = -I\kappa + \gamma \mathcal{F}[g](t),$$

$$Var[M(t_1, t_2)] = \rho^2 \sigma^2 = \rho^2 \int_0^\infty S^2(\nu) \Psi(d\nu).$$

$M(t_1, t_2)$ is, like G, a Gaussian process in the limit as $N \to \infty$ indicating that the presence of even vanishingly small noise causes the Melnikov function to have simple zeros. The state of the system is thus driven from one basin of attraction to that of the competing attractor. Such motion is interpretable as chaotic motion on a single strange attractor.[24]

Of course, the infinitely long tails of the marginal distributions of Gaussian noise are physically unrealistic. Expressions for random forcing with tail-limited marginal distributions can be obtained through nonlinear transformations of Eq. (5). Such tail-limited excitation processes (which may represent, e.g., wave forces whose magnitude is limited by physical constraints) are of interest in engineering applications. From such noise models, the tail-limited marginal distributions of the Melnikov function can be determined, allowing criteria to be developed to guarantee that the Melnikov function has no simple zeros, i.e, that jumps (snap-through dynamics) associated with chaos do not occur.

We now proceed to develop formulae for the average flux factor Φ. Substituting Eq. (10) into Eq. (8) we obtain

$$\Phi = \lim_{T \to \infty} \frac{1}{2T} \int_{-T}^T [\gamma \mathcal{F}[g](\theta_1 - s) + \rho \mathcal{F}[G](\theta_2 - s) - I\kappa]^+ ds. \tag{11}$$

Existence of the limit in Eq. (11) depends on the nature of the excitations g and G and their corresponding convolutions $\mathcal{F}[g] = g * h$ and $\mathcal{F}[G] = G * h$.

To ensure the existence of the limit in Eq. (11), we assume that g is asymptotic mean stationary (AMS): a stochastic process $X(t)$ is defined to be AMS if[9] the limits

$$\mu_X(A) = \lim_{T \to \infty} \frac{1}{2T} \int_{-T}^T E[1_A(X(t))] dt \tag{12}$$

exists for each real Borel set $A \in \mathcal{R}$. Here 1_A is the indicator function, $1_A(x) = 1$ for $x \in A$ and $1_A(x) = 0$ otherwise. This definition applies, in particular, to deterministic functions $X(t)$. If the limits in Eq. (12) exist, then μ_X is a probability measure.[13] μ_X is called the stationary mean (SM) distribution of the process X.

The deterministic forcing function g is assumed to be AMS so, due to the linearity of $\mathcal{F}$, $\mathcal{F}[g]$ is AMS and we denote the SM distribution of $\mathcal{F}[g]$ by $\mu_{\mathcal{F}[g]}$. Assume the spectrum of G is continuous. Then, according to Fact G7, $\mathcal{F}[G]$ is ergodic. Ergodicity implies asymptotic mean stationarity,[9] so $\mathcal{F}[G]$ is AMS also with SM distribution $\mu_{\mathcal{F}[G]}$. All AMS deterministic functions are ergodic so $\mathcal{F}[g]$, like $\mathcal{F}[G]$, is ergodic. Inasmuch as $\mathcal{F}[g]$ is deterministic, $\mathcal{F}[g]$ and $\mathcal{F}[G]$ are jointly ergodic with SM distribution $\mu_{\mathcal{F}[g]} \times \mu_{\mathcal{F}[G]}$.[7] Then the limit Eq. (11) exists and can be expressed in terms of the SM distributions $\mu_{\mathcal{F}[g]}$ and $\mu_{\mathcal{F}[G]}$.

THEOREM 1. (See Frey and Simiu.[7]) Suppose g is AMS and $\mathcal{F}[G]$ is ergodic. Then the limit in Eq. (11) exists, the flux factor Φ is nonrandom and

$$\Phi = E[(\gamma A + \rho B - I\kappa)^+]$$

where A is a random variable with distribution equal to the SM distribution $\mu_{\mathcal{F}[g]}$ of the function $\mathcal{F}[g]$, B is a random variable with distribution equal to the SM distribution $\mu_{\mathcal{F}[G]}$ of the process $\mathcal{F}[G]$ and A and B are independent.

Theorem 1 applies broadly to uniformly bounded and EUC noise processes G with ergodic filtered counterpart $\mathcal{F}[G]$. The modified Shinozuka process (5) belongs to this class provided it is Ψ-admissible with continuous, bandlimited spectrum. Moreover, G in (5) is stationary and $\mathcal{F}[G]$ is asymptotically Gaussian. Hence $\mu_{\mathcal{F}[G]}$ is the marginal distribution of $\mathcal{F}[G]$ and, for large N, B is approximately Gaussian with zero mean and variance σ^2.

THEOREM 2. (See Frey and Simiu.[7]) Suppose g is AMS and G is a Ψ-admissible modified Shinozuka process with continuous bandlimited spectrum. Then the flux factor Φ is approximately

$$\Phi \doteq E[(\gamma A + \rho\sigma Z - I\kappa)^+] \tag{13}$$

where Z is a standard Gaussian random variable. The error in this approximation decreases as N is made larger.

DUFFING OSCILLATOR WITH ADDITIVE NEAR-GAUSSIAN NOISE

The potential energy of the Duffing oscillator is $V(x) = x^4/4 - x^2/2$ with corresponding equation of motion, $\ddot{x} = x - x^3$. The unperturbed Duffing oscillator has a hyperbolic fixed point at the origin $(x, \dot{x}) = (0, 0)$ in phase space connected to itself by symmetric homoclinic orbits. These orbits are given by

$$\begin{pmatrix} x_s(t) \\ \dot{x}_s(t) \end{pmatrix} = \pm \begin{pmatrix} \sqrt{2}\,\mathrm{sech}\,t \\ -\sqrt{2}\,\mathrm{sech}\,t\,\tanh t \end{pmatrix}.$$

The impulse response h of the righthand $(+)$ orbit is $h(t) = \dot{x}_s(-t) = \sqrt{2}\,\mathrm{sech}\,t\,\tanh t$. Thus $I = 4/3$ in Eq. (10).

OSCILLATOR WITH LINEAR DAMPING

We consider first the forced Duffing oscillator with additive noise and linear damping:

$$\ddot{x} = x - x^3 + \varepsilon[\gamma g(t) + \rho G(t) - \kappa \dot{x}]. \tag{14}$$

Here, $\gamma \geq 0$, $\kappa \geq 0$ and $\rho \geq 0$ are constants, g is deterministic and bounded $|g(t)| \leq 1$, and G is the modified Shinozuka noise process reviewed in the previous section.

The flux factor Φ for this system is given exactly in Theorem 1 and approximately in Theorem 2. The approximation in Theorem 2 was obtained by representing the marginal distribution $\mu_{\mathcal{F}[G]}$ of $\mathcal{F}[G]$ by a Gaussian distribution and is appropriate for large N. However, because the Gaussian distribution has infinite tails, Theorem 2 indicates that the flux factor is nonzero for all levels $\rho > 0$ of noise.

Consider the case $\gamma = 0$. According to Theorem 1, $\Phi = E[(\rho\sigma B_N - 4\kappa/3)^+]$. Using *Fact G3*, we take

$$B_N = \sqrt{\frac{2}{N}} \sum_{n=1}^{N} \cos U_n$$

and define

$$\Phi' = \frac{3\Phi}{4\kappa}, \quad \rho' = \frac{3}{\sqrt{2}}\frac{\rho\sigma}{\kappa}, \quad B' = \frac{B_N + \sqrt{2N}}{2\sqrt{2N}}.$$

Then

$$\Phi' = E[(\rho'\sqrt{N}(B' - 1/2) - 1)^+]. \tag{15}$$

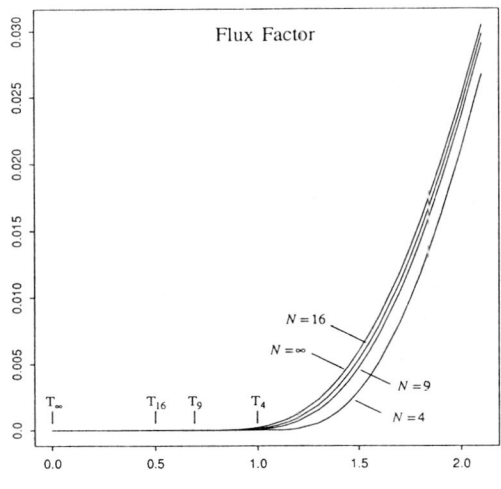

FIGURE 2 The flux factor of Φ' as a function of the noise strength ρ' for various values of N. T_N is the threshold for positive flux.

The support of B' is the interval $(0, 1)$ and is approximately beta-distributed[16] with density

$$\frac{\Gamma(\alpha + \beta)}{\Gamma(\alpha)\Gamma(\beta)} t^{\alpha-1}(1 - t)^{\beta-1}, \quad 0 < t < 1$$

where the parameters $\alpha > 0$ and $\beta > 0$ of the distribution are chosen so that the mean and the variance of the beta distribution are the same as those of B'. $\Phi' = \Phi'(\rho', N)$ is plotted in Figure 2 as a function of ρ' for various values of N using Eq. (16). For comparison, the limiting Gaussian noise case $N \to \infty$ is also plotted using Eq. (13).

OSCILLATOR WITH ATTRACTING HOMOCLINIC ORBIT

If the damping term in Eq. (14) has the form $\epsilon(\kappa - \beta x^2)x$, under suitable conditions on the coefficients κ and β the homoclinic orbits of the unperturbed flow are attracting; i.e., among all orbits passing through a neighborhood of the homoclinic orbit, almost all can be found to be in the vicinity of the homoclinic orbit at times in the far future or past.[5] The dynamics of this oscillator have been considered by Stone and Holmes[22] in the case of white Gaussian noise forcing with spectral intensity ρ. They found from calculations based on the Fokker-Planck equation that the time τ between consecutive returns to a neighborhood of the saddle point is

$$\tau = c - (1/\lambda_u) \ln(\epsilon\rho),$$

where c is a constant and λ_u is the eigenvalue associated with the unstable manifold of the saddle point. τ can be shown to play a prominent role in determining the spectrum of the oscillator dynamics. Using a calculation of the phase space flux with the noise representation in Eq. (5), this expression for τ can be shown to apply not only to white noise, but more generally to the case of colored noise.[21] The phase space flux approach has the added advantage of being simpler to apply.

SYSTEMS WITH MULTIPLICATIVE EXCITATION

We turn now to a more general form for w, the multiplicative excitation model:

$$w(x, \dot{x}, t) = \gamma(x, \dot{x})g(t) + \rho(x, \dot{x})G(t). \tag{16}$$

As in the additive excitation model, the function g represents deterministic forcing while $G(t) = G(t, \omega)$, $\omega \in \Omega$ is a stochastic process representing a random forcing contribution.

The Melnikov function is calculated as in Eq. (9) to be

$$M(t_1, t_2) = \mathcal{M}[g, G] = \int_{-\infty}^{\infty} \dot{x}_s(t)[\gamma(x_s(t), \dot{x}_s(t))g(t+t_1) + \rho(x_s(t), \dot{x}_s(t))G(t+t_2)]dt.$$

We define orbit filters $\mathcal{F}_1$ and $\mathcal{F}_2$ with impulse responses

$$h_1(t) = \dot{x}_s(-t)\gamma(x_s(-t), \dot{x}_s(-t)), \quad h_2(t) = \dot{x}_s(-t)\rho(x_s(-t), \dot{x}_s(-t))$$

and corresponding transfer functions $H_1(\nu)$ and $H_2(\nu)$. Then

$$M(t_1, t_2) = \mathcal{F}_1[g](t_1) + \mathcal{F}_2[G](t_2). \tag{17}$$

Generalizing the additive excitation model Eq. (3) by allowing the coefficients γ and ρ to depend on the state $(x, \dot{x})$ of the system has, according to (17), two significant consequences. First, the orbit filter $\mathcal{F}$ in the additive model is replaced in the multiplicative model by two different orbit filters $\mathcal{F}_1$ and $\mathcal{F}_2$ and, second, the filters $\mathcal{F}_1$ and $\mathcal{F}_2$ are linear, time-invariant and noncausal with impulse responses given solely in terms of the orbit $\vec{x}_s$ of the unperturbed system and the functions γ and ρ.

Substituting Eq. (17) into Eq. (8) gives

$$\Phi = \lim_{T \to \infty} \frac{1}{2T} \int_{-T}^{T} [\rho\mathcal{F}_1[g](\theta_1 - s) + \gamma\mathcal{F}_2[G](\theta_2 - s)]^+ ds. \tag{18}$$

Just as in the case of the additive excitation model, existence of the limit in Eq. (18) hinges on the joint ergodicity of the function $\mathcal{F}_1[g] = g * h_1$ and the process $\mathcal{F}_2[G] = G * h_2$.

THEOREM 3. Consider the system governed by Eq. (2) with perturbation function w as in Eq. (16) such that g is AMS and $\mathcal{F}_2[G]$ is ergodic. Let $\mu_{\mathcal{F}_1[g]}$ and $\mu_{\mathcal{F}_2[G]}$ be the SM distributions of $\mathcal{F}_1[g]$ and $\mathcal{F}_2[G]$, respectively. Then the limit in Eq. (18) exists, the flux factor Φ is nonrandom and

$$\Phi = E[(\gamma A + \rho B)^+]$$

where A is a random variable with distribution $\mu_{\mathcal{F}_1[g]}$, B is a random variable with distribution $\mu_{\mathcal{F}_2[G]}$ and A and B are independent.

Theorem 3 is an extension of Theorem 1 to the general case of multiplicative excitation. Theorem 3 can in turn be extended to systems with more general planar vector fields than that of the system governed by Eq. (2). Only the orbit filters $\mathcal{F}_1$ and $\mathcal{F}_2$ change in these more general cases; the form of the flux factor Φ given in Theorem 3 remains the same.

DUFFING OSCILLATOR WITH SHOT NOISE-LIKE DISSIPATION

As an example of a system with multiplicative shot noise, we consider the Duffing oscillator with weak forcing and non-autonomous damping:

$$\ddot{x} = x - x^3 + \varepsilon[\gamma g(t) - \kappa(K_N(t) + \eta)\dot{x}].$$ (19)

Here $\gamma \geq 0$, $\kappa \geq 0$ and $\eta \geq 0$ are constants, g is deterministic and bounded $|g(t)| \leq 1$, and K_N is a form of shot noise. The perturbation in Eq. (18) is a particular case of the multiplicative excitation model (16) with $\gamma(x, \dot{x}) = \gamma$, $\rho(x, \dot{x}) = -\kappa\dot{x}$, and $G(t) = K_N(t) + \eta$. $\kappa(K_N(t) + \eta)$ in Eq. (18) serves as a time-varying damping factor and plays the same role as the constant κ in Eq. (4). The two terms $\kappa\eta$ and κK_N represent, respectively, viscous and shot noise-like damping forces. We assume for this example that $\eta = 0$. We also assume the shot response r (see below) of K_N to be nonnegative in this example so that the factor κK_N is nonnegative.

The usual model of constant-rate shot noise is a stochastic process of the form[11,23]

$$K(t) = \sum_{k \in \mathcal{Z}} r(t - T_k),$$ (20)

where $\mathcal{Z}$ is the set of integers, $\{T_k, k \in \mathcal{Z}\}$ are the epochs (shots) of a Poisson process with rate $\lambda > 0$ and r is bounded and square-integrable,

$$\int_{-\infty}^{\infty} r^2(t)dt < \infty.$$

r is call the shot response of the process K.

The shot noise model K in Eq. (20) is neither bounded nor EUC and cannot be used in conjunction with Melnikov theory in calculating the phase space flux in chaotic systems. A modification of the model which approximates K and yet has the requisite path properties has been developed[8]:

$$K_N(t) = \sum_{j \in \mathcal{Z}} \sum_{k=1}^{2^N} r(t - T_{jkN} - A_j - T)$$ (21)

where N is a positive integer, $A_j = 2^N(j - 1/2)/\lambda$ and $\{T, T_{jkN}, j \in \mathcal{Z}, k = 1, \ldots, 2^N\}$ are independent random variables such that for each N and j, $\{T_{jkN}, k = 1, 2, \ldots, 2^N\}$ are identically uniformly distributed in the interval $(A_j, A_{j+1}]$ and T is uniformly distributed between 0 and $2^N/\lambda$. As in the usual shot noise model (20), λ is here again the rate of the process; it is the mean number of epochs (shots) T_{jkN} per unit time. We assume just as for K, that r in Eq. (21) is bounded and square-integrable. We further assume that r is UC and that the radial majorant

$$r^*(t) = \sup_{|\tau| \geq |t|} |r(\tau)|$$

of the shot response is integrable; i.e.,

$$\int_{-\infty}^{\infty} r^*(t)dt < \infty.$$

According to this specification of K_N, realizations of the process are obtained by partitioning the real line into the intervals $(A_j, A_{j+1}]$ of length $2^N/\lambda$ with common random phase T and then placing 2^N epochs independently and at random in each interval. The random phase T eliminates the (ensemble) cyclic nonstationarity produced by the partitioning by $(A_j, A_{j+1}]$.

It can be shown that K_N and $\mathcal{F}[K_N]$ are stationary processes, that K_N converges in distribution[4] to the shot noise K with the same shot response r and rate λ, that $\mathcal{F}[K_N]$ is also a shot noise of the form Eq. (21) with shot response $r * h$, and $\mathcal{F}[K_N]$ converges in distribution to the shot noise K with shot response $r * h$ and rate λ, that the variances of K_N and $\mathcal{F}[K_N]$ converge, respectively, to those of K and $\mathcal{F}[K]$, that the spectrum of K_N converges weakly[4] to the spectrum of the shot noise K with the same shot response r and rate λ, and that the spectrum of $\mathcal{F}[K_N]$ converges weakly to the spectrum of the shot noise K with shot response $r * h$ and rate λ. It can also be shown that K_N is uniformly bounded for all N, that K_N is EUC for all N, and that K_N and $\mathcal{F}[K_N]$ are each ergodic for all N. Hence, for large N the shot noise K_N closely approximates the standard shot noise model K in all important respects. Also, K_N, unlike K, can be used in Melnikov's method-type calculations of the flux factor. Additional details can be found in Frey and Simiu.[8]

According to Theorem 3, the Melnikov function for the Duffing oscillator Eq. (19) is

$$M(t_1, t_2) = \mathcal{F}_1[g](t_1) + \mathcal{F}_2[G](t_2)$$

where

$$h_1(t) = \gamma \dot{x}_s(-t) = \gamma\sqrt{2}\mathrm{sech}\, t \tanh t$$

and

$$h_2(t) = -\kappa \dot{x}_s^2(-t) = -2\kappa \mathrm{sech}^2 t \tanh^2 t.$$

The corresponding moduli of the filters $\mathcal{F}_1$ and $\mathcal{F}_2$ are

$$S_1(\nu) = \sqrt{2}\pi\gamma\nu\mathrm{sech}\frac{\pi\nu}{2}$$

and

$$S_2(\nu) = 4\kappa \int_0^{\infty} \mathrm{sech}^2 t \tanh^2 t \cos \nu t\, dt.$$

We have $S_1(0) = 0$ so the d.c. component (if any) of g is completely removed by $\mathcal{F}_1$ and has no effect on the Melnikov function. K_N does have a d.c. component; K_N is ergodic so its d.c. component is $E[K_N] = \lambda R(0)$ where

$$R(0) = \int_{-\infty}^{\infty} r(t)dt > 0.$$

$S_2(0) = 4\kappa/3 > 0$ so the d.c. component of K_N passed by $\mathcal{F}_2$ is

$$E[\mathcal{F}_2[K_N]] = E[K_N]S_2(0) = \frac{4\kappa\lambda R(0)}{3}.$$

The presence of a d.c. component plays a pivotal role in shifting the parametric threshold for chaos. See Frey and Simiu[7] for further discussion.

Assume the deterministic forcing function g is AMS. K_N is uniformly bounded and EUC and $\mathcal{F}_2[K_N]$ is ergodic. Thus $\mathcal{F}_1[g]$ and $\mathcal{F}_2[K_N]$ are jointly ergodic. By Theorem 3, the flux factor Φ exists and

$$\Phi = E[(A - B_N)^+]$$

where the distribution of A is $\mu_{\mathcal{F}_1[g]}$, the distribution of B_N is $\mu_{\mathcal{F}_2[K_N]}$ and A and B_N are independent.

We noted earlier that the distribution of $\mathcal{F}_2[K_N]$ is, for large N, approximately that of the shot noise $\mathcal{F}_2[K]$. This is the basis for the following theorem.

THEOREM 4. The flux factor Φ for the Duffing oscillator (19) with weak forcing and shot noise damping coefficent κK_N is approximately

$$\Phi \doteq E[(A - B)^+]$$

where A is $\mu_{\mathcal{F}_1[g]}$-distributed, B is $\mu_{\mathcal{F}_2[K]}$-distributed, A and B are independent and K is the shot noise (20). This approximation improves as N increases.

Φ can be calculated numerically as follows for given system parameters ν, γ and κ and shot parameters λ and r. Make the following definitions:

$$\Phi' = \frac{\Phi}{\gamma S_1(\nu)}, \quad A' = \frac{A}{\gamma S_1(\nu)}, \quad B' = \frac{B_N}{\gamma S_1(\nu)},$$

$$\lambda' = \frac{16\lambda R^2(0)}{9J}, \quad \kappa' = \frac{3\kappa J}{4\gamma S_1(\nu)R(0)},$$

where

$$J = \int_{-\infty}^{\infty} (r * h)^2(t)dt.$$

Then

$$\Phi' = E[(A' - B')^+]. \tag{22}$$

The random variable B' is approximately gamma-distributed[16] with density

$$\frac{t^{\alpha-1}e^{-t/\beta}}{\beta^\alpha\Gamma(\alpha)}, \quad t > 0$$

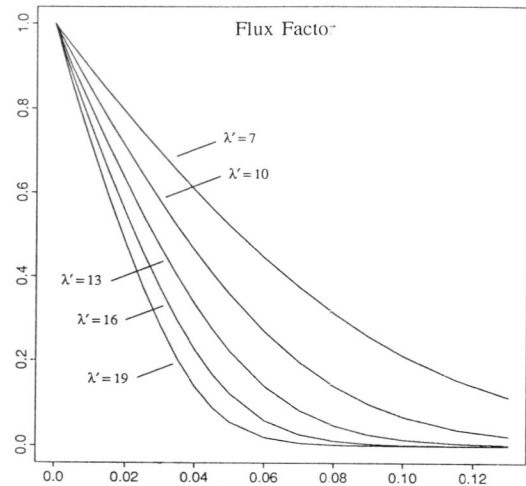

FIGURE 3   The flux factor Φ' as a function of the damping constant κ' for various shot rates λ'.

where the parameters α and β are determined by the condition that $E[B']$ and $Var[B']$ equal the mean and the variance, respectively, of the gamma distribution. $\Phi' = \Phi'(\kappa', \lambda')$ is plotted in Figure 3. Additional details are available from Frey and Simui.[8]

SUMMARY

Noise can cause multistable dynamical systems to exhibit chaotic motion with sensitive dependence upon initial conditions. The theory applicable to noise-induced chaotic dynamics reviewed in this paper rests primarily on the concept of the Melnikov transform and on techniques for approximating noise with any given spectrum and marginal distribution by uniformly bounded, ensemble uniformly continuous processes. The results described here apply to weakly perturbed, one-degree-of-freedom dynamical systems featuring homoclinic or heteroclinic orbits. Results were first given for additive perturbation and then generalized to multiplicative perturbation. Extensions of this work to higher-degree-of-freedom systems and to spatially extended systems are in progress.

ACKNOWLEDGEMENTS

This research was supported in part by the Ocean Engineering Division of the Office of Naval Research, Grant nos. N-00014-93-1-0248 and N-00014-93-F-0028.

REFERENCES

1. Arecchi, F. T., R. Badii, and A. Politi. "Generalized Multistability and Noise-Induced Jumps in a Nonlinear Dynamical System." *Phys. Rev. A* **32(1)** (1985): 402–408.
2. Arrowsmith, D. K., and C. M. Place. *An Introduction to Dynamical Systems.* Cambridge, MA: Cambridge University Press, 1990.
3. Beigie, D., A. Leonard, and S. Wiggins. "Chaotic Transport in the Homoclinic and Heteroclinic Tangle Regions of Quasiperiodically Forced Two-Dimensional Dynamical Systems." *Nonlinearity* **4(3)** (1991): 775–819.
4. Billingsley, P. *Convergence of Probability Measures.* New York: John Wiley and Sons, 1968.
5. Brunsden, V., and P. Holmes. "Power Spectra of Strange Attractors near Homoclinic Orbits." *Phys. Rev. Lett.* **58(17)** (1987): 1699–1702.
6. Brunsden, V., J. Cortell, and P. Holmes. "Power Spectra of Chaotic Vibrations of a Buckled Beam." *J. Sound & Vibra.* **130(1)** (1989): 1–25.
7. Frey, M., and E. Simiu. "Noise-Induced Chaos and Phase Space Flux." *Physica D* (1993).
8. Frey, M., and E. Simiu. "Deterministic and Stochastic Chaos." In *Computational Stochastic Mechanics*, edited by Cheng, A.H-D. and Yang, C.Y. London: Elsevier Applied Science, 1993.
9. Gray, R. M. *Probability, Random Processes and Ergodic Properties.* New York: Springer-Verlag, 1988.
10. Grenander, U. "Stochastic Processes and Statistical Inference." *Arkiv Mat* **1(17)** (1950): 195–277.
11. Iranpour, R. and P. Chacon. *Basic Stochastic Processes: the Mark Kac Lectures.* New York: MacMillan Publishing Co., 1988.
12. Guckenheimer, J., and P. Holmes. *Nonlinear Oscillations, Dynamical Systems, and Bifurcations of Vector Fields* New York: Springer-Verlag, 1983.
13. Jacobs, K. *Measure and Integral.* New York: Academic Press, 1978.
14. Maruyama, G. "The Harmonic Analysis of Stationary Stochastic Processes." *Mem. Fac. Sci., Kyushu Univ., Ser. A, Math.* **4** (1949): 45–106.
15. Meyer, K. R. and G. R.Sell. "Melnikov Transforms, Bernoulli Bundles, and Almost Periodic Perturbations." *Trans. Am. Math. Soc.* **314(1)** (1989).

16. Papoulis, A. *The Fourier Integral and its Applications* New York: McGraw-Hill, 1962.
17. Perko, L. *Differential Equations and Dynamical Systems.* New York: Springer-Verlag, 1991.
18. Shinozuka, M., and Jan, C.-M., Digital Simulation of Random Processes and its Applications,." *J. Sound & Vibra.* **25(1)** (1972): 111–128.
19. Shinozuka, M. "Simulation of Multivariate and Multidimensional Random Processes." *J. Acoust. Soc. Amer.* **49(1)** (1971): 357–367.
20. Simiu, E., and M. Frey. "Spectrum of the Stochastically Forced Duffing-Holmes Oscillator." *Phys. Lett. A*: in press.
21. Simiu, E., and M. Grigoriu. "Non-Gaussian Effects on Reliability of Multistable Systems." In *Proceedings of the 12th Inter. Conf. on Offshore Mechanics and Arctic Engineering*, edited by C.G. Soares. Glasgow, Scotland: American Society of Mechanical Engineers, 1993.
22. Stone, E., and P. Holmes. "Random Perturbations of Heteroclinic Attractors." *SIAM J. Appl. Math.* **53(3)** (1990): 726–743.
23. Snyder, D., and M. Miller. *Random Point Processes in Time and Space.* New York: Springer-Verlag, 1991.
24. Wiggins, S. *Chaotic Transport in Dynamical Systems.* New York: Springer-Verlag, 1991.

C. B. Arnold, B. J. Gluckman, and J. P. Gollub
Physics Department, Haverford College, Haverford, PA 19041 and Physics Department, University of Pennsylvania, Philadelphia, PA 19104

Statistical Averages of Chaotically Evolving Patterns

We briefly summarize some previous work on spatio-temporal chaos, and then discuss novel measurements of time averages of chaotic patterns. These reveal a significant phase rigidity of the patterns even in the presence of large fluctuations.

BACKGROUND

Many investigators concerned with patterns in nonequilibrium systems have recently turned their attention to the problem of *spatio-temporal chaos*. Roughly speaking, the concern here is with situations in which the range ξ of spatial correlations is small compared to at least one dimension L of the system. Given the remarkable progress made in analyzing (a) the chaotic behavior of small systems with only a few degrees of freedom and (b) the steady patterns of extended nonequilibrium systems, it is natural to try to extend the relevant concepts to chaotic extended systems. Indeed, a major body of experimental literature has appeared on a great variety of systems, including Rayleigh-Bénard convection,[2,5,8] rotating convection,[27] Turing patterns,[21] parametrically excited surface waves,[12,25] convection in nematic liquid crystals,[4,19,23] binary mixtures[15,24] rotating films,[22] and Taylor-Couette flow.[13]

In parallel with this experimental work, there has been a considerable numerical and analytical effort, much of it directed at model equations such as the complex Ginzburg-Landau equation, which is believed to provide a plausible approximation to the physics for some systems in which spatio-temporal chaos occurs close to the onset of the basic pattern. The Kuramoto-Sivashinsky equation has provided another approach to modeling, as have coupled map lattices and other discrete models. The theoretical work is now so extensive and diverse that it is impossible to reference adequately; it is thoroughly discussed in a lengthy review by Cross and Hohenberg.[7]

A number of fairly well-defined phenomena have been elucidated, including for example: *spatio-temporal intermittency*, in which "laminar" (or ordered) and "chaotic" (or disordered) domains coexist; *phase turbulence*, in which the pattern amplitude is relatively constant but phase variations are noisy; and *defect-mediated "turbulence"* in which the pattern irregularities are concentrated in localized defects, which are nucleated and annihilated at irregular intervals and locations. In some cases, the initial bifurcations leading to these states have been investigated.

Although the term "chaos" is frequently utilized in connection with deterministic nonlinear phenomena in extended systems, the standard methods of low-dimensional nonlinear dynamics are very difficult to apply experimentally, and have had only limited success so far. These include attempts to reconstruct attractors or to measure the Lyapunov exponents describing the linearized dynamics, using generalizations of the methods developed for nonlinear dynamical systems and recently reviewed by Abarbanel et al.[1] However, any success in generalizing these standard tools to spatially extended systems would undoubtedly have great impact on a variety of disciplines where nonlinear partial differential equations are the standard approach to modeling.

One of the central insights of nonlinear dynamics is the observation that complex and even noisy behavior can sometimes be described in terms of a relatively modest number of degrees of freedom. The dynamical variables, for example the surface displacement field in $\zeta(\vec{r}, t)$ in an interfacial wave experiment, might be written as a linear superposition of a set of suitable eigenfunctions $\phi_n(\vec{r})$ with time varying coefficients $a_n(t)$. It is not obvious how to choose a set of eigenfunctions that minimizes the number required. The Karhunen-Loève procedure (see Aubrey et al.[3] and Lumley[18] for example) for generating empirical eigenfunctions directly from data is one approach that has been used for various types of turbulent flows, and is claimed by its advocates to yield the desired minimal set of modes. The empirical eigenfunctions can be regarded as efficiently representing structures in the flow that are obscured by statistical fluctuations at any particular instant.

An alternate approach to these dynamical systems methods is to acknowledge and utilize the basic statistical character of the phenomena, despite their deterministic origin.[14] For example, one can measure the space-time correlation functions, find characteristic lengths and times, determine how these quantities depend on parameters, and attempt to compare with predictions based on models. Several investigators, including for example Ciliberto and Caponeri,[6] have utilized statistical methods of analysis in novel ways.

STATISTICS: AVERAGE PATTERNS

The simplest statistical quantity to examine is the *time average* of the pattern. Recently, we discovered a novel and interesting phenomenon while conducting experimental studies of parametrically forced surface waves: chaotically evolving patterns can have highly ordered time averages in systems that are considerably larger than the correlation length. While the instantaneous patterns are quite disordered,

(a)   (b)

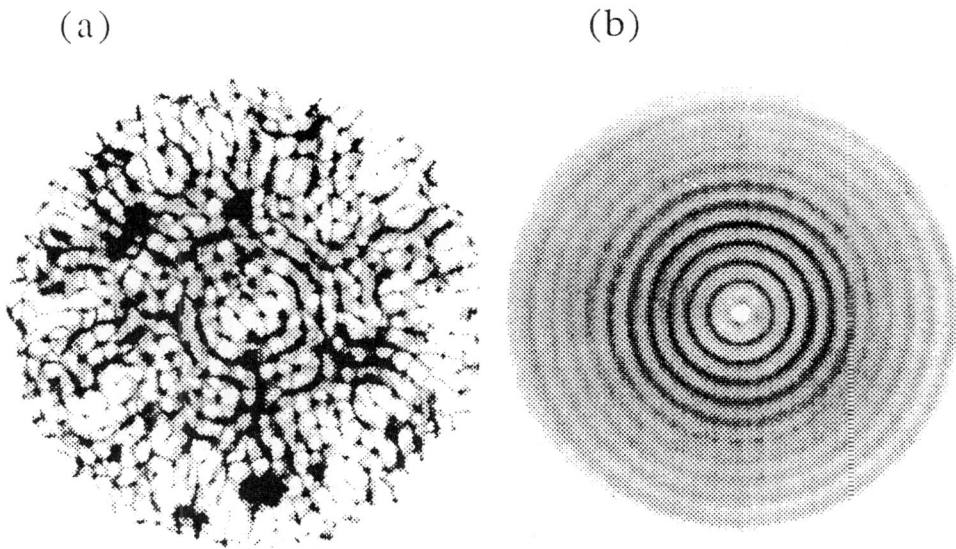

FIGURE 1   (a) Instantaneous and (b) time-averaged shadowgraphs of interfacial wave patterns in a circular cell. These chaotic Faraday wave patterns are produced by a vertical oscillation of the cell at 81 Hz, with a driving amplitude 40% above that required for the wave onset. The structured average pattern indicates a significant degree of phase rigidity despite the spatio-temporal chaos.

fluctuate rapidly, and have a correlation length (relative to the average pattern) that is considerably shorter than the cell dimension, the statistical average over a sufficiently long time is ordered. A preliminary report on this phenomenon[10] has been published, and an example is shown in Figure 1. The top panel shows a snapshot of disordered Faraday waves, obtained using shadowgraph techniques. The system has been excited parametrically by a vertical oscillation of the container at 81 Hz; the amplitude is 40% above the value required to generate standing waves. The bottom panel shows the corresponding time average over about 1000 s, which is far longer than the correlation time of about 1 s.

The geometry of the averaged patterns is related to the spatial symmetry of the boundaries. In the circular case shown in Figure 1, the instantaneous patterns have a tendency to align locally perpendicular to the boundaries, but the time-averaged pattern has rings parallel to the boundary. In the square geometry, the average pattern is close to a product of sinusoidal oscillations in the two orthogonal directions. The structured averages imply a remarkable phase rigidity of the chaotic state.

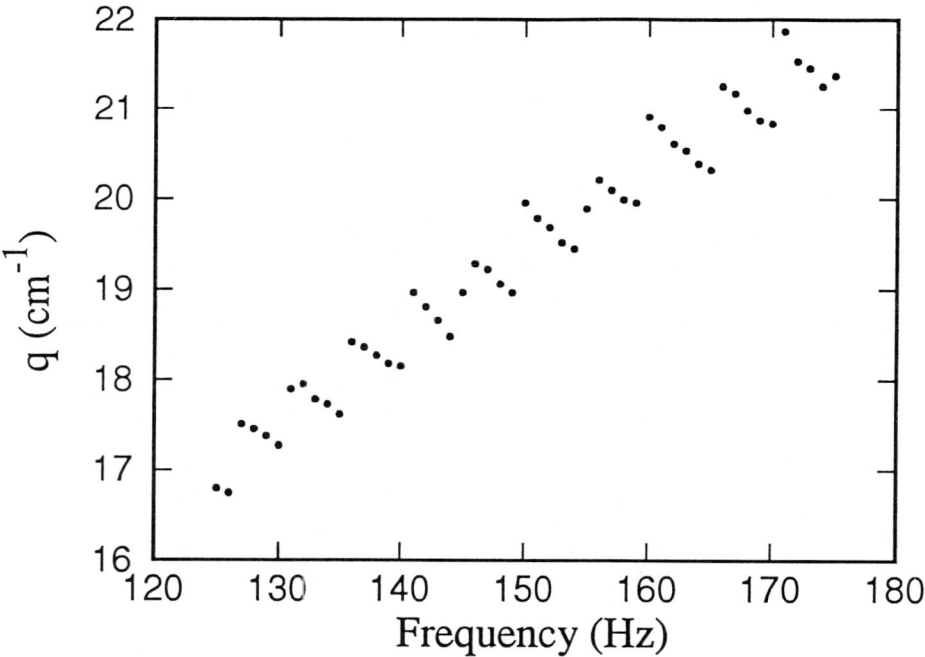

FIGURE 2   Mean wavenumber of the average pattern, showing the phenomenon of statistical quantization.

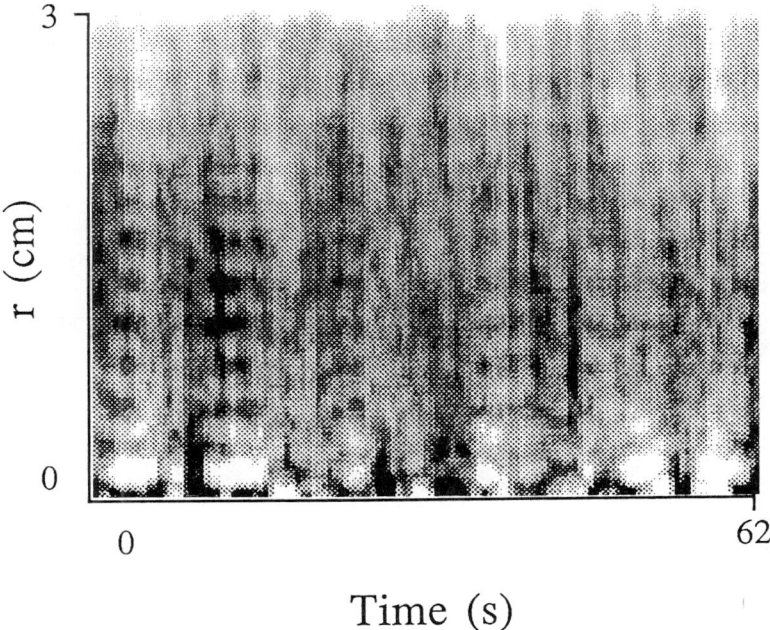

FIGURE 3 Azimuthally averaged instantaneous patterns, showing the radial dependence as a function of time, for an excitation amplitude 50% above the wave onset. The degree of order fluctuates dramatically.

We also noted a phenomenon of *statistical quantization*, in which both the mean wavenumber and the amplitude of the averages are periodic in the cell size. The first of these phenomena is shown for the circular case in Figure 2. Even though the instantaneous patterns have a broad distribution of wavenumbers, the mean wavenumber shows a clear step pattern as the frequency of forcing is increased. Each step corresponds to an additional ring in the pattern.

The amplitude of the ordered average is dependent on the lateral boundary conditions, being stronger when the boundary conditions "pin" the phase of the wave pattern. The evidence to date indicates that the structured average will disappear at large excitation amplitude, and for sufficiently large aspect ratio. The approach to this asymptotic state is described elsewhere.[11]

Our preliminary investigation led to additional interesting results. First, we looked into the form of the average pattern, and found that in the rectangular geometry it is approximately the same as one of the linear modes accessible near onset, but with a different wavenumber. Thus, the average pattern is not simply the persistence of one of the linear modes in the chaotic region. Of course it is not *a priori* obvious why the averages should bear any simple relationship to the onset patterns. In fact, the onset patterns actually realized are often mixed modes rather than individual linear modes. The analogous situation in other geometries is under investigation.

We have also begun to study the time-dependent fluctuations relative to the average pattern. One way to do this in the circular geometry is to average the patterns over angle, and to show the resulting radial dependence as a function of time, using a gray scale. The result, shown in Figure 3, shows that the degree of order in the pattern is highly variable; at certain times in Figure 3 the rings are quite visible, while at other times they are not apparent. When the rings are present, they always have the same spatial phase relative to the boundary; this phase rigidity is responsible for the structured average patterns.

DISCUSSION

These experiments were originally motivated by a suggestion due to M. Golubitsky that there might be an intimate connection between the spatial symmetries of the system and the time-average behavior.[9] However, the observed phenomena are much too rich to be explained by symmetry arguments alone.

It is interesting to note that essentially similar phenomena were subsequently found by Ning, Hu, and Ecke[20] in a study of chaotic roll patterns in rotating Rayleigh-Bénard convection. Thus, it is reasonable to assume that these phenomena will be found to be rather general. In fact, there are several earlier examples that are probably related. For example, turbulent Taylor vortices show a clear mean structure until the Reynolds number becomes very high.[17]

Mean structure has been seen in numerical simulations of several model equations. For example, the Kuramoto-Sivishinsky equation was shown to exhibit mean structure due to boundaries,[26] as does the complex Ginzburg-Landau equation.[9] In the latter case, a transition was noted between asymmetric and symmetric averages as a parameter was varied.

The average patterns provide one useful basis for analyzing chaos in moderately large extended systems. We think that such moderately large systems have the right level of complexity to allow further progress in understanding chaotically evolving patterns. Of course it is also worthwhile to study systems in the "infinite" size limit.

We hope to understand better whether there is a useful analogy between these nonequilibrium macroscopic phenomena and certain microscopic phenomena found in condensed matter. For example, molecular dynamics calculations of ordering in fluids near boundaries at finite temperature[16] show layering in the time average density profile. In both cases, the breaking of translational symmetry provided by boundaries tends to induce order that is surprisingly resistant to the influences that tend to destroy it.

ACKNOWLEDGMENTS

This work was supported by primarily by NSF Grant DMR-8901869, with some additional support from CTS-9115005. We acknowledge helpful discussions with M. Golubitsky, S. Edwards, P. Hohenberg, and B. Shraiman.

REFERENCES

1. Abarbanel, H. D. I., R. Brown, J. J. Sidorowich, and L. S. Tsimring. "The Analysis of Observed Chaotic Data in Physical Systems." *Rev. Mod. Phys.* **65** (1993): 1331.
2. Ahlers, G., and R. P. Behringer. "The Rayleigh-Bénard Instability and the Evolution of Turbulence." *Prog. Theor. Phys. Supp.* **64** (1978): 186.
3. Aubry, N., P. Holmes, J. L. Lumley, and E. Stone. "The Dynamics of Coherent Structures in the Wall Region of a Turbulent Boundary Layer." *J. Fluid Mech.* **192** (1988): 115.
4. Braun, E., S. Rasenat, and V. Steinberg. "The Mechanism of Transition to Weak Turbulence in Extended Anisotropic Systems." *Europhys. Lett.* **15** (1991): 597.
5. Ciliberto, S., and P. Bigazzi. "Spatio-temporal Intermittency in Rayleigh-Bénard Convection." *Phys. Rev. Lett.* **60** (1988): 286.
6. Ciliberto, S., and M. Caponeri. "Thermodynamic Aspects of the Transition to Spatio-temporal Chaos." *Phys. Rev. Lett.* **64** (1990): 2775.
7. Cross, M. C., and P. C. Hohenberg. "Pattern Formation Outside of Equilibrium." *Rev. Mod. Phys.* **65** (1993): 851.
8. Daviaud, F., M. Bonetti, and M. Dubois. "Transition to Turbulence via Spatio-temporal Intermittency in One-Dimensional Rayleigh-Bénard Convection." *Phys. Rev. A* **42** (1990): 3388.
9. Dellnitz, M., M. Golubitsky, and I. Melbourne. "Mechanisms of Symmetry Creation." In *Bifurcation and Symmetry*, edited by E. Allgower, K. Bohmer and M. Golubitsky, 99. Basel: Birkhauser, 1992.
10. Gluckman, B. J., P. Marcq, J. Bridger, and J. P. Gollub. "Time-Averaging of Chaotic Spatio-temporal Wave Patterns." *Phys. Rev. Lett.* **71** (1993): 2034.
11. Gluckman, B. J., C. B. Arnold, and J. P. Gollub. "Time Averages of Chaotic Wave Patterns." To be published.
12. Gollub, J. P., and R. Ramshankar. "Spatio-temporal Chaos in Interfacial Waves." In *New Perspectives in Turbulence*, edited by L. Sirovich, 165. New York: Springer-Verlag, 1991.
13. Hegseth, J. J., C. D. Andereck, F. Hayot, and Y. Pomeau. "Spiral Turbulence and Phase Dynamics." *Phys. Rev. Lett.* **62** (1989): 257.

14. Hohenberg, P. C., and B. Shraiman. "Chaotic Behavior of an Extended System." *Physica D* **37** (1989): 109.

15. Kolodner, P., J. A. Glazier, and H. Williams. "Dispersive Chaos in One-Dimensional Traveling-Wave Convection." *Phys. Rev. Lett.* **65** (1990): 1579.

16. Koplik, J., J. R. Banaver, and J. F. Willemsen. "Molecular Dynamics of Fluid Flow at Solid Surfaces." *Phys. Fluids A* **1** (1989): 781.

17. Lathrop, D. P., J. Fineberg, and H. L. Swinney. "Turbulent Flow Between Concentric Rotating Cylinders at Large Reynolds Number." *Phys. Rev. Lett.* **68** (1992): 1515.

18. Lumley, J. L. "Coherent Structures in Turbulence." In *Transition and Turbulence*, edited by R. E. Meyer, 215. New York: Academic, 1981.

19. Nasuno, S., S. Takeuchi, and Y. Sawada. "Motion and Interaction of Dislocations in Electrohydrodynamic Convection of Nematic Liquid Crystals." *Phys. Rev. A* **40** (1989): 3457.

20. Ning, L., Y. Hu, R. E. Ecke, and G. Ahlers. "Spatial and Temporal Averages in Chaotic Patterns." *Phys. Rev. Lett.* **71** (1993): 2216.

21. Ouyang, Q., and H. L. Swinney. "Transition to Chemical Turbulence." *Chaos* **1** (1991): 411.

22. Rabaud, M., S. Michalland, and Y. Couder. "Dynamical Regimes of Directional Viscous Fingering: Spatio-temporal Chaos and Wave Propagation." *Phys. Rev. Lett.* **64** (1990): 184.

23. Rehberg, I., S. Rasenat, and V. Steinberg. "Traveling Waves and Defect-Initiated Turbulence in Electroconvecting Nematics." *Phys. Rev. Lett.* **62** (1989): 756.

24. Steinberg, V., J. Fineberg, E. Moses, and I. Rehberg. "Pattern Selection and Transition to Turbulence in Propagating Waves." *Physica D* **37** (1989): 359.

25. Tufillaro, N. B., R. Ramshankar, and J. P. Gollub. "Order-Disorder Transition in Capillary Ripples." *Phys. Rev. Lett.* **62** (1989): 422.

26. Zaleski, S., and P. Lallemand. "Scaling Laws and Intermittency in Phase Turbulence." *J. Physique Lett.* **46** (1985): L793.

27. Zhong, F., R. Ecke, and V. Steinberg. "Rotating Rayleigh-Bénard Convection: The Kuppers-Lortz Transition." *Physica D* **51** (1991): 596.

O. Cardoso, D. Marteau, and P. Tabeling
Laboratoire de Physique Statistique, Ecole Normale Supérieure, 24, rue Lhomond, 75231, Paris (France)

Freely Decaying Two-Dimensional Turbulence: A Quantitative Experimental Study

The characterisation of quasi-two-dimensional turbulent flows, produced in thin layers of electrolyte, is presented. The instantaneous vorticity field and the geometrical characteristics of the flows are measured during the decay phase using a pseudo-particle-tracking method. Global quantities such as the energy, the entrophy, and the kurtosis of the vorticity distribution are determined. Unexpected characteristics are obtained, revealing the existence of a new form of quasi-two-dimensional turbulence, which we call "liquid."

The problem of the decay of two-dimensional turbulence has attracted much interest during the last decades[1,3] and, recently, new theoretical approaches have been proposed.[2] According to them, the global characteristics of the flow, both geometrical and dynamical, are governed by power laws, and predictions have been formulated for the values of the corresponding exponents. In a recent experimental study,[5] performed with two of the present authors, we attempted to test such theories; actually, in this work, only the geometrical properties of the flow were determined, and with modest accuracy. In the present work, we perform accurate measurements—both geometrical and dynamical—on the same system. With this improved accuracy, we find somewhat unexpected characteristics for the turbulence, and the purpose of this paper is to report and discuss the corresponding results.

Spatio-Temporal Patterns, Ed. P. E. Cladis and P. Palffy-Muhoray,
SFI Studies in the Sciences of Complexity, Addison-Wesley, 1995   **553**

The experimental system has been described elsewhere,[5] and we briefly summarise the main features: the flow is produced in a thin layer of electrolyte, of thickness b ranging between 2.5 and 4 mm, and the working fluid is a normal solution of sulphuric acid; an electric current I is driven horizontally, from one side of the cell to the other and, just below the flow, arrays of permanent magnets are formed. The interaction of the periodic magnetic field with the electric current produces a system of recirculating flows, whose structure is imposed by the arrangement of the magnets. In most of the experiments, regular arrays are formed, so as to produce ordered lattices of vortices, and thus inject the energy at a given wave number; other patterns have also been considered, such as tripole, quadripole configurations, and disordered lattices.

In the present study, the vortex dimensions are 8×8 mm, and the sizes of the systems that we consider range from 36 up to 182 vortices. Because the layer thickness is small compared to the horizontal extension of the flow, the dynamics of the system is believed to be two-dimensional. The flow is visualised by particles, several tens of microns in size, slightly lighter than the fluid; they are deposited on the free surface, and visualised from above, by using a video camera. The images are sent to a video recorder, and further analysed. To compute the instantaneous velocity, we use a correlation technique: the system is discretised into 30×30 squares, and

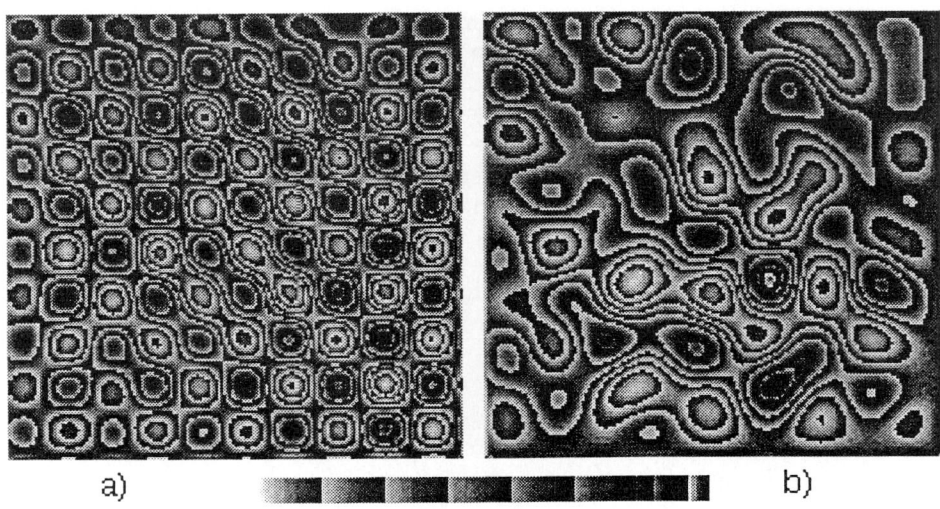

a)   b)

FIGURE 1   Vorticity fields at (a) $t = 0$ and (b) $t = 1$ s for a system of 100 vortices, in an experiment where an electrical current $I = 600$ mA has been driven along the system during a time $t = 0.10$ s. In this case, the fluid thickness b is 3 mm.

maxima of temporal correlation of the light intensity diffused by the particles are calculated; the resulting velocity field is low-pass-filtered at half the Nyquist frequency of the system. The accuracy of the measurement of the velocity field can be estimated to a few percent, and that of the vorticity to about $\pm 10\%$. Global quantities, such as the energy, the entropy, and the kurtosis of the vorticity distribution[1] are also calculated. The number of eddies, and the average distance between them are determined by using the stream function.

All the experiments consist of imposing a steady value of I at $t = -\tau$, quenching the system at $t = 0$, and follow its decay. In Figure 1 we represents the vorticity field at $t = 0$, and 1 s later, for the case of an initially ordered lattice of 10×10 vortices. After the current is quenched, the system undergoes a rapid sequence of merging and aggregations of like-sign vortices, leading to the formation of large structures. Other events, such as mutual advection of vortices, formation and propagation of dipoles, and straining of small vortices by larger ones, are typically observed during this phase.

After a time laps of about 3 s, one seems to reach the final state of the system; the latter is composed by a few vortices, and has no remarkable structure.

In Figure 2 we represent the evolution of the total kinetic energy $E(t)$, and the entropy $Z(t)$ during a typical run. These quantities increase up to $t = 0$, and further decay. The decrease of the energy is exponential, with a time constant of about 5 s; this value is close to the viscous decay time $\tau_\nu = b^2/6\nu$, which represents the viscous damping against the bottom wall for the case of Poiseuilles-type flows. Then, one observes that the energy is essentially burnt by friction against the bottom wall. The entropy also decreases, but, in a first period—i.e., from 0 to about 3 s—the decrease is more rapid than that of the energy. This period coincides with the occurrence of many merging and aggregation processes. After this period, the flow pattern does not change and we find that the two quantities decay at the same rate (see Figure 2). We further focus on the first period, where all the interesting events take place.

In Figure 3 we show the evolution of several quantities: the vortex size $a(t)$, the vortex density $\rho(t)$, the distance between eddies $r(t)$, the renormalised entrophy $Z(t)/E(t)$, the normalised maximum core vorticity $\zeta_{\text{ext}}/\sqrt{E}$, and the kurtosis of the vorticity distribution $K(t)$, for initially ordered conditions, several lattice sizes, various energy levels, and a fluid thickness $b = 3$ mm. The vortex size $a(t)$ is defined as $(E/Z)^{1/2}$, $r(t)$ and $\rho(t)$ are determined from the flow pattern, and ζ_{ext} is determined as the maximum of the absolute value of vorticity in the system. A remarkable result that we obtain is the existence of well-defined power laws for all these quantities, throughout one decade of time variation. In this range, the energy has decreased by about 40%, which is small compared to the decay of the number of vortices, or that of the entropy. Power laws are best defined when the

---

[1]The energy and the entropy are defined respectively as the mean square of velocity and vorticity.

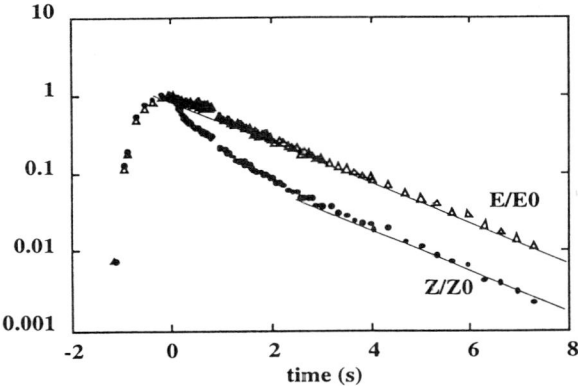

FIGURE 2 Decay of the energy and the entrophy for a system of 100 vortices, with $b = 3$ mm, an initial value of I of 400 mA, and $\tau = 100$ ms. (The ordinates have been renormalized to 1 for $t = 0$.)

lattice is initially ordered, but they are also observed, with comparable exponents, for other experimental conditions such as disordered initial state, systems of tripoles, quadrupoles, various thicknesses (ranging between 2.5 and 4 mm), and various initial energies. Their existence thus appears as an experimental evidence.

A remarkable fact that we observe is that the system remains dense during the decay phase. This is shown in Figure 4, which represents the evolution of ρa^2 (which is a measure of the space occupied by the vortices) and that of the ratio r/a. Both quantities are conserved during the decay phase. The vortices thus occupy all the available space and there is no rarefaction process. The fact that the value of r/a is about 1 indicates that the system is rather compact, and if the vortices were assimilated to molecules, our system would resemble a liquid. Another remarkable fact which is somewhat surprising is that the maximum vorticity is not conserved, but decreases as rapidly as the square root of the entropy (see Figure 4). It is thus difficult to regard ζ_{ext} as an invariant in our system. These two features are observed in a wide range of conditions, and also appear as robust experimental facts.

To describe our turbulent flow, we propose the following laws:

$$a(t) \sim r(t) \sim \sqrt{\frac{1}{\rho(t)}} \sim t^{\nu}, \quad Z \sim \zeta_{ext}^2 \sim E\,t^{-2\nu}, \quad K \sim cst$$

with $\nu = 0.22 \pm 0.04$. This estimate for ν includes all the experiments that we have performed, with ordered and disordered initial conditions, depths varying between 2.5 and 4 mm, and various energy levels. The origin of the scatter in the value of ν is difficult to elucidate, but there is certainly a statistical contribution, due to the fact that we work on a single realisation of the decaying regime.

One can now compare with existing theories. In the Batchelor theory, E is the single invariant of the system as the viscosity decreases to zero.[1] In this theory, the

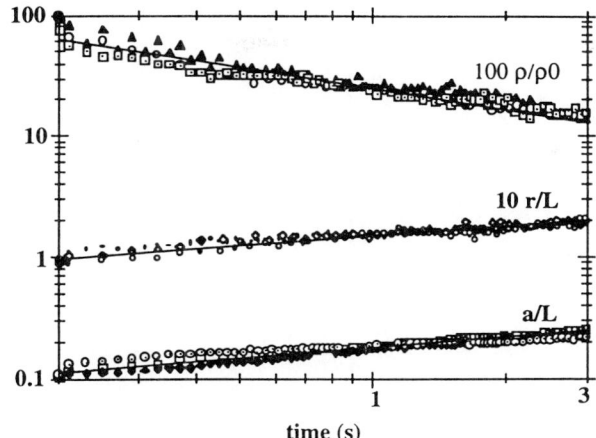

time (s)

FIGURE 3 Temporal evolutions of the vortex densities $\rho/\rho 0$ (where $\rho 0$ is the value for $t = 0$), the mean distance between votices, $r(t)/L$, and the mean vortex size $a(t)/L$ where L is the system size, for systems of 100 vortices, $b = 3$ mm and several initial values of the energy, corresponding to values of the Reynolds number $\sqrt{EL}/\nu$ about 1000.

system remains dense during the decay phase. The corresponding predictions are:

$$r(t) \sim \sqrt{E}\,t, \quad \rho(t) \sim E^{-1}t^{-2}, \quad a(t) \sim \sqrt{E}\,t,$$
$$Z(t) \sim t^{-2}, \quad \text{and} \quad \zeta_{\max} \sim t^{-1}. \tag{1}$$

We do find that the system remains dense, but we do not observe the predicted values of the exponents; our experimental results, thus, disagree with these predictions. Perhaps the strongest discrepancy is obtained on the entrophy, for which we find an exponent about three times smaller than the predicted one. Such a general conclusion was already drawn out in a preceding experimental study.

Another theory was proposed by Carnevale et al.[2]; they assumed the existence of two conserved quantities during the decay phase—the kinetic energy E, and the extremum vorticity ζ_{ext}. These two quantities determine the length scale λ and the time scale τ by the relations:

$$\tau = \frac{1}{\zeta_{\text{ext}}} \quad \text{and} \quad \lambda = \frac{\sqrt{E}}{\zeta_{\text{ext}}}. \tag{2}$$

Using these quantities, Carnevale et al. found power laws for the evolution of $r(t)$, $\rho(t)$, and $a(t)$, in the form:

$$r(t) \sim \lambda \left(\frac{t}{\tau}\right)^{\xi/2}, \quad \rho(t) \sim \lambda^{-2} \left(\frac{t}{\tau}\right)^{-\xi},$$
$$a(t) \sim \lambda \left(\frac{t}{\tau}\right)^{\xi/4}, \quad Z \sim \tau^{-1} \left(\frac{t}{\tau}\right)^{-\xi/2}, \tag{3}$$

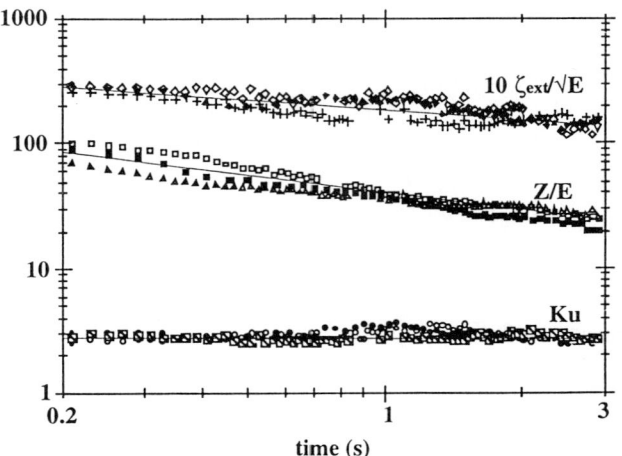

FIGURE 4   Temporal evolutions of the entropy Z (divided by E), the vorticity kurtosis Ku, and the extremal vorticity ζ_{ext} (divided by $\sqrt{E}$), for conditions similar to those of Figure 3.

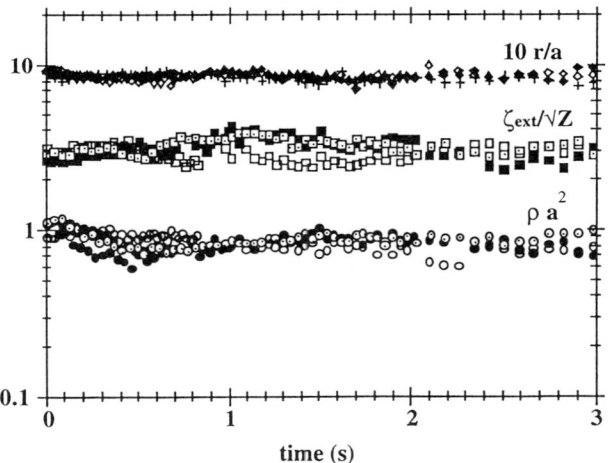

FIGURE 5   Several quantities that appear as invariants in the system ρa^2, ζ_{ext}, and r/a; the experimental conditions are similar to those of Figure 3.

where ξ is a parameter (its numerical estimate is 0.72). In this theory as compared with that of Batchelor, since the average distance between the vortices increases more rapidly than their size, the system tends to rarefy. It turns out that in our experiment, ζ_{ext} is not invariant (see Figures 4 and 5): it decays as rapidly as the square root of the entropy. One of the basic assumption of the theory is, therefore, not consistent with the present experiment. Moreover, the evolution of the kurtosis and that of the ration r/a are qualitatively different from the ones that are predicted. We therefore find a disagreement between this theory and the experiment, despite the fact that some values of the exponents that we find are not far from the ones predicted theoretically.[2]

This study thus reveals that neither the description by Batchelor[1] nor that by Carnevale et al.[2] apply to the particular quasi-two-dimensional turbulence that we have prepared. To the best of our knowledge, no numerical study of freely decaying turbulence has reported characteristics similar to those which we find. Indeed, our experiment lives in a three-dimensional world, but it is not evident that three-dimensionality plays a significant role in our case. A possible explanation is that numerical experiments that have been performed with dilute systems of vortices, compare to our case: for instance, our values of ρa^2 and r/a (which are the space occupied by the vortices and the average distance between eddies) are of order 1, whereas those typically used in numerical experiment are several orders of magnitude smaller (see, for instance, Carnevale et al.[2] and McWilliams[6]). We may, thus, face a "liquid turbulence," whereas the numerical simulations may have examined the gaseous case.

ACKNOWLEDGMENTS

We are grateful to M. E. Brachet, B. Legras, A. Libchaber, Y. Pomeau, and H. Willaime for illuminating discussions.

---

[2] This is the reason why, in a paper by McWilliams,[6] we found good agreement between this theory and the experiment. Such an agreement is now considered as fortuitous.

REFERENCES

1. Batchelor, G. K. "Computation of the Energy Spectrum in Homogeneous Two-Dimensional Turbulence." *Phys. Fluids Suppl. II* **12** (1969): 233.
2. Carnevale, G. F., J. C. McWilliams, Y. Pomeau, J. B. Weiss, and W. R. Young. "Evolution of Vortex Statistics in Two-Dimensional Turbulence." *Phys. Rev. Lett.* **66** (1991): 2735.
3. Couder, Y. "Observation expérimentale de la turbulence bidimensionnelle dans un film liquide mince." *C. R. Acad. Sci.* **297** (1983): 641.
4. Hopfinger, E., R. W. Griffiths, and M. Mory. *J. Mech. Theor. Appl.* (n°Special) (1983): 21.
5. Tabeling, P., S. Burkhart, O. Cardoso, and H. Willaime. "Experimental Study of Freely Decaying Two-Dimensional Turbulence." *Phys. Rev. Lett.* **67** (Dec. 1991): 3772.
6. McWilliams, J. C. *J. Fluid Mech.* **219** (1990): 361.

Michel Decré,† Eric Gailly,† Jean-Marie Buchlin,† and Marc Rabaud‡
†Von Karman Institute, 72 Chaussée de Waterloo, 1640 Rhode-St-Genèse, Belgium
‡Laboratoire de Physique Statistique, Ecole Normale Supérieure, associé au CNRS et aux Universités Paris VI et VII, 24 rue Lhomond, 75231 Paris Cedex 05, France

Spatio-Temporal Intermittency in a Roll-Coating Experiment

We present a statistical description of the transition to chaos by spatio-temporal intermittency for the interface in a forward roll-coating experiment. Similar transition, in the geometry of the Printer's instability, was recently reported (see S. Michalland, et al.[9]). Following their analysis we observed the same general behaviors and a second-order-like phase transition with similar critical exponents.

1. INTRODUCTION

A possible scenario for the transition to chaos in extended one-dimensional systems by spatio-temporal intermittency (STI) was recently described, first in numerical simulations,[1,6] then in experiments.[2,5,7,8,9,11,13] Such transitions are characterized by the possible coexistence of disordered and ordered domains in space and time. A possible analogy with directed percolation model and its exponents was conjectured[10] on the basis of the asymmetric behaviors of ordered and disordered patches: disordered (or chaotic) patches appearing only by a contamination process and not by a spontaneous burst into the ordered (laminar) domain. It appears that critical exponents of the transition were system dependent and no universality was demonstrated. In this chapter, we will describe such a transition for the position of the ribs appearing at the surface of a viscous fluid. These ribs show up at a large enough velocity when the interface is located in the narrow gap between two forward-rotating cylinders.[4] We first describe an experimental bench designed for the industrial study of a roll coating process, as well as the data processing (see Part 2). Then we described the transition to chaotic dynamics in forward coating (see Part 3). In part 4 we compare the results to previous ones of the Printer's instability and conclude.

2. EXPERIMENTAL SETUP AND DATA REDUCTION

The experimental setup consists of two long and horizontal cylinders, one above the other at 45° with vertical axis (see Figure 1). These cylinders are 260 mm long and made of stainless steel. The upper one (radius $R_u = 67.5$ mm) is located such

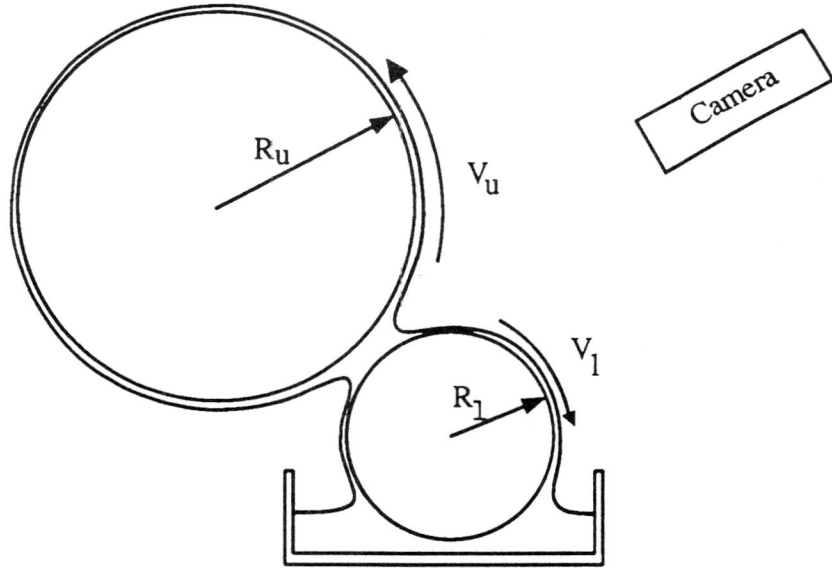

FIGURE 1   Sketch of the experimental setup described in the text.

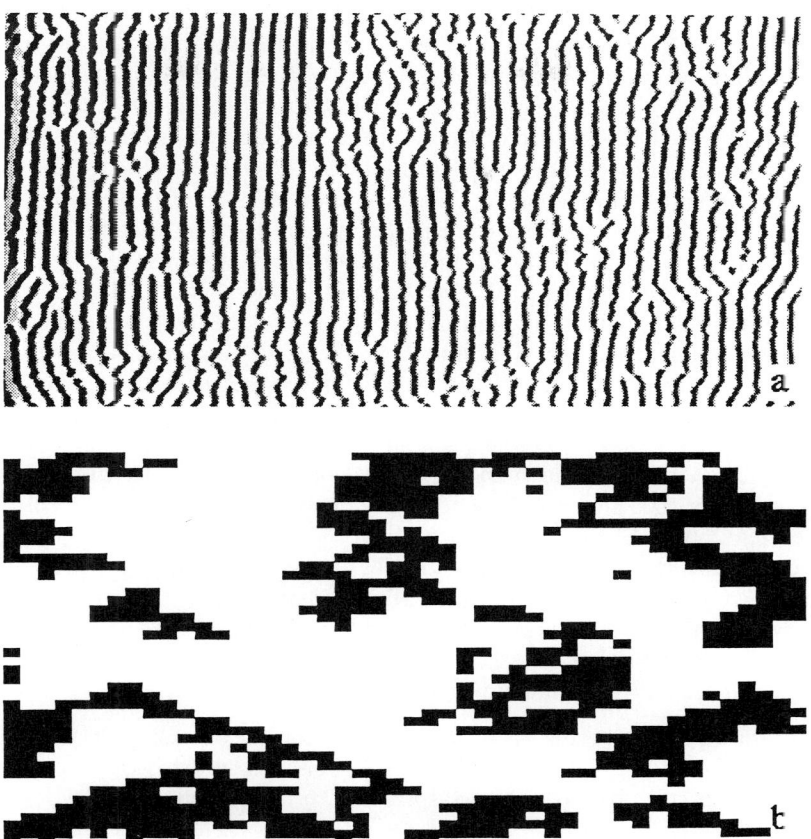

FIGURE 2   Spatio-temporal images of the reflected light by the unstable coated film near the meniscus position. Horizontal axis is along the cylinders axes, time increases downward (2.56 sec. from top to bottom). (a) 64 grey levels image for $Ca_l = 13.02$ and $Ca_u = 4.23$. The mean wavelength $\lambda_{\text{mean}} \approx 5.5$ mm (47 ribs for 260 mm long cylinders). (b) The same image binarized as described in the text.

that the minimum distance with respect to the lower one (radius $R_l = 27.5$ mm) is fixed at the constant value $h = 0.287 \pm 0.015$ mm. An important geometrical parameter is Γ, the ratio of the minimum thickness to the curvature of the gap $\Gamma = h(1/R_l + 1/R_u) \approx 14.7 \times 10^{-3}$. Both can rotate at various angular frequency by the use of two direct current motors. The lower cylinder is partially immersed in the working fluid (latex paint, Levistex code 162) of dynamical viscosity $\mu = 0.741$ Pa·sec and surface tension $T = 0.0382$ N/m. In all the experiments presented here, the velocity V_l of the lower cylinder is kept constant (such that the capillary number, defined as $Ca_l = \mu V_l/T$, is equal to 13.0) and the upper cylinder can rotate in opposite direction (tangential velocity V_u of the same sign as V_l) (see Figure 1).

For a nonrotating upper cylinder, a ribbing of the coated liquid appears on the lower cylinder for a typical capillary number of $Ca_c \approx 0.25$. Such irregularity of the coated film is the signature of the waviness of the meniscus along the cylinder axis. As the two cylinders have exactly the same length, the cellular interface that appears above threshold is usually pinned at the two ends of the cylinders by two larger ribs, as it is the case for the Printer's instability[7] and in some sense similar to what occurs in classical Taylor-Couette experiments.[3] For the chosen velocity of the lower cylinder, of the order of 50 cells form the interface.

The meniscus is lighted at high incidence by a white light lamp and observed almost perpendicular, so that we only observe the ribs between the cells, and not the exact shape of the meniscus. We used a linear C.C.D. camera (IDC131, 1728 pixels, 64 grey levels) to observe the meniscus, and lines of 512 pixels are digitized on a PC computer. Delay time between successive line acquisitions is chosen to be 10 ms. A quite smaller value than the typical period of oscillation of a rib, which is of the order of 0.09 sec. Spatio-temporal video images can also be reconstructed by collecting 256 such lines (see Figure 2(a)).

In order to reduce the description of the dynamics to two states, ordered and disordered, we follow the choice of Michalland et al.,[9] where all static cells belong to ordered domains and moving ones to disordered domains. The algorithm consists then to detect the moving lines shown in Figure 2(a). This is done in three steps as by Michalland[7] and Michalland et al.[9]:

i. First, by subtracting each line to the following one for all the lines of the initial picture. Then, perfectly motionless structure corresponds now to white pixels, grey pixels corresponding to moving ribs.

ii. The resulting image is binarized to black-and-white pixels. In order to reduce the noise, the grey pixels are set to black only if their value is larger than 1/8 of the value of the darker pixel of the image.

iii. The 512×512 pixels image is then replaced by a coarser one consisting of 11×11 boxes. The color (black or white) of each box is decided by counting the number of black pixels in a larger square of 17×17 pixels centered on the box (the pattern wavelength corresponding to 11 pixels in space and the typical period of oscillation of a rib is 9 pixels in time). If this number of black pixels exceeds a threshold value n, the whole box (11×11) is claimed disordered and set to black. This filtering process is then repeated for all boxes.

Figure 2(b) is the final image after such processing of Figure 2(a). We checked that the statistical results are not affected if the value of n remains within a given range. The value of n was then set to the upper bound of this range ($n = 9$) and maintained constant for all the study of the transition. A similar analysis was done in step (ii) for the effect of the thresholding value (1/8) upon the statistical results, and within reason it is not a critical parameter.

---

3. THE SPATIO-TEMPORAL INTERMITTENCY REGIME

For the upper cylinder at rest and rotating lower cylinder at $Ca_l = 13$, spatio-temporal images of the meniscus correspond to an array of vertical black stripes. The paint ribs are motionless. This remains true for small forward rotation of the upper cylinder. After processing, these images become white as all the interface belongs to the same ordered domain. For a larger value of the upper capillary number, disordered patches appear either as bursts in the "bulk" or at the ends of the front. Increasing Ca_u increases the number and the size of these patches, and now the disordered patches can overlap, so that on a typical image as Figure 2(a), the disordered domains are almost completely connected in strong contrast with the ordered ones that are mainly isolated patches. For even larger Ca_u value, the size of the disordered domains increases and finally the whole front is, at all times, disordered. The above description is in agreement with the usual contamination hypothesis for the disordered domains.[1] It is worth noting that the connectivity of the disordered phase is less apparent in the binarized pictures (see Figure 2(b)) than in the original spatio-temporal pictures (see Figure 2(a)). We attribute this to the way we have selected the n value. Note also that the homogeneous nucleation of bursts at intermediate Ca_u values is a new feature that was not observed by Michalland et al.,[9] where the disordered phase always nucleates at the boundaries.

A. CHAOTIC FRACTION. A natural order parameter of the transition from order to disorder is the measure of the proportion of disorder along the interface averaged on long time series to get rid of the temporal intermittency. This is easily done here by measuring the ratio of black pixels on a line over the total number of pixels by line, and taking the average of this value for the 256 lines of 20 successive images as Figure 2(b). The resulting number F_c, called the chaotic fraction, evolves from 0 to 1, as the upper capillary number is increased. In Figure 3, the evolution of F_c appears to be smooth, without any hysteresis, and similar to a slightly imperfect bifurcation with a Ca_u threshold between 2 and 4. On the same curve we also plotted as vertical bars the value of the standard deviation at one σ of the instantaneous chaotic fraction and a best power-law fit.

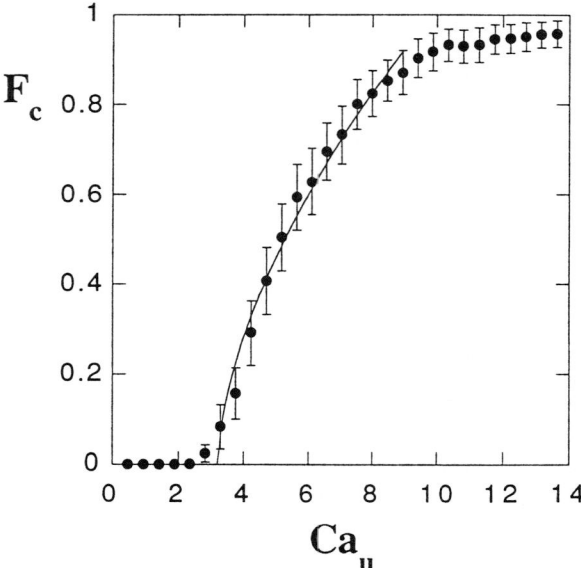

FIGURE 3 Mean chaotic fraction F_c versus Ca_u. Vertical bars are one-sigma-dispersion values for the distribution of instantaneous chaotic fraction. The continuous line is the best fit through the data ($Fc < 0.9$) by a function $F_c = A(Ca_u - Ca_c)^\beta$, best three parameters fit gives $A = 0.33$, $Ca_c = 3.23$ and $\beta = 0.59$.

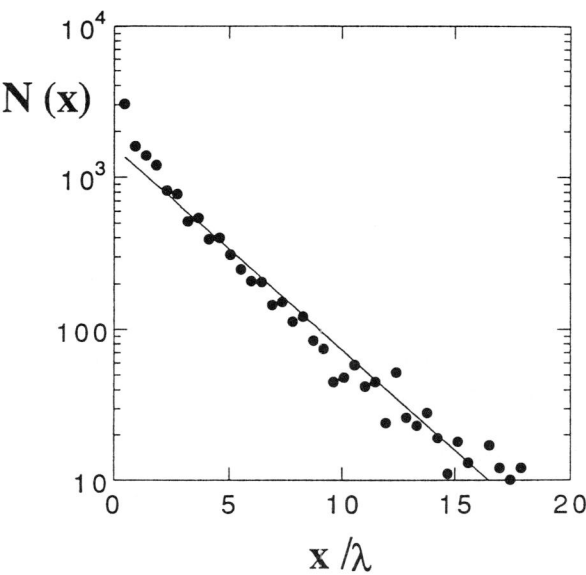

FIGURE 4 Histogram of the number $N(x)$ of laminar patches of dimensionless size x/λ, showing the exponential decay, for $Ca_u = 5.64$. The continuous line is an exponential fit ($\sim e^{-(x/L)}$) corresponding to a typical size for the laminar patches $L/\lambda = 3.27$.

It is worth noting now two quantitative differences between our experiment and the one performed by Michalland et al.[9] First, the standard deviation of Figure 3 is surprisingly much smaller (two to three times for the same statistical sample) than that given by Michalland et al.[9] and, second, the thickness of the STI transition (the typical range of variation of Ca_u) is, in absolute or relative value, much larger. As a consequence, during the transition, the mean wavelength λ increases slowly (8%) with Ca_u and in the following we used an average value of λ to make dimensionless the spatial scale.

B. SPATIAL DISTRIBUTIONS. Another classical function that can be extracted from the black-and-white spatio-temporal images, is the spatial autocorrelation function. It appears that in the present experiment, this function decreases extremely rapidly, even for a very small disorder, giving information only very close to the STI threshold. As Michalland et al.,[9] we choose instead to build the histogram of the width of the laminar domains. For that, on 20 images binarized for the same Ca_u, we count on each line the number of ordered domains of each size. For large enough Ca_u value, such distributions are clearly exponential (see Figure 4), and the exponential decay length is the characteristic width $L(Ca_u)$ of the distribution. For large Ca_u, the laminar patches are few and small, L is then small. For smaller Ca_u, L increases and becomes large near a threshold value (see Figure 5(a)). In Figure 5(b), we plotted the evolution of the square of the inverse of the dimensionless ratio L/λ versus Ca_u. As given by Michalland et al.,[9] the evolution appears to be linear, and the x-intercept of the best linear fit gives a threshold for the STI transition at $Ca_u = 3.2 \pm 0.4$.

Near this value ($2.8 \leq Ca_u \leq 3.8$) the width distribution of the laminar domains cannot be fitted by an exponential, but rather by a power law (see Figure 6). The exponent of this power law is well-defined but evolves with Ca_u. It increases from $\mu_s = 0.60 \pm 0.01$ for $Ca_u = 2.8$ to $\mu_s = 1.0 \pm 0.2$ for $Ca_u = 3.8$.

C. TEMPORAL DISTRIBUTIONS. The same distributions can be built for the duration of the ordered domains: duration is short at high Ca_u, and large at lower Ca_u. Our statistical sample correspond, for each Ca_u values, to 50 vertical lines of 20 successive images. The same general behaviors are obtained, and in particular an exponential decay of the distribution for large enough Ca_u, that permits to determine a typical time scale $T(Ca_u)$ for the ordered domains. These times increase strongly when Ca_u decreases, and a plot of the square of their inverse exhibits a linear behavior that gives a threshold for the STI transition at $Ca_u = 3.6 \pm 0.2$. In the vicinity of this value ($2.8 \leq Ca_u \leq 3.8$) the duration distributions are no more exponential but exhibit a power-law behavior, which exponent evolves from $\mu_t = 0.76 \pm 0.01$ for $Ca_u = 2.8$ to $\mu_t = 1.1 \pm 0.2$ for $Ca_u = 3.8$.

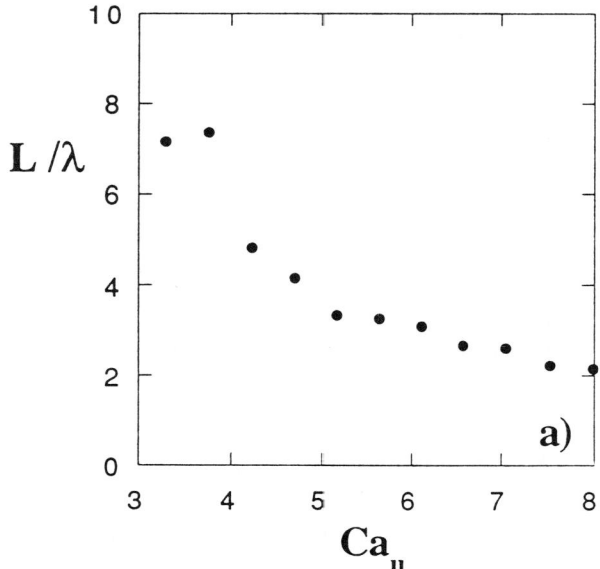

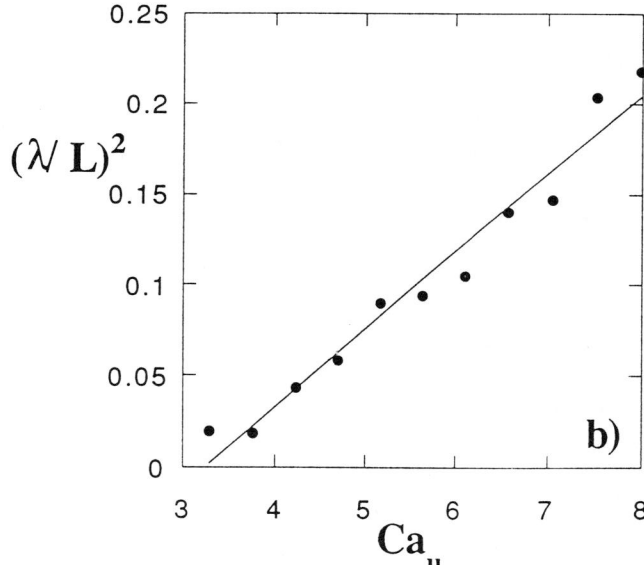

FIGURE 5   Evolution of the typical size, L, of the laminar patches versus Ca_u. (a) L/λ versus Ca_u. (b) $(\lambda/L)^2$ versus Ca_u, with a best linear fit (continuous line) giving a threshold value $Ca_c = 3.24$.

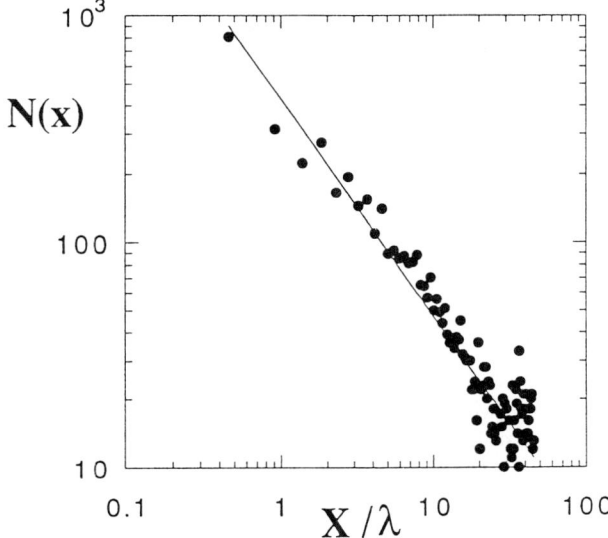

FIGURE 6 Histogram of the number $N(x)$ of laminar patches of dimensionless size x/λ, showing the power-law decay near the STI threshold ($Ca_u = 3.29$). The continuous line is the best fit: $N(x) \sim (x/L)^{-0.96}$.

D. EXPONENTS OF THE TRANSITION. In phase-transition theory, an important exponent is the one that governs the order parameter in the neighboring of the threshold. From the divergence of the characteristic width and duration of the ordered domain, we can estimate a threshold value for the transition by STI. This value is $Ca_u^* \approx 3.4 \pm 0.4$. It corresponds in Figure 3 to a slightly positive value of F_c and it is in agreement with the best fit of Figure 3. Discarding the points below this threshold ascribed to a slightly imperfect transition induced by boundaries, it is possible to estimate directly the exponent for the power-law behavior of F_c versus $(Ca_u - Ca_u^*)$. This best fit corresponds to the equation $F_c \sim (Ca_u - Ca_u^*)^\beta$, with the exponent $\beta \approx 0.46 \pm 0.15$. The low resolution on the β value comes from the strong effect of the choice of Ca_u^*.

4. CONCLUSIONS

In the present work, we reported a transition to disordered pattern by spatio-temporal intermittency. In this forward roll coating geometry, for lower cylinder capillary number $Ca_l = 13$, the transition is similar to the one described for directional viscous fingering.[9] The transition is second-order-like, the chaotic fraction being the order parameter. The exponent β of this transition was found to be

$\beta \approx 0.46 \pm 0.15$, for a threshold value $Ca_u^* \approx 3.4 \pm 0.4$. At this threshold value, the size distributions of the laminar domains obey to power laws which exponents are respectively, $\mu_s = 0.96 \pm 0.02$ for the width distribution, and $\mu_t = 1.0 \pm 0.2$ for the duration distribution.

We must now add few comments.

i. As in some of the other experiments showing a STI transition,[2,5,8] the critical exponents of the histograms μ_t and μ_s in time and space are very close from each other and possibly equal. We have no physical arguments for being so.

ii. The imperfect behavior of the bifurcation below threshold is weak compared to results given by Michalland et al.[9] This is possibly linked to different boundary conditions at the extremity of the front between the two experiments, conditions that make in our case the first contamination by end effect and dilatation waves weaker, and reveals the homogeneous nucleation of the disorder phase at threshold.

iii. Our values of the exponents μ_s and μ_t are slightly larger than the values given by Michalland et al.[9] We cannot rule out the fact that the present binarisation process is slightly different from the one used by Michalland et al.,[9] but more physical reasons can be put forward. One evidence is that the two contamination processes of disorder described in the Printer's instability,[8] dilatation waves and pairs of asymmetrical cells, are weaker in the present experiment. Dilatation waves are damped at long distance, possibly because the capillary number is much larger (4 times) and as shown by Rabaud et al.,[12] dilatation waves are less observed in that case. Asymmetrical cells are also rarely observed, possibly because the aspect ratio Γ is larger too (4 times) and, as shown in that case given by Rabaud et al.,[12] the cells are neither deep enough to observe asymmetrical cells. Then, as suggested in the conclusion given by Michalland et al.,[9] where the non-universality of the STI exponents is ascribed to different contamination processes, the weakening of the contamination processes observed in this work can explain the small changes in the critical exponents.

ACKNOWLEDGMENTS

We thank S. Michalland, P. C. Hohenberg, and P. Wicks for discussions. M. D.'s stays at the Ecole Normale Supérieure was supported by a "bourse de voyage de la Communauté Française de Belgique."

REFERENCES

1. Chaté, H., and P. Manneville. "Transition to Turbulence via Spatiotemporal Intermittency." *Phys. Rev. Lett.* **58** (1987): 112–115.
2. Ciliberto, S., and P. Bigazzi. "Spatiotemporal Intermittency in Rayleigh-Bénard Convection." *Phys. Rev. Lett.* **60** (1988): 286–289.
3. Coles, D. "Transition in Circular Couette Flow." *J. Fluid Mech.* **21** (1965): 385–425.
4. Coyle, D. J, C. W. Macosko, and L. E. Scriven. "Stability of Symmetric Film-Splitting Between Counter-Rotating Cylinders." *J. Fluid Mech.* **216** (1990): 437–458.
5. Daviaud, F., M. Dubois, and P. Bergé. "Spatio-Temporal Intermittency in Quasi One-Dimensional Rayleigh-Bénard Convection" *Europhys. Lett.* **9** (1989): 441–446.
6. Kaneko, K. "Spatiotemporal Intermittency in Coupled Map Lattices." *Prog. Theor. Phys.* **74** (1985): 1033–1044.
7. Michalland, S. "Etude des Différents Régimes Dynamiques de l'Instabilité de l'Imprimeur." Ph.D. Thesis, University of Paris VI, 1992.
8. Michalland, S., and M. Rabaud. "Localized Phenomena During Spatio-Temporal Intermittency in the Printer's Instability" *Phys. D* **61** (1992): 197–204.
9. Michalland, S., M. Rabaud, and Y. Couder. "Transition to Chaos by Spatio-Temporal Intermittency in Directional Viscous Fingering." *Europhys. Lett.* **22** (1993): 17–22.
10. Pomeau, Y. "Front Motion, Metastability and Subcritical Bifurcations in Hydrodynamics." *Phys. D* **23** (1986): 3–11.
11. Rabaud, M., S. Michalland, and Y. Couder. "Dynamical Regimes of Directional Viscous Fingering: Spatiotemporal Chaos and Wave Propagation." *Phys. Rev. Lett.* **64** (1990): 184–187.
12. Rabaud, M., Y. Couder, and S. Michalland. "Wavelength Selection and Transients in the One-Dimensional Array of Cells of the Printer's Instability." *Eur. J. Mech., B/Fluids* **10** (1991): 253–260.
13. Willaime, H., O. Cardoso, and P. Tabeling. "Spatio-Temporal Intermittency in Lines of Vortices." *Phys. Rev. E* **48** (1993): 288–295.

Kim Sneppen and Mogens H. Jensen
Niels Bohr Institute, Blegdamsvej 17, DK-2100 Copenhagen Ø, Denmark

Self-Organized Intermittent Activity for Strings and Interfaces

We study the dynamics of interfaces that through self-organization become critical by a rule similar to that of invasion percolation. In the critical state we observe temporal multiscaling of the interface profile and an activity pattern that exhibits nontrivial power law correlations in both space and time even though only quenched Gaussian noise is applied. One way to characterize this intermittent activity is through avalanches consisting of connected regions of recent local activity. These avalanches display scale invariance, thus reflecting both small- and large-scale properties of the dynamics. An extension of the presented concepts to evolving strings leads to a new universality class with other scalings. Possible systems governed by the investigated algorithms are discussed.

We consider the growth model proposed by Sneppen[11] and Sneppen and Jensen[12] where the motion of the interface is determined by a rule similar to that of invasion percolation.[16] The model is defined on a lattice (x, h). In the one-dimensional version a discrete interface $h(x)$ is defined on $x = 1, 2, 3, \ldots, L$. Along this chain $(x, h(x))$, one initially distributes a sequence of uncorrelated random numbers η. We use periodic boundary conditions. The chain is updated by finding the site with the smallest random number $\eta(x, h(x))$ among all sites on the interface. On this site, one unit is added to h. Then neighbouring sites are adjusted

Spatio-Temporal Patterns, Ed. P. E. Cladis and P. Palffy-Muhoray,
SFI Studies in the Sciences of Complexity, Addison-Wesley, 1995   **573**

upwards ($h \to h + 1$) precisely until all slopes $|h(y) - h(y - 1)| \leq 1$. This induce a local burst of activity that is exponentially bounded. New random η's are assigned to all newly adjusted sites.

One prediction of the above model is the scaling of the saturated interface width $w_q = \langle (h - \langle h \rangle)^q \rangle^{1/q} \propto L^{0.63}$ with system size L, for all moments q.[11] This scaling is, in papers by Sneppen[11] and Tang,[15] connected to the transverse to longitudinal scaling for interfaces glued to directed percolating chains of effectively pinned sites, and is in fact closer to experimentally observed[3,4,9,10] values than KPZ exponents.[5] An interesting consequence of the analogy to directed percolation is that the dynamics induce a self-organized threshold $\eta_{\text{crit}} = 1 - p_c = 0.4615$ below which the corresponding site spontaneously gets activated. Here p_c is the critical density of sites for directed percolation on a square lattice.

The dynamics of the model appear however much richer than expected from this simple analogy alone. Considering the activity along the chain, one observes initially that it is uncorrelated in space, but, as time progresses, subsequent activities gets correlated over larger and larger distances. Finally at saturation a critical state builds up, in the sense that the subsequent spatial activity shown in Figure 1 shows a power law distribution:

$$P(X) \propto X^{-2.25 \pm 0.05}. \tag{1}$$

Therefore the model exhibits self-organized critical behaviour. Notice that the same exponent is observed when one considers the spatial distribution function of distances between the two lowest η along a snapshot of a string.[13] Thus part of the observed dynamical correlation of activity is already an inherent property of static snapshots of the distribution of η in the self-organized critical state. We stress however that other parts of the dynamical correlation do not appear connected to static correlations, for example, the decay of average activity in a given point after there has been activity ($\langle A(t) \rangle \propto t^{-0.6}$) slower than expected from the spatial correlations.

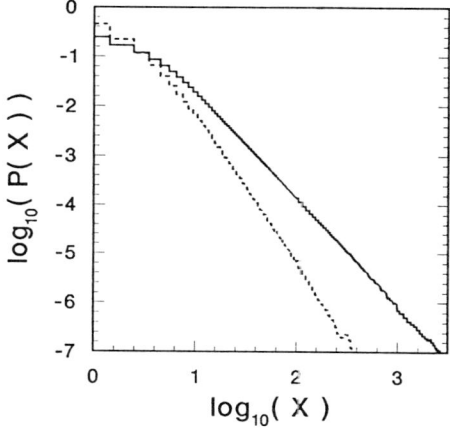

FIGURE 1   Spatial distribution of subsequent activity centers. The solid line is for the interface model, the dashed line for the string model. Activity appears more confined in the string model than in the interface model.

Another more dramatic consequence of the above is the observation of temporal multiscaling.[12] Denoting with τ a time at saturation, we measure the height-height time correlations by

$$W_q(L,t) = \langle (h(x,t+\tau) - h(x,\tau)) - \langle h(x,t+\tau) - h(x,\tau) \rangle)^q \rangle^{1/q} \qquad (2)$$

where average $\langle \rangle$ over $x \in [1, L]$ and members of the ensemble. For not to large times, one observes multiscaling with $W_0 \propto t^{0.60}$, $W_2 \propto t^{0.69}$, and $W_\infty \propto t^{0.40}$.[12] Thus, as is the case for most dynamical systems, the temporal (dynamical) scaling is much richer than the static scaling. For the present model many dynamic scaling exponents can however be derived from the static scalings.[7,6]

In order to characterize the dynamic features of the present model further, we now consider connected avalanches of activity. These are defined by regions of overlapping local bursts of activity, each initiated by localizing a minimal η as described above. When a new burst does not overlap with any of the previous ones, the connected avalanche is considered terminated and a new one is started. Snapshots of interfaces separated by such connected avalanches are displayed in Figure 2(a). In contrast to the locally confined activity at each time step, these

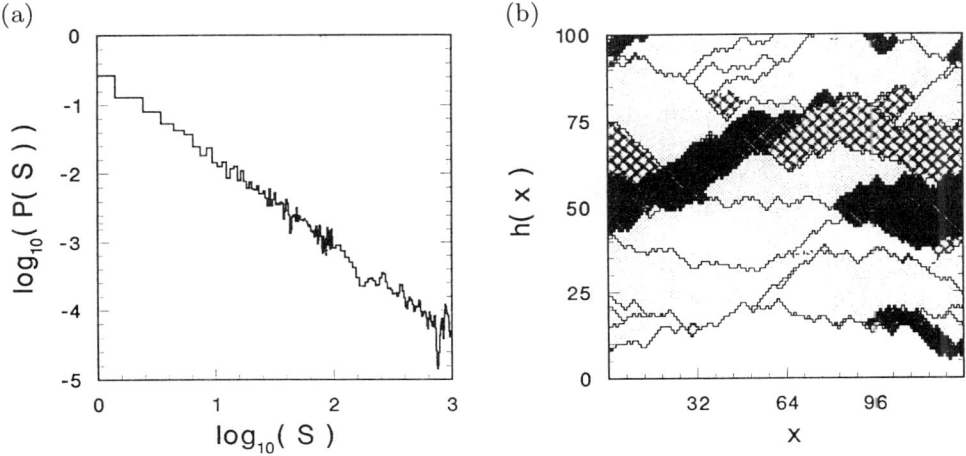

FIGURE 2   (a) Size distribution of connected avalanches of activity, displayed for the interface model. A connected avalanche of activity is defined as a sequence of connected local bursts of activity, stopped by the first local burst that separates spatially from the previous. We observe a power law $P(S) \propto S^{-1.35\pm0.05}$, with small avalanches most probable but with nearly all activity in the big avalanches. (b) Plots of profiles for the interface model, showing snapshots separated by connected avalanches of activity. We see that the activity is dominated by big avalanches.

avalanches appear to display clear power law scalings. In fact, defining the size S of the avalanche by the total number of invaded sites, we numerically observe a power law behaviour $P(S) = S^{-1.35\pm0.05}$ over three decades, as seen in Figure 2(b). Then, as a consequence of all of the above, the space-time plot of the activity shown in Figure 3 exhibits intermittency with big regions of quiescence interrupted by scale-invariant regions of high activity. Thus the algorithm does not only produce a self-organized critical state, but does in fact also predicts self-organization of intermittent bursts of activity and stasis on all scales.

It is interesting that one can define a closely related model similar to the one above, with the exception that the up-down symmetry is restored. The symmetric model is defined on a lattice with sites (x, h). In the one-dimensional version a discrete string $h(x)$ is defined on $x = 1, 2, 3, \ldots, L$. Along the string a sequence of uncorrelated Gaussian random numbers η is distributed. We use periodic boundary conditions. The chain is updated, using a global comparison, by finding the site with

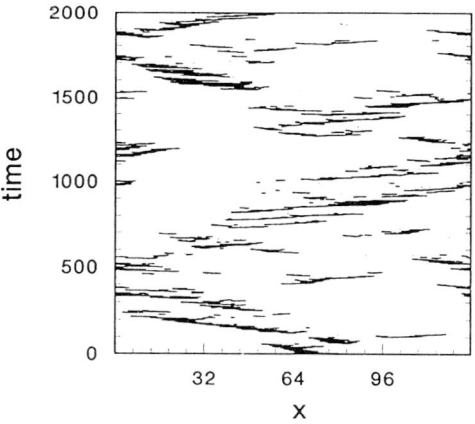

FIGURE 3 Space-time plot of many local bursts of activity for the interface model. Time is counted by the number of local bursts, and measurement starts after saturation is reached. This plot is particularly interesting when one consider, each site as representing a specie in an ecological system, like in the paper by Bak and Sneppen.[2] The eventual application of the interface model for modeling coevolution, instead of the model proposed by Bak and Sneppen,[2] builds on the additional assumption of a specific history dependence for the connectivity between the species. Thus the plot shows mutations in specie-space of an ecology with connectivity given by gradients of an always progressing "evolutionary level" h. In this picture, real time differs from the shown mutation count time by an exponentiated barrier that roughly translate to a stressed exponential in the jump X between subsequent mutations.

the smallest random number η among all sites on the string. On this site, one chose with equal probability to either add or subtract 1 from h. Then neighbouring sites are adjusted in the same direction just until all slopes $|h(y) - h(y - 1)| \leq 1$. This create a local burst of activity that in simulations appear exponentially bounded. New random η's are assigned to all newly adjusted sites.

As is the case for the interface model, this string model also develops a self-organized critical state with long-range correlations of activity

$$P(X) \propto X^{-3.1 \pm 0.2}, \tag{3}$$

and temporal multiscaling, both perpendicular and in fact also parallel to the string. The last is a new phenomena, which we[14] denote multidiffusion. Then, as was the case for the interface model, the intermittent dynamics of this string model contain many more scalings than its static properties indicate.

Finally we would like to mention possible applications of the above-developed models. The asymmetric model for interface propagation appear to mimic, e.g., the interface of oil penetrating a porous medium. Here progress is often observed to occur in avalanches. Other possible and connected areas of interest are depinning of flux lines in type 2 superconductors. More exotic speculations are connected to the observation that snapshots of the interface are closely connected to directed percolating strings at criticality. In this connection it is interesting that Pomeau has conjectured[8] that the division between laminar and turbulent flow may be associated to directed percolation. Thus, if the Pomeau conjecture is correct, the possible existence of dynamical multiscaling perpendicular to the directed percolation could provide new insight.

The application of the string model, on the contrary, is not related to any experiment where one fluid displaces another. Its universality class may, however, turn out to have another and much broader range of applications, connected to an interpretation of the local η as barriers. Bak and Sneppen[2] have proposed to study evolution of a large number of systems, each with many metastable states, and organized such that neighbouring systems can modify each other's barriers. For sufficiently low temperature, this collection of weakly coupled systems is governed by fluctuations connected to passing the overall lowest barrier, like in the present model. The scaling exponents in the evolution model by Bak and Sneppen[2] furthermore appear identical to exponents of the present string model and the models, therefore, belong to the same new universality class.

In any case we emphasize that the dynamic scaling behaviour in this type of models appear much richer than the scalings found in independent snapshots of saturated configurations.

ACKNOWLEDGMENTS

K. Sneppen thanks the Carlsberg Foundation for financial support.

REFERENCES

1. Bak, P., C. Tang, and K. Wiesenfeld. "Self-Organized Criticality: An Explanation of 1/f Noise." *Phys. Rev. Lett.* **59** (1987): 381.
2. Bak, P., and K. Sneppen. "Punctuated Equilibrium and Criticality in a Simple Model of Evolution." *Phys. Rev. Lett.* **71** (1993): 4083. Authors propose a model of coevolutionary dynamics by a sequence of metastable complex systems, each with a minimal random barrier for possible changes. The system evolves by locating the overall lowest barrier and updating this and its two nearest neighbours to new random barriers. The dynamics can be justified by exponential separation of time scales associated to an interpretation of the barriers as energies and the driving performed at very low temperature. Power laws will be observed for distances smaller than a correlation length which increases with decreasing temperature.
3. Buldyrev, S. V., A.-L. Barabási, F. Caserta, S. Havlin, H. E. Stanley, and T. Vicsek. "Anomaleous Interface Roughning in Porous Media: Experiment and Model." *Phys. Rev. A* **45** (1992): R-8313.
4. Havlin, S., A.-L. Barabási, S. V. C. Buldyrev, K. Peng, M. Schwartz, H. E. Stanley, and T. Vicsek. "Anomaleous Surface Roughning: Experiments and Models." In *Proceedings of Granada Conference on "Fractals"*, edited by P. Meakin and L. Sander. New York: Plenum Press, 1992.
5. Kardar, M., G. Parisi, and Y.-C. Zhang. "Dynamic Scaling of Growing Interfaces." *Phys. Rev. Lett.* **56** (1986): 889.
6. Leschhorn, H., and L. H. Tang. "Avalanches and Correlations in Driven Interface Depinning." *Phys. Rev. E* **49** (1994): 1238.
7. Olami, Z., I. Procaccia, and R. Zeitak. "Theory of Self-Organized Interface Depinning." *Phys. Rev. E* **49** (1994): 1232.
8. Pomeau, Y. "Front Motion, Metastability, and Subcritical Bifurcations." *Physica D* **23** (1986): 3.
9. Rubio, M. A., C. Edwards, A. Dougherty, and J. P. Gollup. "Self-Affine Fractal Interfaces From Immiscible Displacement in Porous Media." *Phys. Rev. Lett.* **63** (1990): 1685.
10. Rubio, M. A., C. Edwards, A. Dougherty, and J. P. Gollup. *Phys. Rev. Lett.* **65** (1990): 1389.
11. Sneppen, K. "Self-Organized Pinning and Interface Growth in a Random Medium." *Phys. Rev. Lett.* **69** (1992): 3539.

12. Sneppen, K., and M. H. Jensen. "Colored Activity in Self-Organized Critical Interface Dynamics." *Phys. Rev. Lett.* **71** (1993): 101
13. Sneppen, K., and M. H. Jensen. "Reply." *Phys. Rev. Lett.* **70** (1993): 3833
14. Sneppen, K., and M. H. Jensen. "Multidiffusion in Critical Dynamics of Strings and Membranes." *Phys. Rev. E* **49** (1994): 919
15. Tang, Lei-Han, and H. Leschhorn. "Comment." *Phys. Rev. Lett.* **70** (1993): 3832
16. Wilkinson, D., and J. F. Willemsen. "Invasion Percolation: A New Form of Percolation Theory." *J. Phys.* **A 16** (1983): 3365.

William I. Newman,† **Andrei M. Gabrielov,‡** **S. Leigh Phoenix,*** **and Donald L.Turcotte****

†Departments of Earth and Space Sciences, Astronomy, and Mathematics, University of California, Los Angeles, CA, 90024
‡Department of Geological Sciences, Cornell University, Ithaca, NY, 14853
*Department of Theoretical and Applied Mechanics, Cornell University, NY, 14853
**Department of Geological Sciences, Cornell University, Ithaca, NY, 14853

Failure of Hierarchically Structured Materials: Statics and Dynamics

We consider, by computational and analytic means, the failure properties of hierarchically organized bundles of fibers with specified load sharing among survivors, a problem that may be treated exactly by renormalization methods. We show, *independent of the specific properties of an individual fiber*, that the stress and time thresholds for failure of fiber bundles obey universal, albeit different, scaling laws with respect to the size of the bundles. The application of these results to fracture processes in earthquakes and in engineering practice helps to provide insights into some of the observed patterns and scalings.

1. INTRODUCTION

The failure of both natural and manmade materials constitutes hazards on a wide variety of scales. Natural failures include earthquakes, and all materials are subject to fracture when stressed. In a wide variety of circumstances, failures involve characteristic patterns and scalings. A widely applicable family of models for failure consider lattices in one, two, or three dimensions where the probability of failure of a given lattice point is a function of the applied load σ (essentially instantaneous failure) or of time t (given the load history) is prescribed. The problems considered in this chapter consider the transfer of stress from a failed element to adjacent intact elements. A fundamental feature which distinguishes the nature of failure in such systems is the manner in which the load formerly supported by failed elements is redistributed among the survivors. There is observational evidence in both materials science and in geophysics that universal scaling laws for failure are applicable. One example is the Gutenberg-Richter law relating the frequency of earthquake events to the magnitude (i.e., the logarithm of the energy) of such events. This distribution is self-similar or "fractal" in that the number of earthquakes scales as a power of the rupture area. We wish to understand the pattern and scaling of failure of the macroscopic system given that the microscopic properties are known. To pursue this goal, we will employ renormalization methods, which can be shown to be exact, both in the computational and analytic investigations. Renormalization methods that we apply eliminate *all* redundant operations in the computational investigation (in the same spirit that the FFT optimizes the computation of Fourier transforms), thereby making it possible to explore the behavior over many orders of magnitude. Once the scaling laws are identified from the computation, bounding and asymptotic properties can be analytically derived.

Early efforts to understand this problem utilized as a paradigm the failure properties of bundles of fibers. The question is how the failure of the bundle (e.g., a cable) is related to the failure of the individual fibers (the cable's strands). We will assume that fiber bundles have no bonding or friction between fibers, but are perfectly secured at their ends—in this way, a lattice is indistinguishable from a fiber bundle. While it is easy to see that a tape of fibers and a bundle of fibers are direct analogues of one- and two-dimensional lattices, respectively, fiber bundles can also be employed to explore the behavior of three-dimensional structures. The effective dimensionality of a lattice depends upon the rule for redistributing the load following the failure of individual fibers. There are, essentially, two limiting cases that can be considered: in the first, the stress on a failed group of fibers is transferred equally to *all* surviving fibers, while in the second, the stress on a failed group of fibers is transfered to the nearest surviving group(s) of fibers. The redistribution rule establishes the meaning of "nearest neighbors" and, hence, the effective dimensionality of the system. Although it is useful to visualize fiber bundles as ropes, it is essential to note that the principal mechanism wherein ropes support a load is through the *friction* that exists between different fibers (allowing local stress transfer at fiber breaks), a process that is absent in lattice models and in the idealization we will now pursue.

The first mathematical investigation of the failure properties of bundles of fibers was carried out by Daniels.[4] Following Daniels, we will assume that m such fibers or filaments are placed in parallel along a line to form a planar tape. This tape is then subject to a load σ which is the *force per fiber*, the total force on the bundle being $m\sigma$. [The term *load* is employed in different ways in different disciplines. Among those who work with fibrous, composite materials, it describes the force per unit area or the force per fiber. In the earth sciences, it describes the *total* force acting on or supported by a system. It is most natural for an audience of physicists to employ the former definition, as it invokes the sense of an intrinsic or specific variable, and accordingly we will speak of the load as the stress per fiber.] It is further assumed that these idealized fibers are of equal length, stiffness, and cross-sectional area, and have strengths that are iid (independent and identically distributed) with respect to load or to time (assuming identical load histories).

Daniels considered the static behavior of fibers with cumulative distribution function $P(\sigma)$, $\sigma \geq 0$. In other words, $P(\sigma)$ describes the likelihood that an individual fiber would fail below a load σ. Further, Daniels assumed that the fiber bundle obeyed the so-called "equal-load sharing rule," i.e., all nonfailed fibers share the load equally, and all failed fibers carry no load. Daniels explored the asymptotic distribution for the strength of this ensemble of fibers as m grows large, and also developed recursion relations useful for smaller m. Many other investigators, including Smith and Phoenix,[14] have explored other aspects of this problem, such as the rearrangements of such bundles in series. What becomes clear from these investigations is that the asymptotic behavior depends strongly on the type of load sharing and overall architecture of how such bundles are arranged. As m grows large, the strength of an equal-load-sharing bundle approaches a constant (see Daniels[4]) whereas that for a local-load-sharing bundle (as defined later) approaches zero albeit extremely slowly (see Newman and Gabrielov[11] and Gabrielov and Newman[6]). Moreover, by an insightful application of the central limit theorem, Daniels established that the macroscopic distribution function with respect to the load for the failure of an asymptotically large fiber bundle with equal-load-sharing is normally distributed. It is the nature of the macroscopic distribution function for failure under *local* load sharing that we wish to establish here.

Smalley et al.[13] were the first to appreciate the connection between fiber bundles and lattice models for failure. They considered fiber bundles with $m = 2$ and $m = 4$ as a metaphor for the strength of a fault in the earth's crust, which was considered to be a hierarchically organized system of asperities. In particular, they assumed that m fibers could be assembled to form a bundle that would behave as if it were itself a fiber. Then, they assumed that m of these fiber bundles could themselves be assembled to form a bundle, and so on. This hierarchical assembly process, which we will refer to as hierarchical clustering with coordination number m, can be continued indefinitely. They employed an index n to describe the level within the hierarchy. For example, when $m = 2$, the hierarchical structure corresponds to a successive pairing or twinning of fibers. Then, $n = 0$ implies individual fibers, $n = 1$ implies a pair of such fibers, $n = 2$ implies a pair of pairs of such fibers and so on. Thus, an nth order fiber bundle with coordination number m would contain m^n individual fibers.

The Smalley et al.[13] scheme is a renormalized description of failure and follows in the same spirit as Allègre et al.[1] The description given by Allègre et al., however, does not redistribute the released stress among the surviving members, an essential feature of this problem—consequently, the method given by Smalley et al. is physically more realistic. A main feature of this approach is that the load released by a failed unit of any size in the hierarchy is redistributed onto a regime of comparable spatial extent, so the redistribution of stress is a localized phenomena in contrast to a single Daniels bundle where the load sharing is global. (This represents a specific class of local load sharing models; other classes for local load sharing have also been studied. See, e.g., Harlow and Phoenix.[8]) This feature is the hallmark for the Green's function for the stress field (see Landau and Lifshitz[10]) around a "crack" where the stress field is localized near the failed sites and influences a region whose size is proportional to the crack size and the square of the applied stress. (This is the usual first approximation for the size of the plastic zone around a crack. See Kanninen and Popelar,[9] for example.) A key point that becomes clearer in later discussion is that hierarchical structures become an important concept when catastrophic or weakest link fracture growth stalls as the size scale increases due to some blunting mechanism caused, say, by shear limitations.

Smalley et al.[13] suggested that an additional renormalization procedure could be applied to this problem since the bundles were grouped in a hierarchical fashion. They employed a simplifying approximation in their renormalization calculation and obtained a *critical point*-like behavior. They observed, for a given hierarchical structure defined by the value of m, that a well-defined load would emerge for large assemblies of fibers such that the fiber bundle would survive if the stress was lower than that critical stress and would fail if the stress was higher than that critical stress. The approximation they employed, however, subtly altered the otherwise exact renormalization calculation, and a number of investigators challenged the validity of the implied critical point. Newman and Gabrielov[11] and Gabrielov and Newman[6] explored the problem formulated by Smalley et al. using *exact* computational and analytic means. They employed a renormalization of the computation of the hierarchical system of equations wherein they identified and eliminated all redundant computations. In this way, they were able to simulate the behavior of the hierarchical system *over 90 decades* of system size and to identify the asymptotic nature of the solution. With this information at hand, they were able to derive analytically both bounds and asymptotics for the system, although this required the development of new mathematical techniques. Although a critical point does not formally exist for the static problem, the strength of the bundle declines at an extraordinarily slow rate. Moreover, the width of the distribution function for failure diminishes steadily but remains substantial for any finite-sized realization of a fiber bundle. A simplified version of these developments is presented in the next section.

In this chapter, we present new results we have obtained for the dynamic or time-dependent problem by methods analogous to those used by Newman and

Gabrielov in the static problem. We review the static problem in the next section because what we learn from the static problem in the stress variable σ can in part be transferred to the dynamic problem in terms of the time variable t. In particular, we wish to explore (for a given applied load) how the failure properties in time of fiber bundles vary with respect to the size of the bundle. Coleman[2,3] pioneered the investigation of the lifetime of bundles and these models have been used to explain the failure of yarns, ropes, and cables as well as fibrous composites. (Recall that friction is a major ingredient in some of these situations, and that friction is basically irrelevant to the lattice description of solid materials that concerns us.) For a review of previous work on the failure of chains of bundles, see Smith and Phoenix.[14] Lattice models of material failure have been pursued in the physics literature including dielectric breakdown, failure of superconducting and random resistor networks as well as brittle fracture, largely in the setting of percolation theory. For a review, see Duxbury.[5] Although large cables often involve hierarchical arrangements of bundles, modeling in this vein appears not to have been pursued by others.

In the next section, we will explore the static properties of hierarchical fiber bundles as well as the methods required to investigate these. In the third section, we will extend these methods to develop new computational algorithms and present the results of Monte Carlo simulations which offer a hint as to the nature of dynamic failure in hierarchical systems. The numerical results suggest the possible presence of a *critical point* for failure in time for large systems. We will conclude by discussing the significance of the pattern and scaling properties that are apparent, as well as some of the consequences—in particular the suggestion of a critical point for failure in time could have profound significance in engineering practice—the failure of isolated individual assemblies could predict an avalanche of failure in statistically similar assemblies. In geophysics, this could help explain the success met by some earthquake prediction schemes predicated on premonitory seismicity—the presence of microscopic earthquakes could be an indicator of the catastrophic failure yet to come. Indeed, fiber bundles offer a rich paradigm for the investigation of failure in many varied situations.

2. STATICS

We first consider an assembly of fibers where a load is instantaneously applied and we check to see how each of the fibers responds. Although the sequence of failures is in some sense time-dependent in reality, we consider this to be a static problem since we are considering an instantaneous, in contrast with an evolutionary, response to the loading. In the following section, we will discuss the situation where there is a finite probability of failure during an *interval of time* under a uniform load, in contrast with an instantaneous response. Nevertheless, the conceptual link between these two situations often leads to qualitative similarities in the failure properties with respect to the load in the static situation and to the time in the dynamic one.

Consider an assembly of fibers with no bonding or friction between fibers that has a given cumulative strength distribution $P(\sigma)$ where σ is the applied stress; i.e., a fiber will break under a stress σ with probability $P(\sigma)$. The distribution $P(\sigma)$ is a nondecreasing function with values between 0 and 1 defined for $0 \leq \sigma \leq \infty$ where $P(\sigma) \rightarrow 1$ when $\sigma \rightarrow \infty$. Let us denote individual fibers of the bundle as fibers of order 0. Suppose that these fibers are paired in sequential order (i.e., $m = 2$) and consider each pair of fibers of order 0 as though it were itself a fiber, which we will denote as a fiber or, alternatively, as a fiber bundle of order 1. Suppose that these fibers of order 1 are also paired sequentially and that these pairs then form fibers of order 2, and so on, with fibers of order n consisting of 2^n individual fibers. Equivalently, the order of the system, namely n, varies as $\log_2(\text{mass})$ where the "mass" refers to the total mass of the system. The two fibers of order n that are paired are called neighbours or twins of order n and form a fiber of order $n + 1$. It is natural to equate $P_0(\sigma)$ with $P(\sigma)$, and we denote by $P_n(\sigma)$ the strength distribution for fiber bundles of order n. We will refer to $P_0(\sigma)$ as the primitive or basic cumulative distribution function. The problem here is to determine the properties of $P_n(\sigma)$ as n becomes large. (As we will see later, scaling becomes evident albeit not conclusive when $n \approx 20$, while, for real samples of materials, we typically find $n \approx 25$ to 30. These values of n are generally inaccessible computationally using conventional approaches.)

Since both fibers in a given bundle share equally the load supported by that bundle, a given pair of fibers breaks under an applied stress σ only if both constituent fibers fail under that stress σ, or one fiber fails under its load and then redistributes it to the surviving fiber which then fails under a load between σ and 2σ. This concept can be applied to the failure of fibers of any order and is the *sine qua non* of the renormalization method. The probability that both fibers of order $n - 1$ fail simultaneously under a stress σ is $P_{n-1}^2(\sigma)$. The probability that one fiber fails under the stress σ with the surviving fiber failing under a stress between σ and 2σ is $2P_{n-1}(\sigma)[P_{n-1}(2\sigma) - P_{n-1}(\sigma)]$. The factor of two appears here since failure can occur in two ways, according to which one of the two fibers fails first. Thus, the probability of failure of a fiber bundle of order n is equal to the sum of the previous two terms, namely

$$P_n(\sigma) = P_{n-1}(\sigma)[2P_{n-1}(2\sigma) - P_{n-1}(\sigma)]. \tag{1}$$

The problem can be formulated similarly for the m-fold case when a fiber of order n consists of m fibers of order $n - 1$—however, the elaborate combinatoric nature of the problem for $m > 2$, gives rise to expressions that have the order of 2^m terms!

Smalley et al.[13] explored this transformation *assuming* that $P(2\sigma)$ could be expressed as a simple function of $P(\sigma)$; i.e., $P(2\sigma) = F[P(\sigma)]$ where the functional F was determined to give the correct result when the Weibull distribution $P(\sigma) = 1 - \exp(-a\sigma^2)$ was employed. They continued to use this same functional relationship for $n > 0$, which is not strictly correct as the correspondence between the distribution $P_n(\sigma)$ and $P_n(2\sigma)$ depends on n. Although a useful approximation, it forced the solution to have the character of a critical point—indeed, any functional relationship of this sort necessarily results in a critical point (see Allègre et al.[1] and Newman and Knopoff[12]). (Another way to think of this is to recognize that we are converting a functional iteration of Eq. (1) into a point map.) To provide a sense of how the transformation Eq. (1) develops, consider the case of the general Weibull distribution

$$P(\sigma) = 1 - \exp[-a\sigma^\rho], \tag{2}$$

where ρ defines the *modulus* or *shape parameter* and typically ranges from 1 to 20 and the case of $\rho = 1$ is called the "exponential distribution." In Figure 1, we plot the sequence of cumulative distribution functions $P_n(\sigma)$, for $n = 0, 1, \ldots, 10$,

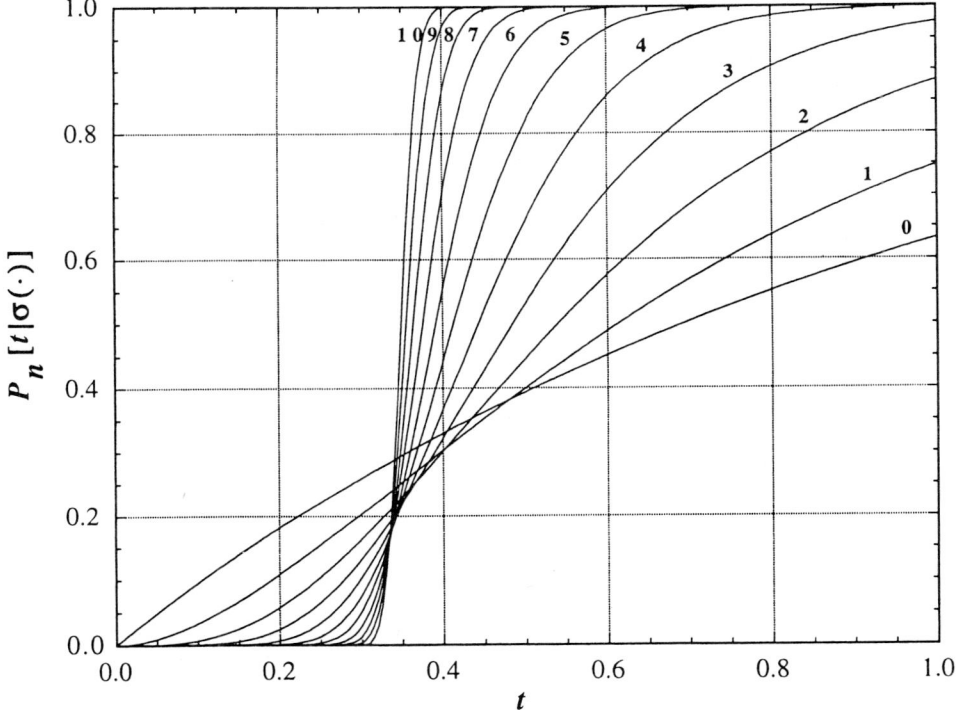

FIGURE 1 Cumulative distribution function $P_n(\sigma)$, $n = 0, 1, \ldots, 10$, for the Weibull exponential distribution $[\rho =]1$.

generated by Eq. (1) using the Weibull exponential distribution ($\rho = 1$ and $a = 1$). The trend that we observe is one where the distribution becomes more and more like a step function that migrates steadily toward the origin as n increases. We observe, however, that the rate of approach toward the origin quickly decelerates and that the width of the step—a measure of the uncertainty in the value of the stress which will precipitate in failure—remains large. A simple argument explaining why weakening occurs can be constructed by considering two representative fibers which are then paired together. Since they are representative of the distribution, it follows that one is likely weaker than the average and that the other fiber is stronger than the average. Hence, failure is initiated in the weaker fiber and the surviving fiber will likely fail under the doubled load (this is the so-called "weakest-link" approximation). The end result is that failure in the paired bundle, on average, will occur at a lower threshold.

To better quantify the observed tendency of the cumulative distribution function to approach the origin, it is useful to define a "failure threshold" σ_n as the stress where $P_n(\sigma) = P_T$ where the value of P_T assigns some assumed threshold for failure. In common engineering practice, this can be the 50% quantile, which we will generally use, or some other criterion for failure. Note that if the iteration Eq. (1) were to result in σ_n converging to a finite nonzero value, we would have a critical point for the iteration. In contrast, suppose that weakest link theory applies, i.e., if one fiber fails, then any surviving fibers in the bundle experiencing an increased load will *always* fail. Thus for $m = 2$ in Eq. (1), this is equivalent to assuming that if one fiber fails, its partner automatically fails and $P_n(2\sigma) = 1$. This conjecture can be employed formally to Eq. (1) to obtain a bound, namely

$$P_{n+1}(\sigma) \leq \begin{cases} P_n(\sigma)[2 - P_n(\sigma)] \\ 1 - [1 - P_0(\sigma)]^{2^n} \end{cases} . \tag{3}$$

(It is worth noting that this expression has the form one would obtain if extreme value statistics were applicable [see Galambos[7]] if 2^n elements were involved—this already provides an indication that something peculiar might emerge from this bound.) Using the example of the Weibull exponential distribution, we find that $P_{n+1} \leq 1 - \exp(-\sigma 2^n)$, leading to the conclusion, if we were to employ the *equality* in Eq. (3), that $\sigma_n \propto 2^{-n} \propto 1/\text{mass}$. This implies that (a lower bound to) the strength of the bundle varies *inversely* as its mass. An alternate interpretation of this bound, given that the total applied force goes as $2^n \sigma_n$, is that a finite force would suffice to break an infinitely massive cable! There is no contradiction here, as this is a lower bound, but it does indicate that this bound is not particularly useful. We will return shortly to the question of determining bounds.

In order to obtain the scaling properties of σ_n, we must be able to efficiently evaluate Eq. (1). The direct evaluation of $P_n(\sigma)$ in Eq. (1) would require 2^n function evaluations. However, we observe that the only stress values that an individual fiber can experience are the $n+1$ values $\sigma, 2\sigma, \ldots, 2^n\sigma$. Assuming that $P_0(x)$ is computed (and stored) for $x = \sigma$, 2σ, $2^2\sigma$, $\ldots$, $2^n\sigma$, we can employ Eq. (1) to calculate $P_1(x)$ for $x = \sigma$, 2σ, $2^2\sigma$, $\ldots$, $2^{n-1}\sigma$. Again, using Eq. (1), we can tabulate $P_2(x)$ for $x = \sigma$, 2σ, $2^2\sigma, \ldots, 2^{n-2}\sigma$. (It follows that this construction can be regarded as a renormalization.) We continue in this way until we are left with $P_n(\sigma)$. The evaluation of $P_n(\sigma)$ required only $n+1$ evaluations of $P_0(\sigma)$ and only the order of n^2 arithmetic operations. Hence, the calculation of $P_n(\sigma)$ is observed to be optimal in the number of computations taken (since the operations taken are now irreducible) and makes feasible the computation of σ_n for very large n. It is worth observing that this procedure for eliminating redundant computations can be developed for any m, including $m = 4$ which is relevant to earthquake faults (see Smalley et al.[13]). The irreducible character of this computational scheme is reminiscent of the Fast Fourier Transform which, using some simple number theoretic results, is able to reduce the computation of an N point discrete Fourier transform from $O(N^2)$ operations to $O(N \log_2 N)$ operations. (This procedure is described in greater detail by Newman and Gabrielov.[11])

Once we had assembled a large array of σ_n, say for $m = 2$ and for $n = 1, \ldots, 100$, we plotted σ_n against a very large number of functions of n until we found a functional relationship that appears to be satisfied. In this way, we discovered that

$$\sigma_n^{-1} \propto \ln(n) \propto \ln[\ln(\text{mass})] \tag{4}$$

for $m = 2$. This is seen in Figure 2 where we plot σ_n^{-1} against $\log_{10} n$ for a variety of Weibull models ($\rho = 1, \ldots, 5$)—the robust character of this relation offers an important indication that this result is a global one valid for almost any physically reasonable choice of $P_0(\sigma)$. With the insight gained from the renormalized computation, Newman and Gabrielov[11] showed that Eq. (4) provides a bound to the behavior for σ_n and Gabrielov and Newman[6] then showed that Eq. (4) was the asymptotic limit for the problem.

The route to this proof is contained within Eq. (1), namely

$$P_n(\sigma) = P_{n-1}(\sigma)[2P_{n-1}(2\sigma) - P_{n-1}(\sigma)]. \tag{1}$$

If we put the first term inside the bracket to the zero power (i.e., we set it to unity), we recover the weakest link theory which is too strong an inequality or lower bound to be useful. If we put the first term inside the bracket to the second power, we can show (see Newman and Gabrielov[11]) that the inequality or upper bound evolves in the wrong direction! The resolution of this problem lies in keeping the first term in the bracket unchanged and ignoring the second term—this shows that the lowest increment of stress to be supported by a surviving fiber dominates the asymptotics including larger values of m. (It is easy to create an inequality, but not all inequalities are created equal.) We found, for general m, that

$$\sigma_n^{-\nu} \propto \ln(n) \propto \ln[\ln(\text{mass})] \tag{5}$$

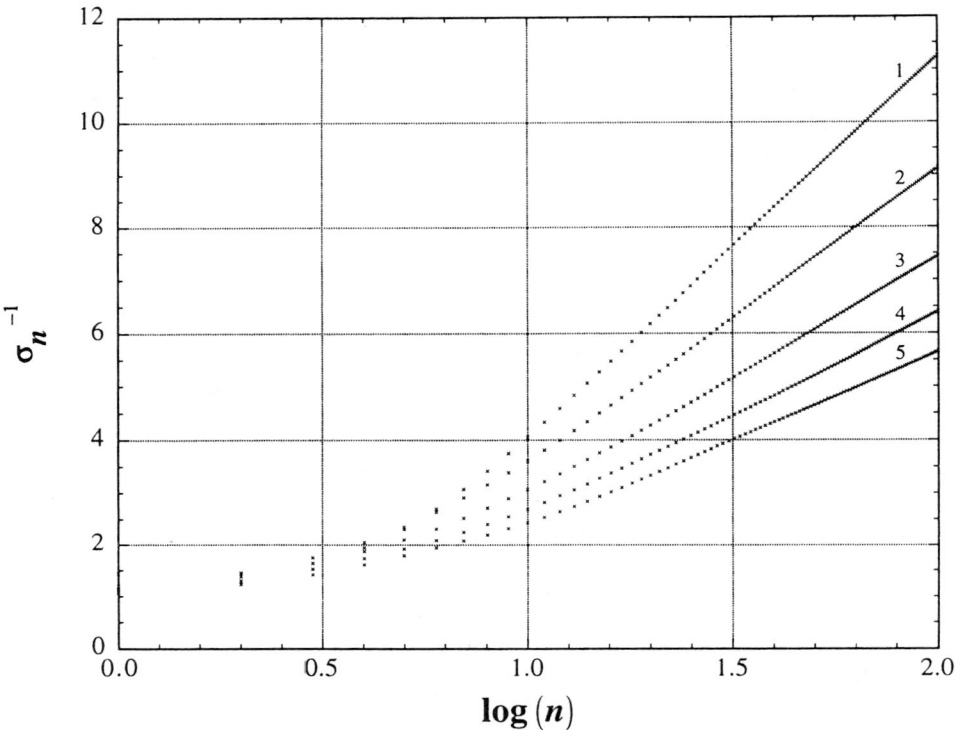

FIGURE 2 Reciprocal of the failure threshold, i.e., σ_n^{-1}, plotted against the logarithm (base 10) of the order index n for Weibull distribution functions with shape parameter $\rho = 1, \ldots, 5$. Coordination number $m = 2$.

where

$$\nu = \frac{\ln 2}{\ln (m/m - 1)}. \tag{6}$$

Figure 3 displays $\sigma_n^{-\nu}$ appropriate to $m = 4$ for over two decades of n—this corresponds to $4^{100} \approx 10^{60}$ fibers!

Thus, our use of renormalization methods and conversion to irreducible form rendered the computational problem both tractable and efficient, and facilitated the determination of the proper pattern of development and scaling for the failure of hierarchical fiber bundles. Based on this insight, we were able to prove analytically[6] that the asymptotic property Eq. (5) is in fact *universal*, valid under very weak restrictions imposed on the original distribution $P_0(\sigma)$. The proof, however, was

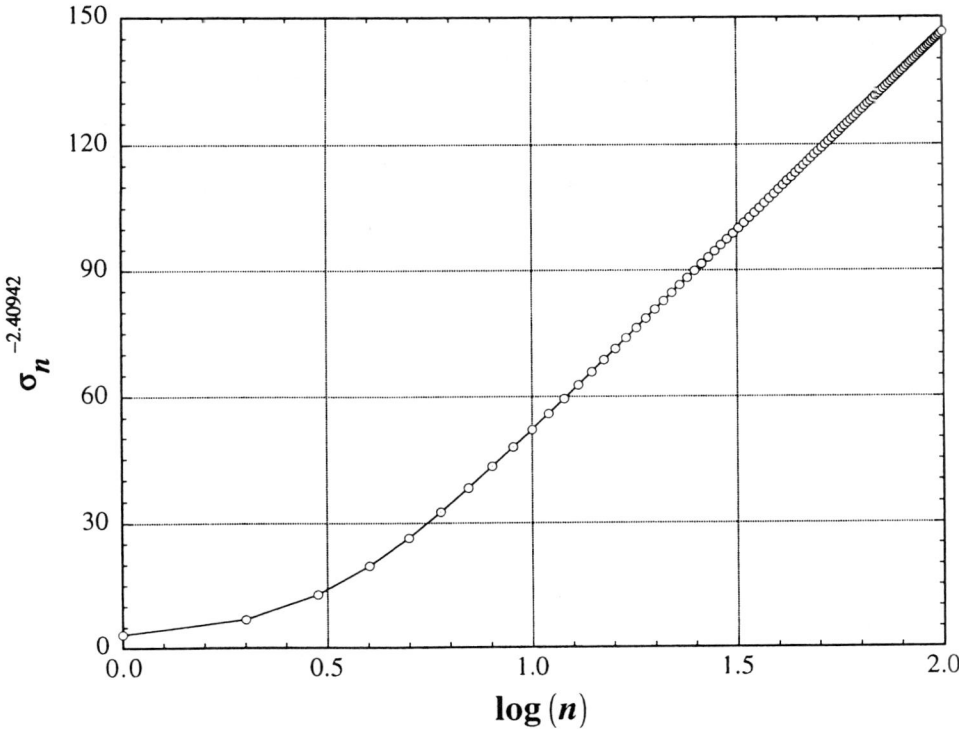

FIGURE 3 The quantity $\sigma_n^{-2.40942}$ plotted against the logarithm (base 10) of the order index n for Weibull exponential distribution functions (shape parameter $\rho = 1$). Coordination number $m = 4$.

difficult and was based on new mathematical techniques. The implication to be drawn from the static case is that, although no critical point for failure exists, the asymptotic rate of decrease of bundle strength is extraordinarily slow—a fractional power of a logarithm of a logarithm. Given the intuitive connection that has developed between static and dynamic problems of this sort, we now turn our attention to the time-dependent problem in order to understand how fiber bundles under a given load fail as a function of time.

3. DYNAMICS

In the time-dependent setting, we assume that fibers are sampled independently from a common source which imparts random physical properties to them. Moreover, we assume that all fibers must break, given sufficient time, and that the fundamental determinant of failure is the integral of time multiplied by some measure of the stress-dependent rate of failure. A single fiber subjected to a *known* load history $\sigma(t')$, $t' \geq 0$ has a random failure time t with a distribution function of the form

$$P[t|\sigma(\cdot)] = 1 - \exp\left\{-\Psi[\int_0^t \kappa\,[\sigma(t')]dt']\right\}, \, t \geq 0. \tag{7}$$

The quantity $\Psi(x)$, $x \geq 0$ is called the *hazard* or *shape function* and basically defines the distribution function in terms of the integrated time. The time integral introduces a hazard rate $\kappa(\sigma)$, $\sigma \geq 0$, which is generally referred to as the *breakdown rule* in terms of the instantaneous stress level σ. Up to this point, our description of the distribution is quite general so long as Ψ is nondecreasing, and both Ψ and κ are nonnegative functions.

Typical examples include the *Weibull shape function*

$$\Psi(x) = \lambda x^\beta, \, x \geq 0, \tag{8}$$

where β and λ are positive constants—this empirical shape function is in good agreement with observations and mimics the static Weibull distribution. (We will, for demonstration purposes, employ the simple exponential shape function where both λ and β are presumed to be unity.) Also, we assume two special forms for the breakdown rule $\kappa(\sigma)$, namely

$$\kappa_e(\sigma) = \phi\exp(\eta\sigma), \, \sigma \geq 0, \tag{9}$$

known as the *exponential breakdown rule* and

$$\kappa_p(\sigma) = \mu\sigma^\rho, \quad \sigma \geq 0, \tag{10}$$

known as the *power-law breakdown rule* where ϕ, η, μ, and ρ are all positive constants.

These forms were initially proposed by Coleman[3] largely on phenomenological grounds and have since accumulated further theoretical and experimental support. The integral form can be thought of as a representation of damage accumulation within the fibers. The two breakdown rules have been associated also with molecular breakdown events. For the exponential breakdown rule, a common interpretation is that thermodynamic considerations require that the hazard rate or breakdown rule contain a Boltzmann factor in the stress. Accordingly, the power-law rule may be regarded as a local approximation to the latter. It is important to note that the power-law breakdown rule is scale free while the exponential case is not—hence, we might expect that the exponential case will not show any simple scaling while the power-law case may have some special scaling properties.

In the case of hierarchical fiber bundles, the essential feature that we must confront is that the time history can be very complicated. Indeed, in a fiber bundle with n levels and coordination number m, there are $(m^n)!$ routes to failure. The general combinatoric problem, even for $m = 2$, appears at first glance to be impossibly difficult—however, we believe there are some grounds for optimism and we are pursuing further this investigation. As in the static case, we want to investigate this problem numerically. Owing to the complicated Bayesian structure of the problem, we choose a Monte Carlo simulation as our first means of investigation. Suppose we select according to some distribution function, say for $m = 2$, a set of 2^n fiber lifetimes. For convenience, we will assume that we have used the power-law breakdown rule $\kappa_p(\sigma) \propto \sigma^\rho$ for, say, $\rho = 2$—this value of ρ is typical of many composite materials. We will use this example to develop a methodology and then a simulation for exploring fiber bundle failure.

Consider now the simplest situation where $m = 2$ and $n = 1$ and, for this realization of the process, the respective lifetimes of the two fibers are t_1 and t_2 where we select $t_1 < t_2$. Suppose T defines the lifetime of the bundle. We can show that

$$T = t_1 + \alpha(t_2 - t_1) = (1 - \alpha)t_1 + \alpha t_2, \tag{11}$$

where $\alpha \equiv 2^{-\rho}$. This equation has the simple explanation that, after the first fiber fails at t_1, the survivor's remaining lifetime $t_2 - t_1$ is reduced by a factor $\alpha < 1$. (For $n > 1$, we can show for the power-law breakdown rule that α is independent of the level in the hierarchy—this is not true in the case of the exponential breakdown rule.) We observe that this latter equation defines a contraction map, offering a hint that the width of the distribution function could be asymptotically decreasing. Moreover, for $\alpha < 1/2$, we note that $T < (t_1 + t_2)/2$—this suggests that the mean of the distribution undergoes a systematic decrease. These features are qualitatively like those obtained in the static problem.

When we look at the general problem of fiber bundle failure for $m = 2$, we see that the phenomenal number of routes to failure obviates the development of a simple simulation procedure. However, "renormalization thinking" does give rise to a simple scheme. In the dynamic case, just as in the static one, a given fiber can experience stresses of $\sigma, 2\sigma, \ldots, 2^n\sigma$. Accordingly, the time history of that fiber is defined completely by the times wherein its load was doubled. With this fact in mind, consider an arbitrary bundle at some arbitrary level in the hierarchy, and assume that this bundle is composed of two sub-bundles identified as a and b. Further, assume that this bundle has had a time history where its external load doubled at times $t_1, t_2, \ldots, t_q$. Thus, we wish to determine the time to failure which we denote by $T_{a+b}(t_1, t_2, \ldots, t_q)$. Suppose we know both $T_a(t_1, t_2, \ldots, t_q)$ and $T_b(t_1, t_2, \ldots, t_q)$. Without loss of generality, we assume that $T_a(t_1, t_2, \ldots, t_q) < T_b(t_1, t_2, \ldots, t_q)$. Then it follows that $T_{a+b}(t_1, t_2, \ldots, t_q)$ is just $T_b[t_1, t_2, \ldots, t_q, T_a(t_1, t_2, \ldots, t_q)]$. This identity can now be converted into an efficient recursive algorithm for computing the time to failure of any bundle given that we know how to compute the lifetime of an individual fiber according to its life history. One can readily show that the number of operations required goes as the total number of elements, say $N = 2^n$ to the power $\ln 3/\ln 2 \approx 1.5849625$. This is in marked contrast with any existing scheme for pursuing this calculation (based upon table look-up for failed fibers) which requires a minimum of $O(N^2)$ operations.

As an illustration of what happens, consider selecting the lifetimes of $2^{15} = 32,768$ fibers according to a Weibull exponential shape function (i.e., $\beta = 1$ and a power-law breakdown rule where $\mu = 1$ and $\rho = 2$. (These parameters are selected,

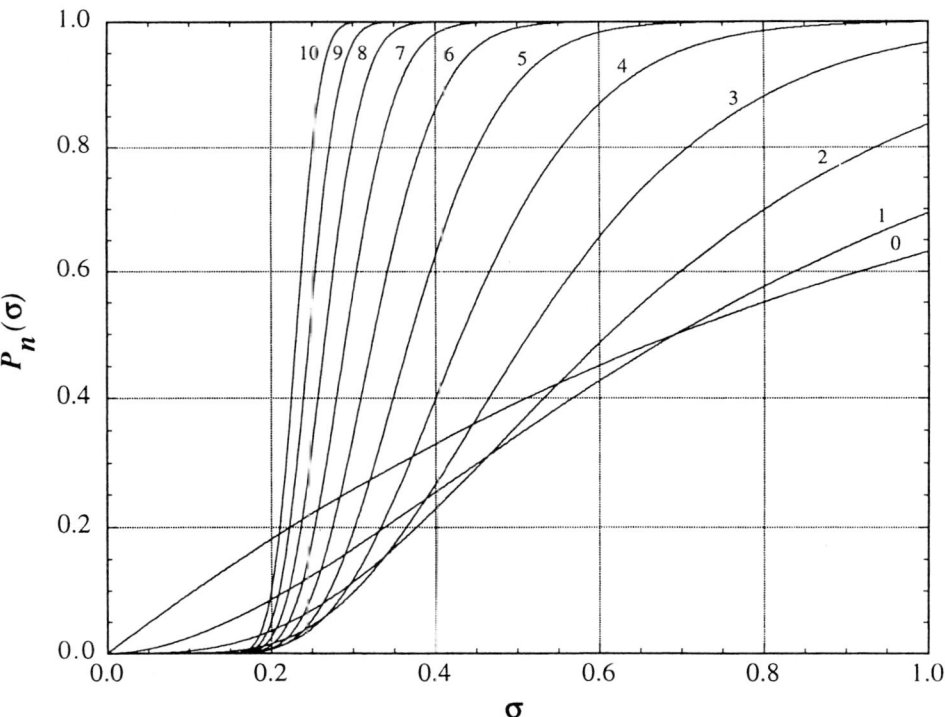

FIGURE 4   Cumulative distribution function $P_n[t; \sigma(\cdot)]$, $n = 0, 1, \ldots, 10$, for a Monte Carlo realization of fiber bundle failure for $m = 2$ with $32,768$ points of the Weibull exponential shape function and power-law breakdown rule $P_0(t|\sigma \cdot) = 1 - \exp\{-\int_0^t \kappa_p[\sigma(t')]dt'\}$, $t \geq 0$ where $\kappa_p(\sigma)$ is the breakdown rule which varies as σ^ρ where $\rho = 2$.

in part, to facilitate comparison with the static case described in Figure 1.) In Figure 4, we illustrate what happens in this situation—we observe a shortlived drop in the median of the distribution (as n increases) followed by a rapid (possibly exponentially fast) reduction in the variance in the distribution. There is some similarity here with the static situation, but an important difference lies in the apparent "critical point" which has now appeared.

In order to develop this problem analytically, several vital questions emerge. First, can we find any essential simplification for this problem? In the general case, the effect of failure events is not commutative, as the stress prevailing at different times can alter the rate of failures in a scale dependent way. Second, can we somehow exploit the contraction properties of the map Eq. (11) to conclusively *prove* that a critical point will emerge? These and related questions will be the subjects of further investigations.

---

4. CONCLUSION

In this chapter, we have described how hierarchically organized models of fiber bundles could provide a useful description for the failure of lattice structures where the stress released by failed units would be redistributed onto regions of comparable size. We considered geometries where the coordination number m described m fibers being grouped together to form a larger bundle, which would then be placed in a group of m similar bundles, and so on. The case $m = 2$ describes a planar tape in the fiber bundle literature, while the case $m = 4$ can readily be associated with a two-dimensional lattice. We employed n to describe the location in the hierarchy, where $n = 0$ describes the root level, $n = 1$ the first combining of fibers, and so on. The failure rule at a given level in the hierarchy is that, whenever a bundle fails, its load is redistributed uniformly among the surviving fibers in that bundle—when all fibers in a given bundle fail, then their load is transferred to a neighboring bundle at the same hierarchical level.

We briefly derived the combinatoric functional mapping which describes the (cumulative) distribution function P_n on increasingly larger spatial scales, established by n, both as a function of the load (per fiber) σ for the static case and the time t for the dynamic case. In the static case, we developed a mechanism for eliminating redundant calculations and, thereby, facilitated the calculation of $P_n(\sigma)$ for large n. Accordingly, we computed the variation of σ_n for $m = 2$ defined by $P_n(\sigma_n) = 1/2$ and observed that $\sigma_n^{-1} \propto \ln[n] \propto \ln[\ln(\text{mass})]$. With this insight, we then developed analytic general bounding and universal asymptotic results for the scaling according to m. In the dynamic case, we developed a mechanism for eliminating redundant calculations and, thereby, facilitated the Monte Carlo simulation of $P_n[t; \sigma(\cdot)]$ for a significant range of n. We observed that the loading history typically manifested in one of two standard representations, one where the "failure rate" varied exponentially in stress (mimicking a thermodynamic dependence which we have not provided here) and one where the rate varied as a power law in stress. In both instances, we observed the development of an apparent critical point with respect to time in the distribution function.

In both the static and dynamic case, we observed that the effective strength of large bundles would systematically decrease but that the rate of decrease would be extraordinarily slow. In the static case, the width of the distribution would remain relatively large while, in the dynamic case, the width of the distribution would shrink to vanishingly small size. The emergence of relatively well defined loads or stresses for the failure of such systems suggests that local load sharing can result in situations where the stress or the time to failure are well established. One important application of this pattern is to the earthquake prediction problem (e.g., $m = 4$) where geophysical folklore has suggested for many years that earthquakes might be predictable based on prevailing stress levels or the time elapsed since a previous earthquake. In engineering applications, the implication here is that, when samples of a product such as a cable having this hierarchical structure begin to fail, we might expect an avalanche of failures in similarly constructed objects. Another important engineering application emerges in situations where a composite material obeys an exponential breakdown rule Eq. (10) with an index $\rho > 1$. In particular, the mean time to failure for the bundle is always less than the mean time to failure for the individual fiber element, consistent with interpretations of lifetime data for fibers in composites. From the standpoint of reliaility, greater caution must be placed on predicting the lifetime of such assemblies. With the insight gained from the Monte Carlo simulations of the time to failure, we will address in a future manuscript the analytic derivation of the emergence of a critical point for failure.

ACKNOWLEDGMENTS

W. I. N. and A. M. G. wish to thank the Center for Nonlinear Studies at Los Alamos National Laboratory where this manuscript was written and several new theorems emerged for its hospitality. W. I. N. wishes to thank the Department of Astronomy at Cornell University for its hospitality while this work was initiated and notes that this research was partly supported by the U.S. Army Research Office through the Mathematical Sciences Institute of Cornell University. A. M. G. wishes to thank the Department of Geological Sciences at Cornell University for its hospitality; his contribution was supported under NSF grant number EAR-91-04624.

REFERENCES

1. Allègre, C. J., J. L. LeMouel, and A. Provost. "Scaling Rules in Rock Fracture and Possible Implications for Earthquake Prediction." *Nature,* **297** (1982) 47–49.
2. Coleman, B. D. "A Stochastic Process Model for Mechanical Breakdown." *Trans. Soc. Rheology* **1** (1957): 862–866.
3. Coleman, B. D. "Statistics and Time-Dependence of Mechanical Breakdown in Fibers." *J. App. Phys.* **2** (1958): 968–983.
4. Daniels, H. E. "The Statistical Theory of Strength of Bundles of Threads." *Proc. Roy. Soc.* Series A **183** (1945): 404–435.
5. Duxbury, P. M. "Breakdown of Diluted and Hierarchical Systems." In *Statistical Models for the Fracture of Disordered Media,* edited by H. Hermann and S. Roux, 189–228. Amsterdam: North Holland, 1990.
6. Gabrielov, A. M., and W. I. Newman. "Failure of Hierarchical Distributions of Fiber Bundles. I." Submitted to *J. App. Prob.,* 1991.
7. Galambos, J. *The Asymptotic Theory of Extreme Order Statistics.* New York: Wiley, 1978.
8. Harlow, D. G., and S. L. Phoenix. "Approximations for the Strength Distribution and Size Effect in an Idealized Lattice Model of Material Breakdown." *J. Mech. Phys. Solids* **39** (1991): 173–200.
9. Kanninen, M. F., and C. H. Popelar. *Advanced Fracture Mechanics.* New York: Oxford University Press, 1985.
10. Landau, L. D., and E. M. Lifshitz. *Theory of Elasticity.* Oxford: Pergamon Press, 1970.
11. Newman, W. I., and A. M. Gabrielov. "Failure of Hierarchical Distributions of Fiber Bundles. II." *Int. J. Fracture* **50** (1991): 1–14.
12. Newman, W. I., and L. Knopoff. "Scale Invariance in Brittle Fracture and the Dynamics of Crack Fusion." *Int. J. Fracture* **43** (1990): 19–24.
13. Smalley, R. F., D. L. Turcotte, and S. A. Solla. "A Renormalization Group Approach to the Stick-Slip Behavior of Faults." *J. Geophys. Res.* **90** (1985): 1894–1900.
14. Smith, R .L., and S. L. Phoenix. "Asymptotic Distributions for the Failure of Fibrous Materials Under Series-Parallel Structure and Equal Load-Sharing." *J. App. Mech.* **103** (1981): 75–82.

Chapter 8
Pattern Formation in Biological Systems

T. Vicsek†‡ A. Czirók,† O. Shochet,* and E. Ben-Jacob*
†Department of Atomic Physics, Eötvös University, Budapest, Puskin u 5-7, 1088 Hungary
‡Institute for Technical Physics, Budapest, P.O.B. 76, 1325 Hungary
*School of Physics, Tel-Aviv University, 69978 Tel-Aviv, Israel

Self-Affine Roughening of Bacteria Colony Surfaces

The surface structure of bacteria colonies grown on nutrient-rich agar is investigated. We find that the surface exhibits anomalous with an exponent $\alpha \simeq 0.85$ over a limited length scale. Our results are interpreted in terms of a stochastic differential equation describing the roughening of interfaces moving in disordered media. The possibility of physically motivated analogies such as the motion of wetting fluid fronts is discussed.

1. INTRODUCTION

The motion of randomly perturbed interfaces typically leads to very complex shapes. If the conditions of a growth process are such that the development of the interface is not unstable, and the fluctuations play a relevant role, the resulting structure has a rough surface which can be well described in terms of self-affine fractals.[4,5,10] In such phenomena, the fluctuations (which usually die out quickly for a stable surface and grow exponentially for an unstable one) survive without drastic changes for a long time leading to surfaces with spatial scaling over many length scales.

The roughening of growing surfaces can successfully be described in terms of *dynamic scaling* based on the general concepts of scale invariance and fractals. Let us consider the $(1 + 1)$-dimensional case corresponding to our experiments to be described later. The growing surface typically can be described by a single-valued function $h(r, t)$ that gives the height (distance) of the interface at position r (along the substrate) at time t measured from the original $d = 1$-dimensional flat surface. We shall concentrate on a part of the surface extending over an interval of length x. During growth, the interface heights fluctuate about their average value $\bar{h}(t) \sim t$ and the extent of these fluctuations characterizes the width or the thickness of the surface. The root mean-square of the height fluctuations $w(x, t)$ over the interval x is a quantitative measure of the surface width and is defined by

$$w(x, t) = [\langle h^2(r, t)\rangle_r - \langle h(r, t)\rangle_r^2]^{1/2}. \tag{1}$$

It has been shown[4,5,10] that in the roughening processes, the dynamic scaling of the *average* surface width has the form

$$w(x, t) = x^\alpha f(t/x^{\alpha/\beta}), \tag{2}$$

where $\alpha/\beta = z$ is the dynamic scaling exponent. In the limit where the argument of the scaling function $f(y)$ is small, $y \ll 1$, the width depends only on t and this implies that for small y the scaling function $f(y)$ has to be of the form $f(y) \sim y^\beta$. Saturation at large times means that $f(y)$ goes to a constant in that limit. From the above it follows that for long times

$$w(x, t \to \infty) \sim x^\alpha, \tag{3}$$

while for fixed x and short times

$$w(x, t) \sim t^\beta. \tag{4}$$

We shall use Eq. (3) to determine the roughness exponent α for bacteria colonies.

Our study of self-affine fractal growth of bacteria colonies was motivated by the rapidly increasing interest in the fractal growth of biological objects (see the paper by Matsushita et al.[7] in this volume and references therein). In particular, it was demonstrated by Matsushita et al.[7] that in the limit when the growth was dominated by the diffusion of the nutrient, the colonies developed a self-similar shape corresponding to the geometry of diffusion-limited aggregates.[11] The reason for self-similarity in these cases is an instability: the bacteria in the most advanced parts of the surface use up most of the nutrient. As a consequence, large empty regions appear within the structure where the nutrient concentration is too small for growth. If the supply of nutrient is not limited, it is the stochastic sequence of the spontaneous multiplication of cells which is expected to play a major role in the development of the surface of a colony.

One of the main points of the present work is to point out that under such conditions the behavior of the colony surface can be described in terms of equations and scaling arguments which are analogous to those widely used for the analysis of physical systems exhibiting kinetic roughening.

2. EXPERIMENTS AND RESULTS

We inoculated plates consisting nutrient-rich agar (10 g/l pepton, 2% agar) that was prepared according to standard methods. All colonies were inoculated from a single culture of a *Bacillus subtilis* strain No 168/Tb2. This strain has previously been used in an extensive series of experiments on morphological transitions during bacteria colony growth.[1] The inoculation was made along a straight-line segment. After this process the dishes were sealed to prevent the agar from drying out.

The growth of bacteria colonies is known to be a rather complicated process (see, e.g., Shapiro[8] and Matsushita et al.[7] and references therein). Under the conditions of our experiment the bacteria could move relatively freely on the agar surface. Under the microscope we could observe their migration ("swimming") which was similar to a relatively random, Brownian-type diffusional motion. The bacteria tended to stay together, forming a smooth surface layer of less active bacteria at the very edge of the colony. This situation is in analogy with the behavior of liquids, with (i) the randomly moving bacteria playing the role of molecules and (ii) their tendency of forming a surface layer corresponding to surface tension. We shall utilize this analogy in our interpretation of the phenomenon. In a previous publication[9] we used dry agar (the bacteria did not swim); here we decided to investigate the situation when the similarity with fluid systems is apparent.

After one day, we took pictures of the colonies (after tinting). The pictures were digitized with a resolution of 3000 × 1500, using a scanner. The surface of the patterns was defined as the set of points that were at the highest position (measured from the line of inoculation) in each column of the array of occupied pixels representing the digitized image of the colonies. A part of a representative colony is shown in Figure 1.

FIGURE 1   Part of a typical colony. We analysed the surface of the patterns which was defined as the set of points belonging to the structure and being at the highest position (measured from the line of inoculation)

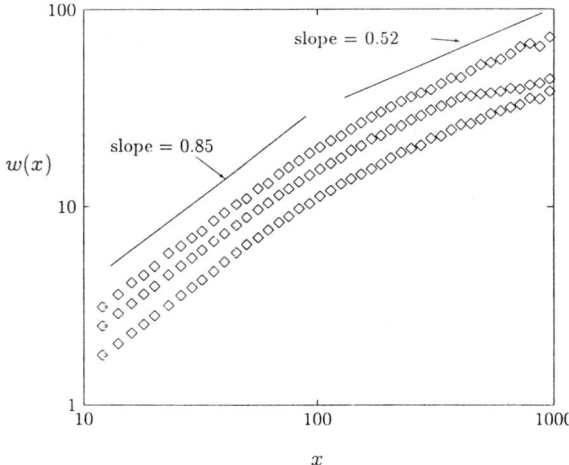

FIGURE 2   The scaling of the average surface width w as a function of the linear size x of the interval over which w was determined.

The roughness exponent α was calculated from the scaling of the width $w(x,t) \sim x^\alpha$. An averaging over many diffrent parts of linear extension x was made to obtain a better estimate of w. In Figure 2, we show how w depends on x for three different colonies. The plots have two major regions. The first part exhibits scaling with an exponent about 0.85 (anomalous value), while the exponent corresponding to the second part is close to the so called universal value 0.5. Both scalings hold over limited length scales only.

3. DISCUSSION

Based on our observations described above, in this section we shall interpret the growing bacteria colony as a "physical system" and use a standard approach to its description. A common method for studying surface roughening in physical systems is the investigation of the corresponding *stochastic differential equations*. In particular, the Kardar-Parisi-Zhang (KPZ) equation[6] has been shown to be very useful in the description of interface dynamics in the presence of temporally uncorrelated (shot) noise. However, we think that the surface roughening in our case is dominated by the irregularities of the agar, therefore, here we have to consider the quenched versions of the KPZ equation (in $1 + 1$ dimensions),

$$\frac{\partial h}{\partial t} = \nu \nabla^2 h + \frac{\lambda}{2}(\nabla h)^2 + v + \eta(r,h),$$
(5)

where ν and λ are constants. In Eq. (5) $\langle \eta(r,h) \rangle = 0$ represents uncorrelated quenched noise with a Gaussian distribution of amplitudes and a correlator which in the continuum limit is formally given by

$$\langle \eta(r_0,h_0)\eta(r_0 + r, h_0 + h') \rangle = D\Delta(h')\delta(r),$$
(6)

where D is a constant and $\Delta(h')$ is only a slowly varying function of h'. It is possible to obtain the exponents α from Eqs. (5) and (6) for the case of d dimensional substrates. Here we give a considerably simpler version of the derivation given by Csahok et al.[3]

From dynamic scaling it follows that the change Δh over an interval Δr is proportional to Δr^α. The scaling of the noise term with Δr can be written in the form $\Delta r^{-d/2}$, since the integral of the correlations $\langle \eta\eta \rangle$ over the interval Δr is normalized to 1. In the $\Delta r \to 0$ limit the relevant terms in the equation should scale the same way. Equating the $(\nabla h)^2$ and the η terms, we get

$$\frac{(\Delta h)^2}{(\Delta r)^2} = \Delta r^{-d/2}.$$
(7)

From the condition $\Delta h \sim \Delta r^\alpha$ we immediately obtain

$$\alpha = \frac{4-d}{4}. \tag{8}$$

For $d = 1$, Eq. (8) predicts $\alpha = 0.75$. This value is somewhat smaller than the experimentally observed value. There are several possible reasons for the discrepancy: (i) for a more complete description further relevant factors should be taken into account, and (ii) the experimental value is upward biased by the small sample size and the digitization procedure.

An alternative way of interpreting our results is to construct a growth model that simulates the development of driven fluid interfaces in inhomogeneous media.[2] This approach leads to α close to the above theoretical value.

In conclusion, we have investigated the roughening of bacteria colony surfaces experimentally, and gave a physically motivated theoretical picture explaining the observed high roughness exponent.

ACKNOWLEDGMENTS

The present research was supported in part by the Hungarian Research Foundation Grant No. T4439; by a grant from the G.I.F., the German-Israeli Foundation for Scientific Research and Development; and by the Program for Alternative Thinking at Tel-Aviv University.

REFERENCES

1. Ben-Jacob E., H. Shmueli, O. Shochet, and A. Tenenbaum. "Adaptive Self-Organization During Growth of Bacterial Colonies." *Physica* **A187** (1992): 378
2. Buldyrev S., A-L. Barabási, F. Caserta, S. Havlin, H. E. Stanley, and T. Vicsek. "Anomalous Interface Roughening in Porous Media: Experiment and Model." *Phys. Rev. A* **45** (1992): R8313
3. Csahók Z., K. Honda, and T. Vicsek. "Dynamics of Surface Roughening in Disordered Media." *J. Phys. A* **26** (1993): L171
4. Family F., and T. Vicsek, eds. *Dynamics of Fractal Surfaces*. Singapore: World Scientific, 1991
5. Jullien, R., J. Kertész, P. Meakin, and D. Wolf, eds. *Surface Disordering*. New York: Nova Science, 1992

6. Kardar M., G. Parisi, and Y.-C. Zhang. "Dynamic Scaling of Growing Interfaces." *Phys. Rev. Lett.* **56** (1986): 889
7. Matsushita, M., J.-I. Wakita, and T. Matsuyama. "Growth and Morphological Changes of Bacterial Colonies." This volume.
8. Shapiro, J. A. "Growth Patterns of Bacteria Colonies Observed by Electronmicroscopy." *J. Bacteriol.* **169** (1987): 142–156
9. Vicsek, T., M. Cserzö, and V. Horváth. "Self-Affine Growth of Bacterial Colonies." *Physica A* **167** (1990): 315
10. Vicsek T. *Fractal Growth Phenomena.* Singapore: World Scientific, 1992
11. Witten, T. A. and L. M. Sander. "Diffusion Limited Aggregation: A Kinetic Critical Phenomenon." *Phys. Rev. Lett.* **47** (1981): 1400

Mitsugu Matsushita,† Jun-ichi Wakita,† and Tohey Matsuyama‡
†Department of Physics, Chuo University, Jasuga, Bunkyo-ku, Tokyo 112, Japan
‡Department of Bacteriology, Niigata University, School of Medicine, Asahimachidori,
Niigata 951, Japan

Growth and Morphological Changes of Bacteria Colonies

Colonies of bacterial species called *Bacillus subtilis* have been grown two-dimensionally on the surface of thin agar plates. It was confirmed that, for hard agar medium with poor nutrient quality, colonies grow self-similarly through diffusion-limited processes in a nutrient concentration field. It was also found that the colony patterns change drastically when varying environmental conditions such as concentrations of nutrient C_n and agar C_a in the agar medium. These patterns are classified into five types: DLA-like, Eden-like, DBM-like, intermediate spreading, and homogeneously spreading. The mechanism of their growth and morphological change is discussed.

1. INTRODUCTION

Pattern formation in biological systems is usually believed to be much more complicated than that in physical and chemical systems. Biological phenomena take place through complicated intertwinement between inherently complex biological factors and environmental (physicochemical) conditions. Setting morphogenesis of individual organisms aside, however, pattern formation of populations of simple biological objects may, in some cases, be dominated by purely physical conditions. Let us here pay attention to the colony formation of bacteria, which are one of the simplest biological objects.

Typical bacteria such as *Escherichia (E.) coli* and *Salmonella (S.) typhimurium* are single-cell organisms. Even a very small number of bacteria (parent cells), once they are inoculated on the surface of an appropriate medium such as semi-solid agar with enough nutrient and are incubated for a while, repeat growth and cell division many times. Eventually the cell number of the progeny bacteria becomes huge, and they form a visible colony. The colony differs in size, form, and color according to bacterial species.[14] It also changes its form sensitively with the variation of environmental conditions. The study of bacterial colonies seems to lead us into the zoo of growing random patterns. For instance, Matsuyama et al.[7] demonstrated that bacteria called *Serratia (S.) marcescens* exerts fractal morphogenesis in the process of spreading growth on an agar surface.

Here we discuss our recent experimental study[4,5,8,12] on a part of this rich-in-variety bacterial colony formation from the viewpoint of the physics of random pattern formation.[1,2,15] In order to study the morphological change due to environmental conditions in the bacterial colony formation, we varied only two parameters: concentrations of nutrient (peptone) C_n and agar C_a in a thin agar plate as the incubation medium. Other parameters such as temperature (35°C) were kept constant. Also, throughout the present experiments, we used only one bacterial species called *Bacillus (B.) subtilis*. Even the strain (OG-01) has been fixed throughout our experiments, unless otherwise stated. This strain is rod-shaped with flagella, and motile in water by collectively rotating these flagella. Otherwise, the experimental procedures are standard and have been described elsewhere.[4,5,8,12]

2. DLA GROWTH OF COLONIES

Present bacterial colonies grow two-dimensionally on an agar surface because the average pore size of the agar-gel network is smaller than the size of bacteria used. Bacterial cells cannot grow or move into the agar medium.

The agar concentration C_a was first fixed at some value in the range of 10 to 15 g/l. Since resultant agar plates are rather hard, bacterial cells cannot move around on the surface of the plates either. They only grow and perform cell division locally by feeding on the nutrient peptone. Under this condition, we observed the effect of nutrient on the colony growth.[4,5] Colonies had been incubated for three weeks after inoculation at various initial nutrient concentrations C_n in the range of 0 to 1 g/l. An increase of C_n within the range was found to enhance the colony growth. In particular, the growth does not occur without nutrient ($C_n = 0$). This implies that the bacterial growth and cell division at the interface of colonies (local growth processes) are governed primarily by the presence of nutrient where they live.

Now we fixed the initial nutrient concentration at $C_n = 1$ g/l. Note that this concentration is much less than usually used, e.g., 15 g/l. We then observed colony patterns with characteristically branched structure after three- or four-week incubation. They are clearly reminiscent of two-dimensional DLA clusters. The DLA (diffusion-limited aggregation) model was first proposed by Witten and Sander[17] as a prototype model for randomly branched patterns grown through diffusion-limited processes.[10] Averaged over about 25 samples of colony patterns from the same strain, we obtained a fractal dimension of $D = 1.72 \pm 0.02$. This is in good agreement with that of two-dimensional DLA clusters.

In addition, we observed clear evidence for the existence of the screening effect in which protruding exterior branches of a colony grow more and prevent interior ones from growing. We also observed repulsion behavior of two neighboring colonies inoculated at two points simultaneously. These behaviors are all characteristic for the pattern formation in a Laplacian field.

Taking all these experimental results into account, it is now clear that the present bacterial colonies grow through DLA processes. There still remains one more problem: what makes the Laplacian field, or what diffuses it? There are two main possibilities: (1) nutrient diffuses in toward a colony, or (2) some waste material excreted by bacteria themselves diffuses out from the colony. Both cases are equally likely to produce the DLA-like pattern, although the effect of nutrient concentration described above strongly supports the first one. In order to determine which possibility takes place mainly, we put the nutrient at a corner of an incubation dish (elsewhere no nutrient initially), and inoculated bacteria at the center. It should be remembered that, in the present case, bacterial cells cannot move around on the surface of an agar plate. Nevertheless, the colony showed a clear tendency to grow toward the area where the nutrient was initially prepared. This result inevitably leads us to support the first possibility. We can, therefore, conclude that for hard agar medium with poor nutrient the present bacterial colonies grow through the *DLA* mechanism in the *nutrient* concentration field.[8]

We have also confirmed the DLA-like colony morphology for common rod-shaped bacteria, *E. coli* and *S. typhimurium*, under similar conditions.[9] This implies that the DLA growth is quite universal in the formation of bacterial colonies.

3. MORPHOLOGICAL CHANGE IN COLONY FORMATION

Let us next discuss the morphological change of colonies under varying environmental conditions. Colony patterns were found to change drastically when varying the concentrations of nutrient C_n and agar C_a. In Figure 1 we show the morphological phase diagram of colonies observed in the present experiment. The abscissa indicates the inverse of agar concentration, which implies the softness of agar medium. Note also that the ordinate is represented by a logarithmic scale. Namely, agar plates become softer as the diagram is followed from left to right, while they become richer in nutrient from bottom to top.

Here we classify colony patterns into five types, each of which was observed in the regions labeled A–E in Figure 1, respectively. Typical patterns observed in these regions are also included in the figure.

In region A, i.e., at low C_n (poor nutrient medium) and high C_a (hard agar plates), we obtained DLA-like colony patterns, as described in detail in the previous section.

Moving from region A to B, i.e., increasing C_n with C_a more or less fixed, colony patterns showed gradual crossover from DLA-like to Eden-like[3] at high C_n (rich nutrient medium). Branches of a colony became gradually thicker as C_n was increased. Eventually they fused together to form a compact pattern. However, an outwardly growing interface looks still rough. Moreover, two colonies inoculated simultaneously at two points grew and came close together in region B, in contrast to region A.[4,5] This implies that the diffusion length ℓ of the nutrient is very small. These properties are characteristic of Eden growth. In order to confirm real Eden growth, however, one has to show that the growing interface is self-affine and obeys characteristic dynamic scaling.[3,16]

In the wide region D with low C_a (soft agar medium), colonies spread homogeneously with smooth, clear-cut, and circular interface and no branching at all. They grew very fast and appeared to be transparent. The growth mechanism will be discussed from the population dynamics viewpoint in the final section.

In the narrow region E between A and D where the nutrient is poor and the medium softness is intermediate, the colony morphology took patterns clearly reminiscent of the so-called dense-branching morphology (DBM).[6,13] In this region, colonies branch densely, while the advancing envelope looks characteristically smooth, compared with DLA-like colonies. In this sense, the colony patterns that we see in region D resemble dense branches fused together to form homogeneously spreading patterns. Up to now, there have been no models that describe the DBM formation unambiguously, although DBM-like patterns have been observed in various systems.[2,6,13] We observed, however, that two DBM-like colonies inoculated at two points simultaneously grew, came close together, and then repelled each other with almost constant gap distance approximately equal to the average distance of neighboring branches.[4,5] This implies that DBM-like patterns seen in region E cannot be described by the diffusion field of nutrient alone. We must look for an additional mechanism that smoothes out the envelope of a growing branched pattern or makes the diffusion length ℓ almost constant. The candidate may be the variation of agar surface tension along the envelope due to some surfactant excreted by bacteria.

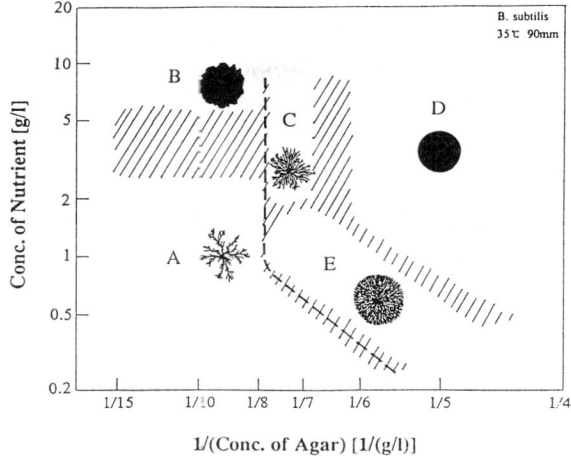

FIGURE 1 Phase diagram of morphological change in colonies of *B. subtilis*.

In region C the morphology took something intermediate. Colonies grew with tip-splitting and narrow branching just as DBM. However, the advancing envelope was rather rough as DLA-like colonies.

4. EFFECT OF BACTERIAL CELL MOVEMENT

We notice in Figure 1 that the morphological change is conspicuous from region A (DLA-like patterns) to region E (DBM-like), compared with gradual change from region A to region B (Eden-like). The growth rate in region E is also much higher than A. It took a colony about a month in region A and a week in region B to grow to the size of about 5 cm (half the diameter of a Petri disk containing agar medium), while it took only a day in regions C and E and half a day in region D. This suggests the essential difference in the way of colony growth. In fact, by microscope observation of colonies' growing zones (tips of outwardly growing branches or growing fronts), we recognized two distinct types of growth processes.

The first one is accompanied by no active movement of individual bacterial cells when the growth is taking place on the colony interface at high C_a (hard agar; regions A and B). This type of growth is relatively "static": only cells composing the outermost part of a colony feed on nutrient and increase their population mass by cell division. It takes more than a few days to develop a specific pattern. Cells in the inner part change to spores and enter into the rest phase or hibernation. They do not grow afterwards unless the nutrient condition improves again. In this region several layers of bacterial cells are piled up to form colonies.

The other type is accompanied by the active movement of individual cells inside colonies for intermediate and low C_a (soft agar; regions C, D, and E). Colonies obtained in these regions are, therefore, made of a monolayer of bacterial cells. This type of colony growth is driven by active cell movement and is remarkably "dynamic": for instance, in the case of region E, the outermost growing front of a colony is enveloped by a thin layer (thickness $\simeq 5$ μm) of bacterial cells whose movement is very dull, while cells inside the layer move around actively and erratically. Sometimes they collide with the layer, break through it, and rush out. However, the rushed-out cells's movement immediately becomes dull, and the cells become part of the layer. As a consequence, the layer or front line expands a little. Although the expansion of the front line is not so speedy as the movement of cells inside it, the growth rate of a colony as a whole is still much higher than that of the first type. Deep inside the growth front the cell movement is again inactive for the colonies grown in regions C and E.

The thick broken line in Figure 1 indicates the boundary beyond which the bacterial cell movement described above was observed. It should be noted that the line coincides with almost-vertical phase boundaries of morphologies between regions A and E, and B and C in the figure. The verticality is understandable because the cell movement is mainly influenced by the softness of agar plates. The bending from verticality in the lowermost part (the boundary between regions A and E) may be attributed to the inactivity of cells due to starvation resulting from poor nutrient quality.

Bacterial cells are known to be able to move actively only by using their flagella. This means that if one can get mutants with no flagellum, the cells cannot show any active movement. Since, as described above, the movement seems to induce the colony's morphological change, we carried out the same experiment described so far except we used immotile mutants with no flagella (OG-01b). We then obtained a surprisingly simple phase diagram consisting of only regions A and B.[12] Namely, the observed morphological change is only a gradual crossover from DLA-like to Eden-like colony patterns in all ranges of the agar concentration examined. We did not observe DBM-like nor homogeneously spreading patterns. In other words, we found that regions A and B in Figure 1 expand laterally to the entire region that was studied in the present experiment, and regions C, D, and E disappear. We can, therefore, conclude that the active movement of bacterial cells observed for the wild type (OG-01) actually triggers the morphological change.

5. VALIDITY OF THE FISHER EQUATION FOR REGION D

Let us here discuss the growth mechanism of homogeneously spreading patterns, seen in region D, from another viewpoint. A colony in this region looks macroscopically like a perfect disk with a clear-cut interface. We noticed by microscopic observations, however, that the growing front is obscure, compared with that in regions C and E. In other words, the interface cannot be defined clearly in this region. We also notice that the population density of bacterial cells inside a colony in region D is low even in a monolayer of cells. This is the reason why colonies in region D looked transparent and were hard to observe. Hence, we assume that the bacterial cell movement in region D can be described in terms of diffusion in two dimensions. (According to close observations, however, their motion does not appear to be completely Brownian but somewhat collective, especially when their population is dense.) The spatio-temporal variation of population density of bacterial cells $b(\mathbf{r}, t)$ and concentration of nutrient $n(\mathbf{r}, t)$ (denoted as C_n in previous sections) is then represented by reaction-diffusion-type equations:

$$\frac{\partial b}{\partial t} = \nabla \cdot (D_b \nabla b) + f(b, n), \tag{1}$$

$$\frac{\partial n}{\partial t} = D_n \nabla^2 n - \nu bn, \tag{2}$$

where D_b and D_n are the diffusion coefficients of bacterial cells and nutrient, respectively; ν is the consumption rate of nutrient by bacteria; and $f(b, n)$ denotes the reaction term due to local bacterial growth. D_b is, in general, dependent on both b and n. In the following, however, we regard D_b as a constant. Moreover, we assume that $f(b, n)$ can be described by the following logisticlike form:

$$f(b, n) = [\varepsilon(n) - \mu b]b, \tag{3}$$

where the term $\varepsilon(n)$ represents the growth rate of individual cells and μ is the coefficient of competition among cells. This form is plausible since too many bacterial cells restrain themselves from increasing their population.

Let us now consider the limiting case in which the nutrient is so rich that n can be spatially constant, i.e., $\varepsilon(n) = \varepsilon$. It should be noted that this does not mean that ε is independent on n. Then Eq. (1) becomes

$$\frac{\partial b}{\partial t} = D_b \nabla^2 b + (\varepsilon - \mu b)b. \tag{4}$$

This is known as the Fisher equation.[11] This equation and its traveling wave solutions have been widely studied. In particular, this equation asymptotically yields isotropically spreadingly, homogeneous solutions with stable traveling wave fronts of constant speed in two dimensions. The internal population density and wave front speed are given, respectively, by $b = \varepsilon/\mu$ and $v = 2(\varepsilon D_b)^{1/2}$. This Fisher equation may describe the simple, homogeneously spreading colony pattern observed in region D of Figure 1.

Let us here discuss experimental confirmation for this conjecture. It turned out, however, that it is very difficult to estimate the value of ε and especially μ

experimentally. We will, therefore, discuss the consistency of experimental data described by the Fisher equation.

We first let a pair of colonies, inoculated simultaneously, collide with each other. They were then observed to fuse together. This is in clear contrast to the case of DLA-like colonies in region A or even DBM-like colonies in region E, which were found to repel each other. Numerical calculations of the Fisher equation (4) that were started from two points in two dimensions were found to yield exactly the same behavior. This is the first (but qualitative) evidence of the conjecture.

Secondly, we estimated the population density b of bacterial cells. We measured the area occupation rate by counting cells in the homogeneously populated region just inside the colony interface for various nutrient concentrations n in region D. We obtained the plausible result that b is approximately proportional to n. This implies that, under the assumption of the Fisher equation with constant μ, ε should be proportional to n, i.e., $\varepsilon = \varepsilon_0 n$.

Next we measured the diffusion coefficient of bacterial cells D_b for various nutrient concentrations n. As described before, the movement of individual cells does not look like Brownian motion either in the shorter periods or for higher nutrient concentrations because of their elongated shape and high population density, respectively. Nevertheless, we tried to estimate D_b from the ensemble average of square displacements of individuals cells $\langle R^2 \rangle$ during time interval t by using the relation $\langle R^2 \rangle = 2D_b t$. We found that D_b does not depend so much on the nutrient concentration n ($D_b \sim n^0$).

Finally, we measured the interface growth speed v for various nutrient concentrations n. The diameter of growing colonies was not quite proportional to the incubation time. There always seems to be a dormant period during which colonies did not grow. After this period, however, colonies grew very fast. Hence, we estimated the growth speed v when the colony diameter reached 5 cm. The results clearly showed that v is proportional to $n^{1/2}$. This is compatible with the speed $v = 2(\varepsilon D_b)^{1/2}$ that the Fisher equation gives.

All these experimental results are consistent with the behaviors derived from the Fisher equation.[11] We can, therefore, conclude that colony formation in region D of Figure 1 can be described by the Fisher equation.

An interesting future problem is whether the set of Eqs. (1) and (2) gives rise to the interfacial instability of traveling wave fronts and eventually to DBM-like patterns observed in region E.

ACKNOWLEDGMENTS

M.M. would like to thank Prof. N. Shigesada and Prof. H. Kawasaki for helpful discussions about the population dynamics approach. We are also grateful to Mr. M. Ohgiwari and Mr. K. Komatsu for their kind assistance. This work was supported in part by Grants-in-Aid for Scientific Research from the Ministry of Education, Science and Culture of Japan (Nos. 02808026, 02804019, and 04452053).

REFERENCES

1. Avnir, D., ed. *The Fractal Approach to Heterogeneous Chemistry*. Chichester: Wiley, 1989.
2. Ben-Jacob, E., and P. Garik. "The Formation of Patterns in Non-Equilibrium Growth." *Nature* **343** (1990): 523.
3. Family, F., and T. Vicsek, eds. *Dynamics of Fractal Surfaces*. Singapore: World Scientific, 1991.
4. Fujikawa, H., and M. Matsushita. "Fractal Growth of *Bacillus subtilis* on Agar Plates." *J. Phys. Soc. Japan* **58** (1989): 3875.
5. Fujikawa, H., and M. Matsushita. "Bacterial Fractal Growth in the Concentration Field of Nutrient." *J. Phys. Soc. Japan* **60** (1991): 88.
6. Grier, D., E. Ben-Jacob, R. Clarke, and L. M. Sander. "Morphology and Microstructure in Electrochemical Deposition of Zinc." *Phys. Rev. Lett.* **56** (1986): 1264.
7. Matsuyama, T., M. Sogawa, and Y. Nakagawa. "Fractal Spreading Growth of *Serratia marcescens* Which Produces Surface Active Exolipids." *FEMS Microbiol. Lett.* **61** (1989): 243.
8. Matsushita, M., and H. Fijikawa. "Diffusion-Limited Growth in Bacterial Colony Formation." *Physica* **A 168** (1990): 498.
9. Matsuyama, T., and M. Matsushita. "Self-Similar Colony Morphogenesis by Gram-Negative Rods as the Experimental Model of Fractal Growth by a Cell Population." *Appl. Environ. Microbiol.* **58** (1992): 1227.
10. Meakin, P. "The Growth of Fractal Aggregates and Their Fractal Measures." In *Phase Transitions and Critical Phenomena*, edited by C. Domb and J. L. Lebowitz, vol. 12, 335. New York: Academic Press, 1988.
11. Murray, J. D. *Mathematical Biology*. Berlin: Springer-Verlag, 1989.
12. Ohgiwari, M., M. Matsushita, and T. Matsuyama. "Morphological Changes in Growth Phenomena of Bacterial Colony Patterns." *J. Phys. Soc. Japan* **16** (1992): 816.
13. Sawada, Y., A. Dougherty, and J. P. Gollub. "Dendritic and Fractal Patterns in Electrolytic Metal Deposits." *Phys. Rev. Lett.* **56** (1986): 1260.
14. Singleton, P., and D. Sainsbury. *Introduction to Bacteria*. New York: Wiley, 1981.
15. Vicsek, T. *Fractal Growth Phenomena*. Singapore: World Scientific, 1989.
16. Vicsek, T., M. Cserzo, and V. K. Horvath. "Self-Affine Growth of Bacterial Colonies." *Physica* **A 167** (1990): 315.

17. Witten, T. A., and L. M. Sander. "Diffusion-Limited Aggregation, a Kinetic Critical Phenomenon." *Phys. Rev. Lett.* **47** (1981): 1400.

Eshel Ben-Jacob†‡, Ofer Shochet†‡, Adam Tenenbaum†* and Orna Avidan†
†School of Physics and Astronomy, Raymond and Beverly Sackler Faculty of Exact Sciences, Tel-Aviv University, 69978 Tel-Aviv, Israel
‡Institute for Theoretical Physics, University of California, Santa Barbara, California, 93106, USA
*Cohn Institute for the History and Philosophy of Science and Ideas, Tel-Aviv University

Evolution of Complexity During Growth of Bacterial Colonies

We present a study of interfacial pattern formation during growth of bacterial colonies. We start from *Bacillus subtilis* that exhibit compact growth. Reinoculation leads to the appearance of a branching mode, that shows tip-splitting growth (phase T). The colonies develop more complex patterns as the growth conditions become more adverse. We observed also phase transitions from phase T to a chiral phase (phase C). The latter exhibits even more complex patterns than phase T. Under adverse growth conditions, there is a competition between the complexity of the response of the phase and the option to perform a phase transition to a more complex phase. We observe also a "vortex" to "turbulent" transition as the growth conditions are varied. We conclude with a comment about the implications of our observations of transitions between modes and phases with different level of complexity on the evolving picture of genome cybernetics.

1. INTRODUCTION

Bacterial colonies can develop patterns[1,8,12,14,16,19] similar to morphologies observed during diffusion-limited growth in azoic (nonliving) systems such as solidification from undercooled melt, solidification from supersaturated solutions, liquid-crystal solidification, and electro-chemical deposition.[4,10,11,13,17,18] The various overall structures of the colony, which we name holoform, are observed as we change the growth conditions. Generally speaking, the colony adopts a more decorated holoform and a more complex structure as the growth conditions become more adverse.

In our previous publication[1] we have demonstrated that the adaptation of the colony requires self-organization on all levels, which may be viewed as the action of a singular interplay between the micro-level (individual bacterium) and the macro-level (the colony) in the determination of the emerging morphologies, like in non-living systems. Furthermore, the observed morphologies can be organized within a morphology diagram,[2,3] indicating the existence of a morphology selection principle similar to the one proposed for azoic systems.

The growth of a bacterial colony presents an inherent additional level of complexity compared to azoic systems, since the building blocks themselves are living systems. Hence, during the growth of the colony, the individual bacterium may go through major changes in response to the growth conditions. Indeed, we have observed[1] morphology transformations due to a reversible change of the bacterium, as well as colony transformations due to irreversible (inheritable) changes in the bacterium.

The observed transformations may be explained according to the conventional view of mutation kinetics. However, as we argued, the situation is simplified if we are willing to extend the conventional view: assume that the genome media can also perform cybernetic processing, and accordingly perform designed changes in the stored information or in its expression. That is, "the genome can be viewed as an adaptive cybernetic unit."[1]

For azoic systems, we know that just as the microscopic dynamics can determine the macroscopic morphology, so can the macroscopic dynamics reach down and affect the micro-structure of the growing phase. In other words, there is a singular feedback mechanism (at present we lack the mathematical description of the mechanism) from the macro-level to the micro-level. The general scenario is: as we drive the system further from equilibrium (in a manner that amplifies the micro-macro interplay), we observe a more complex global structure together with a higher micro-level organization. This is also the phenomenon observed during growth of the bacterial colonies: for more adverse growth conditions, the colonies show a more complex structure and a higher micro-level organization. The complex behavior of bacteria can be viewed from two orthogonal perspectives. One is complexity of the adopted growth pattern and the other is the complexity of the potential strategies for genome reorganizations by mutants of the original bacteria.

As the bacteria were exposed to more adverse conditions, mutations with a higher level of organization and more complex behavior were observed.

In Section 2 we describe the transformation pathway from compact to complex growth and comment about modeling of the growth. In Section 3 we describe the branching modes and their behavior for a wide range of environmental parameters. We show that the branching modes have two distinct, stable and heritable phases. The relative stability of the phases changes according to the growth parameters, each being more stable for different regimes of the parameters space. The faster phase (the one that has faster growing morphology) is the more stable (preferred) one. That is, starting from the slower phase we observe transitions to the faster one. In Section 4 we briefly describe the Vortex mode which is the next level in complexity. It provides dramatic demonstration that bacterial colonies can be regarded as a multicellular organism. Finally, in Section 5 we conclude and comment about a possible interpretation of our observations.

2. FROM *Bacillus subtilis 168* TO BRANCHING MODE

2.1 THE EXPERIMENTAL OBSERVATIONS

Our experimental endeavor started with *Bacillus subtilis* 168. This bacteria is the wild type of *Bacillus subtilis* with an additional genetic marker: it is tryptophan minus (trp^-); i.e., it requires tryptophan in the agar in order to grow. In Color Plate 9 we show colonies of the *Bacillus subtilis* 168 grown on low level of pepton. Although some structure is observed, the growth is rather compact. Seeking for a branching growth similar to the one described by Fujikawa and Matsushita,[8] colonies of the source (*Bacillus subtilis 168*) were grown on low pepton level (1 g/l). After the colonies reached stationary state, we have reinoculated new colonies taken from the edge of the old colonies. After many iterations (several weeks), bursts of branching growth were observed. The new modes of growth, referred to as branching modes, were found to be inheritable, i.e., inoculating bacteria from the new mode (after the burst) leads to a branched colony. We have monitored several events of such mode excitations. Under the microscope, one can clearly observe that while the original *Bacillus subtilis* do not move, the bacteria of the branching mode perform rapid, random walk-like movement within a well-defined envelope.

From the study of diffusive patterning in azoic systems,[4,10,11] we now understand the competition between the diffusion field and the interface kinetics in the determination of the evolved pattern. The diffusion field drives the system towards decorated (on many length scales) and irregular shapes. This effect is a manifestation of the basic diffusive (Mullins-Sekerka) instability. The microscopic effects (surface tension, surface kinetics, etc.), when present, compete with and channel the diffusive instability towards regular and less decorated structures. Generally speaking, the morphology changes from fractal-like, when the diffusive instability dominates the growth, to compact when the interface becomes stable.

In view of the above, one would naively expect that whenever bacterial colonies are grown on low-nutrient agar (diffusion limited) it will lead to a branching growth *regardless of its mobility*. As we have already seen, branching growth is not always observed and the modeling of the bacterial growth is more involved (see Ben-Jacobs et al.[5] for details). For branching growth to occur, the functional form of the growth rate (as function of nutrient concentration and bacterial concentration) is required to have specific properties. We have demonstrated this fact both theoretically and experimentally.

2.2 MODELING OF THE GROWTH—THE DIFFUSIVE-FISHER EQUATION

Generally, the reproduction rate that determines the overall growth rate is limited by the level of nutrients concentration available for the bacteria. The latter is limited by the diffusion of nutrients towards the colony. Hence, the process is similar to a diffusion limited growth in other azoic systems, such as solidification from a supersaturated solution, solidification from undercooled liquid, growth in Hele-Shaw cell, electro-chemical deposition etc. However, there are few essential differences: (1) The concentration field of the nutrient c extends into the colony. (2) Experimentally, the bacteria do not move when their density is too small or too high. This is due to chemicals and fluids that the bacteria secrete. When the density of bacteria is large there is no motion due to sporulation of the older bacteria in the regimes of low density of food. This bacteria–bacteria interaction could affect the dynamics, including the formation of a well-defined envelope. (3) The growth of bacteria can be viewed as the penetration of a stable phase into an unstable phase (unlike the penetration into meta-stable phase for the cases of solidification etc.).

To explain these points, let us first consider the following simplified model of the growth (as it does not include sporulation, wetting effects etc.) which is the Fisher-Kolmogorov model[5,7] coupled to a diffusion field:

$$\frac{\partial \rho}{\partial t} = \nabla(D_\rho(\rho)\nabla\rho) + r \tag{1}$$

$$\frac{\partial c}{\partial t} = D_c\nabla^2 c - \eta r \tag{2}$$

where ρ is the bacteria concentration (projected on two dimensions), c is the nutrients concentration, D_ρ and D_c are the bacterial and nutrients diffusion constant, r is the bacterial growth rate and η is a constant (which describes the nutrients/bacteria ratio). This model leads to a constant propagation in one dimension (Figure 1(a)).

(a)

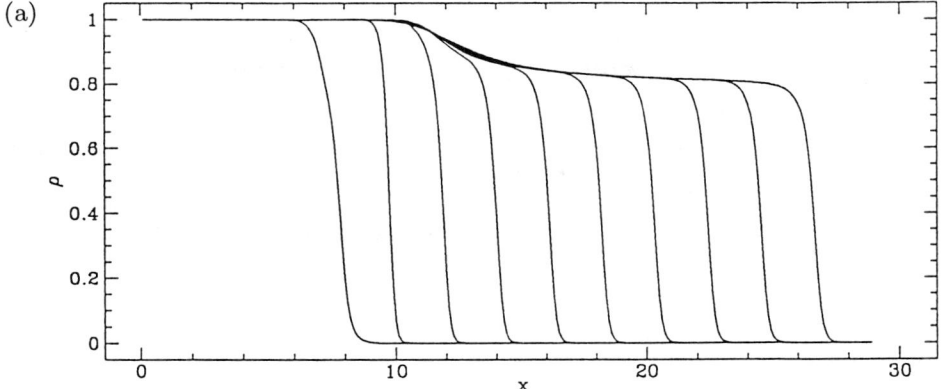

(b)

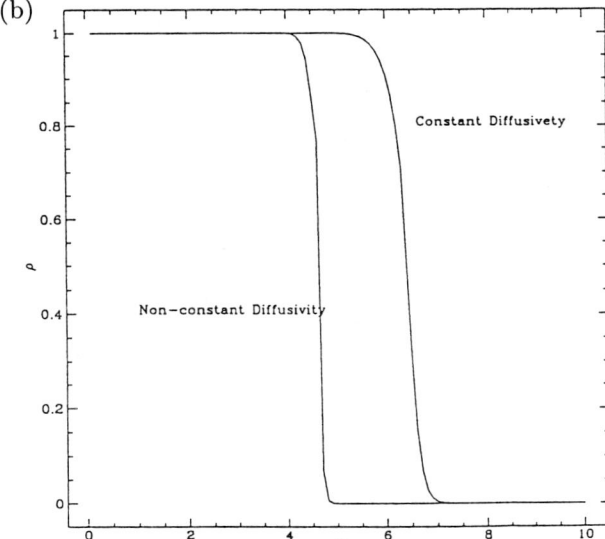

FIGURE 1   Results of one-dimensional numerical simulations of the diffusive Fisher model (Eq. 1). (a) For constant diffusion coefficient. Notice that the asymptotic value of ρ is $\rho_0 = c_\infty/\eta$ and that the front propagates at constant velocity. (b) The effect of nonconstant diffusivity of the bacteria. The front on the right is for constant diffusivity which the profile on the left is of diffusivity which dependes on ρ. The latter profile much sharper than the first one, creating a well-defined boundary for the bacteria colony.

The chemotaxis field is assumed to be proportional to the density field—ρ, and affect the mobility by the functional form of the bacterial diffusion coefficient—$D_\rho(\rho)$. For example, $D_\rho(\rho) \propto e^{-(\rho-0.5)^2}$ ensures low mobility for low bacteria density (due to chemotaxis) and low mobility for high density of bacteria (due to sporulation). We discovered that the growth velocity and stability spectrum are not qualitatively affected by the functional form of D_ρ. Only the profile of growth may change. For the specific D_ρ mentioned above, the growth profile become sharper creating a well-defined envelope for the growing colony (Figure 1(b)).

The simplest possible growth rate is $r = c\rho(1 - \rho)$; i.e. food is required for growth and the dependence on the bacteria density is via the logistic map—the bacteria density is limited to unity and bacteria may grow only in regions where the density does not vanish.

The above diffusive-Fisher model can be solved analytically in one dimension. Only the properties at $\rho = 0$ determines the growth velocity, and no diffusive instability may occur in two dimension. That is, the interface is completely stable and smooth. (Note that for the above growth rate, the model corresponds to Case I—see Ben-Jacobs et al.[5,7]).

It can be shown that if $r \propto c$ and has an unstable fixed point at $\rho = 0$ (as the above example), the interface is, at the most, marginally unstable and no branches will be created. This basic property *does not change* even if we make the model more elaborate, for example, by including sporulation, different mobility, different functional form of r (provided it has an unstable fixed point at $\rho = 0$ and it is proportional to c). The above modifications will only determine whether the interface is smooth or rough. This effect can be understood intuitively when one finds that the bacteria can adjust its local density so that $\rho_0 = c_\infty/\eta$. In this case, as there is no barrier between the fixed points at $\rho = 1$ and at $\rho = 0$, the bacteria will

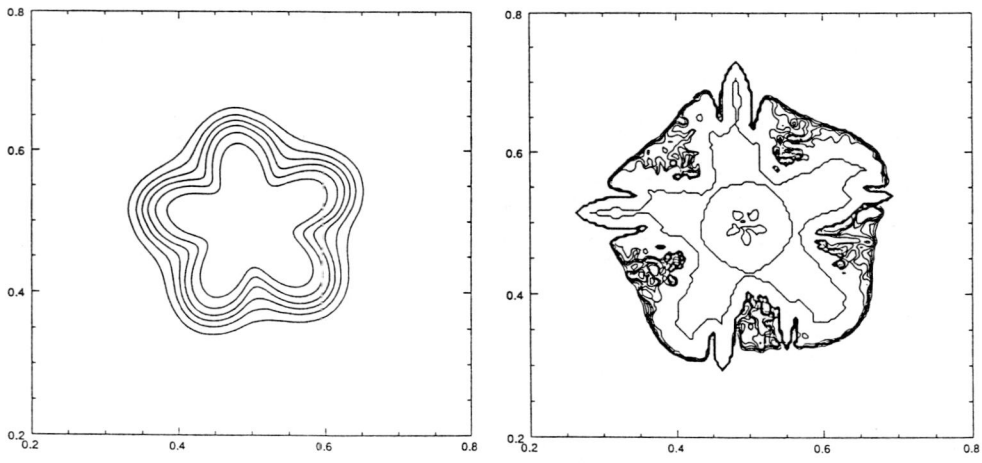

FIGURE 2 Equal density curves of two-dimensional simulations of the diffusive-Fisher model. Both (a) and (b) initiated from a 5-fold seed and show compact growth. (a) case I—the growth velocity depends only on the property of the fixed point at $\rho = 0$ leading to a smooth interface. (b) case II—the velocity depends on the local level of nutrient at the interface leading to a rough interface. The dendritic like tips results from the implicit anisotropy of the lattice.

fill space with density ρ_0 without any branches. The only pattern which is created is due to initial perturbations for smooth growth (Case I) and rough interface (Case II). In Figure 2 we show numerical simulations to demonstrate the above.

We investigated more detailed models which include sporulation. For example, consider the equations of motion:

$$\frac{\partial \rho}{\partial t} = \nabla(D_\rho(\rho)\nabla \rho) + r - \lambda \rho \tag{3}$$

$$\frac{\partial c}{\partial t} = D_c \nabla^2 c - \eta r \tag{4}$$

$$\frac{\partial \rho_s}{\partial t} = \lambda \rho \tag{5}$$

where ρ_s is is the local density of spores and λ is the bacteria sporulation rate. The additional spores field was added in order to ensure no reproduction and no motion for spores. Even for this model no branches will be created. This is understood immediately when one considers the behavior of the combined bacteria density (both active and spores) $\rho_{tot} = \rho + \rho_s$. ρ_{tot} adjusts its value so that $\rho_{tot} = c_\infty/\eta$, leading again to a stable envelope.

Branches are observed as the nature of the fixed point at $\rho = 0$ is changed. In Figure 3 we show a simulation of tip-splitting growth when the nature of the functional form of r was changed (the fixed point at $\rho = 0$ is meta-stable). Moreover, a preliminary study of a phase field model at high driving force indicates patterns of the phase-field which are left behind the growing front. These patterns are similar to the three-dimensional patterns we have observed for the bacterial colonies.

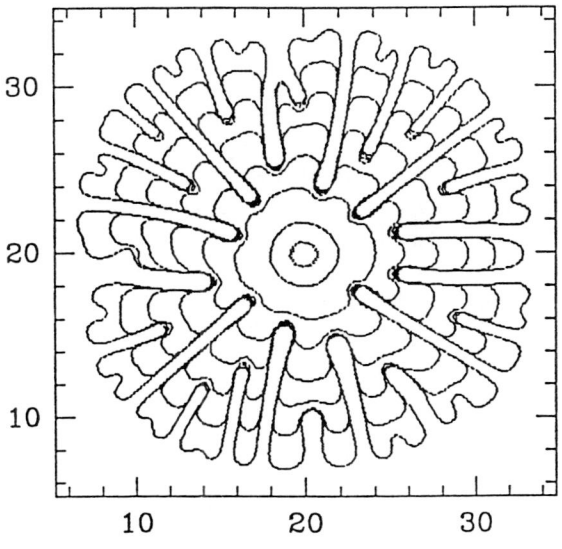

FIGURE 3 Results of two-dimensional simulations when r has a metastable fixed point at $\rho = 0$. The curves shows different time steps.

In view of the above we conclude, that these mode excitations result from the mutation of *Bacillus subtilis 168* to a new strain with a different functional form of the growth rate. We would like to emphasize that the new strain also has a different mobility, but change in the mobility by itself is not sufficient for branching growth, while a change in the functional form can be is sufficient.

We have found that modes B show different growth properties (with respect to the source) on minimal agar plates: In particular, modes B do grow on minimal plates without tryptophan, although with a lower growth rate. The experiments of growth on minimal plates[2,3] also supports our assumption that the mode excitation involves a change in the functional form of the growth rate of the bacteria.

3. THE BRANCHING MODE

In this section we present the branching modes and their behavior as we vary the environmental parameters. Growth of the branching mode under adverse growth conditions led to spontaneous transformations from tip-splitting to chiral growth. Both types of growth are heritable and can be transformed by a single bacterium. Yet the transformations between them differ from ordinary mutations. So we use the term "phases" (borrowed from physics) for their description. The phase that shows chiral growth is referred to as phase $\mathcal{C}$, and the one that shows tip-splitting is referred to as phase $\mathcal{T}$. All the branching modes that we have monitored have both phase $\mathcal{T}$ and phase $\mathcal{C}$. All mode excitations were to phase $\mathcal{T}$, and we have not observed direct transformations from *Bacillus subtilis 168* to a chiral phase.

The phases were separated (see details by Ben-Jacobs et al.[2]) and kept both in fluid environment and on cooled agar plate. They maintained their properties over the entire period that we have monitored (over two years).

3.1 PHASE $\mathcal{T}$

We have studied the morphology diagram as function of agar concentration (from 1% up to 2.5 %) and pepton level (from $0.5g/l$ to $10g/l$). In general, we observe compact growth at high pepton levels and tip-splitting growth at low levels (Color Plate 11). There are two types of branching growth. One that is reminiscent of patterns grown by electrochemical deposition, and the other reminiscent of Hele-Shaw (HS) patterns. The difference between these ECD-like and HS-like tip-splitting is similar to the difference between surface tension and surface kinetic tip-splittings. This observation can provide a hint for modeling of the growth, which we lack at present.

Above a pepton level of $10g/l$, various types of compact growth are observed. These include: compact growth with stable envelope, compact growth with rough or wavy envelope, and compact growth with leading zone of cellular fingers similar to the connective patterns described by Garik et al.[9]

At low pepton levels the patterns become more complex. At high agar concentration the pattern has an overall twist and a pronounced rings structure and a more complex structure in the perpendicular direction. All these phenomena will be described in detail by Ben-Jacobs et al.[2]

3.2 PHASE C

The morphology diagram of phase C was studied for the same range of parameters as in the studies of phase T. There is a clear correspondence with the morphology diagram of phase T: At high pepton level the growth is compact, and becomes less dense for lower pepton levels. At very low levels it looks DLA-like with twisted branches. Phase C shows a richer array of patterns then phase T. It includes a curly structure, HS-like tip-splitting, turbulent-like growth with lower density (self similar structure that is reminiscent of turbulent structure) (Color Plate 10). In general, phase C develops more complex patterns for low agar concentration (below 1.5%) and intermediate pepton levels (about $4g/l$).

3.3 TRANSITION FROM T TO C AND BACKWARDS

Ben-Jacob et al.[1,6] have proposed a new selection principle for azoic systems: "The fastest growing morphology selection principle." This means that if more than one morphology is possible for given growth conditions, the fastest one is dynamically selected.

We have observed coexistence of the two phases (Color Plate 12) and transitions between the phases. These observations indicate that a selection principle indeed exists. In all cases the $T \longrightarrow C$ transitions have been observed when phase C is faster than T. When phase C is much faster then phase T we usually observe a burst of the new phase at one localized point. It might be that the new phase secretes chemicals that inhibit additional bursts. In other cases we observe bursts at a number of localized points. We observe also extended phase transformations, that is, simultaneous transitions over an extended range of the advancing zone of the colony. Usually, the extended phase transformations are observed for values if the parameters that correspond to the boundaries between morphologies of phase T (between ECD-like and HS-like branching growth).

Motivated by the morphology selection principle for azoic systems, we have expected to observe the reverse $C \longrightarrow T$ transformations for values of the parameters where phase T is faster. We found that this is the case for high agar concentration (above 2%). And indeed, starting from phase C at this regime, we have observed transformations to phase T. So far, in all our observations the $C \longrightarrow T$ transformations occur as a burst from a single localized location. Inoculations from locations before and after the transformations demonstrate that both phases maintain their identity, as is the case with the $T \longrightarrow C$ transformations.

We have observed also extended transformations from phase C to a tip-splitting growth. Inoculation from the new growth on 1.5% agar led again to chiral growth. In some cases it seems that the new tip-splitting growth has transformed into phase T after the Petri dishes were kept refrigerated for several weeks.

4. THE VORTEX MODE

In one of the reinoculation experiments we observed the appearance of another mode, which we named vortex for reasons explained bellow. At present it is not clear whether the vortex mode is indeed a mutation of the *Bacillus subtilis 168* or contamination by a different strain of *Bacillus subtilis* .

In Color Plate 13 we show initial stage and late stage growth of the vortex colonies. These colonies spread via leading droplets (the darker dots) that leave behind a clear trail filled with bacteria on the agar. Each droplet consists of many bacteria that spins around a common center (hence named vortex) at typical velocities of 10 μm per second (the velocity is very sensitive to the growth conditions). Depending on the growth conditions, the number of bacteria in a vortex can vary from few to millions. Within a given colony, both clockwise and anti-clockwise vortices are observed. The vortices also vary from being one layered to many layers. In some cases we have observed vortices with an empty core, that is, vortices that have a "bagle" like shape. The latter can either spin in one direction or, in some cases, the inner bacteria and the outer bacteria spin in opposite directions. During some stages of the growth we have observed gliding of droplets which are very reminiscent of the motion of slime mold. Even more exciting are observations of a vortex that starts to glide into a new location and reassembles as a droplet that spins for some time. Then it glides back (almost as playing back a recording of the previous motion) to the original location, spins there and continues the cycle again. We have observed such processes to repeat themselves tens of cycles.

4.1 THE MORPHOLOGY DIAGRAM AND PHASE TRANSITIONS

The vortex mode exhibits a much richer and more complex morphology diagram than the branching modes (for details see Ben-Jacobs et al.[2]). Its special character is mainly expressed at high agar concentration (about 1.5%). At low agar concentration it shows tip-splitting and chiral growth very similar to those developed by phase T and phase C of the branching mode. Even when the vortex nature is expressed (each branch has a leading vortex at the tip), there is much correspondence with the morphology diagrams of phases T and C . At high pepton levels (10 g/l and above) the growth is dense and within a well-defined envelope. It turns more complex for lower pepton levels and adopts a fractal, almost DLA-like shape at very low pepton levels. For some ranges of parameters, the growth has an overall chiral twist similar to that observed for phase T at high agar concentration (about 2.5%).

Another similarity with the branching mode is our observations of different vortex phases and phase transitions, At present we do not have an appropriate geometrical characterization to distinguish between the phases. One of the phases is more remified so we refer to it as the analogue of phase C of the branching mode and denote it as phase γ. Similarly, we denote the other phase to be τ. We have observed phase transitions from phase τ to phase γ under growth conditions where phase γ is the faster one.

4.2 SELF-GENERATED FLOW FIELD

In the introduction we have referred to the vortex mode as higher stage in the evolution of complexity (relative to the branching mode). From the holoform point of view, the reason is the fact that the vortex mode can show both tip-splitting growth and chiral growth in addition to the vortex character.

If we take the micro-level (individual bacterium) view point, we notice the following: The *Bacillus subtilis 168* do not move on the agar unless we mechanically cut the agar surface and form a fluid layer. In phase T, the bacteria produce the necessary conditions on the agar surface so they can move. Yet the motion is possible only above a critical density of the bacteria and the movement is not correlated (very much random walk-like). The correlation is imposed by the advance of the boundaries of the region in which they can move, that is, the boundaries of the excreted chemicals or fluids. In phase C, the critical density needed for motion is much lower (hence the ability to cover large areas with lower densities) and the motion has some correlation,[1,2] mainly in the longitudinal motion along the branches. In the vortex mode there is a very high level of correlation, as all the bacteria in the vortex (or in the advancing "worms") move together. At present, we do not know how to model this collective behavior. We think that the correlated movement of the individual bacterium leads to a self-generated flow field, that is, to a self-generated motion of the excreted chemicals or fluids. In phase γ we have observed a motion of a single (yet very long) bacterium. We even observed a circular motion around a fix circular path of individual bacterium. The velocity oscillates, the bacterium moves for a while, then almost stops, moves again, and so on. In the case of motion of a single bacterium or few bacteria, a clear trail (of yet unknown substances) is observed on the agar. We also noticed propagating waves along the moving single bacterium.

4.3 VORTEX ⟶ TURBULENCE TRANSITIONS

Microscope observations of the tip-splitting growth of the vortex mode reveal that the micro-level differs from that of phase $\mathcal{T}$. While in the latter the bacteria perform random walks, in the former a clear turbulent-like motion is observed. In the studies of turbulence it is presented as a mixture of vortices of different sizes. The vortex mode provides an example for vortex ⟶ turbulence transitions. As we vary the growth conditions we observe changes in size, in the variety of sizes or in the density of the vortices. Under certain conditions the density of vortices increases until we reach a point where it is impossible to follow the individual vortex and the motion becomes turbulent.

---

5. CONCLUSIONS

5.1 NATURAL GENETIC ENGINEERING

Genetic engineering uses processes and mechanisms already available within the genome and within the cells. Shapiro[14] suggested a new perspective for interpreting evolution as controlled by natural genetic engineering. Thus, the integrated operation of DNA reorganization activities can rapidly bring about widespread changes throughout the genome. Now a need arises for hypotheses explaining how biologically appropriate structures result. We believe that here exactly lies the junction where the interlevel transfer of information cannot be viewed just from the "inside" perspective of the genome. A more general cybernetic framework is needed. If the genome is a "smart system," then what is needed are hypotheses on the tasks it has to perform in adverse environments, taking into account the range of simultaneous readjustments and autocatalytic emergence.

5.2 GENOME ADAPTATION AND GENOME LEARNING

Our observations show that the branching modes have two distinct stable and heritable phases. The relative stability of the phases changes according to the growth parameters, each being more stable for different regimes of the parameters space, the faster phase (the one that has a faster growing morphology) being the more stable (preferred) one. That is, starting from the slower phase we observe transitions to the faster one. Clearly, both the potential to enter each of the phases and the tools for "decision making" to perform phase transformations is response to environmental conditions available within the bacteria.

Whether the mechanism is epigenetic our our proposed mechanism based on cybernetors or another one yet to be revealed, the colony has a potential to "choose" the preferred phase for given environmental constraints.

As we have said in Ben-Jacobs et al.[1] to do so "...the colony organization can directly affect the genetic metamorphosis of the individuals." To emphasize the special nature of the phase determination and to distinguish it from ordinary reversible phenotypic adaptation, we refer to it as genome adaptation. The present existence of such a potential for genome adaptation means that it had to be developed (acquired) by the genome in the past. Hence, if we accept genome adaptation, it directly implies the possibility of genome learning. The latter requires the following elements:

1. Exposure of the bacteria to alternating environmental conditions (e.g., alternating between conditions for which phases C and T are preferred).
2. Memory about the environmental variations.
3. Means for problem solving by the genome according to the collected and processed information (both internal and external).
4. Means for genome reorganization (natural genetic engineering) according to the solution to the problem. That is, the genome can be viewed as a learning cybernetic unit.

5.3 COMMENT ABOUT COMPLEXITY

To conclude, we would also like to emphasize the following: In azoic systems at equilibrium we are used to view the system on two levels the micro-level and the macro-level. The interplay between these levels is manifested via the introduction of the entropy as an additional variable (and a functional for isolated systems) on the macro-level. The entropy is a measure of the number of possible microscopic states for a given state of the system. Hence, it can be viewed either as our lack of information about the micro-level (looking from the macro-level) or the freedom of the microdynamics for given imposed macroscopic conditions (looking from the micro-level). During the last decade we have learned that, as an azoic system is driven away from equilibrium, there is a singular interplay between the two levels which determines the self-organization of the entire system. Hence, the interplay can be viewed as balancing between the constraints that each level imposes on the other one. It is tempting to introduce the concept of complexity as a quantitative measure of the interplay. At present we lack a definition of complexity, and it is not clear that indeed such a new real variable does exist. Real in the sense that a measurement of the variable can be defined. Our definition of cybernetics suggests that, to maintain singular interplay between the level of the individual bacterium and the colony, we need a regulating channel to a level below that of the individual bacterium. If correct, it implies that we need more than two levels to describe the interplay in the colony. Hence, instead of applying concepts of statistical mechanics to the genome and the evolution process, we might use the lesson learned from the bacterial colonies as a hint for the development of the concept of complexity in the description of azoic systems out of equilibrium.

ACKNOWLEDGMENTS

We have benefited from conversations with Eliora Ron, Jim Shapiro, Nahum Kaplan, Michael Barber, Herbert Levin, Tamas Vicsek, Menahem Zukerman, and Yuval Shoham. We thank D. Gutnik and A. Drukel for the joint work on the vortex. Figure 3 is part of a study of phase-field model together with Raz Kupferman. We thank Ina Brandis for the technical assistance. E. Ben-Jacob and O. Shochet thank the Institute for theoretical physics and the participants of the program in "Spatially extended nonequilibrium Systems" for hospitality and discussions. This research was supported in part by a grant from the G.I.F., the German-Israeli Foundation for Scientific Research and Development, and by the Program for Alternative Thinking at Tel-Aviv university.

REFERENCES

1. Ben-Jacob, E., H. Shmueli, O. Shochet, and A. Tenenbaum. "Adaptive Self-Organization During Growth of Bacterial Colonies." *Physica A* **187** (1992): 378.

2. Ben-Jacob, E., O. Avidan, A. Tenenbaum, and O. Shochet. "Holo Adaptation, Selection and Transitions During Growth of Bacterial Colonies." In preparation.

3. Ben-Jacob, E., A. Tenenbaum, O. Avidan, and O.Shochet. "Holo Transformations of Bacterial Colonies and Genome Cybernetics." *Physica A* **202** (1993): 7.

4. Ben-Jacob, E., and P. Garik. *Nature* **343** (1990): 523.

5. Ben-Jacob, E., M. Zukerman O. Shochet, O. Avidan, and A. Tenenbaum. "Modeling of Bacterial Growth: The Diffusive Fisher-Kolmogorov Equation." In preparation.

6. Ben-Jacob, E., P Garik, T. Muller, and D. Grier. "Characterization of Morphology Transitions in Diffusion-Controlled Systems." *Phys. Rev A* **38** (1988): 1370.

7. Ben-Jacob, E., H. Brand, G. Dee, L. Kramer, and J. S. Langer, "Pattern Propagation in Nonlinear Dissipative Systems." *Physica D* **14** (1985): 348.

8. Fujikawa, H., and M. Matsushita. "Fractal Growth of *Bacillus subtilis* on Agar Plates." *J. Phys. Soc. Jap.* **58(11)** (1989): 3875.

9. Garik, P., J. Hetrick, B. Orr D. Barkey, and E. Ben-Jacob. "Interfacial Cellular Mixing and a Conjectore in Global Deposit Morphology." *PRL* **66** (1991): 1606.

10. Kessler, D. A., J. Koplik, and H. Levine. *Adv. Phys.* **37** (1988): 255.

11. Langer, J. S. "Dendrites, Viscous Fingering, and the Theory of Pattern Formation." *Science* **243** (1989): 1150.
12. Matsushita, M. "Growth and Morphological Changes of Bacterial Colonies." This volume.
13. Sander, M. *Nature* **322** (1986): 789.
14. Shapiro, J. A. *Genetica* **86** (1992): 99.
15. Shapiro, J. A. *J. Bact.* **169** (1987): 142.
16. Shapiro, J. A., and D. Trubatch. *Physica D* **49** (1991): 214.
17. Stanley, H. E., and N. Ostrowsky, eds. *Random Fluctuations and Pattern Growth: Experiments and Models.* Boston, MA: Martinus nijhoff, 1985.
18. Vicsek, T. *Fractal Growth Phenomena.* New York: World Scientific, 1989.
19. Vicsek, T., A. Czirók, O. Shochet, and E. Ben-Jacob. "Self-Affine Roughening of Bacterial Colony Surfaces." This volume.

J. O. Kessler† and N. A. Hill‡
†Physics Department, University of Arizona, Tucson, AZ 85721, USA
‡Department of Applied Mathematical Studies, The University, Leeds LS2 9JT, U.K.

Microbial Consumption Patterns

Concentrated populations of the aerobic swimming bacteria *Bacillus subtilis* rapidly use up the oxygen dissolved in their culture medium. As a result, oxygen diffuses in from the air interface, creating an upward concentration gradient. The organisms then swim toward the surface where they accumulate. Because this arrangement of mass density is unstable, the entire fluid culture convects. Biological and physical factors thus jointly serve to organize the population, yielding dynamics that improve greatly the transport and mixing of oxygen and the viability of the cells.

Organization that is correlated with consumption can occur also in populations of swimming algal cells. These phototrophic micro-organisms often swim toward or away from a source of illumination, depending on the color and intensity of the light, and the condition of the cells. Since the algae absorb and scatter light, the intensity of the radiation varies within a directionally illuminated population. When that population is appropriately patterned, individual cells generate shade for others and concurrently have access to nearby regions of greater or lesser illumination. Thus, self-shading, together with swimming that is directional within the field of the consumed quantity, generates patterns and makes use of them.

Spatio-Temporal Patterns, Ed. P. E. Cladis and P. Palffy-Muhoray,
SFI Studies in the Sciences of Complexity, Addison-Wesley, 1995   **635**

INTRODUCTION

Many varieties of swimming micro-organisms respond to stimuli which they find attractive or repulsive. When the organisms are members of large populations, the aggregate of individual responses leads to variations in the organisms' concentration. They may then affect the intensity of the stimuli, e.g., by consuming molecules or absorbing light. Most of this paper is concerned with bacteria that, by consumption, create an attractive spatial gradient of oxygen molecules. Self-shading and consumption of radiant energy by algae will also be discussed.

Concentrated fluid cultures of swimming bacterial cells such as motile strains of *Bacillus subtilis* often form patterns.[4,8] They are easy to see when the fluid layer is shallow, and when the illumination provides adequate contrast. These patterns are fairly regular arrays of dots or stripes whose overall diameters and spacings are usually of order millimeters (see Figure 1), i.e., much greater than the size of individual cells, typically micrometers. The pattern dimensions are also much greater than the average spacing between cells ($\approx 10^{-3}$ cm). What we see are marked fluctuations in concentration $n(r, t)$ of bacterial cells, as discussed in the Appendix. Generally, high cell concentrations are associated with, and cause, downward motion of the suspending fluid. However, because of incompressibility the velocity of the fluid, $u(r, t)$ is correlated with both local and remote values of $n(r, t)$. The relations between cell concentration, mean density of the fluid and convection are also discussed in the Appendix.

Physical and biological factors combine to convert an originally static microbial habitat, e.g., a macroscopically quiescent fluid bacterial culture in a Petri dish, into a functional dynamical system. There is a definite sequence of events leading to pattern formation. Oxygen is continuously consumed by the bacteria and supplied from the air, at the surface of the culture. Thus, the oxygen concentration decreases downward from the surface. As consumption progresses the oxygen gradient steepens. The bacterial cells located within this gradient swim upward, toward the air. Oxygen diffusion is usually too slow to reach the lower regions of the fluid. As the cells near the bottom deplete the supply, they stop swimming. The physiological, or biological, components of pattern generation are respiration and oxygen-taxis. Up to this stage, the physical factors are the location of the surface, due to gravity, and the diffusion rate of oxygen.

Initially, there is no general motion of the fluid, except for the slight local fluctuations that accompany the passage of a swimming organism. As the upward swimming of the bacteria toward the air progresses, the mean density of the increasingly cell-laden fluid strata next to the surface surpasses a stability threshold. The dense fluid layer then coalesces into descending plumes. The downward motion of dense fluid is necessarily coupled with the ascent of an exactly equal volume of fluid from below. This convective motion of the fluid and its occupants, driven by them, continues for hours or days, until the culture as a whole becomes senescent and depleted of nutrients.

(a)

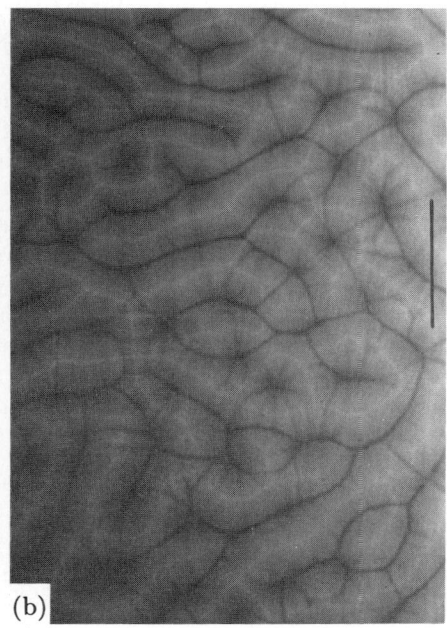

(b)

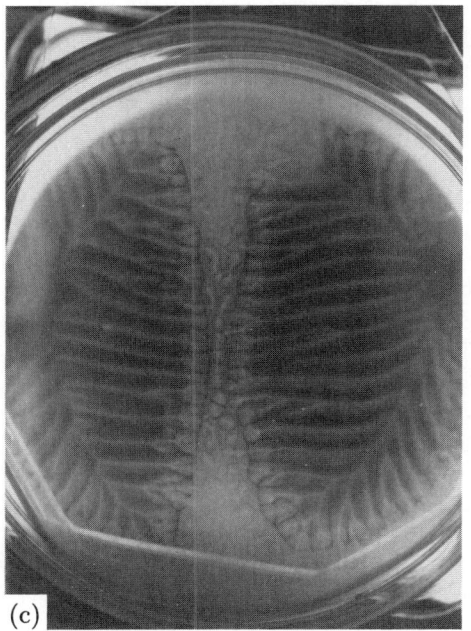

(c)

FIGURE 1 Bacterial concentration convection patterns seen in plan view. Dark field illumination causes regions that contain many cells to appear white; there are fewer cells in the dark regions. The bars are 1 cm long. In (a) the fluid depth is approximately 2 mm; in (b) it is 3 mm. The pattern geometry markedly depends on depth. The diameter of the Petri dish in (c) is 9 cm. Pattern (c) is less common than (a) and (b).

Consumption and upswimming have been recognised as "biological." The rest of the complex dynamics, the convection of fluid and the associated transport mechanisms, are entirely "physical." These specialized physical and biological ingredients generate remarkable and significant biological consequences. There are direct benefits to the cell population in the form of an increased viability of an "average cell." Without convective mixing and transport, the large fraction of the cell population that is well below the surface becomes nonmotile due to lack of oxygen. In contrast, in the early stages of the convection pattern, the stream carries downward a high concentration of well aerated bacteria, together with dissolved oxygen. The displaced fluid that moves upward is oxygen depleted and carries along some of the anoxic cells from below. Eventually, when convection is in full swing, most of the cells are again fully motile, except in the lowest strata where there is almost no movement.

EXPERIMENTAL (BACTERIA)

Two strains of motile wild-type *Bacillus subtilis* were grown in enriched minimal medium, at 37°C, in glass flasks mounted on a shaker. When the cell concentration n was sufficient, usually $n \geq 5 \times 10^8$ cells cm^{-3}, patterns formed spontaneously a few minutes after shaking ceased. The behavior of the two strains was indistinguishable. Generally, patterns were observed by dark field illumination at room temperature, in a shallow layer of fluid culture (see Figure 1). The pattern geometry was rather depth dependent. No patterns occurred for fluid layers shallower than ~ 1 mm. In thick layers the patterns would not be seen in plan view, due to excessive light scatter by the bacterial population. The time lag between first pouring the fluid culture into a diagonally illuminated Petri dish and the initial observation of patterns ranged from 30 seconds to 5 minutes, apparently depending mainly on the cell concentration and the initial aeration of the fluid.

To better understand the relationship between upswimming and the patterns, another geometry was employed. Normally a bacterial culture a few millimeters deep is placed in a flat-bottomed Petri dish, permitting the observation of the patterns that arise in shallow fluid layers. The new geometry consisted of a cuvette constructed of two microscope slides set on edge, separated by 1 mm thick spacers at the bottom and side edges, then sealed with silicone grease. The bacterial culture, approximately 1 cm deep, could now be observed from the side (see Figure 2). It was found that, soon after the cuvette was filled, a region about 1 mm below the air/fluid interface became depleted of bacteria. Because they swam toward the air/water interface, this cell accumulation at the meniscus became gravitationally unstable. Plumes of fluid that contained many more than the average number of cells then slowly began to descend downward from the interface.

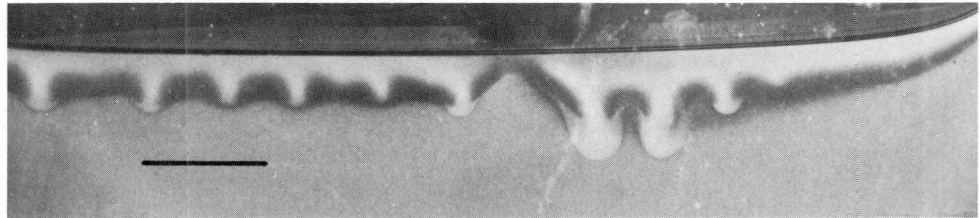

FIGURE 2   Concentration-convection pattern seen from the side. Air over the fluid interface. New plumes are starting while previous plumes penetrate the fluid below. The dark region between the interface and the main body of fluid is depleted of the cells which have swum to the surface. Dark field. Bar is 3 mm long. The uniform grey region below the dark stratum contains anoxic cells that have stopped swimming. The grey stratum is 4 cm deep.

These horizontally observed cell concentration-driven convection plumes (see Figure 2) do not correspond exactly to a side view of the sort of convection rolls seen in plan view (see Figure 1) because of the side wall constraint on the fluid dynamics. The thickness of the cuvette is restricted by the need for visibility and contrast. Nevertheless, it was shown that deep layer plumes do not stir the whole fluid, whereas in shallow layers, e.g. $\leq$ 4 mm, the whole fluid eventually participates in the convective motion.

When a culture in which a pattern has formed is remixed, by shaking or swirling the container, the patterns disappear. The entire fluid system becomes cloudy due to scattering of the illumination from the now uniformly distributed cells. After the container is set down, the patterns reappear. The time for reconstituting the patterns depends on the cell and oxygen concentration, the duration and intensity of the mixing and the depth of the fluid layer. Typically that time varies between 30 and 90 seconds. After the patterns reappear, it takes several minutes before they are nearly steady, and often much longer before a final state is reached. Increase of the cell concentration, by growth, is not a factor in pattern generation since doubling times are $>$ 20 minutes.

Water filters were used and light intensity was kept low, to avoid significant disturbances by thermal convection. It was also necessary to cover the observing dish so as to avoid the generation of an evaporation-induced vertical temperature gradient.

Oxygen-taxis (actually aerotaxis) of the bacteria was demonstrated using a thin layer of fluid bacterial culture between a microscope slide and a coverslip (see Figure 3). Small air bubbles were trapped in the fluid. They eventually served to

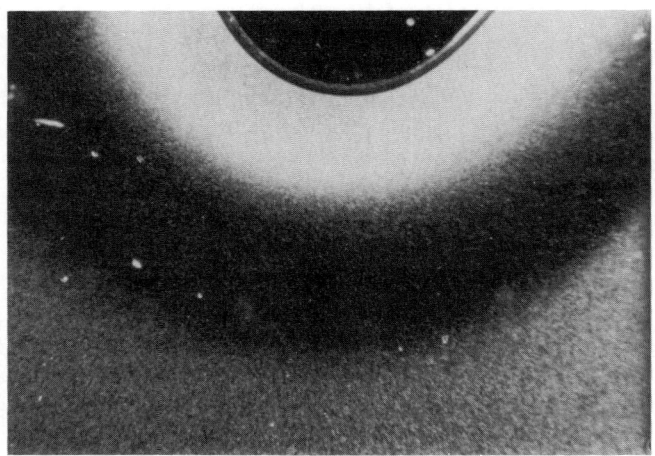

FIGURE 3   Oxygen-taxis of *B. subtilis*. Plan view of bacteria swarming toward an air bubble that is slightly less than 1 mm in diameter. The grey surround shows anoxic nonmotile cells. The dark region contains almost no cells. The former occupants have swum up the oxygen gradient, toward the air bubble, where they show up as a bright white ring.

attract cells as the latter exhausted their oxygen supply and oxygen diffused from the bubbles creating radial gradients. This test did not involve gravity or convection.

Another test of aerotaxis used the patterns themselves. It was found that when gas access to the fluid was occluded by a cover slip floating on the surface, patterns vanished underneath the central region of the coverslip. Some of the patterns from beyond the coverslip steadily persisted approximately 1 mm under the edge of it. That observation indicated that the loss of patterns below the glass cover did not imply that they were driven by surface tension.

In further tests we placed the bacterial culture in a N_2 atmosphere. Initially there were some weak and unusual patterns, presumably due to dissolved oxygen effects. Unlike for the case of an air interface, the patterns vanished after a few minutes.

Patterns generated by motility mutants differed markedly from the normal ones, but not enough data is currently available.

EXPERIMENTAL (ALGAE)

Populations of swimming algae spontaneously generate concentration-convection patterns, even in darkness. The patterns are caused by negative gravitaxis, or upward swimming of the cells. The resultant convective fluid motion[7] is quite analogous to the bacterial dynamics described here. When such an algal bioconvection pattern is unevenly or unilaterally illuminated, the geometry often changes to narrow curtains consisting of a high concentration of cells. The curtain axes are usually oriented approximately in the direction of the illumination.[3] The key to interpreting this phenomenon is the fact that most algal cells appear to seek an optimum luminous ambiance. They swim toward a weak, and away from an intense source of light.

TRANSPORT AND MIXING IN BACTERIAL PATTERNS

A complete mathematical description of consumption-driven bioconvection patterns involves six conservation equations which link the fluid dynamics, the cell's swimming, and the supply, distribution and consumption of oxygen. Some of the required ingredients are unknown, particularly a detailed description of bacterial motility in an oxygen gradient. The fluid dynamical equations provide for calculations of the fluid velocity $\mathbf{u}$, given a driving force. For bioconvection, the driving force is gravity, operating on local departures of the fluid density from the mean. Differences of density are directly proportional to variations of cell concentration. These variations arise from the interaction of the swimming cells' trajectories with guiding influences such as chemical concentration gradients, and with the fluid's velocity. The fluid equations are well known[7] and are not presented here. They have been applied to calculations of algal bioconvection.

The equations that govern oxygen transport and consumption, and cell swimming, can be used to estimate characteristic lengths and times. These calculations are simple in the limiting cases when the fluid velocity is approximately zero, before onset of convection, and when it is vigorous, so that well-mixing becomes a reasonable assumption.

The conservation, consumption and transport of the oxygen concentration c (molecules cm^{-3}) is given by

$$\frac{\partial c}{\partial t} = D\nabla^2 c - \nabla \cdot (\mathbf{u}c) - \gamma n. \tag{1}$$

The bacterial consumption of oxygen is γ (molecules cell^{-1}s^{-1}), the diffusion coefficient is D (cm^{2}s^{-1}), and the cell concentration is n (cells cm^{-3}). If only the vertical direction is of interest, $\nabla \to \partial/\partial z$. The flow, velocity $\mathbf{u}$, transports oxygen, hence, the flux term $\mathbf{u}c$ (molecules $cm^{-2}s^{-1}$). For cells, the analogous equation is

$$\frac{\partial n}{\partial t} = -\nabla \cdot [\mathbf{u} + \mathbf{V}(c)]n. \tag{2}$$

The swimming velocity $\mathbf{V}(c)$ is a function of c, its derivatives, and possibly a function of the history of c in the cells' reference frame. Both deterministic and stochastic motile behaviors are included[6] in $\mathbf{V}(c)$. When $c < c(\text{threshold})$, $V(c) = 0$. When c is large, $V(c) \approx V_b$. In a spatial gradient of c, $\mathbf{V}(c)$ is directed toward higher c, on average.

At the beginning of a typical experiment, the cuvette or Petri dish that contains the cell culture is agitated until the cells are uniformly distributed, $n = n_0$ and the oxygen concentration is $c \approx c_0$, approximately the maximum solubility in equilibrium with air. After the fluid velocity u vanishes due to viscous damping, Eq. (1) can be used to calculate consumption times and oxygen boundary layer thickness.

The bulk consumption time T can be calculated from Eq. (1) by setting $D = 0$. This procedure has validity because, as will be shown, the depth of the boundary layer that develops during T is much smaller than the dimensions of the bulk fluid. This means that after the deep-lying cells consume the initial oxygen concentration they are effectively cut off from supply. Then, without diffusion,

$$c = c_0 - \gamma n_0 t. \tag{3}$$

When $c = c_0/2, T \equiv T(1/2) = c_0/2\gamma n_0$. The approximate penetration depth, i.e., the diffusion boundary layer that develops during T is

$$L(1/2) = [2DT(1/2)]^{1/2} = [Dc_0/\gamma n_0]^{1/2}.$$

Using the estimate[1] $\gamma = 10^6$ molecules cell$^{-1}s^{-1}$, $n_0 = 10^9$ cm^{-3}, $D = 2 \times 10^{-5}$ cm$^2 s^{-1}$, $c_0 = 1.5 \times 10^{17}$ molecules cm^{-3}, one obtains $T(1/2) \approx 75$ s and $L(1/2) \approx 6 \times 10^{-2}$ cm. Thus, by the time half the oxygen is exhausted, the new supply from the air interface has hardly begun to penetrate. The usual depth of the fluid used in the experiments is ≥ 2 mm, which is considerably larger than $L(1/2)$. One may calculate L for other depletions of oxygen. For example, when all the oxygen is used up, $c = 0, T = c_0/\gamma n_0$, and $L(0) = 2^{1/2}L(1/2)$, or about 1 mm. Thus, the thickness of the boundary layer, and hence the penetration of oxygen to the bulk of the fluid, is not sensitive to the assumed total consumption.

Initially all the bacteria are uniformly distributed in the fluid. The cells located near the air interface are the first to sense the oxygen gradient. As a result, they swim upward, toward the interface. When their concentration is sufficient, convection begins in the form of plumes that descend from the interfacial region. The observed time for plumes to start is of order 1 min, and the depth of the layer out of which bacteria swim upward is about 1 mm, both in agreement with the calculated T and L values. The bacteria-depleted layer shows up "black" in dark field illumination, so that the magnitude L can easily be estimated.

The boundary gradient of oxygen can also be calculated from steady conditions, assuming that eventually there is no time dependence of c. The bacteria then consume all the available oxygen which diffuses in from the top surface. Then $\partial c/\partial t = 0$ in Eq. (1). At $z = 0$ the oxygen concentration is approximately in equilibrium with air, $c = c_0$, and at the bottom boundary the oxygen flux vanishes, $\partial c/\partial z = 0$ at $z = H$. Then

$$\frac{c}{c_o} = 1 - \frac{\gamma n}{c_o D}\left[Hz - \frac{z^2}{2}\right]. \tag{4}$$

When $c/c_0 = 1/2$, $z(1/2) = c_0 D/2\gamma nH = L(1/2)^2/2H$; $z(1/2) = 0.5 \times 10^{-2}$ cm for a typical value $H = 3$ mm. The depth of the oxygen gradient is very shallow indeed. This calculation shows that the cell population, except for a thin surface layer, becomes anoxic in the absence of convective distribution of oxygen. The narrowness of the boundary means that it is unimportant to this calculation whether the cells and remaining oxygen in the main bulk of the fluid are stationary or well mixed, by bioconvection, for example. There is, however, a big difference between the two. In the stationary case, the bacteria become anoxic and lose motility. In the convective case, the plumes and convection rolls continually pull pieces of the oxygenated boundary layer from the top surface into the bulk fluid where they exchange places with the upward moving anoxic cell-containing streams. These regions of aerated fluid supply oxygen locally, and relatively slowly, with a characteristic time $T(1/2)$. It is this local aeration of the culture which maintains the observed motility of most of the cells in the population.

For the convective case, it can be argued that H, the depth where the flux is very small, should not be the total height of the cuvette, but some location in the moving fluid. One may set $z = H$ for some particular concentration, e.g., $c/c_0 = 0.1$. Then $z(0.1) = (9/5)L(1/2)$, or about 1 mm, which corresponds to the size scale of some observed features of the convection.

It must be remembered that in the convective case both oxygen and bacteria are transported. Then both Eqs. (1) and (2) are required, with u supplied either from observations or the fluid mechanical equations. The rate of motion of the convection rolls can be experimentally determined by observing suspended particles. Preliminary measurements indicate $u \approx 10^{-2}$ cm s^{-1}.

How do diffusion and convection compete over a distance R? A characteristic time for convective supply is $t_c = R/u$. For diffusion, the time is $t_d = R^2/D$. When the ratio $Pe = t_D/t_c = Ru/D$ is $\gg 1$, the distribution process is dominated by convection, and conversely for $Pe \ll 1$. For $u = 10^{-2}$ and $D \simeq 10^{-5}$, $Pe \equiv R \times 10^3$. For $R > 10^{-2}$ cm the long-range distribution of molecules is convective, but at small distances, e.g., in the gradient boundary layer, diffusion is dominant.

ENERGY BALANCE

The dissipation that accompanies any more or less steady convection pattern must be compensated by some source of energy. Since the system is isothermal, bacterial or algal upswimming, which generates gravitational potential energy, can be the only source of that energy. The gravitational potential energy per unit volume per second that is generated by n organisms per cm^3, each organism having volume v and density ρ_b, swimming in water with density ρ_w is $vn(\rho_b - \rho_w)gV_{bu}$ (ergs/second cm^3). The acceleration of gravity is g and the mean upswimming speed of the bacteria is V_{bu}. The dissipation of energy per time, per volume due to viscosity and shear, which is the spatial variation of the fluid's velocity $\mathbf{u}$, is $\mu(du/dx)^2$erg/second cm^3. The derivative du/dx can be estimated by dividing the observed range of u by a typical wavelength scale of the pattern, determined by measuring the distance between adjacent spots or lines. For bacterial bioconvection, magnitudes of u were determined from the motion of latex spheres suspended in the fluid. This motion was oscillatory, as spheres wandered back and forth in the pattern's convection rolls. A typical order of magnitude was $u = 10^{-2}$ cm s^{-1}.

Does the viscous power dissipation equal the potential energy created per time, i.e., the power generated by upswimming? This equality actually holds at least by order of magnitude: For bacteria, with $v \approx 10^{-12}$ cm^3, $n \approx 10^9$ cm^{-3}, $\rho_b - \rho_w \approx 0.1g$ cm^{-3}, $V_{bu} \approx 10^{-3}$ cms^{-1}, $g = 10^3$cms^{-2}, $\mu = 10^{-2}g$ cm$^{-1}s^{-1}$, $du \approx 10^{-2}$cm s^{-1} and $dx \approx 10^{-1}$cm, one obtains 10^{-4}erg s^{-1}cm^{-3} for both powers. It is actually not necessary that the power balance hold locally, since the laws of fluid mechanics convert locally exerted forces to fluid motions of much greater extent.

As a matter of interest, the upswimming power dissipation is about 10^{-13} erg s^{-1}, per bacterial cell. The power required to just swim, to locomote horizontally, can be estimated from the Stokes drag as $6\pi\mu aV_b^2$. Taking $a \approx 10^{-4}$ cm and $V_b \approx V_{bu} \approx 10^{-3}$ cm s^{-1}, one obtains about 2×10^{-11} erg s^{-1} cell! Thus, upswimming only costs about $1/100$ of the energy to swim altogether. The conservation of energy in the algal convection case was discussed by Kessler.[2]

SHADING PATTERNS OF ALGAE

The intensity of illumination that travels in the x direction within a three-dimensional population of n light absorbing and scattering particles per unit volume is given by Eq. (6) in the Appendix. Let the particles be algae that swim toward the light source when $I > I^*$, and away when $I \leq I^*$ (positive and negative phototaxis). Then $I(x^*) = I^*$ determines a neutral location $x^*(y, z)$ toward which the cells swim. It is probably appropriate to model the light-oriented cell swimming speed by a function such as $V(x) = V_0 \tanh K(x^* - x)$, where the parameters V_0, K, and x^* are governed by distribution functions that depend on the characteristics of particular cell populations. Gravitaxis can affect phototaxis.[5] A complete theory must include that effect.

The initial spatial dependence of $n(x, y, z)$ determines the initial location of x^*. For example, when bioconvection patterns exist before the light is switched on, the initial concentration $n(x, y, z)$ is patterned. The distance x^* is then a function of (y, z). If the cell population's phototactic response to light is fairly uniform, a "curtain" of cells ought to develop. Cells near the "front" of the population swim rearward in the direction of the light beam; in the "rear" cells that are shaded by the population of other cells in front of them swim forward. This dynamic narrowing of the cell distribution generates loci of x^* values. These loci are modulated by the bioconvection which necessarily occurs when a cell distribution becomes locally bunched. Such distributions are actually found experimentally.[3] Narrow light-absorbing curtains, constituted of the swimming cells themselves, make available compact gradients of luminous intensity within which each cell can swim into its optimum location, at least some of the time.

In stationary fluid, i.e., without bio- or other convection, the geometry of the loci x^* is determined by $N(x^*)$, where

$$N(x) = \int_0^x n(\xi)d\xi \tag{5}$$

is the total number of cells per area in the light path, between the source and the position x. In principle, $x^* = f(y, z)$ can therefore be arbitrarily jagged. Swimming of the cells, and bioconvection, can smooth and relocate x^*. This "shimmering" of the pattern is observed. It is a feature of "free-floating" self-organization which relates to a direction, but only weakly to boundaries.

CONCLUSION AND SUMMARY

The spontaneous appearance of spatial and dynamic order, spanning distances great compared to both the size of the organisms and the mean spacing between them, suggests the existence of some form of orchestrated cooperation or direct interaction between the cells. Rather than that, it turns out that signalling is inadvertent, indirect, and interactive with the physical environment. For bacteria, primary organizational factors are the consumption of oxygen by the cell population, and its supply, unilaterally from the interface between fluid and air. The oxygen concentration gradient thus produced elicits directional motility ("aerotaxis" or "oxygen-taxis"). As cells swim upward and accumulate near the top surface, the mean fluid density there increases. Eventually stratification of the heavy fluid above light becomes unstable. Plumes of heavy fluid descend while lighter fluid from below circulates upward. It is this convective dynamics, and the associated variations in cell concentration, that one observes.

The primary causes of algal consumption patterns are again (1) asymmetric supply (of light, rather than oxygen), (2) consumption of light, leading to gradients of luminous intensity, and (3) bidirectional concentrative phototaxis that sharpens these gradients. These consumption patterns can be functional: they generate a chiaroscuro environment within which organisms can readily swim to regions of illumination that are appropriate. Bioconvection driven by concentration fluctuations modulates the cell concentration patterns and mixes the cells.

The spatial and temporal persistence of the self-organised structure arises from the interplay of consumption, behavior, and the long-range nature of the laws of fluid dynamics, which are coupled into the system by the force of gravity. It is especially significant that these consumption-driven dynamic structures function as an effective mechanism for improving molecular and cellular distribution and transport. For bacteria, the entire dynamical system becomes a respiratory engine: Consumption and local depletion of oxygen elicit a complex response that supplies more oxygen. The overall phenomenon provides an example of the auto-constitution of a macroscopic organ, one that develops out of a cell population that organises itself, and is organised, by consumption and the gravitational asymmetry of the physical environment.

APPENDIX

1. LIGHT ABSORPTION AND SCATTERING: VISIBILITY OF PATTERNS

Fluctuations of $n(\mathbf{r}, t)$ generate contrast, since the optical mean path length is $(n\sigma)^{-1}$ where σ is the extinction, or scattering plus absorption cross section. The approximate magnitude of σ is a characteristic cell dimension squared, i.e., $\sigma \approx 10^{-8}$ cm^2 for bacteria, $\sigma \approx 10^{-6}$ cm^2 for algae. The light intensity I, which is transmitted straight through a layer of fluid x cm deep, and neglecting multiple scattering, is

$$I(x) = I(o) \exp\left(- \int_0^x n(\xi)\sigma d\xi \right) = I(o) \exp\left[-\sigma N(x)\right] . \tag{6}$$

Equation (6) relates to the contrasts which are often observed in these patterns, either with dark field or transmitted light. For bacteria, there is some evidence that σ is a function of the cells' orientation. Rate of tumbling and mean orientation are affected by the local fluid shear.

For algae, Eq.(6) also governs the observability of patterns. Furthermore, this equation is central to understanding the generation of the light consumption patterns which result from directional swimming.

2. FLUID DENSITY AND PARTICLE CONCENTRATION.

Why do fluid regions that contain more cells than surrounding regions sink? The mean density $\langle \rho \rangle$ of fluid that contains suspended particles of volume v, concentration n and density ρ_b, where the pure fluids' density is ρ_w, is

$$\langle \rho \rangle = \rho_w - nv(\rho_b - \rho_w).\tag{7}$$

As specified above, for these experimental conditions, $\rho_b - \rho_w \simeq 0.1 g/\mathrm{cm}^3$, $v = 10^{-12}$ cm^3, and $n \simeq 10^9$ cells/cm^3. Then $\langle \rho \rangle - \rho_w \simeq 10^{-4}$ g/cm^3. Although this density offset seems small, it properly estimates the experimentally observed magnitudes of the plumes' speed of descent. The formula used for the speed S (cm s^{-1}) was

$$S = (\langle \rho(n_1) \rangle - \langle \rho(n_2) \rangle) \left(\frac{gr^2}{\mu} \right)\tag{8}$$

where $n_1 - n_2 = 10^9$ cm^{-3}, and the plume radius $r = 10^{-1}$ cm. The formula only provides an estimate of magnitude. An exact formula would also depend on fluid depth, container geometry, and additional factors derived from fluid dynamics. For example, 2/9 is the well known factor for a solid sphere that sediments in an infinite medium.

The expressions for $\langle \rho \rangle$ and S do not depend on the nature of the suspended particles, providing their concentration is sufficient to allow a continuum approach. It is immaterial whether the particles are passive and only undergo thermal motions or whether they swim, as do many algal and bacterial cells.

ACKNOWLEDGMENTS

We should like to acknowledge support from the SERC, U.S. National Science Foundation grant INT 8922466, NASA grant NAG 9442, and the generous support of Ralph and Alice Sheets through the University of Arizona Foundation. We should like to thank T. J. Pedley for many illuminating conversations during J. O. K.'s stay at the University of Leeds, and M. A. Hoelzer and M. F. Wojciechowski for their insights and help with the bacteriology.

REFERENCES

1. Berg, H. C. *Random Walks in Biology.* Princeton: Princeton University Press, 1983.
2. Kessler, J. O. "Cooperative and Concentrative Phenomena of Swimming Micro-organisms." *Contemp. Phys.* **26** (1985): 147–166.
3. Kessler, J. O. "The External Dynamics of Swimming Micro-Organisms." In *Progress in Phycological Research*, edited by F. E. Round and D. J. Chapman, Vol. 4, 257–307. Bristol: Biopress, 1986.
4. Kessler, J. O. "Path and Pattern—The Mutual Dynamics of Swimming Cells and Their Environments." *Comments on Theor. Biol.* **1** (1989): 85–108.
5. Kessler, J. O., N. A. Hill, and D.-P. Häder. "Orientation of Swimming Flagellates by Simultaneously Acting External Factors." *J. Phycol.* **28** (1992): 816–822.
6. Pedley, T. J., and J. O. Kessler. "A New Continuum Model for Suspensions of Gyrotactic Micro-Organisms." *J. Fluid Mech.* **212** (1990): 155–182.
7. Pedley, T. J., and J. O. Kessler. "Hydrodynamic Phenomena in Suspensions of Swimming Micororganisms." *Annl. Rev. Fluid Mech.* **212** (1992): 313–358.
8. Pfennig, N. "Beobachtungen Über das Schwärmen von *Chromatium okenii*." *Archive der Mikrobiologie* **42** (1962): 90–95.

Mark M. Millonas
Complex Systems Group, Theoretical Division and Center for Nonlinear Studies, MS B258, Los Alamos National Laboratory, Los Alamos, NM 87545 and Santa Fe Institute, 1399 Hyde Park Road, Santa Fe, NM 87501

A Nonequilibrium Statistical Field Theory of Swarms and Other Spatially Extended Complex Systems

A class of models with applications to swarm behavior as well as many other types of spatially extended complex biological and physical systems is studied. Internal fluctuations can play an active role in the organization of the phase structure of such systems. Consequently, it is not possible to fully understand the behavior of these systems without explicitly incorporating the fluctuations. In particular, for the class of models studied here the effect of internal fluctuations due to finite size is a renormalized *decrease* in the temperature near the point of spontaneous symmetry breaking. I briefly outline how these models can be applied to the behavior of an ant swarm.

INTRODUCTION

In this chapter, I introduce a class of models which is in line with the basic processes acting in a variety of systems in nature, particularly biological ones. Some systems that fall into this class are insect swarms, swimming bacteria and algae,[6] physical trail formation, the evolution of river networks,[8] diffusive transport in polymeric materials,[1] population distribution models, various types of fractal growth phenomena,[14] and developmental morphogenesis.[12]

Here we study what will be called *stigmergic processes* as a generalization of the concept of stigmergy introduced by Grassé[3] in the context of collective nest building in social insects. The hypothesis of stigmergy, as described by Wilson,[15] is that *it is the work already accomplished, rather than direct communication among nest mates, that induces the insects to perform additional labor.* The concept of stigmergy has also been invoked more recently in regards to swarm behavior.[13]

The more generalized idea of a stigmergic process is realized here in systems composed of three basic ingredients. The first ingredient is a *particle dynamics* which obeys a Markov process on some finite state space $\mathcal{X}$. The particle density $\rho(\mathbf{x}, \tau)$ obeys the Master equation

$$\frac{\partial \rho(\mathbf{x}, \tau)}{\partial \tau} = \int_{\mathcal{X}} \{W_\tau(\mathbf{x}|\mathbf{y})\rho(\mathbf{y}, \tau) - W_\tau(\mathbf{y}|\mathbf{x})\rho(\mathbf{x}, \tau)\} \, d^D\mathbf{y}, \tag{1}$$

where $W_\tau(\mathbf{x}|\mathbf{y})$ is the probability density to go from state $\mathbf{y}$ to $\mathbf{x}$ at time τ. The second element is a *morphogenetic field* $\sigma(\mathbf{x}, \tau)$, representing the environment that the particles both respond to, and act on. We will study one of the simplest situations, a fixed one-component pheromonal field that evolves according to

$$\frac{\partial \sigma(\mathbf{x}, \tau)}{\partial \tau} = -\kappa \sigma + \eta \rho, \tag{2}$$

where κ measures the rate of evaporation, breakdown, or removal of the substance, and η the rate of emission of the pheromone by the organisms. Lastly, some form of *coupling* is made between the particles and the field. This coupling takes the form of a behavioral function that describes how the particles move in response to the morphogenetic field, and in turn, how the particles act back on this field.

As we shall see, small changes in the microscopic behavior of the particles can result in large changes in the global behavior of the swarm, or particle field. This variability has significant implications not only for the behavioral response of the swarm to external stimuli, but also in the evolution of cooperative behavior. Wilson has remarked that an understanding of how this occurs would *constitute a technical breakthrough of exciting proportions, for it will then be possible, by artificially changing the probability matrices, to estimate the true amount of behavioral evolution required to go from [the behavior of] one species to ... that of another.*[15] He has further remarked that such large behavioral changes resulting from small changes in the individual dynamics would provide evidence that social behavior evolves at least as rapidly as morphology in social insects. This could provide an explanation why *behavioral diversity far outstrips morphological diversity at the level of species and higher taxonomic categories* in social insects.

In the region of a nonequilibrium phase transition, the morphogenetic field, and hence the transition matrix, changes very slowly on scales typical of the particle field relaxation time since in this region the unstable modes will exhibit critical slowing down and will relax on a time scale much longer than the time scale of the stable modes. The particle modes are said to be *slaved* to the morphogenetic field, and can be adiabatically eliminated from the picture.[4] We obtain the stochastic order parameter equation

$$\frac{\partial \sigma(\mathbf{x}, \tau)}{\partial \tau} = \kappa \, \sigma + \eta \rho_s[\sigma] + \eta \, g[\sigma] \xi(\mathbf{x}, \tau), \qquad (3)$$

where $\rho_s[\sigma]$ is the quasi-stationary particle density, $g[\sigma]$ is a function describing the fluctuations of the quasi-stationary particle density about its mean value, and $E\{\xi(\mathbf{x}, \tau)\} = 0$, $E\{\xi(\mathbf{x}, \tau)\xi(\mathbf{x}', \tau')\} = \delta(\mathbf{x} - \mathbf{x}')\delta(\tau - \tau')$. Since ρ_s will depend on both the global state of the morphogenetic fields, and on the global boundary conditions, this is a globally coupled set of equations for the evolution of the morphogenetic fields. *Slaving of the particle field therefore allows an explicitly coupled global dynamics to emerge from the strictly local interactions of the model,* providing a key to how a globally integrated response may emerge from a system of locally acting agents.

The fluctuations in the system are state dependent. In addition to amplifying an instability that exists in the absence of noise, this type of fluctuation can also produce transitions and ordered behavior in its own right. One consequence of this fact is that slaved particle field will constructively determine the self-organization properties of the systems *through its fluctuating properties*, as well as through quasi-stationary values. This is a fact that should be constantly borne in mind when studying such models.

For the purposes of this chapter we will consider the case where the transition matrix takes the form $W(\mathbf{x}|\mathbf{y}) \propto f(\sigma(\mathbf{x})) p(|\mathbf{x} - \mathbf{y}|)$, where f is some weighting function describing the effect of the field σ on the motion of the particles, and $p(|\mathbf{x} - \mathbf{y}|)$ is a probability distribution of jumps of length $r = |\mathbf{x} - \mathbf{y}|$. Transition matrices of this type obey detailed balance, $W(\mathbf{x}|\mathbf{y})f(\sigma(\mathbf{y})) = W(\mathbf{y}|\mathbf{x})f(\sigma(\mathbf{x}))$. In this case we can define a partition function

$$Z = \left\{ \frac{1}{V} \int d^D \mathbf{x} \, f(\sigma(\mathbf{x})) \right\}^N,$$

where V is the volume of the state space $\mathcal{X}$, and N is the total number of particles. A one-to-one analogy with a thermodynamic system with energy $U(\sigma(\mathbf{x}))$ and temperature $T = \beta^{-1}$ can be made if we set $f(\sigma(\mathbf{x})) = \exp(-\beta U(\sigma(\mathbf{x})))$, where any parameter T can be regarded as a temperature parameter if $f(\sigma(\mathbf{x}); \alpha \, T) = f^{-\alpha}(c(\mathbf{x}); T)$. Statistical quantities of interest can be calculated from the partition function according to the usual prescriptions. In a closed system the mean particle density and dispersion in the energy state ϵ are given by

$$E\{\rho_\epsilon\} = \frac{N}{VZ} \exp(-\beta\epsilon), \quad E\{(\Delta\rho_\epsilon)^2\} = \frac{E\{\rho_\epsilon\}}{\mu_\epsilon} \left(1 - \frac{\mu_\epsilon}{N} E\{\rho_\epsilon\} \right), \qquad (5)$$

where μ_ϵ is the volume of the system in energy state ϵ. The slaved particle field in energy state ϵ can then be represented, to lowest order in the fluctuations, by $\rho_\epsilon[\sigma] = E\{\rho_\epsilon[\sigma]\} + \sqrt{E\{(\Delta\rho_\epsilon)^2[\sigma]\}}\ \xi(\mathbf{x},t)$.

We introduce the dimensionless parameter $\bar{\rho} = N/V$, the mean density of particles, and $v = \mu^-/\mu^+$, the ratio of the volume of the field $\sigma(\mathbf{x})$ in the σ^- state to the volume in the σ^+ state. We also define the function $R(\sigma^+, \sigma^-) = f(\sigma^+)/f(\sigma^-)$. In the mean-field approximation a Langevin equation

$$\frac{dm}{dt} = -m + F(m) + \frac{1}{\sqrt{N}}Q(m)\ \xi(t) \tag{6}$$

for the order parameter m can be derived,[11] where

$$F = \bar{\rho}(1+v)\frac{R^\beta - 1}{R^\beta + v}, \quad Q^2 = \frac{\bar{\rho}^2(1+v)^4}{v}\frac{R^\beta}{(R^\beta + v)^2}, \tag{7}$$

and where the F and Q are determined as functions of m by

$$R(m) = R\left(\bar{\rho} + \frac{vm}{1+v}, \bar{\rho} - \frac{m}{1+v}\right). \tag{8}$$

The order parameter m is analogous to a gas-liquid order parameter, and represents the difference in the values of the field in the σ^+ and σ^- states after spontaneous symmetry breaking. The behavior of this system is described by the potential function

$$\Phi(m) = \int^m \frac{m - F(m)}{Q^2(m)}dx + \frac{1}{N}\ln Q, \tag{9}$$

where the phases m_i of the system are determined by the conditions $\Phi'(m_i) = 0$, $\Phi''(m_i) > 0$.[5]

In the continuum limit $(N \to \infty)$ it can be shown that the critical value of the mean density $\bar{\rho}_c$ at which spontaneous symmetry breaking occurs is given by the condition $-\bar{\rho}_c\ U'(\bar{\rho}_c) = T$. Generally $\bar{\rho}_c$ will increase with increasing temperature. The relative stability of two phases m_1 and m_2 is determined by the relative potentials $\Phi(m_1)$ and $\Phi(m_2)$ for each phase. Even in the continuum limit the details of the fluctuations cannot be neglected due to the presence of the factor $Q^2(m)$ under the integral in Eq. (9), and the relative stability of the phases will depend on the precise details of the internal fluctuations. Similar observations have been made elsewhere by Landauer and others.[7,9]

When N is finite, the situation is still more complicated. It is clear that the possible values of the order parameter and the phase structure do not remain unchanged under the influence of internal fluctuations. The criterion for spontaneous symmetry breaking in this case is $-\bar{\rho}_c\ U'(\bar{\rho}_c) = \hat{T}$, where $\hat{T}$ is the renormalized temperature $\hat{T} = \gamma(N)T$ where $\gamma(N) = \sqrt{N + (N/2)^2} - N/2$. This is precisely the continuum condition except that the finite size fluctuations have the effect of renormalizing the temperature by the factor $\gamma(N) < 1$. The effect of increasing the internal fluctuations through decreasing the total number of particles has the effect of *decreasing the temperature*. We thus arrive at the seeming paradox that increased internal fluctuations may produce increased order.

I will now briefly outline how the previous analysis can be applied to the example of an ant swarm. More details can be found elsewhere.[10] In this case the individual ants are the particles, and the morphogenetic field is a pheromonal substance which the ants sense with their antennae, and emit from their bodies as they move. The basic measurement the ants make is the quantity of pheromone receive by each antennae. They can therefore respond to difference in the pheromone between the antennae, and move accordingly. A very general model of such motion assume that the particle experiences a *force* that is proportional to the scent gradient at that point multiplied by some nonlinear response function $\chi(\sigma)$ of the scent at that point. The nonlinear response function models the nonlinearities underling the basic physiology of the sensing apparatus, for instance, any nonlinear neural/receptor response to the pheromone, including such effects as saturation of the receptor sites on the antennae by the pheromonal substance. In addition there is an element of randomness due to fluctuations in the external environment as well as internal fluctuations. These are incorporated into an effective random force with a strength in proportion to $\sqrt{T}$ where T is the temperature factor. The motion of a particle can be described by a Langevin equation of the form

$$\frac{d\mathbf{x}(t)}{dt} = \chi(\sigma(\mathbf{x}))\nabla\sigma + \sqrt{2T}\,\xi(t), \tag{10}$$

where $E\{\xi(t)\} = 0$, and $E\{\xi(t)\xi(t')\} = \delta(t - t')$. This can be written in the form

$$\frac{d\mathbf{x}(t)}{dt} = -\nabla U(\mathbf{x}) + \sqrt{2T}\,\xi(t), \tag{11}$$

where $\chi(\sigma(\mathbf{x})) = -U'(\sigma(\mathbf{x}))$. Easy to show that the behavioral function of such a system is given by $f(\sigma) = \exp(-\beta U(\sigma))$.

The microscopic dynamics of the ants, which we will study in the rest of the chapter, is determined by the response function

$$\chi(\sigma) = \alpha + \frac{c\rho}{c + \rho}, \tag{12}$$

where ρ is the pheromone density, and α and c are constants with the units of pheromone density. This function is inspired by the observed behavior of actual ants.[2] The constant α is roughly the threshold where the response of the ants to the pheromone is small unless $\rho > \alpha$. The constant c will be known as the capacity. When ρ approaches c the ants respond less accurately to pheromone gradients. This is because, when the pheromone density is very large, the antennae receptors become saturated and the ant cannot sense the pheromone gradient as accurately.

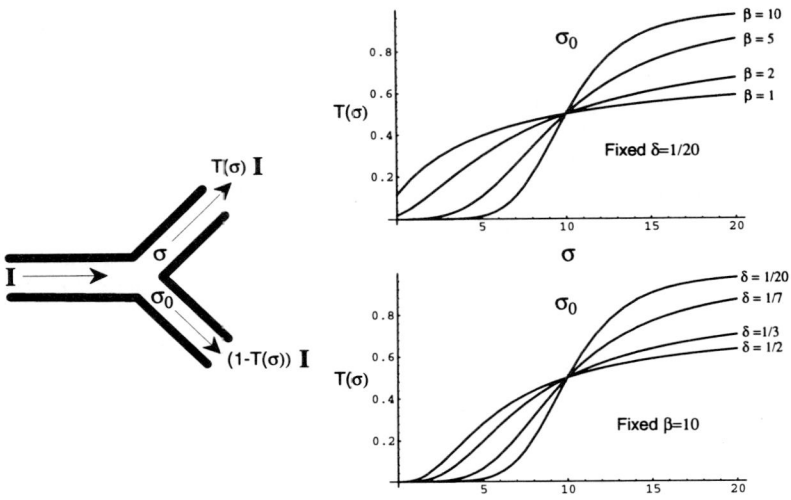

FIGURE 1   Transiton functions for varying β and δ.

For simplicity, we will introduce the dimensionless variable $\sigma = \rho/\alpha$ and the dimensionless parameter $\delta = \alpha/c$, where $1/\delta$ is the dimensionless capacity. The energy function takes the form

$$U(\sigma) = -\ln\left(1 + \frac{\sigma}{1 + \delta\sigma}\right), \tag{13}$$

where we drop off any additive constant term that have no effect on the behavior of the ants. For the case where the density of ants it low and, hence, the pheromone density is low ($\rho << g$), we can make use of the approximate energy function $U_0(\sigma) = -\ln(\alpha + \sigma)$.

An illustration of this effect is shown in Figure 1. A given current of organisms I flows into a junction from the left. On the lower branch the pheromone density is fixed at σ_0, and on the upper branch σ is allowed to vary. $T(\sigma)$, the proportion of the current that flows into the upper branch, is given by the sigmoidal function

$$T(\sigma) = [1 + \exp(\beta U(\sigma)/U(\sigma_0))]^{-1}.$$

The plots on the right of Figure 1 shows $T(\sigma)$ for varying values of β and δ. The upper plot, where δ is fixed, shows the influence of increasing the temperature (lowering β). As the temperature increases the threshold response becomes less and less pronounced. In the opposite limit $\beta \to \infty$, $T(\sigma)$ would be a step function $\Theta(\sigma - \sigma_0)$. In this limit all of the ants would choose the branch with the greatest pheromone density. In the lower plot, the noise level is fixed, and the capacity

$c = \alpha/\delta$ is varied. It is interesting to note that the effects of decreasing the capacity with fixed temperature are similar to the effects of increasing the temperature with fixed capacity. When the density of the ants increases, the pheromone density increases up to and beyond the capacity, the qualitative effects on the behavior of the ants is the same *as if the temperature was increased*. This gives the swarm roughly the ability to modulate its temperature by modulating its numbers.

This can be made more clear by defining an effective temperature factor $\theta(\sigma)$ through the relation $f(\sigma) = \exp(-\beta U_0(\sigma)/\theta(\sigma))$. The variable $\theta(\sigma)$ roughly measures the effective change in temperature as a function of the pheromonal field when compared to the case where $\delta = 0$, which correspond to the energy function U_0. The effective temperature is then given by $\theta(\sigma)T$ where

$$\theta(\sigma) = \frac{\ln\left(1 + \frac{\sigma}{1+\delta\sigma}\right)}{\ln(1 + \sigma)}. \tag{15}$$

In Figure 2, we illustrate the increase in the effective temperature with increasing σ for three different values of δ. Since increasing the temperature tends to decrease stability, we might expect any organized behavior to breakdown when the number of participants grows too large. It is this ability of the swarm to self-modify its temperature which allows it, in a sense, to traverse its various phase-transition boundaries.

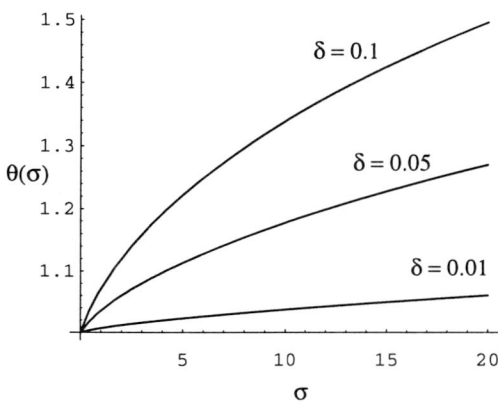

FIGURE 2 Effective temperature factor.

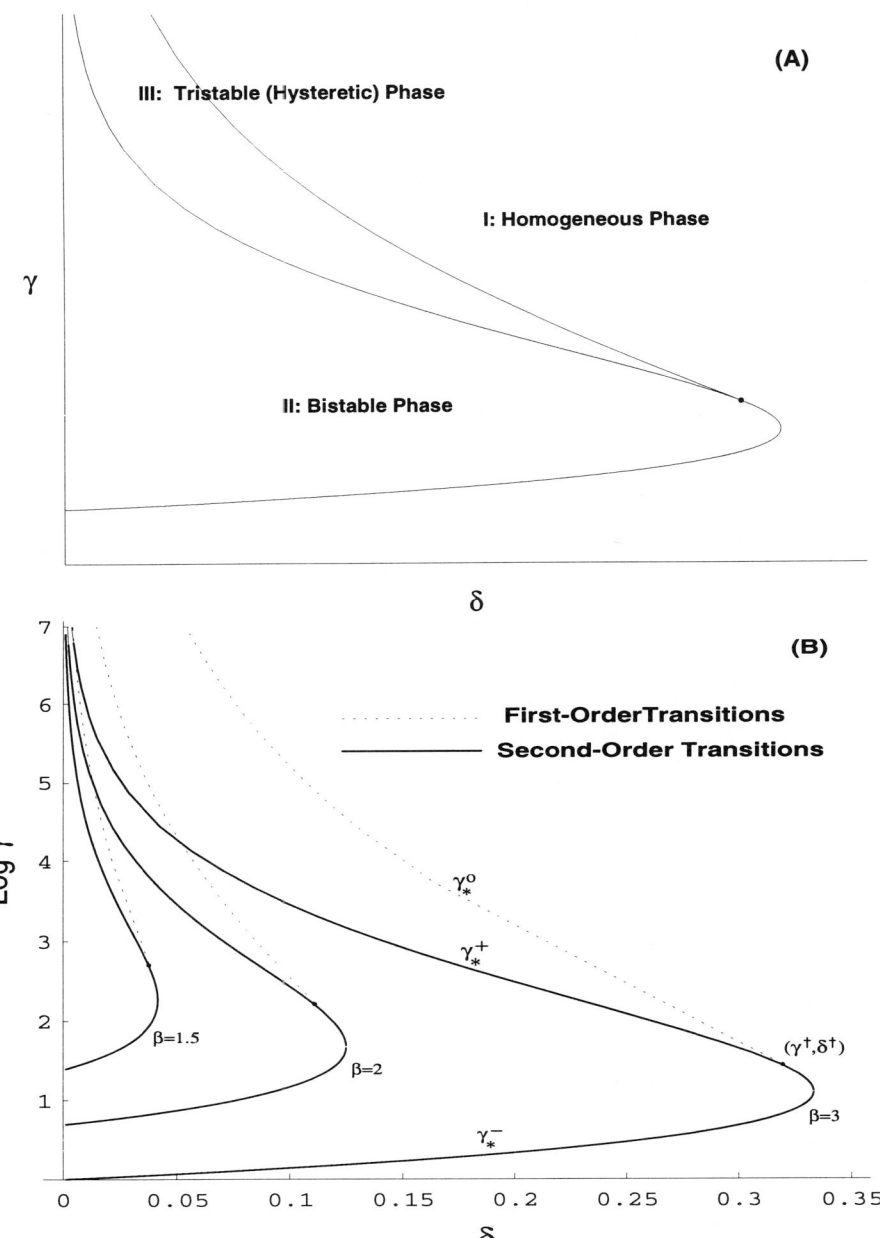

FIGURE 3 $\bar{\rho} - \delta$ phase diagram for the ant swarm.

Figure 3 is a typical phase plot for the ant swarm illustrating regions of homogeneity, bistability, and hysteresis. The plot illustrates the effect of behavioral and swarm parameters on the swarm as a whole. In this case, δ is a behavioral parameter that could be expected to change on the evolutionary time scale, and γ, which is proportional to the number of participants, is a swarm parameter that determines the behavioral "phase" of the swarm. More details may be found in previously published papers[10] where the properties of an ant swarm are analyzed in depth; it is also shown how the collective behavior of real ants[2] can be understood in terms of such models.

REFERENCES

1. Cox, R. W., and D. S. Cohen. *J. Polymer Sci.* B **27** (1989): 589.
2. Deneubourg, J. L., S. Aron, S. Goss, and J. M. Pasteels. *J. Insect Behav.* **3** (1990): 159.
3. Grassé, P. P. *Experientia* **15** (1959): 356.
4. Haken, H. *Synergetics*, 3rd ed. New York: Springer-Verlag, 1983.
5. Horsthemke, W., and R. Lefever. *Noise Induced Transitions.* New York:
   Springer-Verlag, 1984.
6. Kessler, J. O. *Comments Theor. Biol.* **1** (1989): 85.
7. Van Kampen, N. G. *IBM J. Res. Dev.* **32** (1988): 107.
8. Kramer, S., and M. Marder. *Phys. Rev. Lett.* **68** (1992): 205.
9. Landauer, R. *J. Stat. Phys.* **53** (1988): 233.
10. Millonas, M. M. "Swarms, Phase Transitions, and Collective Intelligence." *J. Theor. Biol.* **159** (1992): 529. Also in *Cellular Automata and Cooperative Systems,* edited by N. Bocara, E. Goles, S. Martinez, and P. Pico. New York: Kluwer, 1993. Also in *Artificial Life III*, edited by C. G. Langton, 417–446. Santa Fe Institute Studies in the Sciences of Complexity, Proceedings Volume XVII. Reading, MA City: Addison-Wesley, 1993.
11. Millonas, M. M. *Phys. Rev. E* (1993): to appear.
12. Mittenthal, J. E. "Physical Aspects of the Organization of Development." In *Lectures in the Sciences of Complexity*, edited by D. Stein, 225–274. Santa Fe Institute Studies in the Sciences of Complexity, Lect. Vol. I. Redwood City: Addison-Wesley, 1989.
13. Theraulaz, G., and J. L. Deneubourg. "Swarm Intelligence in Social Insects and the Emergence of Cultural Swarm Patterns." Working Paper 92-09-046, Santa Fe Institute, 1992.

14. Vicsek, T. *Fractal Growth Phenomena.* Singapore: World Scientific, 1989.
15. Wilson, E. O. *The Insect Societies.* Cambridge, MA: Belknap, 1971.

L. Lam, M. C. Veinott, and R. D. Pochy
Nonlinear Physics Group, Department of Physics, San Jose State University, San Jose, CA 95192-0106, U.S.A.

Abnormal Spatio-Temporal Growth

Abnormal spatio-temporal growths in natural or laboratory systems are usually attributed to external causes. However, as exemplified by the boundary probabilistic active walker (BPAW) model studied here we show that at least two kinds of abnormal growth—transformational and irreproducible abnormal growths—could have an *intrinsic* origin coming from the active role of noise. These abnormal growths have been observed in many physical and biological systems. In particular, the transformational abnormal growth pattern obtained in the BPAW model resembles that found in electrodeposit experiments. The idea of intrinsic abnormal growth is compared and contrasted with the two paradigms of deterministic chaos and self-organized criticality.

1. INTRODUCTION

Abnormal growth represents exceptional growth behavior or pattern. In this chapter, we are interested in two particular types of abnormal spatio-temporal growth which manifest themselves in *both* space and time. The first type is what we call *transformational growth*, the occurrence of (usually abrupt) qualitative change in behavior during a growth process. The second type is *irreproducible growth*, the case that unexpected, distinctively different growth forms are obtained from presumably identical samples.

Let us consider the computer generation of the New Mexico state emblem by a growth process as illustrated in Figure 1. Two different algorithms are needed: one algorithm to generate the series of circles in Figure 1(a)–(d), and a second algorithm to grow the needles from the rim of the circle as depicted in Figure 1(e)–(f). This growth process leading to the final pattern in Figure 1(f) is an example of transformational growth, but it is obviously due to an extrinsic cause since the algorithm is changed in the middle of the growth process.

Examples of transformational growth in a real physical system can be found in electrodeposit patterns. As shown in Figure 9 of Lam et al.,[19] the case of $CuSO_4$ with a triangular anode, there are two sharp morphological transformations: the inner one (see Figure 9(c) of Lam et al.[19]) is circular and is insensitive to the shape of the anode; the outer one (see Figure 9(e) of Lam et al.[19]), appearing at about half the distance between the cathode and the anode, assumes the form of the anode. The latter has been interpreted[7] to be caused by the impurities coming from the anode, and is thus of an extrinsic origin. However, the inner one remains unexplained. Is it intrinsic or extrinsic?

Irreproducible growth is common in biological systems.[30] They appear also in electrodeposit experiments. For our purpose here, let us mention two other examples. The first example is the pattern formed by injecting water with a fixed rate into polymer in a radial cell.[31] When the injection rate is low, one always obtains viscous fingers. For high injection rate, one obtains three types of stringy fracture patterns, one of which consists of three branches while the other two have two branches each. The amazing result is that within a region of intermediate injection

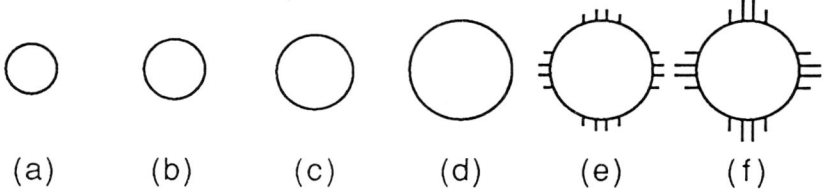

(a)   (b)   (c)   (d)   (e)   (f)

FIGURE 1   The New Mexico state emblem as generated by a growth process.

rate, one sometimes obtain the viscous fingers and sometimes the three-branch fracture pattern. In other words, in this intermediate region the outcome is not unique, i.e., irreproducible.

The second example concerns a sandpile in a semicylindrical drum,[13] the first experiment designed to test the hypothesis of self-organized criticality.[1,2] It is found that as the drum is rotated slowly, when the slope of the sandpile surface θ is below a certain angle θ_r, the angle of repose, the pile is always stable and is stationary. For $\theta > \theta_m$, an angle about 2° above θ_r, the pile is unstable and a global avalanche occurs. Within the narrow range of $\theta_r < \theta < \theta_m$, however, the pile can either flow or be stationary and the result is irreproducible.[13] This observation is attributed to the friction of the sand and the sensitive dependence of the system's behavior on the initial condition.[24] However, is this true? Or, is this the only explanation?

The question we want to raise is that, while extrinsic mechanisms are certainly responsible in some cases, is it also possible that some spatio-temporal growths are *intrinsically* abnormal? For quantum mechanical systems, irreproducible growth can be intrinsic. For example, if one puts ten identical radioactive nuclei on the table and measures them one at a time, some would have decayed and some would not. There is no need of any external cause such as the difference in initial conditions; the process is intrinsic. On the other hand, here we are dealing with complex nonequilibrium systems that are classical in nature. Yet, due perhaps to the noise which is ever present in these open systems, transformational and irreproducible abnormal growths may still arise intrinsically.

Note that it is notoriously difficult to confirm that an experimental abnormal growth is intrinsic, especially in the case of irreproducible growth where unnoticed differences (due, e.g., to contaminants) might exist in the samples. We, therefore, turn to the study of computer experiments.

2. THE ACTIVE WALKER MODEL

The active walker model (AWM) was proposed by Freimuth and Lam[8] to explain in a unified and simple way the abundant appearance of filamentary patterns in natural and laboratory systems,[8,15,17,18] and to provide a systematic method of generating rough surfaces.[17,26] The basic idea of an AWM is very simple. A walker walks on a landscape; at every step it changes the landscape and its next step is determined by the new landscape. The track of the walker forms a filamentary pattern and the resulting landscape forms a rough surface.

The basic AWM has been generalized[17,18] to handle multiwalkers, branching, etc. The AWM is able to produce many patterns as observed in biological, physical, and chemical systems, including those found in retinal neurons, dielectric breakdown of liquids, spirals in bulk and surface chemical reactions. Polymer dynamics[4] such as reptation[3] could be simulated when the distortion due to the active walker assumes a finite lifetime.[18] Furthermore, grooved surfaces as observed in Si film grown by MBE[5] is reproduced by the AWM.[17,18] See Lam and Pochy[18] for a discussion of the relationships between the AWM and other physical concepts such as cellular automata, excitable media, evolutionary biology,[14] self-organized criticality,[1,2] and chaos.[6]

Other recent studies closely related to the AWM include the red queen's walk by Freund and Grassberger,[9] and the behavior of ant swarms by Millonas[21] and by Schweitzer et al.[28] In particular, theoretical discussion has been given by Millonas[22] and by Schweitzer and Schimansky-Geier.[29] In these works—which could be considered as special cases or variants of our AWM—branching, the important ingredient in the formation of many complex patterns[15,17] and in evolutionary biology, is absent. See Lam[16] for further discussion.

2.1 THE BOUNDARY PROBABILISTIC ACTIVE WALKER MODEL

A special case of the AWM capable of producing compact patterns was recently suggested by Lam and Pochy.[18] This so-called boundary probabilistic active walker (BPAW) model is defined by the following steps.

1. One starts with a background landscape represented by a single-valued scalar field—a potential or height—$V_0(i, n)$ defined at every site i of a lattice at time n, where $n = 0, 1, 2, \ldots$
2. At time $n = 0$, a seed (an active walker) is placed at the center of the lattice.
3. The active walker changes the landscape from $V(i, n)$ to $V(i, n + 1)$ such that $V(i, n) = V_0(i, n) + V_1(i, n)$, where

$$V_1(i, n + 1) = V_1(i, n) + W(\mathbf{r}_i - \mathbf{R}(n)) \qquad (1)$$

   with $V_1(i, 0) = 0$; W is called the *landscaping function*; $\mathbf{r}_i$ is the position vector of site i; $\mathbf{R}(n)$ is the location of the walker at time n.
4. One of the perimeter sites is chosen. The probability of perimeter site j being chosen, P_j, is given by $P_j(n) \propto [V(i, n) - V(j, n)]^\eta$ if $V(i, n) > V(j, n)$, and $P_j = 0$ otherwise. Here i is the nearest-neighbor site of j that is occupied; if there are more than one i sites, then the one with the largest $V(i, n) - V(j, n)$ is used. η is a given parameter. The site so chosen is recorded.
5. Step 4 is repeated until one of the perimeter sites is chosen s times first. (Here s is a parameter.) This perimeter site is then filled with a new active walker, which changes the landscape locally according to Eq. (1). The record of perimeter sites chosen in step 4 is then erased.
6. Go to step 4. The process is repeated until the aggregate of walkers touches the boundary of the lattice.

Equivalently, the BPAW model may be viewed as one in which almost every walker at the edge of the aggregate has a certain probability of being chosen to walk one step, followed by a change of the landscape. It is instructive to compare the BPAW model with the dielectric breakdown model (DBM).[25] In the DBM, described in terms of our terminology, one assumes $V = 1$ at the aggregate and $V = 0$ on a large circle surrounding the seed; the Laplace equation $\nabla^2 V = 0$ is solved in the space between the aggregate and the circle. (Due to the boundary conditions of V assumed every perimeter site is lower in potential than the occupied sites.) One of the perimeter sites is chosen in essentially the same manner as in our BPAW. The chosen perimeter site is occupied and the process is repeated. In Figure 2, the change of V in the DBM after the addition of one particle to the aggregate is shown, which is local in character. In comparison, in our BPAW model, this local change of V is replaced by the W function of constant shape (without solving any Laplace equation), which may be taken as an effective potential associated with the walker/particle, similar to that of a polaron or a screened electron.

3. MORPHOGRAM AND INTRINSIC ABNORMAL GROWTHS

In this chapter, $s = 1$, $V_0 = 0$, and a square lattice are assumed. The W function is chosen to be $W_1'(r)$,[17] an isotropic function specified by $W(0) = W_0$, $W(r_1) = -1$, $W = 0$ for $r \geq r_2$, and linear in the two intervals $[0, r_1]$ and $[r_1, r_2]$. By varying the three parameters in W_1 (W_0, r_2 and $\rho \equiv r_1/r_2$), and η, growth patterns ranging from very compact to purely filamentary are generated. In particular, for $W_0 = 20$, $r_2 = 20$, $0.01 \leq \rho \leq 0.9$, and $0 \leq \eta \leq 9$, the patterns obtained from the BPAW belong to one of five different classes, which are named blob, jellyfish, diamond, lollipop, and needle, respectively (see Figure 3). Note that the blob obtained with $\eta = 0$ is similar but not identical to the Eden pattern. In the Eden model, *every* perimeter site can be chosen while in the BPAW only those perimeter sites with

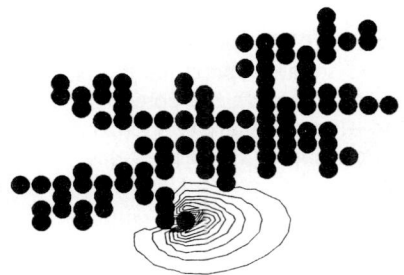

FIGURE 2 Change of potential ΔV calculated from the Laplace equation after the addition of one particle to the aggregate in the DBM. Number of particles present is 101. The change of ΔV is 84×10^{-4} between two adjacent contour lines.

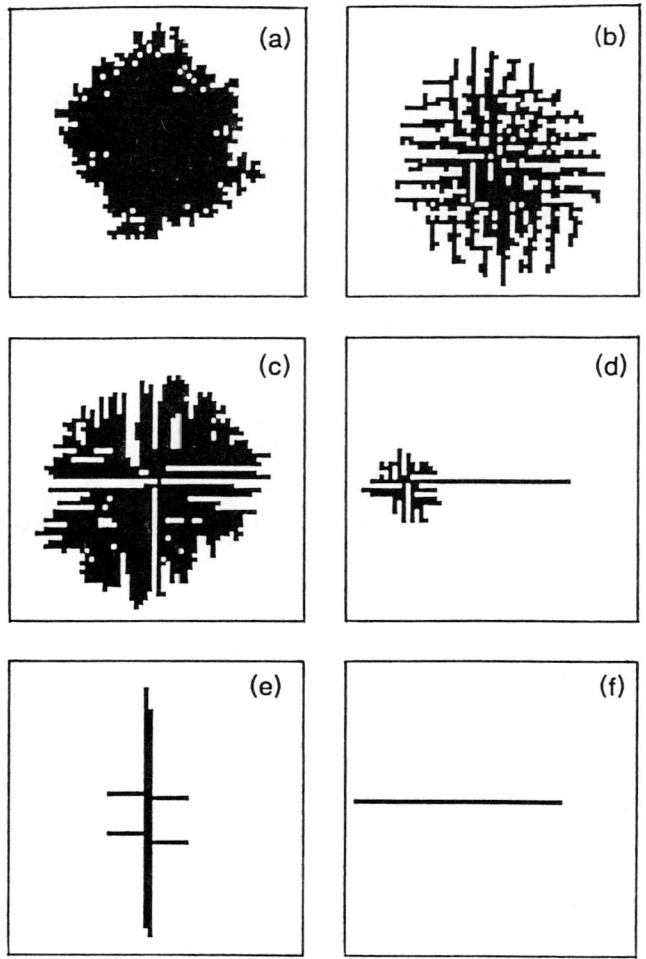

FIGURE 3   The five classes of patterns obtained from the BPAW model. $W = W_1$, $W_0 = 20$, $r_2 = 20$, $0.01 \leq \rho \leq 0.9$, and $0 \leq \eta \leq 9$. The square represents the 64×64 lattice boundary. (a) Blob. (b) Jellyfish—the boundary is filamentary; the core can be either compact or ramified (see Figure 5). (c) Diamond—the boundary is approximately a parallelogram; the core has vertical or horizontal narrow gaps. (d) Lollipop—the "candy" at the end of the "stick" can be jellyfish or diamond. (e) and (f) Needle.

potential lower than the neighboring occupied sites are available; in both models, the available sites have equal probability of being chosen.

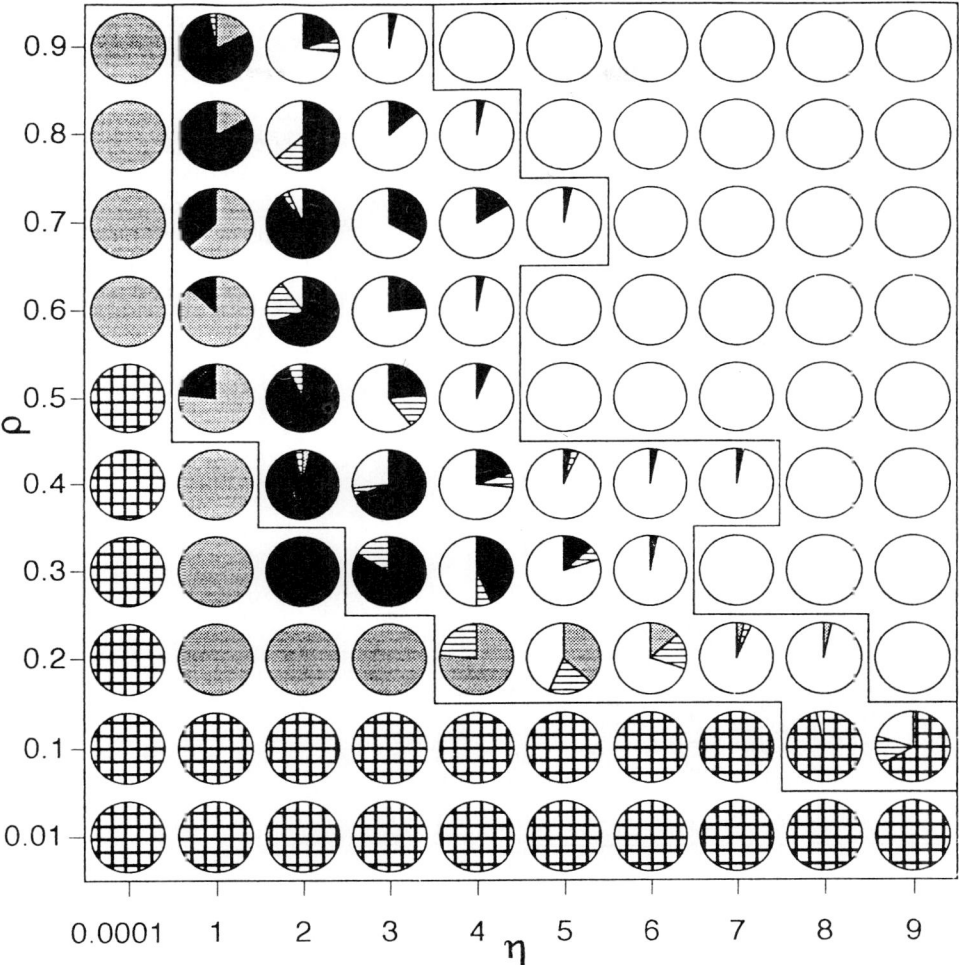

FIGURE 4 Morphogram in the (η, ρ) plane from the BPAW. Parameters are the same as in Figure 3. The pie chart at each data point represents the percentage of each pattern obtained from 30 runs of the computer program. The grid corresponds to blob, grey to jellyfish, black to diamond, horizontal shades to lollipop, and white to needle.

The morphogram in the (η, ρ) plane is given in Figure 4. For each point shown, the computer program written in Microsoft C language is run 30 times with different seed numbers in the random number sequence; a PC486 computer is used. The same 30 seed numbers are used for each data point. The region enclosed by the two solid lines is a *sensitive zone* in which more than one kind of pattern are obtained out of the 30 runs. In other words, in the sensitive zone, the resulting growth pattern depends sensitively on the seed number and is *intrinsically* irreproducible since the seed number is randomly chosen. (For nonequilibrium systems morphogram is a better word than morphological diagram, phase diagram, or state diagram.)

In Figure 4, note the existence of a reentrant sequence of needles at $\eta = 5$ and $\eta = 7$, respectively, as ρ is varied. Incidentally, the lower boundary of the sensitive zone is approximately parabolic in shape, similar to the one obseved in electrodeposits.[11] Before one can identify the sensitive zone with the transition region of crossover reported in Figure 1 of Grier et al.,[11] it should be checked whether the patterns in the transition region are irreproducible (this point is not mentioned by Grier et al.[11]).

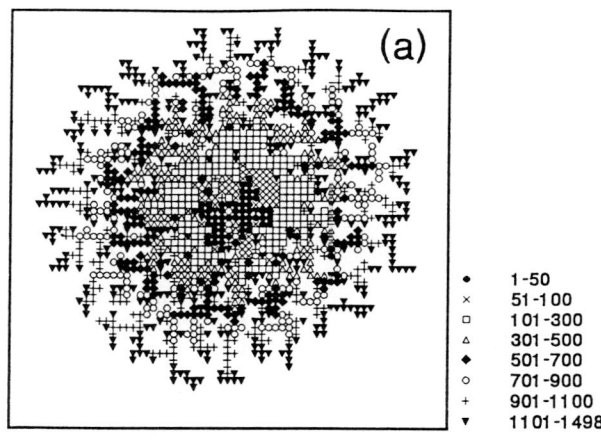

•	1-50
×	51-100
□	101-300
△	301-500
◆	501-700
○	701-900
+	901-1100
▼	1101-1498

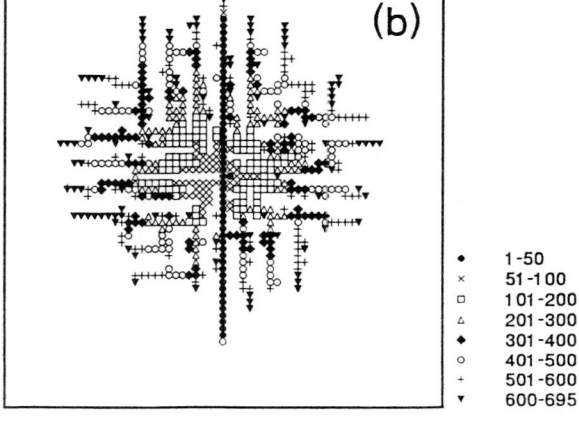

•	1-50
×	51-100
□	101-200
△	201-300
◆	301-400
○	401-500
+	501-600
▼	600-695

FIGURE 5   Two types of time evolution of the jellyfish pattern. $W_0 = 20$, $r_2 = 20$, and $\rho = 0.2$. The square represents the 64×64 lattice boundary. The numbers shown are the pixel numbers with the initial seed counted as one. (a) $\eta = 2$. (b) $\eta = 5$.

Two types of time evolution of the jellyfish pattern are shown in Figure 5. In Figure 5(a), the pattern grows from a compact core to filaments at the rim (resembling that in Figure 9(c) of Lam et al.,[19] mentioned in Section 1). In Figure 5(b), the pattern starts from a needle and grows into radial filaments. Both these two cases are examples of what we called transformational growth and are *intrinsic* in nature.

4. DISCUSSIONS

As shown in Section 3, transformational and irreproducible abnormal growths can occur naturally due to the order the perimeter sites are chosen, which is dictated by the sequence of random numbers involved. In real systems, the random numbers correspond to the noise coming from the environment and are not controlled externally; these abnormal growths are thus intrinsic. The fact that noise can play an important and active role in nonequilibrium systems has been recognized before,[10,12,23] but the examples presented here are the first ones involving spatio-temporal growth patterns.

The morphogram in Figure 4 will change slightly when different numbers of runs or seed numbers are used (see Lam and Pochy[18]), but the qualitative feature is expected to remain intact. The size of the lattice used will also affect the morphogram, but this is a separate issue since growth patterns observed and compared are always grown in a limited space after a finite amount of time. We have tested only those data points shown in Figure 4; when more data points were tested one could find out whether the sensitive zone is actually a multilayered object, like the fractal basins in chaos. (The five classes of pattern may be viewed as attractors in the space of configuration states.)

We expect that in any kind of pattern-forming nonequilibrium system, it is very likely that between two regions of very different morphologies there is a sensitive zone in which these two morphologies and something intermediate between them will appear irreproducibly; i.e., each type will appear with a certain probability. This possibility should be examined experimentally in electrodeposit,[11,27] biological,[20] and other systems. Note that within the sensitive zone the different morphologies appear separately; they coexist with each other only if they are grown from different points in the sample which are far from each other. In this sense, they differ from the coexistent phases in a first-order phase transition in equilibrium systems. Mathematically, the sensitive zone may correspond to the "hysteretic phase" deduced from the bifurcation diagrams.[21]

The origins of cancer cell growth[30] are not yet understood. However, the growth morphology of cancer cells is distinctively different from that of normal cells. For

TABLE 1 Three paradigms concerning exceptional phenomena.

	Chaos	SOC	IAG
Phenomena	Randomness	Catastrophes	Transformational and irreproducible growths
Old wisdom	Noise	Extrinsic	Extrinsic
New explanation	Deterministic nonlinearity	Self-organization	Intrinsic noise
Signature	Sensitive dependence on initial condition (strange attractor)	Power laws in space and time	Sensitive dependence on noise (sensitive zone, sharp morphological change)

example, on a solid substrate normal cells divide until they form a monolayer, while cancer cells form a rough surface of several layers deep.[30] Whether there is an intrinsic mechanism behind these abnormal spatio-temporal cancerous growth is a topic still to be explored.

Finally, as shown in Table 1, intrinsic abnormal growth (IAG) presents a paradigm that—like those in chaos[6] and self-organized criticality (SOC)[1]—some interesting, exceptional phenomena could be intrinsic in origin, in contrary to the conventional wisdom.

ACKNOWLEDGMENTS

One of us (LL) thanks James Maher, Mitsugu Matsushita, Mark Millonas, Frank Schweitzer, Emil Simiu, and Chao Tang for helpful discussions; Walter Fontana for bringing work by Freund and Grassberger[9] to his attention; and Pat Cladis and Peter Palffy-Muhoray for the invitation to this wonderful workshop. We are grateful to Daniela Kayser and Daniel Ratoff for their assistance in producing some of the diagrams presented in this paper. This work is supported by the NSF-REU program and a Syntex Corporation grant from the Research Corporation.

REFERENCES

1. Bak, P., C. Tang, and K. Wiesenfeld. "Self-Organized Criticality: An Explanation of $1/f$ Noise." *Phys. Rev. Lett.* **59** (1987): 381–384.
2. Bak, P. "Catastrophes and Self-Organized Criticality." *Comput. Phys.* **5** (1991): 430–433.
3. de Gennes, P. G. "Reptation of a Polymer Chain in the Presence of Fixed Obstacles." *J. Chem. Phys.* **55** (1971): 572–579.
4. Doi, M., and S. F. Edwards. *The Theory of Polymer Dynamics.* Oxford: Clarendon, 1988.
5. Eaglesham, D. J., H. L. Gossmann, and M. Cerullo. "Limiting Thickness h_{epi} for Epitaxial Growth and Room-Temperature Si Growth on Si(100)." *Phys. Rev. Lett.* **65** (1990): 1227–1230.
6. Eubank, S. G., and J. D. Farmer. "Chaos and Randomness" In *Introduction to Nonlinear Physics*, edited by L. Lam, Part II. New York: Springer, to appear.
7. Fleury, V., M. Rosso, and J. N. Chazalviel. "Geometrical Aspect of Electrodeposition: The Hecker Effect." *Phys. Rev. A* **43** (1991): 6908–6919.
8. Freimuth, R. D., and L. Lam. "Active Walker Models for Filamentary Growth Patterns." In *Modeling Complex Phenomena,* edited by L. Lam and V. Naroditsky, 302–313. New York: Springer, 1992.
9. Freund, H., and P. Grassberger. "The Red Queen's Walk." *Physica A* **190** (1992): 218–237.
10. Frey, M., and E. Simiu. "Noise-Induced Transitions to Chaos." This volume.
11. Grier, D., E. Ben-Jacob, R. Clarke, and L. M. Sander. "Morphology and Microstructure in Electrochemical Deposition of Zinc." *Phys. Rev. Lett.* **56** (1986): 1264–1267.
12. Horsthemke, W., and R. Lefever. *Noise-Induced Transitions.* New York: Springer, 1984.
13. Jaeger, H. M., C. H. Liu, and S. R. Nagel. "Relaxation at the Angle of Repose." *Phys. Rev. Lett.* **62** (1989): 40-43.
14. Kauffman, S. A. *The Origins of Order: Self-Organization and Selection in Evolution.* New York: Oxford University Press, 1993.
15. Kayser, D. R., L. K. Aberle, R. D. Pochy, and L. Lam. "Active Walker Models: Tracks and Landscapes." *Physica A* **191** (1992): 17–24.
16. Lam, L. "Active Walks, Landscapes and Complex Systems." In *Introduction to Nonlinear Physics,* edited by L. Lam, Ch. 15. New York: Springer, to appear.
17. Lam, L., R. D. Freimuth, M. K. Pon, D. R. Kayser, J. T. Fredrick, and R. D. Pochy. "Filamentary Patterns and Rough Surfaces." In *Pattern Formation in Complex Dissipative Systems,* edited by S. Kai, 34–46. River Edge, NJ: World Scientific, 1992.

18. Lam, L., and R. D. Pochy. "Active Walker Models: Growth and Form in Nonequilibrium Systems." *Comput. Phys.* **7** (1993): 534–541.
19. Lam, L., R. D. Pochy, and V. M. Castillo. "Pattern Formation in Electrodeposits." In *Nonlinear Structures in Physical Systems,* edited by L. Lam and H. C. Morris, 11–31. New York: Springer, 1990.
20. Matsushita, M., J. I. Wakita, and T. Matsuyama. "Growth and Morphological Changes of Bacteria Colonies." This volume.
21. Millonas, M. M. "A Connectionist Type Model of Self-Organized Foraging and Emergent Behavior in Ant Swarms." *J. Theor. Biol.* **159** (1992):529–552.
22. Millonas, M. M. "A Nonequilibrium Statistical Field Theory of Swarms and Other Spatially Extended Complex Systems." This volume.
23. Moss, F. "Stochastic Resonance: From the Ice Ages to the Monkey's Ear." In *Some Problems in Statistical Physics,* edited by G. H. Weiss. Philadelphia, PA: SIAM, 1992.
24. Nagel, S. R. "Instabilities in a Sandpile." *Rev. Mod. Phys.* **64** (1992): 321–325.
25. Niemeyer, L., L. Pietronero, and H. J. Wiesmann. "Fractal Dimension of Dielectric Breakdown." *Phys. Rev. Lett.* **52** (1984): 1033–1036.
26. Pochy, R. D., D. R. Kayser, L. K. Aberle, and L. Lam. "Boltzmann Active Walkers and Rough Surfaces." *Physica D* **66** (1993): 166–171.
27. Sawada, Y., A. Dougherty, and J. P. Gollub. "Dendritic and Fractal Patterns in Electrolyte Metal Deposits." *Phys. Rev. Lett.* **56** (1986): 1260–1263.
28. Schweitzer, F., K. Lao, and F. Family. "Active Brownian Walkers Simulate Trunk Trail Formation by Ants." Preprint, 1994.
29. Schweitzer, F. and L. Schimansky-Geier. "Clustering of Active Walkers in a Two-Component System." *Physica A* in press.
30. Watson, J. D., N. H. Hopkins, J. W. Roberts, J. A. Steitz, and A. M. Weiner. *Molecular Biology of the Gene.* Menlo Park, CA: Benjamin/Cummings, 1987.
31. Zhao, H., and J. V. Maher. "The Fracture Transition in Swollen Polymer Networks." This volume.

Index